Drug Metabolism and Pharmacokinetics

Drug Metabolism and Pharmacokinetics

Frontiers, Strategies, and Applications

Edited by

Liang Shen
Vice President, Head of DMPK Department
WuXi AppTec

WILEY

Published by John Wiley & Sons, Inc., Hoboken, New Jersey.
Published simultaneously in Canada.

Library of Congress Cataloging-in-Publication Data applied for:
Hardback ISBN: 9781394300129
ePDF ISBN: 9781394300143
epub ISBN:9781394300136

Cover Design: Wiley
Cover Image: Created by and Courtesy of WuXi AppTec

Set in 9.5/12.5pt STIXTwoText by Straive, Pondicherry, India

SKY10135614_123025

Contents

About the Editors

Liang Shen

Ph.D., Vice President, Head of DMPK Department, WuXi AppTec

Dr. Liang Shen has more than 20 years of experience in drug metabolism and pharmacokinetics (DMPK), toxicology, clinical pharmacology, and precision medicine. Dr. Shen has published or co-authored over 50 research articles and has led or participated in the compilation or translation of four books in drug discovery or DMPK field. His extensive experience in drug discovery, preclinical candidate (PCC) profiling, and investigational new drug (IND) submissions has successfully supported multiple pipelines in the fields of oncology, immuno-oncology, immunology, metabolic diseases, infectious diseases, central nervous system (CNS), and cardiovascular diseases to proof-of-concept at the early clinical stage. He received his PhD degree from the University of Georgia in the United States.

WuXi AppTec DMPK

As a global company with operations across Asia, Europe, and North America, WuXi AppTec provides a broad portfolio of R&D and manufacturing services that enable the global pharmaceutical and life sciences industry to advance discoveries and deliver groundbreaking treatments to patients. The Department of DMPK of WuXi AppTec was established in 2006 and offers *in vivo* and *in vitro* DMPK services covering stages from discovery screening to preclinical development and clinical studies. With research facilities located in the United States (New Jersey) and China (Shanghai, Suzhou, Nanjing, and Nantong), the department has supported thousands of investigational new drug (IND) and new drug applications (NDA). The services of WuXi AppTec DMPK include *in vitro* absorption, distribution, metabolism, excretion (ADME) studies, *in vivo* pharmacokinetic studies, metabolite profiling and identification studies, synthesis of radiolabeled compounds, radiolabeled ADME including human absorption, metabolism, and excretion (AME) studies, and bioanalysis. For more information, visit dmpkservice.wuxiapptec.com.

Preface

Drug metabolism and pharmacokinetics (DMPK) studies involve characterizing the absorption, distribution, metabolism, and excretion (ADME) properties of compounds. This aids in identifying drugs with enhanced safety and potency, minimizing the risk of drug-drug interactions, and establishing a foundation for determining clinical dosage and frequency. The advancement of the pharmaceutical industry and the emergence of novel drug modalities have presented new challenges and opportunities for drug development, driving continuous innovation and progress in DMPK research.

This book is written by a team of scientists specializing in DMPK research from the DMPK Department of WuXi AppTec. The team was established in 2006 and has successfully supported thousands of investigational new drug (IND) applications globally. Drawing on a wealth of extensive practical experience and theoretical research, this book encapsulates the most recent advancements and illustrative applications. It is intended for pharmacy students, pharmaceutical industry researchers, and DMPK scientists, especially those exploring novel drug modalities.

Sixty-eight relatively independent yet interconnected articles compose this book, each offering a unique perspective and providing in-depth interpretation. Readers can either read systematically or select specific topics of interest from the table of contents. Basic concepts, frontier advancements, DMPK research strategies, and technical methods are covered in the book for novel drug modalities, and therapeutics in different disease areas. Furthermore, the book encompasses a wide range of application and validation cases for DMPK research, including studies in *in vitro* ADME, *in vivo* pharmacokinetics, metabolite profiling and identification, radiolabeled ADME, and bioanalysis.

The book is divided into three sections. Section I explores the pharmacokinetic strategies and research advancements in novel drug modalities, including proteolysis-targeting chimeras (PROTACs), antibody-drug conjugates (ADCs), peptide-drug conjugates (PDCs), peptide drugs, oligonucleotide (OLIGO) drugs, and mRNA-based vaccines and therapeutics. The mechanisms of these new drug modalities impose higher requirements on their DMPK studies. For example, ADC drugs are composed of antibodies and small molecule toxins and exhibit both large molecule and small molecule characteristics, making their *in vivo* process complicated and dynamic with biotransformation and drug-to-antibody ratio (DAR) changes. This complexity poses challenges to ADC studies such as stability, metabolite identification, and bioanalysis.

Section II focuses on the drug metabolism and pharmacokinetic studies of therapeutics in specific disease areas. With an aging population and an increasing number of patients with chronic diseases, there is a growing demand for drug development in disease areas with unique pharmacokinetic characteristics, such as ophthalmic drugs, respiratory medications, transdermal and topical drugs, and central nervous system drugs. Specific preclinical study models or methods need to be developed to address these characteristics resulting from different pathogenesis and therapeutic modes of action. For example, developing ophthalmic drugs will encounter difficulties such as multiple biological barriers, analytical complexities, and intricate experimental operations.

Section III delves into front-edge metabolism and pharmacokinetic research strategies, methods, and applications, covering topics such as *in vitro* ADME, *in vivo* pharmacokinetics, metabolite profiling and identification, radiolabeled ADME studies, and advanced bioanalysis techniques and platforms. These approaches enable more comprehensive and accurate drug evaluation in DMPK studies.

Scientists are devoted to drug metabolism and pharmacokinetic studies, and expand the breadth and depth of their research by summarizing and optimizing the methodologies and insights from massive experience. Nevertheless, it is essential to be aware that the DMPK research will be full of opportunities and challenges with technological advancements. Continuous investment of time and resources is essential to keep pace and add to clinical breakthroughs and health sciences.

The publication of this book aims to bring together the knowledge and experience from WuXi AppTec DMPK's forefront scientists, and contribute to new drug development in this era of industrial transformation. I sincerely thank the authors for their efforts and contributions, as well as Wiley for their review and support. I hope readers can enjoy reading while finding useful tactics in this book.

Dr. Liang Shen
Vice President, Head of DMPK Department
WuXi AppTec

Acknowledgments

The compilation and publication of this book have been a joint effort. Credit should be given to the authors and editors, who have delivered profound insights by drawing upon their vast DMPK expertise. Their hard work, research, and commitment in crafting this book deserve special recognition. We would like to thank all the authors and editors, listed here in alphabetical order.

Peiyun An, Weiqun Cao, Dr. Genfu Chen, Dr. Jacob Zhi Chen, Dr. Shiyan Chen, Song Chen, Yu Chen, Liwei Cheng, Dr. Qigan Cheng, Hongfang Cui, Qian Cui, Tiantian Dang, Ya Ding, Xuan Dong, Jian Fang, Quanli Feng, Flora Fu, Xinfa Fu, Li Gao, Lijia Gao, Yafei Gao, Yang Gao, Lian Guo, Qiqi Han, Huan He, Lijuan Hou, Jie Hu, Weimin Hu, Dr. Jiaming Jiang, Dr. Lifang Jiang, Dr. Furong Jiao, Dr. Jing Jin, Wei Lei, Chengyuan Li, Hongye Li, Huan Li, Huayu Li, Dr. Huihui Li, Jie Li, Lin Li, Dr. Peng Li, Qian Li, Ruixing Li, Tingting Li, Xiaotong Li, Xin Li, Yingying Li, Zhihai Li, Dr. Zhiyu Li, Danqing Lin, Haijuan Liu, Dr. Huan Liu, Huan Liu, Jian Liu, Mengjie Liu, Qing Liu, Shoutao Liu, Siyu Liu, Yanfeng Liu, Chunhong Lu, Jinlian Lu, Yikuan Lu, Dr. Dongke Ma, Dr. Liping Ma, Chen Ning, Jie Pan, Dr. Yan Pan, Linlin Pei, Hongfeng Qian, Huijuan Qian, Gengyao Qin, Dr. Li Qu, Xiang Ren, Yanfu Ren, Tingting Ruan, Dr. Liang Shen, Dr. Liqi Shi, Xinmeng Shi, Miaomiao Song, Yuanqiang Su, Hong Sun, Jianping Sun, Ziqian Sun, Cheng Tang, Liping Tang, Xue Tang, Dr. Yi Tao, Binbin Tian, Dr. Hongmei Wang, Huijuan Wang, Dr. Jie Wang, Junjie Wang, Dr. Xiangling Wang, Dr. Xiaoqi Wang, Yu Wang, Yuxi Wang, Xinxin Wen, Lili Xing, Haiwei Xiong, Tao Xiong, Dr. Mingcheng Xu, Ye Xu, Jia Xue, Dr. Huan Yan, Jinyu Yan, Dr. Yunyi Yan, Guang Yang, Rong Yang, Xianqing Yu, Dr. Jiaqi Yuan, Chao Zhang, Hefeng Zhang, Dr. Hong Zhang, Honghong Zhang, Dr. Lingling Zhang, Xuan Zhang, Yingying Zhang, Yu Zhang, Hang Zhao, Dr. Nan Zhao, Lijin Zheng, Dr. Maotian Zhou, Xinyun Zhou, Ying Zhou, Wenjun Zhu, and Zhangpei Zhu.

Section I

Novel Drug Modalities: Advancements and Their Pharmacokinetic Strategies

1

Proteolysis-Targeting Chimeras (PROTACs)

1.1 An Overview of PROTAC Technology and Drug Metabolism and Pharmacokinetic (DMPK) Research Strategies

Chengyuan Li, Yu Wang, Jing Jin

1.1.1 Introduction

In recent decades, significant advances have been achieved in the development of new therapeutic approaches. Proteolysis-targeting chimera (PROTAC) offers a new approach to some disease treatments, which has increasingly attracted the attention of drug developers in the last five years. Additionally, PROTAC, as a new therapeutic technology, has facilitated substantial investment worldwide. In this chapter, we summarized the PROTAC's structure and mechanism of action, its global landscape, the drug metabolism and PK (DMPK) challenges in PROTAC research, and corresponding potential solutions.

1.1.2 Structure and Mechanism of Action of PROTACs

PROTAC utilizes the body's natural protein degradation mechanism, known as the ubiquitin–proteasome system [1]. PROTAC is a bifunctional molecule that consists of three key structural parts: a ligand that binds the target protein, a ligand that binds the ubiquitin–protein ligase (E3), and a linker that connects the two ligands (Figure 1.1).

Because of its unique structure, PROTAC brings the E3 ligase and the target protein to proximity, which induces the E3 ligase to label the target protein with ubiquitin. This causes the target protein to degrade [2]. The PROTAC molecules that detach after the degradation of the target protein can be recycled (Figure 1.2). In contrast to the occupancy-driven pharmacology of most small-molecule drugs, PROTAC can access proteins that were previously inaccessible without occupying an active pocket or relying on target occupancy to disrupt the function of the target protein. This is known as event-driven pharmacology.

Therefore, PROTAC offers many advantages in drug discovery:

- It has the potential to target many undruggable protein targets that lack active sites.
- While retaining the advantage of directly targeting intracellular proteins, it has the potential to overcome some of the disadvantages of small molecules such as drug resistance.
- To improve patient compliance, it can be administered orally.

Drug Metabolism and Pharmacokinetics: Frontiers, Strategies, and Applications, First Edition.
Edited by Liang Shen and WuXi AppTec DMPK.

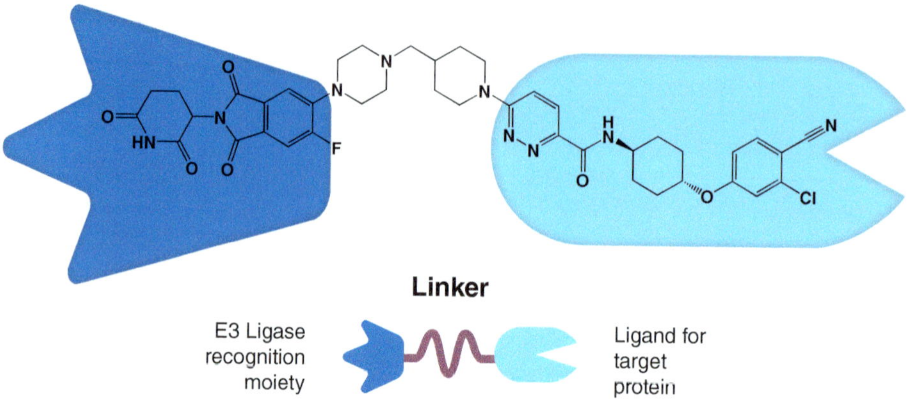

Figure 1.1 Schematic of typical PROTAC structure.

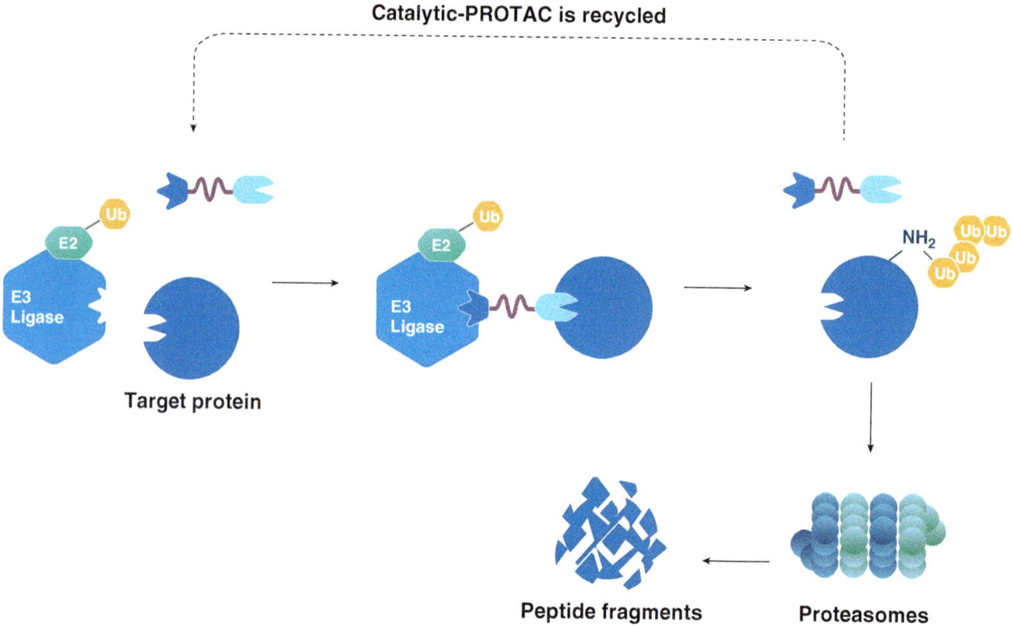

Figure 1.2 Mechanism of action of PROTAC.

1.1.3 PROTAC Landscape Analysis

By using the event-driven pharmacology, PROTAC offers an attractive therapeutic concept to control target protein levels. Although this is a relatively new modality, it has made rapid progress in the drug discovery pipeline.

1.1.3.1 PROTAC Research and Development Progress: International Pharmaceutical Companies

PROTAC has boomed from 2019 to 2022 with Arvinas' two candidate molecules, ARV-110 and ARV-471, being the first to demonstrate positive clinical data. In addition, ARV-766, the company's

Table 1.1 PROTAC research and development progress of some international pharmaceutical companies.

Company	PROTAC	Indications	Target	Study phase
Arvinas, Inc.	ARV-471	Breast cancer (ER+/ HER2-breast cancer)	Estrogen receptor (ER)	Clinical phase III
	ARV-393	B-Cell Malignancies	B-cell lymphoma 6 protein (BCL6)	Clinical phase I
	ARV-102	Parkinson's Disease Progressive Supranuclear Palsy	Leucine-rich repeat kinase 2 (LRRK2)	Clinical phase I
Nurix Therapeutics, Inc.	NX-2127	B cell malignancies	BTK + IMiD activity	Clinical phase I
	NX-1607	Immuno-oncology	CBL-B	Clinical phase I
	NX-5948	B cell malignancies and autoimmune diseases	BTK	Clinical phase I
Kymera Therapeutics, Inc.	KT-474	Allergic dermatitis, hidradenitis suppurativa, rheumatoid arthritis	Interleukin-1 receptor-associated kinase 4 (IRAK4)	Clinical phase II
	KT-621	Immunology	STAT6	Clinical phase I
	KT-333	Solid tumors	STAT3	Clinical phase I
	KT-253	Solid tumors	MDM2	Clinical phase I
C4 Therapeutics, Inc.	CFT7455	Relapsed/refractory non-Hodgkin's lymphoma or multiple myeloma	IKZF1/3	Clinical phase I/II
	CFT8919	Non-small cell lung cancer (NSCLC) with drug-resistant EGFR mutation	EGFR L858R	Clinical phase I
	CFT1946	V600 mutant cancers	BRAF V600	Clinical phase I/II
Bristol-Myers Squibb	AR-LDD	Prostate cancer	AR	Clinical phase I
Dialectic Therapeutics, Inc.	DT-2216	Tumor	BCL-XL	Clinical phase I
Accutar Biotechnology Inc.	AC0699	Breast cancer	ER	Clinical phase I
	AC0682	Breast cancer	ER	Clinical phase I
	Ac176	Prostate cancer	AR	Clinical phase I
	AC0676	Autoimmunity	BTK WT & C481S	Clinical phase I

Note: **The information was obtained from the official websites of the companies** mentioned in the table **and the deadline for information collection is 15 December 2024.**
(Only research pipelines with compounds are listed.) Vividion Therapeutics, ERASCA, and other PROTAC companies do not disclose pipeline information.

third PROTAC molecule, is set to enter the clinical phase. Other international companies, such as Kymera Therapeutics, C4 Therapeutics, and Nurix Therapeutics, that focus on developing protein degraders have a molecule in clinical phase I as well. In addition, other pharmaceutical companies such as Bristol-Myers Squibb, Dialectic Therapeutics, and Accutar Biotechnology also own a PROTAC molecule in clinical phase I (Table 1.1).

Table 1.2 PROTAC R&D progress of Chinese pharmaceutical companies.

Company	PROTAC	Indications	Target	Study phase
Lynk Pharmaceuticals Co., Ltd.	LNK01002	Hematological tumor	—	Clinical phase I
Kintor Pharmaceutical Limited	GT20029	Androgenetic alopecia, acne	AR	Clinical phase II
Haisco Pharmaceutical Group Co., Ltd.	HSK29116	B-cell malignancies	BTK	Clinical phase I
BeiGene Ltd.	BGB-16673	B-cell malignancies	BTK	Clinical phase I
Cullgen Inc.	CG001419	Solid tumor	TRK	Clinical phase I/II
Hinova Pharmaceuticals, Inc.	HP518	mCRPC for standard treatment failure	AR	Clinical phase I/II
	HP568	Breast cancer	ER	Preclinical
	HC-X029	Last-line treatment for mCRPC that has failed standard treatment	AR-sv	Preclinical
	HC-X035	KRAS mutated cancers	Protein tyrosine phosphatase (SHP2) containing two SH2 (Src homology 2) domains	Preclinical

***Note:* The information was obtained from the official websites of the above companies and the deadline for information collection is 15 December 2024.**
Hangzhou Polymed Biopharma, Fendi Technology, Five Elements Biotechnology, Seed Therapeutics, and other PROTAC companies have not disclosed their pipeline information.

1.1.3.2 PROTAC Research and Development Progress: Chinese Pharmaceutical Companies

In China, PROTAC molecules are also developed by many companies, such as Lynk Pharmaceuticals, Kintor Pharmaceuticals, Haisco, Meizer Pharma, Cullgen, Hinova Pharmaceuticals, BeiGene, and Seed Therapeutics, a subsidiary of BeyondSpring, and all these companies have laid out in the PROTAC field. Among them, Lynk Pharmaceuticals, Kintor Pharmaceuticals, BeiGene, and Haisco own a PROTAC molecule in clinical phase I (Table 1.2).

1.1.4 Why Is DMPK Research Critical to PROTAC Drug Development?

Despite the popularity of PROTAC technology because of its unique mechanism of action, the development of PROTAC molecules still faces multiple challenges (Figure 1.3).

- Oral administration is the ideal drug delivery route for the treatment of most diseases, but because of its unique structure, PROTAC has a high molecular weight and poor solubility, which makes it difficult to meet the classical Lipinski's Rule of Five.
- It has poor *in vivo* and *in vitro* permeability, leading to poor absorption and low bioavailability.
- Several studies of PROTAC have shown that at appropriate or low concentrations, a normal ternary complex is formed. However, at high concentrations, a target protein-PROTAC or

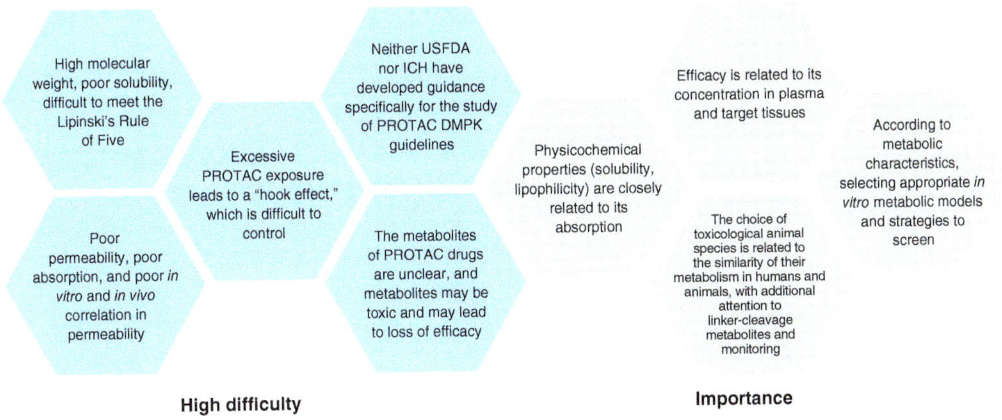

Figure 1.3 Difficulties and importance of PROTAC research.

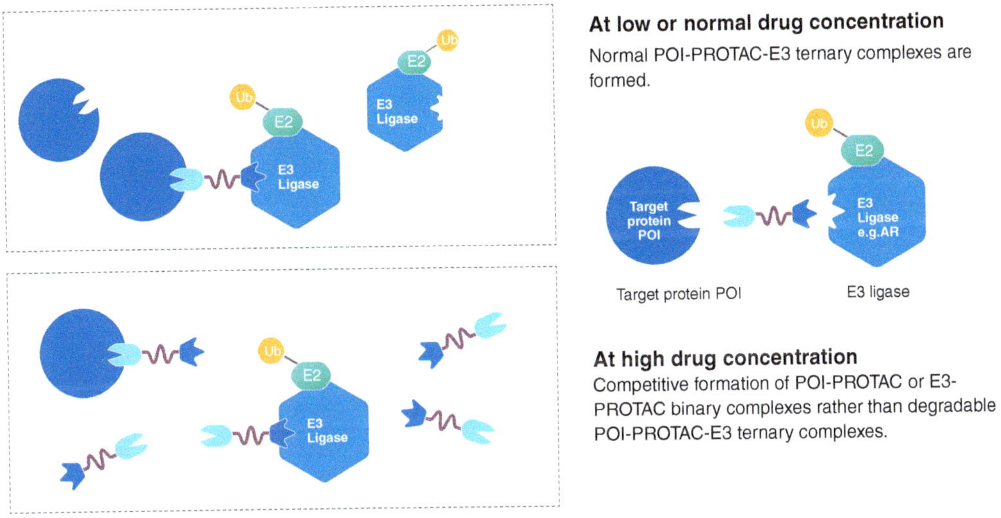

Figure 1.4 Schematic representation of the hook effect.

E3 ligase-PROTAC binary complex is formed competitively, leading to reduced efficacy or toxic reactions, also known as the "hook effect" (Figure 1.4). Therefore, it is essential to fully understand the PKPD relationship of PROTAC, to timely modify the dosages.

- These molecules show high plasma protein binding in various species, and their plasma protein binding ratio may not be accurately measured when using the routine plasma protein binding evaluation method. Furthermore, the metabolism of PROTAC compounds in plasma might be affected by the high plasma protein binding.
- Metabolites of PROTACs may result in toxic reactions. Therefore, *in vivo* metabolite monitoring is a crucial part of its preclinical screening, with particular attention to linker-cleavage metabolites.
- There are no guidelines specifically for PROTAC drugs, hence, the guideline for small-molecule drugs is generally used as a reference.

1.1.5 The Strategy of the DMPK Study for PROTAC Drugs

The preclinical optimization of PROTAC drugs is mainly performed through the cascade optimization of physicochemical and DMPK properties [3].

- In the preliminary screening stage, the *in vivo* and *in vitro* characterization of PROTAC molecules is required, including physicochemical properties, permeability, protein binding, and drug-drug interaction (DDI).
- In the optimization stage, to better understand the DMPK properties of PROTAC, such as absorption and metabolism, the focus should be on improving the metabolic clearance and solubility of PROTAC molecules combined with the PK properties of extravascular drug delivery. The structure of PROTACs poses a challenge in improving their permeability. Therefore, at this phase, more attention should be paid to solubility and metabolic stability when administered orally.
- At the PCC stage, in order to obtain a more in-depth exposure–response relationship, PROTAC molecules with potent and better oral bioavailability can be used in further pharmacokinetics/pharmacodynamics (PK/PD) studies.

1.1.6 Summary

Although there is no marketed drug for PROTAC until July 2024, it is attracting more and more biopharmaceutical innovators and entrepreneurs to compete on this brand-new track. Over the course of more than 20 years of development, PROTAC has redefined small molecules with its unique mode of action. It has shattered the conventional rules of drug discovery and ushered in a new era of drug development. We anticipate a growing number of designed and developed PROTAC molecules, which will unlock fresh opportunities in various disease areas and give hope to a greater number of patients.

References

1 Domanski, P. Recycling the cell: autophagy and the ubiquitin-proteasome processes. Biomol blog resources biomol GmbH - life science shop.
2 Churcher, I. (2018). Protac-induced protein degradation in drug discovery: breaking the rules or just making new ones? *Journal of Medicinal Chemistry* 61 (2): 444–452.
3 Cantrill, C., Chaturvedi, P., Rynn, C. et al. (2020). Fundamental aspects of DMPK optimization of targeted protein degraders. *Drug Discovery Today* 25 (6): 969–982.

1.2 Strategies to Improve the Oral Bioavailability of PROTACs

Liping Ma, Jie Hu, Chengyuan Li, Jing Jin

1.2.1 Introduction

Because of its unique mechanism of action, PROTAC has gradually become one of the most popular small molecules over the last five years. However, there are many obstacles to overcome due to its druggability. One of the most discussed challenges is that, unlike traditional small-molecule drugs, PROTAC does not adhere to the classical Lipinski's Rule of Five. PROTAC is considered to violate the tenets of medicinal chemistry and functions as a redefined small-molecule compound. Consequently, poor oral absorption remains a common issue for PROTAC compounds. Here, we discuss the PROTAC molecules' properties that are closely related to oral absorption and bioavailability, including solubility, lipophilicity, permeability, metabolic stability, etc., to provide ideas for further research and development of PROTAC molecules.

1.2.2 PROTAC Molecules Hardly Meet Lipinski's Rule of Five

Oral small-molecule drugs with high bioavailability and desired pharmacokinetic properties generally meet Lipinski's Rule of Five:

a) The molecular weight is less than 500.
b) No more than 5 hydrogen bond donors.
c) No more than 10 hydrogen bond acceptors.
d) Octanol–water partition coefficient log P not greater than 5.
e) No more than 10 rotatable bonds.

PROTAC consists of three key structural parts, a ligand binding with the target protein, a ligand binding with the E3 ubiquitin ligases, and a linker connecting the two ligands, which results in a large molecular weight of PROTAC, typically more than 700 Da [1]. PROTAC is also poorly soluble and permeable.

The two PROTAC molecules, ARV-110 and ARV-471, which have entered Phase II clinical trials, exhibit physical and chemical characteristics that are closer to Lipinski's Rule of Five (Figure 1.5). This may indicate that PROTACs that adhere more closely to Lipinski's Rule of Five may possess better pharmacokinetic properties [2].

1.2.3 Strategies to Improve PROTAC's Bioavailability

1.2.3.1 Improve the PROTAC's Solubility

The solubility of compounds in a pH 7.4 phosphate buffer solution (PBS) is typically measured by conventional solubility studies. However, since drug absorption occurs in the gastrointestinal tract, it is important to consider the solubility of PROTAC in physiological media as well. It has been reported that the solubility of PROTAC molecules significantly improves in fasted-state simulated intestinal fluid (FaSSIF) and fed-state simulating intestinal fluid (FeSSIF), with optimal solubility observed in FeSSIF (Figure 1.6). The clinical trial design of ARV-110 and ARV-471 published by Arvinas disclosed that the phase I clinical administration modes of these two PROTAC molecules

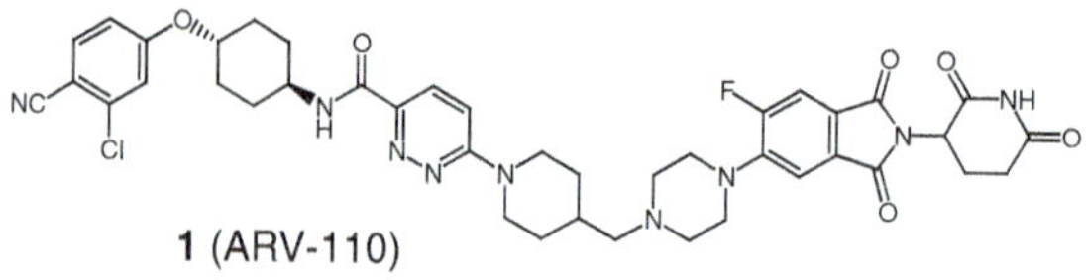

1 (ARV-110)

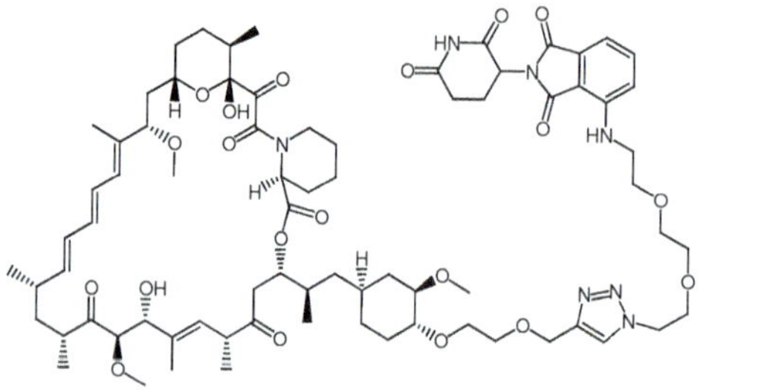

3

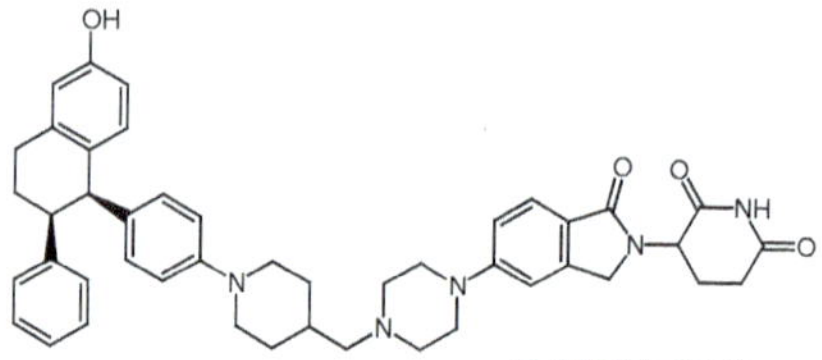

2 (ARV-471)

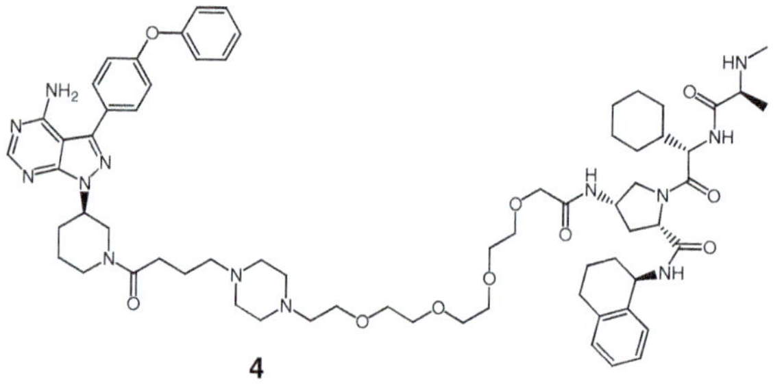

4

	MW (Da)	cLogP	HBD	HBA	TPSA	NRotB
bRo5 outer	1000	10	6	15	250	20
1	812	4.5	2	10	181	11
2	724	6.0	2	5	96	7
3	1426	5.3	4	21	338	23
4	1241	3.5	5	15	262	35

Figure 1.5 Structure and physicochemical properties of ARV-110, ARV-471, and two other PROTAC molecules. *Source:* [2] / with permission of Taylor & Francis Group.

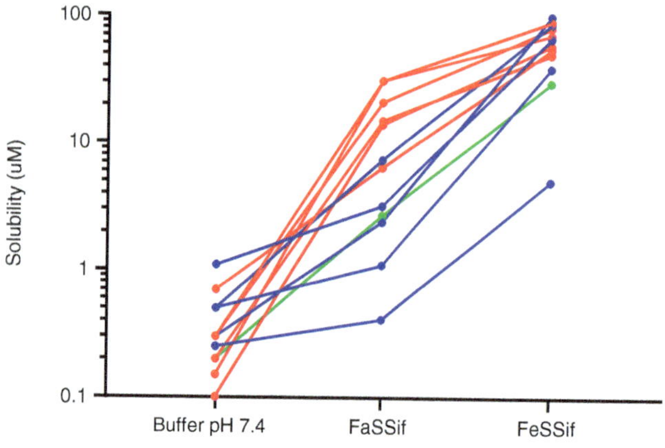

Figure 1.6 Solubility of PROTAC molecules in different buffer solutions. *Source:* [3] / ELSEVIER.

are both "once daily with food." Our study also found that postprandial administration increased the exposure of some PROTAC molecules in animals. This suggests that the *in vivo* pharmacokinetics of PROTAC may improve its *in vivo* drug exposure via postprandial administration.

Another way to solve the solubility problem of the PROTAC is to use optimized formulations. Studies have shown that the *in vivo* exposure can be increased by employing amorphous solid dispersions (such spray drying and hot melt extrusion), nano delivery systems, self-emulsifying delivery methods, or different solvents to make the formulation clear.

1.2.3.2 Improve the PROTAC's Permeability

Permeability is another crucial factor that affects oral bioavailability. On the one hand, drugs need to pass through the membrane barrier of small intestinal epithelial cells for oral absorption. On the other hand, for intracellular protein degradation, PROTACs also need to enter target cells. It has been demonstrated that the cell permeability of PROTACs can be significantly increased by substituting the PEG linker with a 1,4-disubstituted benzene ring [4].

In the preclinical compound screening stage, it is critical to select an appropriate *in vitro* permeability model for PROTAC. Conventional models for *in vitro* permeation research include Caco-2, MDCK, LLC-PK1, and PAMPA. When researching PROTAC molecules, PAMPA is not recommended as a suitable *in vitro* permeability model because of the limitations of its noncellular structure. Caco-2, MDR1-MDCK, or LLC-PK1 cells can be used to evaluate the *in vitro* permeability of PROTAC, whereas PROTAC molecules showed low permeability in most of the models (Figure 1.7a–d).

In addition to low permeability, the low solubility and high nonspecific binding (NSB) would also lead to the insufficient recovery of PROTAC molecules in permeability models. Low recovery makes it hard to determine whether the permeability data are accurate, thus greatly reducing the credibility of the data. We carried out additional validation studies to solve this problem. The recovery of PROTAC molecules in the Caco-2 cell models can be greatly improved by using a transfer buffer with BSA or a physiological medium as the transfer buffer. The optimized Caco-2 cell systems have been recommended for *in vitro* permeability evaluations of PROTAC.

1.2.3.3 Improve the PROTAC's Metabolic Stability

When absorbed by the intestine, the compound undergoes metabolism by the liver or intestine before it enters the circulatory system. This process is known as "first-pass" metabolism, which restricts the oral absorption of many drugs. One strategy to enhance the oral bioavailability of PROTACs is to improve their metabolic stability and reduce the first-pass metabolism. Various approaches have been explored to achieve this, including altering the length of the linker, modifying the anchor point of the linker, employing cyclic linkers, and changing the attachment site of the linker [5].

1.2.3.4 Choose Smaller E3 Ligand

PROTACs' properties are closely related to the type of E3 ligase ligands. The most commonly utilized E3 ligases for PROTAC mainly include CRBN, VHL, cIAP, and MDM2, with CRBN and VHL being the most frequently employed. As shown in Figure 1.8, CRBN has a smaller volume compared to VHL ligands. It has been reported that VHL ligand-containing PROTACs usually have poor oral absorption due to larger molecular weights [2]. Vasanthanathan et al. compared the chemical properties of PROTAC molecules based on different E3 ligases with oral drugs included in Drugbank and found that the PROTAC based on VHL **(green)** is far away from the oral drug region **(blue)** and it has been reported that the bioavailability of such PROTAC is indeed low. In contrast, the PROTAC based on CRBN E3 ligase **(red)** is closer to the oral drug region [2]. The two PROTAC molecules (ARV-110 and ARV-471) that have entered the clinical phase II are with CRBN E3 ligase. Therefore, searching for new E3 ligands with smaller molecular weights is worth exploring.

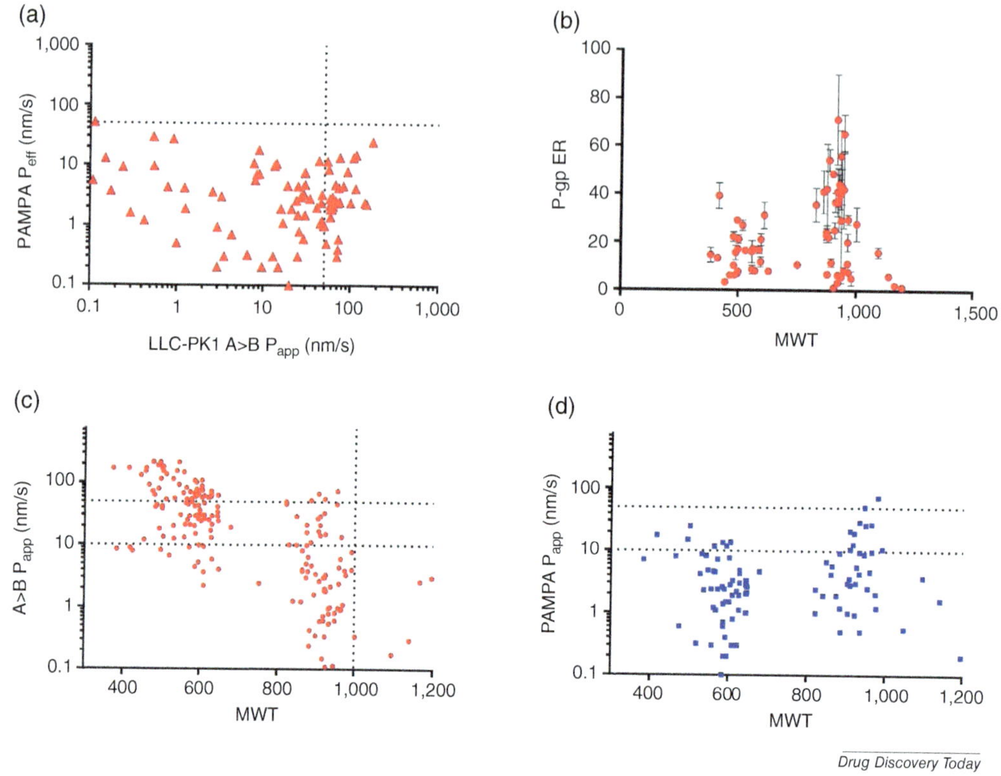

Figure 1.7 Molecular weight and permeability of PROTAC. *Source:* [1] / ELSEVIER. (a) The correlation between PAMPA and LLC-PK1, the area under the dotted line represents low permeability; (b) relations between P-gp efflux ratio and molecular weight; (c) the permeability of compounds with different molecular weights in LLC-PK1 cell model, molecular weight less than 650 is a single ligand, and molecular weight more than 650 is a complete PROTAC; (d) the permeability of compounds with different molecular weights in the PAMPA model, molecular weight less than 650 is an individual ligand, and molecular weight more than 650 is an intact PROTAC.

1.2.3.5 Introduce Intramolecular Hydrogen Bonds

PROTACs usually have high polarity and many rotatable bonds, and these structures generally make it difficult to penetrate the lipid bilayer of the cell membrane. Recent research found that the formation of intramolecular hydrogen bonds can reduce the polar molecular surface area of PROTACs and then improve their permeability. Under the action of intramolecular hydrogen bonds, an original strip-type molecule will be transformed into a "ball" form, which makes it easier to penetrate the lipid bilayer of the cell membrane [6].

1.2.3.6 Select Appropriate Solutions

PROTAC compounds generally have long structures, large molecular weight, high polarity, and a large number of rotatable bonds. Atilaw et al. showed that a PROTAC acts like a chameleon in that its conformation changes with the environment [6]. In solutions that mimic extra- (dimethyl sulfoxide, DMSO) or intracellular (DMSO mixed with water by 10:1), PROTAC molecules present an elongated shape and have a high molecular polarity. However, in the solution that mimics a cell membrane interior (chloroform), PROTAC molecules are folded by forming intramolecular hydrogen bonds and π–π interaction, thus becoming a molecule with a smaller polar surface area (Figure 1.9). In other words, it shifts from the conformation of a polar molecule to the conformation of a nonpolar molecule. This finding suggests that the permeability of PROTAC may be related to whether it can form a conformation with a small polar surface area during the permeation process.

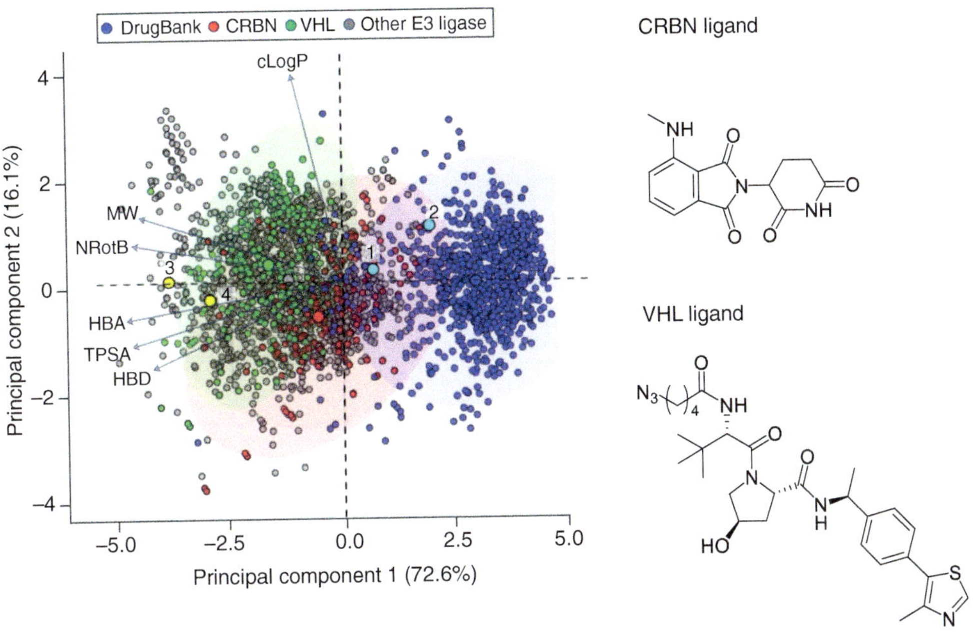

Figure 1.8 Structures of CRBN, VHL ligands, and comparison of chemical properties of PROTAC molecules with oral drugs included in the Drugbank. Blue circles represent oral drugs included in the Drugbank (n = 888), and red, green, and grey circles represent PROTAC of CRBN, VHL, and other E3 ligases (n = 2,082), respectively. *Source:* [2] / with permission of Taylor & Francis Group.

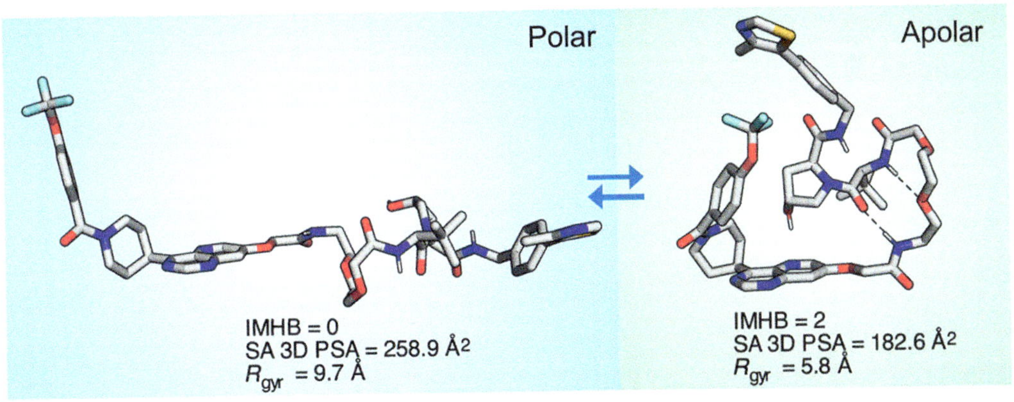

Figure 1.9 Conformation change of PROTAC in different solutions. *Source:* [6] / American Chemical Society.

1.2.3.7 Use Prodrug Strategy

Prodrug is a common approach to improving the oral bioavailability of drugs. A prodrug is obtained by structural modification of pharmacologically active compounds. Prodrugs themselves have little or no activity, and pharmacologically active parent drugs will be released *in vivo* by enzymatic catalysis. Chemists designed a prodrug from a PROTAC by adding a lipophilic group to the CRBN ligand. The results showed that the bioavailability of a PROTAC was significantly increased by prodrug design [7]. To improve the oral bioavailability of different PROTACs with similar E3 ligands, CRBN ligand-based prodrug design can also be used (Figure 1.10). However, increasing the molecular weights of PROTACs is a potential problem when designing prodrugs for PROTACs.

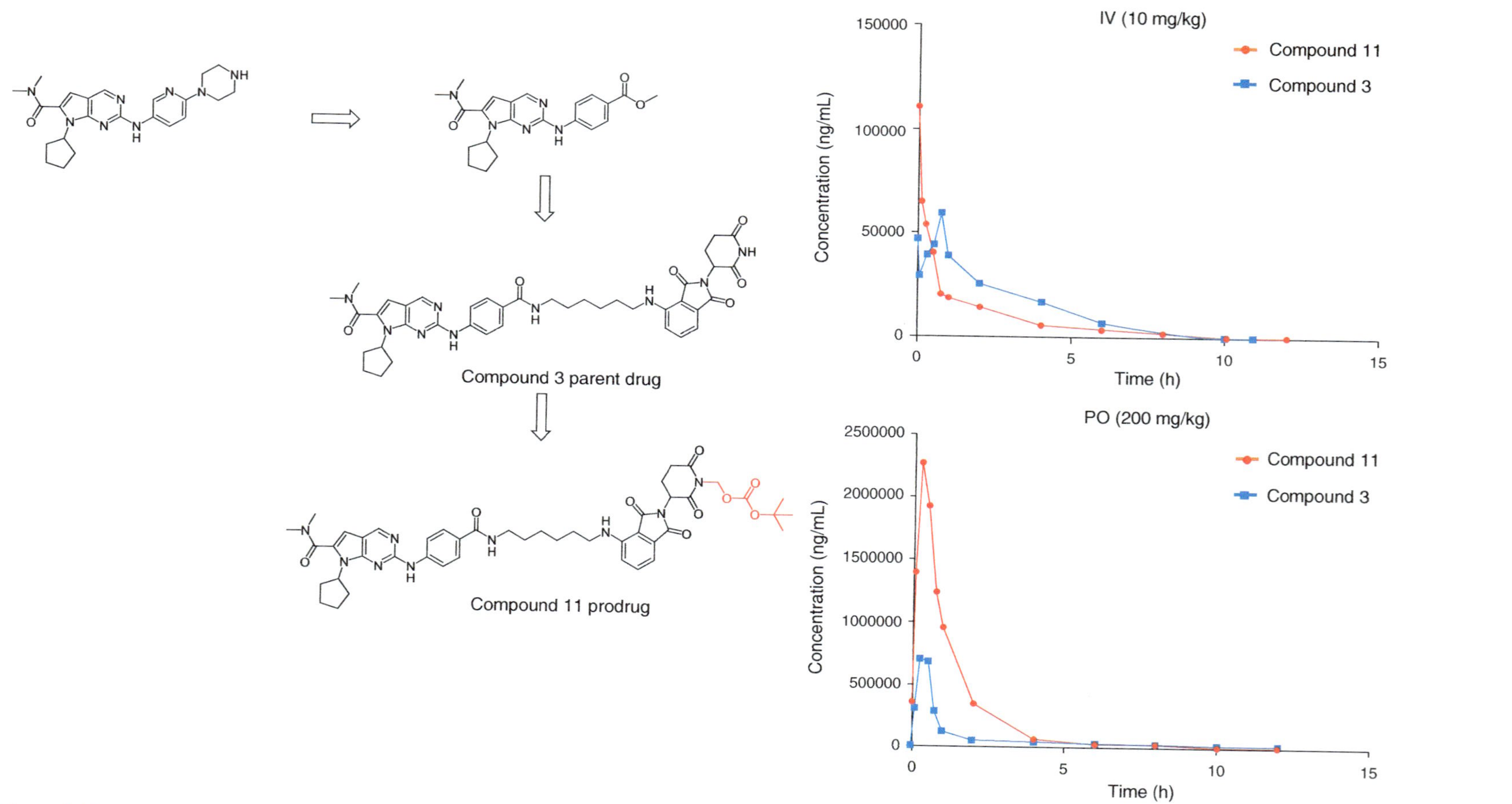

Figure 1.10 Oral bioavailability of compound 3 was significantly improved after its prodrug design as compound 11. *Source:* [7] / ELSEVIER.

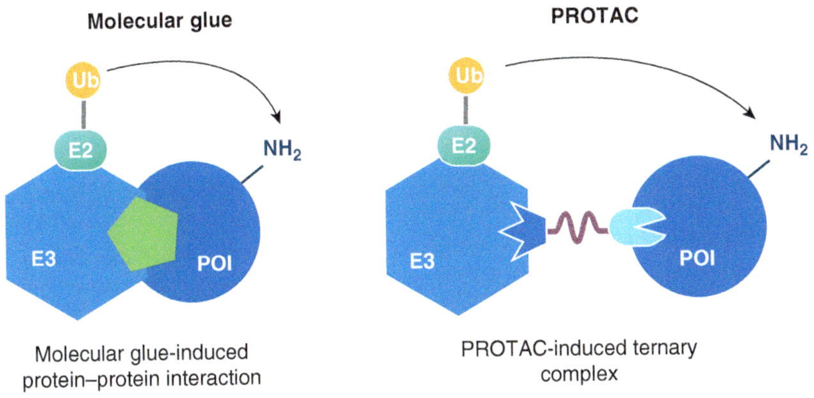

Figure 1.11 Comparison of the mechanism of action between molecular glues and PROTACs.

1.2.3.8 Use Molecular Glues

Because of the large molecular weight of PROTACs, chemists continue to explore new strategies to reduce molecular weight to make PROTACs more "drug-like." Molecular glues are considered to be a more "dense molecule" than PROTACs and they can also trigger a ternary complex like that of PROTACs (Figure 1.11). For these reasons, the molecular glues have better drug-like properties than PROTACs [8].

1.2.4 Summary

The development of PROTAC technology has provided unprecedented therapeutic options for drug development, with the most exciting potential ability to target "undruggable" targets. Despite the industry's optimism toward some late-stage clinical studies, researchers still face challenges in optimizing the pharmacokinetic behavior of PROTACs without compromising their efficacy. By modifying the structure of PROTAC molecules, such as modifying the linker, introducing intramolecular hydrogen bonds, designing as prodrugs, as well as selecting appropriate administration methods, such as administering with meals, the oral bioavailability of PROTAC molecules can be enhanced. It is hoped that academic and industrial researchers in the field of drug development will collaborate to address the challenges in pharmacokinetics faced by PROTACs and unleash their enormous potential.

References

1 Cantrill, C., Chaturvedi, P., Rynn, C. et al. (2020). Fundamental aspects of DMPK optimization of targeted protein degraders. *Drug Discovery Today* 25 (6): 969–982.

2 Poongavanam, V. and Kihlberg, J. (2022). PROTAC cell permeability and oral bioavailability: a journey into uncharted territory. *Future Medicinal Chemistry* 14 (3): 123–126.

3 Pike, A., Williamson, B., Harlfinger, S. et al. (2020). Optimising proteolysis-targeting chimeras (PROTACs) for oral drug delivery: a drug metabolism and pharmacokinetics perspective. *Drug Discovery Today* 25 (10): 1793–1800.

4 Farnaby, W., Koegl, M., Roy, M.J. et al. (2019). Publisher correction: BAF complex vulnerabilities in cancer demonstrated via structure-based PROTAC design. *Nature Chemical Biology* 15 (8): 846.

5 Goracci, L., Desantis, J., Valeri, A. et al. (2020). Understanding the metabolism of proteolysis targeting chimeras (PROTACs): the next step toward pharmaceutical applications. *Journal of Medicinal Chemistry* 63 (20): 11615–11638.

6 Atilaw, Y., Poongavanam, V., Svensson Nilsson, C. et al. (2020). Solution conformations shed light on PROTAC cell permeability. *ACS Medicinal Chemistry Letters* 12 (1): 107–114.

7 Wei, M., Zhao, R., Cao, Y. et al. (2021). First orally bioavailable prodrug of proteolysis targeting chimera (PROTAC) degrades cyclin-dependent kinases 2/4/6 *in vivo*. *European Journal of Medicinal Chemistry* 1 (209): 112903.

8 Maneiro, M., De Vita, E., Conole, D. et al. (2021). PROTACs, molecular glues and bifunctionals from bench to bedside: unlocking the clinical potential of catalytic drugs. *Progress in Medicinal Chemistry* 1 (60): 67–190.

1.3 Research on PROTAC Metabolism: Strategies and Main Approaches

Liping Ma, Chengyuan Li, Jing Jin

1.3.1 Introduction

Drug metabolism refers to the process in which a drug undergoes structural changes under the action of various drug-metabolizing enzymes (primarily in the liver), and this process is also named biotransformation (Figure 1.12). Notably, the drugs are metabolized into non-pharmacologically active metabolites or pharmacologically active metabolites, or even toxic metabolites. The activity or safety concerns arising from these metabolites can significantly impact the success or failure of drug development to some extent.

Despite over two decades of research on PROTAC, there are no PROTAC drugs that have been approved for the market, and little literature on PROTAC metabolism. This chapter summarizes the studies on the metabolism of PROTAC molecules according to the literature and the practical experience to assist researchers in studying PROTAC molecules. The main parts of this chapter are as follows: (i) understanding the key metabolic enzymes of PROTACs, (ii) selecting an appropriate *in vitro* metabolism model, and (iii) facilitating structural optimization through studies on metabolic soft spots.

1.3.2 Understanding the Key Metabolic Enzymes of PROTACs

First, it is critical to identify the enzymes that may participate in the PROTACs' metabolism. The molecular weight of PROTAC compounds is about 600–1,200, which is slightly larger than that of conventional small molecules (the molecular weight of compounds meeting Lipinski's Rule is generally less than 500). Its metabolism may be more complex because the PROTAC is a ternary

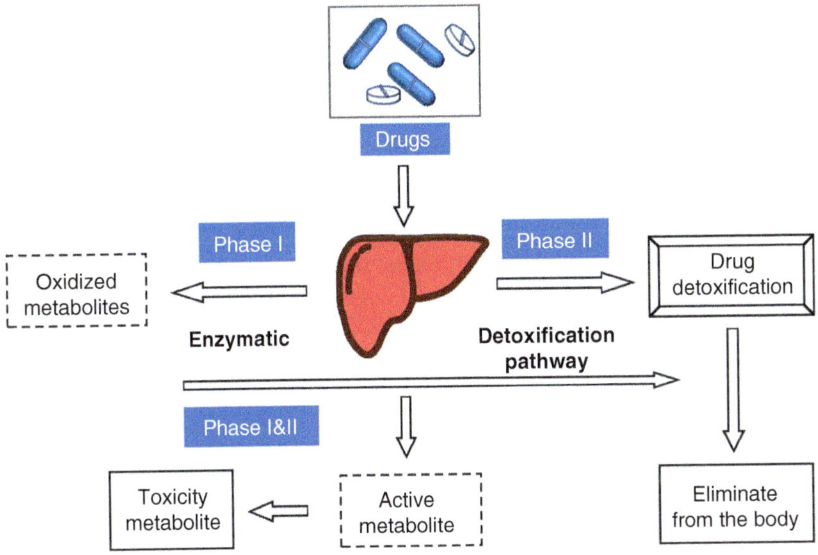

Figure 1.12 General pathway of drug metabolism. *Source:* [1] / Frontiers Media S.A. / CC BY 4.0.

complex made up of two small molecular compounds linked through linkers. Common drug-metabolizing enzymes involved in small-molecule metabolism mainly include phase I metabolizing enzymes (CYP1A2, CYP2B6, CYP2C8, CYP2C9, CYP2C19, CYP2D6, CYP3A, etc.) and phase II metabolizing enzymes (uridine diphosphate (UDP)-glucuronosyltransferases (UGTs), sulfotransferases (SULTs), etc.). The enzyme system of phase I metabolism, represented by cytochrome P450 (CYP450) enzymes, is considered to be involved in the metabolism of about 50% of clinical drugs [2].

Goracci et al. [3] reported that CYP3A4 and human aldehyde oxidase (hAOX) are involved in PROTAC metabolism. CYP3A4 is the most major phase I metabolizing enzyme among the CYP450 enzymes. However, as a cytosolic drug-metabolizing enzyme that is mainly expressed in the liver, aldehyde oxidase (AO) is a non-CYP drug-metabolizing enzyme. It is worth noting that there are marked species differences in the metabolism by AO, with large individual differences in rats, almost no expression in dogs, and the most expression in humans [4]. Therefore, although compounds that are mainly metabolized by AO are metabolically stable in animals, they may be metabolically unstable in humans. As the metabolism by AO is easily neglected in early development, some candidate compounds are terminated due to rapid clearance and low bioavailability in humans. From the perspective of drug screening, the role of AO in the metabolism of PROTACs should be identified as early as possible.

Additionally, *in vitro* plasma or whole blood stability is another important metabolic system of PROTAC screening at an early stage. According to the experimental data, we found that PROTAC molecules have various degrees of stability problems in the plasma or blood. The plasma or blood contains rich hydrolases that can catalyze the hydrolysis of drugs. Compounds with functional groups, such as ester bonds, amide bonds, sulfonamide bonds, and peptide bonds, may be unstable in the blood or plasma. Besides, some PROTACs may be unstable in the linker or linker binding sites.

Overall, the drug-metabolizing enzymes involved in PROTAC metabolism are phase I or phase II metabolizing enzymes involved in the metabolism of conventional small molecules. For the types of metabolizing enzymes, attention needs to be paid to PROTAC metabolism by CYP3A4, aldehyde oxidase, and hydrolase. Aldehyde oxidase, in particular, is not a common metabolizing enzyme. As shown in Figure 1.13c,f, aldehyde oxidase and xanthine oxidase (XO) are only involved in the metabolism of about 3% of the drugs in the market. In addition, further study on identifying particular enzymes that may cause the instability of PROTAC in the plasma or blood needs to be conducted.

1.3.3 Selecting an Appropriate *In Vitro* Metabolic Model

Some metabolic models, including liver microsomes, hepatocytes, and liver S9, are usually used in the *in vitro* metabolism of small molecules and are also applicable to PROTAC molecules (Figure 1.14). The appropriate *in vitro* metabolic model can be selected according to the PROTAC metabolism characteristics. Studies found that the metabolic rates of some PROTAC compounds using liver microsomes or hepatocytes, respectively, may differ significantly. This has also been confirmed in our studies. Liver microsomes mainly contain phase I metabolizing enzymes and their metabolism can better reflect the *in vitro* metabolism when compounds are mainly metabolized by CYP enzymes. However, when there is strong non-phase I metabolism, the hepatocyte system is more appropriate. The mentioned aldehyde oxidase is a drug-metabolizing enzyme in the cytosol, and its content in liver microsomes is low. The hepatocyte contains a membrane barrier and a complete metabolizing enzyme system, so it is an appropriate system for studying the *in vitro* metabolism of PROTAC.

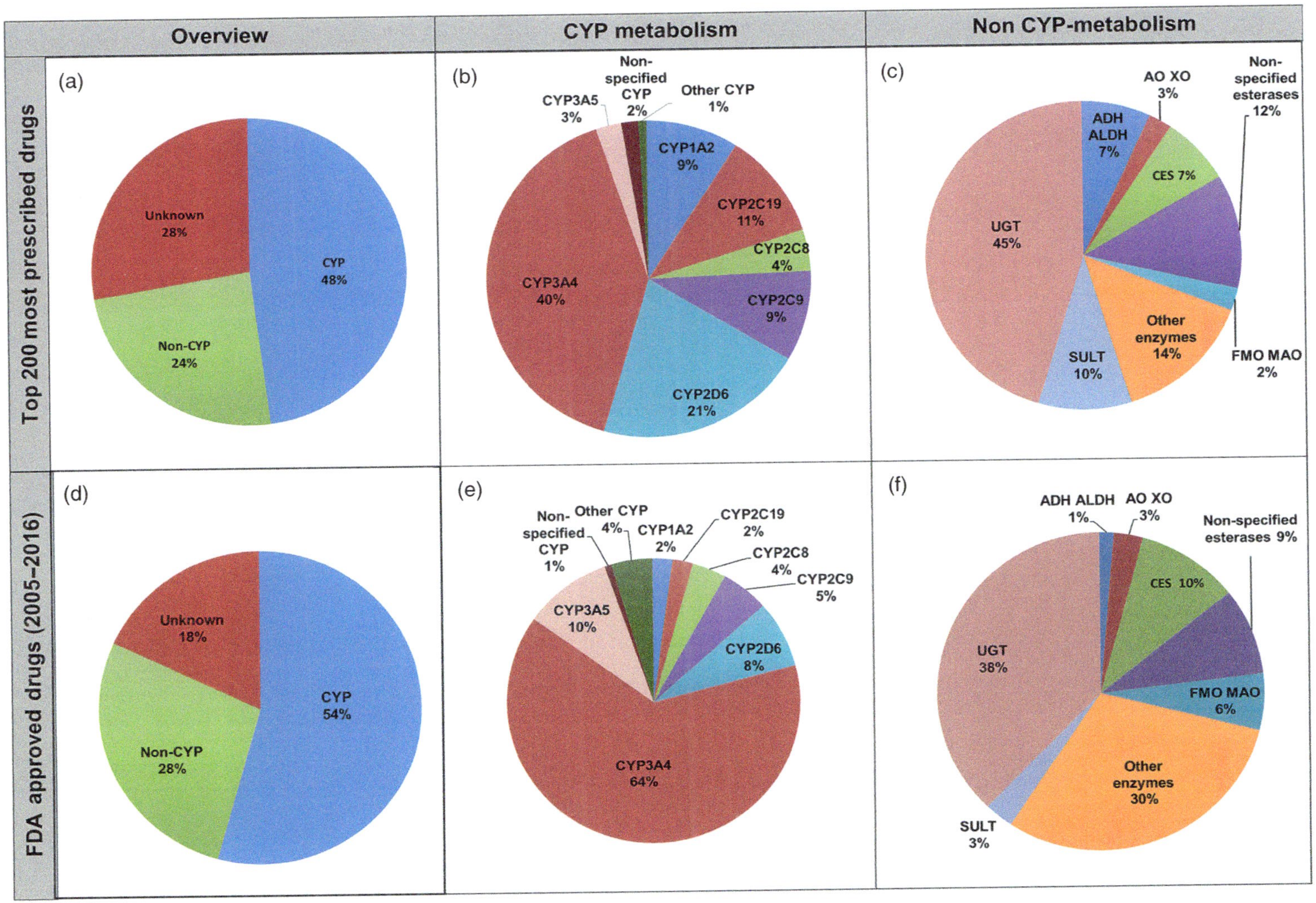

Figure 1.13 Drug metabolism involved in CYP and non-CYP [2]. Note: The contribution of major enzymes (a,d), major cytochrome P450s (CYPs) (b,e), and major non-CYP enzymes (c,f) in the metabolism of "top 200 most prescribed" and "2005–2016" FDA-approved drugs.

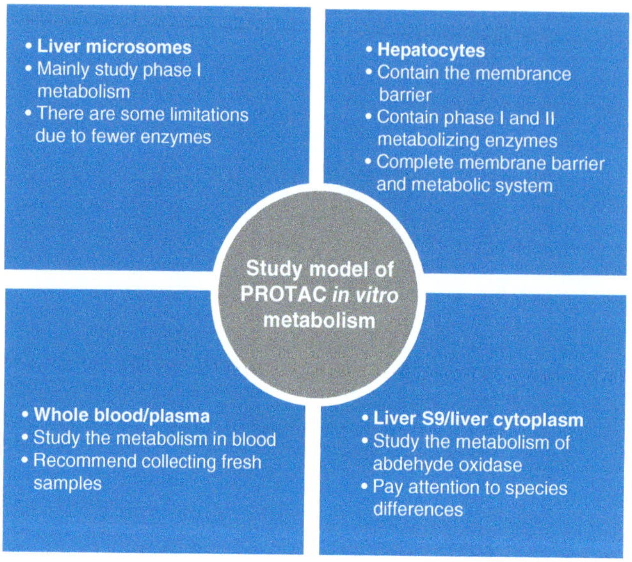

Figure 1.14 Study model of PROTAC *in vitro* metabolism.

It is necessary to study the metabolism of aldehyde oxidase specifically if aldehyde oxidase may be involved in PROTAC metabolism. Because aldehyde oxidase is found in the cytosol, it can be studied via liver S9 or the liver cytosol system. At the same time, the effects on metabolism with or without aldehyde oxidase inhibitors need to be investigated. If aldehyde oxidase participates in metabolism, the metabolic rate with or without aldehyde oxidase inhibitors may be different.

There may be stability problems with PROTAC in the blood or plasma. The most direct assessment method is to investigate the stability of PROTAC compounds in the whole blood/plasma *in vitro*. The conventional incubation time (about two hours at most) could meet the needs of the *in vitro* study of PROTAC molecules. Our study found that the stability data obtained from the whole blood and plasma may be quite different. It is speculated that the enzyme activities of frozen plasma may be different from those of freshly collected whole blood. Therefore, it is recommended to use freshly collected whole blood to study the metabolic stability.

It is worth noting that the protein binding rate of PROTAC molecules is generally high, and the free concentrations in the incubation system need to be fully considered when the interspecies comparison or *in vitro–in vivo* extrapolation (IVIVE) is conducted. It is recommended to conduct IVIVE after the protein binding rate has been accurately determined. It was reported that *in vitro–in vivo* correlation was good by using the *in vitro* hepatocyte data of mice, with the theoretical *in vivo* clearance of most PROTAC molecules within threefold of the actual value (Figure 1.15).

1.3.4 Facilitating Structural Optimization Through Studies on Metabolic Soft Spots

The PROTAC molecule is a ternary complex formed by connecting two ligands via a linker. The cleavage of the linker or the metabolism of the two ligand are both potential metabolic sites. The following conclusions were drawn by Goracci et al. after comparing the metabolic soft spots and metabolic rates of the PROTAC molecule and its ligands [3]:

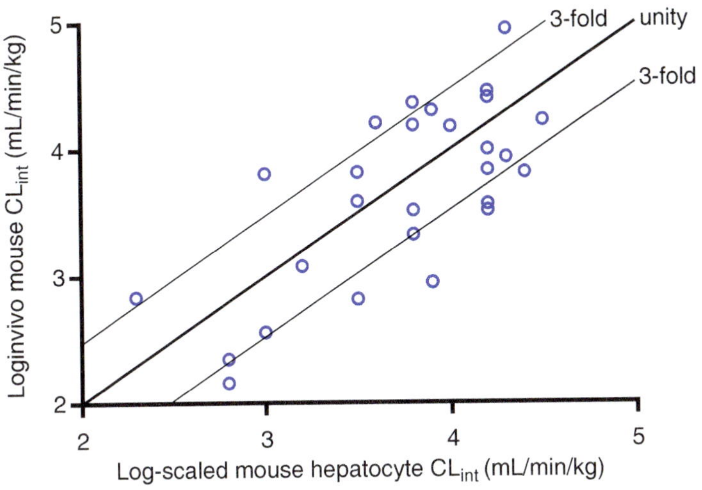

Figure 1.15 Analysis of *in vitro–in vivo* correlation of PROTAC metabolism via hepatocytes. *Source:* [5] / ELSEVIER.

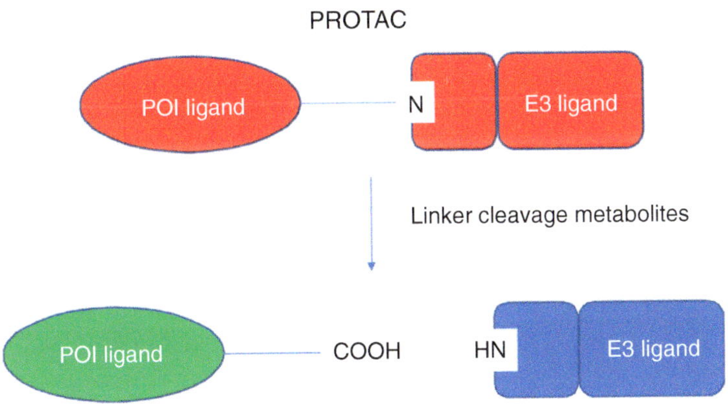

Figure 1.16 Linker cleavage metabolites produced by PROTAC. *Source:* [5] / ELSEVIER.

a) The metabolic soft spot of PROTAC might be completely different from that of ligands.
b) The linker cleavage site of PROTAC is a common metabolic site.
c) The metabolic rate of PROTAC might be completely different from that of ligands.

Therefore, it was difficult to predict the metabolism of PROTAC molecules according to the metabolism of ligands. PROTAC metabolism shall be studied as a whole. Furthermore, the linker cleavage site is a metabolic type of PROTAC that is different from conventional small molecules (Figure 1.16). Linker cleavage can produce two ligand-related metabolites, which may have the following characteristics:

a) The metabolic clearance rates are completely different from that of PROTAC.
b) The protein binding rates are lower than that of PROTAC due to the small molecular weight.
c) Two ligands are intact in their structure so that the target protein or E3 ligase can be bound.

If there are obvious linker cleavage metabolites, the effect of metabolites on drug efficacy needs to be fully considered in the PK/PD study of PROTAC. In addition, possible safety problems associated with metabolites shall be studied as needed.

1.3.5 Summary

PROTAC metabolism can be evaluated and predicted by various *in vitro* metabolic models. At present, it is difficult to identify which models are completely appropriate for PROTAC metabolism. It has been reported that good prediction results can be obtained by IVIVE via hepatocyte metabolic data, but this does not mean that the method is suitable for all PROTAC molecules. During the screening process, it is recommended to summarize the metabolic characteristics of the same series of PROTAC compounds, selectively conduct metabolic studies, and find an *in vitro* metabolic model appropriate for PROTAC molecules to predict *in vivo* clearance or evaluate the early screening of compounds. In view of the study on PROTAC metabolism, we expect more experience to be summarized to gain an in-depth understanding of the metabolic characteristics of PROTAC molecules.

References

1 Wang, D., Liu, W., Shen, Z. et al. (2020). Deep learning-based drug metabolites prediction. *Frontiers in Pharmacology* 30 (10): 501593.

2 Saravanakumar, A., Sadighi, A., Ryu, R. et al. (2019). Physicochemical properties, biotransformation, and transport pathways of established and newly approved medications: a systematic review of the top 200 most prescribed drugs vs. the FDA-approved drugs between 2005 and 2016. *Clinical Pharmacokinetics* 58: 1281–1294.

3 Goracci, L., Desantis, J., Valeri, A. et al. (2020). Understanding the metabolism of proteolysis targeting chimeras (PROTACs): the next step toward pharmaceutical applications. *Journal of Medicinal Chemistry* 63 (20): 11615–11638.

4 Dalvie, D., Xiang, C., Kang, P. et al. (2013). Interspecies variation in the metabolism of zoniporide by aldehyde oxidase. *Xenobiotica* 43 (5): 399–408.

5 Pike, A., Williamson, B., Harlfinger, S. et al. (2020). Optimising proteolysis-targeting chimeras (PROTACs) for oral drug delivery: a drug metabolism and pharmacokinetics perspective. *Drug Discovery Today* 25 (10): 1793–1800.

1.4 Metabolic Characteristics and Research Strategies for Metabolite Identification of PROTACs

Junjie Wang, Hong Zhang, Ruixing Li, Weiqun Cao

1.4.1 Introduction

PROTAC is an emerging class of therapeutic agents that has attracted much attention in academia, industry, and investment in the field of biomedicine because it expands the coverage of druggable targets and has the potential to solve the problem of drug resistance of small-molecule inhibitors. In October 2021, the Health and Environmental Sciences Institute (HESI) brought together more than 150 pharmaceutical scientists, regulators, and academic researchers to deeply explore important issues, such as the safety evaluation of targeted protein degradants [1]. Participants generally agreed that the safety challenges of PROTAC are mainly involved in off-target protein degradation, differences in PK/PD from conventional small-molecule drugs, and the selection of appropriate species for safety assessment in nonclinical trials. Therefore, we urgently need to study the metabolites of PROTAC in depth as early as possible, which is essential for analyzing the distribution of drugs in human circulation, assessing their safety, comparing metabolic differences between different species, and selecting appropriate toxicological species. In this section, we initially discuss the importance of metabolite identification (MetID) at various stages of PROTAC drug development. Subsequently, we summarize the structural and metabolic characteristics of PROTAC. Finally, we outline the challenges encountered during PROTAC MetID research and suggest our corresponding strategies according to our experience. We aim to lay a foundation for developing robust MetID research strategies, help medicinal chemists design PROTAC molecules with favorable pharmacokinetic (PK) attributes, and provide effective direction for the development of PROTAC drugs.

1.4.2 Importance of Metabolite Identification Studies in the Development Process of PROTAC Drugs

Given the relatively large molecular weight of PROTACs, enhancing permeability and solubility often necessitates the incorporation of heteroatoms into the structure during the designing of the drug. However, this approach will escalate the risk of more metabolic soft spots being introduced. Specifically, the linker component of PROTACs can undergo metabolic reactions, such as dealkylation or hydrolysis, leading to the production of cleavage products [2]. These metabolites have the potential to competitively bind with the target of PROTACs, thereby inhibiting the pharmacodynamic effects of these drugs or leading to off-target toxicities. For example, E3 ligase ligand fragments obtained by linker cleavage can bind to nontarget proteins as molecular gels and produce off-target effects [1]. Results obtained from *in vitro* MetID studies can help identify the metabolic soft spots of PROTAC and screen linker-cleaved products. This information further guides the structural optimization of PROTAC to obtain more metabolically stable compounds. In addition, because the metabolic characteristics of PROTAC cannot be predicted from one of the ligands used for their design and synthesis, PROTACs represent independent chemical entities [2].

During the preclinical development phase, conducting *in vitro* MetID studies in different species can help assess the metabolic differences of PROTAC between species and select species suitable for toxicological studies. At the same time, the *in vitro* and the *in vivo* metabolic correlation of PROTAC can be assessed by comparative analysis with the results of *in vivo* MetID studies, and then the

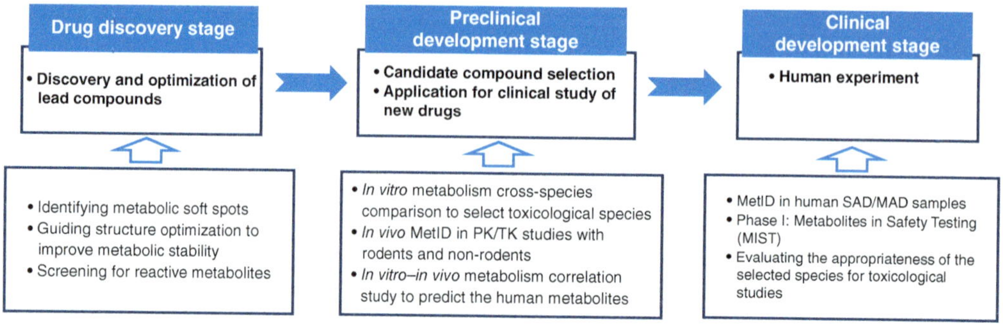

Figure 1.17 The role of metabolite identification studies in various stages of PROTAC drug development.

PROTAC metabolism in humans can be speculated. Additionally, by analyzing the concentration of PROTAC and its metabolites in plasma and tissues such as the liver, kidney, and tumor, we can evaluate the *in vivo* efficacy and toxicity of PROTAC. This is crucial to understand the potential off-target risks associated with PROTAC [3]. In clinical phase I, MetID studies can help analyze the metabolic pathways and clearance of PROTAC in the human bodies and can be used to assess whether the selection of toxicological species for preclinical studies is appropriate (Figure 1.17).

Hence, it is crucial to undertake MetID studies of PROTAC as early as feasible to aid its drug discovery and safety evaluations. The experimental methods and analytical techniques required for PROTAC MetID are more demanding than those for traditional small-molecule drugs. While PROTAC falls under the category of small-molecule compounds, its metabolic characteristics significantly diverge from traditional small-molecule drugs, attributable to its larger molecular weight and complex structural features.

1.4.3 Structural Characteristics of PROTAC Drugs

PROTAC drugs bring significant therapeutic advantages with their unique structure and mechanism of action. However, this particular structure also endows it with unique physicochemical properties as well as different metabolic characteristics from traditional small-molecule drugs. For a deeper understanding of its metabolic profile, first, we will introduce the structural characteristics of PROTAC, providing a base to summarize and investigate its metabolic characteristics. PROTAC primarily consists of three critical structures: one region is the ligand structure that targets the protein of interest (POI); another region is the ligand structure for E3 ligase; and situated between these two is the linker, which connects both ligands (Figure 1.18). In the following sections, we will individually introduce these three components based on their structural characteristics.

1.4.3.1 POI Ligand

Among the reported PROTAC molecules as of 2024, the number of POI ligands exceeds 360 [4]. The target proteins include androgen receptor (AR), estrogen receptor (ER), bromodomain-containing protein (BRD), and Bruton tyrosine kinase (BTK). The POI ligand component of PROTAC typically utilizes active inhibitors that are either marketed or reported in scholarly literature [3]. For instance, Gray's group, in 2019, developed BSJ-03-204 [5], deriving from the CDK4/6 inhibitor Palbociclib. In 2021, Jinlei's team developed B03 [6], which is based on the CDK9 inhibitor BAY-1143572.

PROTAC drugs target not only conventional small-molecule drugs but also extend to non-druggable proteins, scaffold function proteins, protein polymers, drug-resistant mutant proteins,

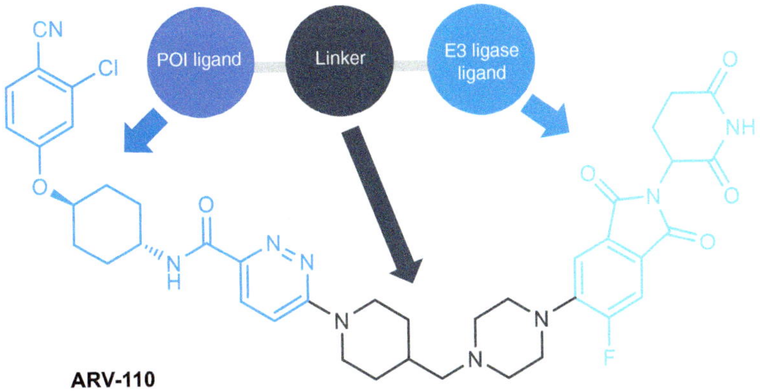

ARV-110

Figure 1.18 Structural diagram of PROTAC and representative structure of PROTAC: ARV-110.

Figure 1.19 STAT3 degradant SD-36.

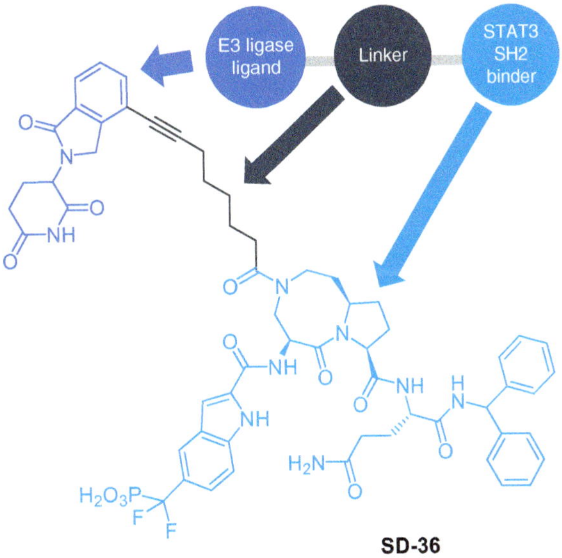

SD-36

and specific isoform proteins. For example, signal transducer and activator of transcription 3 (STAT3) is a transcription factor closely related to the development of cancer and is considered to be a non-druggable protein. Based on SI-109, a binder of the SH2 domain in STAT3, Wang's group [7] developed the PROTAC molecule SD-36 which has high degradation activity for STAT3 protein in both *in vitro* and *in vivo* experiments (Figure 1.19). KT-333, also a PROTAC drug targeting STAT3, developed by Kymera, has entered phase I clinical trials to assess its safety in patients with large granular lymphocytic leukemia.

1.4.3.2 E3 Ligand

There are over 600 known types of E3 ligases encoded by the human genome. However, in PROTAC-related drug design, the primary E3 ubiquitin ligases include cereblon (CRBN, with 71 subtypes), von Hippel-Lindau (VHL, with 47 subtypes), an inhibitor of apoptosis protein (cIAP, with 7 subtypes), and murine double minute gene (MDM2, with 4 subtypes) (Figure 1.20) [3].

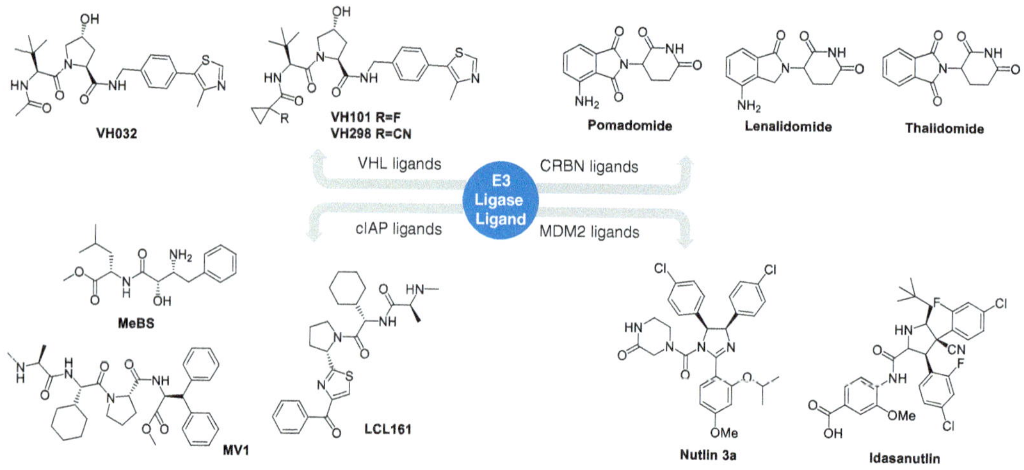

Figure 1.20 Representative E3 ligands.

Among PROTACs which have entered the clinical stage and published structures, more than 39% of the molecules have CRBN ligands, and only DT2216 contains VHL-containing ligand.

Investigating novel E3 ubiquitin ligase targets is a prominent area of interest in PROTAC research. In recent years, several structurally unique E3 ligases have been discovered, including DCAF11, DCAF15, DCAF16, RNF114, RNF4, AhR, FEM1B, and KEAP1. Additionally, over 80 novel E3 ligase ligands have been reported. Moving forward, it is anticipated that researchers will identify a more diverse range of E3 ligases, as well as E3 ligase ligands with smaller molecular weight and higher activity.

1.4.3.3 Linker

The linker, an important part, connects the two active groups of PROTAC drugs, playing an essential role in the activity, metabolic stability, and kinetic properties of PROTAC molecules. Over 1,500 different linker structures have been reported [4], and from a molecular variability perspective, linkers can be divided into two types: flexible and rigid (Figure 1.21). Relevant literature [2]

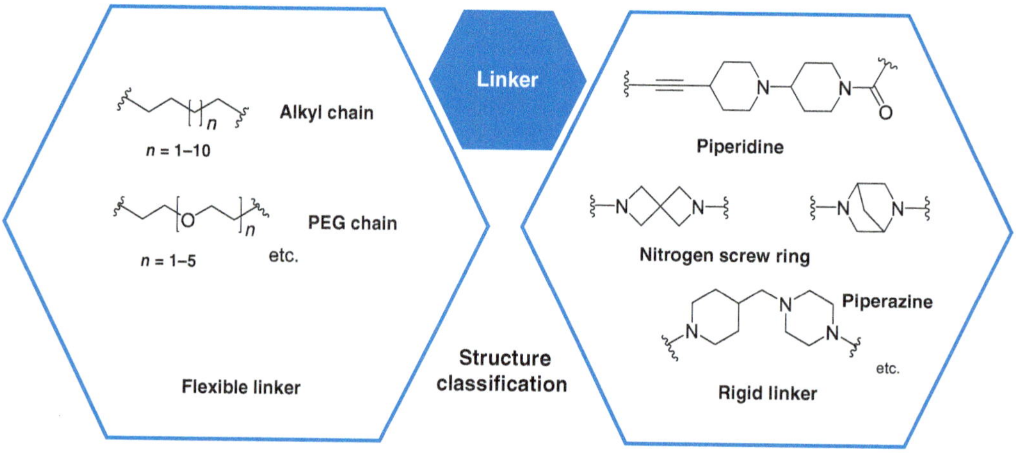

Figure 1.21 Representative structures of linkers used in PROTACs.

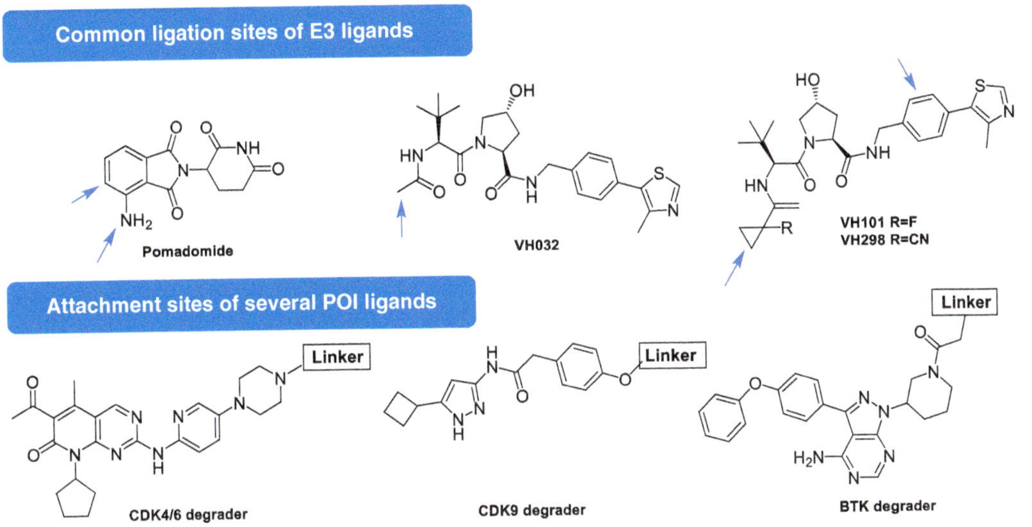

Figure 1.22 Common attachment sites of linkers with E3 and POI ligands.

indicates that neither short nor long linkers are beneficial for the efficacy of PROTAC. If the linker is too short, it can impede the binding of ligands on both sides to the corresponding protein due to steric hindrance, making the formation of the ternary complex difficult. Conversely, when the linker is too long, the E3 ligase cannot sufficiently approach the target protein, preventing the ubiquitination process from occurring, and the excessive molecular weight reduces PROTAC's membrane permeability. Studies suggest that linkers possessing rigid elements (e.g., a piperidine moiety) or polar groups (e.g., polyethylene glycol (PEG)) can enhance the pharmacokinetic properties of PROTAC.

Besides the structure of the linker, its attachment site can also influence the selectivity and activity of PROTAC. Generally, the principles for selecting ligation sites are as follows: (i) do not diminish the binding capacity between E3 ligase ligands or POI ligands and their receptors; (ii) select the solvent-exposed region of the ligand binding pocket; (iii) ensure the integrity of POI ligands as much as possible when attaching the linker to prevent any impact on their binding capacity. Figure 1.22 illustrates the common ligation sites of E3 ligase ligands (indicated by the red arrow section) as well as the ligation sites of several POI ligands (3).

1.4.4 Metabolic Characteristics of PROTAC Drugs

Due to its unique structural characteristics, the molecular weight and physicochemical properties of PROTAC often exceed the limitations of Lipinski's Rule of Five, and its metabolic process is also significantly different from that of traditional small-molecule drugs. An in-depth understanding of their metabolic properties will help researchers to more effectively cope with the challenges encountered in the study of their metabolites. Therefore, this part will thoroughly summarize the metabolic characteristics of PROTAC drugs based on their structural features.

1.4.4.1 The Influence of PROTAC's E3 Ligand on PROTAC Metabolism

Cruciani's team [2] discovered, after comparing three datasets, that for PROTACs targeting BET, CK2, and PARP proteins, when the POI ligands remain constant and linkers are similar, the

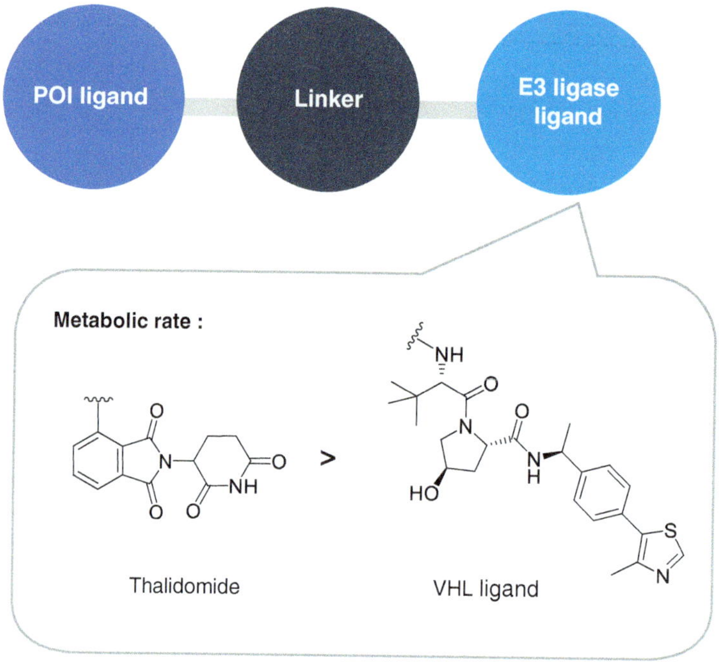

Figure 1.23 Effect of E3 ligands on the metabolic stability of PROTACs.

metabolic stability of PROTACs with thalidomide as the E3 ligase ligand is significantly lower than those with VHL ligand (as depicted in Figure 1.23). The team hypothesizes that the plausible reason could be that thalidomide and its analogs contain multiple amide structures, which might undergo non-enzyme-catalyzed hydrolysis reactions. Another reason is that due to the larger molecular weight of the VHL ligand, the corresponding PROTAC has reduced permeability [3], thereby not providing an accurate estimation of its half-life. Therefore, for metabolic stability studies of PROTACs containing VHL ligands, liver microsomes may be a more appropriate *in vitro* metabolic study system [8].

1.4.4.2 The Influence of PROTAC's POI Ligand on PROTAC Metabolism

Cruciani's team [2] discovered that the metabolic stability of PROTAC molecules changes as the POI ligand varies, keeping the linker and E3 ligase ligand constant. Figure 1.24 illustrates that for PROTACs using the AR receptor antagonist as the POI ligand, their metabolic stability is lower compared to other POI ligands, regardless of changes in the linker and E3 ligase ligand. MetID studies identified that the primary soft metabolic spots of these compounds are on the POI ligand, not on the E3 ligase ligand. Given the fast metabolic rate of the POI ligand itself (a half-life of 18.3 minutes), this suggests that the metabolic stability of the POI ligand needs to be considered when designing PROTAC drugs to avoid significant weaknesses. For example, the PROTAC drug ARV-110 uses an AR receptor antagonist with better metabolic stability than the POI ligand, boasting a half-life of up to 110 hours.

1.4.4.3 The Influence of Linker on PROTAC Metabolism

In PROTAC molecules, the linker is the part that is most likely to be metabolized. The main metabolic sites are where the linker attaches to the ligand. Variations in the linker's length, attachment

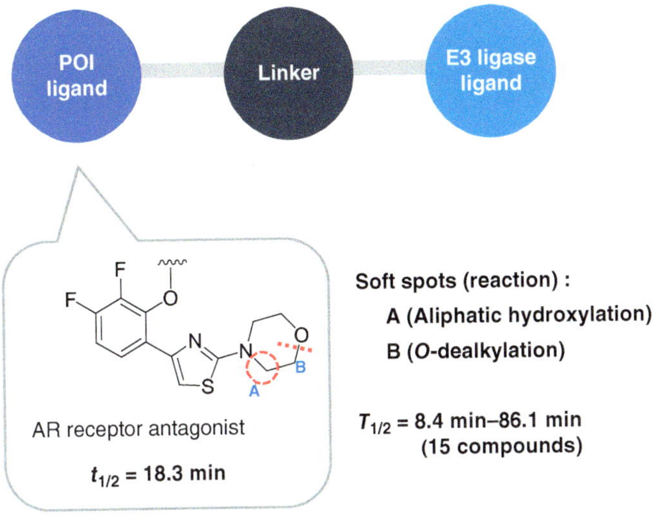

Figure 1.24 Effect of POI ligands on metabolic stability.

Figure 1.25 Effect of linker length on the metabolic stability of PROTACs.

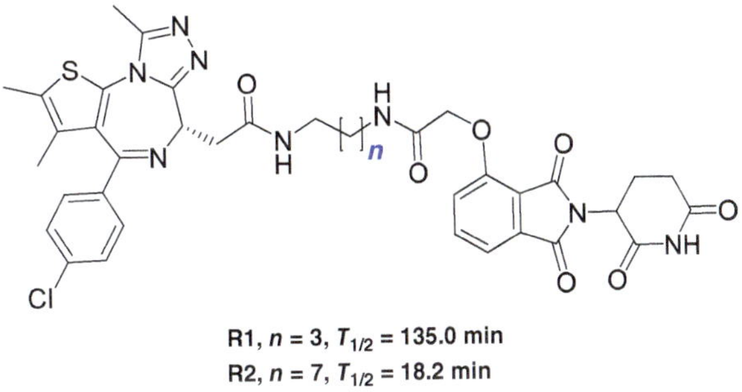

points, and the degree of rigidity or flexibility can significantly influence the overall metabolic stability of the molecule.

1.4.4.3.1 Effect of Linker's Length and Ligation Site on the Metabolic Stability of PROTAC

In the majority of PROTACs, the metabolic stability of the molecule tends to diminish as the linker length extends. As depicted in Figure 1.25, R1 represents a PROTAC, ingeniously designed using JQ1 (a BET inhibitor) and Thalidomide. When the straight-chain alkyl linker in R1 is elongated from four methylene units to eight, forming R2, the half-life dramatically contracts from 135 minutes to just 18.2 minutes. This phenomenon can likely be attributed to the increased steric hindrance of the shorter linker, which effectively prevents the PROTAC from entering the catalytic site of the metabolic enzyme.

It has also been found that the changes in the ligation site of linker and E3 ligase ligand may also affect the metabolic stability of PROTAC, as shown in Figure 1.26.

1.4.4.3.2 Effect of Linker Rigid Flexibility on the Metabolic Stability of PROTAC

Studies have shown that when both the POI ligand and the E3 ligase ligand are kept constant, the assembly of the PROTAC molecule with different linkers can significantly impact its metabolic stability (as depicted in Figure 1.27). Specifically, PROTACs featuring piperidine or triazole ring

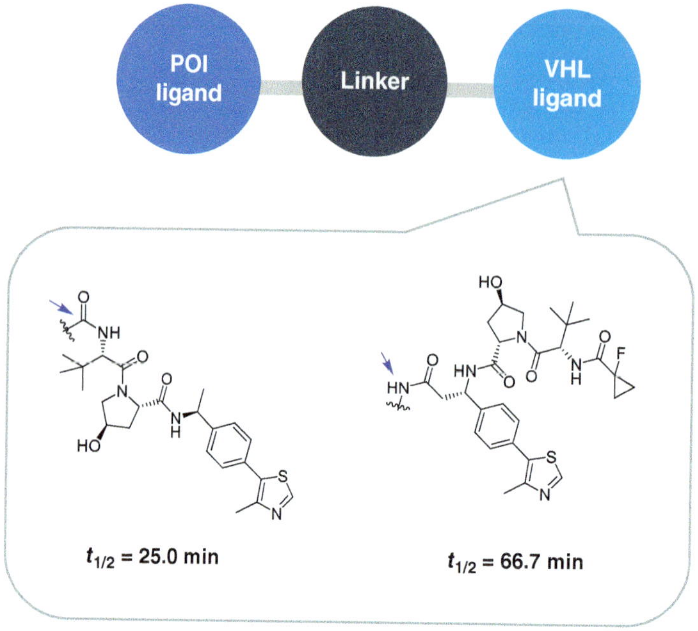

Figure 1.26 Impact of the linker attachment site on the metabolic stability of PROTACs.

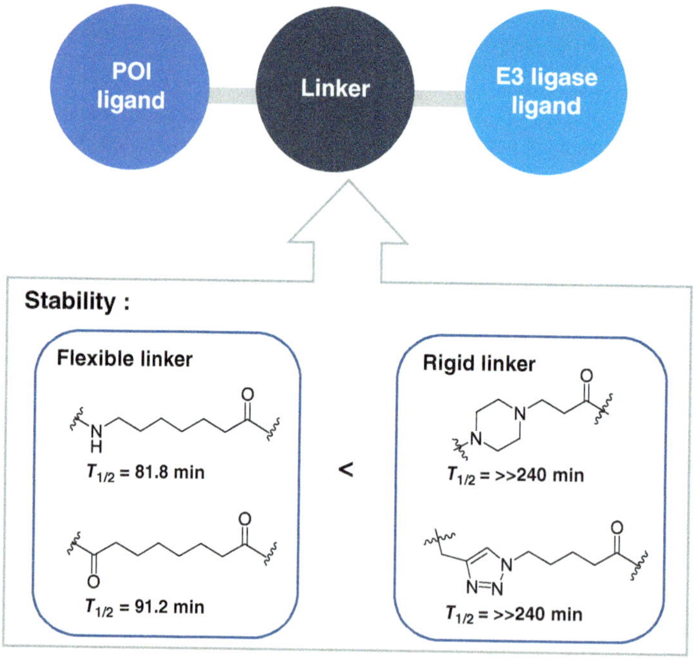

Figure 1.27 Effect of linker rigidity on metabolic stability.

linkers demonstrate a prolonged half-life compared to those with straightforward linear linkers. This evidence suggests that PROTACs constructed with rigid linkers tend to exhibit enhanced metabolic stability.

Incorporating rigid structures into the linker presents a superior approach to bolster metabolic stability. For instance, in the development process of ARV-110 (as depicted in Figure 1.28), the

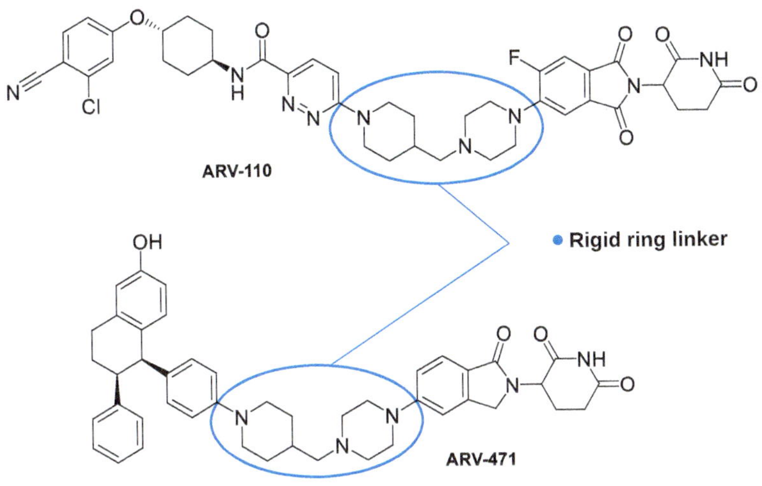

Figure 1.28 Structure of ARV-110 and ARV-471 with rigid cyclic linker.

initial flexible chain-like linker was refined into a stiff cyclic linker, bridged by piperidine and piperazine rings. This optimization notably enhanced its metabolic stability and therapeutic efficacy. Arvinas, the company that developed ARV-110, implemented the same linker design in another PROTAC drug, ARV-471 [2].

In 2021, Wang Yonghui's research team at Fudan University [9] unveiled the PROTAC molecule 6e. Despite its high degradation activity against BTK protein, 6e demonstrated relatively low metabolic stability, with a half-life of just 1.3 minutes in mouse liver microsomes. However, in 2023, the team significantly enhanced the metabolic stability of the molecule by implementing a strategy to increase the rigidity of its linker. Through conducting structure–activity relationship studies on a series of rigid linkers and CRBN ligands, they replaced the linear PEG linker with a more rigid one, featuring two piperidine rings. This modification led to the creation of molecule 3e, which not only displayed a greatly improved metabolic stability, with a half-life exceeding 145 minutes in mouse liver microsomes, but also outperformed its predecessor, 6e, in terms of activity (as depicted in Figure 1.29).

1.4.4.4 Relationship Between the Metabolic Characteristics of Various PROTAC Components and the Whole Molecule

Considering that a PROTAC molecule is composed of three crucial structures, one might wonder if it is possible to predict the metabolic characteristics of PROTAC based on the metabolic traits of each ligand. To address this, Cruciani's team [2] embarked on a comprehensive study. As demonstrated in Figure 1.30, the POI ligand and CRBN ligand individually undergo metabolic reactions of alkyl hydroxylation and amide hydrolysis, respectively. However, when these ligands are conjugated via a linker into a PROTAC molecule, neither ligand exhibits the respective metabolism, with the primary metabolic activity occurring on the linker itself. Moreover, the metabolic rate of the entire PROTAC molecule varies from that of its separate ligands. Thus, as a novel chemical entity, PROTAC's metabolic characteristics cannot be extrapolated merely from the metabolic attributes of its constituent parts, underscoring the need for a holistic investigation of its metabolites.

1.4.4.5 Summary of Metabolic Characteristics of PROTACs

From the above analysis, it can be seen that the type of E3 ligase, the stability of the POI ligand, and the length, ligation site, and rigidity of the linker all affect the metabolism of PROTAC. The

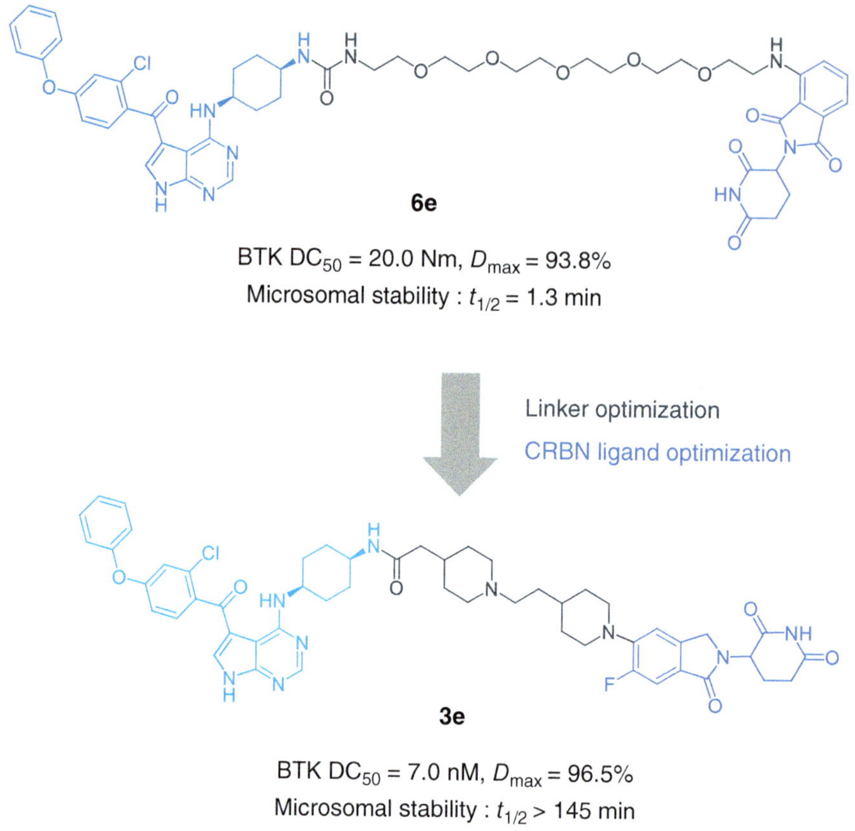

Figure 1.29 Study case: enhancing PROTAC metabolic stability and activity through structural optimization.

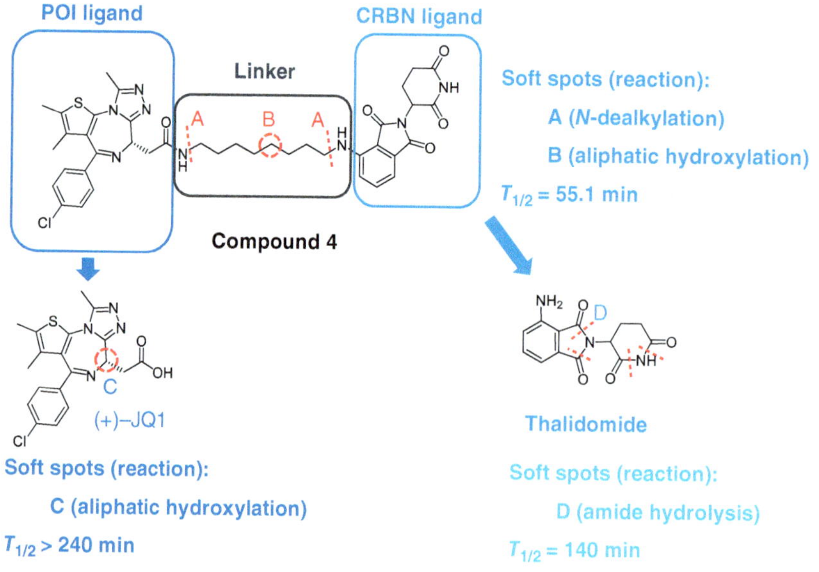

Figure 1.30 Comparison of PROTAC's metabolic soft spots with those of its individual ligands.

metabolic characteristics of PROTAC molecules are different from the sum of metabolic characteristics of their various partitions, and the metabolic rate may also be completely different.

Metabolic pathways of PROTAC include hydroxylation of its ligand's moiety, amide hydrolysis, and *O*-dealkylation and hydroxylation of the linker's moiety, *N*-dealkylation, and amide hydrolysis. If the drug contains a PEG-like linker, there are a large number of *O*-dealkylation reactions. Enzymes involved in the metabolism of PROTAC mainly include phase I metabolizing enzymes (CYP enzymes) and common phase II metabolic enzymes (UGTs, SULTs, etc.). In addition, Cruciani's team [2] found that for PROTACs containing VHL ligands, aldehyde oxidase (hAOX) is involved in metabolism and catalyzes the hydroxylation of its thiazole ring. Since aldehyde oxidase is more expressed in humans, metabolic studies targeting this enzyme need to be paid attention to.

1.4.5 Challenges and Strategies for PROTAC Metabolite Identification Studies

The process of identifying metabolites for a PROTAC molecule aligns closely with that associated with traditional small-molecule drugs (Figure 1.31). Initially, an appropriate metabolic study

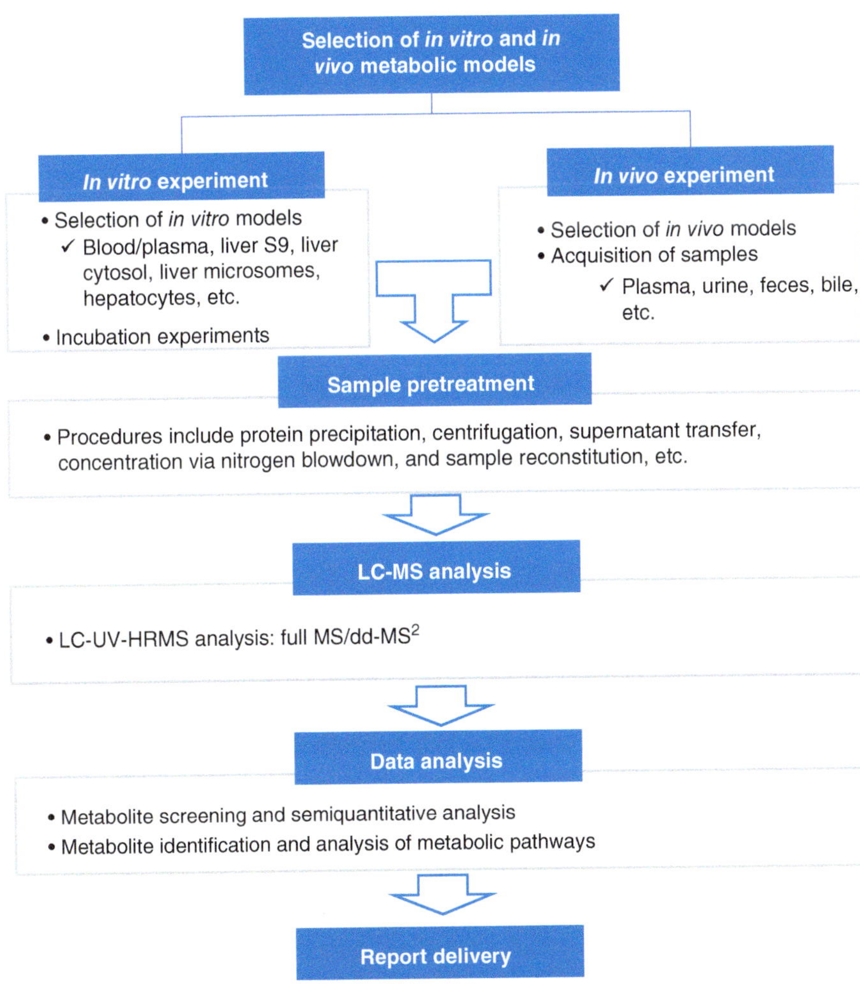

Figure 1.31 Workflow for metabolite identification of PROTACs.

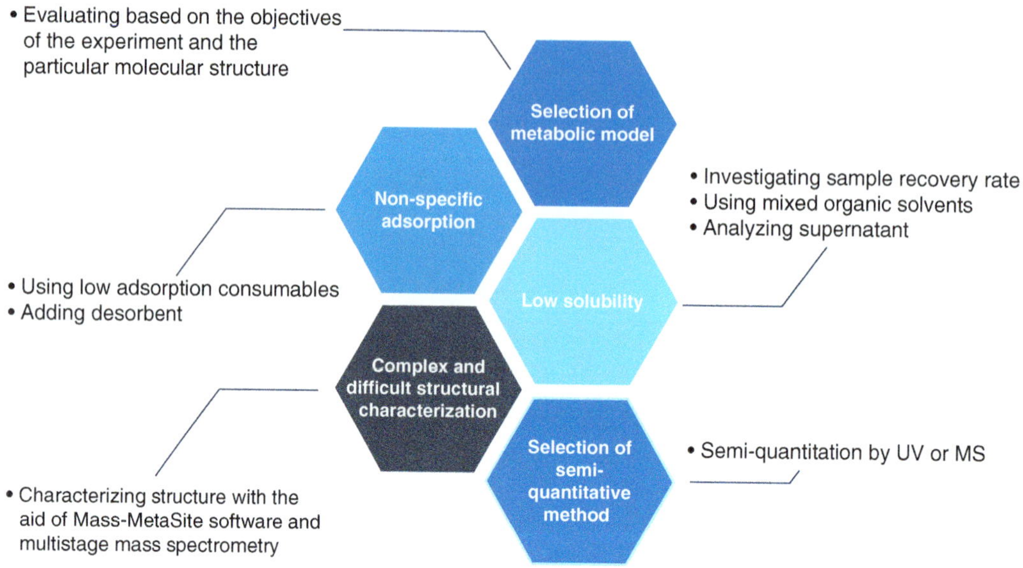

Figure 1.32 Diagram of challenges and strategic solutions for PROTAC metabolite identification studies.

system is selected for either *in vitro* incubation or *in vivo* administration in animals. Following this, samples, either from incubation or collected *in vivo*, undergo a process of liquid chromatography-ultraviolet-high-resolution mass spectrometry (LC-UV-HRMS) analysis after protein precipitation, centrifugation, supernatant transfer, nitrogen blowing, concentration, and redissolution. The parent drug and its metabolites are separated through chromatography, and a comparison of the information between the drug substance and its metabolites based on the MS and MS^2 signals from high-resolution mass spectrometry, the metabolic pathways, and reaction sites of PROTAC is analyzed.

However, due to the distinct structural and metabolic characteristics of PROTAC drugs compared to traditional small molecules, the study of their metabolites presents additional challenges. These include the selection of appropriate *in vitro* and *in vivo* models, addressing issues of nonspecific adsorption and solubility of parent drugs during sample preparation, and efficiently acquiring the relative abundance of metabolites. Furthermore, in the data analysis phase, accurately completing the structural characterization also poses a challenge.

Drawing from the extensive research expertise of the WuXi AppTec DMPK (drug metabolism and pharmacokinetics) MetID team, we have devised a comprehensive research strategy to effectively address the challenges encountered during PROTAC MetID studies (Figure 1.32). Our approach aims to deliver high-quality research reports to our clients, thus expediting the drug development process for PROTAC molecules. We delve into these strategies in greater detail in the following sections.

1.4.5.1 Selection of Metabolic Models

The selection of an appropriate metabolic model is key for successful *in vitro* and *in vivo* metabolite research. For the study of PROTAC molecules, from which dimensions should appropriate metabolic models be selected?

The selection of an appropriate *in vitro* metabolic system for studying PROTAC molecules depends on the experiment's objective and the molecule's structural characteristics (Table 1.3). If

Table 1.3 The *in vitro* and *in vivo* systems for PROTAC metabolism and their respective purposes.

Study subjects	Matrix	Purpose
PROTAC	*In vitro* blood/plasma	Investigating hydrolases-mediated metabolism in blood/plasma
	Liver S9/liver cytosol (with or without aldehyde oxidase inhibitor)	Investigating aldehyde oxidase-mediated metabolism
	Liver microsomes	Investigating CYP enzyme-mediated metabolism
	Hepatocytes	Investigating both phase I and II enzyme-mediated metabolism
	In vivo samples (plasma, blood, urine, feces, and bile)	Investigating the correlation between *in vitro* and *in vivo* metabolism to predict metabolic processes in the human body
	In vivo samples obtained from the human body	Evaluating the exposure of PROTACs in the human body and the appropriateness of species selection for toxicology studies

the study involves the metabolic stability of PROTAC under the action of blood hydrolytic enzymes, *in vitro* plasma or incubation can be utilized. Liver S9 or liver cytosol is selected as the metabolic system when aldehyde oxidase is involved in metabolism. This is particularly critical for PROTACs containing a VHL ligand, as its thiazole ring may be hydroxylated under aldehyde oxidase catalysis [2]. Hepatocyte systems are used to study the effects of CYP enzymes and non-phase I metabolizing enzymes on compounds. For larger molecular weight PROTAC molecules with poor cell permeability, the liver microsomal system is more suitable than the hepatocyte system for studying reactions involving only CYP enzymes [8].

For *in vitro–in vivo* metabolism correlation, plasma, urine, feces, bile, and other *in vivo* samples can be used to evaluate the correlation between *in vitro* and *in vivo* metabolism of PROTAC, helping predict human metabolism. MetID studies of human *in vivo* samples can assess the intra-body exposure scenarios of PROTAC and determine the appropriateness of the selected toxicological species.

1.4.5.2 Method Development

During the method development of PROTAC molecules, nonspecific adsorption and solubility issues are common challenges. Improper handling of these issues can significantly impact the comprehensiveness and accuracy of metabolite analysis results. Hence, it is crucial to develop methods and strategies in advance to effectively address these challenges.

For nonspecific adsorption issues, the large molecular weight of PROTAC can result in easy NSB to the surface of consumables [10]. To mitigate this, it is recommended to use low-adsorption consumables and assess compound adsorption during the method development stage. If a compound is found to bind nonspecifically to consumables, the addition of an appropriate amount of desorbent may be required. For further information on this topic, refer to Section 1.6 PROTAC Bioanalysis: Challenges and Strategies.

To address solubility issues, the solubility can be first evaluated by investigating the recovery of PROTAC and selecting the suitable solvent. If the recovery rate is insufficient, it may result in

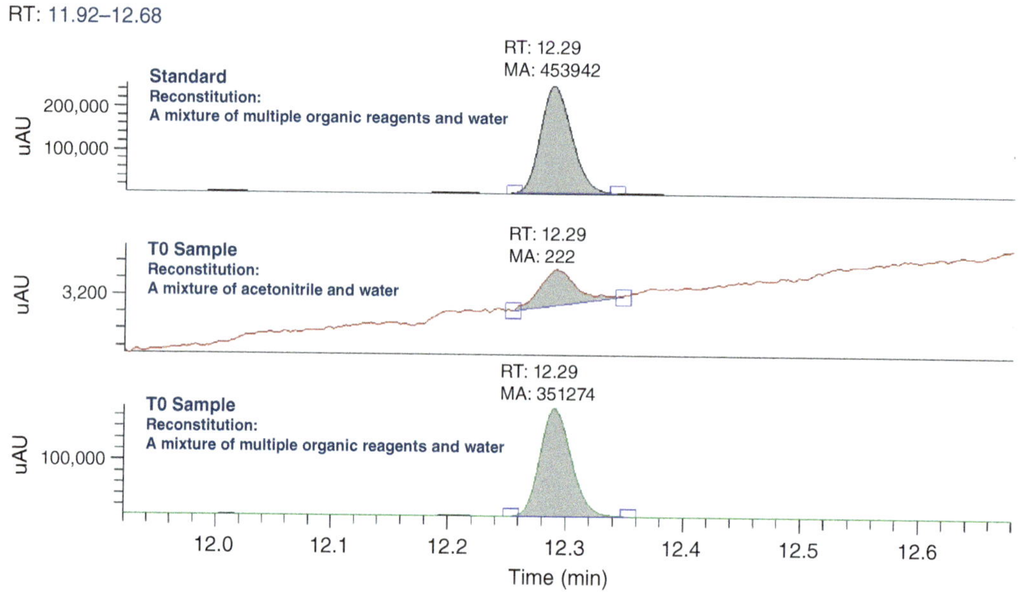

Figure 1.33 LC-UV chromatograms of a PROTAC in different reconstitution solvents.

missed metabolites, inaccurate semiquantitative results, and ultimately unreliable MetID. For instance, a PROTAC molecule in a mixed acetonitrile/water solvent may show low recovery at 0 minutes of hepatocyte incubation (T0 sample) and cause system stratification. Adjusting the proportion of acetonitrile might not improve this situation, suggesting the need for other organic reagents in the reconstitution solution. In one case, the highest recovery rate was achieved when the reconstitution solution was a mixed solvent made up of several organic reagents and water, fulfilling the analytical requirements (as depicted in Figure 1.33).

To solve PROTAC's solubility problem, strategies may include using solutions containing mixed organic reagents for compound dissolution, dilution, and pretreatment, or directly using the supernatant after protein precipitation and centrifugation for LC-UV-HRMS analysis. By adopting these strategies, the problems of nonspecific adsorption and solubility of PROTAC samples can be effectively overcome to ensure the comprehensiveness and accuracy of metabolite analysis and identification results.

1.4.5.3 Structural Characterization of Metabolites

Structural characterization is a critical step in MetID. For PROTAC molecules, their large molecular weight and the presence of multiple reaction sites pose significant challenges for efficient and high-quality structural characterization, a crucial aspect of metabolite research. To solve these challenges, we've developed a set of strategies specifically for PROTAC metabolite profiling and structure identification (Figure 1.34). These strategies, derived from extensive experimental explorations and experiences, leverage the strengths of the MetID software Mass-MetaSite, the expertise of senior scientists, and high-resolution mass spectrometry platforms.

In this strategy, our experienced professionals characterize metabolite structures by interpreting the UV and mass spectral information from LC-UV-HRMS analysis data. This includes elucidating the metabolic pathways and metabolic sites of PROTAC metabolites through chromatographic

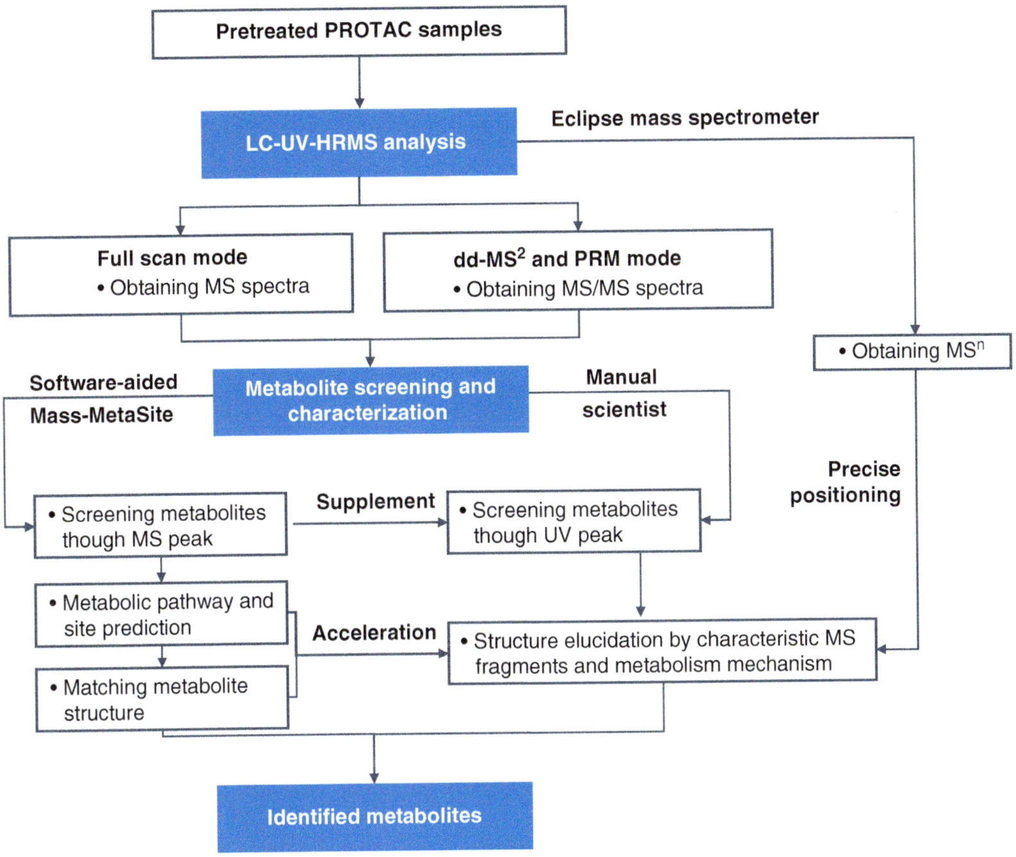

Figure 1.34 Workflow for the metabolite identification of PROTACs.

peak searching and mass spectral fragment information interpretation. Simultaneously, we employ Mass-MetaSite software for automated screening and identification of metabolites, calculating metabolic pathways and potential metabolic sites, and structurally aligning them with experimental results. This integrated approach that combines expert interpretation with automated software analysis speeds up the process of characterizing PROTAC metabolite structures. This assists researchers and developers in their work, making the process more efficient and reliable.

Additionally, WuXi AppTec DMPK platform features an industry-leading Thermo Scientific™ Orbitrap Eclipse™ Tribrid™ mass spectrometer capable of multistage (third grade and above) mass spectrometry data. During the structural characterization process, R&D scientists utilize this multistage mass spectrometry data to deeply analyze the fragmentation behavior of drug candidates and their metabolites. This allows for the acquisition of more refined fragment information and precise localization, ensuring high-quality structural characterization.

Through our strategy that combines professional software, experienced expert teams, and industry-leading high-resolution mass spectrometry platforms, structural characterization in MetID can be efficiently carry out. This approach helps us navigate the challenges posed by the complexity of PROTAC molecules, speeds up PROTAC metabolite research, and provides crucial support for the design and development of PROTAC drugs.

1.4.5.4 Semi-quantification of Metabolites

Through semiquantitative analysis of characterized metabolites, we can estimate the relative abundance of the parent drug and its metabolites. This allows for an initial understanding of the proportion of each metabolite and helps identify the main metabolites. This process is crucial for identifying metabolic soft spots, comparing *in vitro* metabolic differences across species, and comparing *in vitro* and *in vivo* metabolic differences. The information generated provides a basis for further structural optimization of subsequent drugs, prediction of drug metabolism in the human body, and safety evaluation of metabolites.

Given the structural properties of PROTAC, it is vital to ensure the accuracy of semiquantitative results [11]. In semi-quantification using MS, the difference in ionization efficiency of the parent drug and its products in the ion source can impact the semiquantitative data accuracy. Moreover, PROTAC and its metabolites often produce not only single-charged ions but also multi-charged ions and the ions cleaved in the ion source. These factors need to be fully considered when performing semi-quantification by mass spectrometry integration. This involves synchronously extracting and integrating single-charged ions, multi-charged ions, and ions cleaved in the source, which enhances the accuracy of semi-quantification. For instance, in a PROTAC study, we integrated the extracted ion current for a metabolite. The peak area after superimposing the extracted single- and multi-charged ions was approximately eight times that of the single-charged extracted ion.

When employing a UV detector for semi-quantification, it's crucial to note that if the E3 ligand of PROTAC has a different UV chromophore compared to the POI ligand, the metabolite of linker breakage might differ from the parent drug in terms of the absorbance coefficient. This discrepancy could significantly reduce the accuracy of UV semiquantitative data. In such scenarios, other detectors, such as charged aerosol detector (CAD), can be considered for more accurate results.

Therefore, the approach to semi-quantifying metabolites should be chosen comprehensively, taking into account their structural characteristics, UV spectra, and mass spectra.

1.4.6 Summary

After over two decades of development, PROTAC has become popular among leading pharmaceutical companies and research institutions worldwide due to its unique mechanism of action and advantages. Metabolite studies are crucial during the discovery, preclinical, and clinical stages of PROTAC drugs, making the experimental data for MetID increasingly critical for new drug applications (NDA). Leveraging the advanced Waters ACQUITY UPLC liquid-phase system and mass spectrometry systems such as Orbitrap Eclipse Tribrid and Q-Exactive™ HF, in conjunction with data processing software like Mass-MetaSite, Compound Discoverer, and Metabolynx, WuXi AppTec DMPK MetID team has developed a research platform for PROTAC metabolite profiling and identification. This platform, based on ultra-performance ultraviolet-liquid chromatography-high-resolution mass spectrometry (UPLC-UV-HRMS) technology, has established a series of robust solutions to overcome the challenges in the experimental process, thereby aiding the metabolic study of PROTAC molecules in numerous drug development companies globally. Looking forward, we anticipate an increasing number of PROTAC drugs transitioning into clinical studies and, ultimately, achieving a successful market launch, bringing hope to more patients.

References

1 Jones, L.H., Mitchell, C.A., Loberg, L. et al. (2022). Targeted protein degraders: a call for collective action to advance safety assessment. *Nature Reviews Drug Discovery* 21 (6): 401–402.

2 Goracci, L., Desantis, J., Valeri, A. et al. (2020). Understanding the metabolism of proteolysis targeting chimeras (PROTACs): the next step toward pharmaceutical applications. *Journal of Medicinal Chemistry* 63 (20): 11615–11638.

3 Cao, C., He, M., Wang, L. et al. (2022). Chemistries of bifunctional PROTAC degraders. *Chemical Society Reviews* 51 (16): 7066–7114.

4 Weng, G., Cai, X., Cao, D. et al. (2023). PROTAC-DB 2.0: an updated database of PROTACs. *Nucleic Acids Research* 51 (D1): D1367–D1372.

5 Jiang, B., Wang, E.S., Donovan, K.A. et al. (2019). Development of dual and selective degraders of cyclin-dependent kinases 4 and 6. *Angewandte Chemie International Edition England* 58 (19): 6321–6326.

6 Qiu, X., Li, Y., Yu, B. et al. (2021). Discovery of selective CDK9 degraders with enhancing antiproliferative activity through PROTAC conversion. *European Journal of Medicinal Chemistry* 211: 113091.

7 Bai, L., Zhou, H., Xu, R. et al. (2019). A potent and selective small-molecule degrader of STAT3 achieves complete tumor regression *in vivo*. *Cancer Cell* 36 (5): 498–511.

8 Khojasteh, S.C., Argikar, U.A., Driscoll, J.P. et al. (2021). Novel advances in biotransformation and bioactivation research - 2020 year in review. *Drug Metabolism Reviews* 53 (3): 384–433.

9 Chen, S., Chen, Z., Lu, L. et al. (2023). Discovery of novel BTK PROTACs with improved metabolic stability via linker rigidification strategy. *European Journal of Medicinal Chemistry* 255: 115403.

10 Fandozzi, C., Evans, C., Wilson, A. et al. (2019). 2019 white paper on recent issues in bioanalysis: chromatographic assays (part 1 - innovation in small molecules and Oligonucleotides & Mass Spectrometric Method Development Strategies for large molecule bioanalysis). *Bioanalysis* 11 (22): 2029–2048.

11 Pike, A., Williamson, B., Harlfinger, S. et al. (2020). Optimising proteolysis-targeting chimeras (PROTACs) for oral drug delivery: a drug metabolism and pharmacokinetics perspective. *Drug Discovery Today* 25 (10): 1793–1800.

1.5 Evaluation of Drug-Drug Interactions of PROTACs with Efflux Transporters

Liping Ma, Chengyuan Li, Jing Jin

1.5.1 Introduction

Drug transporters can be major determinants of drug pharmacokinetics. To evaluate the drug interactions with these drug transporters, suitable *in vitro* methods or models should be applied. The literature has revealed the drug interactions between a PROTAC drug and rosuvastatin. The elevation of aspartate aminotransferase and alanine aminotransferase (AST/ALT) was observed in subjects taking both the rosuvastatin and a PROTAC drug; symptoms alleviated after discontinuation of rosuvastatin, and treatment with the PROTAC drug could be continued. The possible reason is that the PROTAC drug inhibits the breast cancer resistance protein (BCRP) transporter, for which rosuvastatin is a known BCRP transporter substrate [1]. The combination of rosuvastatin with BCRP inhibitors may inhibit its efflux in the intestine and its biliary excretion in the liver, thus resulting in elevating the plasma concentrations of rosuvastatin and triggering corresponding adverse effects [2]. Considering the potential DDI risk in the clinic, it is necessary to assess the DDI of PROTAC with BCRP transporters at the early screening stage. Based on our research experience, *in vitro* transporter studies of PROTAC molecules present some challenges. Improper model selection may lead to erroneous results and mislead the assessment of DDI risk.

1.5.2 What Is a BCRP Transporter?

BCRP is a major ATP-binding cassette (ABC) transporter that plays an important role in drug transport. BCRP is predominantly expressed in the apical membrane of the intestinal epithelia, in liver hepatocytes, and in the basolateral membrane of the blood–brain barrier, which can exert a significant impact on restricting the entry of BCRP substrates into intestinal epithelia, mediating drug biliary excretion and the blood–brain barrier [3]. Many therapeutic agents commonly used in clinical practice are substrates for BCRP, such as the statins (rosuvastatin and pitavastatin) and the antineoplastic agents (axitinib, canatinib, imatinib, lapatinib, sunitinib, irinotecan, topotecan, etoposide, SN-38, and methotrexate) [2]. The latest Food and Drug Administration guidelines for *in vitro* DDI studies also clarify that BCRP is one of the nine drug transporters recommended for evaluation [4].

1.5.3 How to Evaluate the Inhibition of BCRP by PROTACs

The risk of DDI can be assessed at an early stage using *in vitro* transporter models. If *in vitro* studies indicate that a drug is a BCRP inhibitor, the sponsor should consider whether to conduct a clinical drug interaction study. With regard to the selection of transporter models, Caco-2, BCRP-overexpressed cells, or membrane vesicles with a high expression of BCRP can be used to determine if an investigational drug can inhibit BCRP. Of these, Caco-2 cells are common models for evaluation due to their easy access. First, we selected the most frequently used Caco-2 model to evaluate the BCRP inhibition potential of PROTACs. But here's the lesson we learned: improper model selection may lead to erroneous conclusions. We would like to share our experiences on this study and emphasize the model selection for PROTAC molecules.

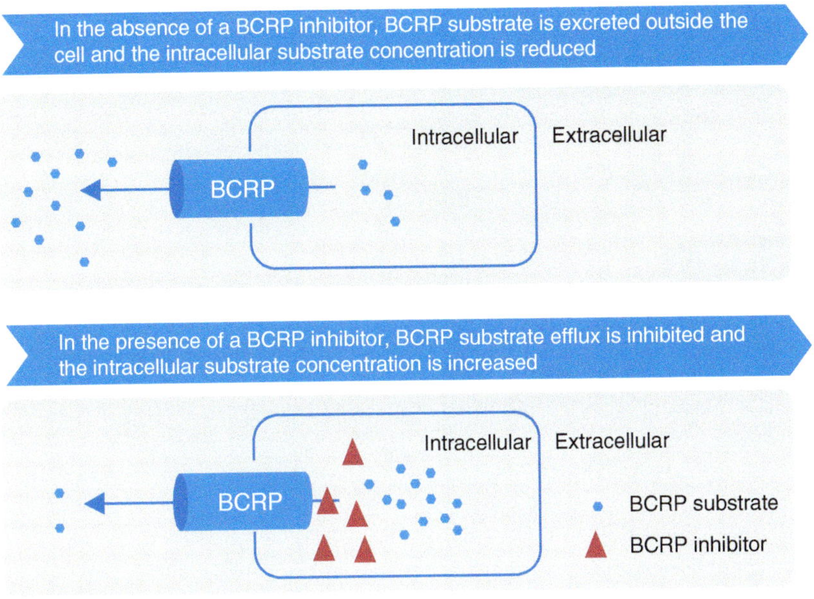

Figure 1.35 Breast cancer resistance protein (BCRP) efflux and transport inhibition.

1.5.4 Case Study and Analysis on the Inhibition of BCRP by PROTACs

The mechanism of the BCRP inhibition assay is shown in Figure 1.35. BCRP substrate can be pumped out of cells via BCRP, and the direction is from inside to outside. In the presence of the BCRP inhibitor, the BCRP-mediated efflux of the substrate can be inhibited. The inhibitory effect of a PROTAC molecule on BCRP was determined by using a Caco-2 cell model. The results of the assay showed no inhibitory effect of the PROTAC molecule on BCRP, which is inconsistent with the literature. We ruled out solubility and adsorption issues and tried to find the underlying reasons.

The molecular mechanisms of BCRP inhibition are relatively complicated: (i) one class of inhibitors, such as fumitremorgin C and Ko143, inhibits the ATPase activity of BCRP; (ii) another class of inhibitors has the same binding site as the substrate, affecting the binding of the substrate to BCRP; and (iii) the third class of inhibitors binds outside of the substrate binding site, causing a conformational change in BCRP and thus affect substrate transport [5]. Reports on the third mechanism are relatively limited, for which the inhibitor binding site may be intracellular or extracellular, whereas the other two inhibition mechanisms require the inhibitor to enter the cell first.

A general characteristic of PROTAC molecules is poor permeability. In the Caco-2 assay, inhibitors in the incubation system need to enter the cells to work. Although the concentration of PROTAC outside Caco-2 cells is high, the intracellular drug concentrations may be low due to poor permeability. Therefore, we speculate that it is likely that the concentration of PROTAC in Caco-2 cells is too low to exert an inhibitory effect (Figure 1.36).

1.5.5 Customized Assay on PROTACs

Based on the analysis of the above influence factors, we switched to a membrane vesicle model to evaluate BCRP inhibition by PROTAC. Membrane vesicles expose the substrate binding sites of ABC transporters to extracellular buffers, which is conducive to studying the interaction of drug

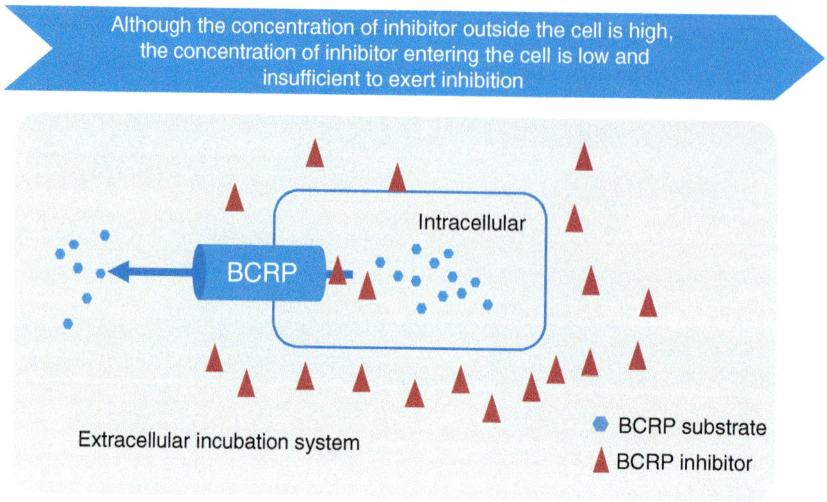

Figure 1.36 A low intracellular inhibitor concentration fails to exert an inhibitory effect.

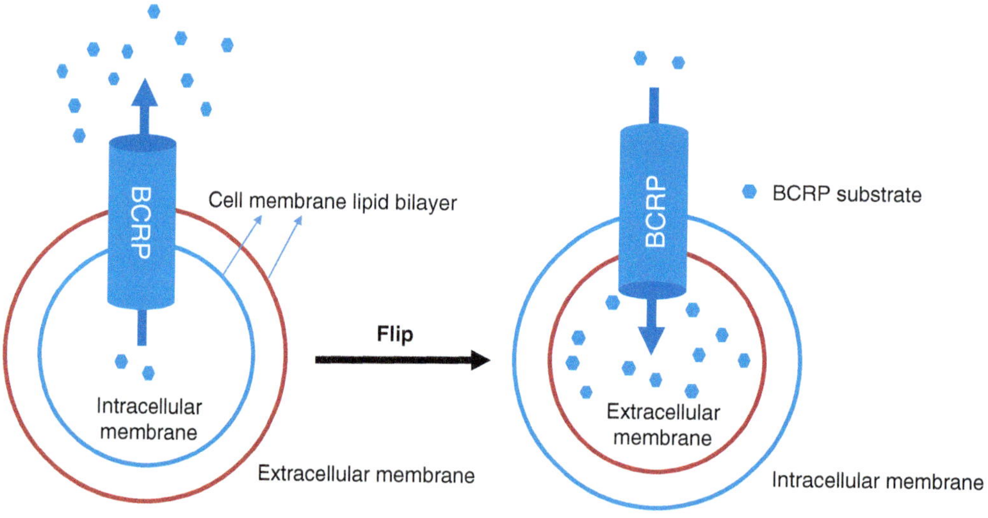

Figure 1.37 Mechanism of eversion of epithelial membrane vesicles.

candidates as substrates and inhibitors with ABC transporters. Figure 1.37 illustrates the mechanism of everted membrane vesicles. Originally, the substrate transport direction mediated by efflux transporters is from intracellular to extracellular, but after eversion, the substrate transport direction is reversed (extracellular to intracellular). In this model, the inhibitor can be in direct contact with the transporter substrate binding site outside the membrane.

1.5.6 Results

The membrane vesicle method showed that the PROTAC molecule had a strong inhibitory effect on BCRP, which was consistent with the previous literature reports.

1.5.7 Summary

The vesicle model is more sensitive in assessing efflux transporter inhibition than conventional methods that fail to obtain inhibition. This case caused us to pay attention to the model selection in the study of the inhibition of efflux transporters by PROTAC molecules. In addition to BCRP, the same problem may exist in the inhibition of other efflux transporters by PROTAC molecules. The poor permeability of PROTAC may lead to an underestimation of the risk of PROTAC inhibition of efflux transporters when assessing the inhibition of efflux transporters by PROTAC molecules using an intact cell model. The membrane vesicle model is more suitable to study the inhibitory effect of low-permeability PROTAC molecules on the efflux transporter.

References

1 Petrylak, D.P., Gao, X., Vogelzang, N.J. et al. (2020). First-in-human phase I study of ARV-110, an androgen receptor (AR) PROTAC degrader in patients (pts) with metastatic castrate-resistant prostate cancer (mCRPC) following enzalutamide (ENZ) and/or abiraterone (ABI). *Journal of Clinical Oncology* 38 (15): 3500.

2 Lee, C.A., O'Connor, M.A., Ritchie, T.K. et al. (2015). Breast cancer resistance protein (ABCG2) in clinical pharmacokinetics and drug interactions: practical recommendations for clinical victim and perpetrator drug-drug interaction study design. *Drug Metabolism and Disposition* 43 (4): 490–509.

3 Giacomini, K.M., Huang, S.M., and Tweedie, D. (2010). Membrane transporters in drug development. *Nature Reviews Drug Discovery* 3: 9.

4 US Department of Health and Human Services Food and Drug Administration Center for Drug Evaluation and Research. (2020). Studies—cytochrome IV. P450 enzyme-and transporter-mediated drug interactions guidance for industry. https://www. fda. gov/drugs/guidance-compliance-regulatory-information/guidances-drugs (accessed 10 March 2020).

5 Mao, Q. (2005). Role of the breast cancer resistance protein (ABCG2) in drug transport. *The AAPS Journal* 7: E118–E133.

1.6 PROTAC Bioanalysis: Challenges and Strategies

Tiantian Dang, Weimin Hu, Weiqun Cao, Lili Xing

1.6.1 Introduction

In order to exert their therapeutic effects, traditional small-molecule drugs or large-molecule antibodies generally must be bound to the active sites of an enzyme or receptor; however, 80% of the targets in human cells lack such sites. As an emerging strategy for new small-molecule drug development, PROTAC significantly combines the advantages of small-molecule compounds and small-molecule nucleic acids. They could capture target proteins through any binding site anywhere in a cell, making it easier to enter cells and exert the desired effect. PROTAC technology has embraced constant iterations and updates and has become a hot research area in recent years.

However, the mechanism involved in PROTAC's crossing of the cell membrane remains unclear. Therefore, to support the research on its absorption, distribution, metabolism, excretion, and toxicity, further theoretical and practical studies are necessary. Despite PROTAC having substantial potential, their complex structures may result in poor druggability, posing huge challenges to clinical trials, bioanalysis, and PK/PD studies. This chapter focuses on some challenges and practical experiences in PROTAC bioanalysis.

1.6.2 Key Points of PROTAC Bioanalysis

According to bioanalysis, a PROTAC molecule consists of a target protein–ligand, linker, and E3 ligand, resulting in a relatively large molecular weight, typically around 1,000 Da. Multi-charged ions in mass spectrometry could cause signal dispersion, potentially leading to a loss of sensitivity. In addition, the linker structure in a PROTAC molecule is fragile and prone to breakage [1]. Based on our practical experience, the optimization of MS parameters is crucial in the LC-MS/MS analysis process to avoid in-source fragmentation. Furthermore, PROTAC compounds often have multiple chiral centers, which can result in peak splitting during liquid chromatography. Therefore, adjusting the liquid-phase method to optimize peak shape is necessary for achieving smoother analysis. The exposed ions may also nonspecifically bind to the experiment material, such as glass or plastic [2]. Hence, it is essential to eliminate nonspecific adsorption to ensure no compound loss during the operation. Although the lyophilized powder of such compounds is stable, some hidden stability issues may arise when they exist in a biological matrix. Therefore, timely detection and preventive measures are necessary to ensure data accuracy. The following are the key points for PROTAC bioanalysis.

1.6.2.1 Optimal MS Parameters

The fragile linker structure in PROTAC makes it prone to in-source fragmentation, which can be reduced by optimizing the low ionizing energy and ion source temperature. In addition, for compounds with high molecular weights, the MS conditions are further optimized to enhance sensitivity by seeking out multiple charges. The current lower limit of quantification (LLOQ) can reach 1 ng/mL.

1.6.2.2 Peak Shape Optimization – Peak Splitting

The existence of multiple chiral centers [3] in the structure of PROTAC leads to peak splitting during liquid chromatography. By optimizing the method of liquid chromatography (such as mobile phase replacement, chromatographic column screening, gradient regulation, and ion source

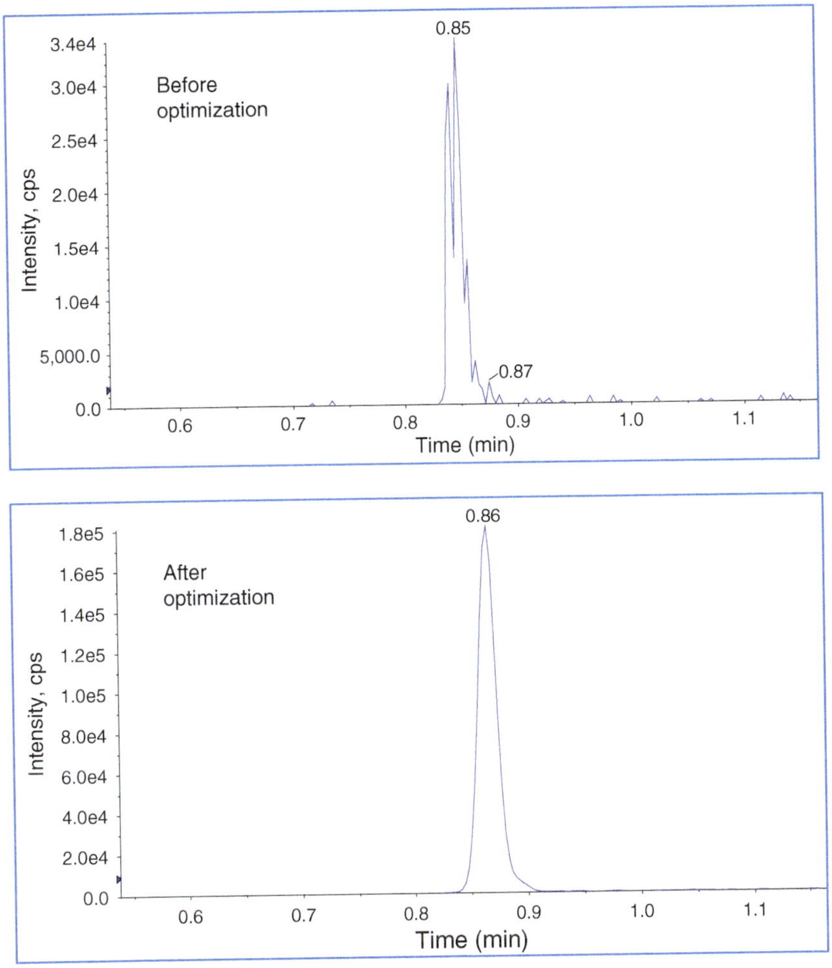

Figure 1.38 Comparison of compound A chromatographic peaks before and after optimization.

temperature reduction), a symmetrical and good peak shape can be achieved, and the compound's signal can be enhanced (Figure 1.38).

1.6.2.3 Elimination of Nonspecific Binding

PROTAC is a small-molecule compound with a large molecular weight (~1,000 Da) and exposed ions tend to nonspecifically bind to the surface of experiment material such as glass or plastic. Therefore, it is necessary to use appropriate container materials to collect a sample after dosing, use low-binding materials as much as possible, and investigate the NSB of compounds during the development of analytical methods. If NSB was found, a certain proportion of desorbent (such as Triton X-100 and Tween 20) should be added to block the NSB sites, ensuring that the compound remains intact during operation (Table 1.4).

1.6.2.4 Controllable Matrix Stability

PROTAC exhibits distinctive and covert plasma stability issues. Typically, enzyme activity in the fresh matrix is higher than that in the frozen matrix, making enzyme-sensitive compounds less

Table 1.4 Comparison of compound B transferred three times in blank plasma before and after adding a desorbent.

Sample	Compound B remaining (before adding the desorbent)	Compound B remaining (after adding the desorbent)
T0	100%	100%
T1	90%	87%
T2	85%	90%
T3	65%	85%

Note: T1: compound B was transferred once in the blank plasma; T2: compound B was transferred twice in the blank plasma; T3: compound B was transferred three times in the blank plasma.

Table 1.5 Comparison of stability of compound C in fresh and frozen plasma at room temperature for two hours.

Matrix	Compound C remaining
Fresh plasma	95%
Frozen plasma	35%

stable in the fresh matrix or samples. Some PROTAC compounds are stable when placed at room temperature in the fresh matrix for two hours, but they become highly unstable in the frozen matrix, and no inhibitors have been identified to effectively stabilize them (see Table 1.5). Therefore, a fresh blank matrix should be used to prepare calibration standards and quality control samples.

Furthermore, it is necessary to avoid repetitive freezing of samples to ensure the accuracy of the unknown sample's concentration. After collecting biological samples, protein precipitation is directly carried out to obtain the preserved supernatant for testing. In this case, the stability of the treated supernatant is the main concern. In the case of compound C, a standard solution was quantitatively added to the fresh matrix to simulate the real sample. The supernatant, after protein precipitation, was placed in a sample injector at 4°C and a refrigerator at −40°C for 24 hours, respectively. Next, the supernatant of 4°C and −40°C were injected alongside the 0-hour sample, indicating that the supernatant was stable after storage in the sample injector at 4°C and refrigerator at −40°C for 24 hours (Table 1.6). This shows that it is feasible to obtain and store the supernatant samples for preservation in the analysis of a DMPK project by immediately performing protein precipitation on biological samples after collection.

Table 1.6 Stability of compound C in supernatant of fresh plasma.

Sample	Remaining of the substance to be tested
0-hour sample	100%
Storage in sample injector at 4°C for 24 hours	95.4%
Storage in the refrigerator at −40°C for 24 hours	99%

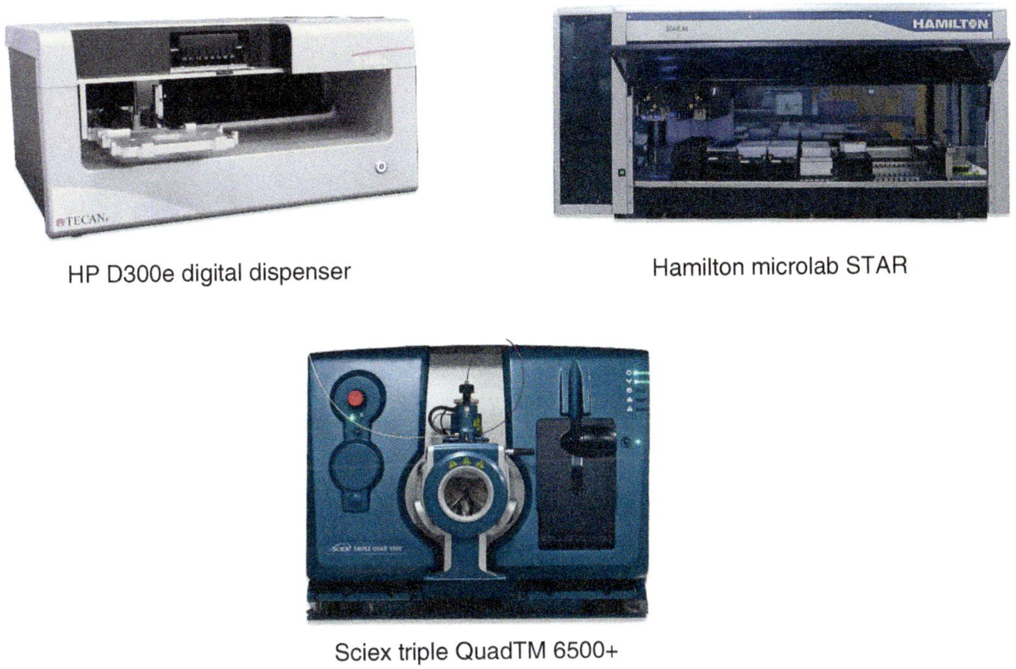

Figure 1.39 Biological sample analysis platform.

Each step of the bioanalysis operation is interlinked. During the sample pretreatment stage, the use of HP D300e digital titration equipment allows for swift preparation of calibration standards and quality control samples within eight seconds, eliminating the need for multiple transfers of compound solutions and mitigating NSB. Biological samples are precipitated immediately after collection, and the whole sample pretreatment process is carried out on wet ice conditions to ensure the accuracy and reliability of sample concentration detection, which provides reliable data support for the absorption, distribution, metabolism, and excretion (ADME) experiments of the PROTAC *in vivo* study.

1.6.3 PROTAC Bioanalysis Process

Based on different characteristics of PROTAC and the above-mentioned analysis experience, the PROTAC bioanalysis process is summarized as follows (Figure 1.39):

Optimization of MS parameters: Adjusting ionizing energy and ion source temperature; exploring multiple charges to improve sensitivity.

Optimization of liquid chromatography conditions: Screening for a suitable chromatographic column, mobile phase, and mobile phase gradient to obtain a better peak shape.

Elimination of nonspecific binding: Carry out the method development initially and screen suitable desorbents if adsorption occurs.

Optimization of sample processing procedure: Rapid operation at low temperatures, precipitation immediately after sample collection, and using the fresh matrix to prepare calibration standards and quality control samples.

Method reproducibility: Further optimizing the method to meet analysis requirements (reproducibility, LLOQ, matrix effects, etc.).

References

1 Goracci, L., Desantis, J., Valeri, A. et al. (2020). Understanding the metabolism of proteolysis targeting chimeras (PROTACs): the next step toward pharmaceutical applications. *Journal of Medicinal Chemistry* 63 (20): 11615–11638.

2 Fandozzi, C., Evans, C., Wilson, A. et al. (2019). 2019 white paper on recent issues in bioanalysis: chromatographic assays (part 1–innovation in small molecules and oligonucleotides & mass spectrometric method development strategies for large molecule bioanalysis). *Bioanalysis* 22: 2029–2048.

3 Xiao, Y., Wang, J., Zhao, L.Y. et al. (2020). Discovery of histone deacetylase 3 (HDAC3)-specific PROTACs. *Chemical Communications* 56 (68): 9866–9869.

2

Antibody–Drug Conjugates (ADCs)

2.1 General Considerations for Pharmacokinetic Studies of ADCs

Liping Ma, Jie Wang, Nan Zhao, Ruixing Li

2.1.1 Introduction

As a novel antitumor therapeutic drug, antibody–drug conjugates (ADCs) have achieved substantial progress in recent years. By the end of 2022, 15 ADCs have been approved globally; among them, 10 ADCs have been approved since 2019, triggering a new boom in the development of ADC. ADCs consist of three components: the antibody, the toxic payload, and the linker connecting the antibody with the payload. There are two key factors in the successful development of novel ADCs. First, an ADC should remain stable in blood circulation and attack tumor cells selectively, leading to the rapid release of the payload after being endocytosed by tumor cells. Poor stability in blood circulation or excessive distribution to other tissues can result in severe toxicity of ADCs. Second, to kill tumor cells quickly, the payload released by an ADC in tumor cells should be highly cytotoxic (Figure 2.1).

2.1.2 ADC Drug PK Evaluation

From a PK perspective, **ADC development exhibits two characteristics**. First, the ADC PK study is challenging. As a novel drug, there is limited understanding of the human PK properties of ADCs within the industry. As of 2022, no radiolabeled mass balance studies have been conducted for ADCs in humans. ADC is a combination of large and small molecules; this combination makes it necessary to study the PK of these two molecules. Second, the study of payload release is essential for the development of ADC. Specifically, the released concentration of payload in target organs is directly related to ADC efficacy, while its release in nontarget tissues is directly related to ADC toxicity. Additionally, the DDI of ADCs is also associated with the exposure, metabolism, and excretion of payloads (Figure 2.2).

As of 2022, the most authoritative document in the literature concerning the ADC PK study is "Current Approaches for Absorption, Distribution, Metabolism, and Excretion Characterization of Antibody-Drug Conjugates: An Industry White Paper" published by the American Society for Pharmacology and Experimental Therapeutics in 2016 [1]. The PK studies proposed in the white paper include human and animal PK studies after ADC administration, animal ADME studies, and the identification of small-molecule metabolites in humans and animals. However, this white

Drug Metabolism and Pharmacokinetics: Frontiers, Strategies, and Applications, First Edition.
Edited by Liang Shen and WuXi AppTec DMPK.

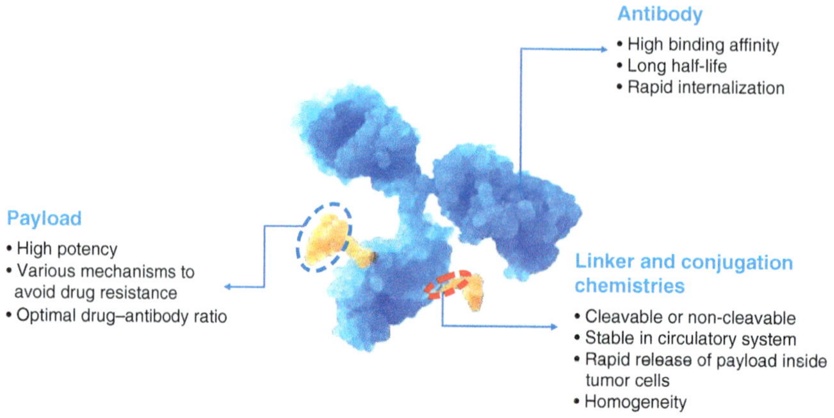

Figure 2.1 ADC structure.

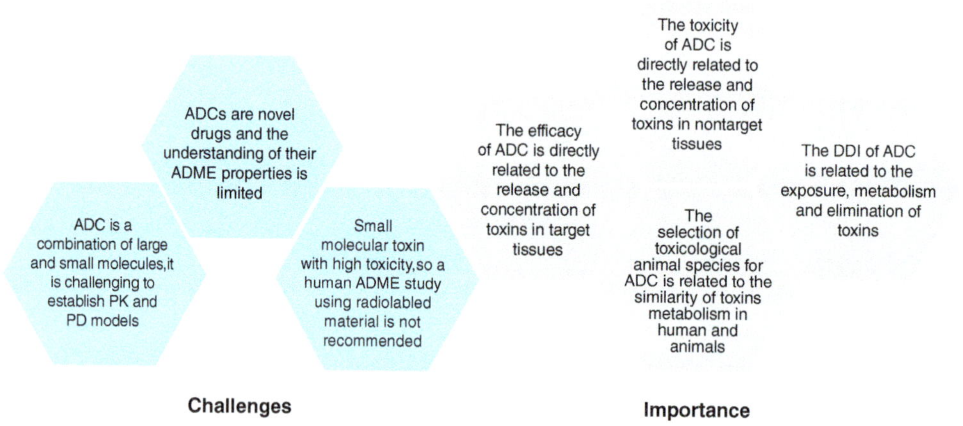

Figure 2.2 Challenges and importance of the ADC PK study.

paper lacks detailed descriptions of the strategies, and the methods and scope of ADC PK studies have been evolving in recent years. Therefore, ADC PK studies should be tailored to meet the evolving requirements.

In September 2023, the Chinese National Medical Products Administration (NMPA) published the "Guidelines for Non-clinical Studies of Antibody-Drug Conjugates" [2]. According to this guideline, the preclinical PK of ADC drugs can be mainly divided into three parts, as shown in Table 2.1: the corresponding research on ADCs, new free small molecules, and new linkers. Free small molecules refer to free payloads and other free payload-related components released from the ADC. To determine what kinds of free small molecules need to be evaluated, payload-related metabolite profiling and identification studies for ADCs need to be conducted.

2.1.3 ADC Drug PK Study Strategies

The ADC PK studies have been categorized into three main aspects: ADME, *in vitro* DDI, and PK. Each of these aspects holds particular significance in drug discovery and preclinical development stages (Table 2.2).

Table 2.1 Preclinical ADC PK studies.

Testing article	Study contents
ADCs	*In vitro* stability Plasma concentration–time curve Tissue distribution of ADCs intended to target specific tissue
New free small molecules	Plasma protein binding Absorption, distribution, metabolism, and excretion Drug-drug interaction with drug metabolism enzymes and drug transporters
New linkers	*In vivo* metabolism

Table 2.2 Pharmacokinetic studies of ADCs at different stages of research.

	Preclinical discovery	Preclinical development
DMPK studies	**Lead optimization**	**ADME characterization and Investigational New Drug (IND) application**
ADME	• Plasma/serum/blood stability of ADC • Identification of metabolites related to payload released by ADC *in vitro*	• Microsomal/hepatocyte metabolite identification of payload • Plasma protein binding of payload • ADME study of radiolabeled payload in animals • *In vivo* metabolite identification of ADC
DDI	• Study on the permeability and efflux transports of payload	• Study on inhibition and induction of P450 enzyme by payload • Metabolic enzyme phenotyping of payload • Study on the inhibition of drug transporters by payload
PK	• PK study of ADC or payload in rodents • Screening PKPD study of ADC	• PK studies of ADC or payload in non-rodents • PKPD study of ADC

Based on the literature, analysis of PK data from approved ADC drugs, and a summary of research experience, we have identified four specific requirements for ADC PK research:

- The ability to conduct PK studies of small and large molecules.
- Comprehensive capabilities and facilities for ADME and bioanalysis, including study capacity for small and large animals, bioanalysis of large and small molecules, *in vitro* DDI, *in vitro* and *in vivo* metabolite identification, and radiolabeled ADME.
- Specifically design the overall strategies for PK studies and the specific protocols according to the structure of the ADC and the demand for the effectiveness and safety of new drugs.
- Excellent project management capability and the ability to coordinate and collaborate across multiple departments.

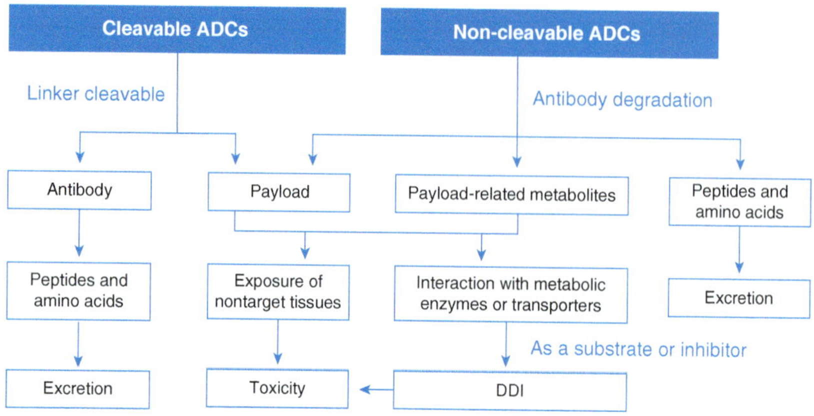

Figure 2.3 Payloads released from ADCs.

2.1.4 Relationship Between ADC Payload Release and ADC Toxicity as well as DDI

It is worth noting that the release, metabolism, and clearance of payloads in nontarget tissues are closely related to ADC toxicity and DDI. ADC can be divided into two types based on their structure: one is **cleavable ADCs** that release payloads under the catalysis of lysosomal enzymes or acidic conditions, and the released payloads have a relatively clear structure. The other is **non-cleavable ADCs** that release payloads that are bound to partial peptide chains or amino acids through the degradation of antibodies, and the structures of payload-related metabolites are unknown. The exposure of the payloads released from ADCs to nontarget tissues may be toxic. If the metabolism or clearance of these cytotoxic drugs is affected by metabolic enzyme or drug transporter inhibitors, their systemic exposure will increase and thus result in toxic effects (Figure 2.3).

To support the research and development (R&D) of ADC drugs, our ADC team has specifically developed strategies, processes, and methods for the study of the release and metabolism of payloads. The *in vitro* metabolic system of ADCs is used to determine whether payloads are released according to the ADC design or to identify unknown payload-related metabolites through proteolysis.

If the ADC is found to release new payload or payload-related metabolites, a series of studies need to be conducted for the small molecules. Such studies mainly focus on two aspects: the first is the metabolite identification, including the comparison of *in vitro* metabolites in multiple species, animal ADME studies, and payload-related metabolite identification in human plasma, and the second is the *in vitro* DDI studies, including CYP450 enzyme induction and inhibition, metabolic enzyme phenotyping, and inhibition or substrate evaluation of drug transporters. The results of these studies provide a scientific basis for the later-stage development of ADCs, such as guidance for the selection of toxic species, genetic polymorphism study of metabolic enzymes, clinical DDI study design, and studies for specific populations.

References

1 Kraynov, E., Kamath, A.V., Walles, M. et al. (2016). Current approaches for absorption, distribution, metabolism, and excretion characterization of antibody-drug conjugates: an industry white paper. *Drug Metabolism & Disposition* 44 (5): 617–623.

2 The National Medical Products Administration (NMPA). (2023). Guidelines for non-clinical studies of antibody-drug conjugates. https://www.cde.org.cn/zdyz/domesticinfopage?zdyzIdCODE=42d891ebccbcece7dd757d6429acff33.

2.2 Application of Integrated Bioanalysis in ADC DMPK Studies

Maotian Zhou, Huan Yan, Li Qu, Zhiyu Li, Lili Xing

2.2.1 Introduction

DMPK studies are conducted throughout all stages of ADC drug development, including screening discovery and the preclinical and clinical phases. There are different research focuses at different stages. Generally, each stage will include the PK evaluation of total antibody, conjugated antibody (ADC), free payload, and possible metabolites. *In vitro* stability, drug–antibody ratio (DAR) evaluation, and biotransformation characterization are usually added during the ADC discovery phase, and the immunogenicity evaluation is usually included in preclinical and clinical phase studies.

Bioanalysis is indispensable in helping scientists understand ADCs' properties at different stages of development. Regardless of category or phase, investigating and developing ADCs require a combination of small- and large-molecule approaches or even hybrid approaches to monitor the drug's PK and biotransformation. This chapter will detail bioanalytical approaches to investigating ADCs and provide recommendations for drug sponsors developing them.

2.2.2 Overall Strategy for ADC Bioanalysis

ADCs have the characteristics of both small-molecule and large-molecule therapeutics. Therefore, the ADC bioanalysis methods must consider these unique characteristics. Specifically, the ADC bioanalysis strategies can be divided into three following aspects that cover the discovery, preclinical development, and clinical development stages.

2.2.2.1 ADC DAR Characterization in Animal Plasma/Serum

ADCs with different DAR values may have completely different PK and toxicological characteristics and could even be considered as different drugs. Inaccurate preclinical study results or unpredictable toxicity may occur when ADCs are roughly studied as a mixture. Therefore, accurate ADC DAR characterization is necessary.

Developing a bioanalytical method for characterizing the molecular structure of ADCs in plasma/serum is crucial to identify analytes with varying DARs in circulation. This method can provide abundant data in *in vitro*/*in vivo* processes of ADCs (e.g., whether the antibody still carries most of the covalently bound drug over time) (see Section 2.3).

2.2.2.2 Quantitative Analysis of ADC-Related Components

According to their structure and *in vivo* transformation characteristics, ADC analytes undergoing PK analysis are usually divided into total antibody, ADC, conjugated payload, unconjugated payload, and payload-related metabolites.

Combined with traditional ligand binding assays (LBAs), liquid chromatography–mass spectrometry (LC-MS/MS) assays, affinity capture LC-MS/MS, affinity capture LC–high-resolution MS (LC-HRMS), and other new analytical methods, feasible quantitative analysis methods were developed to conduct quantitative analysis of ADC components in animal plasma or serum. Among these approaches, the LC-MS method offers the advantages of having a short time cycle and good versatility and has also demonstrated utility in large-molecule quantitative analysis.

Table 2.3 Definitions of ADC-related components and methods of bioanalysis.

Analytes	Definition	Analytical methods	Special abilities
Total antibody	DAR ≥ 0	ELISA and LC-MS	LC-HRMS intact protein analysis and DAR analysis
ADC	DAR ≥ 1	ELISA and LC-MS	
Conjugated payload	Small-molecule drugs conjugated to antibodies	LC-MS	
Unconjugated payload	Released small-molecule drug	LC-MS	
Payload-related metabolites	Payload-related components with medicinal properties generated after metabolism	LC-MS	
ATA	Anti-therapeutic antibody	ELISA	

2.2.2.3 Anti-therapeutic Antibody (ATA) Investigation of ADCs

Appropriate methods should be developed to determine the anti-therapeutic antibody (ATA) of ADCs. This part is frequently completed with a platform of LBAs (see Table 2.3).

2.2.3 ADC DAR Characterization

2.2.3.1 What Is DAR?

DAR (the number of payloads conjugated to one antibody) is an important parameter used to describe the physiochemical properties of ADCs. It indicates the payload carried by a single ADC molecule, which in turn relates to ADC safety and effectiveness.

Most ADCs approved until 2024 use heterogeneous conjugation technology (e.g., amine-based lysine conjugation or thiol-based cysteine conjugation). Since each antibody has multiple Lys/Cys loci, and the conjugation process is random, the number and location of the payload formed by conjugation are uncertain, and the ADC eventually obtained is often a mixture of multiple components with different DARs. For example, trastuzumab emtansine (T-DM1) has a DAR that ranges from 0 to 7, with an average DAR of ~3.5 (Figure 2.4).

DAR will affect the stability of ADCs in *in vivo* circulation. Typically, an ideal DAR is 3–4. An overly high DAR will increase drug clearance, decrease half-life, and may lead to drug aggregation and increased toxicity.

2.2.3.2 How to Determine ADC DAR?

DAR analysis can be divided into two aspects: first is DAR distribution, which characterizes the proportion of ADC molecules with different DARs in the total ADC molecules, and second is the average DAR, which is the ratio of the molar concentration of the total conjugated payload to that of the antibody molecule in the system. We have developed an LC-HRMS intact protein analysis to characterize the ADC DAR. Compared to hydrophobic interaction chromatography–ultraviolet (HIC-UV) and traditional DAR characterization methods, the LC-MS method can greatly reduce sample consumption and provide more accurate DAR results [1].

For ADCs in a biomatrix, the general approach is to first use the antihuman IgG, idiotype antibody, or antigen that can specifically bind to the target ADC to conduct immunoaffinity capture for the target ADC. Next, the LC-HRMS platform is used to conduct direct analysis on the purified

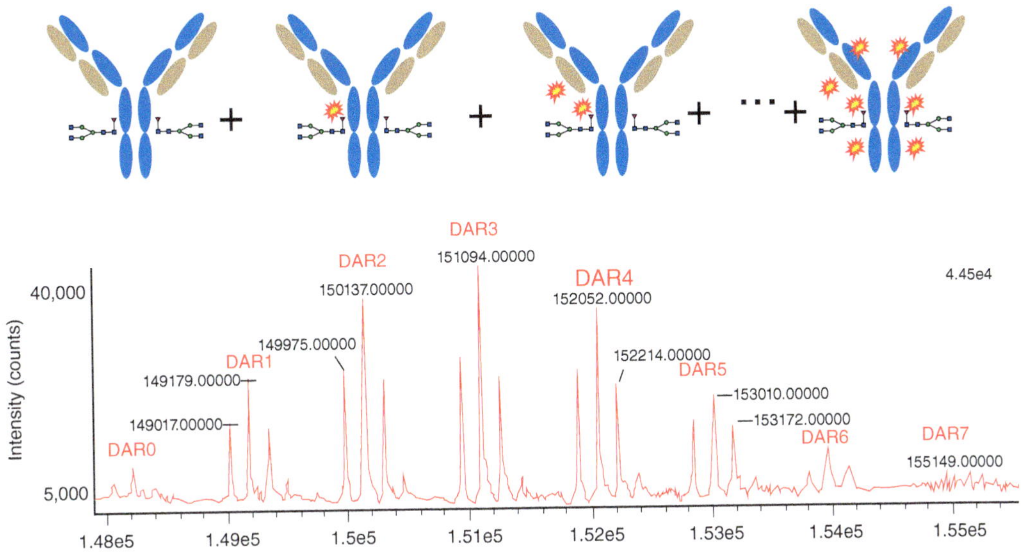

Figure 2.4 T-DM1 DAR characterization results. T-DM1 is a mixture of ADCs with different DARs. With the LC-HRMS platform, the characterization of ADCs with different DARs can be realized.

Figure 2.5 ADC DAR characterization process.

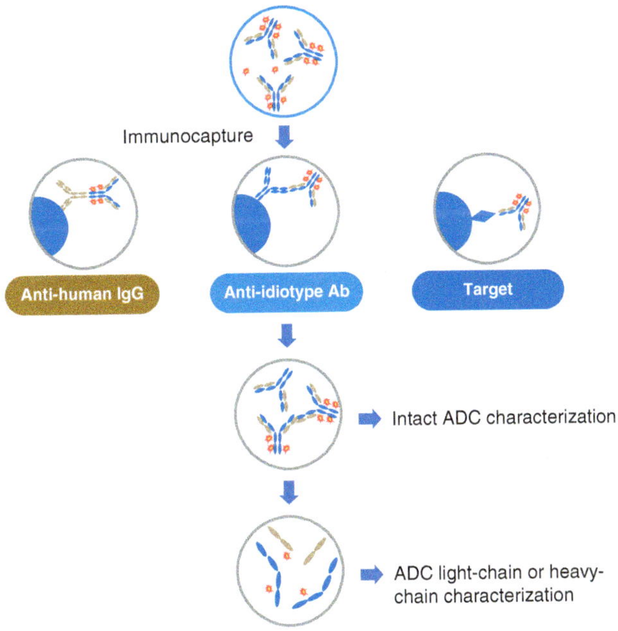

ADC, and finally, a deconvolution software is utilized for data processing (see Figure 2.5). To complete the DAR analysis of ADCs, the specific peaks of ADC molecules with different absolute molecular weights and different DARs are obtained.

In addition, purified ADC can be broken down to separate the light and heavy chains, and then it is subjected to an LC-HRMS assay. Once the deconvolution of the spectrogram is acquired, DARs of light and heavy chains can be obtained. The DAR analysis of ADCs can then be performed indirectly.

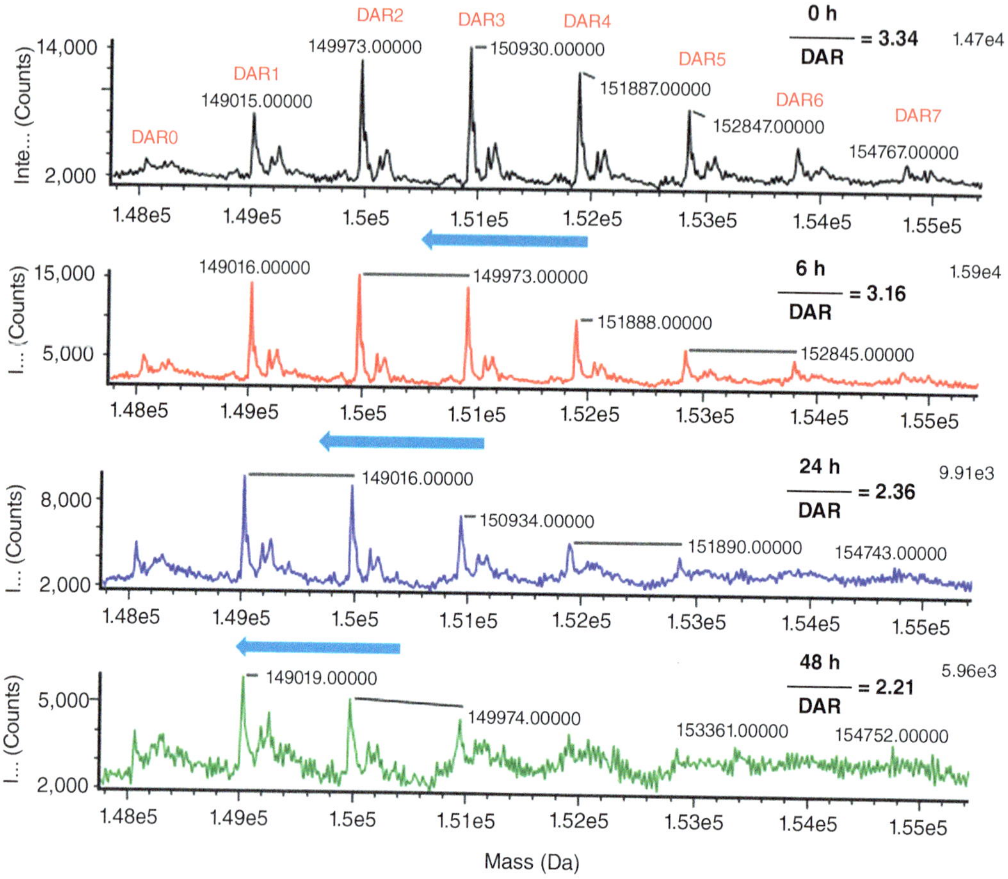

Figure 2.6 *In vitro* plasma stability of T-DM1. With the increase in incubation time, the content of high DAR ADCs (DAR6 in the figure disappeared after 24 hours) gradually decreased, and the average DAR of ADC saw a corresponding decrease from 3.34 at T_0 to 2.21 after 48 hours.

This platform is used to verify the *in vitro/in vivo* stability of T-DM1. It provides a DAR analysis method with ultra-high sensitivity (LLOQ: 0.5–1 μg/mL), specificity, and accuracy. The *in vitro* stability study results are shown in Figure 2.6. The intact protein analysis results revealed the variation in DARs of T-DM1.

2.2.4 Quantitative Bioanalysis of ADC Drug Components in Serum/Plasma or Other Biological Matrices

2.2.4.1 Ligand Binding Assays (LBAs)

The LBA method is the most commonly used strategy to measure the antibody part of the ADC; it measures both total antibodies and conjugated antibodies. This method offers adequate sensitivity to analyze large molecules. It also promises high throughput and robustness at a relatively low cost of implementation and instrumentation, and there is well-established guidance that can be referenced for the LBA method. To meet different assay requirements, various LBA formats can be designed (Figure 2.7).

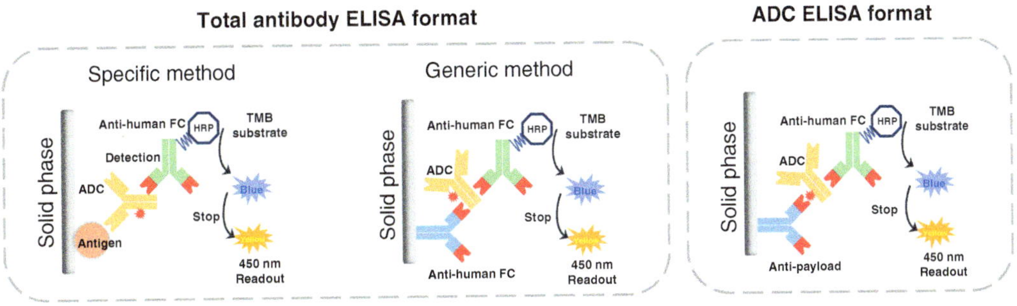

Figure 2.7 Common ELISA assay formats for total antibody and conjugated antibody bioanalysis.

The LBA platform has several challenges specific to ADC bioanalysis. The conjugated antibody quantification generally cannot provide the DAR information, and it is not sensitive to the biotransformation of ADC. The specificity and the sensitivity of LBA are highly dependent on the selection of critical reagents. The development of critical reagents for LBA can be costly and time consuming, especially for anti-payload antibodies. Perhaps the most significant disadvantage in using LBA for ADC bioanalysis is its complete inability to capture information about the ADC's structure or possible biotransformation.

2.2.4.2 Liquid Chromatography–Mass Spectrometry (LC-MS)

In recent years, with the continuous development of MS, LC-MS/MS and LC-HRMS have been increasingly used for large-molecule characterization and quantification [2–4]. Compared with the traditional LBA method, LC-MS offers a shorter development cycle and better standardization. In addition, LC-HRMS not only quantifies large-molecules but also determines the DARs of ADCs, thus facilitating the improvement of ADCs' biological characterization.

ADCs are often present *in vivo* as complex and dynamically changing mixtures caused by biotransformation, DAR changes, or a combination of these processes. In this case, the existing PK is no longer clear with regard to the expression of "therapeutic concentration" and time, which presents a unique challenge to quantitatively analyzing ADCs. As shown in Table 2.3, the components for analysis available to characterize ADC PK include total antibody (an antibody that is conjugated or unconjugated to the payload), conjugated antibody (ADC, an antibody that is conjugated to at least one payload and is a prototype drug of ADC), unconjugated payload, conjugated payload, and payload-related metabolites (effector formed after ADC cleavage or decomposition) [5].

The PK of different analytes reflects different content and significance, presenting the complete picture of ADC metabolism *in vivo*. Therefore, a comprehensive LC-MS platform is needed for the PK analysis of total antibodies, intact ADCs, unconjugated payloads, and conjugated payloads.

2.2.4.2.1 Quantitative Analysis of Total Antibody by LC-MS

For total antibody analysis of ADCs, sample preparation and analysis remain consistent with other antibody drugs. The process can be divided into three steps (Figure 2.8):

1) Immuno-affinity capture, which involves antibodies or antigens that are specifically bound to the ADC antibody to purify the target total antibody.
2) Proteolysis, which generates the signature peptides of the antibody.
3) LC-MS/MS, which is used for quantitative determination.

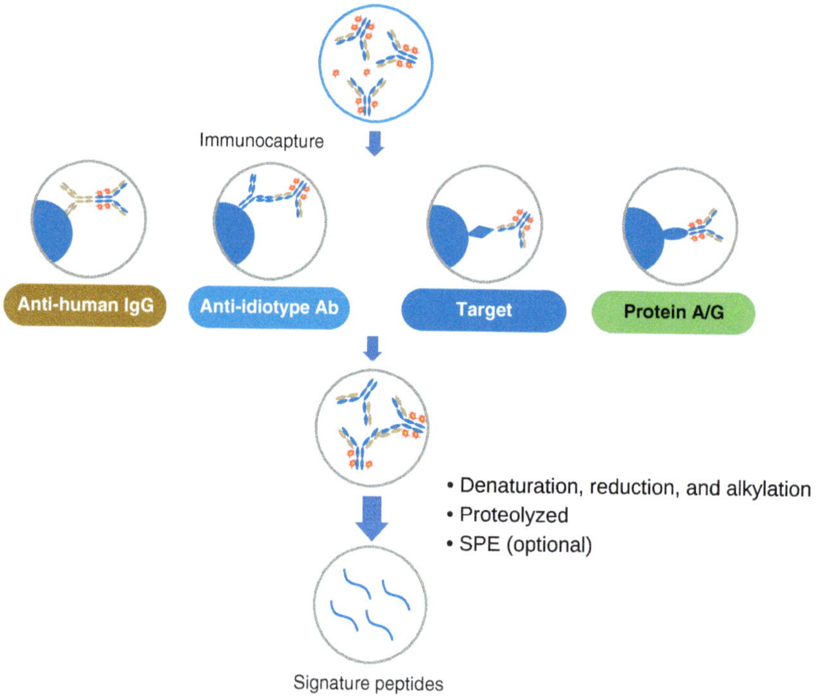

Figure 2.8 LC-MS/MS quantification process of total antibody.

The determination of total antibody is quite consistent with the first step of DAR analysis. The only difference is that when the LC-MS/MS platform is used to determine the total antibody, the requirements for the specificity of the capture reagents are reduced, meaning that a wide range of options are available to reduce the cost. Beyond antihuman IgG, idiotype antibody, and antigen, we can also use protein A/G, which has better universality, poorer specificity, and lower price.

2.2.4.2.2 Quantitative Analysis of ADC by LC-MS

This section begins by defining the target analyte of ADC quantitative analysis. There are generally two ways to define the PK target analyte of ADCs:

1) From the perspective of antibodies, it can be defined as antibody molecules bound to at least one drug.
2) From the perspective of drug load, it can be defined as the total concentration of the drug bound to the antibody.

From the two definitions, we have developed two different bioanalysis strategies. If an antibody molecule bound to more than one drug is used as the analyte, we can use a method similar to total antibody analysis. The difference is that the capture reagent used in the first step must be an antidrug antibody (ADA). Therefore, the antibody molecules conjugated to drugs can be extracted and quantified using LC-MS/MS [2]. If the drug conjugated to the antibody is used as the analyte, and the linker can be cleaved through sample processing to release the drug, then the conjugated drug can be released after the purified total antibody is captured (Figure 2.9). Furthermore, the concentration of the conjugated drug can be determined to achieve the quantitative ADC bioanalysis.

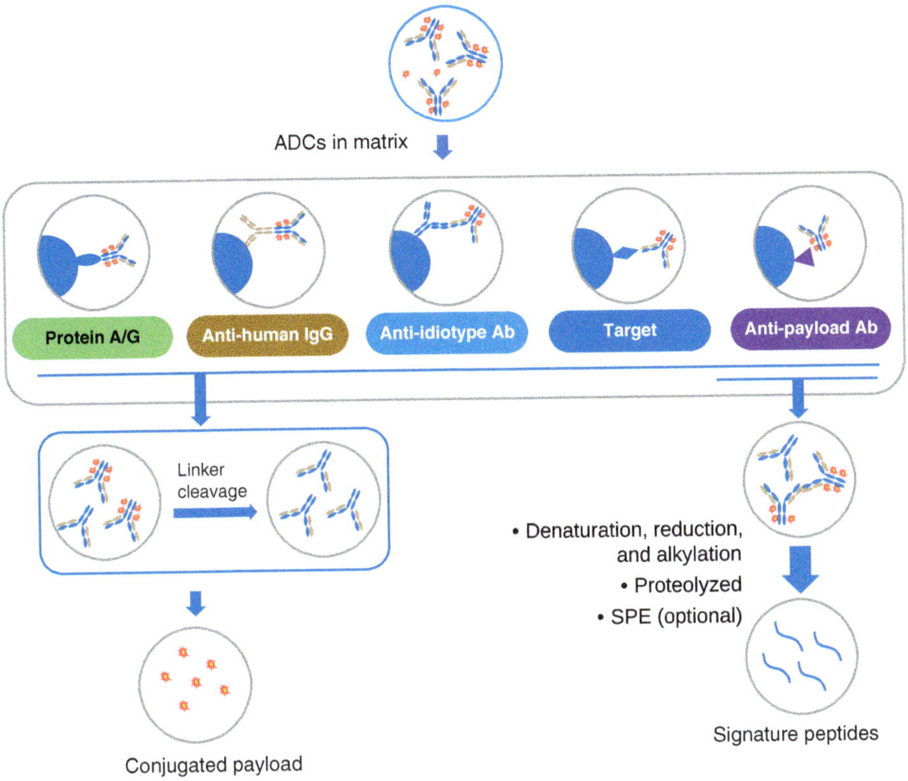

Figure 2.9 Quantitative strategy for ADC: (i) for non-cleavable linkers, the ADC can be captured by using an anti-payload antibody, and the antibody molecules conjugated to the drug can be quantified; (ii) for cleavable linkers, the total concentration of the drug bound to the antibody can be quantified by releasing the conjugated payload.

In addition to the two situations noted earlier, through the steady development of site-specific conjugation [6], the "payload–linker–amino acid" generated by proteolysis has also been used as the analyte for ADC quantitative analysis in recent years. This makes it possible to achieve the quantitative analysis of conjugated payload. Theoretically, the average DAR of ADCs can be indirectly detected through the analysis of the quantitative results of conjugated payload and total antibody.

2.2.4.2.3 Quantitative Analysis of Free Payload by LC-MS

An important factor affecting ADC toxicity is the free payload. The free payload may mediate unexpected off-target toxicity, so monitoring the free payload is necessary and has become a very important part of ADC research. The very low concentration of free payload in plasma is a greater challenge to sensitivity. We have established quantitative bioanalytical methods using the highly sensitive AB 6500 and AB 7500 for the quantification of free payload. Figure 2.10 shows the LC-MS/MS chromatogram of monomethyl auristatin E (MMAE) at 20 pg/mL in the rat plasma sample. Based on the signal-to-noise ratio, the LLOQ can be lower than 20 pg/mL. The linearity meets the method requirements. LC-MS/MS quantitative analysis of free payload is often applied in experiments to determine whether the free payload is released during ADC drug storage, plasma stability of the linker, stability of toxin in the biological matrix, small-molecule PK, and tissue distribution.

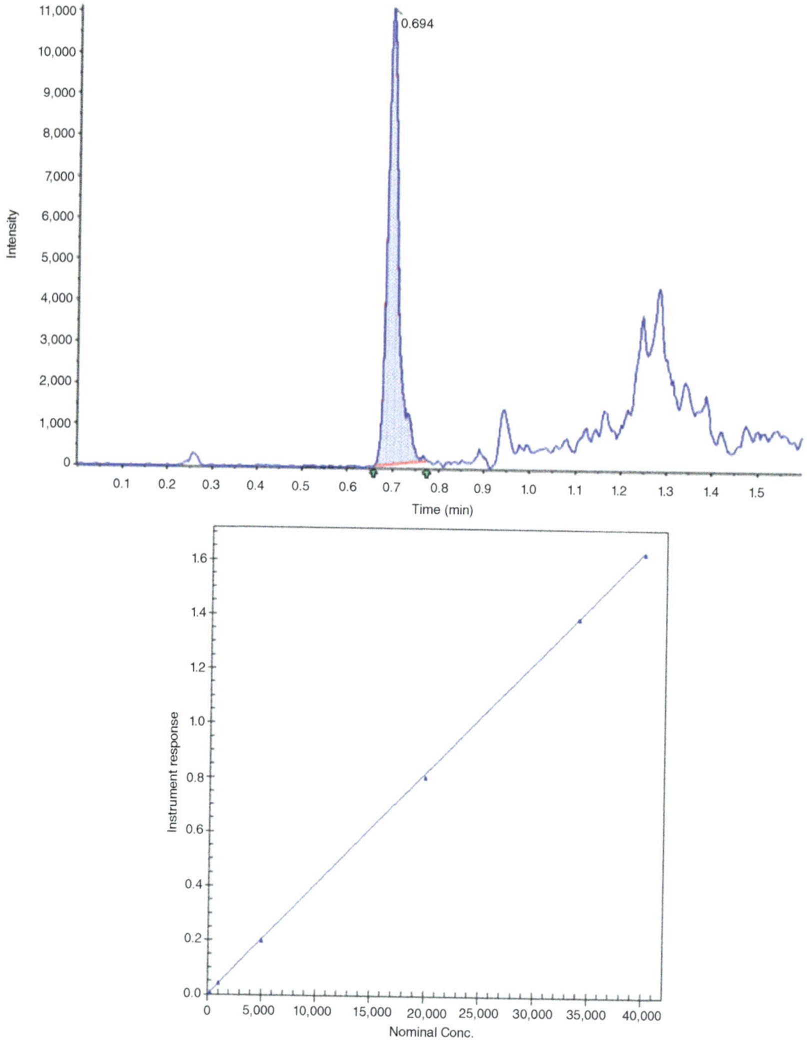

Figure 2.10 LC-MS/MS chromatogram and standard curve of MMAE at 20 pg/mL in rat plasma sample.

2.2.5 Summary

Our ADC analysis team has established a qualitative/semiquantitative analysis method for intact ADCs from plasma/serum or other matrices, which can detect the change in *in vivo* and *in vitro* ADC DAR values and possible biotransformation. In addition, there is a mature and complete quantitative analysis protocol for various ADC components that can partially represent the PK characteristics, including total antibody, ADC, unconjugated payload, and conjugated payload (Figure 2.11). The intact ADC analysis method based on HRMS can even provide extremely extensive PK characteristic parameters of ADCs in just a single detection, including antibody concentration and DAR. All these methods collectively form a complete set of ADC bioanalysis strategies.

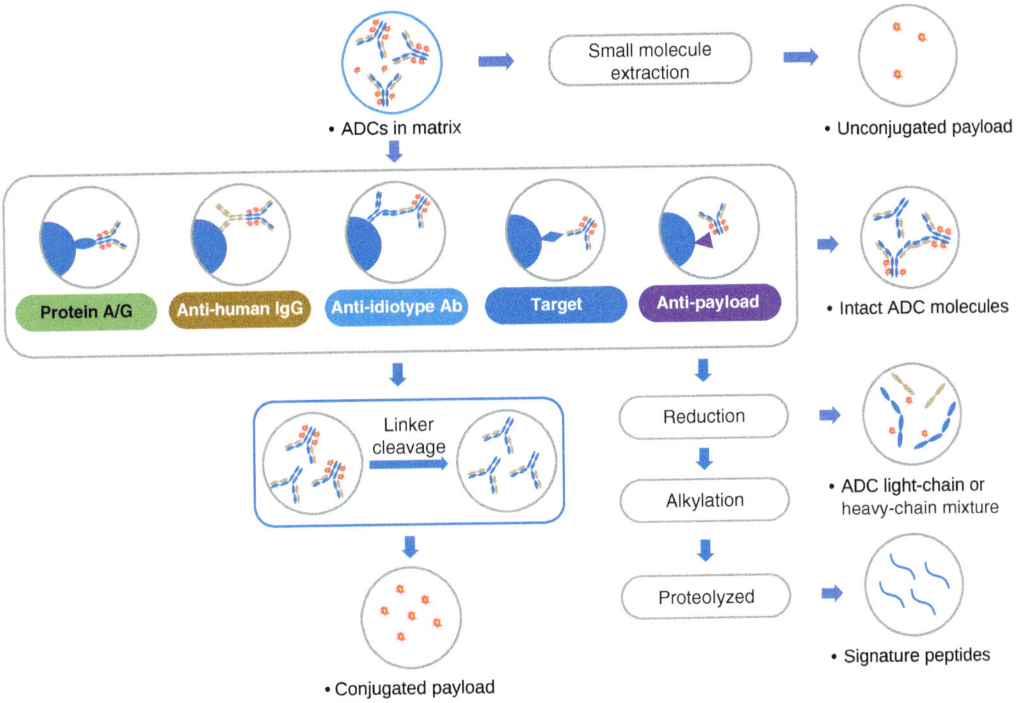

Figure 2.11 Summary diagram of LC-MS assay strategy for ADC-related components.

References

1 Källsten, M., Hartmann, R., Artemenko, K. et al. (2018). Qualitative analysis of antibody–drug conjugates (ADCs): an experimental comparison of analytical techniques of cysteine-linked ADCs. *Analyst* 143 (22): 5487–5496.

2 Huang, Y., Mou, S., Wang, Y. et al. (2021). Characterization of antibody–drug conjugate pharmacokinetics and *in vivo* biotransformation using quantitative intact LC-HRMS and surrogate analyte LC-MRM. *Analytical Chemistry* 93 (15): 6135–6144.

3 Jin, W., Burton, L., and Moore, I. (2018). LC–HRMS quantitation of intact antibody drug conjugate trastuzumab emtansine from rat plasma. *Bioanalysis* 10 (11): 851–862.

4 Kellie, J.F., Pannullo, K.E., Li, Y. et al. (2020). Antibody subunit LC-MS analysis for pharmacokinetic and biotransformation determination from in-life studies for complex biotherapeutics. *Analytical Chemistry* 92 (12): 8268–8277.

5 Zhu, X., Huo, S., Xue, C. et al. (2020). Current LC-MS-based strategies for characterization and quantification of antibody-drug conjugates. *Journal of Pharmaceutical Analysis* 10 (3): 209–220.

6 Lee, B.I., Park, M.H., Byeon, J.J. et al. (2020). Quantification of an antibody-conjugated drug in fat plasma by an affinity capture LC-MS/MS method for a novel prenyl transferase-mediated site-specific antibody–drug conjugate. *Molecules* 25 (7): 1515.

2.3 Study Models and Bioanalysis Methods for the Payload-Related Metabolite Identification Studies of ADCs

Liqi Shi, Weiqun Cao

2.3.1 Introduction

ADC drugs can be divided into cleavable and non-cleavable types: cleavable ADCs will release toxin molecules by the hydrolysis of the linker, while non-cleavable ADCs will release toxin molecules through the degradation of the antibody part. As the main components with pharmacological activity, the relevant structures of toxin molecules (payload-related metabolites) released by ADC should be determined as early as possible. These compounds are also the object of subsequent ADME studies, and their concentrations need to be simultaneously detected in PK or toxicokinetic (TK) studies.

2.3.2 Study Models

A variety of study models are available to investigate the payload release of ADC drugs. According to the mechanism of action of ADC drugs, liver lysosome and liver S9 can be selected to study the release of payload-related metabolites, and the tumor cell incubation system can be used as a PD model to clarify the effective substances. After the major payload-related metabolites are determined, studies of the metabolic pathways of the small molecules can be conducted using hepatocytes, liver microsomes, or liver S9, and species differences can be compared to provide a basis for the selection of toxicological species. The metabolic study of ADC *in vivo* mainly focuses on the identification of payload-related metabolites in the plasma, which can verify the stability of ADC drugs in systemic circulation (Figure 2.12) [1]. Radiolabeled ADCs can be used to study tissue distribution in tumor-bearing mice to assess the extent of ADC enrichment in target tissues.

2.3.3 Analytical Procedures

There are many challenges in the detection of payload-related metabolites due to the low concentrations released from ADCs and the difficulty in prediction of released products of non-cleavable ADCs. To this end, an LC-HRMS method was established for the discovery and identification of payload-related metabolites released by ADC drugs. This method uses both targeted and nontargeted analytical techniques in data processing to quickly and comprehensively discover payload-related metabolites with unknown conjugation sites or connected with unknown amino acid sequences (Figures 2.13 and 2.14).

LC-HRMS can accurately distinguish the metabolite signals from the background signals within several mDa mass tolerance. The use of background subtraction technology can highlight the mass spectrum signals of the products from complex background signals, and it is a typical nontargeted data analysis technology.

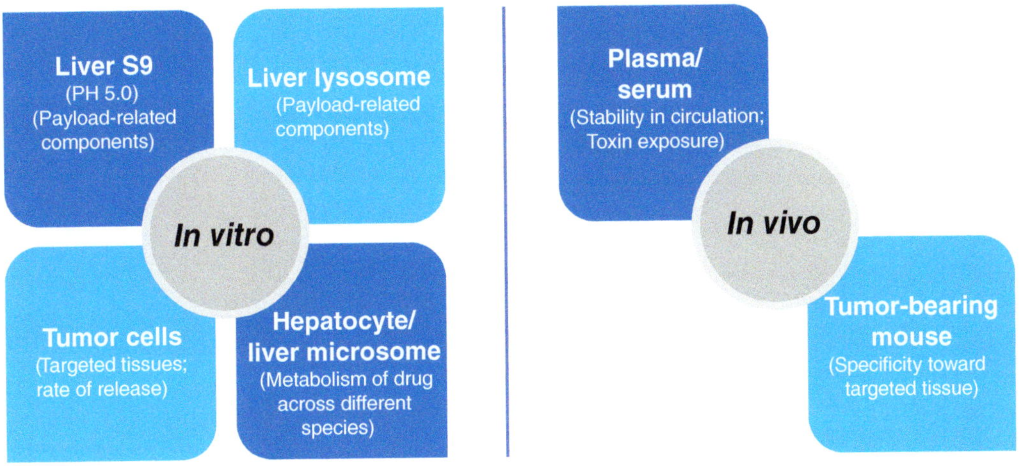

Figure 2.12 ADC drug metabolism study strategy.

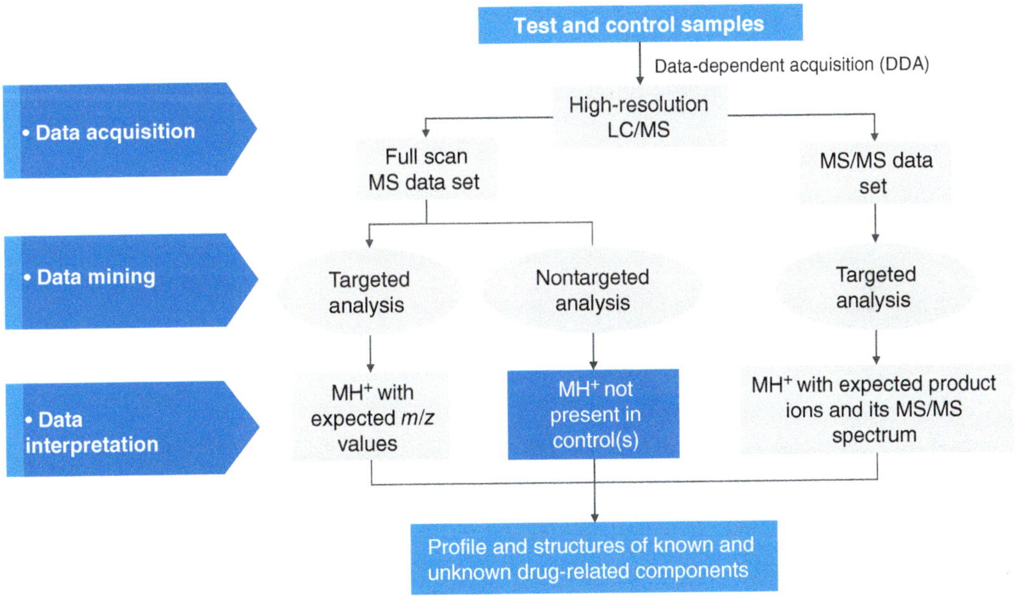

Figure 2.13 LC-HRMS method for ADC metabolites.

2.3.4 Case Analysis

Example 1: Nontargeted high-resolution mass spectrometry method for finding payload-related metabolites released by ADCs in *in vitro* lysosomes.

A schematic diagram of the structure of ADC-1 is shown in Figure 2.15.

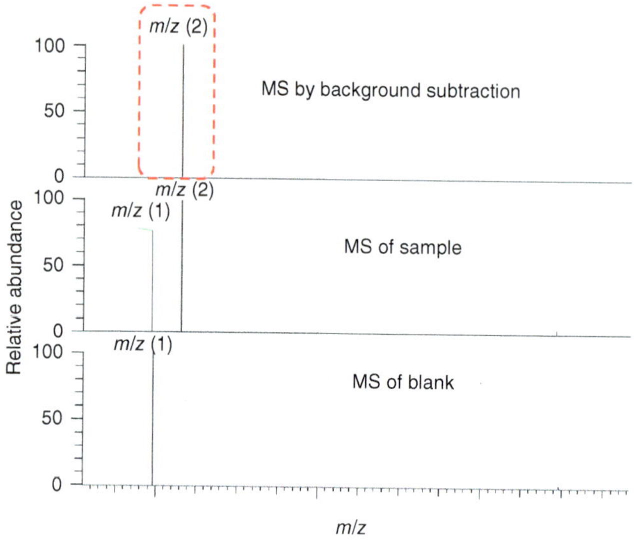

Figure 2.14 LC-HRMS after background subtraction.

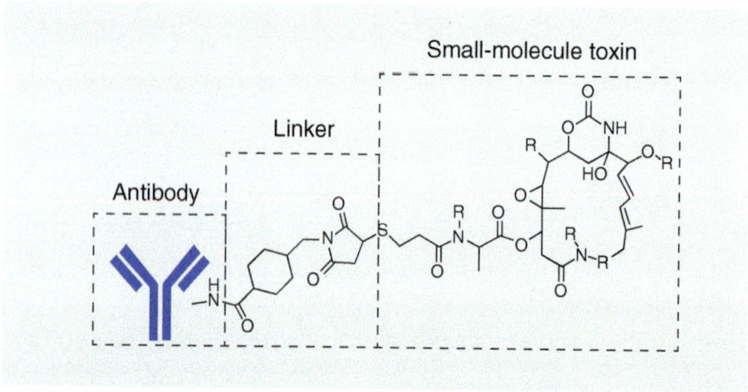

Figure 2.15 Structural diagram of ADC-1.

ADC-1 was a non-cleavable ADC. In ADC-1, the payload is coupled to the antibody on the lysine residues of the antibody through a linker. ADC-1 and the corresponding antibody were incubated in an acidic human lysosomal incubation system for 48 hours. Primary and secondary MS data were collected by an HRMS. After incubation of ADC-1 in acidified lysosomes, LC-UV, LC-HRMS, and background-subtracted sample chromatograms were obtained as shown in Figure 2.16.

The major degradation product released by ADC-1, shown in Figure 2.16, is lysine residue–linker–payload, which is consistent with the results reported in the literature and confirms the reliability of the *in vitro* metabolic system and detection method. In addition to major degradation products, minor metabolites with unknown amino acid sequences can also be observed after

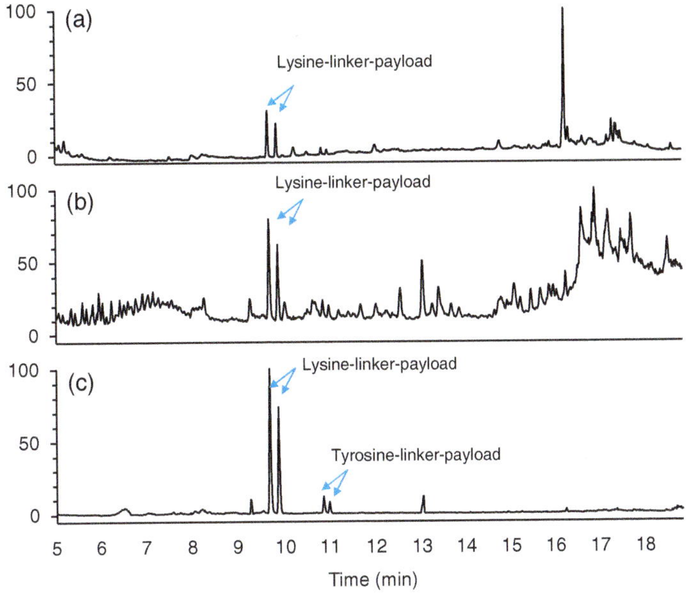

Figure 2.16 LC-UV (a), LC-MS (b), and background-subtracted LC-MS (c) of ADC-1 after acidified lysosomal incubation.

processing using the background subtraction method, which indicates that untargeted data processing technology can help discover unknown degradation products.

Example 2: High-resolution mass spectrometry analysis of small molecules released by ADC-2 in acidified liver S9.

The structural diagram of ADC-2 is shown in Figure 2.17.

We applied the established assay to analyze the unknown drug ADC-2, which consists of a known payload and new antibodies. ADC-2 was incubated with liver S9 for 48 hours. After that, full-scan chromatography was performed by LC-HRMS (Figure 2.18a). However, due to the low concentrations of small molecules released from ADC-2, which were more severely interfered with by the matrix, Panel A did not show small molecules. Background subtraction data processing effectively removed the matrix interference signal and obtained full-scan LC-HRMS chromatograms (Figure 2.18b) displaying Cys-mc-MMAF. It is the main degradation product. According to the parent ion of Cys-mc-MMAF, we obtained the LC-HRMS spectrum of Cys-mc-MMAF in the MS/MS data, which obtained the structure of Cys-mc-MMAF and confirmed the site connecting the linker–payload in ADC-2 and the release mechanism.

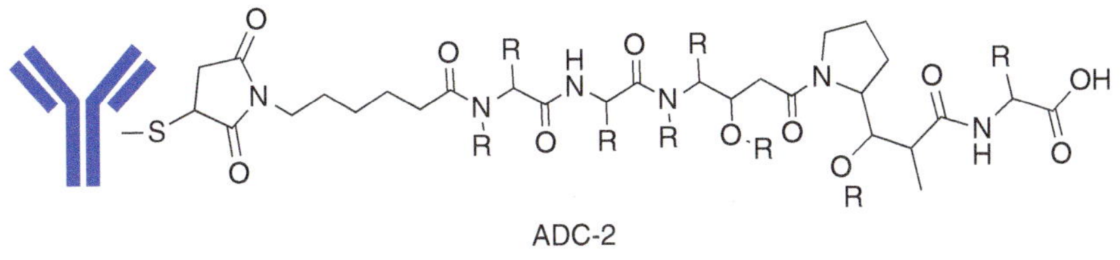

Figure 2.17 Structural diagram of ADC-2.

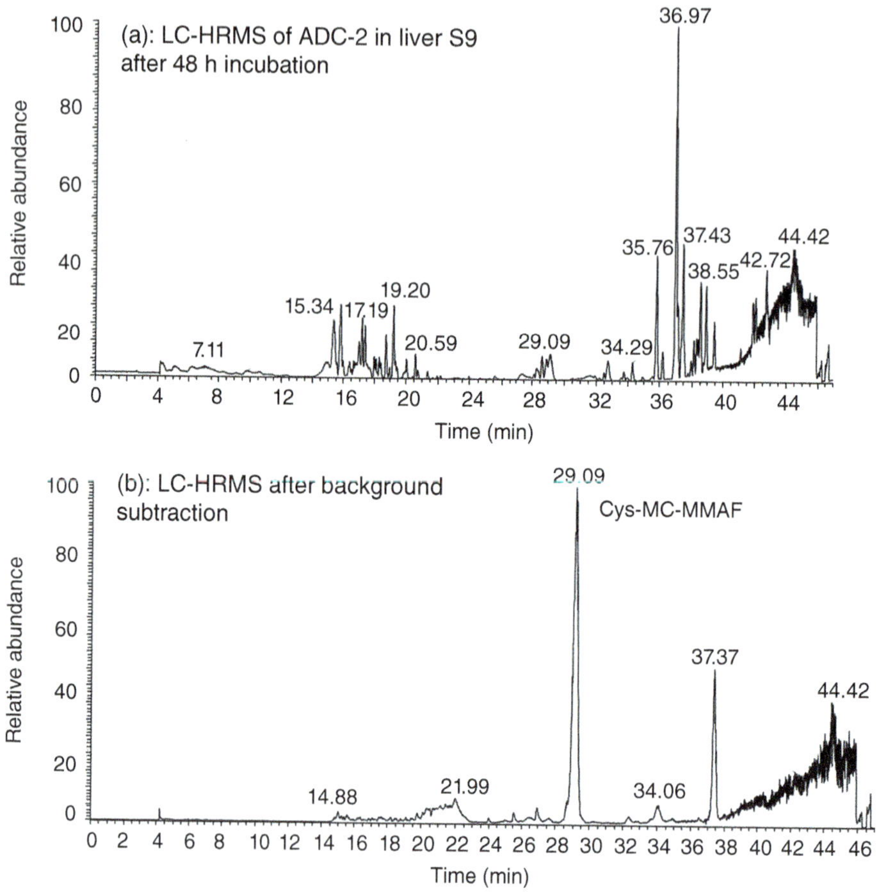

Figure 2.18 LC-HRMS (a) and LC-HRMS with background subtraction (b) after ADC-2 incubation in liver S9.

2.3.5 Conclusion

Two examples used nontargeted HRMS to search for and identify payload-related metabolites in ADC incubation with liver lysosomes and S9 *in vitro* illustrate that liver lysosomes and S9 are effective *in vitro* metabolic models to study the release of toxin molecules by the ADC and that nontargeted HRMS is an effective method to discover and identify unknown payload-related metabolites released by the ADC.

Reference

1 Tingting Cai, Liqi Shi, Huihui Guo, Ruixing Li, Weiqun Cao, Liang Shen, Mingshe Zhu, Yi Tao, Detection and Characterization of In Vitro Payload-Containing Catabolites of Noncleavable Antibody-Drug Conjugates by High-Resolution Mass Spectrometry and Multiple Data Mining Tools, Drug Metabolism and Disposition,Volume 51, Issue 5,2023, Pages 591–598.

2.4 Strategies and Challenges in the Determination of Drug-to-Antibody Ratio (DAR) Values for ADCs

Liqi Shi, Weiqun Cao

2.4.1 Introduction

ADCs have entered the stage of rapid and explosive growth in recent years. According to incomplete statistics, the number of ADC preclinical/clinical R&D pipelines worldwide has also been accelerating annually (listed in Table 2.4).

Different antibodies, linkers, or payloads can affect the PK and TK characteristics of ADCs. Even the distribution of conjugated payloads within the same ADC may also lead to different PK and TK properties. Therefore, it is important to evaluate ADC through the analysis of the distribution of DAR values (the average number of drug molecules conjugated to an antibody). This analysis, whether from a pharmacology, PK, or safety evaluation perspective, can help optimize ADCs to ensure the optimal therapeutic effect while minimizing potential adverse effects. It holds significant guidance for R&D, clinical trial design, and treatment plans of ADC drugs.

Determining payload distribution and average DAR value requires consideration of several factors, including coupling method, linker type, payload, and analysis scenario (synthesis characterization or PK analysis). The previous chapter explained the methodology for DAR value analysis mainly using T-DM1 (Kadcyla®), a representative lysine-conjugated ADC, as an example. This chapter uses DS-8201 (trastuzumab deruxtecan [T-Dxd], Enhertu®) as an example to highlight the differences in qualitative and semiquantitative DAR value detection methods between cysteine- and lysine-conjugated ADC. By analyzing the changes of DAR values of DS-8201 in *in vitro* incubation system and *in vivo* plasma post-administration, the considerations, difficulties, and countermeasures for detecting the distribution and changes of ADC's DAR values in DMPK studies are explored in this chapter.

2.4.2 Common Conjugation Strategies for ADCs

ADCs are primarily differentiated by their conjugation sites and mainly include lysine-conjugated, cysteine-conjugated, genetically engineered unnatural amino acid-conjugated, and N-glycosylated conjugated ADCs [1]. The drug structures are summarized in Table 2.5. Among the marketed ADC drugs until 2023, lysine-conjugated ADCs (four), exemplified by T-DM1 (Kadcyla®), and cysteine-conjugated ADCs (nine), represented by DS-8201 (Enhertu®), dominate the market.

Table 2.4 The number of global preclinical/clinical development pipelines for ADCs from 2018 to 2023.

Year	2018	2019	2020	2021	2022	2023
Preclinical	171	154	154	182	220	315
Phase I	48	49	46	48	56	95
Phase II	27	29	29	30	42	53
Phase III	6	8	6	8	6	13

Table 2.5 Summary of conjugation method of approved ADCs up to 2023.

Trade name	Year approved	Target	Antibody	Conjugation site	Payload	Linker
Mylotarg®	2000/ September 2017	CD33	Gemtuzumab	Lysine	Calicheamicin	Hydrazone
Adcetris®	August 2011	CD30	Brentuximab	Cysteine	MMAE	MC-VC-pABC
Kadcyla® (T-DM1)	February 2013	HER2	Trastuzumab	Lysine	DM1	Non-cleavable
Besponsa®	August 2017	CD22	Inotuzumab	Lysine	Calicheamicin	Hydrazone
Lumoxiti®	September 2018	CD22	Moxetumomab	Fusion protein	Pseudomonas exotoxin A	NA
Polivy®	June 2019	CD79β	Polatuzumab	Cysteine	MMAE	MC-VC-pABC
Padcev®	December 2019	Nectin-4	Enfortumab	Cysteine	MMAE	MC-VC-pABC
Enhertu® (T-Dxd, DS-8201)	December 2019	HER2	Trastuzumab	Cysteine	Dxd	GGFG
Trodelvy®	April 2020	Trop-2	Sacituzumab	Cysteine	SN38	CL2A
Blenrep®	August 2020	BCMA	Belantamab	Cysteine	MMAF	MC-VV
Zynlonta®	April 2021	CD19	Loncastuximab	Cysteine	PBD	MC-PEG-VA-pABC
Aidixi®	June 2021	HER2	Disitamab	Cysteine	MMAE	MC-VC-pABC
Tivdak®	September 2021	TF	Tisotumab	Cysteine	MMAE	MC VC-pABC
Elahere®	November 2022	FRα	Mirvetuximab	Lysine	DM4	Disulfide bond

Cysteine-conjugated ADCs also include Seagen's series of drugs (Adcetris®, Padcev®, Polivy®, and Tivdak®) based on the vedotin platform with MC linker and MMAE payload. Between 2019 and 2022, cysteine-conjugated ADCs accounted for 90% of the marketed ADC drugs.

2.4.3 Structural Characteristics of Cysteine-Conjugated ADCs

The immunoglobulin G1 (IgG1)-type antibody has four pairs of interchain disulfide bonds composed of cysteine residues that could be chemically reduced to yield eight free thiol groups. These free thiol groups can undergo a conjugation process with linker–payload via the sulfhydryl groups located through a chemical reaction with the linker (e.g., Michael addition reaction with a nucleophilic reaction substrate such as maleimide), and this can be considered as a form of site-specific conjugation. Figure 2.19 shows that ADCs with a maximum conjugated number, DAR8, can be obtained when saturated conjugation is used. The following discussion provides a deep understanding of cysteine-conjugated ADC drugs by examining the structure of the classical drug DS-8201 (Figure 2.20).

DS-8201 uses trastuzumab as the antibody, a camptothecin analog Dxd as the payload, and enzymolysis-cleavable GGFG tetrapeptide as the linker [2]. The conjugation of antibodies to the drug is achieved through the addition of the thiol group on a monoclonal antibody (mAb) to the maleimide group on the linker, resulting in an average DAR of 7–8. The linker optimization of DS-8201 can improve its enzymatic hydrolysis stability to reduce the untargeted payload release; it can

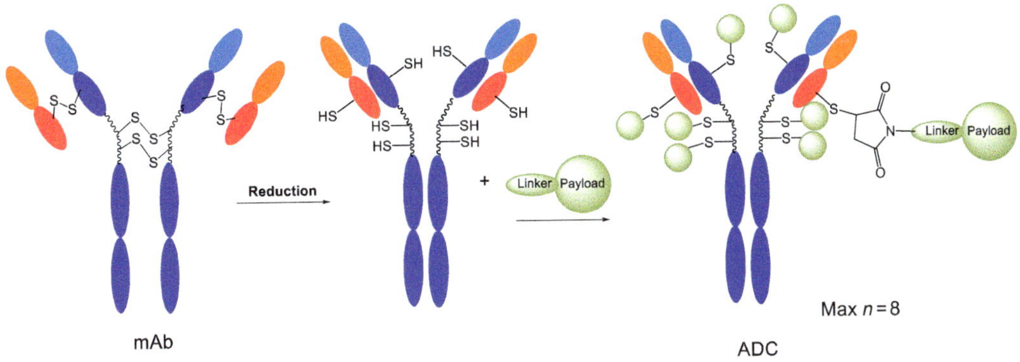

Figure 2.19 Structural representation of cysteine-conjugated ADC.

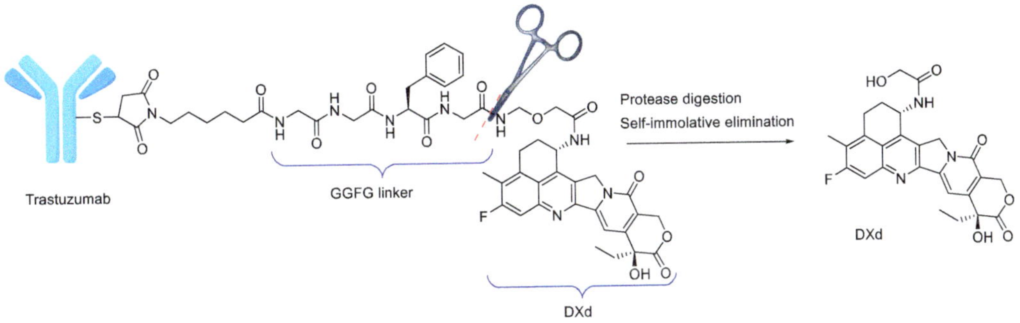

Figure 2.20 Structural diagram and release mechanism of DS-8201.

also avoid physicochemical property disadvantages such as hydrophobic payload aggregation at high DAR values [3]. This allows maximized benefits of payload release quantity and site-directed conjugation uniformity.

The success of DS-8201 brings more confidence in cysteine site-specific conjugation techniques for ADC development. According to public information [4–7], other ADCs have been developed in different clinical stages, such as patritumab deruxtecan (U3-1402) [4] developed by Daiichi Sankyo targeting HER3 and SGN-CD228A [5] developed by Seagen. Patritumab deruxtecan (U3-1402) uses the same GGFG linker and Dxd payload as DS-8201, and SGN-CD228A uses glucuronic acid as the linker and MMAE as the payload. In China, cysteine site-specific conjugate ADCs, such as SHR-A1811 [6] developed by Hengrui and 23C2 HER2-2-DDDXD [7] developed by Chia Tai Tianqing, are under development.

2.4.4 Application of High-Resolution Mass Spectrometry (HRMS) in the Detection of DAR Values of ADC Drugs in Biological Samples

HRMS is an effective technique for determining the DAR values of ADC. Immunocapture enrichment is required before detecting the DAR values of ADC in biological samples (e.g., serum/plasma) by HRMS. As shown in Figure 2.21, the immunocapture method usually uses a streptavidin-modified carrier (such as magnetic beads or extraction columns) to form a complex with biotin-labeled capture protein, which binds to the ADC analyte in the samples for antigen–antibody

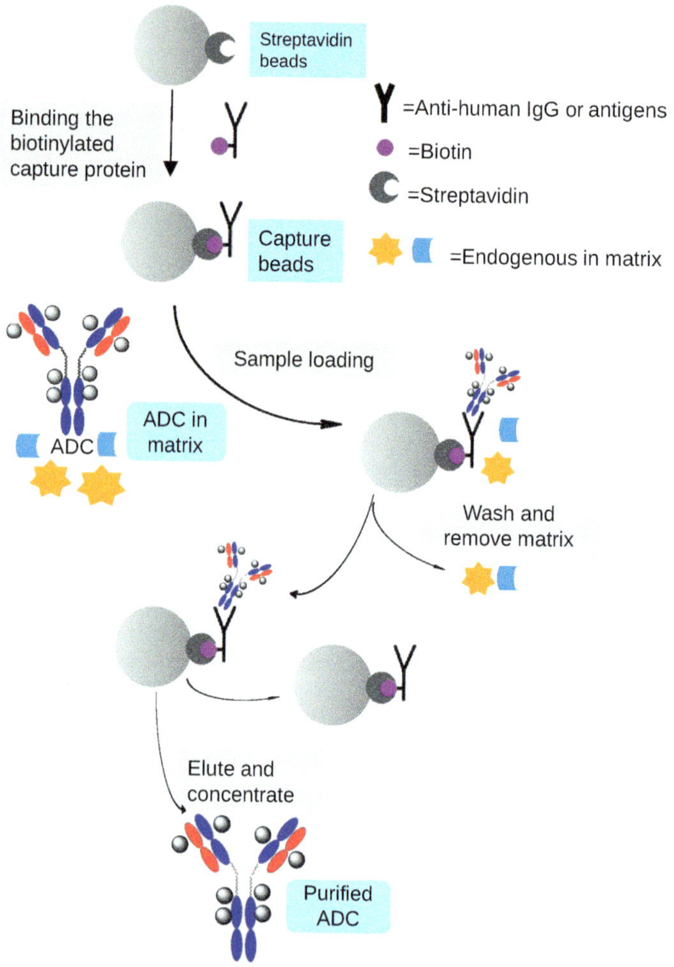

Figure 2.21 Method workflow of immunocapture of ADC in biological samples by magnetic beads.

reaction and retains on the carrier. After discarding the solution, the carrier is eluted to obtain purified ADC and antibody protein.

ADCs with different DAR values have distinct molecular weights. The intact protein or protein subunits (light chain/heavy chain) can be separated by LC and detected by HRMS. The mass–charge ratio (m/z) of each component in the samples can be obtained. As shown in Figure 2.22, the proteins with different numbers of charges present the distribution of the valence mass–charge ratio in the mass spectrum. The deconvoluted mass spectrum of different components needs to be converted by software processing (Figure 2.23). The DAR values of different components are characterized according to the mass to obtain the DAR value distribution information of ADC drugs, and then, the average DAR value can be calculated by normalization according to the MS intensity of components with different DAR values.

As shown in Figure 2.23a, for lysine-conjugated ADCs, the molecular weight of the intact protein is usually directly detected, while the mass of light-chain/heavy-chain substructures is analyzed to obtain DAR values for cysteine-conjugated ADCs (depicted in Figure 2.23b).

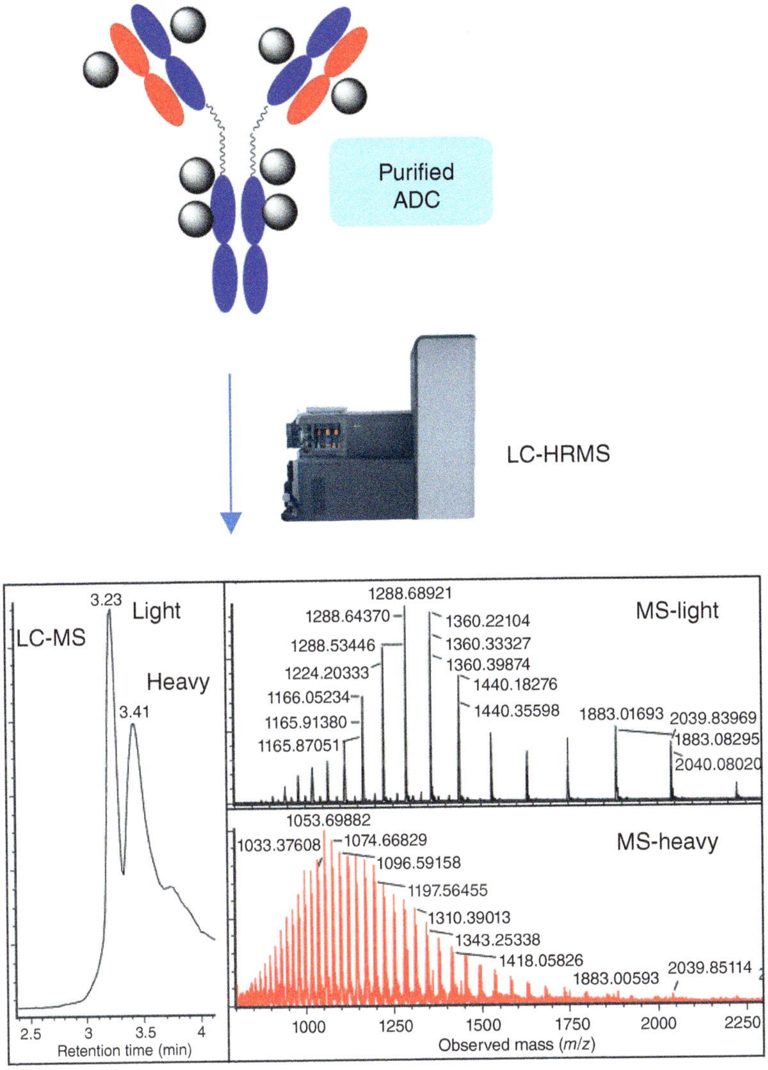

Figure 2.22 LC-MS profiles and mass spectra of ADC by liquid chromatography–high-resolution mass spectrometry (LC-HRMS).

When all interchain disulfide bonds are reduced and conjugated (DAR8), the antibodies of cysteine-conjugated ADCs will present a dissociation state so that two light chains and two heavy chains will appear in the reverse-phase chromatography of the acidic mobile phase due to the absence of interchain disulfide bonds (shown in Figure 2.24a). What can be detected in the HRMS is the light chain conjugated with one payload (L1) and the heavy chain conjugated with three payloads (H3).

When incomplete conjugation is used for synthesis (such as DAR4), only part of disulfide bonds is reduced to thiols and conjugated, and there may be various interchain disulfide bond combination morphologies such as light–heavy chain, heavy–heavy chain, and light–heavy–heavy chain. This leads to complex MS peaks, complicating the calculation of the DAR value and even hindering the calculation of the accurate DAR values. However, even in the complete conjugation mode, the

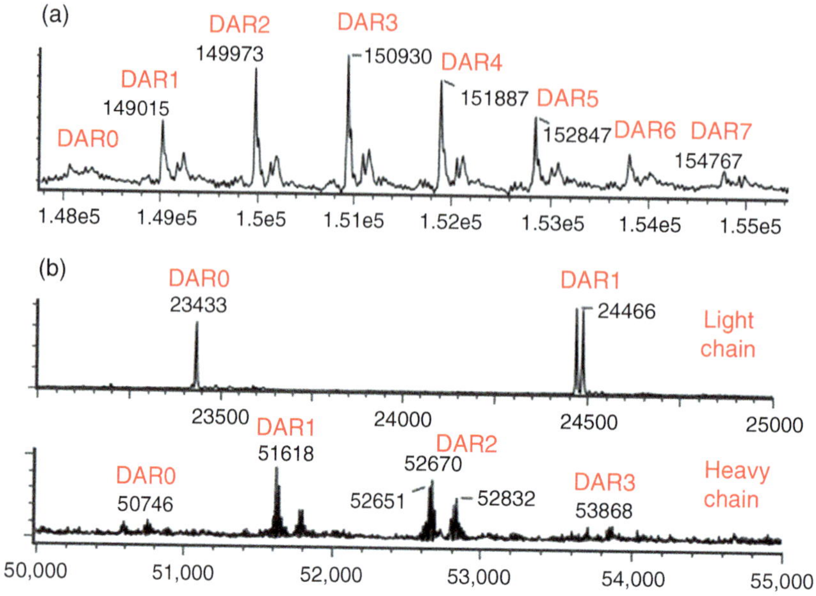

Figure 2.23 Deconvoluted mass spectrum and DAR value distribution of ADC. (a) Intact protein mass of the lysine-conjugated ADC; (b) mass of the light-chain/heavy-chain substructure of the cysteine-conjugated ADC.

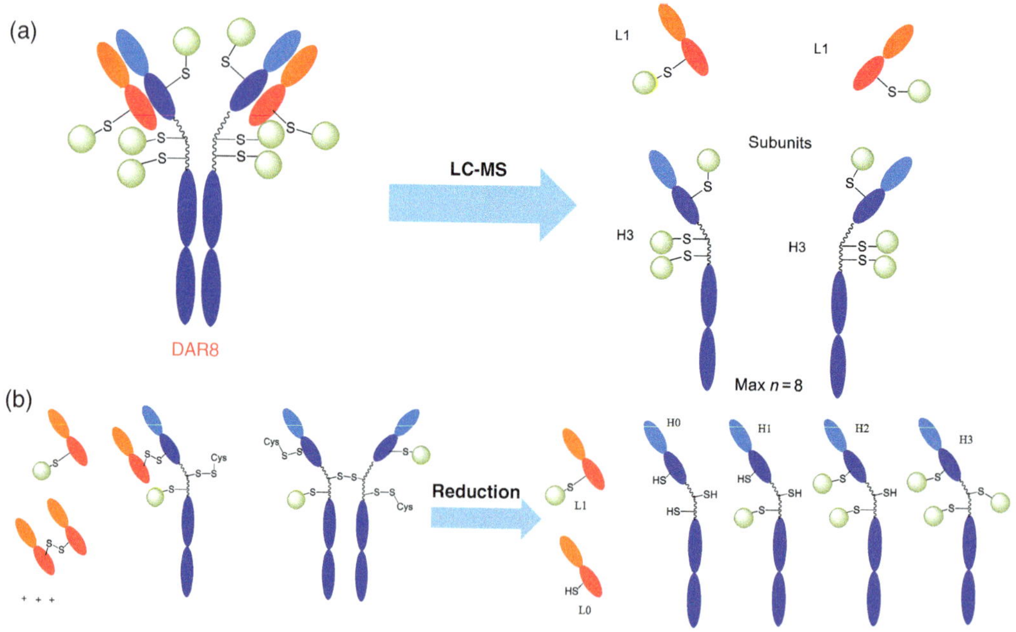

Figure 2.24 Schematic diagram of cysteine-conjugated ADC melting in chromatography (a) and reduction pretreatment (b).

ADC with a high DAR value may still have linker deconjugation, and then, the exposed thiols may form oxidized interchain disulfide bonds, which complicates the calculation of the DAR value.

For the interchain disulfide bond combination in these cases, the reduction step can be used post-immunocapture to ensure that the interchain disulfide bonds are opened (demonstrated in Figure 2.24b). The test objects degenerate into two forms of DAR = 1 (L1) and DAR = 0 (L0) for the light chain and four forms of DAR = 3 (H3), DAR = 2 (H2), DAR = 1 (H1), and DAR = 0 (H0) for the heavy chain. After obtaining the mass spectrometric intensities of the light chains and heavy chains, the overall average DAR value of the ADC can be obtained.

For non-site-specific lysine-conjugated ADCs, the conjugation sites include 80–90 lysine residues distributed on the antibody without changes in interchain disulfide bonds. If the disulfide bonds of these ADCs are also reduced, the distribution of DAR values of their light and heavy chains will be more complex, and the difficulty of data processing will increase, so the non-reduced intact proteins will be selected as analytes in this case.

2.4.5 Case Study: Analysis of DAR Values for DS-8201

As shown in Figure 2.25, according to the results from the monkey *in vivo* PK study of DS-8201 reported in the literature [8], the half-life of total antibody and total ADC was about 4 days, and the plasma concentrations decreased from about 100 μg/mL to below 10 μg/mL within 0–14 days. From a DMPK perspective, the incubation conditions of *in vitro* plasma stability experiments can be set up according to the C_{max} (100–200 μg/mL) of *in vivo* studies after administration, while for the detection of DAR values of *in vivo* samples from animals and clinical trials (the clinical administration

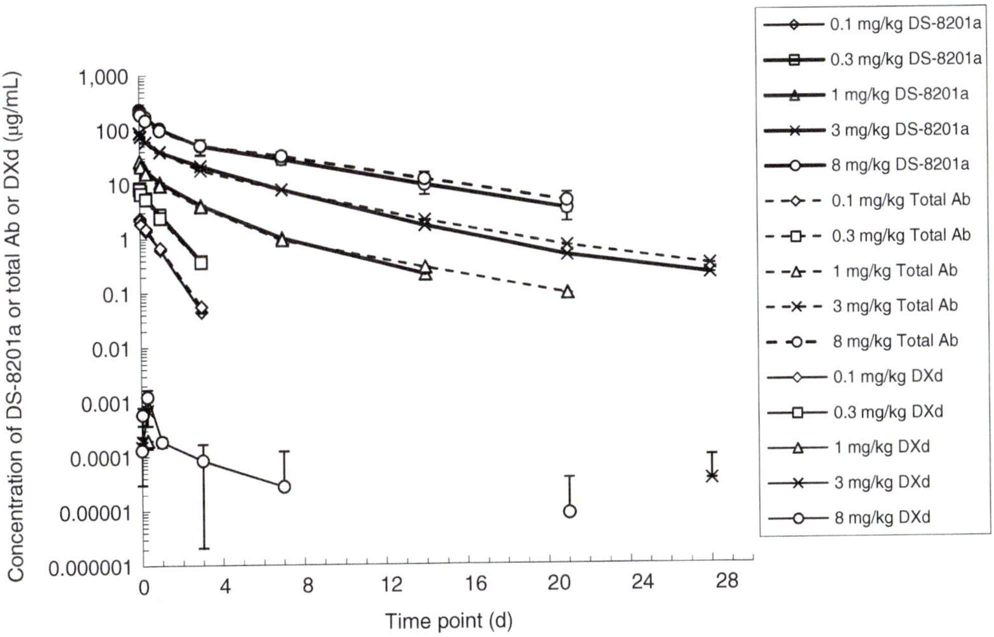

Figure 2.25 *In vivo* pharmacokinetic data after injection of DS-8201 in monkeys (dosage of 0.1–8 mg/kg; —: total ADC; ---: total antibody; and —: payload Dxd) [8] / with permission of Taylor & Francis Group.

dose of DS-8201 is 5.4 mg/kg), the limit of detection of nearly 1 μg/mL level needs to be met, making the analysis more challenging.

Referring to the literature data, we performed *in vivo* PK experiments in SD rats at a dose of 3 mg/kg of DS-8201 (T-Dxd) and performed *in vitro* stability studies in rat plasma at an incubation concentration of 100 μg/mL of DS-8201 (T-Dxd). The obtained *in vitro* and *in vivo* plasma samples were subjected to acid elution after affinity immunocapture, followed by neutralization and treatment with a reducing agent. The final processed samples were detected by ACQUITY I-Class liquid chromatogram (Waters) tandem Vion Q-Tof MS (Waters), and the data were processed using UNIFI software (Waters).

Figure 2.26 shows the deconvoluted mass spectra of T-Dxd in plasma samples at different time points through the above analysis process. The molecular weight of the linker–payload of T-Dxd was 1,034. Figure 2.26a-1 and b-1 shows the mass of the 0-payload light chain (L0) and 1-payload light chain ($L1 = L0 + 1034$ Da). As the plasma incubation time increases, the signal of L0 gradually rises as the signal of L1 decreases, which represents that the light chain loses the mass of an intact linker–payload, that is, the reverse Michael elimination reaction occurs. At the same time, the light chain gradually generates a peak of $L1 + 18$ Da, which represents the hydrolytic ring opening of succinimide according to the structure and mechanism of the linker, and this hydrolytic ring opening will competitively inhibit the decoupling reaction of reverse Michael elimination. Since no other mass was detected between L0 and L1, no cleavage of linker–payload occurred except for the loss of intact linker–payload, reflecting the high stability of the tetrapeptide linker design of T-Dxd and the payload Dxd.

For the heavy chains shown in Figure 2.26a-2 and b-2, the mass species distribution in the mass spectrum is much more complex than that of the light chain. The mass spectrum shows that in addition to the main peak H3 (heavy chain containing three payloads), there is a mass peak with a difference of 162 Da, representing the different glycoform distribution on the heavy chain inherited from the mAb. With the extension of incubation time, when the entire linker–payload deconjugation occurs, the 2-payload heavy chain (H2), the 1-payload heavy chain (H1), and the 0-payload heavy chain (H0) are gradually produced with corresponding glycoform peaks. In addition, the succinimide of each linker can undergo hydrolytic ring opening, and then the heavy chain Hn ($n = 1, 2, 3$) can undergo $n + 18$ Da mass changes in addition to the number of removed masses of 1,034 Da. When calculating the DAR values, each species with different degrees of hydrolysis needs to be counted because the degree of hydrolysis is dynamically changed, and the degree of hydrolysis varies for mass species with different DAR values.

2.4.6 Difficulties and Countermeasures of DAR Value Detection by LC-HRMS

Due to the conjugation with the linker and payload, there are some differences in the structure and properties between ADC and mAb. Some challenges will be encountered in DAR value detection by LC-HRMS. The following sections discuss two main difficulties and provide possible countermeasures.

2.4.6.1 Immunocapture Efficiency

When the steric hindrance of the payload is significant, the conjugated ADC may easily form immune binding compared with the mAb. To obtain better immune capture efficiency, it is critical to select the appropriate capture reagent. Immunocapture usually targets the Fc region of the antibody, using antihuman Fc antibody capture, or targets the variable region using antigen protein capture. Since ADCs usually use human-derived IgG-based antibodies, when the matrix is human plasma, various human-derived IgGs dominate the matrix, and extraction with anti-Fc antibody protein lacks selectivity. At this time, only target antigen protein can be used for capture, with a relatively high cost, while in the animal matrices, the anti-Fc antibody is usually used to obtain good extraction selectivity. However, when the capture efficiency is low or the matrix interference is

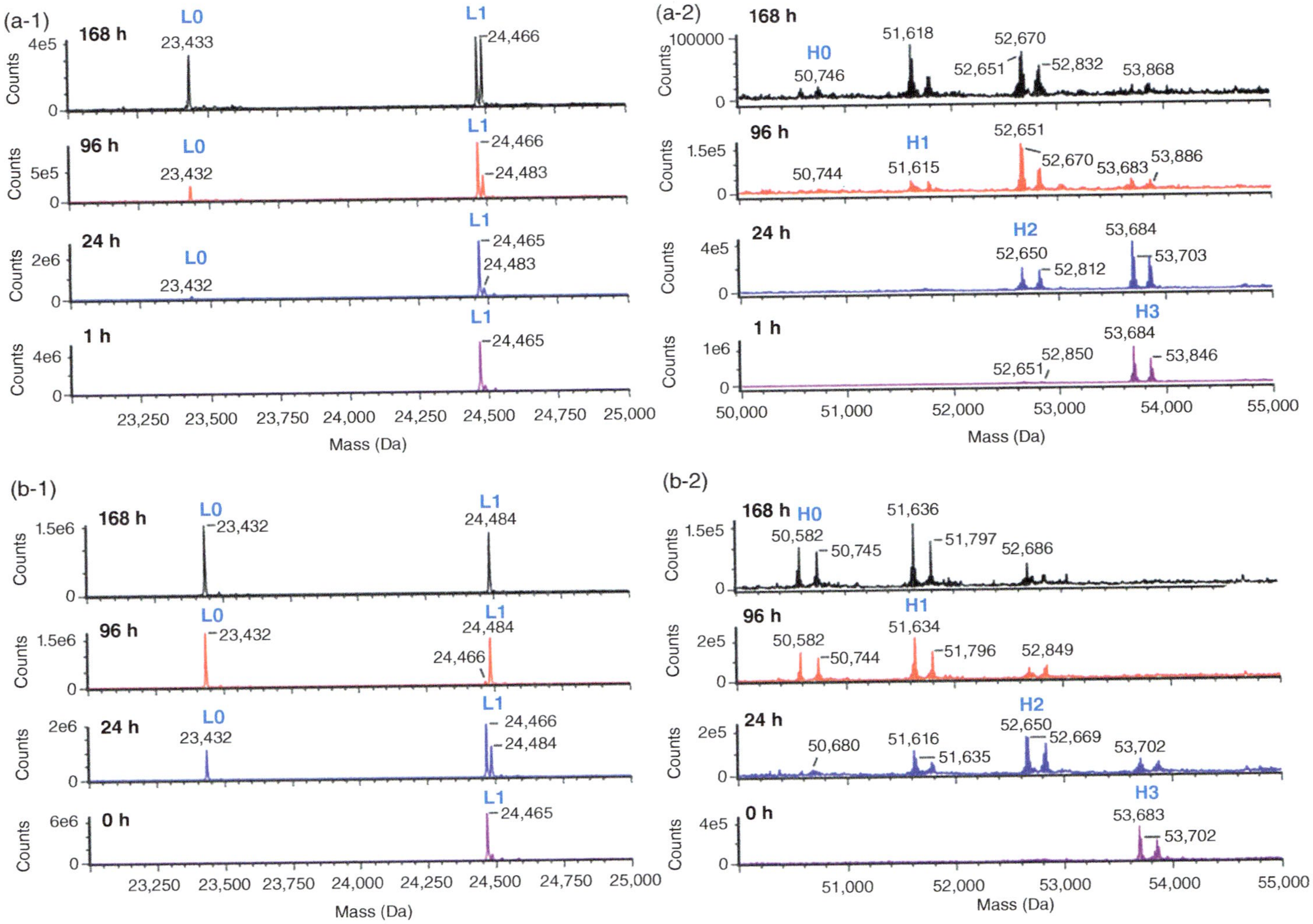

Figure 2.26 Deconvoluted mass spectra of T-Dxd in plasma samples (0, 24, 96, and 168 hours after incubation *in vitro*; 1, 24, 96, and 168 hours after administration *in vivo*). (a-1) *In vivo* light chain; (a-2) *in vivo* heavy chain; (b-1) *in vitro* light chain; and (b-2) *in vitro* heavy chain.

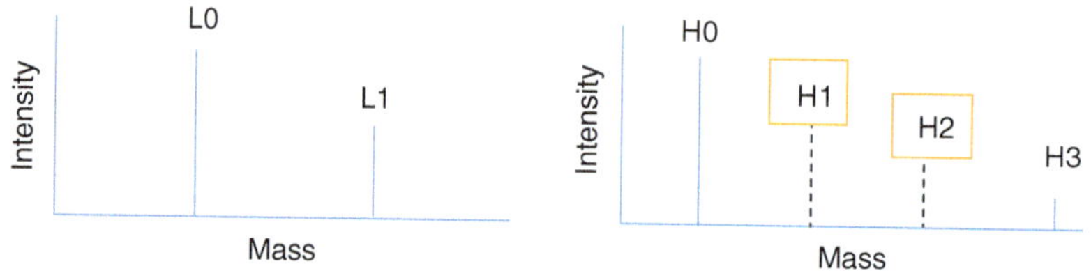

Figure 2.27 Schematic intensity of different DAR species of antibody in the mass spectrum.

large, it may still be necessary to consider the use of target antigen protein as the capture reagent to enhance the selectivity and improve the extraction efficiency.

Optimization of extraction recovery by immunocapture can also be achieved by adjusting the amount of solid-phase, the amount of capture protein and the volume of plasma sample. The loading volume of the plasma sample is evaluated according to the *in vitro* incubation concentration, *in vivo* administration dosage, and the limit of detection of the instrument. The amount of capture protein is slightly excessive depending on the estimated drug concentration, and saturation of the surface site of the solid phase needs to be considered to reduce sample loss due to non-specific adsorption or excessive matrix impurity residue.

2.4.6.2 Mass Spectrometry Response for Different DAR Values

The polarity and charge distribution of the payload affect the ionization efficiency of ADCs with different DAR values. These factors cause variations in the mass spectrometry response signal of ADCs with different DAR values, leading to overestimation or underestimation (Figure 2.27). In this case, the correction curve of response intensity can be obtained by proportionally mixing the naked antibody and ADC with the maximum DAR values. The coefficient of relative response/relative concentration of the species with a high DAR value relative to the fragment with a low DAR value can be calculated, and then, the weight coefficient can be added when calculating the DAR value. When the payload not only affects the capture efficiency but also inhibits the MS response, the response sensitivity of the components with high DAR value will be severely limited, especially considering that the heavy chain will simultaneously superimpose the effects of glycoform distribution, DAR distribution, and hydrolysis peak distribution, and its signal will be significantly lower than that of the light chain. These issues need to be taken into full consideration for the prediction of effective test concentrations; deglycosylation operations can be performed, and mass spectrometric detection modes and ionization parameters can be adjusted during method development. If the effective signal-to-noise ratio cannot be obtained, the effective qualitative and semiquantitative analysis of DAR values cannot be performed for the samples of the PK experiments.

2.4.7 Summary

As part of the continuous development of ADC technology, engineered cysteine conjugation technology has been developed. These methods mutate the specific antibody sites into cysteine, allowing the generation of the cysteine-conjugated ADC with a determined conjugation position and DAR without opening the interchain disulfide bond, such as the THIOMAB technology developed by Genentech [9]. In addition, the retro-Michael elimination reaction can be prevented by hydrolytic ring-opening pretreatment to the maleimide or other covalent binding methods that resist decoupling, such as SKB264/BT001035 developed by Kelun-Biotech [10], which uses a methanesulfonylpyrimidine–thiol coupling method, to minimize decoupling.

For the MS analysis technology of antibodies, researchers are also continuously developing, adjusting, and optimizing to better meet specific needs, such as microfluidic and capillary electrophoresis (CE)-based separation, native analysis, and other methods.

From the large-molecule protein perspective, biotransformation can be qualitatively measured, and DAR value changes can be calculated based on molecular weight. Combined with the metabolite identification of payload-related small-molecule compounds released from ADCs, this can help comprehensively evaluate the biotransformation of ADCs.

References

1 Fu, Z., Li, S., Han, S. et al. (2022). Antibody-drug conjugate: the "biological missile" for targeted cancer therapy. *Signal Transduction and Targeted Therapy* 7 (1): 93.
2 Nakada, T., Sugihara, K., Jikoh, T. et al. (2019). The latest research and development into the antibody–drug conjugate, [fam-] trastuzumab deruxtecan (DS-8201a), for HER2 cancer therapy. *Chemical and Pharmaceutical Bulletin* 67 (3): 173–185.
3 Ross, P.L. and Wolfe, J.L. (2016). Physical and chemical stability of antibody drug conjugates: current status. *Journal of Pharmaceutical Sciences* 105 (2): 391–397.
4 Hashimoto, Y., Koyama, K., Kamai, Y. et al. (2019). A novel HER3-targeting antibody–drug conjugate, U3-1402, exhibits potent therapeutic efficacy through the delivery of cytotoxic payload by efficient internalization. *Clinical Cancer Research* 25 (23): 7151–7161.
5 Patnaik, A., Meric-Bernstam, F., Rocha Lima, C.M.S.P. et al. (2020). SGN228-001: a phase I open-label dose-escalation, and expansion study of SGN-CD228A in select advanced solid tumors. *Journal of Clinical Oncology* 38 (15): TPS3652.
6 Zhang, T., Xu, J., Yin, J., et al. (2023). SHR-A1811, a Novel Anti-HER2 Antibody–Drug Conjugate with Optimal Drug-to-Antibody Ratio, Superior Bystander Killing Effect and Favorable Safety Profiles. Research Square.
7 Zhang, X. WO2022033578. Antibody drug conjugate.
8 Nagai, Y., Oitate, M., Shiozawa, H. et al. (2019). Comprehensive preclinical pharmacokinetic evaluations of trastuzumab deruxtecan (DS-8201a), a HER2-targeting antibody-drug conjugate, in cynomolgus monkeys. *Xenobiotica* 49 (9): 1086–1096.
9 Sadowsky, J.D., Pillow, T.H., Chen, J. et al. (2017). Development of efficient chemistry to generate site-specific disulfide-linked protein–and peptide–payload conjugates: application to THIOMAB antibody–drug conjugates. *Bioconjugate Chemistry* 28 (8): 2086–2098.
10 Cheng, Y., Yuan, X., Tian, Q. et al. (2022). Preclinical profiles of SKB264, a novel anti-TROP2 antibody conjugated to topoisomerase inhibitor, demonstrated promising antitumor efficacy compared to IMMU-132. *Frontiers in Oncology* 12: 951589.

2.5 *In Vitro* Stability of ADCs in Blood Matrices

Xinmeng Shi, Yang Gao, Liping Ma, Jing Jin

2.5.1 Introduction

ADCs are typically composed of an mAb covalently attached to a cytotoxic drug *via* a chemical linker [1]. The antibody helps to deliver a payload to the disease site *via* targeting a specific antigen expressed on the surface of the cancer cells. Therefore, the ADC can decrease the side effects on normal cells. It combines both the advantages of highly specific targeting ability and highly potent activity to achieve accurate and efficient elimination of cancer cells, which has become one of the hotspots for the R&D of anticancer drugs [2].

The stability of ADCs in circulation is critical for maximum efficacy and minimal toxicity. An intact ADC reaching the intended target can deliver the highest possible drug load to the tumor and reduce off-target toxicity caused by the premature release of payload in the blood [2]. Several factors may impact the stability of ADCs, including the linker [3], conjugation [4], physicochemical properties of the payload [5], and DAR value [6]. It is a relatively simple and direct method to assess the *in vitro* stability of ADCs using blood matrices. The NMPA issued *Guidelines for Non-clinical Studies of Antibody-Drug Conjugates* in September 2023 [7], which stated that "Before starting the clinical trials, the *in vitro* stability of ADCs should be investigated in plasma and/or blood matrices in humans and animals (pharmacology and/or toxicology species). This study is required to select appropriate analytes based on the release characteristics of ADCs, such as ADC remaining and/or the generation of free payload."

This chapter introduced various *in vitro* incubation systems and stability results and summarized the *in vitro* plasma and serum stability studies of marketed ADCs. We hope this information will provide a reference for the *in vitro* stability evaluation in ADC development.

2.5.2 *In Vitro* Stability of ADCs in Plasma

Plasma is the most frequently used matrix for evaluating the *in vitro* stability of ADCs. It is well known that plasma is obtained by the centrifugation of blood containing anticoagulant, in which the blood cells are removed, and plasma contains many kinds of hydrolytic enzymes [8].

Plasma has been widely used in various preclinical and clinical studies due to its advantages, such as simple and quick preparation, high yield, and convenience of frozen storage and transportation. In fact, the anticoagulant selection requires attention in the plasma preparation. Ethylenediaminetetraacetic acid (EDTA), a commonly used anticoagulant, is an inhibitor of protease and nuclease, and plasma prepared with EDTA may mislead the stability results of ADCs that are sensitive to protease. To mimic *in vivo* conditions, the candidate molecule is typically incubated at 37°C. The remaining ADC and/or the generation of free payload are determined at different time points. The incubation time generally varies from 4 to 21 days, and sterility should be maintained during the incubation. The selected species should include human and intended pharmacology and toxicology species, and the incubated ADC concentrations are recommended to be close to the observed or predicted *in vivo* C_{max} [9]. In addition, negative control would typically include appropriate buffers (e.g., phosphate-buffered saline [PBS]) with protein (e.g., bovine serum albumin [BSA]) to determine the stability of the ADC under buffer systems [10].

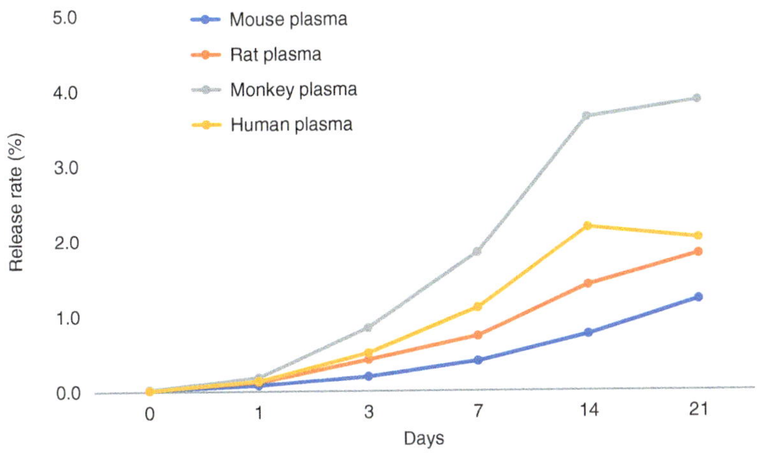

Figure 2.28 *In vitro* stability of DS-8201a in plasma. *Source:* Adapted from reference [11].

So far, literature data indicated that the *in vitro* plasma stability of many ADCs translates well with *in vivo* results. In the study of Enhertu (DS-8201a), 1.2%–3.9% of free Dxd was detected in mouse, rat, monkey, and human plasma on day 21 (Figure 2.28), indicating that Enhertu is stable in plasma. For another marketed ADC of Polivy, the concentrations of total antibody significantly decreased within 24 hours, while the decrease was much slower after 24 hours (Figure 2.29a). At 96 hours, the conjugated payload decreased to around 60% (Figure 2.29b) [12].

In the plasma stability study of maleimide-conjugated ADC of Tras-mcVC-Auristatin, Han et al. [13] observed a discrepancy between the measured free payload concentrations and loss of DAR (Figure 2.30a–d). The changes in DAR values suggested that the ADC had lost 30%–50% of its payload, while free payload concentrations measured in plasma showed that very little loss of payload occurred. Further studies suggested that the deconjugated maleimide linker–payload is transferred to serum albumin to form adducts, causing a decrease in DAR. However, this part of the payload cannot be measured as the form of free payload, which also indicates that there is a certain risk in assessing the stability of the ADC by only detecting free payload generation.

It has been reported that the frozen plasma will reduce the activity of enzymes involved in linker–payload modification [14]. Therefore, caution is needed for studies using frozen plasma. Inconsistency in stability *in vitro* and *in vivo* has also been reported; for example, Dorywalska et al. found that the valine-citrulline-p-aminocarbamate (VC-PABC) linker appears to be cleaved more efficiently *in vivo* compared with an *in vitro* plasma stability assay in mice. The researchers rationalized that this could be due to a continuous enzyme source *in vivo* or the contribution of intracellular degradation [15].

2.5.3 *In Vitro* Stability of ADCs in Serum

Serum is the clear liquid that separates from blood when blood is allowed to clot completely and then centrifuged. Therefore, the serum contains no fibrinogen and no anticoagulants. Compared with plasma, using serum can circumvent the potential risks associated with inhibition of enzyme activity by anticoagulants, as well as the effect on protein binding and stability in serum [8]. Serum is also commonly used in *in vitro* stability studies of ADCs. The study design of incubation conditions and detection is similar to plasma stability.

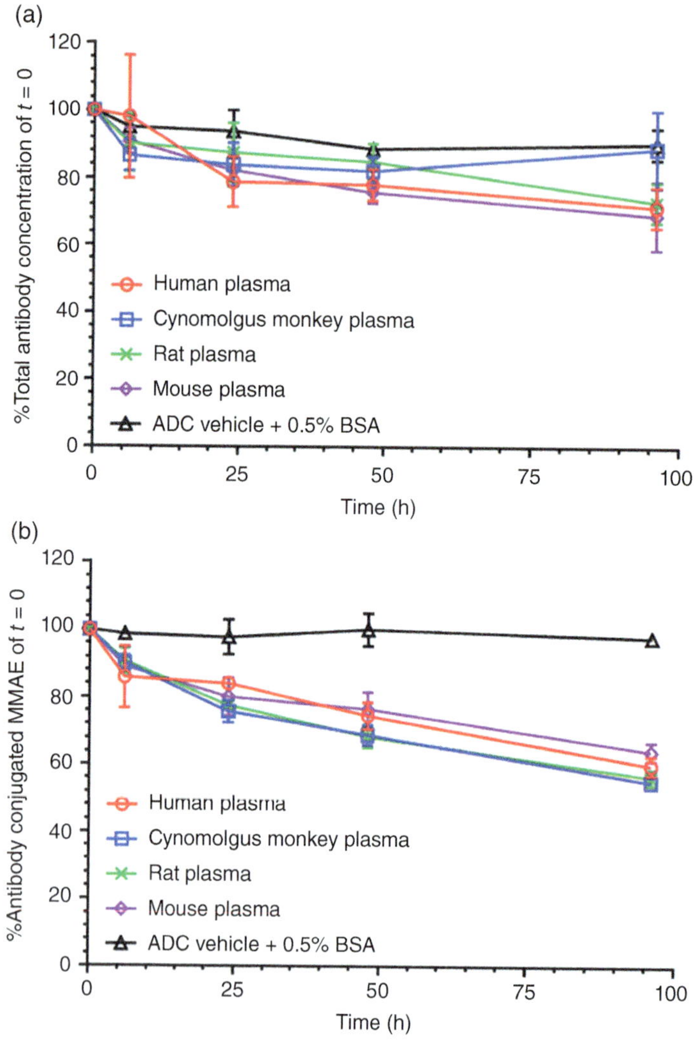

Figure 2.29 Total antibody (a) and antibody-conjugated MMAE (b) levels of Polivy following 96 hours of incubation in human, cynomolgus monkey, rat, and mouse plasma at 37°C. *Source:* [12].

Aidixi was incubated in human, rat, and mouse serum for 24 hours at 4°C and 37°C, separately, and the release of MMAE was detected. The results showed that the release of MMAE in mouse serum was 4.3% (at 37°C) and less than 1% under other conditions. Aidixi was most stable in human serum, followed by rats and mice, according to their MMAE release, indicating species differences. Overall, Aidixi showed good serum stability [16]. In the study of Trodelvy, free SN-38 concentrations were measured after 120 hours of incubation in human and monkey serum. The results showed that the half-life of SN-38 release in serum *in vitro* was approximately 1 day [17].

2.5.4 *In Vitro* Stability of ADCs in Whole Blood

Whole blood is considered to better mimic the *in vivo* circulatory environment, which includes blood cells and plasma. Whole blood has been included as one of the recommended matrices for

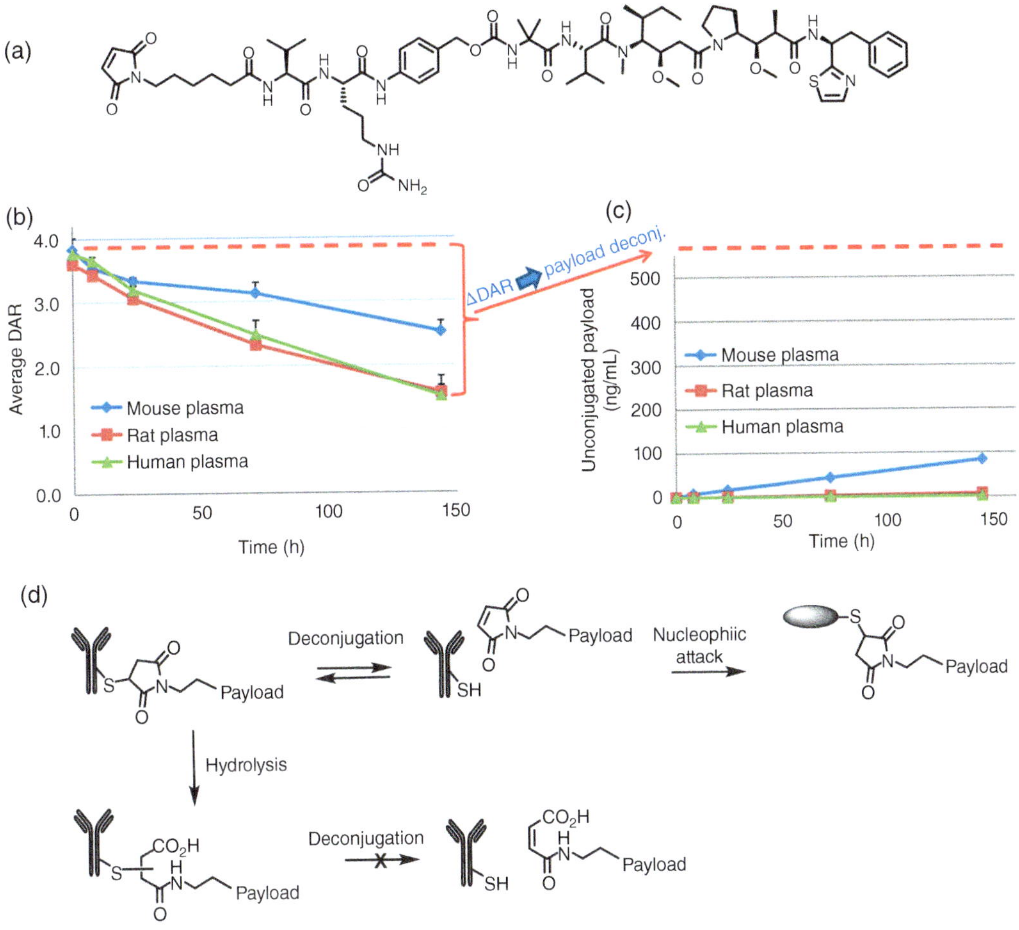

Figure 2.30 (a) Chemical structure of linker–payload of Tras-mcVC-Auristatin ADC; (b) average DAR profiles of the Tras-mcVC-Auristatin ADC over the time course of the *in vitro* plasma incubation; (c) unconjugated/free payload levels of Tras-mcVC-Auristatin over the time course of the *in vitro* plasma incubation; and (d) scheme of the maleimide exchange from the ADC. *Source:* [13] / American Chemical Society.

the *in vitro* stability study of ADCs in the NMPA guidelines. Different from plasma and serum, whole blood cannot be frozen at −80°C and is inconvenient for storage and transportation. The stability study requires a fresh matrix to be performed. As of 2023, there are only a few published reports on the *in vitro* stability of ADCs using whole blood as the matrix.

Aimee et al. used a variety of THIOMAB™ ADCs (TDCs) to evaluate the stability of the ADC in both whole blood and plasma and to correlate *in vitro* data to *in vivo* data [14]. Results of whole blood showed an improved correlation to *in vivo* stability ($R^2 = 0.87$, coefficient of determination) compared with those of unfrozen or frozen mouse plasma ($R^2 = 0.34$ and 0.01, respectively). Further studies showed that whole blood assay can also predict *in vivo* stability in other preclinical species (e.g., rat and cynomolgus monkey) and in humans. Based on our research experience, red blood cells may be broken if blood is incubated *in vitro* for over 24 hours. For ADCs that require more than 24 hours of incubation, the applicability of whole blood needs to be further evaluated (Figure 2.31a–c).

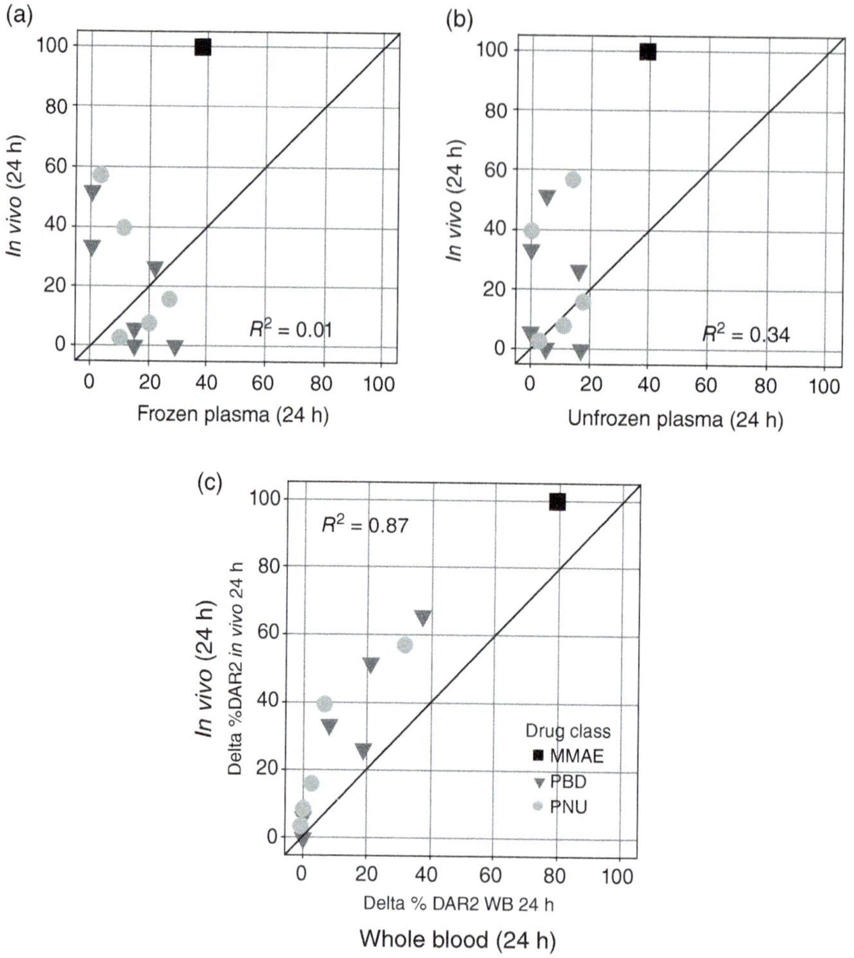

Figure 2.31 Correlation of *in vivo* with *in vitro* stability for 13 TDCs in frozen mouse plasma (a), unfrozen mouse plasma (b), mouse whole blood (c) at 24 hours, as all the tests were performed at 24 hours for the three matrices (c). *Source:* [14]/ with permission of Taylor & Francis Group. *Note:* Axes represent the percent loss of drug relative to 0 hour.

2.5.5 Analyte Selection for ADC *In Vitro* Stability Study

There are multiple options for the analytes in the *in vitro* stability study of ADCs. Various analytes and their characteristics were summarized as follows:

a) Free payload: The *in vitro* stability of ADC is evaluated by detecting the free payload released at different time points. However, it is difficult to determine payload release when the free payload is unstable in matrices or easily covalently bound to proteins. In addition, the sensitivity of detection is required to be high enough, and quantitative analysis is recommended.
b) ADC: ADC concentration is generally measured by LBA. Since conjugated antibodies with one or more payloads will be treated as the ADC in LBA, measuring the ADC concentrations by LBA cannot sensitively evaluate the payload release. Besides, anti-payload antibodies, which are necessary for ADC detection by the LBA method, are usually not available at an early stage for ADCs conjugated with new payloads.

c) DAR value: ADC stability is assessed by detecting changes in DAR over time. The average DAR values can sensitively evaluate the stability of ADC. Payload release from ADC will cause the DAR value changes, no matter what components are being released or whether the payload is stable or not, DAR detection can effectively avoid those issues, and it is more recommended to evaluate ADC stability.
d) Total antibody: This method is mainly used to evaluate the stability of antibodies. By comparing the differences between the conjugated antibody and the total antibody, we can further analyze whether the instability of ADC is caused by payload release or the instability of the antibody itself.
e) Conjugated payload: ADC stability is assessed by detecting the concentration of payload conjugated to ADC. First, it requires purification of ADC using immunocapture, followed by hydrolyzing the payload from ADC, which is applied to cleavable ADCs but not non-cleavable ADCs. Hydrolyzing the payload effectively from ADC is the most challenging part of the sample pretreatment, which requires case-by-case optimization of the hydrolysis conditions.

Table 2.6 Summary of *in vitro* stability study design for marketed ADCs.

	Drug	Matrix	Species	Incubation time	Incubation drug	Analyte
1	Adcetris	Plasma, 1% BSA buffer	R, C, H	21 d	ADC	MMAE
2	Kadcyla	Plasma, 0.5% BSA buffer	R, C, H	4 d	Tmab-MCC-[3H] DM1	Radioactivity, ADC/ total antibodies in acetonitrile-treated supernatant and precipitate
3	Besponsa	Plasma, 1% BSA buffer	R, C, H	4 d	[3H] Inotuzumab ozogamicin [3H] N-Ac-γ-calicheamicin DMH	Calicheamicin
4	Polivy	Plasma, vehicle	C, M, R, H	4 d	ADC	Total antibodies, MMAE
5	Enhertu	Plasma, 1% BSA buffer	C, M, R, H	21 d	ADC	Dxd
6	Trodelvy	Serum	C, H		ADC	SN-38
7	Blenrep	Plasma	C, R, H	4 d	ADC	McMMAF
8	Zynlonta	Plasma	C, H	7 d	ADC	
9	Aidixi	Serum	H, R, M	1 d	ADC	MMAE
10	Tivdak	Plasma, 1% BSA buffer	M, C, H	14 d	ADC	MMAE
11	Mylotarg	NA				
12	Lumoxiti	NA				
13	Padcev	NA				
14	Akalux	NA				
15	Elahere	NA				

Note: C: monkey; M: mouse; R: rat; H: human.

2.5.6 Summary of *In Vitro* Stability Studies of Marketed ADCs

Among the 15 ADCs approved worldwide, *in vitro* stability studies have been conducted on many of them. Here, we summarized the information in Table 2.6. For most marketed ADCs, plasma was selected to evaluate the *in vitro* stability, and a buffer group containing BSA was also included. Compared with conventional small molecules, the incubation time in the ADC plasma stability study is longer, varying from 1 to 21 days. By detecting the concentration of free payload and/or total antibody, the payload release or the instability of the antibody part is evaluated. Among all the approved ADCs, serum was only used to evaluate *in vitro* stability for Trodelvy and Aidixi.

2.5.7 Summary

In ADC drug development, except for the selection of antibody and payload, it is also important to incorporate an appropriate linker to combine both parts and comprehensively consider the effects of various factors on the stability, toxicity, PK, and PD of ADCs. The *in vitro* stability study in blood matrices can help to assess ADC stability in the circulatory system, off-target risk, and species differences.

References

1 Samantasinghar, A., Sunildutt, N.P., Ahmed, F. et al. (2023). A comprehensive review of key factors affecting the efficacy of antibody-drug conjugates. *Biomedicine & Pharmacotherapy* 161: 114408.

2 Tsuchikama, K. and An, Z. (2018). Antibody-drug conjugates: recent advances in conjugation and linker chemistries. *Protein Cell* 9 (1): 33–46.

3 Su, Z., Xiao, D., Xie, F. et al. (2021). Antibody-drug conjugates: recent advances in linker chemistry. *Acta Pharmaceutica Sinica B* 11 (12): 3889–3907.

4 Zhou, Q. (2017). Site-specific antibody conjugation for ADC and beyond. *Biomedicines* 5 (4): 64.

5 Shen, B.Q., Xu, K., Liu, L. et al. (2012). Conjugation site modulates the *in vivo* stability and therapeutic activity of antibody-drug conjugates. *Nature Biotechnology* 30 (2): 184–189.

6 Sun, X., Ponte, J.F., Yoder, N.C. et al. (2017). Effects of drug-antibody ratio on pharmacokinetics, biodistribution, efficacy, and tolerability of antibody-maytansinoid conjugates. *Bioconjugate Chemistry* 28 (5): 1371–1381.

7 The National Medical Products Administration (NMPA). (2023). Guidelines for non-clinical studies of antibody-drug conjugates. https://www.cde.org.cn/zdyz/domesticinfopage?zdyzIdCODE=42d891ebccbcece7dd757d6429acff33.

8 Uges, D.R. (1988). Plasma or serum in therapeutic drug monitoring and clinical toxicology. *Pharmaceutisch Weekblad Scientific Edition* 10 (5): 185–188.

9 Kraynov, E., Kamath, A.V., Walles, M. et al. (2016). Current approaches for absorption, distribution, metabolism, and excretion characterization of antibody-drug [1] conjugates: an industry white paper. *Drug Metabolism & Disposition* 44 (5): 617–623.

10 Fisher, J.E. Jr. (2021). Considerations for the nonclinical safety evaluation of antibody-drug conjugates. *Antibodies (Basel)* 10 (2): 15.

11 FDA (2020). NDA/BLA multi-disciplinary review and evaluation (BLA 761139).

12 Li, D., Lee, D., Dere, R.C. et al. (2019). Evaluation and use of an anti-cynomolgus monkey CD79b surrogate antibody-drug conjugate to enable clinical development of polatuzumab vedotin. *British Journal of Pharmacology* 176 (19): 3805–3818. And its supplement.

13 Wei, C., Zhang, G., Clark, T. et al. (2016). Where did the linker-payload go? A quantitative investigation on the destination of the released linker-payload from an antibody-drug conjugate with a maleimide linker in plasma. *Analytical Chemistry Journal* 88 (9): 4979–4986.

14 Fourie-O'Donohue, A., Chu, P.Y., Dela Cruz Chuh, J. et al. (2020). Improved translation of stability for conjugated antibodies using an *in vitro* whole blood assay. *MAbs* 12 (1): 1715705.

15 Dorywalska, M., Strop, P., Melton-Witt, J.A. et al. (2015). Site-dependent degradation of a non-cleavable auristatin-based linker-payload in rodent plasma and its effect on ADC efficacy. *PLoS One* 10 (7): e0132282.

16 Li, L., Xu, M.Z., Wang, L. et al. (2020). Conjugating MMAE to a novel anti-HER2 antibody for selective targeted delivery. *European Review for Medical and Pharmacological Sciences* 24 (24): 12929–12937.

17 Goldenberg, D.M., Cardillo, T.M., Govindan, S.V. et al. (2015). Trop-2 is a novel target for solid cancer therapy with sacituzumab govitecan (IMMU-132), an antibody-drug conjugate (ADC). *Oncotarget* 6 (26): 22496–22512.

2.6 Application of Radiolabeling Techniques in ADC Pharmacokinetic Studies

Dongke Ma, Huan Li, Wei Lei, Lingling Zhang

2.6.1 Introduction

ADCs are like precision-guided missiles, with payloads as the explosive warhead. The therapy success of ADCs depends on how many payloads are delivered to and released in the target tissues. Meanwhile, the toxicities of ADCs may be associated with the exposure of payloads in nontarget tissues.

Although some payloads of the ADCs were approved drugs, attention should still be paid to the formation of new payload-related components arising from different release mechanisms. It is necessary to study plasma protein binding, tissue distribution, metabolism, excretion/mass balance, and cytochrome P450 enzyme- and transporter-mediated drug interaction studies, which are commonly conducted for new chemical entities.

Due to low systemic exposure of payload-related components, severe matrix interferences, and the difficulty in predicting the cleavage products of non-cleavable ADCs, there are many challenges in tissue distribution, *in vivo* metabolism, and excretion studies of ADCs. The "Guidelines for Non-clinical Studies of Antibody-Drug Conjugates" published by the NMPA suggest that tissue distribution studies are needed for ADCs targeting specific tissues, which can be carried out via radiolabeling on the payload or radiolabeling on the antibody and payload to obtain more comprehensive tissue distribution characteristics [1].

This chapter summarizes the current radionuclide labeling technique for ADCs and provides examples of how to screen ADC with optimal DAR value in the drug discovery and development stage and apply radiolabeling technology to achieve comprehensive *in vivo* characterization of ADCs at the drug application stage.

2.6.2 Radiolabeling Strategies for ADC Drugs

Before performing radioactivity detection, appropriate radionuclides should be selected for the synthesis of radiolabeled payload or antibody of an ADC (Figure 2.32).

Payloads are usually radiolabeled with low-energy radionuclides [^{14}C] or [^{3}H], which have the highest specific activities of ~62.4 mCi/mmol and ~28.6 Ci/mmol, respectively. It is challenging in

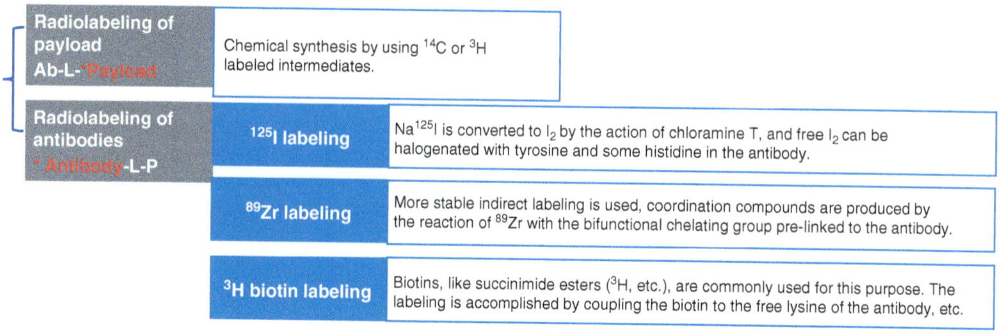

Figure 2.32 Common methods for radiolabeling. *Indicates radionuclide labeling position.

radioactivity detection due to low concentrations of payload released *in vivo*, so it is recommended to increase the administered radioactivity dose to meet the detection requirements.

Generally, a higher DAR value or more radionuclides labeled on a single payload can achieve a higher specific activity after synthesis. The [^{3}H]-labeled specific activity is significantly higher than that of the [^{14}C]-labeled specific activity. However, considering that ^{3}H labeling is prone to hydrogen–tritium exchange *in vivo* and both samples before and after volatile drying may need to be tested, the difficulty and cost of the study will be increased.

Antibodies are usually radiolabeled with high-energy radionuclides such as ^{125}I or ^{89}Zr. Because of their high specific activity, the detection limit is no longer a limiting factor. Therefore, factors considered during synthesis include the types of amino acids contained in the antibody (different nuclides are correspondingly labeled with specific amino acids), radionuclides decay half-life (not suggested to be shorter than the biological half-life of the antibody *in vivo*), and stability and activity of the labeled antibody. ^{125}I is a more common choice due to its accessibility, simple labeling process, and minimal impact on the activity of the antibody.

Marketed ADC drugs, such as Adcetris® and Padcev® from Takeda Pharmaceuticals, Kadcyla® and Polivy® from Roche, Besponsa® from Pfizer, Enhertu® from AstraZeneca, Zynlonta® from ADC Therapeutics, and Tivdak® from Seagen/Genmab, use low-energy radionuclides (^{14}C or ^{3}H) to label payloads for tissue distribution, mass balance, and metabolite identification studies. Meanwhile, radionuclides (^{125}I, ^{111}In, ^{3}H, and ^{89}Zr) were also used to label antibodies in Kadcyla®, Polivy®, Enhertu®, Trodelvy®, and Tivdak® to reveal the differences in tissue distribution in comparison with those observed by using only payload-labeled ADC drugs.

2.6.3 Screening ADCs with Optimal DAR Value at the Drug Discovery and Development Stages

At the development stage, highly sensitive radiolabeling technology allows for the clear determination of ADC PK behaviors *in vivo*. Comparing the differences in PK characteristics caused by different labeling positions and conjugation methods through separately labeling naked antibodies, antibodies and payloads of ADCs could contribute to screening ADCs that meet the expectations at the earliest possible stages.

SAR3419 is an ADC comprising the antibody-targeting CD19 (huB4) and microtubule inhibitor DM4 (maytansine, shortened to May) (Figure 2.33). [^{125}I]-labeled antibody ([^{125}I]-Ab) and conjugates ([^{125}I]Ab-L-May) were obtained and administered to mice separately (Ab: antibody; L: linker). The results indicated that their tissue distributions were identical (Figure 2.34), illustrating that the *in vivo* tissue distribution of the naked antibody was not changed after the huB4 was conjugated to the payload DM4 (May) with a DAR value between 3.5 and 4 [3].

Figure 2.35 evaluates whether the PK of the payload could be changed for ADCs with the same antibody but different DAR values. After the synthesis of [^{3}H]-labeled payload, cleavable ADCs (M9346A-sulfo-SPDB-DM4) and non-cleavable ADCs (J2898A-SMCC-DM1) were prepared with DAR values ranging from low (mean ~2, range: 0–4) to very high (mean ~10, range: 7–14). PK results in mice showed that clearance was slow for ADCs with a mean DAR value below six and became significantly faster for ADCs with a mean DAR value higher than nine. Tissue distribution results indicated that the tissue distribution characteristics of ADCs with lower DAR were similar to those of naked antibodies, while tissue distribution characteristics of ADCs with DAR above nine differed significantly from those of naked antibodies. Payload concentrations declined rapidly in whole blood after dosing, accumulated rapidly in the liver, and were rapidly excreted via the hepatic routes, resulting in limited *in vivo* exposure and subsequently suboptimal PD performance (Figure 2.36) [4].

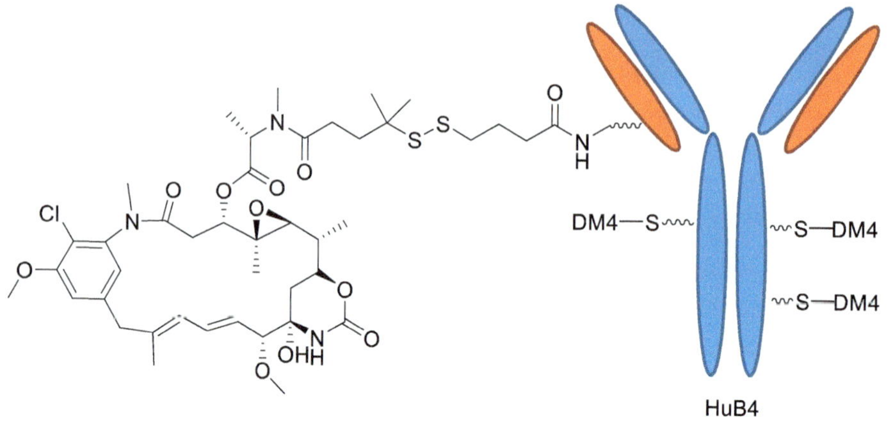

Figure 2.33 Structure of SAR3419. *Source:* Adapted from reference [2].

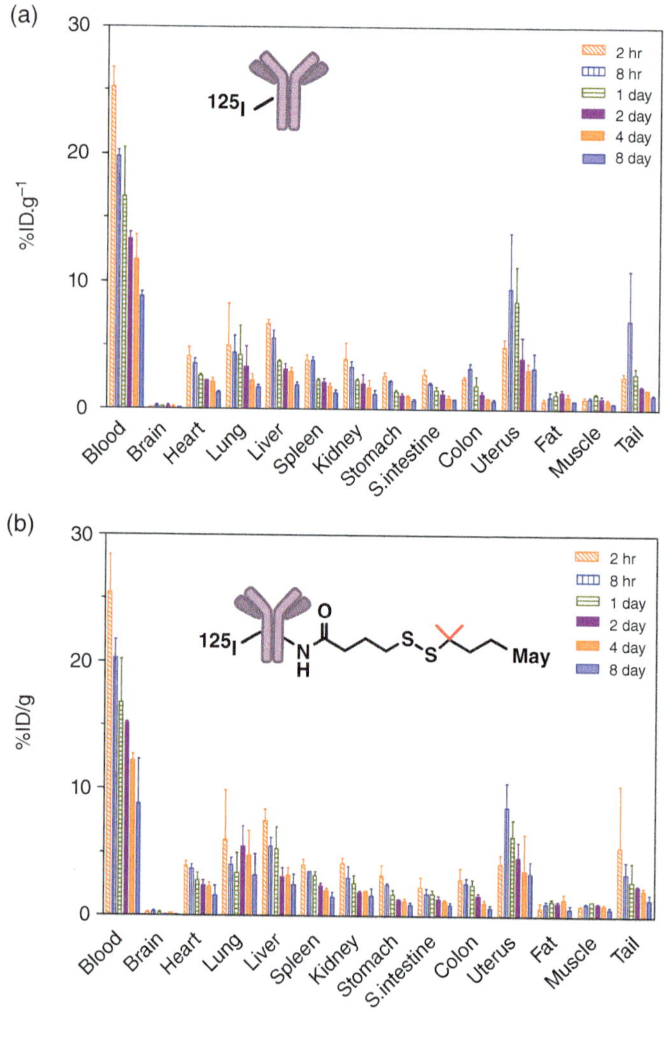

Figure 2.34 Tissue distribution of [^{125}I]Ab (a) and [^{125}I]Ab-L-May (b) in mice following a single intravenous administration. *Source:* [3] / with permission of Springer Nature.

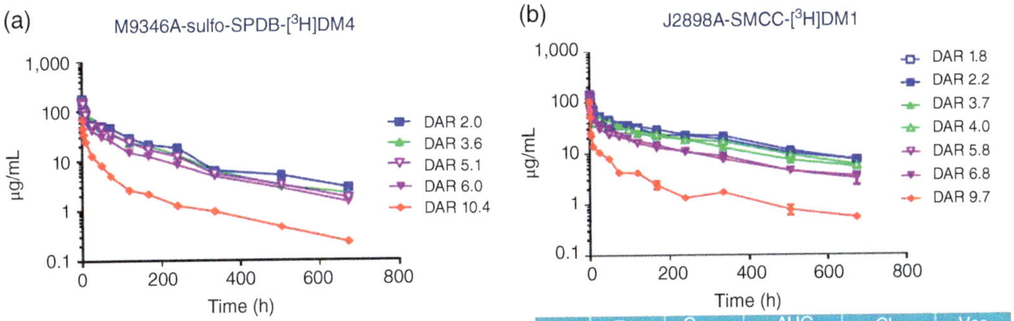

(a)

DAR	$T_{1/2}$ (Days)	C_{max} (μg/mL)	AUC (h*μg/mL)	CL (mL/h/kg)	Vss (mL/kg)
2.0	6	180	12791	0.78	147
3.6	5.5	185	11152	0.90	149
5.1	5.4	160	10132	0.99	163
6.0	5.8	152	7530	1.33	234
10.4	5.1	118	1812	5.49	690

(b)

DAR	$T_{1/2}$ (Days)	C_{max} (μg/mL)	AUC (h*μg/mL)	CL (mL/h/kg)	Vss (mL/kg)
1.8	11.0	144	13926	0.60	217
2.2	10.0	147	15745	0.55	181
3.7	9.9	146	11478	0.75	239
4.0	10.3	144	12988	0.66	217
5.8	10.2	433	7383	1.16	366
6.8	9.0	141	8158	1.09	298
9.7	8.7	103	1730	5.29	1199

Figure 2.35 Pharmacokinetic study in CD1 mice after intravenous injection of M9346A-sulfo-SPDB-[^{3}H]DM4 (a) or J2898A-SMCC-[^{3}H]DM1 (b). *Source:* [4]] / American Chemical Society.

2.6.4 Using the Radiolabeling Technique to Profile the *In Vivo* PK Characteristics of ADCs at the Application Stage

Tissue distribution studies are necessary for ADCs targeting specific tissues and are recommended to be conducted in PD animal models.

In recent years, quantitative whole-body autoradiography (QWBA) has been frequently used to evaluate the target tissue specificity of ADCs. It can provide average concentrations within entire tissues and can distinguish concentration differences among regions within the tissue, and especially, it can offer incomparable advantages in assessing dose–response relationships between the ADC distribution depth in tumor tissues and drug efficacy, compared with conventional harvesting methods.

Trastuzumab deruxtecan (code name, DS-8201a or T-Dxd; US brand name, Enhertu®), launched in 2019, is a combination of a HER2-targeting antibody and Dxd (a DNA topoisomerase I inhibitor) for treating adult patients with HER2-positive unresectable or metastatic breast cancer. Its PK studies are summarized based on the published literature (Figure 2.37).

To distinguish different radiolabeling positions, the ADC is represented as Ab-L-Dxd. [^{3}H]Ab-L-Dxd was denoted after labeling ^{3}H on the antibody of ADC, and Ab-L-[^{14}C]Dxd was denoted after labeling ^{14}C on the payload. Additionally, [^{14}C]Dxd was denoted after only labeling ^{14}C on the payload.

Two groups of cynomolgus monkeys were administered with [^{3}H]Ab-L-Dxd or Ab-L-[^{14}C]Dxd, and the differences in tissue distribution between them were assessed using QWBA. The results showed that the distribution characteristics of the two ADCs with different radiolabeling locations were similar in whole blood and important tissues and organs. The only difference was that Ab-L-[^{14}C]Dxd exhibited a high distribution in the large intestine, while [^{3}H]Ab-L-Dxd exhibited a low distribution in the intestinal tract. Ab-L-[^{14}C]Dxd was mainly excreted via feces (67.3%), followed by urine (18.7%). No metabolites other than [^{14}C]Dxd were detected in feces and urine. In addition,

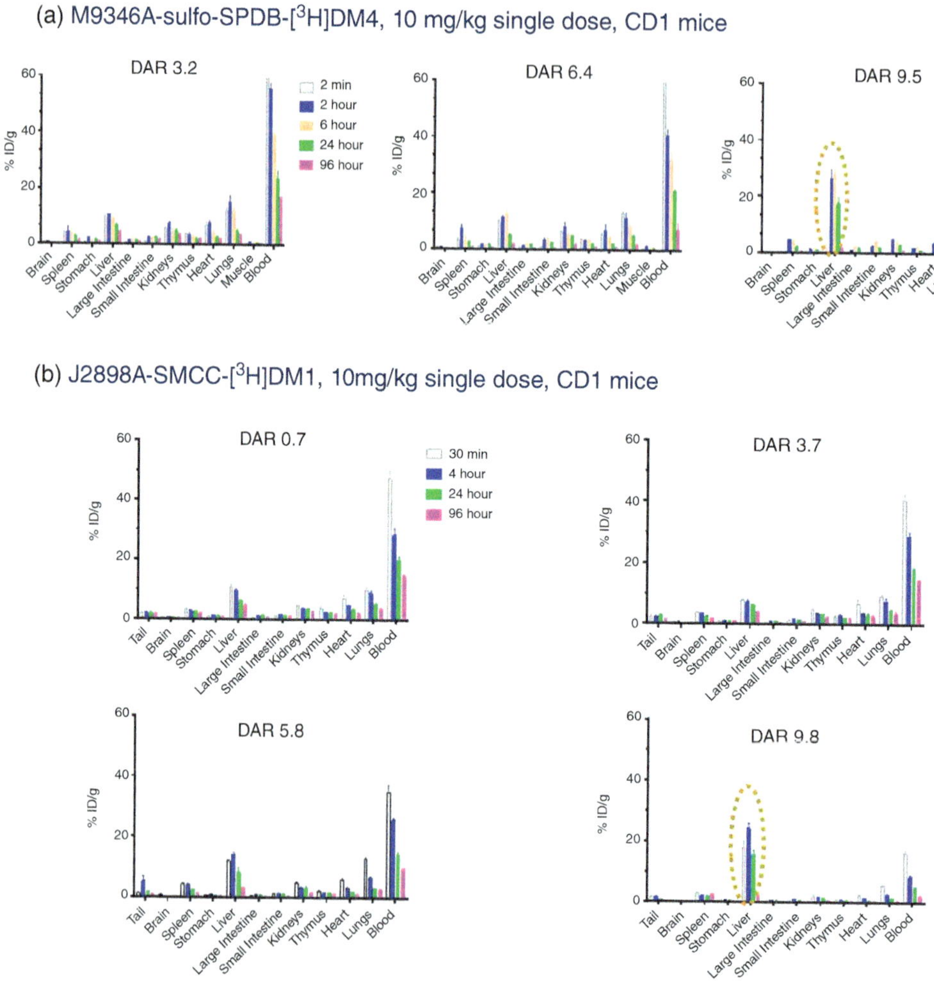

Figure 2.36 Tissue distribution study in CD1 mice after intravenous injection of 10 mg/kg M9346A-sulfo-SPDB-[^{3}H]DM4 (a) or J2898A-SMCC-[^{3}H]DM1 (b). *Source:* [4] / American Chemical Society.

the results of bile duct cannulation studies in rats showed that 71.5% of the unconjugated payload [^{14}C]Dxd was excreted via the bile (Figures 2.38 and 2.39) [5].

The results from these animal models provided a comprehensive understanding of the *in vivo* PK process of Ab-L-Dxd: after Ab-L-Dxd is injected into the body, it distributes throughout the body and the levels of radioactivity in the blood are the highest, and it does not accumulate in normal tissues, which considerably reduces its safety risk. After the release of Dxd in the body, Dxd is not metabolized and most of it flows into the intestinal lumen via the bile and is eventually excreted via the feces [4].

The QWBA technique can also be used to investigate the tissue distribution of ADC in tumor-bearing rats. Figure 2.40 shows the structure of an ADC named [^{3}H]DM1-LNL897.

The ADC [^{3}H]DM1-LNL897 was radiolabeled with [3H] on the payload. After intravenous administration to tumor-bearing rats, QWBA was used to analyze the tissue distribution of the

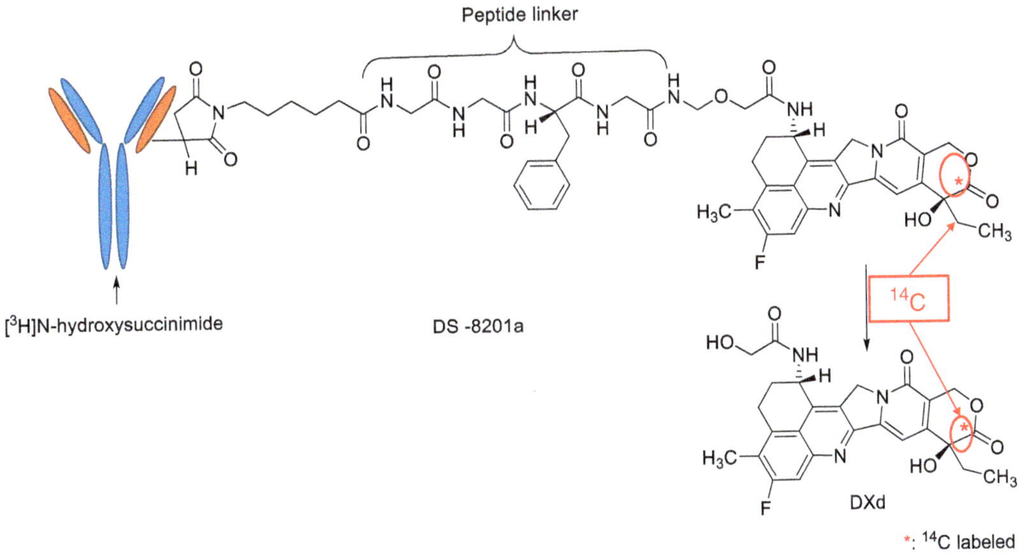

Figure 2.37 Constitutional formula and radiolabeling position in DS-8201a. *Source:* [5] / with permission of Taylor & Francis Group.

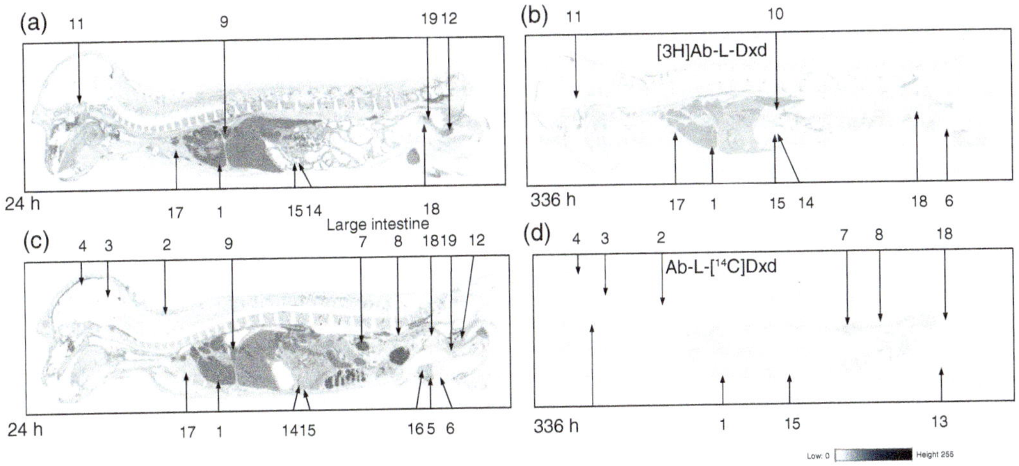

Figure 2.38 Representative quantitative whole-body autoradiography images at 24 hours (a,c) and 336 hours (b,d) after intravenous injection of [^{3}H]Ab-L-Dxd (a,b) and Ab-L-[^{14}C]Dxd (c,d) in cynomolgus monkeys. *Source:* [5] / with permission of Taylor & Francis Group.

payload at different time points [5]. Representative images were selected to illustrate the tissue distribution of radioactivity-related components in this study.

At 1 hour post-dose, radioactivity was mainly concentrated in the blood-rich tissues, such as blood, heart, and liver, and radioactivity was also distributed in the tumor tissues. At 24 hours post-dose, the concentration of radioactivity in the tumor increased significantly, which far exceeded the concentrations in other tissues. Only little radioactivity is distributed in the skin, and almost no radioactivity is distributed in the brain or muscle tissues (Figure 2.41).

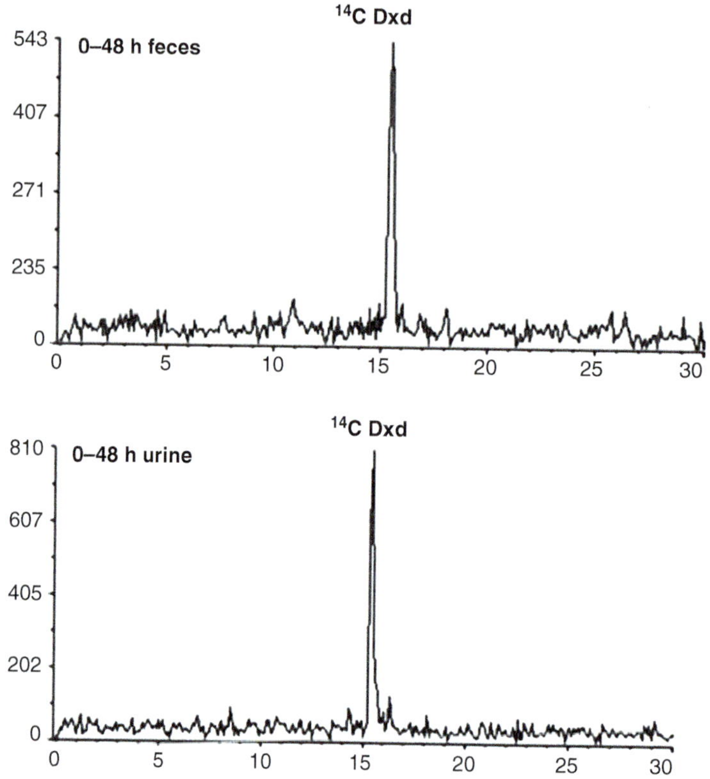

Figure 2.39 Radioactive metabolite profiles in feces and urine after intravenous injection of Ab-L-[^{14}C]Dxd in Cynomolgus monkeys. *Source:* [5] / with permission of Taylor & Francis Group.

At 72 hours post-dose, the payload-related components (including intact ADC and free payload) were mainly concentrated in the tumor and rarely distributed to other tissues, which demonstrated the target of the ADC. The radioactivity exposure in nontarget tissues decreased continuously. At 168 hours post-dose, the payload-related radioactivity could still be detected in the tumor (Figure 2.42).

2.6.5 Summary

Combining application documents and relevant literature on the PK of approved ADCs, preliminary determination of payload-related components released from ADCs based on the results of *in vitro* assay, plasma concentration, and tissue distribution results, is essential for ADC efficacy and safety assessments in animal studies. It is recommended to adopt the radiolabeling technique to track the tissue distribution *in vivo* and to examine whether the tissue distribution characteristics of radiolabeled payloads and antibodies are similar. If a significant difference is observed in a particular tissue, the specific reason can be inferred by the *in vivo* behavior of the unconjugated payload. Subsequently, based on metabolite identification results from radioactive plasma, urine, and feces, it can be further confirmed whether other payload-related components are released *in vivo*, and the supplementary plasma protein binding and DDI studies *in vitro* of the payload-related components need to be considered.

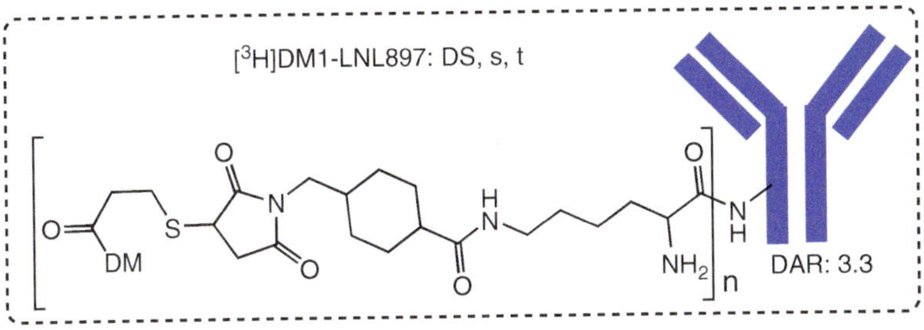

Figure 2.40 Structure of [^{3}H]DM1-LNL897. *Source:* [6] / with permission of The American Society for Pharmacology and Experimental Therapeutics.

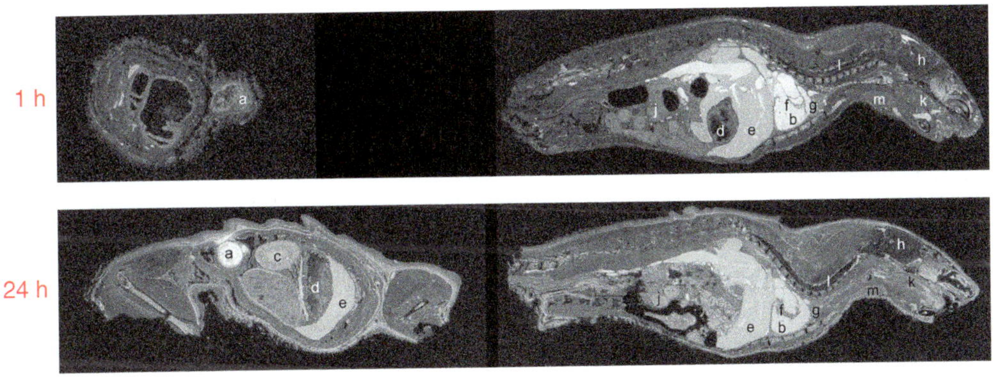

Figure 2.41 Tissue distribution of radioactivity in tumor-bearing rats at 1 hour and 24 hours post-dose after administration of [^{3}H]DM1-LNL897. *Source:* [6] / The American Society for Pharmacology and Experimental Therapeutics.

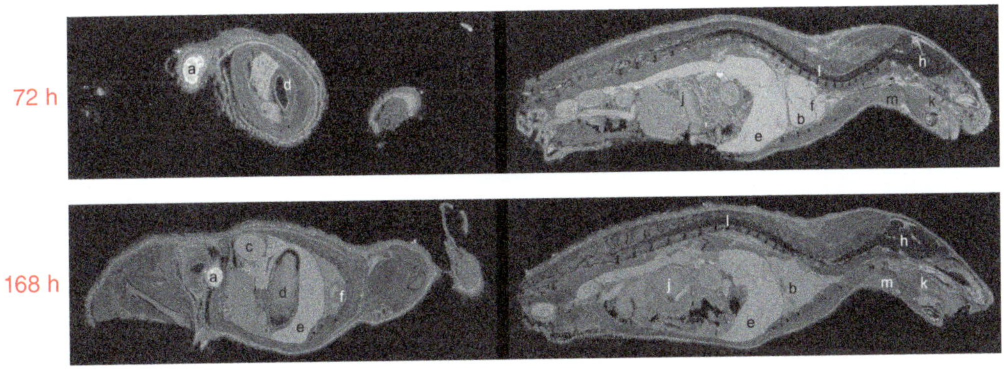

Figure 2.42 Tissue distribution of radioactivity in tumor-bearing rats at 72 hours and 168 hours post-dose after administration of [^{3}H]DM1-LNL897. *Source:* [6] / The American Society for Pharmacology and Experimental Therapeutics.

References

1 The National Medical Products Administration (NMPA). (2023). Guidelines for non-clinical studies of antibody-drug conjugates. https://www.cde.org.cn/zdyz/domesticinfopage?zdyzIdCODE=42d891ebccbcece7dd757d6429acff33.

2 Blanc, V., Bousseau, A., Caron, A. et al. (2011). SAR3419: an anti-CD19-Maytansinoid Immunoconjugate for the treatment of B-cell malignancies. *Clinical Cancer Research* 17 (20): 6448–6458.

3 Erickson, H.K. and Lambert, J.M. (2012). ADME of antibody–maytansinoid conjugates. *The AAPS Journal* 14: 799–805.

4 Sun, X., Ponte, J.F., Yoder, N.C. et al. (2017). Effects of drug–antibody ratio on pharmacokinetics, biodistribution, efficacy, and tolerability of antibody–maytansinoid conjugates. *Bioconjugate Chemistry* 28 (5): 1371–1381.

5 Nagai, Y., Oitate, M., Shiozawa, H. et al. (2019). Comprehensive preclinical pharmacokinetic evaluations of trastuzumab deruxtecan (DS-8201a), a HER2-targeting antibody-drug conjugate, in cynomolgus monkeys. *Xenobiotica* 49 (9): 1086–1096.

6 Walles, M., Rudolph, B., Wolf, T. et al. (2016). New insights in tissue distribution, metabolism, and excretion of [3H]-labeled antibody maytansinoid conjugates in female tumor-bearing nude rats. *Drug Metabolism and Disposition* 44 (7): 897–910.

3

Peptide–Drug Conjugates (PDCs)

3.1 Introduction to PDCs and DMPK Research Strategies

Qigan Cheng, Liping Ma, Jing Jin

3.1.1 Introduction

Traditional chemotherapy drugs, such as paclitaxel and doxorubicin, still play an important role in cancer treatment and have achieved significant therapeutic effects, but the side effects are also obvious. The reason is that chemotherapy unspecifically poisons both tumor cells and normal cells. As early as 1913, German scientist Paul Ehrlich proposed a "biological bullet" – if toxic drugs were installed on carriers specifically targeting tumor cells, it would be possible to kill tumor cells accurately without harming normal cells. In 2000, the first ADC went on the market, which marked the successful implementation of the concept of "biological bullet". So far, the total number of ADCs approved by the FDA has reached more than 10. Although the treatment with ADCs continues to make progress, there are still some problems with these treatments. For example, the complex structure of the ADC leads to high production costs. With a long half-life in the circulatory system, the ADC is prone to exert systemic toxicity. This has led to the emergence of PDC, which consists of a homing peptide, linker, and payload (Figure 3.1). The PDC takes advantage of the high affinity of the homing peptide with tumor surface receptors to deliver a payload to the target.

3.1.2 Types of PDC Payloads

The following are the types of PDC payloads:

1) Cytotoxic drugs, such as paclitaxel and gemcitabine.
2) Radionuclides, which have been used for accurate diagnosis of diseases and targeted radiotherapy of tumors. Radioisotopes used for diagnosis include fluorine-18 (^{18}F), copper-64 (^{64}Cu), gallium-68 (^{68}Ga), and iodine-123 (^{123}I). These radioisotopes are coupled with targeted peptides, and by combining them with targeted receptors on tumor cells, the positron released by radioisotopes can be detected to accurately locate malignant tissues. When the targeted peptide is linked to radionuclides such as indium-111 (^{111}In), yttrium-90 (^{90}Y), and lutetium-177 (^{177}Lu), the PDC is used as a targeted radiotherapy drug.
3) Other types include proteins, peptides [1], nucleic acids, and PROTAC [2].

Drug Metabolism and Pharmacokinetics: Frontiers, Strategies, and Applications, First Edition.
Edited by Liang Shen and WuXi AppTec DMPK.

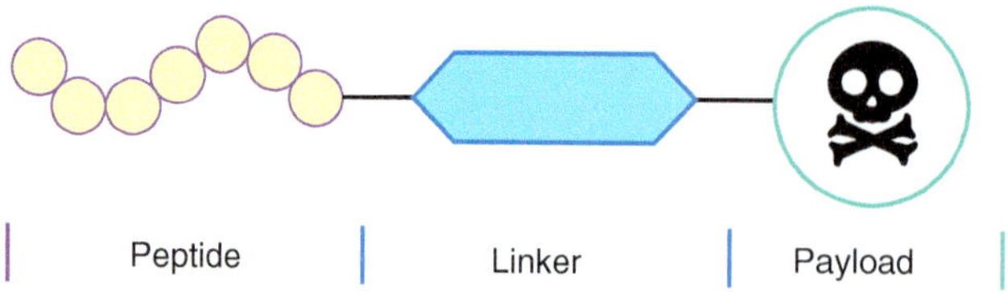

Figure 3.1 Structure of PDCs.

Table 3.1 Comparison of the PDC with chemotherapy and the ADC.

Item	Chemotherapy	ADC	PDC
CMC cost	Low	High	Low
Pharmacokinetic (PK)	Widely distributed; prone to resistance	Long half-life in circulation; poor tissue penetration	Short half-life in circulation; good tissue penetration
Targeting	None	Good	Moderate to good
Immunogenicity	None	Yes	No or low

3.1.3 Advantages of PDCs

The PDC shares the same advantage as the ADC. To some extent, the PDC overcomes some shortcomings of the ADC. As shown in Table 3.1, compared with the ADC, the PDC has a smaller molecular weight, better tissue penetration, and lower or no immunogenicity. Compared with the complicated chemistry, manufacturing, and control (CMC) process of the ADC, the PDC is easier to synthesize, purify, and characterize, and the cost is much lower. In recent years, with the application of cyclization technology, phage display technology, and mRNA display technology in the screening of targeted peptides, the development of the PDC has been accelerated. The PDC is expected to become the next generation of targeted drugs after small molecule, antibody, and ADC.

3.1.4 PDCs in the Clinical Stage and on the Market

As of 2024, a variety of PDCs are in the clinical stage, and selected examples are listed in Table 3.2. The approved PDCs are summarized in Table 3.3.

3.1.5 Pharmacokinetic Characteristics of PDCs

3.1.5.1 Absorption

Compared with traditional small molecules, the PDC has a relatively large molecular weight and the permeability is not so superior. Due to the instability in the gastrointestinal tract, PDCs on the market and in the clinical stage are administered by intravenous injection, which does not involve the absorption process.

The obstacle of the PDC is its poor stability *in vivo* and short half-life. Peptides can be rapidly degraded by protease and easily cleared in the kidneys. In the treatment of solid tumors, it is necessary to prolong the circulation of the PDC so that the PDC can fully penetrate the tumor tissue. Currently, the modification of peptides is an important strategy to enhance stability and prolong half-life [3].

Table 3.2 Selected PDCs in clinical stages.

PDC	Company	Stage	Target	Indication
AEZS-108 (zoptarelin doxorubincin)	Aeterna Zentaris, Sinopharm	Phase III	GnRH	Endometrial carcinoma
SNG1005/ ANG1005	Shenogen Pharma, Angiochem	Phase III	LRP-1	Brain metastatic non-small cell lung cancer (NSCLC), breast cancer, and recurrent glioma
BT1718	Bicycle Therapeutics	Phase I/II	MMP14	Esophageal carcinoma and NSCLC
BT5528	Bicycle Therapeutics	Phase I/II	EphA2	Solid tumor
BT8009	Bicycle Therapeutics	Phase I/II	Nectin-4	Esophageal carcinoma and NSCLC
PEN-221	Tarveda Therapeutics	Phase I/II	SSTR	NSCLC and neuroendocrine tumor
CBX-12	Cybrexa Therapeutics	Phase I/II	TOPO1	Solid tumor
EP-100	Esperance Pharmaceuticals	Phase II	GnRH	Ovarian cancer
TH1902	Theratechnologies	Phase I	SORT1	Three-negative breast cancer
CBP-1008	Coherent Biopharma	Phase I	FRα, TRPV6	Solid tumor

Table 3.3 Approved PDCs.

PDC	Approval time	Company	Target	Payload	Indication
^{68}Ga-DOTATATE	June 2016	Advanced Accelerator Applications S.A./Novartis	SSTR	^{68}Ga	Diagnosis of neuroendocrine tumor
^{177}Lu-dotatate	January 2018	Advanced Accelerator Applications S.A./Novartis	SSTR	^{177}Lu	Gastrointestinal pancreatic neuroendocrine tumor
^{68}Ga-DOTATOC	August 2019	Advanced Accelerator Applications S.A./Novartis	SSTR	^{68}Ga	Diagnosis of neuroendocrine tumor
^{64}Cu-dotatate	September 2020	RadioMedix Inc.	SSTR	^{64}Cu	Diagnosis of neuroendocrine tumors
Pepaxto	February 2021[a]	Oncopeptides	Aminopeptidase	Melflufen	Multiple myeloma
TLX591-CDx	December 2021	Telix Pharmaceuticals	PSMA	^{68}Ga	Diagnosis of prostate cancer

[a] Withdraw in October 2021.

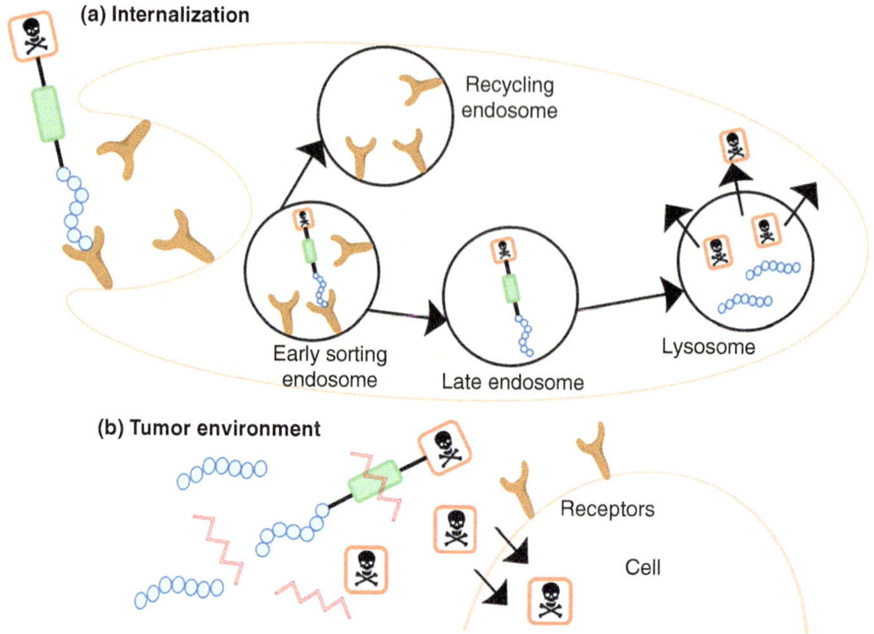

Figure 3.2 Targeting mode of the PDC to tumor. *Source:* [3] / Royal Society of Chemistry / CC BY 4.0.

3.1.5.2 Distribution

The peptides of PDCs are divided into two categories: cell-targeting peptides (CTPs) and cell-penetrating peptides (CPPs). CPPs are not used as homing peptides of PDC because of their poor specificity to cells. There are two approaches for PDC to target tumors.

1) PDC binds to the receptor and is endocytosed into the cells, is captured by lysosomes in the cells, and it releases toxins in the lysosomes (Figure 3.2a).
2) PDC releases toxins when it reaches the tumor environment, and the toxins enter the tumor cells to kill the tumor cells (Figure 3.2b).

3.1.5.3 Metabolism

There are three main metabolic pathways for PDCs *in vivo* [4].

1) The first one is the target-mediated drug disposal (TMDD) approach that is specific. The PDC binds to the receptor on the target cells followed by endocytosis into the cells, and then, it is degraded into peptides and amino acids in lysosomes. However, the number of receptors on the cell surface is usually limited, so TMDD could be saturated, which results in a nonlinear pharmacokinetic profile.
2) The second one is the nonspecific metabolic pathway. Nontarget-mediated and nonspecific metabolic processes, such as protease hydrolysis and nonspecific endocytosis, lead to the release of small molecular cytotoxins to untargeted tissues and exert off-target side effects.
3) The third one is the formation of antidrug antibodies, which will be removed by the immune system. It is generally believed that the immunogenicity or antidrug antibody reaction found in preclinical animals cannot predict the antidrug antibody reaction in humans.

3.1.5.4 Excretion

The molecular weight of the PDC is less than the filtration threshold of the glomerulus (60 kDa), so it can be quickly cleared by the kidney, which makes it impossible for cytotoxins to accumulate effectively in tumors.

Renal excretion may be the main clearance route of PDCs *in vivo*. In this case, *in vivo* clearance is equal to the glomerular filtration rate (f_u * GFR) of PDCs without considering renal tubular reabsorption and active secretion [5].

3.1.6 Challenges in the Development of PDCs

Despite many experiences that have been recorded for the ADC, the research and development of the PDC is not regarded to be smooth. The PDC faces both opportunities and challenges. Compared with the ADC, the tissue specificity and tumor targeting of peptides are inferior for the PDC, which brings great challenges to the screening of targeted peptides. At the same time, oral administration has become more challenging because of the poor gastrointestinal stability of peptides. Additionally, the PDC also has potential challenges related to bioanalysis and potential off-target effects.

3.1.7 DMPK Strategy for PDCs

In essence, the PDC uses a targeting strategy to deliver drugs, which share the properties of both the peptide and small molecules. The stability and targeting of the PDC are relatively poor compared to the ADC, so special attention should be paid to the toxicity of the payload. Table 3.4 summarizes the DMPK strategy for the PDC.

Table 3.4 DMPK strategy for PDC[a].

DMPK	Screening	PCC	IND-enabling
In vitro	Metabolic stability and MetID of homing peptide and PDCs in whole blood, plasma, and renal homogenate *in vitro* Stability of PDCs in cathepsin (or lysosome, acid buffer, or glutathione (GSH)) (depending on the type of linker) Blood/plasma ratio of payload PPB and tissue protein binding of PDCs and payload	MetID in tumor cells Stability and MetID in kidney/ liver homogenate or S9	PPB of PDCs Stability in plasma and kidney homogenate of PDCs and the corresponding MetID studies CYP inhibition, induction, and transport inhibition of PDCs *In vitro* absorption, distribution, metabolism, and excretion (ADME) studies for payload
In vivo	MetID of peptide and PDCs in plasma and urine from *in vivo* studies. PK of peptide PK and tissue distribution of PDCs	Non-rodent PK of PDCs Tissue distribution of PDCs in PD models	Bioanalytical method qualification of PDCs and payload anti-drug antibody (ADA) method qualification Dose linearity *In vitro* MetID Tissue distribution and mass balance

[a] Applicable when the payload is a cytotoxic drug.

In the screening stage, the main content of DMPK research is to improve the stability of homing peptide and PDCs in whole blood, plasma, and kidney homogenate and continue to optimize the structure of PDCs through metabolite identification (MetID) *in vitro* and *in vivo*. According to the different types of linkers (non-cleavable or cleavable; if cleavable, refer to cleavage mode), a suitable *in vitro* model is selected to evaluate the release of payload. To assess the potential toxicity of the payload in different organs, it is suggested to conduct a tissue distribution study in rodents using PDCs in the early stage of the project.

In the preclinical candidate (PCC) stage, a suitable *in vitro* model is recommended to evaluate the release of payload and the metabolism of PDC. It is necessary to simultaneously detect the concentration of the PDC and the payload in the plasma. In the case of repeated administration, it is recommended to monitor antidrug antibodies. A tissue distribution study could be performed in pharmacodynamics (PD)-relevant species to evaluate the targeting of PDCs.

In the IND-enabling stage, if the payload is brand new, a comprehensive *in vitro* evaluation of the payload is needed. It is suggested to use radiolabeled PDC to study the tissue distribution and mass balance in rodents.

3.1.8 Desired Capabilities for PDCs

3.1.8.1 Plasma Protein Binding of PDCs

The molecular weight of the PDC is generally large with potential nonspecific binding, so the commonly used equilibrium dialysis may not be suitable. Plasma protein binding of PDCs can be carried out using the ultracentrifugation method or, if needed, in conjunction with plasma dilution.

3.1.8.2 Bioanalysis of PDCs

PDCs suffer from stability and nonspecific binding issues during bioanalysis. Other issues, such as multi-charge state, low sensitivity, low extraction recovery, and endogenous interference, bring more challenges to the bioanalysis of PDCs.

For some PDCs with cleavable linkers, it is suggested that the plasma from *in vivo* PK should be precipitated immediately as a precaution after collection to get supernatant, because these PDCs may have stability issues in plasma.

In addition, due to the high toxicity of payload, the generally required concentration in normal tissues is at a very low level. This brings about higher requirements for the lower detection limit of the payload, and it will be great to reach the level of pg/mL. As a high-sensitivity mass spectrometry was set up, combined with the rich experience of homogenate of various tissues, the detection limit of the payload can be further lowered.

3.1.8.3 Bioanalysis of Immunogenicity

NMPA requests to test immunogenicity for peptide or their conjugates, so its immunogenicity from PDCs should be evaluated [6]. MSD® electrochemiluminescence (MSD ECL) can be used to determine the ADA of PDCs, including screening tests, confirmation tests, and titer tests.

3.1.8.4 MetID of PDCs

For evaluating the release of PDC payloads, different *in vitro* models could be used, such as liver S9, lysosome, plasma, whole blood, and tumor cells. *In vivo* models include healthy animals, tumor-bearing animals, and so on. In these models, the identification of PDC metabolites can be carried out to evaluate the payload release of PDCs (Figure 3.3).

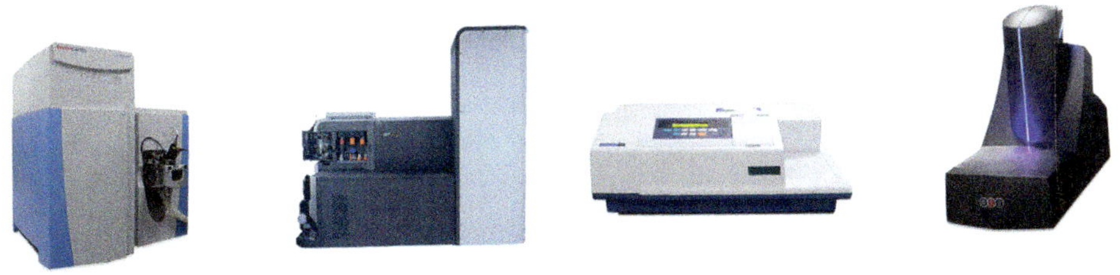

Figure 3.3 PDC-related instruments for bioanalysis.

3.1.9 Summary

The design of PDCs combines the advantages of targeted peptides and different types of drugs. In recent years, with the regulatory approval of radionuclide therapy and radionuclide diagnosis drugs [7], the advantages of these drugs have been proven. The clinical pipelines also show encouraging results. With the coupling of various molecular entities and targeted peptides, the future of PDCs is promising.

References

1 Hurov, K., Lahdenranta, J., Upadhyaya, P. et al. (2021). BT7480, a novel fully synthetic Bicycle tumor-targeted immune cell agonist™ (Bicycle TICA™) induces tumor-localized CD137 agonism. *Journal of Immunotherapy Cancer* 9 (11): e002883.

2 Jin, J., Wu, Y., Chen, J. et al. (2020). The peptide PROTAC modality: a novel strategy for targeted protein ubiquitination. *Theranostics* 10 (22): 10141–10153.

3 Cooper, B.M., Iegre, J., O'Donovan, D.H. et al. (2021). Peptides as a platform for targeted therapeutics for cancer: peptide–drug conjugates (PDCs). *Chemical Society Reviews* 50 (3): 1480–1494.

4 Haoze, Z., Zilong, S., and Hanmei, X. (2017). Recent progress in pharmacokinetic analysis of protein and peptide drugs. *Progress in Pharmaceutical Sciences* 41 (08): 592–599.

5 Jobin, J. and Bonjour, J.P. (1985). Measurement of glomerular filtration rate in conscious unrestrained rats with inulin infused by implanted osmotic pumps. *American Journal of Physiology-Renal Physiology* 248 (5): F734–F738.

6 NMPA. (2021). Guidance on drug immunogenicity. https://www.cde.org.cn/zdyz/domesticinfopage?zdyzIdCODE=7e37810560d2acae0277334400aa3320.

7 Peltek, O.O., Muslimov, A.R., Zyuzin, M.V. et al. (2019). Current outlook on radionuclide delivery systems: from design consideration to translation into clinics. *Journal of Nanobiotechnology* 17: 1–34.

3.2 Targeted Release Principle and Metabolite Identification Studies of PDCs

Liqi Shi, Gengyao Qin, Weiqun Cao

3.2.1 Introduction

The ideal drug delivery system can improve efficacy, improve the therapeutical window, reduce the dosage, and minimize the toxic and side effects through targeted and controlled release. Especially in the field of cancer therapy, the precise and efficient killing of tumor cells using targeted delivery drugs has significant advantages over the indiscriminate use of radiotherapy or chemotherapy, which causes damage to both tumor cells and normal cells (Figure 3.4).

By the end of 2022, 15 ADCs with antibodies as targeted delivery systems have been approved for marketing worldwide. PDCs, which are also emerging as a new modality, are composed of peptides as drug delivery carriers and small-molecule compounds as payload connected through linkers. Compared with ADCs, PDCs have a smaller molecular mass (generally < 8 kDa) and feature low immunogenicity, adjustable physical and chemical properties, and low manufacturing cost.

Biologics are generally not metabolized by CYP450 enzymes but are mainly cleared through glomerular filtration (< 60 kDa), enzymatic hydrolysis, and catabolism. Compared with ADCs, PDCs with a relatively small molecular weight are cleared more rapidly with a shorter plasma exposure time. Nevertheless, targeting remains one of the most important design principles in PDC drug development from a safety and efficacy perspective, while the study of metabolite identification (MetID) is essential for drug release and targeting assessment in PDC.

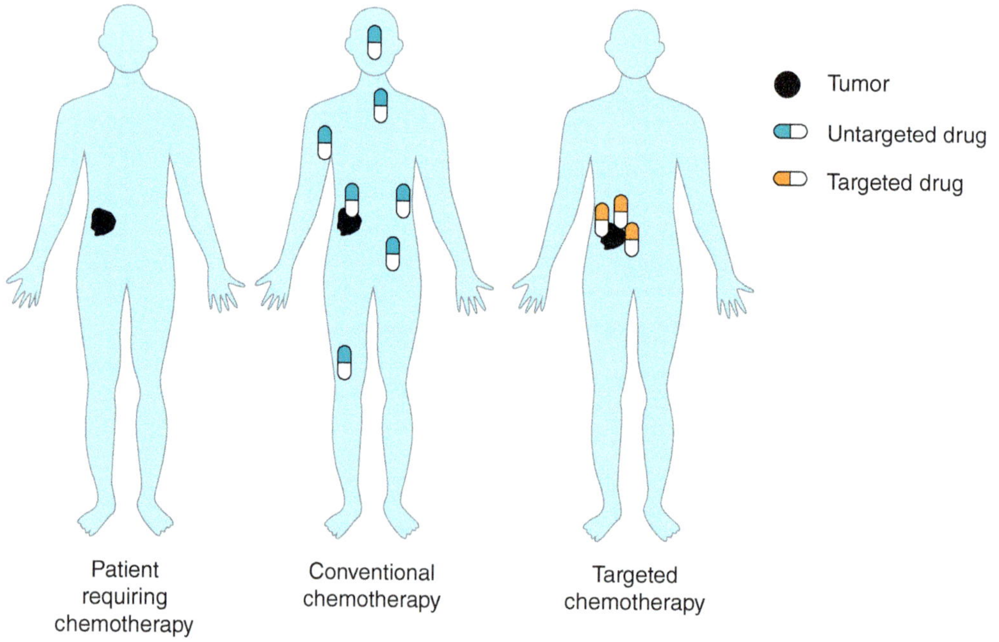

Figure 3.4 Tumor nontargeted and targeted therapy. *Source:* Adapted from reference [1].

3.2.2 Metabolic Characteristics of Marketed PDCs

The approved PDC Pepaxto® (melphalan flufenamide) improves bioavailability by coupling the toxin molecule melphalan with a fluorobenzamide through peptide bonds to improve lipophilicity. The metabolic release of toxin molecules occurs through passive transport (active transport of non-protein targets) into the cell, followed by hydrolysis by aminopeptidases which is highly expressed in tumor cells (Figure 3.5) [2]. The design idea is more similar to the design of the prodrug in this case.

Another class of marketed PDCs is radionuclide drug conjugates (RDCs) using peptides as targeting ligands. For example, Lutathera® [4] is a drug composed of octreotide coupled with radionuclide ^{177}Lu. Taking advantage of the targeting of octreotide to the SSTR2 receptor, targeted radiotherapy of radionuclides can be achieved without the need to release the loaded nuclide through metabolism after being captured by tumor cells. Lutathera is not metabolized by the liver and is mainly cleared by the kidney (Figure 3.6).

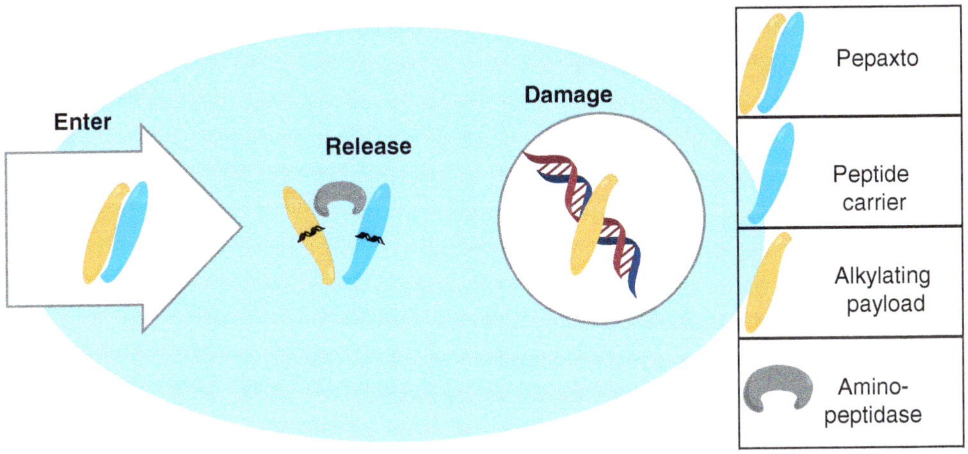

Figure 3.5 Mechanism of action (MOA) of Pepaxto®. *Source:* Adapted from reference [3].

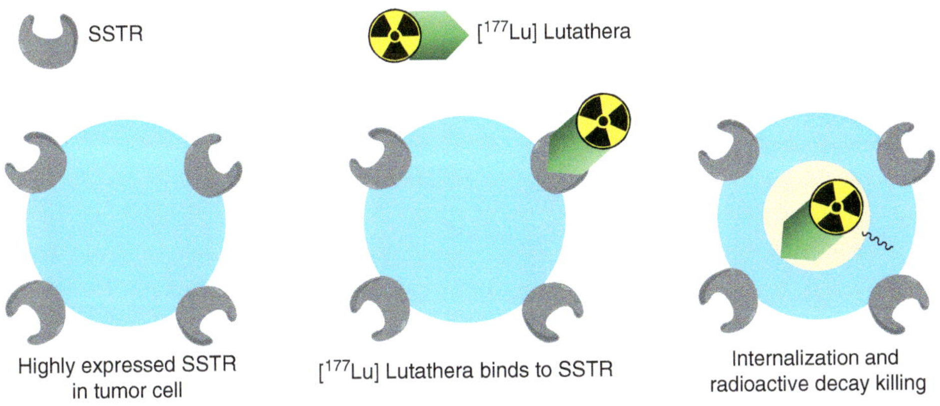

Figure 3.6 MOA of Lutathera®. *Source:* Adapted from reference [4].

3.2.3 Design Principle of Current Mainstream PDCs and PDCs Under Development

Currently, most of the mainstream PDCs in the clinical/preclinical stage use chemotherapeutic drugs and toxin molecules (cellular microtubule inhibitors or nucleic acid inhibitors such as paclitaxel, camptothecin, maytansine, and aplysia toxins) as payloads. Macromolecules in the coupled state have steric inhibition, and their payloads need to be released by metabolic degradation before they can be effective. Similar to ADCs, it is expected to maintain structural integrity and stability in the circulatory system for PDCs before entering the target cells, while they can effectively release the payloads after entering the target cells. The design idea is as follows:

1) The effective delivery of the payload depends on the targeted design of the peptide.
2) The precise release of the payload depends on the targeted release mechanism.

3.2.3.1 Targeted Delivery: Target Receptor Versus Targeted Peptide (Table 3.5)

Targeted peptides targeting specific receptors on tumor cells can be obtained by receptor target screening using phage display technology or other approaches, The common receptors on tumor cells are shown in Figure 3.7.

3.2.3.2 Targeted Release: Payload Release Mechanism

After PDCs are targeted into tumor cells, there can be different payload release methods according to different designs of linkers.

Table 3.5 Partial targets and targeted peptide sequences.

Receptors	Targeted peptide sequence
αvβ3	Cilengitide [cyclo (RGDf-N(Me)V-)] LXW7 (cGRGDdvc)
αvβ6	RTDLXXL
APN/CD13	NGR
PEPT	Ser-Glu
EGFR	GE11 (YHWYGYTPQNVI) Disruptin (SVDNPH)
HER2	KCCYSL LTVSPWY
PSMA	KYLAYPDSVHIW WQPDTAHHWATL
uPAR	VSNKYFSNIHW
GRPR	EQRLGNQWAVGHLM
SSTRs	OCT (FCFWKTCT)
VEGFR	Peptide SP5.2 (NGYEIEWYSWVTHGMT)
Insulin	MCR (RRLFYKKVGLFYKKVRR)
Ephrin receptors	EWLSPNLAPSV TNYLFSPNGPIA YSA peptide (YSAYPDSVPMMS)

Source: Adapted from reference [5].

A) Non-cleavable linkers: The proteolytic enzymes of lysosomes release the payload after degradation of the peptide moiety.
B) Cleavable linkers: The breakage of linkers occurs in pathways such as redox and hydrolysis (including hydrolysis in lysosomes) in the microenvironment of tumor cells, thereby releasing the payloads (Figure 3.8).

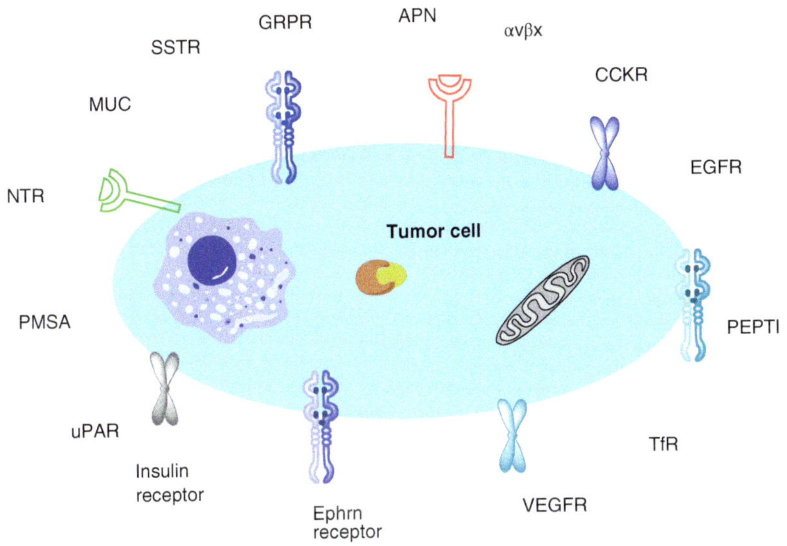

Figure 3.7 Common receptors on tumor cells. *Source:* Adapted from reference [5].

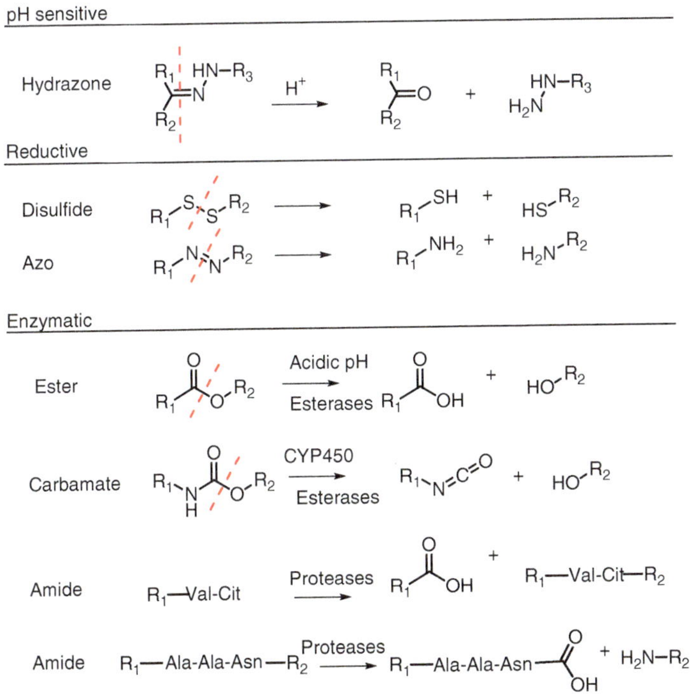

Figure 3.8 Common linker-cleavage mechanism. *Source:* [6] / Frontiers Media S.A. / CC BY 4.0.

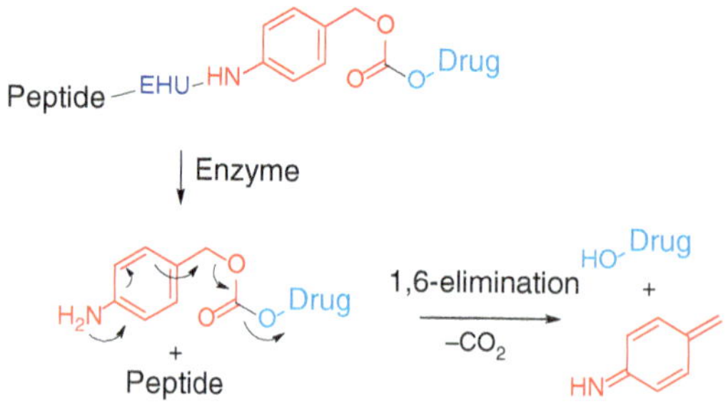

Figure 3.9 Mechanism of digestion – elimination of self-degradable groups. *Source:* [1] / Beilstein Journal of Nanotechnology / CC BY 4.0.

Acid-cleavable linkers utilize the pH difference between tumor cell lysosomes (pH <5) and plasma (pH 7.4), selectively metabolizing within tumor cells to release the active payload, such as hydrazone, that can be decomposed into aldehydes/ketones and amines under acidic conditions.

Disulfide linkers can be released by glutathione-mediated disulfide reduction/exchange to form cleavage, and the glutathione content in tumor cells exceeds that in blood by 1,000 times. Azo linkers can be cleaved under the catalysis of reduced coenzyme II (NADPH)-azoreductase, which is overexpressed in the hypoxic environment of tumor cells.

Cathepsin B is a proteolytic enzyme highly expressed in tumor cells/tissues. Cathepsin B could recognize hydrolytic-specific peptide sequences, such as Val-Cit and GFLG, and these short peptides can be used as directed enzyme cleavable linkers. In addition, when the enzyme cleavable linker is used, the self-degrading group is often used at the same time. For example, a para-aminobenzyl carbamate (PABC) linkage was used as a self-immolative spacer. When the linker is hydrolyzed, the self-immolation group undergoes spontaneous chemical elimination to release the payload (Figure 3.9).

3.2.3.3 Examples of PDCs in the Clinical Stage

Currently, the mainstream companies that focus on PDCs include Angiochem, Bicycle Therapeutics, Cybrexa Therapeutics, Oncopeptides, PeptiDream, Theratechnologies, and Borui Pharmaceutical. The following are some PDCs that have entered the clinical stage, which are composed of different peptides, linkers, and payloads, and their targeting properties and release mechanisms are typically representative.

A) AEZS-108 [7]: Adriamycin (doxorubicin) is used as the payload and the luteinizing hormone-releasing hormone (LHRH) peptide is used as the targeting peptide, which is catabolized and released under the action of lysosomes after entering the cell. It has entered the phase III clinical stage.
B) ANG1005 [8]: It contains peptide-coupled multiple paclitaxel targeting the low-density lipoprotein (LDL) LRP-1 receptor, which is highly expressed on tumor cells and mediates drug transmembrane entry into tumor cells, followed by the release of paclitaxel through hydrolysis in lysosomes. It has entered the clinical phase II/III.
C) CBX-12 [9]: It is an antigen-independent targeted drug composed of pHLIP® peptide (in cooperation with pHLIP, Inc.), linker, and small molecule anti-cancer drug based on the alphalex™ platform developed by Cybrexa Therapeutics, Inc. The principle is to target the specific low-pH environment of the tumor, which could form the alpha helix, allowing for the insertion of the peptide within the cell membrane. After delivery of the C-terminal toxin across the membrane, the payload would be released through the cleavage of disulfide bonds. CBX-12, as a candidate lead compound, has entered clinical phase I/II.

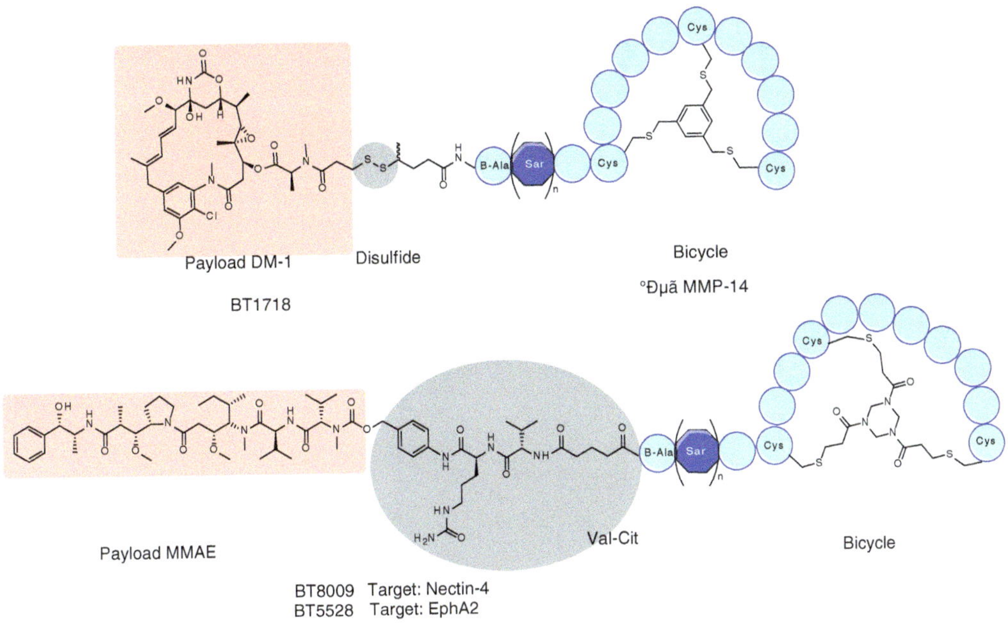

Figure 3.10 Structure of BT1718, BT8009, and BT5528. *Source:* [10] / Royal Society of Chemistry / CC BY 4.0.

D) Bicyclic peptide series PDC [10]: Bicyclic peptides, developed by Bicycle Therapeutics, Inc., have high targeting, stability, and affinity. BT1718 is composed of the cleavable disulfide-linked payload DM1 and bicyclic peptides, whereas BT8009 and BT5528 use MMAE as the payload and the cleavable dipeptide Val-Cit as the linker (Figure 3.10), which has entered clinical phase I/II status.

3.2.3.4 Other Research Hotspots of PDCs

Using the combination of different functional peptides and multiple site-cleavage patterns, the advantages of multiple structures can be combined from functionality and targeting. Deng [11] and others designed a multifunctional peptide combinatorial PDC consisting of targeting peptides (TTP), linked peptides, CPPs, and toxin molecules, targeting the LHRH receptor of tumor cells with LHRH peptides by using peptides (PLGLAG) that can be specifically degraded by MMP2 enzymes highly expressed between tumor tissues as the linker arm. After degradation of the linker peptide, toxin molecules are introduced into the cell along with the transmembrane peptide, and toxin molecules are released by rapid hydrolysis of the penetrating peptide or cleavage of the cleavable linker (Figure 3.11). The design simultaneously uses targeted peptides that promote the accumulation of drugs in tumor cells and transmembrane peptides that facilitate transport into cells, as well as two environmentally targeted lysis mechanisms, intracellular and extracellular, to improve targeting and drug properties.

3.2.4 Study Strategy and Case Sharing for Metabolite Profiling and Identification of PDCs

3.2.4.1 *In Vitro* and *In Vivo* Metabolism Research System of PDCs

At the early stage of drug development, *in vitro* metabolism research systems, such as acidified liver S9 (simulating lysosomal degradation in an acidic environment of tumor cells), plasma/blood,

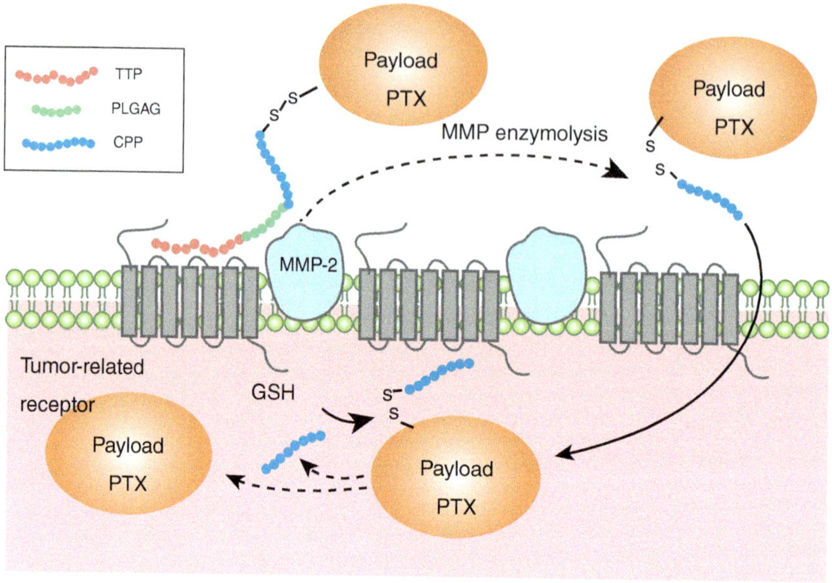

Figure 3.11 PDC of targeted peptide (TTP)-linked peptide-penetrating peptide-toxin molecules. *Source:* [11] / ELSEVIER.

Table 3.6 *In vitro* metabolism study system and purpose of PDCs and their payloads.

In vitro test article	Incubation system	Purpose
PDCs	Plasma/blood and hepatocytes	To study the stability and off-target possibility in its circulatory system
	Acidified liver S9/tumor cells	To evaluate the payload release and analysis of its associated metabolite
	Kidney S9	To assess the possible metabolic pathway payload during its renal clearance
Payloads	Hepatocytes/ microsomes/plasma	Metabolic pathways to study payload and metabolic enzymes

hepatocyte, and kidney S9, can help to determine the stability and targeting of PDC, as well as the release mechanism of payloads. Distinguishing the actual form of payload-related products released by PDCs as early as possible can help optimize and improve the structural design and can also provide accurate information on expected metabolites for subsequent pharmacokinetics and toxicokinetic studies. Preclinical and clinical *in vivo* studies, on the other hand, confirmed the metabolism of PDCs and released payloads *in vivo* by MetID in plasma and excreta, such as urine (Tables 3.6 and 3.7).

Like conventional small molecules, the payload itself, or the identified main release products of PDC (such as payload linkers and other main released moieties), is also required to determine the metabolism and its possible metabolic pathway (Figure 3.12, the metabolic pathways of commonly used payload MMAE), which is then helpful for the related drug interactions and toxicology studies.

Table 3.7 Purpose of *in vivo* metabolism for PDC.

In vivo test article	Sampling matrix	Purpose
PDCs	Plasma/blood	To assess the safety by confirming the exposure of payloads and their associated metabolites in the circulatory system
	Urine/feces	To identify the routes of elimination for excretion and assess recovery
	Tissue homogenate	To investigate the targeting or accumulation of drugs or metabolites *in vivo*

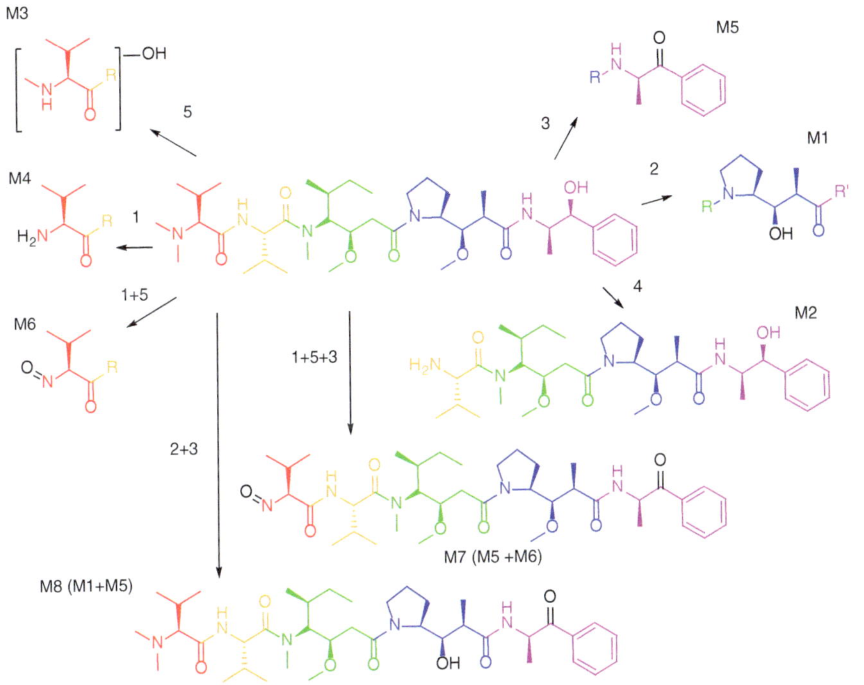

- N-demethylation (CYP3A)
- O-demethylation (CYP3A)
- Dehydrogenation (CYP3A)
- Amide hydrolysis
- Oxygenation

Figure 3.12 Metabolites and metabolic pathways of MMAE in humans. *Source:* Adapted from reference [12].

3.2.4.2 Experimental Method for MetID of PDCs

The MetID of PDCs mainly employed a LC-UV-HRMS. The analytical samples obtained by incubation with the above-mentioned *in vitro* study system or collected *in vivo* samples are treated with protein precipitation. After centrifuging, the supernatant is analyzed by LC-UV-HRMS. If the compounds are chemically stable during N_2 drying, they can also be analyzed after concentration. If the peptide is neutral, the reverse liquid chromatographic system with water/acetonitrile can be used for gradient elution. If the peptide is strongly alkaline (such as transmembrane peptide trans-activating transcriptional activator (TAT) and other peptides containing more basic amino acid residues), an appropriate amount of trifluoroacetic acid can also be added to the mobile phase to improve the peak shape.

The components were separated by liquid chromatography and the chromatographic peaks of the metabolites were found. Then, the *m/z* information of the mass spectrum (MS) was compared with that of the parent drug. The metabolic changes and metabolic sites were analyzed according to the primary MS and secondary MS (MS2) signals of the high-resolution MS.

3.2.4.3 Difficulties and Solutions in Metabolite Analysis

The analysis of metabolites by LC-MS/MS focuses on chromatographic tracer localization and structure elucidation. Radioisotopic and nonisotopic methods are used for chromatographic tracer localization. The radioisotopic method uses radiolabeled compounds for metabolic experiments, but the synthesis of radiolabeled compounds is complicated with high costs and laboratory qualification requirements are high, which create obstacles in the screening stage. Nonisotopic methods typically rely on additional peaks compared to the blank matrix for tracer localization, such as LC-UV or LC-MS/MS.

Due to the structural characteristics, PDCs and its main metabolites may have weak UV absorption; thus, UV detection cannot effectively trace and locate them. Hence, it is necessary to obtain the secondary chromatography by data processing methods such as LC-MS/MS background subtraction to locate the possible exogenous new peaks on the chromatography (Figures 3.13 and 3.14, D-E-F in the following cases).

Compound molecular size can affect the complexity of structure elucidation. Similar to ADCs, due to the steric hindrance, their payloads are usually metabolized by CYP enzymes after release. However, generally, only the release mode and metabolites of payloads are considered for ADCs. The antibody part is mainly degraded through the catabolic mechanism in the body, while the peptide and linker of PDCs may be metabolized by hydrolases. Therefore, when analyzing the possible metabolites and judging the sites of metabolic changes, the payload, the peptide, and the linker need to be taken into consideration. Currently, the commonly used data processing software is mainly through the mass changes of metabolic characteristics for library matching and combined with exhaustive matching based on chemical bond breakage. Therefore, both manual analysis and software calculation are used in the data processing of MetID for PDCs. With the increase of molecular weight (number of chemical bonds), its computational complexity increases by geometric orders compared with small molecular compounds. Therefore, upgrading the software and algorithm is important to improve the efficiency of MetID for PDCs.

3.2.4.4 Case Study of PDC Metabolite Identification

The following is a case study based on the *in vitro* metabolism of a PDC with cyclic peptide and a cleavable linker. The instrument used is shown in Figure 3.16.

The metabolism of this PDC in acidified liver S9 (pH = 5) and hepatocyte incubation systems were compared. In the acidified S9 incubation system, only the UV peaks of the parent drug and one metabolite M1 could be found in the LC-UV spectrum, and the MS peak of another major metabolite M2 could be found by background subtraction (sample minus blank matrix) in the LC-MS/MS (Figure 3.14). By comparing the signals of MS and MS2, it was confirmed that the cyclic peptide M1 and the payload M2 were obtained after hydrolysis of the linker. Because the characteristic UV absorption is mainly contributed by some amino acids of cyclic peptides, the UV absorption of M2 is weak and not effectively detected on LC-UV, which can only be detected by LC-MS/MS identification.

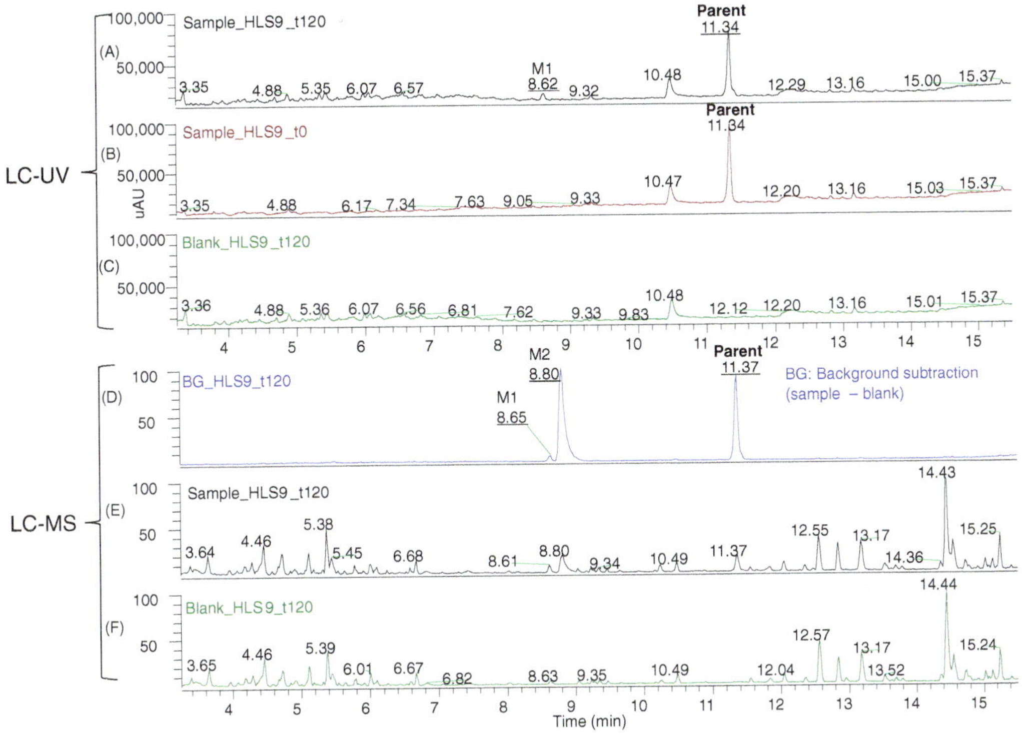

Figure 3.13 LC-UV (A: 120 minutes sample; B: 0 minute sample; C: blank sample) and LC-MS/MS spectrum (D: after background subtraction; E: 120 minutes sample; F: blank sample) of a PDC after incubation with human acidified liver S9 system.

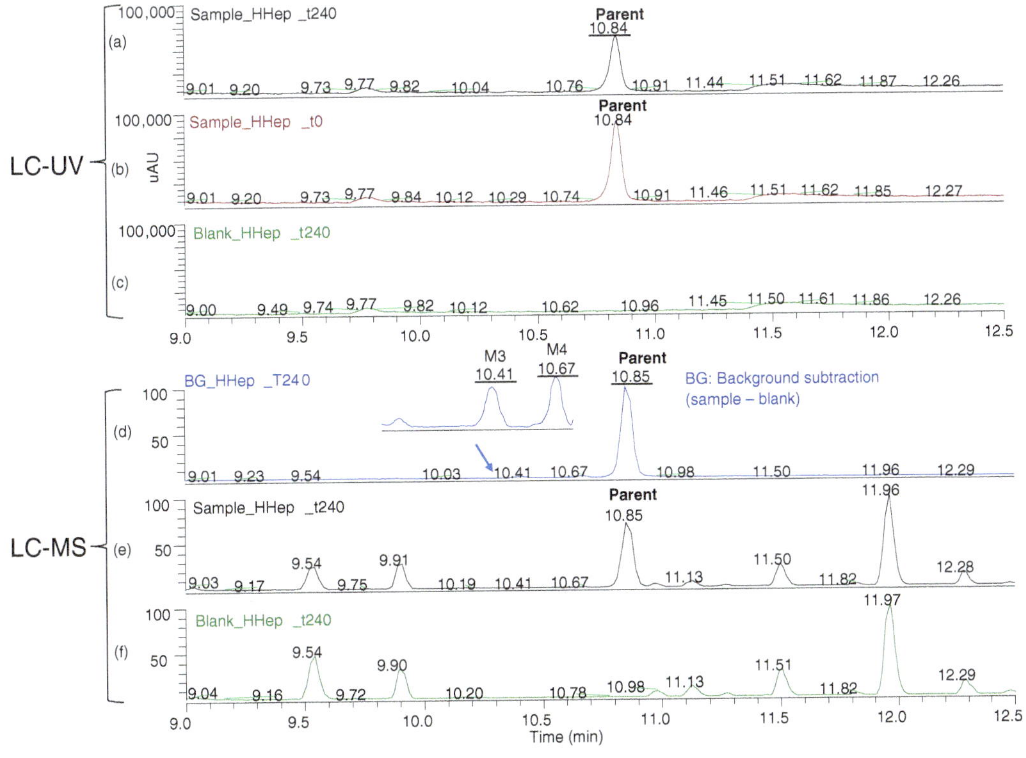

Figure 3.14 LC-UV (A: 240 minutes sample; B: 0 minute sample; C: blank sample) and LC-MS/MS spectrum (D: after background subtraction; E: 240 minutes sample, F: blank sample) of a PDC after incubation with acidified S9 and hepatocytes.

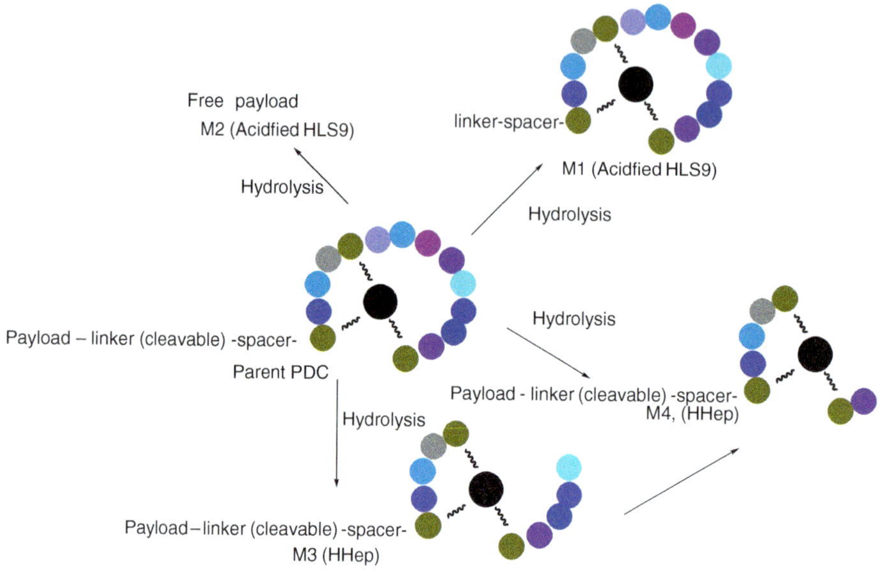

Figure 3.15 Difference in metabolic pathways of a PDC after incubation with acidified S9 and hepatocytes.

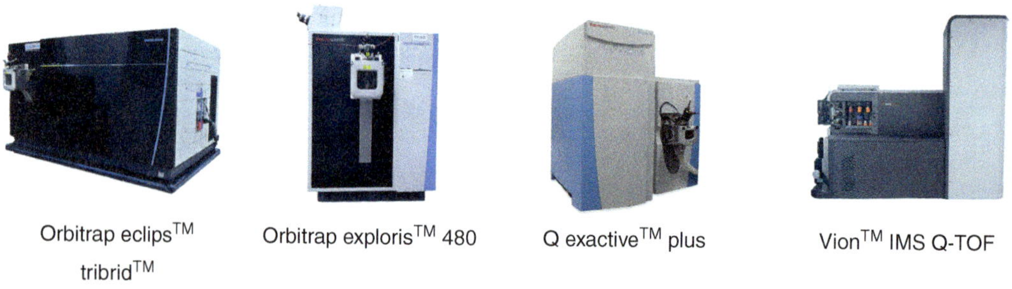

Figure 3.16 High-resolution LC-MS/MS analysis and product structure elucidation for identification of PDC metabolites.

In the hepatocyte incubation system, because the metabolite level was low, no significant metabolite peak was found in the LC-UV, and two low-abundancy metabolites M3 and M4 were found after background subtraction processing in the LC-MS/MS (Figure 3.15). By comparing the signals of MS and MS2, it was judged that the payload-linker part did not change, while the metabolic change was mainly the hydrolysis of the peptide part. The molecular weight and mass changes of each amino acid residue were compared, and the generated peptide after hydrolysis was concluded.

This PDC was found to be mainly hydrolyzed through linkers to release payload M2 and cyclic peptide M1 in acidified S9; partial hydrolysis on cyclic peptides occurred in the MS to produce M3 and M4, whereas linkers remained stable. From the difference in metabolic pathways (Figure 3.16), it can be seen that the metabolic release mode of the payload also has tumor environmental targeting (relative to hepatic metabolism).

In addition, Val-Cit-type linkers commonly used in conjugated drugs are unstable in mouse plasma, which is due to the high expression of Ces1c enzyme in mouse plasma that can hydrolyze the linkers; although it is very low in human plasma, there will be a large variation when using mice as an animal model to study the pharmacokinetics, which needs more attention. In the study of such compounds, we also confirmed that the free payload generated in the mouse plasma is much higher than that generated in the human plasma by the *in vitro* plasma incubation system.

3.2.5 Summary and Prospect

With the innovative development of peptide and linker technology, PDCs have a broad development prospect. Metabolic studies of PDCs simultaneously include assessments of the metabolic stability, targeting, and payload release. This section summarizes the metabolic characteristics and research strategies of PDCs, and an MetID case is shared. More PDCs are expected to be designed and developed to bring good news to patients.

References

1 Vrettos, E.I. (2018). On the design principles of peptide–drug conjugates for targeted drug delivery to the malignant tumor site. *Beilstein Journal of Organic Chemistry* 14: 930–954.

2 Drug Approval Package: PEPAXTO, FDA. (2021). https://www.accessdata.fda.gov/drugsatfda_docs/nda/2021/214383Orig1s000TOC.cfm.

3 Drug Approval Package: PEPAXTO, FDA. (2021). https://www.oncopeptides.com/contentassets/a6783cb36a194c2fa467a8202605f5d3/fda-pepaxto-approval-webcast-presentation.pdf. Archived July 04, 2022.

4 Hennrich, U. and Kopka, K. (2019). Lutathera®: the first FDA- and EMA-approved radiopharmaceutical for eptide receptor radionuclide therapy. *Pharmaceuticals (Basel).* 12 (3): 114.

5 Zhao, N. (2018). Tumor-targeting peptides: ligands for molecular imaging and therapy. *Anti-Cancer Agents in Medicinal Chemistry* 18: 74–78.

6 Hoppenz, P. (2020). Peptide-drug conjugates and their targets in advanced cancer therapies. *Frontiers in Chemistry* 8: 571. https://doi.org/10.3389/fchem.2020.00571.

7 Engel, J., Emons, G., Pinski, J., and Schally, A.V. (2012). AEZS-108: a targeted cytotoxic analog of LHRH for the treatment of cancers positive for LHRH receptors. *Expert Opinion Investigational Drugs* 21 (6): 891–899.

8 Alas, M., Saghaeidehkordi, A., and Kaur, K. (2021). Peptide-drug conjugates with different linkers for cancer therapy. *Journal of Medicinal Chemistry* 64 (1): 216–232.

9 Gayle, S., Aiello, R., Leelatian, N. et al. (2021). Tumor-selective, antigen-independent delivery of a pH sensitive peptide-topoisomerase inhibitor conjugate suppresses tumor growth without systemic toxicity. *NAR Cancer* 3 (2): zcab021.

10 Cooper, B.M. (2021). Peptides as a platform for targeted therapeutics for cancer: peptide–drug conjugates (PDCs). *Chemical Society Reviews* 50: 1480.

11 Deng, X. (2021). Discovery of novel cell-penetrating and tumor-targeting peptide-drug conjugate (PDC) for programmable delivery of paclitaxel and cancer treatment. *European Journal of Medicinal Chemistry* 213: 113050.

12 Han, T.H. (2013). CYP3A-mediated drug–drug interaction potential and excretion of brentuximab vedotin, an antibody–drug conjugate, in patients with CD30-positive hematologic malignancies. *Journal ofClinical Pharmacology* 53 (8): 866–877.

4

Peptide Drugs

4.1 Development, Characteristics, and Pharmacokinetic Research Strategies of Peptide Drugs

Jianping Sun, Yan Pan, Yu Wang, Yanfeng Liu, Jing Jin

4.1.1 Introduction

Peptides are compounds formed by linking α-amino acids together through peptide bonds. Peptides composed of less than 2–20 amino acids are called oligopeptides, and those composed of more than 10 amino acids and up to 40 amino acids are called polypeptides or peptides. The molecular weight (MW) is generally below 10 kDa, which has no clear boundary with the protein. According to the Further Consolidated Appropriations Act of 2020 (FCA), which revised the Biologics Price Competition and Innovation Act of 2009 (BPCIA), the FDA defines a "Protein" as "Any α-Amino acid polymer with a well-defined sequence and a size greater than 40 amino acids. Such protein products are subject to biological license application (BLA)" [1]. Therefore, the regulatory requirements for small molecules can still be referred to for the development of peptide drugs comprising 40 amino acids or less.

We know that interactions between proteins are essential for life process, and changes in protein structure or function can potentially initiate disease onset. As structural units of proteins, peptides certainly play a pivotal role in disease treatment.

Nature Review Drug Discovery published a review article "Trends in peptide drug discovery" [2], in which the research history and trend of peptide drugs were introduced (Figure 4.1). In 1922, F.G. Banting and C.H. Best extracted insulin from the bovine pancreas, marking it the first peptide drug. This revolutionized the treatment of type 1 diabetes and heralded the beginning of peptide drug development. In 1954, Vincent du Vigneaud's team first chemically synthesized peptides, published "The total synthesis of Oxytocin and Vasopressin," and won the Nobel Prize in Chemistry in 1955. The next leap was Bruce Merrifield's idea of assembling amino acids on the solid phase to automatically synthesize peptides, which led to the invention of solid-phase peptide synthesis (SPPS) in 1963 and won the Nobel Prize in Chemistry in 1984. The advent of recombinant technology in the 1980s made it possible for the clean production of peptides with greater MWs.

However, compared with small-molecule drugs, limitations of peptides such as low oral bioavailability, poor plasma stability, and short circulation time were also exposed. In addition, during the 1970s and 1980s, for pharmaceutical companies, the high cost associated with peptide production led to a stagnant period in the development of peptide drugs.

Drug Metabolism and Pharmacokinetics: Frontiers, Strategies, and Applications, First Edition.
Edited by Liang Shen and WuXi AppTec DMPK.

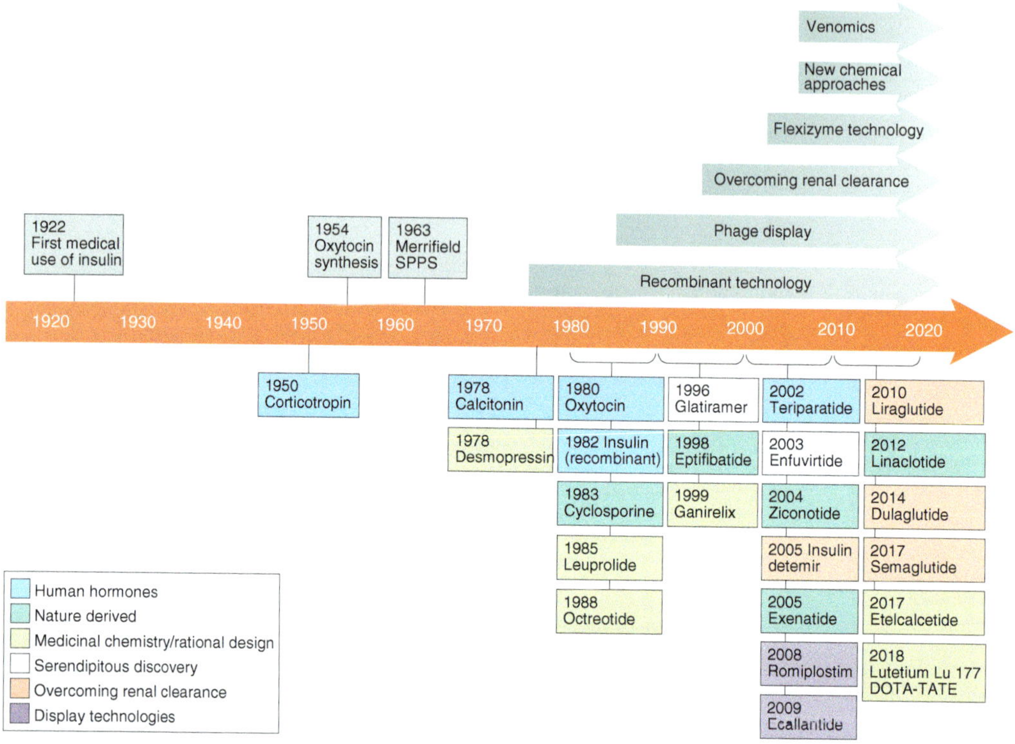

Figure 4.1 Development and milestones of peptide drugs. *Source:* [2] / with permission of Springer Nature.

Until the late 1980s, it was found that peptides could be used as selective probes of subtype receptors, as well as their enduring presence as lead compounds, laying the foundation for a second wave of peptide drug development and drawing the attention of venture capital and biotechnology companies. Since then, drug delivery technology, formulations, and synthesis technology have also made great progress, which has created a good prospect in the peptide drug market.

After the milestones of this peptide research, the average annual growth rate of approved peptide drugs was about 7.7% from 1959 to 2019, and more than 80 peptide drugs have been marketed by 2020. These drugs were mostly agonists and were used for the treatment of a variety of diseases, including diabetes, cancer, osteoporosis, multiple sclerosis, acquired immune deficiency syndrome (AIDS), and chronic pain. In 2019, peptide drugs accounted for 5% of the $1.2 trillion drug market, and the $25 billion sales of insulin and its analogs accounted for approximately 50% of the peptide drug market.

The peptide pipeline in clinical research doubled from 2000 to 2010, with more than 150 peptides in clinical research and 400–600 peptide compounds in preclinical research. Peptide drugs have emerged as one of the areas with the most substantial market potential in the pharmaceutical industry.

4.1.2 Characteristics and Advantages of Peptide Drugs

Peptide drugs combine the benefits of both small molecules and protein drugs. Small molecules have small binding surfaces, low activity, and poor specificity to target proteins, which often lead to toxic side effects. Peptides exhibit strong affinity to targets, high specificity, and few side

Table 4.1 Comparison of advantages and disadvantages of various drugs [5] / MDPI / CC BY 4.0.

	Advantages	Disadvantages
Small molecule	High oral bioavailability Metabolically stable Numerous targets Small size	High toxicity Many side effects Potential low solubility Poor selectivity
Peptide	High activity High selectivity Broad target Low toxicity Low tissue accumulation High chemical and biological diversity	Poor metabolic stability Weak membrane permeability Low oral bioavailability Fast clearance Sometimes, low solubility
Macromolecule	High activity High selectivity Low toxicity Low tissue accumulation High chemical and biological diversity Difficult for biosimilar	Poor membrane permeability Poor oral bioavailability The target is mainly extracellular High cost of manufacturing

effects [3]. Compared to macromolecular drugs, peptide drugs offer distinct advantages such as low immunogenicity and low production cost. However, peptide drug development faces challenges such as low permeability, poor stability, and low oral bioavailability. A comparison of the advantages and disadvantages of various drugs is shown in Table 4.1 [4].

4.1.3 Drug Metabolism and Pharmacokinetics Properties of Peptides

4.1.3.1 Absorption

The main modes of peptide absorption are convection of fluid, passive diffusion, and receptor-mediated active transport. Peptides, except for oligopeptides or small peptides, typically have MWs greater than 1,000 Da and exhibit stronger polarity, resulting in slower diffusion rate compared to smaller molecules. Peptides with larger MWs (5,000–12,000 Da) are mainly absorbed and transported through the lymph. The time for the drug to reach peak concentration will be significantly extended because of the slow lymph circulation. The flip-flop phenomenon can also occur *in vivo* due to absorption retardation (when the absorption constant is much smaller than the elimination constant), resulting in a longer half-life and pharmacodynamic effect [6]. For instance, the gonadotropin-releasing hormone (GnRH) agonist leuprorelin, the monthly long-acting intramuscular injection, gradually releases the drug into the systemic circulation to inhibit follicle-stimulating hormone (FSH) and luteinizing hormone (LH) secretion in the long term for the treatment of prostate cancer and endometriosis.

Orally administered peptide drugs encounter the mucus layer of the gastrointestinal tract, which secretes certain mucinous proteins that, through intermolecular interactions and disulfide bonds, impede the diffusion of peptide drugs into the cells. Kovalainen et al. reported that the penetration of intestinal epithelial cells is restricted to molecules with a MW of no more than 3,500 Da [7]. Therefore, poor gastrointestinal absorption has become a challenging aspect in the development of oral peptide drugs, and the following methods are often adopted to enhance the absorption of peptides.

1) Identify the appropriate absorption site based on drug characteristics such as acidity/alkalinity, MW, and LogP and take corresponding formulation strategies to extend the residence time at the absorption site.
2) Evaluate suitable absorption enhancers (e.g., surfactants, bile salts, phospholipids, fatty acids, and glycerides) to increase the permeability across the gastrointestinal tract. In addition, sodium caprate (C10, also known as decanoic acid) and sodium N-[8-(2-hydroxybenzoyl) amino] octanoate (SNAC, salcaprozate sodium) are also commonly used penetration enhancers.
3) Utilize bio-adhesive formulations, such as polycarbophil, to enhance the absorption of peptides in the mucus layer.

4.1.3.2 Distribution

Human serum albumin (HSA) has a strong binding affinity to peptides, and it has a greater impact on the distribution of peptides because of its high physiological concentration. Peptides have weak lipophilicity and low permeability, and it is difficult to directly cross the cell membrane into cells. It primarily uses endocytosis to enter cells, including pinocytosis or phagocytosis, and occasionally binds to certain cell membrane surface receptors for cell entry. Given the volume of distribution (Vd), values of peptides are typically small and generally do not exceed the volume of extracellular fluid; utilizing various nanocarrier systems, including liposomes, polymers, protein complexes, and inorganic materials, etc., is considered a more effective strategy to enhance the intracellular delivery and targeting of peptides.

There also exists a specialized peptide called the CPP, which is a class of short peptides that can carry molecules or proteins across membrane permeation, derived mainly from viral peptide sequences, nonviral proteins, or smaller molecules. Its ability to penetrate membranes is independent of classical endocytosis and can directly cross the cell membrane into cells.

4.1.3.3 Metabolism

Peptides are mainly degraded by proteases or peptidases. Proteases are widely distributed in the body, including the liver, kidney, and gastrointestinal tissues, lungs, blood and vascular endothelium, skin, and other tissues and organs. Proteases are divided into endopeptidases and exopeptidases (including aminopeptidases and carboxypeptidases) [8]. The hydrolysis of peptides and proteins typically begins with endopeptidases, which act on the inner part of the protein, and oligopeptides produced by enzymatic hydrolysis can be further degraded by exopeptidases. Additionally, the well-known CYP450 enzyme can metabolize prodrugs or metabolites of some peptides. Proteases involved in peptide metabolism and their metabolic sites are listed in Table 4.2.

In addition to direct metabolism by proteases in plasma, gastrointestinal tract, etc., endocytic elimination and target-mediated elimination (TMDD) (Table 4.3) are also the main routes of elimination of peptides [9].

To improve the stability of peptide compounds, optimization can be carried out in the following aspects [10]:

1) N-terminal or C-terminal modifications such as methylation and N-terminal acetylation. The half-life of glucose-dependent insulinotropic polypeptide (GIP) (1-42) is very short only ranging from two to five minutes. Mabilleau et al. developed an enzymatically stabilized GIP analog by acetylating Tyr1 in GIP (N-AcGIP), resulting in a circulating half-life exceeding 24 hours.
2) Replacing natural amino acids with D-amino acids and artificial amino acids to reduce affinity to proteases.

Table 4.2 Proteases and metabolic sites.

Protease	Metabolic site
Pepsin	Peptide bonds are formed by amino groups of aromatic amino acids and carboxyl groups of acidic amino acids
Trypsin	Peptide bonds are formed by carboxyl groups of basic amino acids
Chymotrypsin	Hydrolysis of aromatic amino acids
Elastase	Peptide bonds are formed by carboxyl groups of aliphatic amino acids
Aminopeptidase	Amino-terminal of oligopeptide
Carboxypeptidase A, carboxypeptidase B	Carboxy-terminal of neutral and basic amino acids
Dipeptidyl peptidase IV (DPPIV)	The degradation of Ala or Pro as the second amino acid residue at the N-terminus

Table 4.3 Endocytic elimination and target-mediated elimination pathways.

Endocytic elimination	Target-mediated elimination (TMDD)
Peptides with relatively larger molecular weight (e.g., insulin) are endocytosed into cells and then degraded by lysosomes.	It specifically binds to the target/antigen on the cell surface and is eliminated by internalization.
Small peptides with molecular weights <500 endocytose through the membrane into cells and are first localized in the endosome and then eliminated by the endosome–lysosome pathway.	Common peptide drugs such as GLP-1 receptor agonists bind to the G Protein-Coupled Receptors (GPCRs) of the target cell, and nonlinear elimination of drugs occurs when the drug concentration saturates the receptor.

3) Implementing cyclization or bicyclization of peptides to improve compound rigidity.
4) Increasing MW, such as the addition of aliphatic chains, and PEGylation.
5) The addition of specific ligands to improve binding to albumin can effectively extend the half-life of peptide drugs.

4.1.3.4 Excretion

Glomerular pores are approximately 8 nm in size. It is generally believed that molecules with a MW smaller than 50 kDa can be eliminated by renal filtration [11]. The smaller the peptide molecule, the greater the renal clearance. The filtration rate is decreased with renal impairment, which may lead to increased drug concentration and half-life in the body.

Larger peptides filtered by the glomerulus are eliminated by endocytosis and lysosomal degradation and finally hydrolyzed into small peptide fragments and amino acids. Smaller peptides filtered through the glomerulus are hydrolyzed to amino acids by exopeptidases in the brush-border membrane of the tubular lumen and then reabsorbed into the systemic circulation by specific amino acid transport systems, or they are first broken into small peptides and then transported into the proximal tubular epithelial cells and hydrolyzed intracellularly. The differences in DMPK properties between peptides and other types of drugs are listed in Table 4.4.

Table 4.4 DMPK property differences between peptides and other types of drugs.

	Traditional chemical drugs	Unmodified peptides	Monoclonal antibody	Antibody–drug conjugate (ADC)
Molecular weight	Generally, not >1 kDa	1–10 kDa	~150 kDa	~150 kDa
Biological activity and specificity	Low	High	High	High
Immunogenicity	None	None or low	Yes	Yes
Purity and cost	High purity and low cost	High purity and cost	Low purity and high cost	Low purity and high cost
Half-life	Short	Short	Long	Long
Oral bioavailability	Acceptable	Very low	No	No
Routes of administration	Oral typically	Intravenous, subcutaneous, intramuscular	Intravenous, subcutaneous, intramuscular	Intravenous
Distribution	Wide V_d range, may be transporter substrate	V_d approximates the extracellular fluid volume	V_d approximates plasma volume	ADC is similar to monoclonal antibodies, and unbound drugs are similar to small molecules
Metabolism	Phase I and II enzyme metabolism	Peptide hydrolases, Proteases	Antibody catabolism	
Excretion	Biliary and renal excretion	Renal excretion	Body recirculation	
PK linearity	General linear with low dose may be nonlinear with high dose	For small molecular weights similar to chemical drugs, and for large molecular weights, similar to those of macromolecules	General linear with high dose; nonlinear with low dose	General linear with high dose, nonlinear with low dose
PK analyte	Parent drug and metabolites	Peptides	Antibody	ADC or bound drug, total antibody, unbound drug
Analytical methods	Liquid chromatography–mass spectrometry	LC-MS, radiolabel, or ELISA analysis	LC-MS, radiolabel, or ELISA analysis	LC-MS, radiolabel, or ELISA analysis

4.1.4 DMPK Research Strategy for Peptide Drugs

In the screening stage, the plasma stability, plasma protein binding, and kidney homogenate stability studies should be investigated. Radiolabeling is commonly employed for excretion experiments of peptide drugs. In the IND application stage, the species differences of drugs in various metabolic systems should be fully considered, and the animal species with similar metabolic characteristics to humans should be selected; the appropriate administration route, dosage, and detection time points should be determined based on the pharmacological mechanism and metabolic characteristics, especially the potential nonlinear kinetics. Given that peptide drugs may be immunogenic, the concentration of anti-drug antibodies needs to be measured to assess potential pharmacological impacts. The components of PK studies on peptide drugs vary at different stages of research and development (Table 4.5).

In terms of clinical data prediction, the processing of peptides across different mammals is relatively conserved. This implies that it is rather reliable to extrapolate human data based on the PK

Table 4.5 Pharmacokinetics studies at different stages of peptide drug development.

Development phase	Pharmacokinetics studies
Screening phase	Plasma protein binding (high albumin binding leads to low clearance) Plasma stability Stability of simulated gastrointestinal fluid (for oral peptide drugs) Kidney S9/homogenate stability Brush-border membrane vesicle (BBMV) stability Liver/intestine S9/cytoplasmic metabolic stability MDCK, Caco-2 permeability P450 enzyme inhibition (optional) PEPT1, PEPT2 transporter substrate and inhibition Vehicle screening *In vivo* PK studies
Preclinical candidate phase	Metabolites identification in plasma, liver/kidney S9/homogenate Hepatocyte metabolite identification (optional) Metabolites identification in *in vivo* plasma samples Dose escalation pharmacokinetics (PK) studies (support pilot toxicology)
IND phase	*In vivo* analytical method validation *In vitro* ADME assay *In vivo* PK in rats and large animals (including anti-drug antibody detection) Tissue distribution and mass balance in rats *In vivo* metabolite identification

data obtained from animal studies. Therefore, employing allometric scaling for prediction tends to yield more accurate results. However, for peptides with a relatively small MW, when hepatic metabolism is the primary clearance pathway or there are metabolic differences between species, it may be more appropriate to employ *in vitro–in vivo* extrapolation.

4.1.5 Summary

Peptide drugs, with their abundant targets, high biological activity, strong specificity, and minimal side effects, have been garnering increasing attention in recent years. Despite many challenges in peptide drug development, emerging methods such as new biosynthesis, formulations, and delivery carriers are likely to bring significant breakthroughs in the application. The combination of peptides with other types of drugs, such as PDCs, also holds promising prospects. Occupying a distinct niche in the pharmaceutical field, peptide drugs are poised to deliver health benefits to patients globally leveraging their unique advantages.

References

1 Food and Drug Administration. ANDAs for certain highly purified synthetic peptide drug products that refer to listed drugs of rDNA origin, guidance for industry. https://www.fda.gov/regulatory-information/search-fda-guidance-documents/andas-certain-highly-purified-synthetic-peptide-drug-products-refer-listed-drugs-rdna-origin.
2 Muttenthaler, M., King, G.F., Adams, D.J. et al. (2021). Trends in peptide drug discovery. *Nature Review Drug Discovery* 20 (4): 309–325. https://doi.org/10.1038/s41573-020-00135-8.

3 Srivastava, V. (2017). *Peptide-based Drug Discovery: Challenges and New Therapeutics*. The Royal Society of Chemistry https://doi.org/10.1039/9781788011532.

4 Lau, J.L. and Dunn, M.K. (2018). Therapeutic peptides: historical perspectives, current development trends, and future directions. *Bioorganic & Medicinal Chemistry* 26 (10): 2700–2707.

5 Manna, S.L., Natale, C.D., Florio, D. et al. (2018). Peptides as therapeutic agents for inflammatory-related diseases. *International Journal of Molecular Sciences* 19 (9): 2714.

6 Di, L. (2015). Strategic approaches to optimizing peptide ADME properties. *AAPS J* 17 (1): 134–143. https://doi.org/10.1208/s12248-014-9687-3.

7 Haibo, D. et al. (2018). Progress in the kinetic characteristics and metabolic mechanism of polypeptide drugs. *Chinese Journal of Pharmacology & Toxicology* 32 (3): https://doi.org/10.3867/j.issn.1000-3002.2018.03.010.

8 Jinfeng, Y. et al. (2013). Progress in the metabolism of peptide drugs. *Chinese Pharmacological Bulletin* 29 (7): 895–899.

9 Peletier, L.A. and Gabrielsson, J. (2012). Dynamics of target-mediated drug disposition profiles and parameter identification. *Journal of Pharmacokinetics and Pharmacodynamics* 39: 429. https://doi.org/10.1007/s10928-012-9260-6.

10 Patel, G. and Misra, A. (2011). Oral delivery of proteins and peptides: concepts and applications. In: *Challenges in Delivery of Therapeutic Genomics and Proteomics* (ed. A. Misra), 481–529. London: Elsevier https://doi.org/10.1016/B978-0-12-384964-9.00010-4.

11 Datta-Mannan, A. (2019). Mechanisms influencing the pharmacokinetics and disposition of monoclonal antibodies and peptides. *Drug Metabolism and Disposition* 47 (10): 1100–1110. https://doi.org/10.1124/dmd.119.086488.

4.2 Metabolic Characteristics and Optimization Strategies of Peptide Drugs

Jianping Sun, Jing Jin

4.2.1 Introduction

Since insulin was used as the first peptide drug for the treatment of type 1 diabetes in 1922, peptides have gained significance in the field of drug research and development due to their diverse physiological regulatory effects, notable activity, high selectivity, and low toxicity. Over the past decade, with the surge in biomedical R&D investment, peptide drugs have gained increased attention and development.

Peptide drugs also have non-negligible weaknesses; their chemical and physiological stability are relatively poor and vulnerable to the degradation by proteolytic enzymes in the body, resulting in poor oral absorption, rapid metabolism, short half-life, and other problems, which bring many challenges to the research and development of peptide drugs. DMPK studies can help optimize the structure of peptide drugs, evaluate the appropriate route of administration, and determine metabolism-related species, which are important processes in peptide drug research and development. In the previous chapter, the overall idea of peptide research is introduced. In Section 4.2, the metabolic mechanism of peptide drugs and how to carry out a metabolic investigation and improve metabolic stability will be introduced and discussed.

4.2.2 Metabolic Enzymes and Sites of Peptide Drugs

Peptide bonds are the natural metabolic sites of proteolytic enzymes, and peptide drugs will be metabolized by proteolytic enzymes into active or inactive metabolites and excreted out of the body. There are approximately 560 human and 640 mouse proteases known, and these metabolic processes occur widely and rapidly, resulting in the short half-life of peptide drugs *in vivo*. Figure 4.2 shows the classification of proteolytic enzymes and the mode of catalysis [1].

Endopeptidases break peptide bonds within the molecule. Their catalytic mechanism is related to the chemical groups involved in the hydrolysis process and can be subdivided into five different

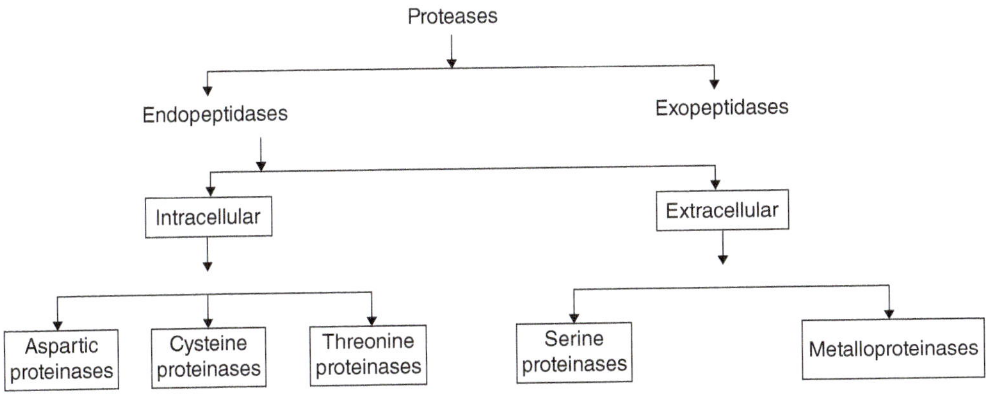

Figure 4.2 Protease main type. *Source:* [1] / with permission of Taylor & Francis Group.

Table 4.6 Major protease species and associated amino acid substrates.

Protease	Relevant amino acid substrates
Pepsin	Leu, Phe, Trp, Tyr (N-terminal residues)
Trypsin	Lys, Arg (C-terminal residues)
Chymotrypsin	Phe, Tyr, Trp (C-terminal residues)
Elastase	Ala, Gly, Ser, Val (C-terminal residues)
Aminopeptidase	—
Carboxypeptidase A, carboxypeptidase B	Phe, Tyr, Trp or Leu, Lys or Arg
Dipeptidyl peptidase IV (DPP IV or DPP-4)	Ala, Pro

Leu, leucine; Phe, phenylalanine; Trp, tryptophan; Tyr, tyrosine; Lys, lysine; Arg, arginine; Ala, alanine; Gly, glycine; Ser, serine; Val, valine; Pro, proline.

categories: aspartic proteases, cysteine proteases, threonine proteases, serine proteases, and metalloproteases. Exopeptidases specifically cleave substrates at the N- or C-terminus of peptides and can therefore be subdivided into aminopeptidases and carboxypeptidases.

The metabolism of peptides by peptidases is nonspecific and involves multiple steps. A peptidase can metabolize various types of substrates, or a substrate can be metabolized by multiple peptidases and is a sequential multistep degradation process. Peptides are usually metabolized in the body by endopeptidases to oligopeptides, which are then further degraded to amino acids by exopeptidases and subsequently involved in various biochemical processes such as protein synthesis in the body [2].

Proteases are mainly distributed in the plasma, gastrointestinal tract, blood–brain barrier, liver, spleen, kidney, and other tissues, resulting in poor absorption and tissue distribution of peptide drugs and thus inconvenience of administration and inability to reach the target site and other defects. Table 4.6 shows the major protease categories and mode of metabolism.

DPP IV is abundant in the kidney, small intestine, submaxillary gland, and liver, and some are present in the circulating blood system in a soluble form. The activity is optimal at a pH of ~7.8 and maintains similar activity even over a wide range of pH 5–10. DPP IV is very sensitive to the protease inhibitors diisopropyl fluorophosphate (DFP), but much less sensitive to other common serine protease inhibitors, such as diethyl p-nitrophenyl phosphate and phenylmethanesulfonyl fluoride (PMSF) [3].

Glucagon-like peptide-1 (GLP-1) is an important hormone that promotes insulin secretion and controls blood glucose. It is rapidly degraded by DPP IV *in vivo*, with a half-life of 0.9 minute in humans [4]. GLP-1 receptor agonists such as exenatide, liraglutide, dulaglutide, and lixisenatide have effects similar to native GLP-1, while they are used to treat type 2 diabetes and weight management because structural optimization reduces the metabolism caused by DPP IV and prolongs the duration of action. DPP IV inhibitors such as sitagliptin are also used as drugs for the treatment of diabetes. Figure 4.3 shows the mechanism of action of DPP IV inhibitors.

4.2.2.1 Gastrointestinal Metabolism

The gastrointestinal tract contains abundant proteases, which become the main barrier to peptide absorption. Figure 4.4 shows the secretion and activation of proteases in the gastrointestinal tract. These proteases come from three sources:

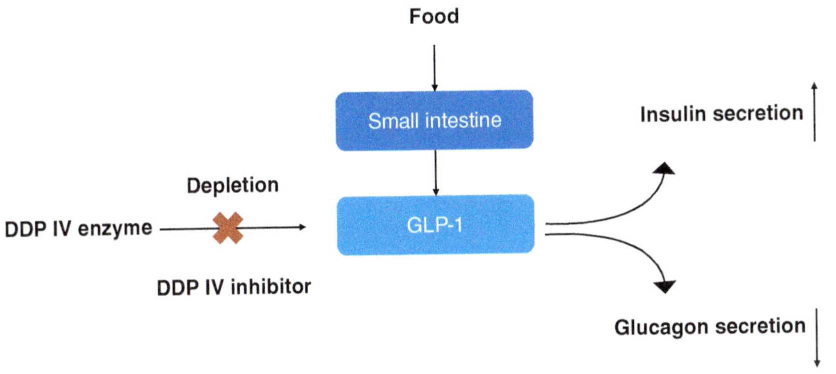

Figure 4.3 Mechanism of action of DPP IV inhibitors.

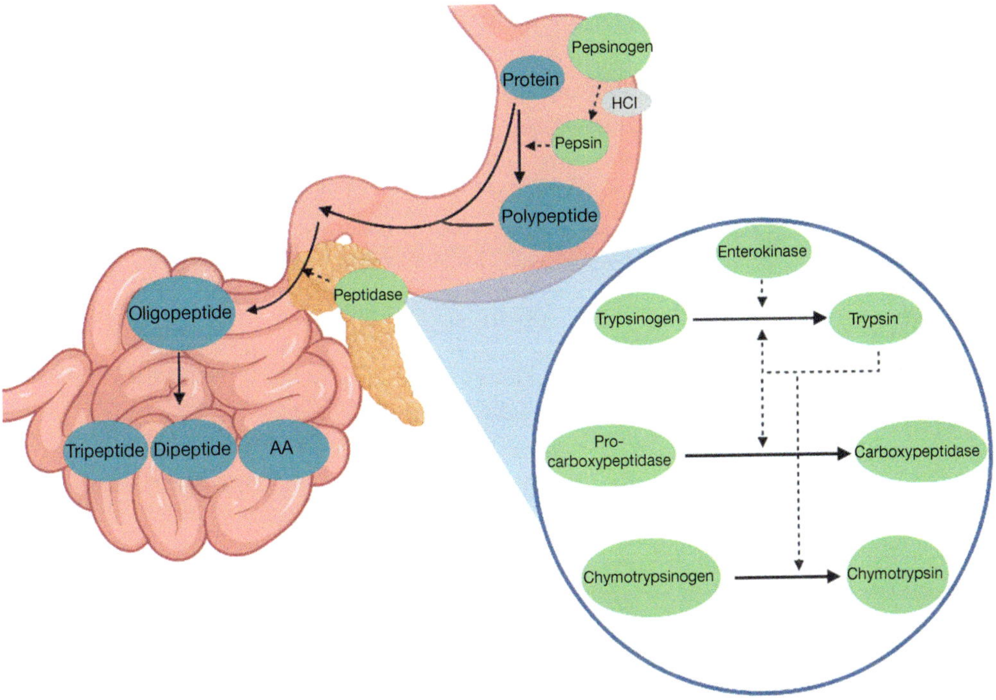

Figure 4.4 Secretion and activation of proteases in the gastrointestinal tract. *Source:* [6] / MDPI. / CC BY 4.0.

1) Secreted enzymes, including pepsin, trypsin, chymotrypsin, elastase, carboxypeptidase A, and carboxypeptidase B.
2) Brush-border membrane-bound enzymes, including various carboxypeptidases and aminopeptidases.
3) Proteases and other oxidases in the cytoplasm of gastrointestinal epithelial cells.

The stability of 17 kinds of peptide drugs in the gastrointestinal tract was studied in a paper [5], and the peptide drugs were divided into three classes according to the number of amino acids.

1) Five drugs with less than 12 amino acids and with partial or all cyclic structures.
2) Seven linear peptide drugs of less than 12 amino acids.
3) Five drugs greater than 12 amino acids.

The findings suggest that in human gastric fluid, larger peptides, including somatostatin, calcitonin, secretin, glucagon, and insulin, are metabolized rapidly, whereas smaller ones showed good stability. However, in human small intestinal fluid, both smaller and larger peptides are rapidly degraded except for cyclic peptide cyclosporin and disulfide-bridged peptide octreotide and desmopressin. The stability of peptides in simulated and porcine gastric fluid correlated well with the stability in human gastric fluid. The rapid degradation of larger polypeptides in gastric fluid may be due to multiple factors. For example, larger peptides have a large number of pepsin-sensitive peptide bonds, are structurally flexible, have more hydrogen bond acceptors/donors, and have a higher polar surface area, thereby increasing the interaction with pepsin.

4.2.2.2 Hepatic Metabolism

Peptide drugs are mostly less hydrophobic and cannot enter hepatocytes by passive diffusion. It will be metabolized after entering hepatocytes via carrier-mediated transmembrane transport or endocytosis. Receptors associated with LDL receptors found on hepatocyte membranes may be involved in the metabolism of plasminogen activators [7]. A study on the metabolism of two GLP-1 metabolites in human and mouse hepatocytes showed that both metabolites are rapidly metabolized at the N-terminus but not at the C-terminus [8]. Salmon calcitonin is another peptide metabolized by the liver, and when incubated *in vitro* with rat liver homogenates, salmon calcitonin is first cleaved at the His^{17}-Lys^{18} and Val^{8}-Leu^{9} bonds and then these cleavage products are further degraded by exopeptidases (aminopeptidase/carboxypeptidase) [9].

CYP450 enzymes in the liver can also metabolize a variety of polypeptide drugs. For example, the prodrug of the opioid peptide H-Tyrd-AlaGlyPhed-LeuOH (DADLE) is sensitive to CYP450 oxidation [10].

4.2.2.3 Renal Metabolism

Depending on the MW, polypeptides are metabolized in the kidney by two main mechanisms. After glomerular filtration, macromolecular peptides will be removed by endocytosis and lysosomal degradation and finally hydrolyzed into small peptides and amino acids. Peptides with smaller MWs will be hydrolyzed to amino acids by exopeptidase on the brush-border membrane of the tubule after glomerular filtration and then reabsorbed into the systemic circulation or degraded to small peptides and transported to the proximal tubule epithelial cells through specific amino acid transport systems. Peptide transporter 1 (PEPT1) and PEPT2 are involved in the reabsorption mechanism of oligopeptides.

After subcutaneous injection of teriparatide acetate in rats, kinetic data show that the kidney is the main organ of its distribution and degradation, but excretion is not significant [11]. The metabolism of natriuretic peptides (NPs) is also an example, and D-type NPs (DNPs) are more stable in rabbit plasma compared with other NPs (e.g., atrial NPs); further investigating the metabolism of DNP in other organs and the results indicate that the kidney is the main organ for its degradation [12].

4.2.2.4 Blood Metabolism

Peptide drugs are hydrophilic and are more likely to be degraded by proteases in plasma than in tissues and organs. For example, bradykinin is rapidly degraded by proteases found in human plasma [13]. Neuropeptide Y (NPY) is also metabolized by proteolytic cleavage when incubated in human serum. NPY1-36 can be rapidly converted to three metabolites, and this degradation process can be prevented by inhibiting DPP IV, aminopeptidase P, and kinin-releasing enzyme [14].

4.2.2.5 Metabolism in Other Tissues

Because of the poor oral bioavailability of peptide drugs, the most common route of administration in clinical practice is intravenous or subcutaneous administration. When administered subcutaneously, the drug is also metabolized in the subcutaneous tissue before entering the systemic circulation. Subcutaneous interstitial fluid (SIF) contains proteases similar to plasma, but the content is only one-third of the plasma, similar to the lymph [15]. The albumin content of subcutaneous tissue is less than that of plasma. For the polypeptide bound to albumin to prolong the half-life, the free ratio in subcutaneous tissue is higher than that in plasma; therefore, the subcutaneous tissue is its main metabolic site. Ito et al. [16] incubated leuprorelin in rat skin tissue samples to account for the low subcutaneous bioavailability observed *in vivo*.

4.2.2.6 Receptor-Mediated Metabolism

Because peptide drugs have better target affinity, a large proportion of drug molecules will bind to target receptors. Peptide drugs are usually internalized into cells after binding to target receptors on the cell membrane surface and are degraded by intracellular endosomes or lysosomes. Due to the limited number of target receptors, this route of elimination is easily saturated. Therefore, this pathway is one of the important reasons for the nonlinear PK properties of peptides or macromolecular drugs. Target receptors expressed in the blood generally have a more significant effect on the elimination of peptide drugs than target receptors expressed in tissue cells.

4.2.3 Stability Test of Peptides and Investigation of Metabolic Sites

As mentioned above, peptides are easily metabolized by proteases in tissues such as blood, liver, kidney, gastrointestinal tract, and skin, so *in vitro* incubation at 37°C in blood or plasma is often used to investigate stability. In addition, liver S9, kidney S9, simulated gastrointestinal fluid, and skin homogenate are also recommended stability test systems. It should be noted that because of the low permeability of peptides, *in vitro* S9, homogenate, and other systems may overestimate the extent of the metabolism of peptide drugs in tissues *in vivo*. Metabolites of peptides are usually also biologically active, and HRMS can be used to study the metabolic sites and products of peptides in preclinical practice to help guide structure optimization to improve stability and focus on the activity and toxicity of major metabolites.

If the main types of metabolic enzymes of drugs are understood, stability testing using recombinant proteases will provide more direct results. However, there are few types of proteases commercially available, and specific protease inhibitors can also be considered to investigate the metabolic phenotype.

4.2.4 Methods to Improve the Metabolic Stability of Peptides

In response to the problems of rapid metabolism and short duration of action of peptide drugs, medicinal chemists have explored a variety of methods to improve the metabolic stability of peptides.

4.2.4.1 Substitution of L-Amino Acids for D-Amino Acids or Unnatural Amino Acids

The primary recognition site for protease in organisms is the natural L-amino acids. Thus, substituting L-amino acids with D-amino acids at the metabolic site can reduce the affinity toward protease while preserving the physiological activity of the polypeptide.

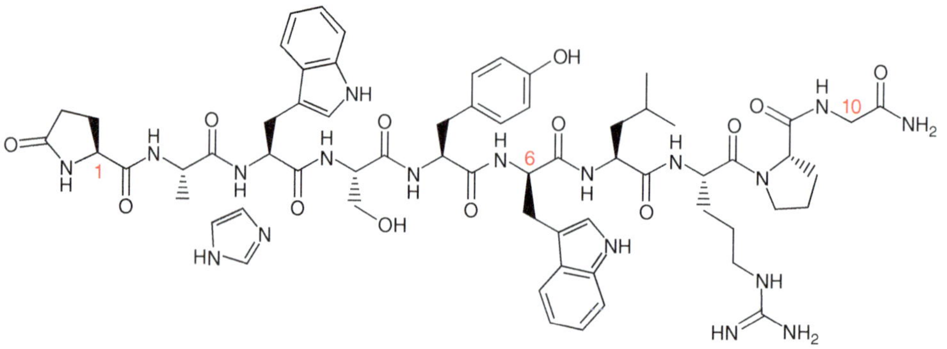

Figure 4.5 Structural formula of triptorelin (unnatural amino acid residues in red). *Source:* [17] / John Wiley & Sons.

Triptorelin (10-amino acid peptide), whose structure is shown in Figure 4.5, is a GnRH analog. The amino acid residues at positions 1, 6, and 10 of native GnRH are substituted with three unnatural amino acid residues, and its half-life after intravenous injection is prolonged to 2.8 hours compared with five minutes of native GnRH, which is mainly cleared by the liver and kidney [17].

4.2.4.2 *N*- or *C*-Terminal Modification

Amino acid residues at both ends of linear peptides are susceptible to degradation by exopeptidases, so modification of terminal residues by methylation or N-acetylation can play a role in metabolic protection.

The half-life of the GIP (1-42) is only two to five minutes, and Mabilleau et al. developed an enzymatically stable GIP analog by acetylating Tyr1 (*N*-AcGIP) in GIP with an *in vivo* half-life of greater than 24 hours.

4.2.4.3 Cyclization or Bicyclization

The configuration of the linear peptide is unstable and easily transformed to bind to the protease catalytic center. The method of cyclization can stabilize the peptide configuration, reduce the binding constant (K_a) to the protease, and reduce the affinity to the protease. It can also protect the amino acid residues at both ends or metabolic sites, which is an effective way to enhance the stability of peptides. Currently, more than two-thirds of the peptide drugs that have been marketed are cyclic peptide compounds [18].

Many antimicrobial agents from natural sources are cyclic peptides, such as cyclosporin and daptomycin. Cyclosporin is highly stable to gastrointestinal peptidases (>90% remains intact after two hours of incubation) [19]. Sandimmune® is one of its lipid-based formulations, whereas Neoral® is a microemulsified formulation with improved bioavailability. Its terminal half-life ranges from 5 to 18 hours after oral administration.

Lanreotide is an 8-residue cyclic peptide analog of 14-amino acid peptide somatostatin for the treatment of acromegaly (gigantism) and unresectable advanced or metastatic gastroenteropancreatic neuroendocrine tumors (GEP-NETs) caused by excessive hormone release from the pituitary gland. The investigators developed a one-month sustained-release subcutaneous depot formulation (Somatuline Autogel® and Somatuline Depot®) to avoid daily injections. Lanreotide contains two unnatural amino acid residues at positions 1 and 4, followed by amidated Thr8, extending its elimination half-life to 23–30 days, compared with three minutes for endogenous somatostatin [20].

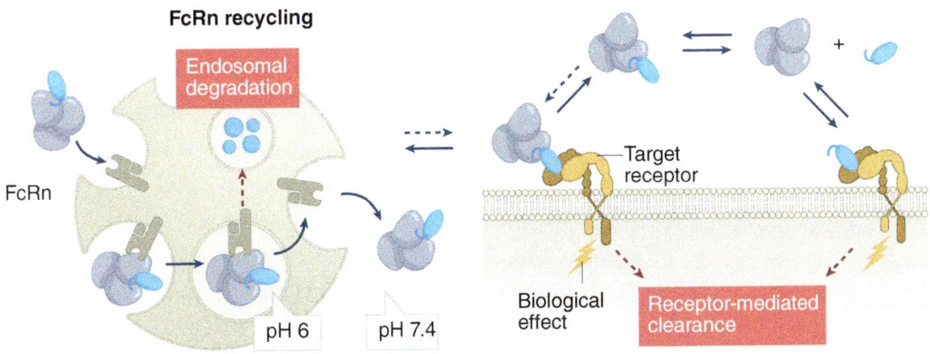

Figure 4.6 FcRn cycle. *Source:* [22] / with permission of Springer Nature.

4.2.4.4 PEGylation

PEG has a variety of drug-related properties, such as high water solubility, high fluidity, low toxicity, and immunogenicity, and is easily removed from the body. When the drug is PEGylated, it acquires certain properties, and the transmission of these properties positively correlates with the MW of PEG. PEGylation modification at the end of the peptide chain not only protects the amino acid residues but also significantly increases the MW and steric hindrance. Lee et al's research demonstrated that site-specific PEGylation of GLP-1 resulted in a 16-fold increase in plasma half-life in rats [21].

Another significant function of PEGylation with large MW is to decrease glomerular filtration rate and reduce renal clearance. The combination of both mechanisms can greatly prolong the elimination half-life *in vivo*.

4.2.4.5 Addition of Ligand Improves Binding to Albumin or IgG

Albumin or IgG in the organism uses the neonatal Fc receptor (FcRn) cycle (Figure 4.6) pathway to avoid degradation. When albumin is internalized into cells, it binds to FcRn at pH 6 in the endosome, protecting albumin from lysosomal degradation. The conjugates are then released into the extracellular environment at a physiological pH of 7.4 to release albumin. In relation to this pathway, researchers have modified the peptide structure with the aliphatic chain to bind albumin, and both liraglutide and semaglutide have significantly extended the half-life in this way. Other research approaches also include the covalent binding of peptides to albumin and recombinant expression [22].

In addition to the above ways, the *in vivo* half-life of the drug can also be prolonged by co-administration with enzyme inhibitors and by using sustained-release formulations such as liposomes to protect free peptides.

4.2.5 Summary

Deepening the understanding of peptide metabolic stability and early identification of metabolic mechanisms and soft spots can aid researchers in conducting structural modification to improve drug stability, which is crucial for successful clinical development. Nowadays, several long-acting peptide drugs and sustained-release delivery formulations have successfully entered the market. With the advancement of more research, including the exploration of new molecular types such as multifunctional peptides and PDCs, peptides stand a chance of being utilized in a wider array of disease areas.

References

1 Pernot, M., Vanderesse, R., Frochot, C. et al. (2011). Stability of peptides and therapeutic success in cancer. *Expert Opinion Drug Metabolism and Toxicology* 7: 793–802.

2 Yao, J.F., Yang, H., Zhao, Y.Z., and Xue, M. (2018). Metabolism of peptide drugs and strategies to improve their metabolic stability. *Current Drug Metabolism* 19 (11): 892–901.

3 Rawlings, N.D. and Salvesen, G. (ed.) (2013). *Handbook of Proteolytic Enzymes*, 3e. Amsterdam: Elsevier/AP.

4 Zhu, L., Tamvakopoulos, C., Xie, D. et al. (2003). The role of dipeptidyl peptidase IV in the cleavage of glucagon family peptides *in vivo* metabolism of pituitary adenylate cyclase-activating polypeptide- (1-38). *Journal of Biological Chemistry* 278: 22418–22423.

5 Wang, J., Yadav, V., Smart, A.L. et al. (2015). Toward oral delivery of biopharmaceuticals: an assessment of the gastrointestinal stability of 17 peptide drugs. *Molecular Pharmaceutics Journal* 12 (3): 966–973.

6 Kurz, A. and Seifert, J. (2021). Factors influencing proteolysis and protein utilization in the intestine of pigs: a review. *Animals* 11 (12): 3551. https://doi.org/10.3390/ani11123551.

7 Authier, F., Posner, B.I., and Bergeron, J.J.M. (1996). Endosomal proteolysis of internalized proteins. *FEBS Letters* 389: 55–60.

8 Sharma, R., McDonald, T.S., Eng, H. et al. (2013). *In vitro* metabolism of the glucagon-like peptide-1 (GLP-1) -derived metabolites GLP-1 (9-36) amide and GLP-1 (28-36) amide in mouse and human hepatocytes. *Drug Metabolism & Disposition* 41: 2148–2157.

9 Liao, S., Qie, J.K., Xue, M. et al. (2010). Metabolic stability of human parathyroid hormone peptide hPTH (1-34) in rat tissue homogenates: kinetics and products of proteolytic degradation. *Amino Acids* 38: 1595–1605.

10 Nofsinger, R., Fuchs-Knotts, T., and Borchardt, R.T. (2012). Factors that restrict the cell permeation of cyclic prodrugs of an opioid peptide, part 3: synthesis of analogs designed to have improved stability to oxidative metabolism. *Journal of Pharmacological Sciences* 101: 3486–3499.

11 Serada, M., Sakurai-Tanikawa, A., Igarashi, M. et al. (2012). The role of the liver and kidneys in the pharmacokinetics of subcutaneously administered teriparatide acetate in rats. *Xenobiotica* 42 (4): 398–407.

12 Kim, S.M., Kim, S.Y., Kim, S.H., and Kim, S.Z. (2014). Dendroaspis natriuretic peptide is degraded by a metalloproteinase in the rat kidney. *Molecular Medicine Reports* 9: 1037–1043.

13 Laine, J., Murphey, D.L., Hachey, J.A. et al. (2000). Metabolism of bradykinin *in vivo* in humans: Identification of BK1-5 as a stable plasma peptide metabolite. *Journal of Pharmacology and Experimental Therapeutics* 294: 263–269.

14 Abid, K., Rochat, B., Lassahn, P.G. et al. (2009). Kinetic study of neuropeptide Y (NPY) proteolysis in blood and identification of NPY3-35 a new peptide generated by plasma kallikrein. *Journal of Biological Chemistry* 284: 24715–24724.

15 Esposito, S., Orsatti, L., and Pucci, V. (2022). Subcutaneous catabolism of peptide therapeutics: bioanalytical approaches and ADME considerations. *Xenobiotica* 52 (8): 828–839. https://doi.org/10.1080/00498254.2022.2119180.

16 Ito, Y., Murano, H., Hamasaki, N. et al. (2011). Incidence of low bioavailability of leuprolide acetate after percutaneous administration to rats by dissolving microneedles. *International Journal of Pharmaceutics* 407 (1–2): 126–131.

17 Muller, F.O., Terblanchè, J., Schall, R. et al. (1997). Pharmacokinetics of triptorelin after intravenous bolus administration in healthy and in males with renal or hepatic insufficiency. *British Journal of Clinical Pharmacology* 44 (4): 335–341.

18 Zorzi, A., Deyle, K., and Heinis, C. (2017). Cyclic peptide therapeutics: past, present and future. *Current Opinion of Chemical Biology* 38: 24–29.

19 Wang, J., Yadav, V., Smart, A.L. et al. (2015). Toward oral delivery of biopharmaceuticals: an assessment of the gastrointestinal stability of 17 peptide drugs. *Molecular Pharmaceutics Journal* 12 (3): 966–973.

20 Brian Chia, C.S. (2021). A Review on the metabolism of 25 peptide drugs. *International Journal of Peptide Research and Therapeutics* 27: 1397–1418. https://doi.org/10.1007/s10989-021-10177-0.

21 Lee, S.H., Lee, S., Youn, Y.S. et al. (2005). Synthesis, characterization, and pharmacokinetic studies of PEGylated glucagon-like peptide-1. *Bioconjugate Chemistry* 16: 377–382.

22 Kurtzhals, P., Østergaard, S., Nishimura, E. et al. (2022). Derivatization with fatty acids in peptide and protein drug discovery. *Nature Reviews Drug Discovery.* 22 (1): 59–80.

4.3 Insights into ADME Studies of 37 Approved Therapeutic Peptides: Strategies and Radiolabeling Considerations

Jia Xue, Peng Li, Lian Guo, Weiqun Cao, Lingling Zhang

4.3.1 Introduction

Since the clinical introduction of insulin dating back to 1922, therapeutic peptides have undergone over a century of development. Typically comprising 10–40 amino acids, peptides exhibit a MW of less than 10 kDa. Positioning between small molecules and biological macromolecules in terms of MW, therapeutic peptides stand out due to their high specificity and potency, low dosage, and toxicity. In recent years, with the development of many blockbuster drugs such as dulaglutide, semaglutide, and tirzepatide, therapeutic peptides have become one of the most emerging fields in drug research and development globally. By November 2023, around 90 therapeutic peptides have been marketed worldwide. Notably, these encompass diverse targets, including GLP-1R, insulin, HER 2, and others, among which GLP-1R is emerging as a prominent focus.

In this chapter, we will focus on an industry white paper about ADME research of therapeutic peptides published in 2023: Metabolism and Excretion of Therapeutic Peptides: Current Industry Practices, Perspectives, and Recommendations [1]. This white paper explores the current ADME research landscape and research strategy of therapeutic peptides *in vivo* and shares the key considerations of radiolabeling technology for the PK study of therapeutic peptides.

4.3.2 Current Landscape of Radiolabeled ADME Studies for Therapeutic Peptides

Therapeutic peptides can be classified into three types based on the structure: TPepA, linear peptides composed of only natural amino acids; TPepB, linear peptides containing unnatural amino acids and/or organic linkers with lipophilic side chains; and TPepC, cyclic peptides.

The white paper summarizes 31 therapeutic peptides approved by the FDA from 2011 to 2022, and we supplemented six new therapeutic peptides approved by the FDA in 2023, which are trofinetide, rezafungin, paxlovid, flotufolastat F 18, motixafortide, and zilucoplan (listed by approval order in Figure 4.7). These 37 therapeutic peptides include 8 TPepA (natural amino acid linear peptides), 19 TPepB (unnatural amino acid linear peptides), and 10 TPepC (cyclic peptides).

The *in vitro* metabolism studies of 37 therapeutic peptides approved by the FDA from 2011 to 2023 are summarized in Table 4.7, with additional details provided in Table 4.A1. The metabolism of the three types of therapeutic peptides was studied *in vitro* in a variety of biological materials, whereas the liver-associated matrices were commonly used to study TPepB and TPepC. Some of these drugs, such as difelikefalin and odevixibat, have been studied *in vitro* with a variety of biological substrates using ^{14}C and/or ^{3}H.

Seventy-three percent (mainly TPepB and TPepC) of these 37 therapeutic peptides have been researched using radiolabeled animal and/or human ADME studies (summarized in Table 4.8) (metabolism and excretion studies summarized in Table 4.A1). Among these approved therapeutic peptides, the structures of nine drugs (boceprevir, telaprevir, ombitasvir, paritaprevir, grazoprevir, voxilaprevir, difelikefalin, odevixibat, and trofinetide) are similar to small molecules (MW is less than 1,000), and all contain unnatural amino acids. Human radiolabeled ADME studies were conducted for all of them, elucidating the pattern of metabolism and elimination of therapeutic peptides in humans.

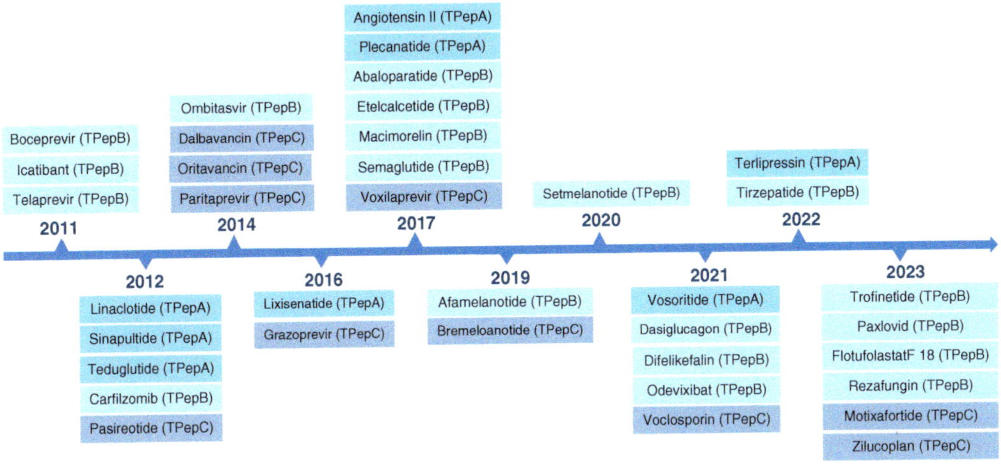

Figure 4.7 Summary of therapeutic peptides approved by the FDA from 2011 to 2023.

Table 4.7 Summary of *in vitro* metabolism studies of therapeutic peptides approved by the FDA from 2011 to 2023.

In vitro metabolism study matrix	Mouse	Rat	Rabbit	Dog	Monkey	Human
Liver homogenate/liver S9/liver microsomes/hepatocytes/liver cytosol	TPepA (1)	TPepA (1)	NS	TPepA (1)	NS	TPepA (1)
	TPepB (5)	TPepB (14)	TPepB (1)	TPepB (10)	TPepB (7)	TPepB (15)
	TPepC (3)	TPepC (6)	TPepC (3)	TPepC (5)	TPepC (5)	TPepC (8)
Kidney homogenate/kidney S9/kidney microsomes	NS	TPepB (4)	NS	NS	TPepB (2)	TPepB (4)
		TPepC (1)			NS	TPepC (1)
Intestinal S9/intestinal contents/intestinal fluid/intestinal microsomes	TPepA (1)	TPepA (2)	NS	NS	NS	TPepA (2)
	NS	TPepB (1)			TPepB (1)	TPepB (1)
Skin microsomes/skin S9/skin cells	NS	NS	NS	NS	NS	TPepB (2)
Blood/serum/plasma	TPepA (1)	TPepA (1)	NS	TPepA (1)	NS	TPepA (1)
	NS	TPepB (7)		TPepB (3)	TPepB (2)	TPepB (8)
	NS	TPepC (1)		TPepC (1)	NS	TPepC (2)

Note: NS, no studies mentioned in submission document(s).
Figures in parentheses indicate the number of peptides utilized for *in vitro* metabolism studies in this matrix type among 37 therapeutic peptides approved by the FDA.

4.3.3 Industry Recommendations for Conducting *In Vivo* ADME Studies of Therapeutic Peptides

There is no regulatory guidance on conducting metabolism and excretion studies for therapeutic peptides. Based on industry status and regulatory requirements, the white paper provides a decision tree delineating the necessity of radiolabeled ADME experiments to support the marketing of therapeutic peptides (Figure 4.8).

Table 4.8 Summary of *in vivo* radiolabeled metabolism and excretion studies of therapeutic peptides approved by the FDA from 2011 to 2023.

Species/Radionuclide	Mouse	Rat	Rabbit	Dog	Monkey	Human
^{14}C	NS	TPepA (1)	NS	NS	NS	NS
	TPepB (1)	TPepB (9)		TPepB (1)	TPepB (3)	TPepB (9)
	TPepC (1)	TPepC (4)		TPepC (4)	TPepC (2)	TPepC (4)
^{3}H	TPepB (1)	TPepB (3)	NS	TPepB (1)	TPepB (1)	TPepB (1)
$^{3}H/^{14}C$	NS	TPepA (1)	NS	NS	NS	NS
	TPepC (1)	TPepC (3)	TPepC (2)	TPepC (2)		TPepC (2)
^{125}I	NS	TPepB (1)	NS	NS	NS	NS
^{18}F	NS	NS	NS	NS	NS	TPepB (1)

Note: NS, no studies mentioned in submission document(s).
Figures in parentheses indicate the number of peptides utilized for *in vivo* radiolabeled studies among 37 therapeutic peptides approved by the FDA.

For a peptide containing only natural amino acids, *in vitro* MetID and *in vivo* metabolism/excretion studies are not mandated, following ICH-S6(R1) [2] for regulatory filing. However, these studies are required when metabolic insights are critical to understanding the PK/PD relationship.

In contrast, for peptides containing unnatural amino acids and/or organic linker moieties (TPepB and TPepC), there are three distinct scenarios meriting considerations:

- If the unnatural amino acids or the organic linker moiety have been used in other marketed therapeutic peptides and shown safety in humans, it may not be necessary to conduct radiolabeled ADME studies. Nonetheless, metabolism studies are still recommended to evaluate major circulating and potential active metabolites if needed.
- If the therapeutic peptides contain novel unnatural amino acids and/or a novel organic linker moiety that has not been incorporated in any marketed drugs, radiolabeled ADME studies are generally recommended to characterize its metabolism and elimination pathways. It is worth noting that if a therapeutic peptide contains multiple (≥ 3) novel unnatural amino acids and/or organic linker moieties, it is not obligatory to label all those novel moieties.
- Radiolabeled metabolism and excretion studies may not be performed if sufficient *in vitro* and/or *in vivo* ADME data or literature knowledge is provided to demonstrate the safety of specific unnatural amino acids or organic linker moieties.

4.3.4 Utilization of Radioisotopes in *In Vivo* ADME Studies of Therapeutic Peptides and Considerations for Labeling Sites

Radioactive tracer labeling is a widely used technique to study the *in vivo* ADME properties of small-molecule drugs, as well as drugs such as peptides, oligonucleotides, and ADCs. Similar to small molecules, peptides are commonly labeled with either ^{14}C or ^{3}H [3–6], with occasional application with other radioisotopes (e.g., ^{18}F or ^{125}I). Among the 37 peptide drugs approved by the FDA between 2011 and 2023, 17 drugs were labeled with ^{14}C for *in vivo* ADME studies, 4 drugs were labeled with ^{3}H, and 4 drugs were developed with the initial application of ^{3}H labeling; before transitioning to ^{14}C-labeled compounds, 1 drug was labeled with ^{125}I, 1 drug

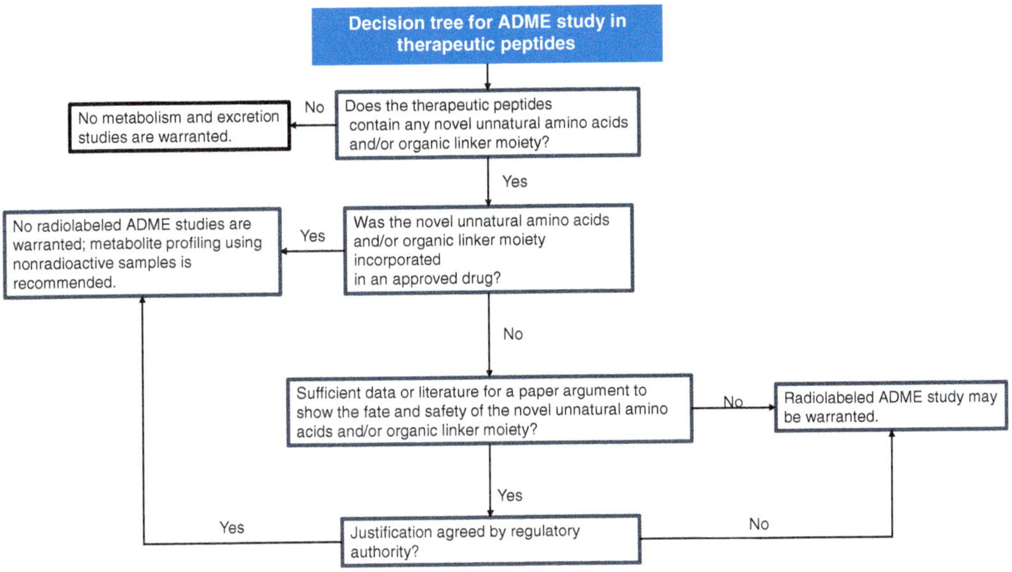

Figure 4.8 Industry recommendations for ADME studies of therapeutic peptides. Adapted from Reference [1].

(diagnostic drug) was labeled with ^{18}F, and other drugs were linear peptides containing only natural amino acids (TPepA) or had sufficient safety support data, and no radiolabeled ADME studies were conducted.

^{14}C-labeling is a more prevalent choice than other radioisotopes [7]: ^{3}H can be lost from the molecule due to hydrogen–tritium exchange [8], thus affecting the accuracy of the ADME study; ^{125}I-labeling may alter the structure and composition of the molecule and ADME properties, particularly for small peptides.

Although ^{14}C is a commonly used radioisotope, it has limitations compared to ^{3}H labeling. The specific activity of ^{14}C is much lower than ^{3} H [7] which can make detection of the ^{14}C labeling difficult for low-dose drugs. The specific activity could be increased by introducing multiple ^{14}C atoms to the same molecule. However, analytical challenges with ^{14}C labeling persist for low-dose peptides with a large MW. The traditional radioactivity detection methods may lack the requisite sensitivity. To overcome this limitation, accelerator mass spectrometry can be used to analyze low radioactivity concentrations of ^{14}C samples [9].

The selection of a radiolabeling site is critical and warrants careful deliberation. The radioisotope should be introduced in a position to monitor significant biotransformation pathways or to track the long-lived circulating metabolites. Prior to radiolabeled ADME studies, a qualitative metabolic profile for the therapeutic peptides from *in vitro* and *in vivo* nonradiolabeled metabolism studies can assist in determining the optimal labeling site [10].

Additionally, consideration must be given to the potential reincorporation of radiolabeled natural amino acids into the proteome of test species, which is unethical in the case of humans, and may potentially affect the total recovery of radioactivity in the excreta [2]. As a rule of thumb, xenobiotic moieties like novel unnatural amino acids and organic linkers with lipophilic side chains are preferred labeling positions over the natural amino acids within a peptide. In cases where multiple parts of a therapeutic peptide need to be monitored, efforts may be made to label a therapeutic peptide at two or more distinct positions.

4.3.5 Case Analysis of Radioactive Tracer Technique in the *In Vivo* ADME Study of Therapeutic Peptides

Semaglutide is a GLP-1 receptor agonist developed by Novo Nordisk for the treatment of type 2 diabetes. It has a MW of 4113.6 and uses radioisotope ^3H. As shown in Figure 4.9, the labels are located at the α and β positions of the carbonyl group in the amide bond of the octadecanedioic acid group of the side chain of the semaglutide (two labeling positions in total) [11]. The investigators conducted separate radiolabeled ADME studies of ^3H-labeled semaglutide in rats, monkeys, and humans using a liquid scintillation counter (LSC) to determine total radioactivity. The results showed that the recovery of total radioactivity in rats was 94%, much higher than 75% in humans and 58% in monkeys, as detailed in Table 4.9 [11]. The total radioactivity of rat carcasses accounted for 22%, which was the main reason why its total recovery was higher than that of humans and monkeys. Lower recovery of total radioactivity in humans and monkeys presumably may be due to *in vivo* hydrogen–tritium exchange resulting in the loss of ^3H from the molecule. It may also result

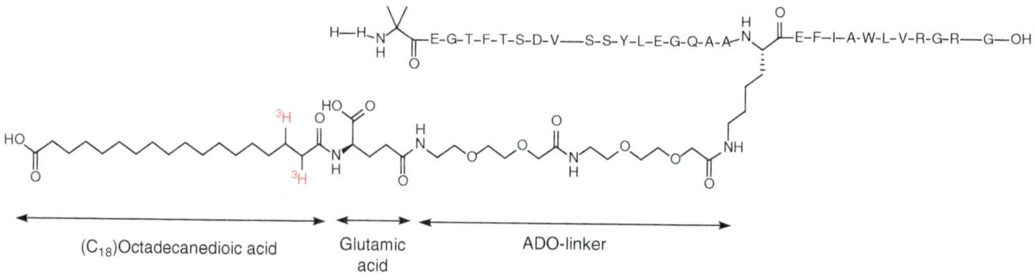

Figure 4.9 [^3H]-Semaglutide structure.

Table 4.9 Summary of radioactivity excretion recovery following subcutaneous administration of [^3H]-semaglutide [11].

Species	Human (% of Dose [% CV]) *n* = 7	Rat (% of Dose [% CV]) *n* = 3	Monkey (% of Dose [% CV]) *n* = 3
Dosage	0.5 mg 16.7 MBq	0.3 mg/kg 10 MBq/kg	0.03 mg/kg 14 MBq/kg
Collection time	0–64 d	0–1 wk	0–2 wk
Sex	M	M	M
Urine	53.0 (8.2)	35.6 (27.7)	30.3 (25.3)
Feces	18.6 (19.9)	32.6 (9.7)	20.7 (3.0)
Expired air	3.2 (9.0)	< 0.2	NA
Carcass	NA	22.4 (27.0)	NA
Cage bedding	NA	3.7 (22.2)	7.2 (28.4)
Total excretion	75.1 (5.2)	72.1 (8.8)	58.2 (10.2)
Total recovery	75.1	94.5 (0.5)	58.2

from the hydrolysis of ^{3}H-semaglutide to octadecanedioic acid, which enters into the body's circulation. To improve the recovery of total radioactivity in human mass balance and reduce the effect of hydrogen–tritium exchange, a ^{14}C-labeled compound for therapeutic peptides ADME studies is recommended, but there is still a need to comprehensively consider the administered dosage of therapeutic peptides, MW, and ^{14}C-labeled radioactivity.

4.3.6 Summary

In conclusion, radiolabeling technology provides a powerful tool for *in vivo* ADME studies of therapeutic peptides. In applications, it is necessary to select the appropriate isotope and labeling position based on the specific situation to ensure the accuracy and reliability of the study results. By leveraging radiolabeling techniques, invaluable ADME insights can be obtained, thereby advancing drug development and therapeutic optimization.

References

1. He, M.M., Zhu, S.X., Cannon, J.R. et al. (2023). Metabolism and excretion of therapeutic peptides: current industry practices, perspectives, and recommendations. *Drug Metabolism and Disposition* 51 (11): 1436–1450.
2. International Council for Harmonisation of Technical Requirements for Pharmaceuticals for Human Use (2011). *S6 (R1) Preclinical Safety Evaluation of Biotechnology - Derived Pharmaceuticals*. London, UK: ICH (International Council for Harmonisation of Technical Requirements for Pharmaceuticals for Human Use).
3. Subramanian, R., Zhu, X., Kerr, S.J. et al. (2016). Nonclinical pharmacokinetics, disposition, and drug-drug interaction potential of a novel d-amino acid peptide agonist of the calcium-sensing receptor AMG 416 (etelcalcetide). *Drug Metabolism and Disposition* 44 (8): 1319–1331.
4. Jensen, L., Helleberg, H., Roffel, A.D. et al. (2017). Absorption, metabolism, and excretion of the GLP-1 analogue semaglutide in humans and nonclinical species. *European Journal of Pharmaceutical Sciences* 104: 31–41.
5. Lee, S. and Lee, D.Y. (2017). Glucagon-like peptide-1 and glucagon-like peptide-1 receptor agonists in the treatment of type 2 diabetes. *Annals of Pediatric Endocrinology & Metabolism* 22 (1): 15.
6. Subramanian, R., Zhu, X., Hock, M.B. et al. (2017). Pharmacokinetics, biotransformation, and excretion of [^{14} C] etelcalcetide (AMG 416) following a single microtracer intravenous dose in patients with chronic kidney disease on hemodialysis. *Clinical Pharmacokinetics* 56: 179–192.
7. Dell'isola, A., Brown, R.T., Jones, S. et al. (2019). Synthesis of carbon - 14 – labelled peptides. *Journal of Labelled Compounds and Radiopharmaceuticals* 62 (11): 713–717.
8. Knadler, M.P., Ackermann, B.L., Coutant, J.E. et al. (1992). Metabolism of the anticoagulant peptide, MDL 28, 050, in rats. *Drug Metabolism and Disposition* 20 (1): 89–95.
9. Fifield, L.K. (1999). Accelerator mass spectrometry and its applications. *Reports on Progress in Physics* 62 (8): 1223.
10. Knadler, M.P., Ackermann, B.L., Coutant, J.E. et al. (1992). Metabolism of the anticoagulant peptide, MDL 28, 050, in rats. *Drug Metabolism and Disposition* 20 (1): 89–95.
11. Jensen, L., Helleberg, H., Roffel, A.D. et al. (2017). Absorption, metabolism and excretion of the GLP-1 analogue semaglutide in humans and nonclinical species. *European Journal of Pharmaceutical Sciences* 104: 31–41.

4.4 Challenges and Strategies in the Bioanalysis of Peptide Drugs

Yanfeng Liu, Weimin Hu, Weiqun Cao, Lili Xing

4.4.1 Introduction

Peptide drugs, known for their strong target affinity and minimal side effects, have become one of the hotspots in new drug research. Compared to protein drugs, peptide drugs have simpler spatial structures and lower immunogenicity. They are widely used in medical fields such as vaccines, oncology, endocrine, and cardiovascular, offering numerous advantages such as high specificity, potent bioactivity, remarkable efficacy, low therapeutic dosage, and less likelihood of developing drug resistance. However, they also have shortcomings such as short half-life, instability, susceptibility to degradation, and low oral absorption rate [1]. This section outlines the various challenges in the bioanalysis of peptide drugs and proposes targeted strategies for peptide analysis.

4.4.2 Advantages of LC-MS/MS in the Bioanalysis of Peptide Drugs

Peptide drugs, offering numerous advantages, have emerged as a leading choice for drug development. Nevertheless, poor stability, prone to enzymatic degradation, and pH sensitivity present more challenges in their bioanalysis.

Common bioanalytical techniques for peptides mainly include ligand binding assay (LBA) and LC-MS/MS. Compared with LBA, LC-MS/MS offers the advantages of lower cost, highly specific at molecular level, and the capacity for multiple component analysis (Table 4.10), providing a significant advantage in PK research.

4.4.3 Challenges and Strategies in LC-MS/MS Analysis of Peptide Drugs

When using LC-MS/MS technology to analyze peptide drugs, challenges such as low response, nonspecific binding, endogenous interference, and poor stability may arise. The following strategies can be implemented to solve these issues:

Table 4.10 Comparison of LBA and LC-MS/MS [2].

Category	LBA	LC-MS/MS
Basis of measurement	Biological binding reactions	Physicochemical properties of analyte
Detection	Indirect	Direct
Concentration response	Nonlinear	Linear
Standard curve range	Narrow range, ≤2 orders of magnitude	Wide range, ≥6 orders of magnitude
Specificity	Comparatively less specific for discriminating drug from its metabolites	Highly specific at molecular level and have the capacity for multiple component analysis
Sensitivity	Higher sensitivity	Comparatively less sensitive
Recurring costs	Comparatively high	Low

4.4.3.1 Low Response

Optimize sample pretreatment methods to improve compound extraction recovery; optimize chromatographic conditions and mass spectrometry methods to increase compound response.

4.4.3.2 Nonspecific Binding

Use low-binding consumables, incorporate appropriate surfactants, and screen organic solvents and protein-rich solutions to facilitate desorption.

4.4.3.3 Endogenous Interference

Optimize liquid chromatographic conditions and mass spectrometry methods to remove or separate the compound from endogenous interferences, or choose an interference-free surrogate matrix alternatively.

4.4.3.4 Poor Stability

Conduct the bench work at low temperatures or screen for an appropriate stabilizer to add.

Figure 4.10 summarizes the process based on years of experience in analyzing peptide drugs using LC-MS/MS. It mainly summarizes method optimization from three aspects: sample pretreatment, chromatographic conditions, and mass spectrometry conditions, to meet the expected quantitative analysis requirements. The stability issue is related to the specific structure and needs to be evaluated specifically and will not be further discussed here.

4.4.4 Case Studies of Peptide Analysis Using LC-MS/MS

Case 1: LC-MS/MS analysis of exenatide

Exenatide, a synthetic analog of the GLP-1 Exendin-4, is composed of 39 amino acids and is primarily used for the treatment of type 2 diabetes. We've established a bioanalytical method for exenatide with an LLOQ at 41.8 pg/mL (Figure 4.11a,b). As shown in Table 4.11, our results are comparable to those achieved with radioimmunoassay (RIA).

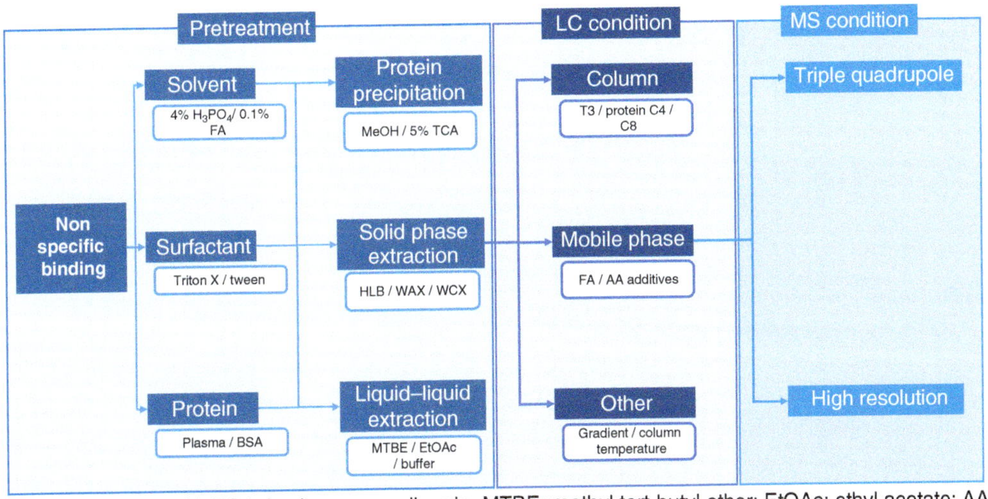

Note: FA: formic acid; BSA: bovine serum albumin; MTBE: methyl tert-butyl ether; EtOAc: ethyl acetate; AA: acetic acid.

Figure 4.10 Workflow for LC-MS/MS analysis of conventional peptide drugs.

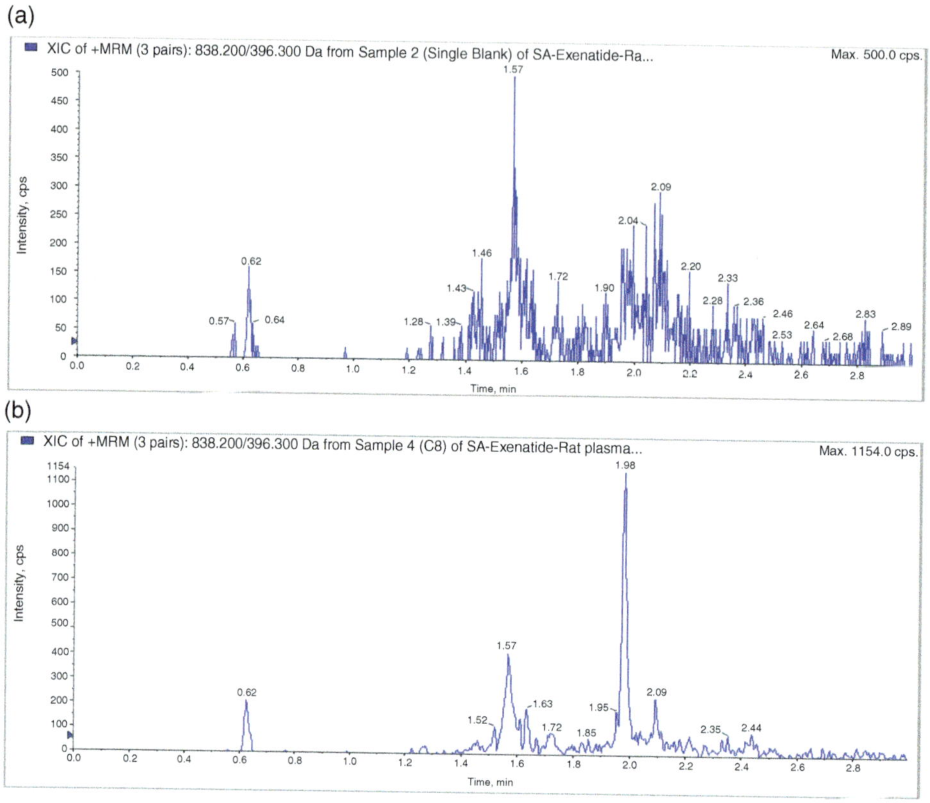

Figure 4.11 (a) LC-MS/MS chromatogram of blank sample and (b) LC-MS/MS chromatogram of LLOQ (41.8 pg/mL).

Table 4.11 Comparison of exenatide analysis methods.

Category	Literature 1 [3]	Literature 2 [4]	WuXi AppTec DMPK
Analysis method	RIA	LC-MS/MS	LC-MS/MS
LLOQ	25 pg/mL	10 ng/mL	41.8 pg/mL
Sample preparation	Solid-phase extraction	Solid-phase extraction	Solid-phase extraction

4.4.5 Summary

The PK characteristics of peptides are closely related to their physicochemical properties, chemical modifications, drug delivery systems, and properties of the conjugated drugs. Good bioanalytical methods are crucial for studying the PK of peptide drugs. This section summarizes the challenges and solutions in the bioanalysis of peptide drugs based on the LC-MS/MS technology, which can better utilize the LC-MS/MS technology to provide accurate results rapidly for peptide analysis, thereby enhancing PK research.

References

1 Markus, M., Glenn, F.K., David, J.A. et al. (2021). Trends in peptide drug discovery. *Nature Reviews* 20: 309–325.

2 Londhe, V. and Rajadhyaksha, M. (2019). Review of recommendations for bioanalytical method validation: chromatographic assays and ligand binding assays. *Chromatographia* 82: 523–535. https://doi.org/10.1007/s10337-018-3677-z.

3 Ai, G., Zhen, Z., Shan, C. et al. (2008). Single- and multiple-dose pharmacokinetics of exendin-4 in rhesus monkeys. *International Journal of Pharmaceutics* 353: 56–64.

4 Kehler, J.R., Bowen, C.L., Boram, S.L. et al. (2010). Application of DBS for quantitative assessment of the peptide Exendin-4; comparison of plasma and DBS method by UHPLC-MS/MS. *Bioanalysis* 2: 1461–1468.

Appendix

Table 4.A1 Summary of metabolism and excretion studies of therapeutic peptides approved by the FDA from 2011 to 2023.

Name	Molecular weight	Peptide category	*In vitro* metabolism matrix	*In vivo* metabolism and excretion studies	Mechanism of action/indication
oceprevir	519.7	TPepB	Mouse, rat, monkey, and human liver S9; human liver microsomes, liver cytosol, liver mitochondrial fraction, and plasma	Rat and human ^{14}C-ADME study	A hepatitis C virus (HCV) NS3/4A protease inhibitor
Icatibant	1304.5	TPepB	Dog liver microsomes and liver S9; human liver microsomes, liver S9, and hepatocytes	Mouse, rat, and dog ^{3}H-ADME study	A potent and selective bradykinin B2 receptor and antagonist
Telaprevir	679.9	TPepB	Rat, dog, monkey, and human liver microsomes and liver S9; human hepatocytes, skin microsomes, and skin S9	Human ^{14}C-ADME study	A hepatitis C virus (HCV) NS3/4A protease inhibitor
Carfilzomib	719.9	TPepB	Rat and human hepatocytes, plasma, and plasma; human liver microsomes	Rat ^{3}H-ADME study	A protease inhibitor for the treatment of relapsed and refractory multiple myeloma
Linaclotide	1526.7	TPepA	Rodent and human intestinal contents; human intestinal fluid and microsomes	Excretion (nonradiolabeled) study in rats	A guanylate cyclase C (GC-C) agonist
Pasireotide	1047.2	TPepC	Rat and human liver microsomes, hepatocytes, and kidney microsomes; monkey liver microsomes	^{14}C-ADME studies in rats, monkeys, and humans	A somatostatin analog that is a second-generation analog of somatostatin inhibitors
Sinapultide	2469.5	TPepA	NS	NS	A peptide that mimics the function of human surfactant protein B (SP-B)
Teduglutide	3752.1	TPepA	NS	NS	A GLP-2 analog for the treatment of short bowel syndrome (SBS)
Dalbavancin	1816.7	TPepC	Rat and dog liver microsomes, hepatocytes, and plasma; human liver microsomes and hepatocytes	$^{3}H/^{14}C$-ADME studies in rats, New Zealand rabbits, and dogs; human (nonradiolabeled) ADME study	For the treatment of acute bacterial skin or skin structure infections (ABSSSI) caused by Gram-positive bacterial infections, including MRSA (methicillin-resistant *Staphylococcus aureus*)
Ombitasvir	894.1	TPepB	Rat, dog, monkey, and human liver microsomes and hepatocytes; recombinant enzymes	^{14}C-ADME studies in mice, rats, dogs, and humans	Treatment of patients with chronic hepatitis C virus (HCV) genotype 1 infection, often in combination

Oritavancin	1793.1	TPepC	Monkey and human liver microsomes	^{14}C-ADME studies in mice, rats, and dogs; human excretion (nonradiolabeled) studies in urine and feces	A lipopeptide antibacterial used to treat acute bacterial skin and skin structure infections (ABSSSI) caused by certain susceptible bacteria
Paritaprevir	765.9	TPepC	Rat, dog, monkey, and human liver microsomes and hepatocytes	Dog and human ^{14}C-ADME study	Treatment of patients with chronic hepatitis C virus (HCV) genotype 1 infection, often in combination
Grazoprevir	766.9	TPepC	Mouse liver microsomes; rat, rabbit, dog, and human liver microsomes and hepatocytes	Rat, rabbit, dog, and human ^{3}H- or ^{14}C-ADME studies	An NS3/4A protease inhibitor, an oral hepatitis C drug, often in combination
Lixisenatide	4858.6	TPepA	Mouse, rat, dog, and human liver S9 and plasma	$^{3}H/^{14}C$-ADME study in rats; excretion in milk (radiolabeled) study	A GLP-1 receptor agonist for use in adults with type 2 diabetes
Abaloparatide	3960.7	TPepB	Rat and human kidney and liver homogenates; chymotrypsin and cathepsin B	Rat ^{125}I-ADME study	A parathyroid hormone-related peptide [PTHrP (1-34)] analog used in the treatment of postmenopausal women with osteoporosis
Angiotensin II	1046.2	TPepA	NS	NS	A vasoconstrictor that is the major bioactive peptide of the renin/angiotensin system
Etelcalcetide	1048.3	TPepB	Rat and human liver microsomes, liver S9, liver cytosol, hepatocytes, kidney S9, and liver cytosol; dog liver microsomes, liver S9, and liver cytosol	Rat and human ^{14}C-ADME study	A calcimimetic used in the treatment of secondary hyperparathyroidism in adult hemodialysis patients
Macimorelin	474.6	TPepB	Mouse liver microsomes; rat and dog liver microsomes and plasma; human liver microsomes, liver S9, and plasma	NS	An oral ghrelin agonist
Plecanatide	1681.9	TPepA	Intestinal fluid from rats and humans	Rat ^{14}C-ADME study	An orally administered guanylate cyclase C (GC-C) receptor agonist
Semaglutide	4113.6	TPepB	Mouse hepatocytes; rat, monkey, and human hepatocytes and plasma	Rat, monkey, and human ^{3}H-ADME studies	A GLP-1 receptor agonist for glycemic control in adults with type 2 diabetes mellitus

(*Continued*)

Table 4.A1 (Continued)

Name	Molecular weight	Peptide category	*In vitro* metabolism matrix	*In vivo* metabolism and excretion studies	Mechanism of action/indication
Voxilaprevir	868.9	TPepC	NS	Rat, dog, and human ^{14}C-ADME studies	A noncovalent, reversible HCV NS3/4A protease inhibitor
Afamelanotide	1646.8	TPepB	NS	NS	A melanin 1 receptor (MC1-R) agonist for erythropoietic porphyria (orphan drug)
Bremelanotide	1025.2	TPepC	Rat, dog, and monkey liver S9; mouse, rat, rabbit, and dog hepatocytes; human liver microsomes, liver S9, hepatocytes, and plasma	Monkey and human ^{14}C-ADME Study	A melanocortin receptor-4 (MC4R) agonist for acquired generalized loss of libido in premenopausal women
Setmelanotide	1117.3	TPepB	Rat, monkey, and human liver microsomes, hepatocytes, and kidney microsomes; human dermal fibroblast cell line (Hs68)	^{14}C-ADME study in rats; excretion (nonradiolabeled) urine study in humans	A melanocortin receptor-4 (MC4R) agonist for chronic weight management in MC4R pathway impairment
Dasiglucagon	3381.6	TPepB	Mouse, rat, rabbit, dog, and human hepatocytes; rat, dog, and human serum and plasma	Rat and dog plasma metabolism (nonradiolabeled) studies	Glucagon analog for the treatment of severe hypoglycemia in pediatric and adult diabetic patients 6 yr of age and older
Difelikefalin	679.8	TPepB	Rat, dog, monkey, and human hepatocytes (nonradiolabeled); rat, monkey, and human hepatocytes, intestine S9, and kidney S9 (^{3}H-radiolabeled); rat, monkey, and human hepatocytes (^{14}C radiolabeled)	Monkey, rat ^{14}C-excretion study; rat milk excretion study (nonradiolabeled); ^{14}C-ADME in healthy volunteers and hemodialysis patients	A K-chain opioid receptor agonist for the treatment of moderate to severe pruritus associated with chronic kidney disease in adults
Odevixibat	740.9	TPepB	Mouse, rat, dog, and human hepatocytes (nonradiolabeled and ^{14}C-radiolabeled)	Rat ^{14}C-excretion study and milk excretion study (nonradiolabeled); human ^{14}C-radiolabeled ADME	An ileal bile acid transporter (IBAT) inhibitor for the treatment of pruritus in patients with progressive familial cholestasis (PFIC) for 3 mo and beyond
Voclosporin	1214.6	TPepC	Mouse, rat, rabbit, dog, monkey, and human liver microsomes	Rat, dog, monkey, and human metabolism (nonradiolabeled) studies; mice, rat, and human $^{3}H/^{14}C$-ADME study	For the treatment of active lupus nephritis in adults

Vosoritide	4102.8	TPepA	Neprilysin	NS	For the treatment of pediatric patients with achondroplasia starting at 2 yr of age until the growth plate is closed
Terlipressin	1227.4	TPepA	NS, literature references	NS	For improvement of renal function in adult patients with hepatorenal syndrome with rapidly declining renal function
Tirzepatide	4813.5	TPepB	NS	^{14}C-ADME studies in rats, monkeys, and humans	GIP and GLP-1 receptor agonists for the control of type 2 diabetes in adults
Trofinetide	315.3	TPepB	Rat, dog, and human plasma; rat, dog, and human liver microsomes; recombinant human cytochrome P450 isoenzymes	Rat, human ^{14}C-ADME study	Drugs approved for the treatment of Reiter's syndrome (orphan drugs)
Rezafungin	1226.4	TPepB	Rat, monkey, and human liver microsomes; rat, dog, monkey, and human hepatocytes	^{14}C-ADME studies in rats, monkeys, and humans	Treatment of adult candidemia and invasive candidiasis (orphan drug)
Paxlovid (Nirmatrelvir/ Ritonavir)	499.5/720.9	TPepB	Rat, monkey, human liver microsomes and hepatocytes	Human (nonradiolabeled) ADME study, ^{19}F NMR ADME study	Treatment of adult patients with mild to moderate novel coronavirus pneumonia at high risk for progression to severe disease (COVID-19).
Flotufolastat F 18	1470.6	TPepB	NS	Human ^{18}F-ADME study	Radioactive targeted contrast agents for PET imaging of PSMA-positive lesions in men with prostate cancer
Motixafortide	2159.5	TPepC	Human liver microsome	Dog, human (nonradiolabeled) ADME studies; rat and dog 14C-ADME studies	Chemokine receptor CXCR4 antagonist, in combination, mobilizes hematopoietic stem cells to peripheral blood
Zilucoplan	3562.2	TPepC	NS	Human (nonradiolabeled) ADME study	Generalized myasthenia gravis (gMG) targeted C5 complement inhibitor

5

Oligonucleotide (OLIGO) Drugs

5.1 Oligonucleotide Drugs: An Introduction and Overview of Pharmacokinetic Strategies

Nan Zhao, Zhiyu Li, Lili Xing

5.1.1 Introduction

In recent years, nucleic acid drugs have gained significant attention as a highly promising area of new drug development. With their unique technical features, they are expected to become the third wave of pharmaceutical innovation, following small-molecule and antibody drugs. Among the known disease-related pathogenic proteins in the human body, over 80% cannot be targeted by conventional small-molecule drugs or biological agents, making them "undruggable" protein targets. Some nucleic acid drugs can act directly on disease-causing target genes, recognize the target messenger RNA (mRNA), and inhibit translation through base complementarity, thus regulating protein expression to achieve therapeutic goals. Consequently, nucleic acid drugs are capable of overcoming the limitations of existing druggable targets and offer tremendous potential for treating previously "undruggable" diseases. Nucleic acid drugs with structure modification are also less prone to drug resistance, exhibiting high specificity and long-term efficacy, greatly improving their safety and drug efficacy.

Nucleic acid drugs include antisense oligonucleotides (ASOs), small interfering RNA (siRNA), microRNA (miRNA), small activating RNA (saRNA), mRNA, aptamer, ribozyme, and antibody–oligonucleotide conjugates (AOCs). They represent a gene therapy modality and a new generation of pharmaceutical technology following small-molecule drugs and protein-based drugs. This chapter focuses on a class of nucleic acid drugs (hereafter referred to as oligonucleotides) that are shorter than 30 nucleotides (nt) in length. At present, 19 oligonucleotides have been approved by the FDA or the European Medicines Agency (EMA), of which 11 are ASOs and 6 are siRNAs. Two to four oligonucleotides have been approved each year since 2018. As of 2022, oligonucleotides (marketed and under clinic trials) have been approved or tested in 102 different indications across 14 therapeutic fields with the main therapeutic fields being metabolic disorders, cancer, neurology, and ophthalmology [1].

5.1.1.1 Introduction to Oligonucleotide Drugs

Oligonucleotides, consisting of synthetic single or double-stranded modified ribonucleotides (RNA) or deoxyribonucleotides (DNA) less than 100 nt in length, act on target genes based on

Drug Metabolism and Pharmacokinetics: Frontiers, Strategies, and Applications, First Edition.
Edited by Liang Shen and WuXi AppTec DMPK.

Watson–Crick base pairing and can modulate gene expression via a variety of mechanisms. Since oligonucleotides target selected sequences through Watson–Crick base pairing with DNA or RNA, they can theoretically target any gene of interest by designing an appropriate nucleotide sequence that is highly complementary to the target DNA or RNA sequence. Therefore, oligonucleotide drugs are expected to have better drug-like properties compared to small-molecule drugs.

Exogenous oligonucleotides need to overcome multiple obstacles to enter the body and exert pharmacological function: being structurally unstable and susceptible to degradation by endogenous nucleases; having large and negatively charged molecular structure, which makes them difficult to penetrate cell membranes and reach the intracellular space; having ineffective endosomal escape into the cytoplasm; and being immunogenic, potentially activating immune responses in the body. With technological advancements, many of these challenges have been addressed through chemical modification and delivery system technology breakthroughs, which have played a crucial role in the development of oligonucleotide drugs.

- Oligonucleotide Drug Chemical Structure Modifications

Chemical modification of oligonucleotides can increase their stability against nucleases and improve their target binding affinity. The first generation of chemical modifications focused on modifying the phosphate backbone, such as phosphorothioate (PS) modification. The second generation of chemical modifications involved modifications at the 2'-O position of RNA and the 2′ position of DNA. Commonly used modifications include 2'-O-methyl (2-OMe), 2′-methoxyethyl (2-MOE), and 2′-fluoro (2-F). The third generation of chemical modifications included modification of the ribose and base. Common chemical modifications of oligonucleotides [2, 3] are shown in Figure 5.1.

- Delivery of Oligonucleotide Drugs

Oligonucleotides require the aid of delivery systems to improve delivery efficiency, reduce dose, and improve tolerability, thereby improving drug safety. Over the years, various delivery systems have been developed. The successful delivery systems are lipid nanoparticle (LNP) technologies and targeted delivery achieved through specific ligand binding (e.g., N-acetylgalactosamine, GalNAc) [4].

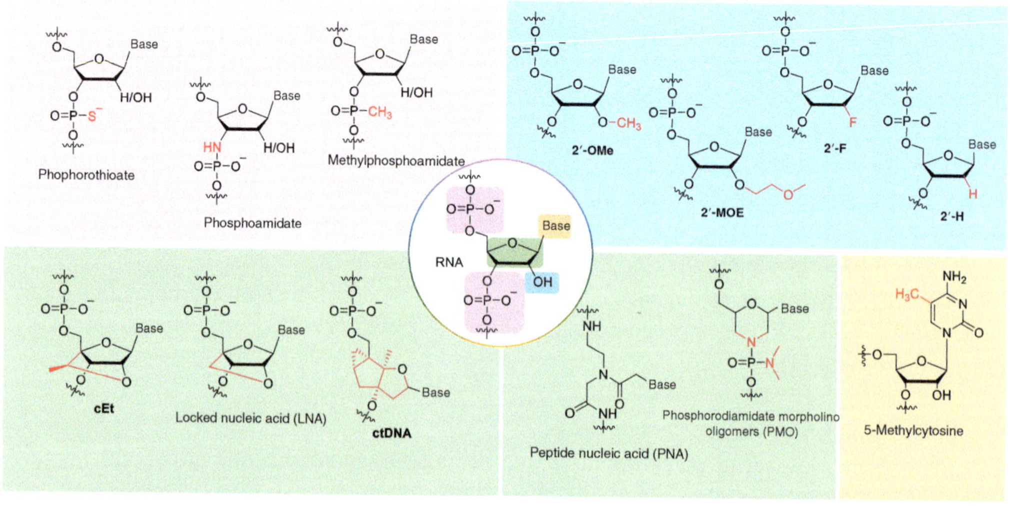

Figure 5.1 Common chemical modifications of oligonucleotides. *Source:* [2]. Suzan M Hammond, et al. 2021; [3]. Anna Kilanpwska, et al., 2020.

- Mechanisms of Action of Oligonucleotide Drugs

Oligonucleotide drugs include various types such as ASOs, siRNA, miRNA, saRNA, mRNA, aptamer, ribozyme, and AOCs. Mechanisms and sites of action of various oligonucleotides are illustrated in Figure 5.2. Depending on their mechanism of action, oligonucleotide drugs can decrease, increase, or restore protein expression [2].

Due to the distinct mechanisms of action of each oligonucleotide, they exert inhibitory effects at different stages of pathogenesis. As shown in Figure 5.3, decoy targets DNA-encoded transcription factors in the upstream process of expression; ASO, siRNA, and miRNA act at the mRNA level; and aptamers directly inhibit the activity of proteins involved in pathogenesis. Their common feature is their ability to inhibit disease progression by interfering with gene expression [5].

5.1.2 Advantages of Oligonucleotide Drug Development

Oligonucleotide drugs represent an emerging field of global interest due to their advantages of short development time, lower production costs compared to biologics, and high success rate in research and development.

5.1.2.1 Short Development Time

Oligonucleotide drugs are designed based on the principle of base complementarity, where the majority of nucleic acid drugs rely on the pairing of bases. Therefore, designing drugs based on the target gene's base sequence is relatively simple. The design of chemical modifications and delivery

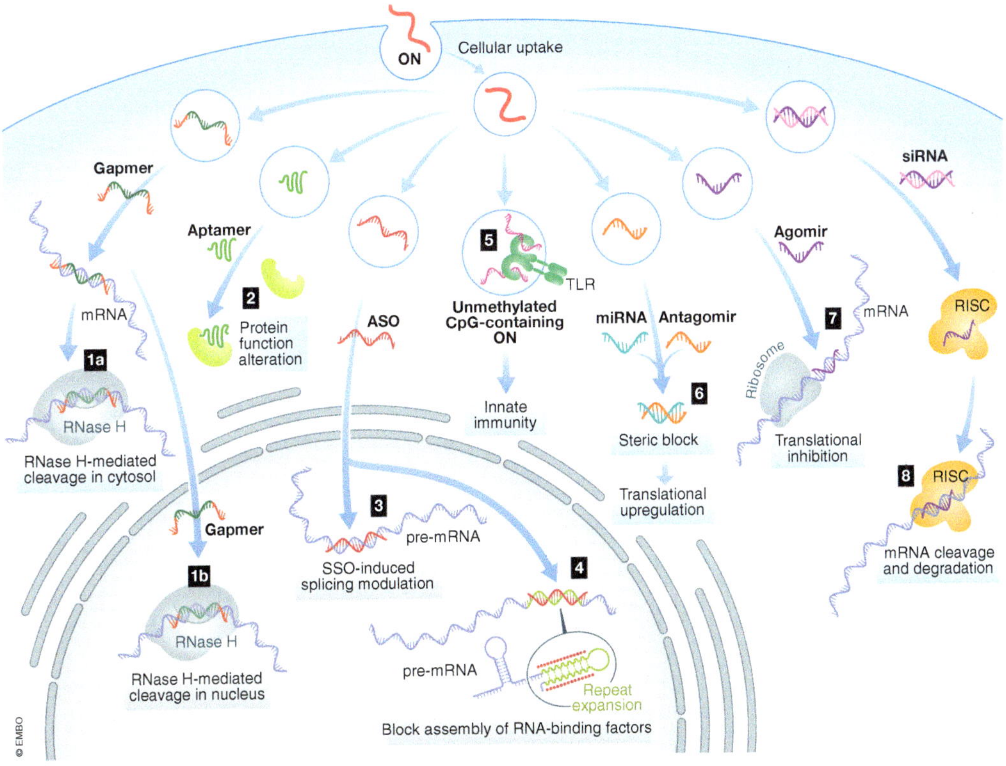

Figure 5.2 Mechanisms and sites of action of oligonucleotides. *Source:* [2]/EMBO Press/CC BY 4.0.

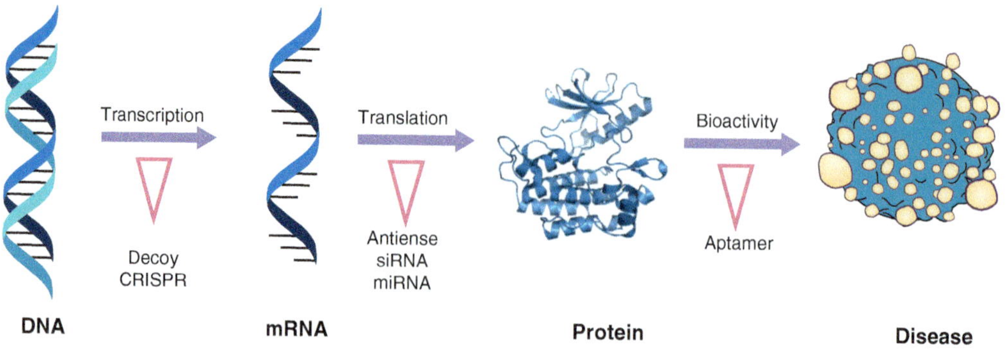

Figure 5.3 Schematic of oligonucleotide drugs acting on different stages of pathological gene expression to ameliorate disease pathogenesis and progression. *Source:* [5]/MDPI./CC BY 4.0.

systems is independent of sequence design. There are multiple nucleic acid sequence design software available for analyzing various parameters, including sequence conservation, homology, immunogenicity, off-target effects, etc. These software tools enable rapid cross-species sequence alignment and homology analysis, allowing for the selection of oligonucleotide sequences with high specificity and potential activity.

5.1.2.2 Lower Production Cost than Biologics

Oligonucleotides are chemically synthesized and can be scaled up to industrial levels, resulting in lower production costs compared to biological molecules. The current technology for oligonucleotide synthesis is well established and supports production from laboratory scale to commercial scale. These advantages significantly shorten the oligonucleotide drug research and development cycle. For example, Milasen, an oligonucleotide drug developed for a single patient, took only 10 months from drug design to first clinical administration.

5.1.2.3 High Success Rate in Research and Development

Oligonucleotide drugs have revolutionary advantages in terms of success rates in research and development. Taking Alnylam, a leading company in the siRNA field, as an example, it has achieved a 64.3% success rate in progressing from phase 1 clinical trials to positive results in phase 3 clinical trials, far surpassing the industry average of 5.7% for new drug development.

5.1.3 Challenges in the Pharmacokinetic (PK) Evaluation of Oligonucleotide Drugs

The regulatory guidelines and filing requirements for oligonucleotide drugs are still relatively vague. For example, the FDA proposes that although the physicochemical properties of oligonucleotides are similar to biological products, they are still regulated as new chemical entities because they are usually chemically synthesized. However, marketed oligonucleotide drugs until 2024 are usually referenced to the examination of the immunogenicity of the drug by large biological molecules. Therefore, both new chemical entity rules and the aspects involving biologics often need to be considered for PK studies of oligonucleotides.

To improve nucleotide drug stability and targeting capabilities, new technologies involving the chemical modification of oligonucleotides and delivery systems are constantly emerging. The chemical modifications of oligonucleotide drugs that have been developed include modification of the nucleic acid backbone, the sugar portion of ribose, and nucleobases. Delivery systems mainly include LNP delivery systems, polymer delivery systems, conjugate delivery systems, and viral vectors [6]. Preclinical PK properties of oligonucleotides and their delivery systems should be evaluated simultaneously.

PK studies involve a variety of complex bioanalytical tools. Different oligonucleotides require different bioanalytical methods. For the same oligonucleotide drug, different bioanalytical methods may be required at different stages of development as well as for different sample types at the same stage. This means that PK analysis often involves collaboration among multidisciplinary teams, using multiple technical testing platforms.

5.1.4 Summary

The PK study of oligonucleotide drugs requires the establishment of a comprehensive evaluation system by combining the nature of oligonucleotides themselves, chemical modifications, and principles and characteristics of delivery technologies. We look forward to the development of more oligonucleotide drugs to fill the current unmet clinical needs.

References

1 Moumné, L., Marie, A.C., and Crouvezier, N. (2022). Oligonucleotide therapeutics: from discovery and development to patentability. *Pharmaceutics* 14 (2): 260.

2 Hammond, S.M. et al. (2021). Delivery of oligonucleotide-based therapeutics: challenges and opportunities. *EMBO Molecular Medicine* 13 (4): e13243.

3 Kilanpwska, A. et al. (2020). *In vivo* and *in vitro* studies of antisense oligonucleotides – a review. *RSC Advances* 10: 34501–34516.

4 Roberts, T.C. et al. (2020). Advances in oligonucleotide drug delivery. *Nature Reviews. Drug Discovery* 19: 673–694.

5 Takakura, K. et al. (2019). The clinical potential of oligonucleotide therapeutics against pancreatic cancer. *International Journal of Molecular Sciences* 20. 3331.

6 Roberts, T.C., Langer, R., and Wood, M.J.A. (2020). Advances in oligonucleotide drug delivery. *Nature Reviews. Drug Discovery* 19: 673–694.

5.2 Delivery Systems of siRNA Drugs and Their Effects on the Pharmacokinetic Profile

Huan Liu, Jing Jin, Maotian Zhou, Miaomiao Song, Liping Ma, Lili Xing

5.2.1 Introduction

With the increasing number of approvals of oligonucleotide drugs, they attract greater attention in the pharmaceutical industry. Although several oligonucleotide drugs have been successfully approved, their target tissue delivery remains challenging. Due to their physicochemical characteristics, unmodified free oligonucleotide drugs may be rapidly cleared by the body after dosing and pose a risk of off-target and toxic side effects. Therefore, nucleic acid-based drugs usually require chemical modifications and appropriate delivery systems to exert therapeutic effects. Although chemical modification plays an essential role in the delivery of oligonucleotide drugs, it is challenging for siRNAs drugs to reach the target sites by chemical modifications alone, and it is similar to the delivery of ASOs. Therefore, an effective delivery system is vital to successful oligonucleotide drug development. At present, the leading companies in siRNA drug development all possess licensed delivery technology platforms. In the future, a breakthrough in delivery systems will undoubtedly become the key factor for companies to dominate the market, facilitating the development of various nucleic acid drugs. This chapter focuses on nucleic acid drug delivery technologies with the beginning of the delivery system of siRNA drugs.

5.2.2 Delivery Systems for Approved siRNA Drugs

In vivo delivery of siRNA drugs could be divided into three main phases, and the body sets different biological barriers for siRNA drugs in each stage:

- **Reaching target tissues:** After dosing, the siRNA first needs to evade both degradation by nucleases in plasma and tissues and capture by the immune system, before successfully entering the target tissue.
- **Entering cells:** Due to their high molecular weights and negative charges, siRNA drugs do not readily cross cell membranes through passive diffusion and thus require cellular internalization through endocytosis to enter the cells.
- **Lysosomal escape:** siRNA needs to escape from endosomes before they fuse with lysosomes, entering the cytoplasm where they bind to target mRNA to induce gene silencing [1].

The main task of a delivery system is to protect the siRNA against all biological barriers to reach the cytoplasm and bind to target mRNA, thus exerting drug efficacy.

5.2.2.1 Lipid Nanoparticle (LNP) Delivery System

Due to the charge characteristics of siRNAs, cationic liposomes appear to be the best candidate delivery systems. However, the toxicity of traditional cationic liposomes impedes significantly the development of siRNA drugs. Alnylam has achieved a breakthrough in the successful delivery of oligonucleotides using Arbutus' ionizable cationic lipid materials. In 2018, we saw the approval of Onpattro, the first siRNA drug using LNPs as a delivery system, which is used for the treatment of polyneuropathies induced by hereditary transthyretin amyloidosis (hATTR).

The major success factor of LNP delivery systems is using ionizable cationic lipids. These lipids exhibit different charging characteristics under different pH conditions, being positively charged in acidic pH environments and nearly neutral charged at physiological pH. The ionizable properties of lipids provide intelligent protection for siRNAs across biological barriers (see Figure 5.4) [2].

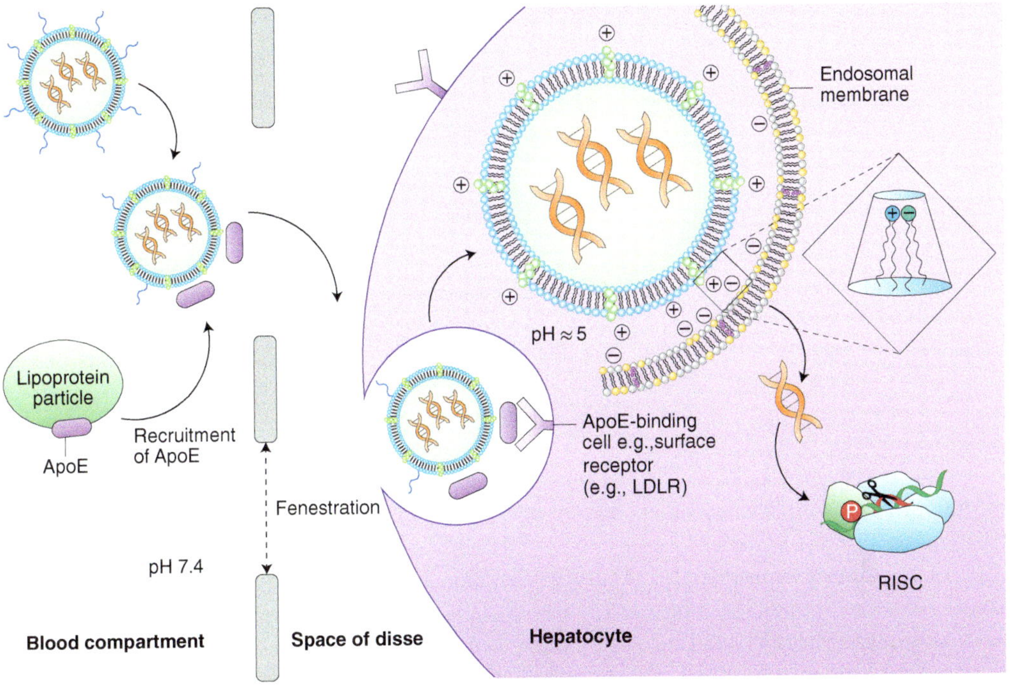

Figure 5.4 Lipid nanoparticle (LNP) mediated delivery of siRNAs *in vivo*. *Source:* [2]/EMBO Press,/ CC BY 4.0.

First, siRNA drugs must be produced under low pH conditions. During production, the ionizable lipids are positively charged to achieve stable and optimal siRNA drug encapsulation rates. PEG lipids are added to the liposomes to obtain liposomes of 100 nm or smaller sizes. Liposome size (20–100 nm) can be regulated by adjusting the proportion of PEG lipids to other lipids.

After administration, the LNP delivery system presents an essentially neutral charge surface in plasma, which helps the siRNA drug to evade capture by the immune system and degradation by nucleases in conjunction with the PEG lipids.

As PEG lipids dissociate from the liposome surface, the siRNA-carrying liposomes enter cells via endocytosis. Due to the low pH in endosomes, the ionizable lipids protonated and their positive charge destabilizes the endosome phospholipid bilayer structure, facilitating siRNA escape from endosomes. This releases siRNA into the cytoplasm where it binds to target mRNA to downregulate the target gene expression.

From among more than 300 candidate ionizable lipid materials, Onpattro selected the ionizable lipid material N,N-dilinoleoyl-monomethylamine-3-(N,N-dimethylamino)-2-hydroxypropylamine (Dlin-MC3-DMA) with the most desirable pKa characteristics. In clinical trials, Dlin-MC3-DMA-based liposomes showed superior therapeutic efficacy compared to first-generation ionizable lipids, with over 80% reduction in thyroxine transporter protein in serum.

5.2.2.2 N-Acetylgalactosamine (GalNAc) Delivery System

A delivery system based on GalNAc–siRNA conjugation has received attention from nucleic acid drug development organizations in recent years for its efficient liver targeting and good safety profile. GalNAc is a ligand for the asialoglycoprotein receptor (ASGPR), which is predominantly expressed on hepatocyte cell membranes (~500,000 ASGPR per cell). GalNAc–siRNA conjugates transport siRNA from the cell surface into the cell through the specific binding of ASGPR followed

Table 5.1 EMA/FDA-approved GalNac–siRNA conjugate drugs.

Drug name	R&D company	Target genes	Indications	Administration	First approval year
Givlaari, Givosiran	Alnylam Pharmaceuticals	ALAS1	Acute hepatic porphyria	Subcutaneous injection	2019
Leqvio, Inclisiran	Novartis/Alnylam Pharmaceuticals	PCSK9	Hyperlipidemia	Subcutaneous injection	2020
Oxlumo, Lumasiran	Alnylam Pharmaceuticals	HAO1	Primary hyperoxaluria type 1	Subcutaneous injection	2020
Amvuttra, Vutrisiran	Alnylam Pharmaceuticals	hATTR	Hereditary transthyretin amyloidosis with polyneuropathy (hATTR-PN) in adults	Subcutaneous injection	2022

by endocytosis. Consequently, the GalNAc–siRNA conjugates dissociate from ASGPR. ASGPR recycles back to the cell surface, while GalNAc–siRNA conjugates further dissociate, releasing free siRNAs that exert pharmacological effects by silencing genes in the cytoplasm [3, 4]. The 3′ end of the siRNA sense strand (SS) chemically bonds to GalNAc with a trivalent structure, and GalNAc–siRNA conjugate stability is increased by enhanced stabilization chemistry (ESC) technology to achieve increased liver exposure and prolonged gene silencing time [4, 5]. Several GalNac–siRNA drugs have received FDA and EMA approval (Table 5.1).

5.2.3 Influence of Delivery Systems on siRNA Drugs' PK Profile

The delivery system assists siRNA drugs transporting from the administration site to the target cell, and the siRNA exists mostly as a complex with the delivery system during this process. Therefore, the delivery system influences the *in vivo* processes and distribution of siRNAs. Some researchers determined the tissue distributions of ^{3}H-labeled free siRNAs and LNP-encapsulated siRNAs using quantitative whole-body autoradiography (QWBA) and matrix-assisted laser desorption/ionization mass spectrometry (MS) imaging (MALDI-MSI) techniques, respectively. The QWBA study revealed that the animal kidneys and salivary glands exhibited the highest radioactivity after a single intravenous dosing of free siRNAs. However, the highest radioactivity levels were found in the livers and spleens after the dosing with LNP-encapsulated siRNA, with radioactivity lasting >168 hours. MALDI-MSI imaging results showed that the distribution of cationic lipid material was consistent with that of radiolabeled siRNAs, suggesting that siRNA may circulate together with lipid material and distribute to tissues in the form of nanoparticle complexes [6]. By altering free siRNA's tissue distribution characteristics, the delivery system achieves targeting and effectively prolongs the *in vivo* retention time of siRNA drugs.

LNP delivery systems primarily exhibit passive hepatic targeting. However, GalNac–siRNA conjugate delivery systems achieve active hepatic targeting by binding ligand GalNac to the ASGPR. Adjusting the structure, proportions, or particle size of nanoparticle components or modifying oligonucleotide structures may result in significant changes in tissue distribution and gene-silencing efficacy of siRNA drugs. In Onpattro, the optimal pKa value for ionizable lipids is around 6.4, and a deviation in pKa of only 0.5 may lead to a 100-fold reduction in its efficacy. To enable the rapid distribution of Onpattro to target tissues in the liver, scientists have developed C14 alkyl chain PEG lipids, which are short-chain PEG lipids that can dissociate more rapidly from the liposome surface *in vivo* than conventional PEG lipids. The C14 alkyl chain PEG lipids allow liposomes that are not encapsulated by PEG to rapidly adsorb endogenous apolipoprotein E (ApoE), triggering ApoE receptors on the surface of hepatocytes to endocytose these liposomes. In GalNac–siRNA delivery systems, trivalent and tetravalent ligands exhibit a higher affinity for ASGPR than monovalent or divalent ligands. Therefore, trivalent GalNAc was selected to conjugate with siRNA for this delivery system. Exposure of GalNac–siRNA conjugates in the liver was significantly increased after structural modification of siRNA by Advanced ESC technology [7]. Refinement of the structure and proportion of each component is key to optimizing the targeted distribution and efficacy of oligonucleotide drugs.

5.2.4 Prospects for the Development of Oligonucleotide Drug Delivery Systems

The indications for siRNA drugs that have been approved or entered clinical stages mainly concentrate on the liver or liver-related disease areas, which is closely associated with the successful development of liver-targeted delivery systems. Expanding indications to target other tissues and organs has become the focus of commercial oligonucleotide drug R&D at this stage.

Studies indicate that siRNAs can be distributed and released in tissues including the heart, lungs, fat, and muscle by conjugating with various lipids (saturated fatty acids, unsaturated fatty acids, cholesterol, vitamins, etc.). Arrowhead's TRiM™ (Targeted RNAi Molecule) platform [8] may facilitate this expansion, targeting multiple tissues, including the liver, lung, and tumors. This platform utilizes ligand-mediated delivery for tissue-specific targeting. The TRiM™ platform includes a highly potent RNA trigger that can optimize high affinity targeting ligands for each drug candidate, enhancing drug structure and PK. It also contains specific nucleic acid sequences that assist in achieving targeted drug delivery and maintaining optimal pharmacological activity and safety. Another startup, utilizing cyclic cell-penetrating peptides as core technology, should also be paid attention to. The Endosomal Escape Vehicle (EEV™) platform of Entrada Therapeutics [9] aims to develop intracellular therapeutics. The EEV™ technology platform conjugates and delivers various drugs, including oligonucleotides and antibodies, to the cytoplasm of target cells through chemical modifications. Preclinical studies have shown that more than 90% of EEV conjugates can be absorbed by the body, and those conjugates that enter a target cell can escape from early endosomes efficiently and rapidly. Scientists have observed that approximately 50% of EEV conjugates reach intracellular targets, which shows a significant improvement compared to the less than 2% escape rate of many current biologics. siRNAs targeting tumors and various diseased tissues via peptide nanoparticles, viral vectors, or binding antibody fragments are also current R&D hotspots.

AOC (antibody–oligonucleotide conjugates) therapeutics are designed to combine the proven technology of specific targeting antibodies or antibody fragments (e.g., targeting transferrin receptor for muscle and central nervous system specificity) with the precision and potency of oligonucleotide therapies to access extrahepatic tissues and provide new potential treatments for patients in recent years. There are many biotechs focused on the development of AOCs, such as Avidity Biosciences, Tallac Therapeutics, Dyne Therapeutics, and Denali Therapeutics. In 2019, Avidity launched the first AOC drug AOC 1001 (Tfr1-targeted mAB for DM1 disease with anti-DMPK siRNA) worldwide, which has entered clinical phase I [10]. The AOC 1044 (Tfr1-targeted mAB for DMD disease with anti-DMD PMO) has recently been approved to be an orphan drug and has obtained FDA fast track, driving further enthusiasm for the development of AOC drugs. In addition to the targeting determined by the antibody part, the conjugation strategy is the key to affecting the efficacy of AOC drugs by influencing the release of oligonucleotides in lysosomes. The primary conjugation techniques are charge-based affinity, streptavidin/biotin binding, direct conjugation, and complementary base pairing.

In the development of delivery systems, quantifying the effects of changes in the delivery system on tissue distribution using PK studies combined with pharmacodynamic validation is key to unlocking the "smart" oligonucleotide delivery. This involves targeting different diseased tissues and treating multiple indications by adjusting the structure or proportion of delivery system components [11, 12].

In recent years, siRNA drug delivery systems have also attracted the attention of companies involved in the development of other nucleic acid drugs. By conjugating with delivery systems, the efficacy of ASOs can be significantly enhanced, and several GalNac-ASO conjugate drugs are already in clinical trials. A lipid-based nanoparticle delivery system has also been successfully adopted by Pfizer and Moderna for the development of an mRNA vaccine against the coronavirus disease [13].

References

1 Wang, J. et al. (2010). Delivery of siRNA therapeutics barriers and carriers. *The AAPS Journal*.

2 Akinc, A. et al. (2019). The Onpattro story and the clinical translation of nanomedicines containing nucleic acid-based drugs. *Nature Nanotechnology*.

3 Springer, A.D. et al. (2018). GalNAc-siRNA conjugates: leading the way for delivery of RNAi therapeutics. *Nucleic Acid Therapeutics* 28 (3): 109–118.

4 Kulkarni, J.A. et al. (2021). The current landscape of nucleic acid therapeutics. *Nature Nanotechnology* 16(6): 630–643.

5 Zimmermann, T.S. et al. (2017). Clinical proof of concept for a novel hepatocyte-targeting GalNAc-siRNA conjugate. *Molecular Therapy* 25 (1): 71–78.

6 Park, J. et al. (2016). Pharmacokinetics and biodistribution of recently developed siRNA nanomedicines. *Advanced Drug Delivery Reviews* 104: 93–109.

7 Brown, C.R. et al. (2018). *Mechanistic Insights and Progress on the GalNAc-siRNA Conjugate Platform for Targeted Delivery of RNAi Therapeutics to the Liver*. Alnylam Pharmaceuticals.

8 arrowheadpharma.com/science

9 entradatx.com

10 Mullard, A. (2022). Antibody-oligonucleotide conjugates enter the clinic. *Nature Reviews. Drug Discovery* 21 (1): 6–8.

11 Benizri, S. et al. (2019). Bioconjugated oligonucleotides recent developments and therapeutic applications. *Bioconjugate Chemistry* 30 (2): 366–383.

12 Biscans, A. et al. (2019). Diverse lipid conjugates for functional extra-hepatic siRNA delivery *in vivo*. *Nucleic Acids Research* 47 (3): 1082–1096.

13 Mulligan, M.J. et al. (2020). Phase I/II study of COVID-19 RNA vaccine BNT162b1 in adults. *Nature* 586(7830): 589–593.

5.3 Bioanalytical Strategies for Oligonucleotide Drugs

Zhiyu Li, Nan Zhao, Siyu Liu, Hefeng Zhang, Hongmei Wang, Hong Sun, Yafei Gao, Qian Cui, Jiaming Jiang, Lili Xing

5.3.1 Introduction

As of July 2023, there are five approved siRNA drugs on the market, and an increasing number of siRNA drugs are entering clinical research stages. To improve nucleotide drug stability and targeting capabilities, new technologies involving the chemical modification of oligonucleotides and delivery systems are constantly emerging. Hence, it includes evaluations of siRNA plasma PK, tissue PK, RNA-induced silencing complex (RISC) PK, the degradation efficiency of target gene mRNA, and the inhibition efficiency of the target protein.

Due to diverse chemical modifications, delivery systems, and the involvement of multiple types of sample analysis, multiple analysis platforms are needed to support the quantitative detection of different analytes.

5.3.1.1 Diverse Chemical Modifications and Delivery Systems

Oligonucleotides are basically negatively charged, highly polar, and highly hydrophilic; however, different chemical modifications and delivery systems can significantly affect their physicochemical properties. Therefore, laboratory technicians need to select different analytical platforms based on the physicochemical properties, chemical modifications, and delivery systems.

5.3.1.2 Involvement of Multiple Bioanalytical Methods

Different stages of oligonucleotide drug development and different sample types require different levels of detection sensitivity; thus, the selection of an appropriate analytical platform is necessary. Cross-validation is also required when multiple analytical methods are used to analyze samples from different stages of the same oligonucleotide drug.

5.3.1.3 Involvement of Multiple Types of Sample Analysis

Oligonucleotide drugs rapidly distribute from plasma into tissues within minutes to hours, followed by a slower elimination phases with half-lives of several weeks to months. Oligonucleotide drugs are distributed into more than 80% of systemic tissues, mainly in the kidney and liver, where exposure is hundreds to thousands of times higher than that in the plasma, making oligonucleotide exposure in individual tissue important data for PK and toxicological studies. Usually, the tissue distribution of oligonucleotide drugs involves more than a dozen different tissues, and the sample collection, tissue sample homogenization, and matrix effects present challenges that need to be addressed in the development of bioanalysis methods.

5.3.1.4 Immunogenicity

Oligonucleotide drugs could induce immunogenicity in the body; however, the detection of antidrug antibodies against oligonucleotides has always been a challenge in bioanalysis. Because oligonucleotides have a relatively smaller molecular weight and carry a large negative charge compared to antibody-based drugs and can easily bind to positively charged biomolecules *in vivo* with relatively high affinities. Therefore, they exhibit poor specific binding to antidrug antibodies and require analytical methods with high specificity and selectivity. In addition, the presence of abundant DNA or RNA fragments in the body can also interfere with the experiments.

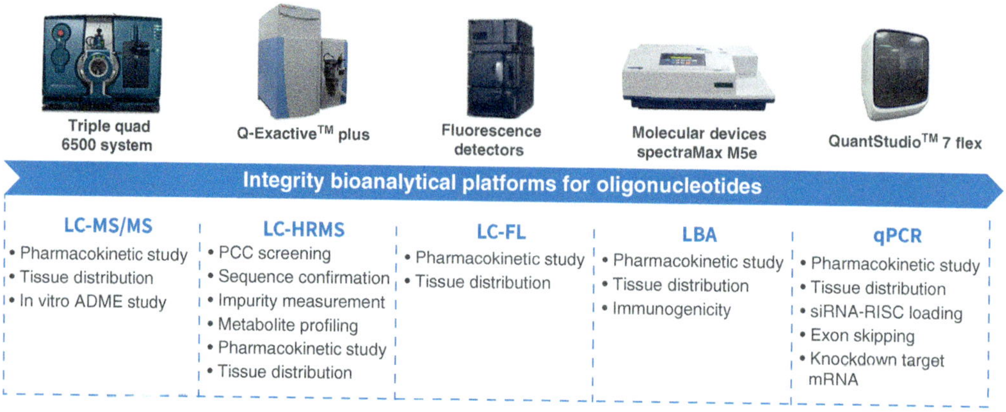

Figure 5.5 Oligonucleotide bioanalysis platforms.

There are four major bioanalysis platforms: LC-MS/MS and LC-HRMS, a fluorescence detection platform (LC-FL), an LBA platform, and a quantitative polymerase chain reaction (qPCR) platform (Figure 5.5). These platforms enable us to obtain high-quality *in vitro* and *in vivo* data, conduct diverse biological analyses, and accelerate the drug development process.

Mass spectrometry platforms: MS platforms offer advantages in the development of quantitative analysis methods, including short development cycles and good specificity due to no requirement for assay-specific probe design and synthesis. These platforms also allow qualitative and quantitative analyses of metabolites. It has been widely used in the field of nucleic acid drug detection, but there are still many challenges, such as sample preparation, extraction, and suppression of MS signals by ion-pairing reagents. We have established an oligonucleotide MS platform including high-resolution MS, with a detection sensitivity of 1–10 ng/mL to support early screening and preclinical PK studies.

Fluorescence detection platform: For long (25–60 nt) oligonucleotides or oligonucleotides with specific structures, fluorescent probes complementary to the target oligonucleotides can be used for target identification/capture. Hybridized oligonucleotides are separated using LC and used for indirect quantitative analysis of the nucleic acid drug and its known metabolites using fluorescence chromatography. For this platform, we can perform the design and screening of probes with a detection sensitivity of 1–5 ng/mL.

LBA platform: LBA methods can also be used for the quantitative analysis of longer oligonucleotides in biological matrices with a detection sensitivity of up to 50 pg/mL. These techniques offer strong inclusivity for chemical modifications and conjugation of oligonucleotides but are susceptible to interference by metabolites. We have developed a hybridization-based enzyme-linked immunosorbent assay (HELISA) and MSD® electrochemiluminescence assay for the quantitative analysis of oligonucleotides in biological matrices, with a detection sensitivity of 50 pg/mL.

qPCR platform: The qPCR method has high sensitivity, but oligonucleotide drugs are usually short, requiring the use of stem-loop PCR or overlap extension PCR. Ligand binding platforms, fluorescence detection platforms, and MS platforms all have corresponding regulatory guidelines; however, there is no specific regulatory guidance for qPCR platforms until July 2024 used in the quantitative analysis of oligonucleotides. Only suggestions made in industry white papers are available. Based on our experience in oligonucleotide drug research, the precautions for oligonucleotide drug development issued by different regulatory agencies, and cutting-edge research, we have established a comprehensive process for the development and validation of bioanalytical methods, with a platform detection sensitivity of 10 pg/mL or lower.

5.3.2 Application of Liquid Chromatography-Mass Spectrometry (LC-MS) in Quantitative Analysis of Oligonucleotides

5.3.2.1 LC-MS/MS for Oligonucleotide Quantification

The quantitative analysis of oligonucleotides in biological samples using LC-MS/MS (Figure 5.6) can be summarized in the following three steps: sample collection (tissue samples require homogenization), sample pretreatment, and sample analysis using LC-MS/MS.

5.3.2.2 Advantages of LC-MS/MS Technology

Quantitative analysis methods for oligonucleotides include hybridization-based LBA, LC-FL, LC-ultraviolet (UV), and LC-MS/MS. The development of LBA and LC-FL involves the design and synthesis of probes complementary to the target oligonucleotide, resulting in longer development cycles and higher costs. Therefore, it has become a bottleneck in nucleic acid drug development, especially in early drug screening. HELISA has been widely used in PK studies of oligonucleotides because of its significant advantages of high sensitivity and high throughput. However, full-length oligonucleotides and nucleolytic metabolites cannot be distinguished, leading to the overestimation of prodrug concentrations due to cross-hybridization. LC-UV cannot be used to analyze low-concentration biological samples because of its low detection sensitivity.

LC-MS/MS technology combines high separation capacity liquid chromatography with high sensitivity and selective MS. In addition to accurately quantifying nucleic acids in biological samples, this method can provide information on base composition, sequence structure, and other information for MetID. The MS platform does not require special reagents or probes, and the method development cycle is short, which can significantly accelerate the speed of drug development. Therefore, LC-MS/MS has advantages in the quantitative analysis of oligonucleotide biological samples in early *in vitro* and *in vivo* drug screening and nonclinical studies, providing more effective, sensitive, rapid, and accurate assay data.

5.3.2.3 Challenges of LC-MS/MS Technology

The quantitative analysis of oligonucleotides using the LC-MS/MS method faces several challenges, primarily due to the strong hydrophilicity of oligonucleotides, low ionization efficiency, and tendency to generate a large number of metal adduct ions, resulting in poor peak shape, low recovery, and low MS response. To improve the sensitivity and reproducibility of the method, the following three methods can be employed:

1) Sample pretreatment was optimized to improve the extraction recovery.
2) Improvement of the multicharged distribution of oligonucleotides and optimization of MS parameters for more abundant precursor ions.
3) Reduction of metal adduct ion formation in the MS analysis.

5.3.2.4 Development of Sample Pretreatment Methods

When analyzing oligonucleotides in biological samples, the development of sample pretreatment methods is as important as establishing LC-MS/MS conditions. The following points should be considered for sample pretreatment:

1) Common oligonucleotide extraction methods include liquid–liquid extraction (LLE) and/or solid-phase extraction (SPE). For the LLE method, thorough mixing and complete sample separation from the organic layer are important steps for removing proteins from biological samples. The SPE method requires strict pH control to ensure high extraction recoveries. Oligonucleotides exhibit nonspecific binding in polypropylene and glass containers. The use of

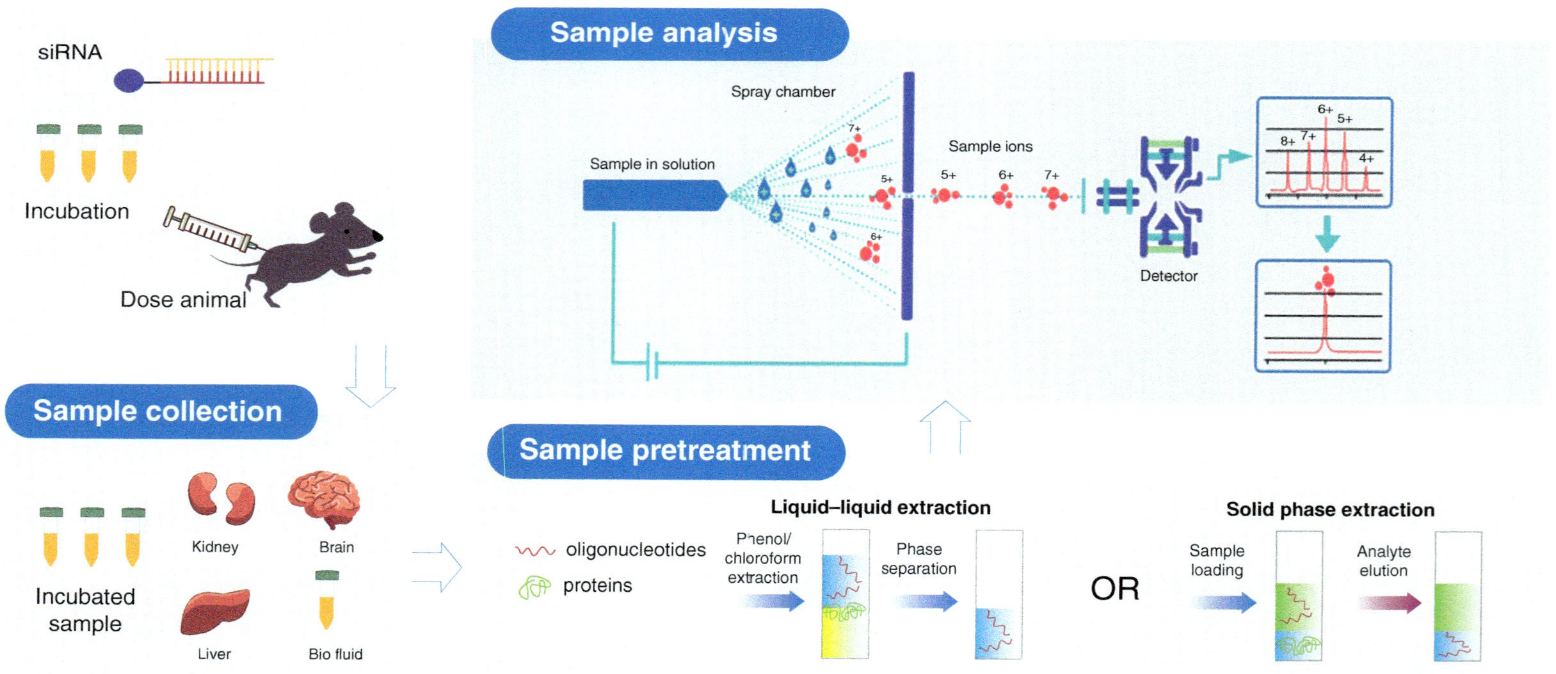

Figure 5.6 The flow of quantitative analysis using LC-MS/MS.

low-adsorption consumables or alkylating treatment of the containers, along with sample dissolution in a solvent containing triethylamine (TEA), can prevent nonspecific binding. The process of nitrogen blow-drying followed by reconstitution is frequently employed to concentrate oligonucleotides prior to injection. However, complete drying of the sample through evaporation can sometimes affect reconstitution. In such cases, the addition of a solubilizing agent or evaporation of a portion of the sample can be considered.

2) In method development, it is important to fully consider the impact of chemical modifications and delivery systems of oligonucleotides on their chemical properties and polarity. Therefore, flexibility in optimizing the existing extraction methods is critical.

5.3.2.5 LC-MS/MS Method Development

LC-MS/MS method development is mainly concerned with the selection of columns and optimization of the mobile phases and MS parameters. There are several liquid chromatography columns available for oligonucleotide separation, such as Phenomenex Clarity Oligo-RP and Thermo DNAPac™ PA200 RS, which will not be further discussed here. Ion-pairing reagents such as hexafluoroisopropanol (HFIP)/TEA can be used to retain highly polar and anionic oligonucleotides and are widely used for oligonucleotide LC-MS/MS analysis. However, TEA can form "pseudoneutral molecules" with oligonucleotides through ionic bonding, thereby shielding oligonucleotide charges and suppressing the MS response. Therefore, the retention provided by ion-pairing reagents must be balanced with their ion suppression to maximize the sensitivity of bioanalysis using MS. TEA is a nonvolatile base that tends to bind to MS detectors, causing contamination and significantly decreasing MS detection sensitivity. Hydrophilic interaction liquid chromatography (HILIC) has recently been used for oligonucleotide chromatographic separations as an alternative to reverse-phase separation to reduce the impact on MS detection sensitivity [1].

Oligonucleotides respond well to negative ion mode electrospray ionization (ESI) sources and can easily form multiple charged ions. Furthermore, the multiple reaction monitoring (MRM) optimization steps are similar to those used for small molecules after selecting the ion transition. In addition, HRMS, especially full-scan MS, does not require specific optimization of MS parameters for each oligonucleotide, which greatly simplifies this optimization process and shortens the method development time.

5.3.2.6 Application of HRMS

In the analysis of oligonucleotides, triple quadrupole MS or ion trap MS is commonly used for MS.

- Triple quadrupole MS, with high ion transmission efficiency, wide dynamic range, and high sensitivity, offers significant advantages in quantitative analysis. It is widely used for oligonucleotides in biological samples and has a sensitivity of 5–10 ng/mL. Oligonucleotides in triple quadrupole MS mainly produce fragment ions of m/z 125 thymine and m/z 95 phosphate through collision-induced dissociation, but they cannot be used for sequence analysis.
- Ion trap MS can generate diagnostic ions related to the sequence of oligonucleotides. These ions can be used not only for quantification but also for identification of their metabolites.

In recent years, the application of HRMS in full-scan mode has attracted great interest in quantitative oligonucleotide analysis. It provides quantitative information on target oligonucleotides and information related to impurities and metabolites. The detection sensitivity of Orbitrap high-resolution MS (Orbitrap HRMS) is similar to that of triple quadrupole MS, with a wide linear range and a mass deviation of less than 3 ppm. It has a high-energy collision (HDC) dissociation fragmentation mode, which provides abundant ion fragmentation for MetID. As shown in Figure 5.7,

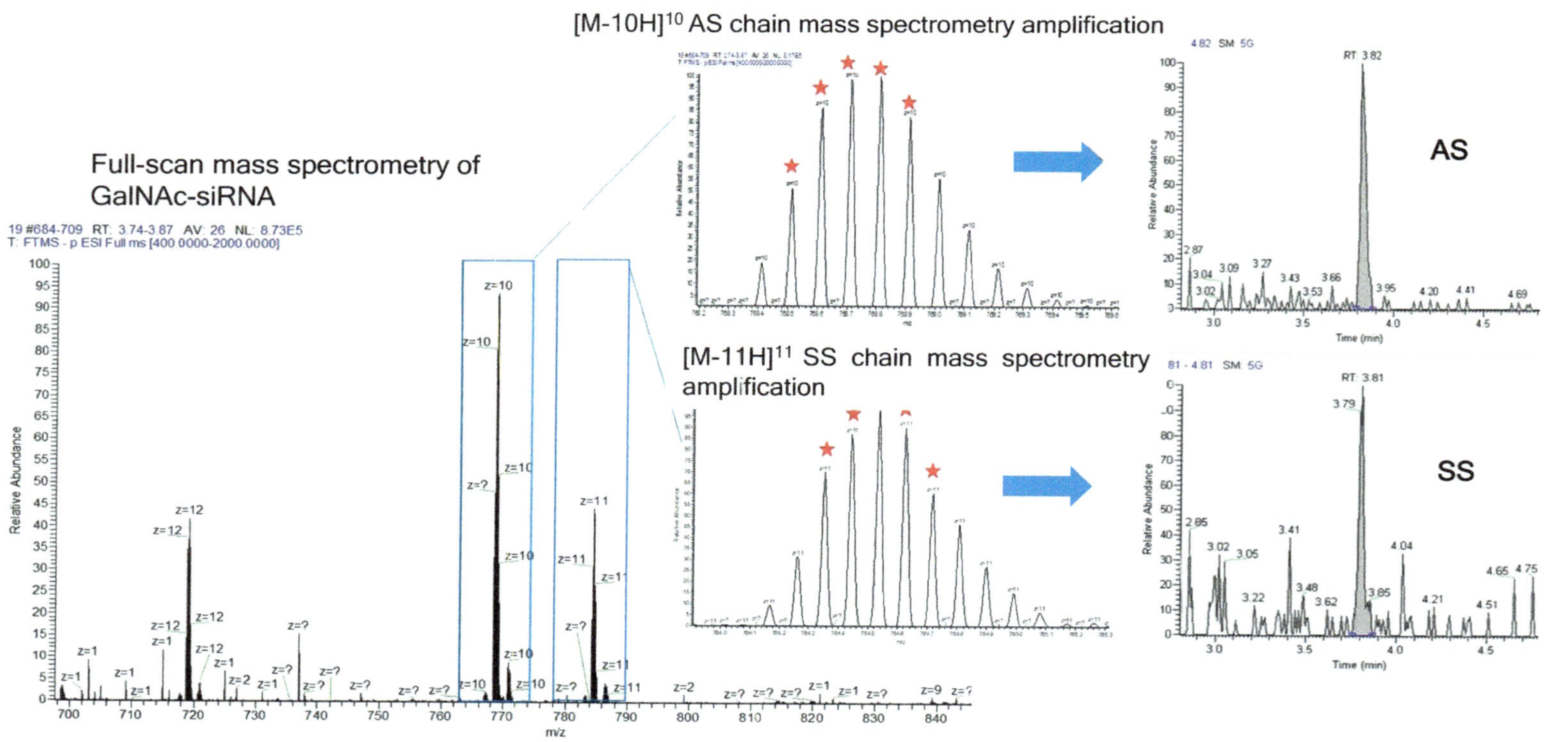

Figure 5.7 Quantitative analysis of siRNAs in the rat plasma using Orbitrap HRMS full scan (linear range: 10–2500 ng/mL).

the siRNA level was quantified in the rat plasma using Orbitrap HRMS, with a detection sensitivity of up to 10 ng/mL in the plasma.

In addition, quadrupole time-of-flight MS (QTOF MS) combines quadrupole and time-of-flight mass analyzers and can be used for *in vitro* metabolism studies and sequence confirmation of oligonucleotides. MetID of oligonucleotides and corroboration of modification sites require specialized software to provide compositional analysis methods and identify new peaks. BioPharma Finder 4.0 software (Thermo Scientific) and ProMass (Novatia, LLC) are commonly used oligonucleotide sequence confirmation software. With the continuous advancement of MS technology, various techniques, such as CE-MS, matrix-assisted laser desorption/ionization-time of flight MS (MALDI-TOF-MS), and ion mobility MS (IM-MS), have been employed for quantitative analysis of oligonucleotide in biological samples.

The MS platforms used for quantitative oligonucleotide analysis are shown in Figure 5.8.

5.3.2.7 Process for Establishing LC-MS/MS Bioanalytical Methods for Oligonucleotides

Based on our experience, we have summarized the process of establishing a bioanalytical method for oligonucleotides in the following five steps:

1) **Optimization of MS conditions**
   a) Confirmation of the charge distribution of test articles.
   b) Optimization of the mobile phase to minimize metal adduct ions.
   c) Optimization of MRM conditions and the use of multiple MRMs to improve the sensitivity.
2) **Optimization of chromatographic conditions**
   a) The use of different mobile phases and columns.
   b) After further optimization of the chromatographic conditions, the gradient and column temperature were adjusted to separate the analyte and interferences.

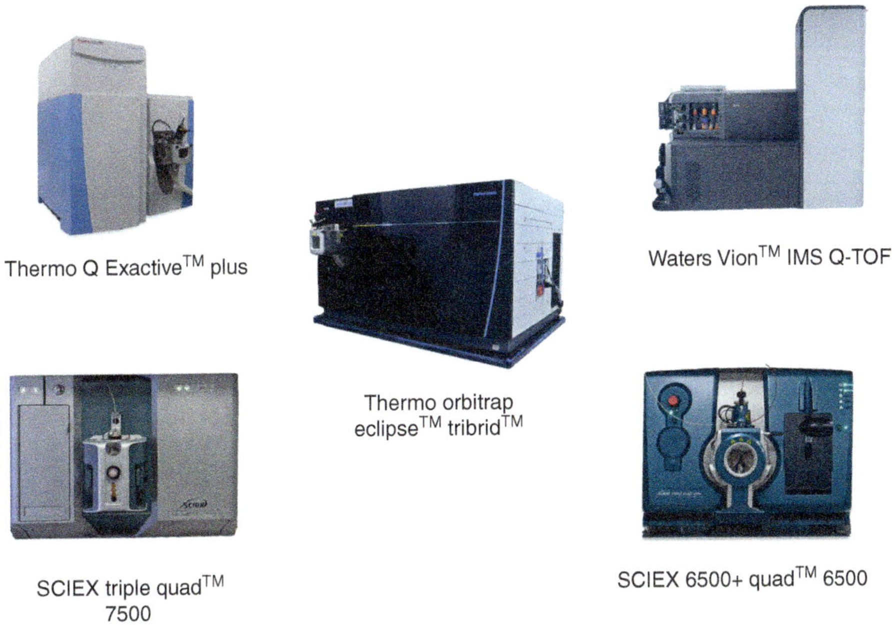

Figure 5.8 Mass spectrometry (MS) platform for quantitative oligonucleotide analysis.

3) **Optimization of sample storage conditions**
   a) The appropriate containers and solvents were used to avoid nonspecific binding.
   b) Confirmation of sample storage stability (temperature).
4) **Optimization of sample extraction conditions**
   a) Extraction recovery optimization.
   b) Evaluation of the stability of the blow-drying and reconstitution processes.
5) **Method qualification**
   a) Further optimization of methods is needed to meet the sensitivity requirements.
   b) Assessment of the selectivity, reproducibility, and matrix effects of the method.
   c) The stability of the samples was confirmed during processing.

5.3.3 Application of Liquid Chromatography Fluorescence (LC–FL) Detection Methods for Quantitative Bioanalysis of Oligonucleotides

5.3.3.1 Principle of the Hybridization-Based LC-FL Assay Method

Hybridization-based LC-FL combines probe hybridization and an LC-FL assay. Quantitative oligonucleotide bioanalysis (Figure 5.9) using LC-FL consists of three main steps:

1) Binding of fluorescently labeled peptide nucleic acid (PNA) probes to the target oligonucleotides through Watson–Crick base pairing to form fluorescently labeled complexes.
2) Separation of fluorescently labeled complexes from excess PNA probes by ion-exchange LC.
3) Detection of fluorescently labeled complexes using a fluorescence detector.

PNA is a DNA analog with a peptide backbone replacing the phosphate backbone (see Figure 5.10). Due to the absence of charged phosphate groups, PNA exhibits electrical neutrality and displays high affinity and specificity for target DNA or RNA through complementary hybridization. PNA is very stable both *in vivo* and *in vitro* and is resistant to nucleases and proteases [2]. The unique physicochemical properties of PNA allow it to be widely used in nucleic acid sequence determination, genetic testing, and disease diagnosis.

ATTO fluorescent dyes are usually derivatives of coumarins, rhodamine, or oxazine, which have high solubility, and their wavelengths range from ultraviolet to near-infrared, making them the most useful full-band fluorescent markers.

Before adding the PNA fluorescent probe, biological samples are digested with proteinase K to remove most of the matrix proteins, and the oligonucleotide analytes are combined with the ATTO-labeled PNA probe to form a fluorescently labeled complex. Next, the excess PNA fluorescent probe was separated from the fluorescently labeled oligonucleotide complex sample using exchange chromatography. The resolution of oligonucleotides and their metabolites and the resolution of oligonucleotides with secondary structures are selectively controlled by adjusting the pH,

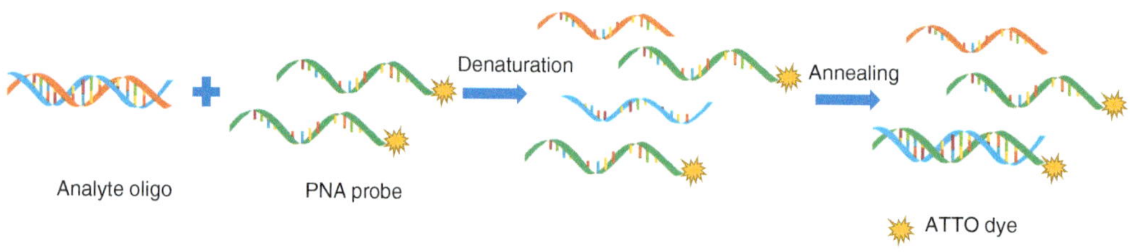

Figure 5.9 Scheme of the liquid chromatography-fluorescence analysis method.

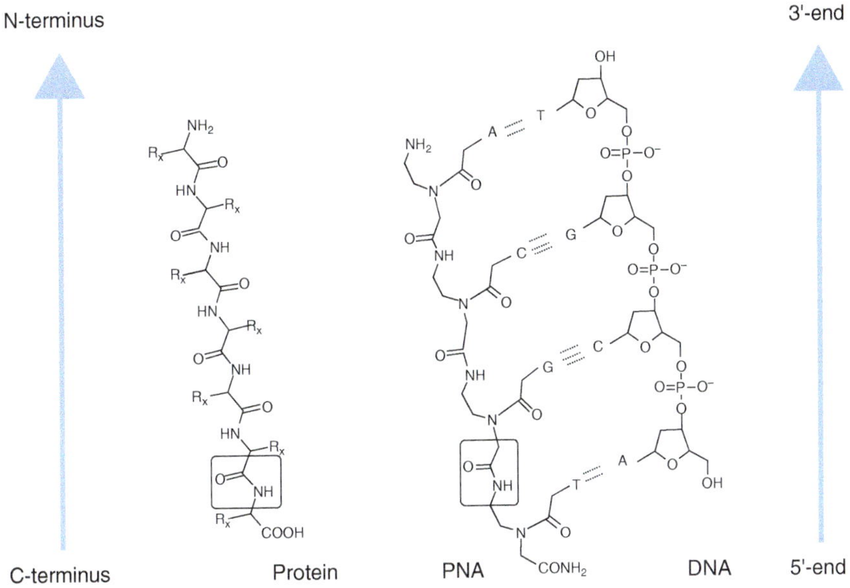

Figure 5.10 Comparison of peptide nucleic acid (PNA) and DNA structures. *Source:* [2]/Springer Nature.

salt, and solvent of the eluent. After absorbing a certain wavelength, fluorescent dyes can emit light at wavelengths greater than the wavelength of the absorbed light. A fluorescence detector can quantify the emitted light by detecting the energy of the emitted wave.

5.3.3.2 Advantages of the LC-FL Method

1) **High detection sensitivity for preclinical and clinical sample analysis**

Similar to the hybridization-based ELISA method, the hybridization-based LC-FL method can easily achieve a limit of detection of 1 ng/mL or less and can quantify oligonucleotides in the plasma or tissue homogenates without an extraction step after sample pretreatment [3]. The fluorescence detector has a stable response, high accuracy and precision, and a wider linear range than the ELISA method. Approved oligonucleotides, such as Patisiran (an RNAi therapeutic agent), are used for PK analysis via the LC-FL method to determine the concentration of Patisiran in preclinical and clinical samples [4].

2) **Good selectivity for quantitative analysis of metabolites**

The hybridization-based LC-FL method avoids interference by other components in biological samples that are not fluorescently labeled and have good selectivity. In addition to relying on high-performance LC, the hybridization-based LC-FL method enables the quantitative analysis of oligonucleotide metabolites. Some truncated oligonucleotide metabolites (e.g., $n-1$ and $n-2$) can efficiently hybridize with fluorescently labeled PNA probes and be separated from oligonucleotides by ion-exchange chromatography.

3) **Unaffected by oligonucleotide structural modifications and delivery systems**

Compared to the qPCR method, chemical modifications do not affect the detection sensitivity or specificity of the hybridization-based LC-FL method. This method can be used for the detection of GalNAc-conjugated oligonucleotides, LNP-delivered oligonucleotides, and oligonucleotides from other delivery systems in preclinical and clinical studies of biological samples.

5.3.3.3 Challenges of Hybridization-Based LC-Fluorescence Detection Methods

1) **Probe design and screening**

The main challenge in developing hybridization-based LC-FL methods is the design and screening of probes, which is critical to the sensitivity and specificity of the assay, and the probes need to be progressively optimized and screened during method development. The design and screening of PNA probes can be a critical factor in determining the cost and time consumption of LC-FL methods. The PNA probe must be complementarily paired with the target oligonucleotide, with a sequence length of 12–21 bases. Because PNA/PNA interactions are stronger and have a greater affinity than PNA/RNA interactions, they can affect the formation of fluorescently labeled complexes. Therefore, any self-complementary sequences, such as inverted repeats, hairpins, and palindromic sequences, should be avoided during probe design. Purine-rich sequences should also be avoided because the proportion of purines affects the solubility of PNA probes in aqueous solutions. Therefore, the proportion of purines should be limited as much as possible, and purine extensions should be avoided. Solubility enhancers, such as O linkers, E linkers, X linkers, or amino acids, can also be added to improve the solubility of PNA probes.

2) **Quantitative analysis of metabolites**

Since the LC-FL method isolates the target oligonucleotides and their truncated metabolites (e.g., *n*−1, *n*−2) by ion-exchange chromatography, metabolite standards are required to confirm the retention times of the metabolites. For double-stranded siRNAs, the metabolites of the SS and antisense strand (AS) in the siRNA can only be detected separately since the PNA probe can only detect one strand at a time. In addition, each P=S modification introduces a chiral center, making chromatographic separation of PS-modified oligonucleotides and their metabolites more difficult [5].

The column temperature and mobile-phase conditions are also essential for the sensitivity and the specificity of the method. Therefore, it is necessary to specifically optimize each specific probe and/or target oligonucleotide. LC-FL methods are often used to support late-stage preclinical or clinical studies, whereas LC-tandem MS (MS/MS) or LC-HRAM-MS assays are more commonly used in drug discovery or early development stages.

5.3.3.4 Case Study of the Quantitative Analysis of GalNAc–siRNA Conjugates Using an LC-FL Method

The hybridization-based LC-FL method has high sensitivity, accuracy, and precision in quantitative assays. We developed a hybridization-based LC-FL method to quantify GalNAc–siRNA conjugates and validated the method for specificity, selectivity, sensitivity, linear range, and accuracy. This method used an ATTO425-labeled PNA fluorescent probe to complementarily hybridize to the AS in the GalNAc–siRNA duplex to detect the concentration of the AS. The concentration of GalNAc–siRNA was reported based on the concentration of the AS.

- **Calibration curve and quantitative range**

Figure 5.11 shows the quantitative performance of spiked rat plasma samples after sample treatment. The lower limit of quantitation (LLOQ) was as low as 1 ng/mL, with a linear range between 1 and 200 ng/mL (Figure 5.12).

- **Accuracy and precision**

The within-batch accuracy of the quality control samples ranged from 87.1% to 119.7%, and the between-batch accuracy ranged from 97.4% to 103.2%. The within-batch precision (% relative standard deviation [RSD]) ranged from 2.1% to 20.0%, and the between-batch precision (% RSD) ranged from 7.7% to 19.5%. These data met the requirements for accuracy (±20%) and precision (±20%) dictated by regulatory agencies.

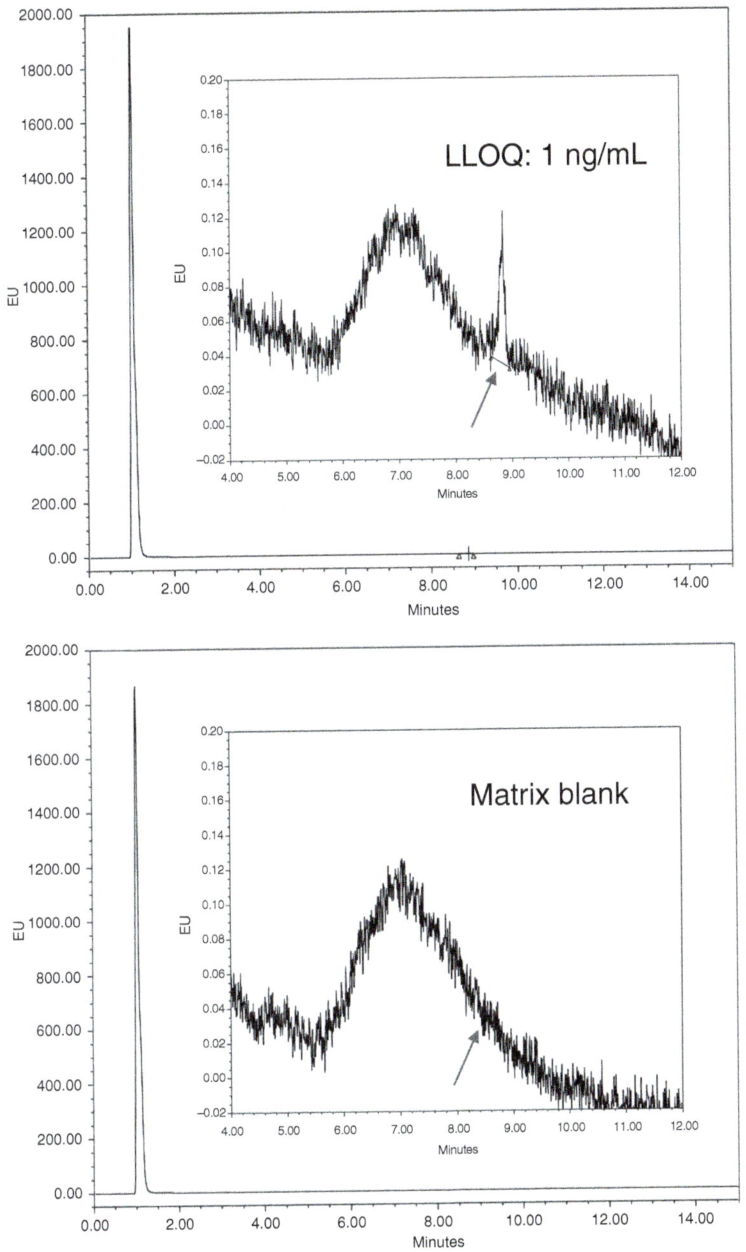

Figure 5.11 Representative chromatograms of the lower limit of quantitation (LLOQ; 1 ng/mL) and control blank samples.

- **Separation of metabolites**

Due to the very limited volume of the biological samples, the simultaneous separation of the oligonucleotides and their antisense chain metabolites (3′*n*−1 and 3′*n*−2) was conducted using ion-exchange high-performance LC and fluorescence detection. To ensure detection sensitivity, the separation achieved excellent results. Figure 5.13 shows the separation of the tested oligonucleotides and their antisense chain metabolites, 3′*n*−1 and 3′*n*−2, by ion-exchange high-performance LC.

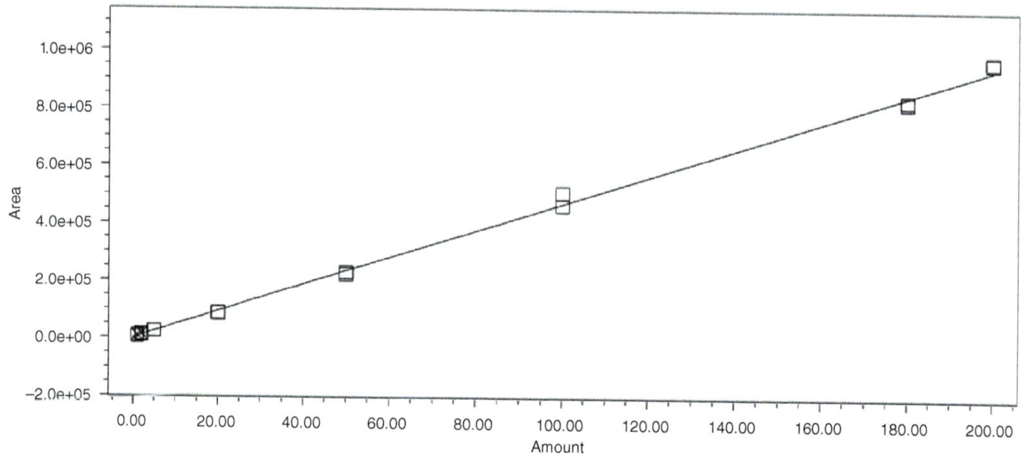

Figure 5.12 Calibration curve of GalNAc–siRNA hybridized with a PNA fluorescent probe in the rat plasma (1–200 ng/mL; linear $1/x^2$, $R^2 = 0.99$).

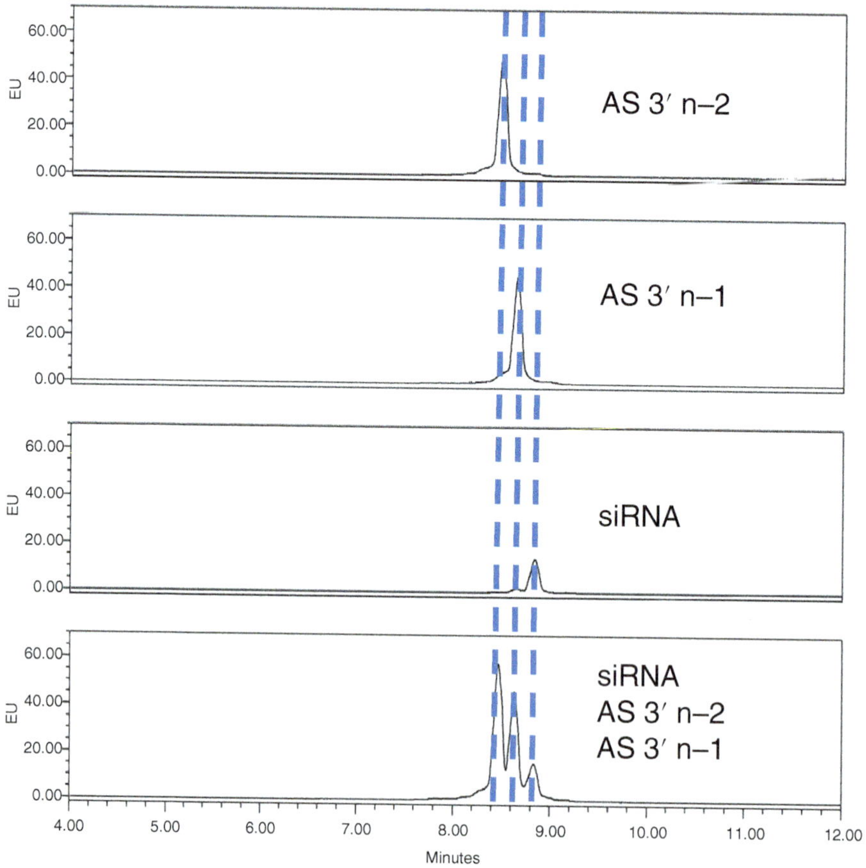

Figure 5.13 Separation of GalNAc–siRNA and its metabolites using ion-exchange high-performance liquid chromatography.

5.3.4 Application of LBA Detection Methods for Quantitative Bioanalysis of Oligonucleotides

5.3.4.1 Principle of Ligand Binding Assays (LBAs)

LBA, which is primarily referred to as HELISA and MSD ECL, is used to detect oligonucleotides. HELISA is an enzyme-linked immunosorbent assay (ELISA) based on complementary hybridization that directly or indirectly hybridizes enzyme-labeled nucleic acid probes with target oligonucleotides adsorbed on a solid carrier. After the addition of a substrate solution, the substrate's hydrogen donor is transformed from a colorless reduced form to a colored oxidized form under the enzyme's action, resulting in a colored substance. By qualitatively or quantitatively detecting the colored product, the concentration of the target oligonucleotide in the sample can be determined. Similar to the HELISA workflow, the MSD®ECL method uses electrochemiluminescent labels to mark the nucleic acid probe and quantitatively measures the target oligonucleotides in the sample by detecting the intensity of the emitted light during electrochemical reactions. Based on the analytical principles, LBA can be categorized into one-step, two-step, sandwich, double-bind, and competitive methods (Figure 5.14).

- **One-step method**

The one-step method involves the use of a single-stranded nucleic acid probe to form a double-stranded body through complementary hybridization with the target oligonucleotide. One end of the nucleic acid probe is labeled with biotin and affixed to a streptavidin plate, whereas the other end is labeled with a detection tag. The one-step method quantifies double-stranded structures labeled with the detection tag using S1 nuclease to remove unhybridized single-stranded probes. The one-step method is unable to differentiate between full-length oligonucleotides and metabolites generated through enzymatic degradation. This susceptibility to cross-hybridization can result in an overestimation of the concentration of the target oligonucleotide.

- **Two-step method**

This two-step method involves the use of capture and detection probes. The capture probe comprises one nucleotide sequence complementary to the target oligonucleotide and an additional nucleotide sequence complementary to the detection probe. One end of the capture probe was labeled with biotin and affixed to a streptavidin plate. The target oligonucleotide complementarily hybridizes with the capture probe and binds to the detection probe under the action of T4 DNA ligase. Simultaneously, the detection probe is complementary to the capture probe, followed by the addition of a substrate and detection of the reaction signal. The two-step method is not affected by nucleic acid metabolite degradation at the 3′ end; however, it cannot distinguish between full-length oligonucleotides and nucleic acid metabolites degraded at the 5′ end.

- **Sandwich method**

The sandwich method also requires capture and detection probes. The capture probe is labeled with biotin at one end and immobilized on a streptavidin plate, whereas the detection probe is labeled with a detection label. The capture and detection probes complementarily hybridize with sequences on the target oligonucleotide, forming a double-stranded structure for quantitative detection. Similar to the one-step method, the sandwich method is also affected by metabolites, leading to an overestimation of the concentration of the target oligonucleotide.

- **Double-bind method**

The double-bind method, employing capture, detection, and template probes, exhibits good specificity and remains unaffected by metabolites of nucleic acid degradation. The capture probe is

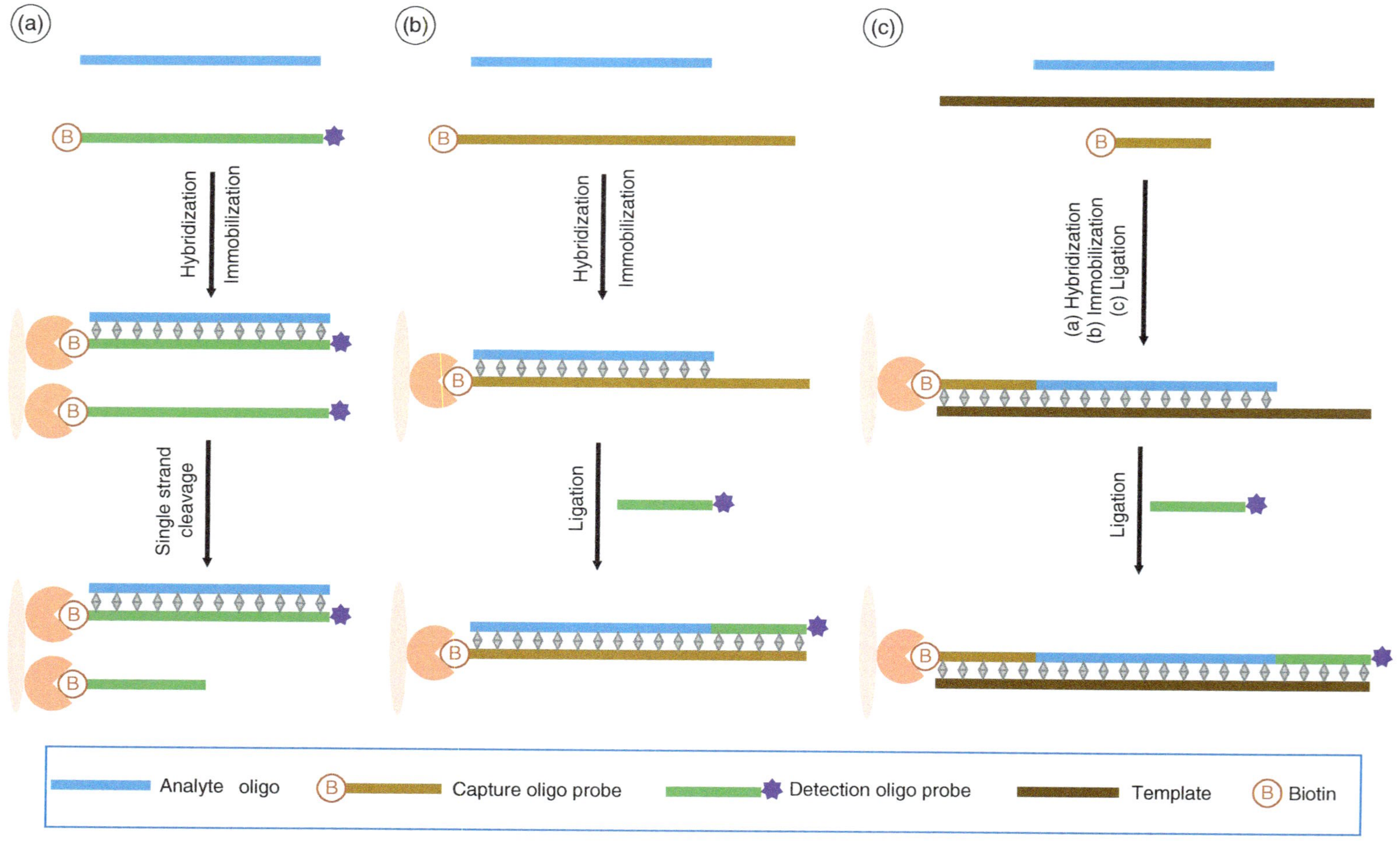

Figure 5.14 Principles of ELISA: (a) One-step hybridization ELISA, (b) two-step hybridization ELISA and (c) dual ligation ELISA. *Source:* [6] / with permission of Taylor & Francis Group.

labeled with biotin at one end and immobilized on a streptavidin-coated plate, whereas the detection probe is labeled with a detection tag at one end. The template probe can simultaneously hybridize with the capture probe, detection probe, and target oligonucleotide, forming an intact double-stranded structure in the presence of T4 DNA ligase. Subsequently, a substrate solution was added to detect the reaction signal.

- **Competitive method**

The competitive method involves the use of a capture probe with the same sequence as that of the target oligonucleotide and a detection probe that is complementary to the target oligonucleotide. The capture probe is labeled with biotin at one end and immobilized on a streptavidin plate, while the detection probe is labeled with a detection tag at the other end. The target oligonucleotide competes with the capture probe for complementary hybridization with the detection probe. By quantifying the interference of the target oligonucleotide on the expected signal, the concentration of the oligonucleotide in the sample can be detected. However, this method cannot distinguish between full-length oligonucleotides and metabolites of nucleolytic degradation.

When selecting an analytical method, in addition to sensitivity, the selectivity and specificity of the method should be considered. If there is potential interference from metabolic products, the two-step or double-bind method can be chosen.

5.3.4.2 Advantages of Ligand Binding Assays (LBAs)

- **High sensitivity of the LBA**

Resembling LC-FL methods based on hybridization principles, HELISA easily offers low detection limits of 0.5 ng/mL and even lower limits of detection at 0.05 ng/mL with MSD®ECL. This makes it suitable for preclinical, late-phase, and clinical PK studies of oligonucleotides and the detection of biological samples in low-dose groups or long experimental periods. MSD®ECL utilizes SULFO-TAG™ as the chemiluminescent substrate. SULFO-TAG™, which has a relatively small molecular weight, has minimal steric hindrance, which makes it easy to label antibodies without interfering with their capacity to bind to the analyte. SULFO-TAG™ generates highly efficient, stable, and continuous electrochemiluminescent signals, thereby substantially improving the detection sensitivity. MSD®ECL can simultaneously detect oligonucleotides at both high and low abundance levels, with a dynamic range spanning 4–5 orders of magnitude.

- **High-throughput LBA**

Compared to chromatographic methods, the HELISA method hardly requires sample pretreatment or chromatography-related linear injection methods, thus significantly reducing the detection time. Moreover, the use of automated equipment such as TECAN (Männedorf, Switzerland) and Hamilton (Nevada, USA) liquid handling workstations, automated plate washers, and HP D300e digital dispensers could reduce handling time and manual pipetting errors and increase sample throughput and the success rate of sample analysis.

- **Independence from chemical modifications and delivery systems**

The HELISA method demonstrated better tolerance to chemical modifications of oligonucleotides than the qPCR methods, with its detection sensitivity remaining unaffected. This method can be used to detect biological samples of GalNAc-conjugated oligonucleotides, LNP-delivered oligonucleotides, and other delivery system-based oligonucleotides in preclinical and clinical research. Additionally, the MSD®ECL method reduces sample consumption and enables multiplex detection, providing more data for studying rare samples.

5.3.4.3 Challenges of Ligand Binding Assays (LBAs)

- **Probe design**

The capture and detection design of probes should essentially ensure proper affinity for the oligonucleotide analyte and minimize potential self-annealing. During the method development stage, screening of capture and detection probes is necessary. Locked nucleic acid (LNA) is a novel nucleic acid analog that contains a 2'-O,4'-C methylene bridge (Figure 5.15). This bridging moiety is locked in the 3′-endo conformation, restricting the flexibility of the furanose ring and locking the structure into a rigid bicyclic form. LNA exhibits a high affinity toward complementary DNA (cDNA) or RNA, which enhances the thermal stability of the hybridized double-stranded structure and the specificity of probe-target sequence hybridization [7]. The MSD®ECL method using LNA capture and detection probes has been used to measure short interfering RNA (siRNA) in the serum and tissue homogenates of preclinical biological samples [8].

Additionally, optimizing the design of probes also contributes to improving the selectivity of the HELISA method. For example, if 3′- or 5′-end metabolites are expected, optimizing the length of the capture and detection probe sequences can reduce hybridization between the metabolites and probes, thereby enhancing the selectivity of the method.

- **Metabolite**

Compared to LC-MS methods, HELISA lacks specificity for the target oligonucleotide in biological samples, making it difficult to distinguish between the full-length oligonucleotide and long-chain metabolites such as $n-1$ and $n-2$. This could lead to cross-hybridization and overestimation of the concentration of the target oligonucleotide. Therefore, it is necessary to prioritize the use of LC-MS methods for metabolic stability studies and develop probes and methods specifically targeting oligonucleotide analytes and major metabolites for preclinical and clinical research purposes. In addition, it can be used to synthesize potential metabolites and evaluate their quantitative impact on test oligonucleotides during method development and validation. Additionally, selecting an appropriate HELISA method can reduce interference from metabolites on oligonucleotide analytes. As previously mentioned, two-step and double-bind methods, which introduce T4 DNA ligase into ASSAY, can further enhance the specificity of the HELISA method.

5.3.4.4 Case Study of the Quantitative Analysis of GalNAc–siRNA-Conjugated Oligonucleotides Using LBA Methods

Two bioanalytical methods have been developed, based on HELISA and MSD®ECL, for quantitatively analyzing GalNAc–siRNA-conjugated oligonucleotides in monkey serum samples. These methods have been methodologically validated to assess their specificity, selectivity, sensitivity, linear range, and accuracy. The method uses a sandwich method with a biotin-labeled LNA capture probe and a detection-labeled LNA detection probe to hybridize with the AS of the GalNAc–siRNA duplex. The concentration of AS was measured, and the concentration of GalNAc–siRNA

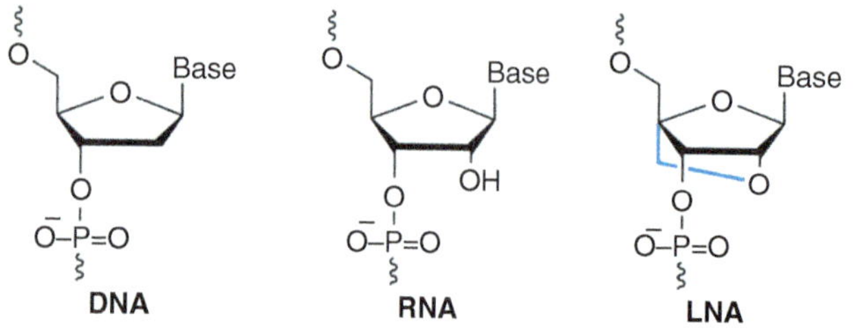

Figure 5.15 Structural comparison of DNA, RNA, and LNA nucleotides. *Source:* [9]/Microsynth.

was reported based on the AS concentration. The sensitivities of the HELISA and MSD®ECL methods were 0.5 and 0.05 ng/mL, respectively.

- **Standard curves and quantification range**

Figure 5.16 illustrates the quantitative performance of the HELISA method in analyzing spiked monkey serum samples after sample processing, with a lower limit of quantification (LLOQ) as low as 0.5 ng/mL and a linear range from 0.5 to 50.0 ng/mL.

Figure 5.17 illustrates the quantitative performance of the MSD®ECL method in the analysis of spiked monkey serum samples after sample processing. The LLOQ was as low as 0.05 ng/mL, and the linear range was from 0.05 to 10.0 ng/mL.

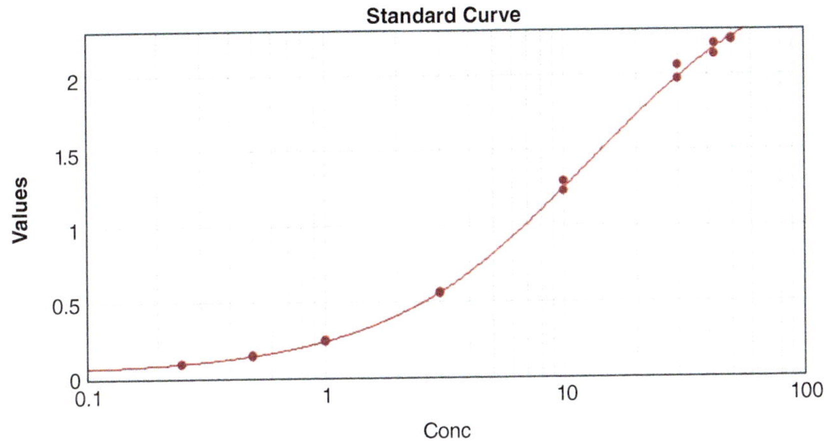

Figure 5.16 Standard curve of GalNAc–siRNA in monkey serum using the HELISA method. (0.5–50.0 ng/mL, curve fit: 4-parameter logistic, weighting factor: 1/values^2, R^2 = 1.000).

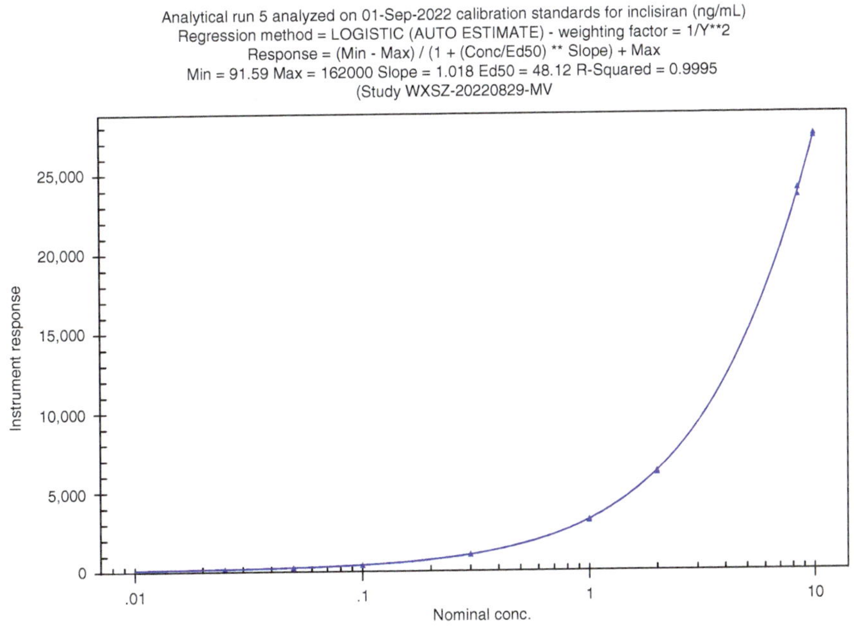

Figure 5.17 Standard curve of GalNAc–siRNA in monkey serum using the MSD® method. (linear range: 0.05–10.0 ng/mL, curve fit: 4-parameter logistic, weighting factor: 1/Y^2, R^2 = 0.9995).

- **Accuracy and precision**

In the HELISA method, the within- and between-batch accuracy of the quality control samples ranged from 84.2% to 112.8% and 98.5% to 101.5%, respectively. The within- and between-batch precision percentages of the %RSD ranged from 1.7% to 15.1% and from 6.7% to 13.7%, respectively. In the MSD®ECL method, the within-batch accuracy of the quality control samples ranged from 92.3% to 103.9%, and the between-batch accuracy ranged from 95.4% to 101.6%. The within-batch precision %RSD ranged from 0.9% to 7.0%, while the between-batch precision %RSD ranged from 2.5% to 6.0%. These data meet the regulatory requirements for CV% (±20%) and bias% (±20%).

- **Interference detection of metabolites**

During the method development process, the potential interference of two metabolites (AS 3′*n*−1 and AS 3′*n*−2) and the SS of GalNAc–siRNA double-stranded structures on the MSD®ECL sandwich method was evaluated. Monkey serum samples were prepared with high and low concentrations of GalNAc–siRNA, AS 3′*n*−1, AS 3′*n*−2, and SS and detected using the MSD®ECL sandwich method. Compared to those of GalNAc–siRNA, the two potential metabolites (AS 3′*n*−1 and AS 3′*n*−2) showed a significant signal response during detection, which suggests a lower specificity for GalNAc–siRNA in monkey serum samples. To address this issue, it is recommended to prioritize metabolic stability studies using LC-MS methods before sample analysis. The results demonstrated that the quantification of AS in GalNAc–siRNA double-stranded structures using the MSD sandwich method is not affected by the presence of the SS.

Figure 5.18 shows the signal response of the two potential metabolites (AS 3′*n*−1 and AS 3′*n*−2) and the SS chain detected in monkey serum using the MSD®ECL method.

5.3.5 Application of RT-qPCR Detection Methods for Quantitative Bioanalysis of Oligonucleotides

5.3.5.1 Principles of Oligonucleotide Quantitative Analysis Using RT-qPCR

qPCR refers to the introduction of fluorophore-labeled probes or nucleic acid dyes in PCR, facilitating real-time analysis of the fluorescence signal at each reaction moment. Furthermore, the

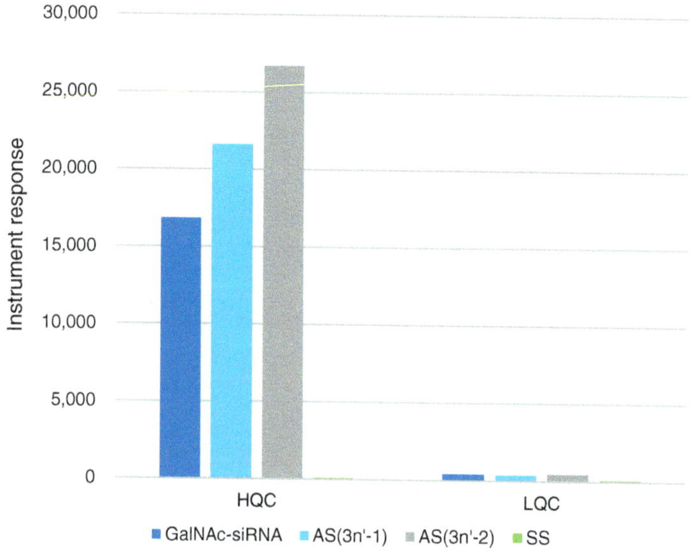

Figure 5.18 Detection of GalNAc–siRNA, AS 3′*n*−1, AS 3′*n*−2, and SS in monkey serum using the MSD® ECL method.

fluorescence signal intensity is proportional to the number of PCR products, enabling the calculation of the amount of initial template and thus facilitating the quantitative detection of nucleic acids. qPCR is recognized as one of the standard methods for quantifying nucleic acids harboring known sequences. RT-qPCR is usually used for the quantitative analysis of RNA. This method involves converting the RNA template to cDNA using reverse transcriptase, followed by the amplification of the cDNA via PCR. In addition, some qPCR technologies, such as ligase-based qPCR, do not require reverse transcription.

Oligonucleotides typically consist of 19–25 bases, which are too short for designing primers and probes using conventional methods to amplify nucleotide sequences. Therefore, it is necessary to extend the length of the test article and construct a long PCR template for further quantitative analysis using PCR technology [10]. To address this challenge, several methods for lengthening nucleic acid sequences have been developed, including stem-loop RT-qPCR, poly(A) polymerase tailing-based RT-qPCR, ligation-based RT-qPCR, and primer-extension qPCR [11]. Among these methods, stem-loop RT-qPCR and poly(A) polymerase tailing-based RT-qPCR are more commonly used [12].

- **Quantitative analysis of oligonucleotides using stem-loop RT-qPCR**

Chen et al. [13] first described stem-loop reverse transcription-qPCR (SL-RT-qPCR) in 2005. Stem-loop RT-qPCR differs from conventional RT-qPCR because it involves designing a special reverse transcription primer to directly extend the template in the reverse transcription reaction. Figure 5.19 shows that this reverse transcription primer is composed of a "stem-loop" structure and a sequence that is complementary to the target oligonucleotide. The reverse transcription primer extends the template chain by pairing with 3′ ~6 nt of target oligonucleotides and synthesizing cDNA under the action of reverse transcriptase. Subsequently, forward and reverse primers and Taqman probes can be designed for the extension of cDNA templates for amplification and detection [12]. The unique structure of stem-loop primers increases the stability of reverse transcription primers and their target oligonucleotides, thereby reducing nonspecific binding to endogenous RNA or DNA. A specially designed 6–8 specific base pair can also increase sequence specificity and reduce background noise signals. This method can achieve a wide linear range spanning 7–8 orders of magnitude, with the lower limit of quantification (LLOQ) being as low as ~2 pg/mL.

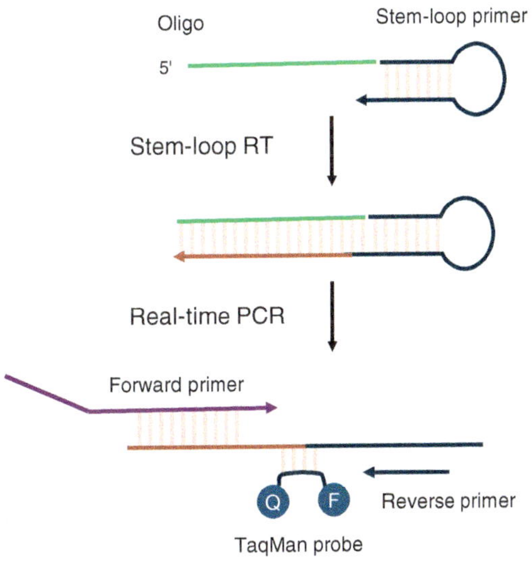

Figure 5.19 Principle of stem-loop qPCR. *Source:* [14]/Oxford University Press.

- **Quantitative analysis of oligonucleotides using poly(A) polymerase tailing RT-qPCR**

Poly(A) polymerase tailing refers to the addition of an oligoadenylate tail to the 3′ end of target oligonucleotides under the action of poly(A) polymerase (Figure 5.20), followed by reverse transcription of cDNA with a universal reverse transcription primer to extend the template sequence. This reverse transcription primer comprises three essential structures: (i) a poly(T) sequence complementarily paired with the poly(A) sequence; (ii) a universal sequence at the 5′ end for binding of the reverse primer to the cDNA during PCR; and (iii) one or two anchor bases at the 3′ end, allowing the reverse transcription primer to specifically bind to the segment of the poly(A) sequence adjacent to target oligonucleotides. Subsequently, PCR amplification can be performed using a specific forward and universal reverse primer, and quantitative detection can be achieved using SYBR Green. The greatest strength of this method is its low cost because it is relatively easy to perform and does not require the design of special reverse transcription primers. Additionally, multiple target sequences can be reverse transcribed simultaneously in a single reaction. However, the disadvantage of this method is its poor specificity, which may decrease its accuracy.

- **Quantitative analysis of oligonucleotides using primer-extension qPCR**

Primer-extension qPCR involves a primer that contains a 3′-end gene-specific sequence and a universal 5′-end sequence. It can be used to specifically convert a target oligonucleotide into cDNA. Subsequently, amplification and detection were performed using paired universal forward and specific reverse primers with the SYBR Green dye method. The key to this method is an LNA-modified reverse primer of approximately 15 nt. Two to three of the first eight bases at the 5′ end of this reverse primer were replaced with specially modified nucleic acid analogs. LNA nucleotides contain a methylene bridge that connects the 2′-oxygen of ribose with the 4′-carbon. This bridge

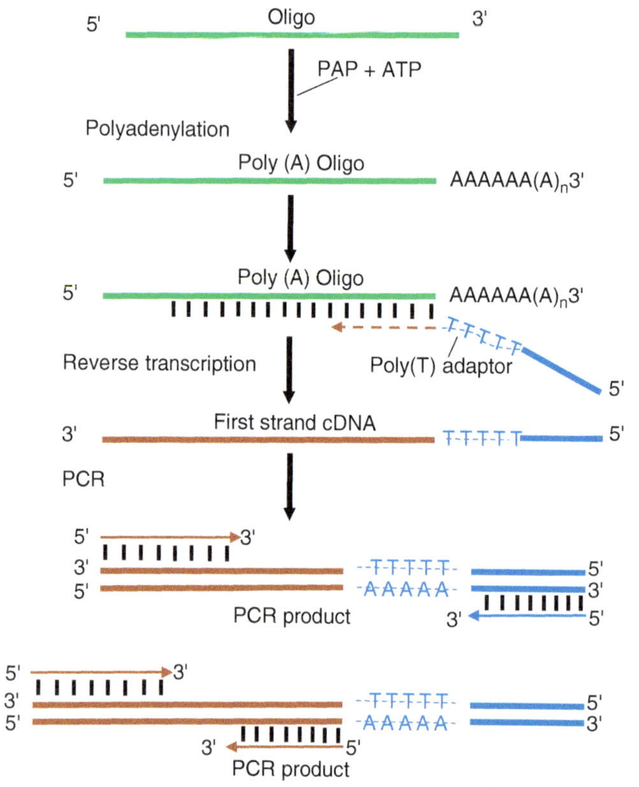

Figure 5.20 Principle of PolyA polymerase tailing.

results in a locked 3′-endo conformation (Figure 5.21), reducing the conformational flexibility of the ribose and increasing the local organization of the phosphate backbone without affecting its adherence to the Watson–Crick base-pairing principle; LNA modification enhances binding affinity, sequence specificity, thermal stability, and nuclease resistance. This method shows a high dynamic range of 6–7 orders of magnitude with sensitivity in the FM range. However, the results may be affected by nonspecific binding via the use of SYBR Green-based qPCR.

- **Quantitative analysis of oligonucleotides using ligation-based qPCR**

The method preserves the qPCR process but changes the synthesis process of cDNA. Instead of using reverse transcription, cDNA can be synthesized using DNA ligases, such as T4 DNA ligase or SplintR ligase, which can fill gaps and construct complete cDNA without reverse transcription [12]. This method requires two DNA probes with a target oligonucleotide recognition sequence and a pair of forward and reverse PCR primers. In the presence of the RNA sequence (Figure 5.22), the two probes hybridize to the RNA sequence through base pairing, which is followed by the formation

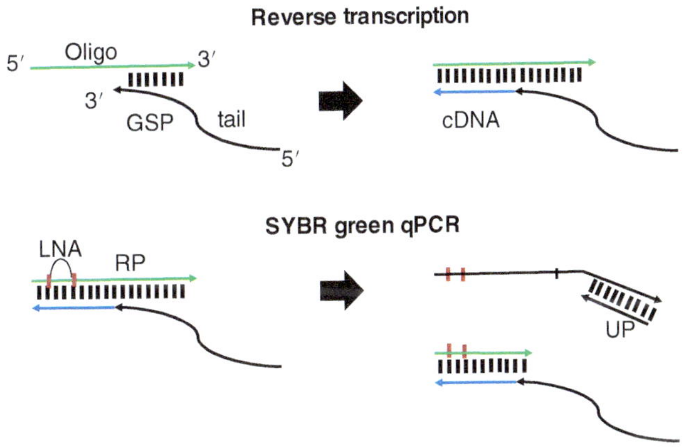

Figure 5.21 Principle of primer extension. Adapted from reference [10].

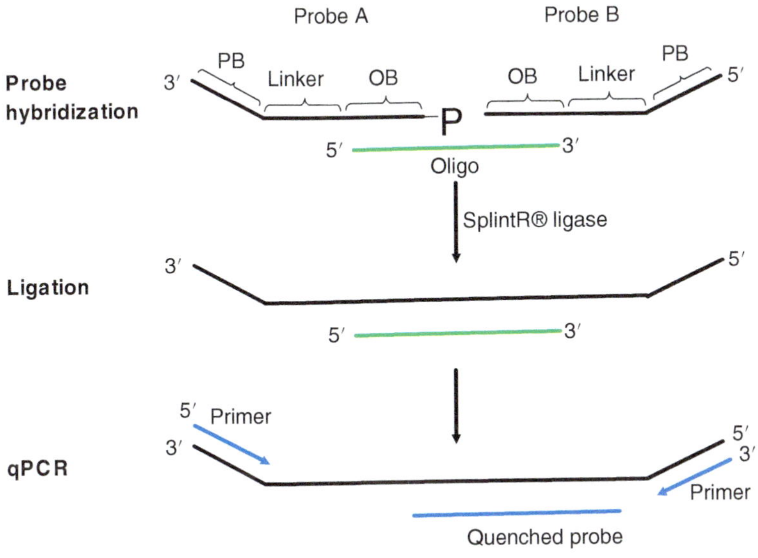

Figure 5.22 Principle of the DNA probe binding method.

of a complete, extended-length cDNA template for subsequent PCR quantification using DNA ligase activity. However, DNA ligases may nonspecifically bind two DNA probes to form a nonspecific amplification signal. Additionally, the efficiency of DNA ligase-linked DNA probes using RNA as a template is relatively low, which may affect specificity and sensitivity.

5.3.5.2 Advantages of RT-qPCR for Oligonucleotide Quantitative Analysis

- **Ultrahigh detection sensitivity and wide dynamic linear range**

RT-qPCR is a molecular biology technique for amplifying specific nucleic acid fragments. It can amplify low-copy-number targets millions of times in a short period, thereby providing ultra-high sensitivity. In the detection of oligonucleotide drugs, the LLOQ can reach the pg/mL level, which is much lower than the ng/mL level of MS detection. Oligonucleotides exhibit high efficiency and long duration, with rapid elimination in plasma and long half-lives in target tissues. For the detection of plasma samples in low-dose groups or long experimental periods, a highly sensitive detection method is necessary. RT-qPCR offers a wide dynamic linear range for quantification, typically over 7–8 orders of magnitude, facilitating the bioanalysis of high- and low-abundance samples.

Key indicators of oligonucleotide pharmacodynamics (e.g., changes in gene silencing effects/associated protein levels in target cells) correlate with exposure to siRNAs conjugated with RISC complexes. However, the concentrations of RISC-loaded siRNAs are usually extremely low, and ultrasensitive detection methods are needed (detection limit <50 pg/mL). The ultrahigh sensitivity of RT-qPCR enables the detection of RISC-loaded siRNA, providing information for the intracellular delivery of siRNA and the mechanism of RNA silencing [15]. Moreover, PD/PK models for key pharmacodynamic indicators and RISC-loaded siRNAs could be established to help design and optimize drugs and predict effective human doses.

- **Enabling high-throughput sample analysis and reducing the use of biological samples**

When preparing an RT-qPCR sample, especially when configuring 96- or 384-well plates, the application of high-throughput and high-precision automated liquid handling systems to construct experimental systems and reaction systems allows the avoidance of inhomogeneity caused by manual fatigue during operation in a high-pressure environment, the reduction of experimental errors, and the improvement of sample processing efficiency. Additionally, RT-PCR requires only approximately 10 μL of biological sample, thereby reducing the use of biological samples compared to other assays and facilitating multiple analyses of each sample.

5.3.5.3 Challenges and Solutions in Using RT-qPCR for Oligonucleotide Quantification

- **Primer and probe design**

One difficulty in the development of RT-qPCR methods for oligonucleotide detection lies in primer and probe design. The quality of primers and probes significantly affects assay sensitivity. In terms of SL-RT-qPCR, although universal reverse transcription primers, reverse primers for PCR, and probe sequences are available, forward primers need to be designed specifically based on different oligonucleotide sequences. Short native template sequences limit the selection and length of primers, making it difficult to meet the T_m-value requirements. In this case, a segment of the G or C sequence can be added to the 5′ end to bring the Tm value close to 60°C. Furthermore, it is necessary to prevent the formation of primer–primer and primer–probe dimers, which may lead to nonspecific amplification or abnormal results. In addition, it is important to ensure that the Tm values of the primer pairs are as close as possible, and that the Tm value of the probe is 5–10°C higher than that of the primer. If generic primers and probes cannot meet these requirements, it is vital to design primers and probes based on the template sequence. In addition, due to the long probe synthesis cycle, two to three primer–probe sets need to be designed for screening during the method development phase, leading to increased costs.

- **Harsh operating environment and consumable requirements**

Since RT-qPCR assays are very sensitive and can detect only a few copies of the target sequence, they are strongly influenced by the cleanliness of the environment and consumables. Therefore, it is necessary to strictly avoid aerosol contamination and interference from nucleases during the experimental process.

- **Metabolites may affect method accuracy**

Although qPCR is sensitive, it lacks perfect specificity for oligonucleotide analytes in biological samples and cannot distinguish full-length oligonucleotides from long-chain metabolites, such as n–1 and n–2. This shortcoming may lead to nonspecific amplification owing to the binding of reverse transcription primers to metabolites and the overestimation of the concentration of the target oligonucleotides, thus affecting the accuracy.

- **Chemical modifications or carriers may affect sensitivity and accuracy**

Studies have shown that several chemically modified sequences, including glycol nucleic acid (GNA)-modified siRNA, reduce the detection sensitivity of RT-qPCR. This is possible because such acyclic hydrocarbon bases with three carbon atoms are not the natural substrates for reverse transcriptase, leading to a decrease in reverse transcription efficiency. Therefore, unique optimization steps, such as changing the ion concentration in the reaction system, are necessary for this type of oligonucleotide.

- **Lack of guidance from regulatory agencies**

Validation of qPCR/RT-qPCR assays is not currently required by the FDA or NMPA, and there are no uniform standards for experimental design, evaluation processes, or acceptance criteria for sample analysis. Therefore, the acceptance criteria for qPCR/RT-qPCR method validation and sample analysis in clinical and nonclinical studies will vary depending on the scientific understanding of the bioanalytical scientist.

5.3.5.4 Case Study of the Quantitative Analysis of GalNAc–siRNA Using SL-RT-qPCR

The author developed an SL-RT-qPCR method with ultra-high sensitivity for GalNAc–siRNA quantification in rat plasma and conducted methodological validations of specificity, selectivity, sensitivity, linear range, accuracy, and precision. The method uses a specific stem-loop primer to complementarily hybridize to the AS in the double strand of GalNAc–siRNA. Subsequently, the extended cDNA was obtained via reverse transcription followed by a TaqMan probe-based qPCR to determine the concentration of the AS. The concentration of GalNAc–siRNA will be determined based on the AS concentration. This suggests that the method displays good accuracy and precision (Figures 5.23 and 5.24).

- **Standard curve and quantification range**

The method exhibits excellent sensitivity and a wide dynamic range, with an LLOQ as low as 0.01 ng/mL and a linear range from 0.01 to 50 ng/mL. The amplification efficiency of the method ranged from 90% to 110%.

- **Accuracy and precision**

The ranges of within- and between-batch accuracy for the quality control (QC) samples were 78.2–115.2% and 91.8–100.7%, respectively. The ranges of within- and between-batch precision (%CV) were 2.7–15.0% and 11.6–12.8%, respectively.

5.3.6 Summary

Qualitative and quantitative analysis of oligonucleotide drugs and their metabolites is essential for drug development and evaluation, and robust bioanalytical methods are critical for PK and toxicological studies. Oligonucleotide molecular weights fall in the range of 7,000–15,000 Da between those of large and small molecules. The reported analytical techniques for oligonucleotide analysis

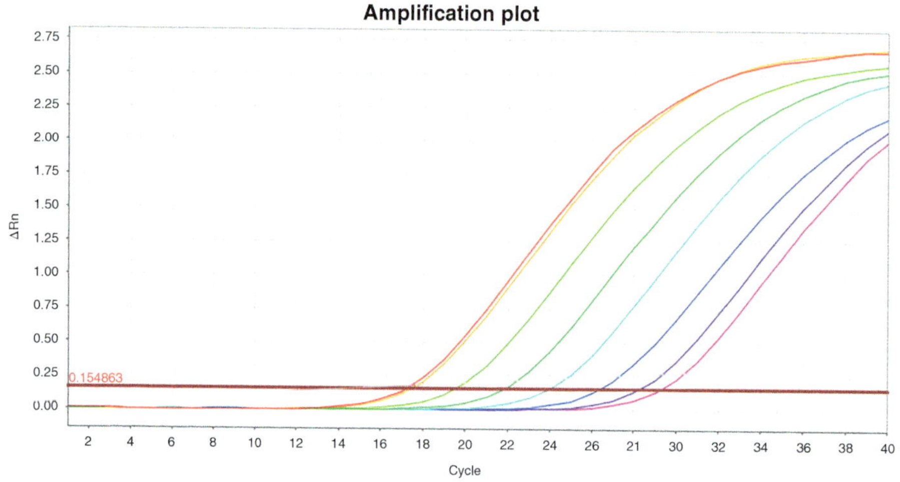

Figure 5.23 Amplification curves of GalNAc–siRNA in rat plasma samples.

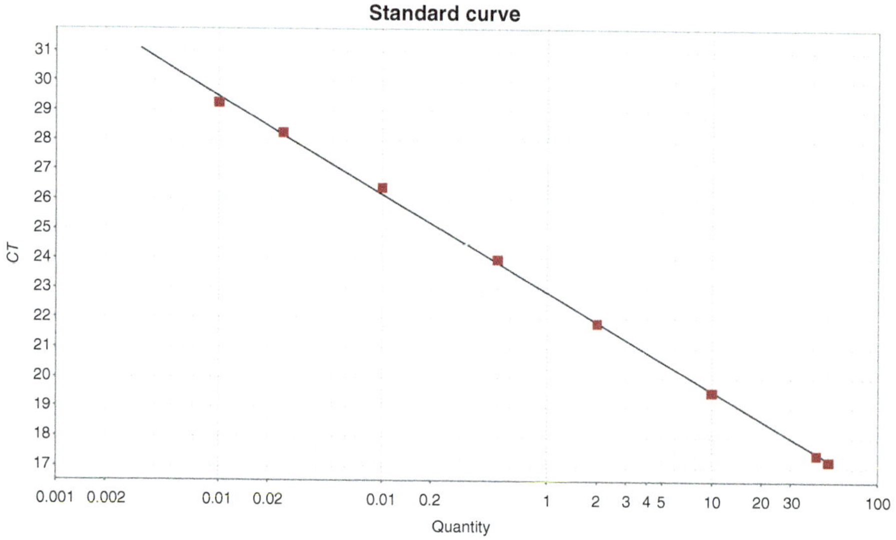

Figure 5.24 Calibration curve of GalNAc–siRNA in rat plasma (0.01–50 ng/mL; R^2 = 0.999).

can be broadly categorized into five types (Figure 5.25), relying on charge, mass, hybridization, imaging, or amplification/extension for quantitative analysis. Commonly used bioanalytical platforms include LC-MS/MS, LC-HRMS, hybridization-based LC-FL, LBA, and qPCR. The inherent polyanionic nature of oligonucleotides greatly affects the sensitivity of their bioanalytical measurements, which are dependent on the length and chemical modification of the oligonucleotide molecule. In general, increasing oligonucleotide length/size decreases the detection sensitivity of LC-MS/MS and LC-HRMS methods but increases the sensitivity of hybridization-based analysis. Additionally, individual variations in each oligonucleotide drug and the sensitivity of the method need to be considered, as well as the different stages of drug development.

Here, we compare different bioanalytical methods to help choose the right method for a specific purpose at the right time (Table 5.2).

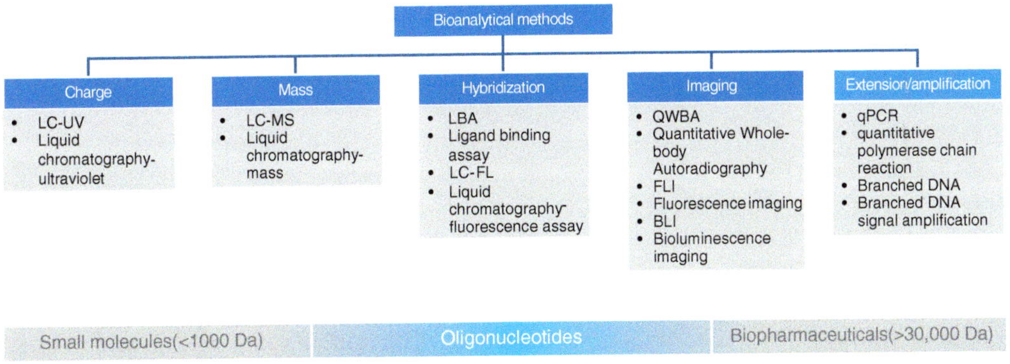

Figure 5.25 Bioanalytical methods for oligonucleotides.

Table 5.2 Comparison of four bioanalytical methods.

Method	Mass spectrometry	Fluorescence detector	LBA	RT-qPCR
Instrument	LC-MS	LC-FL	Plate reader; MSD for hybridized MSD	Stem loop reverse transcriptase PCR
Oligonucleotide detection size and tolerance of chemical modifications	Analyzes small oligonucleotides <25 mer, tolerates chemical modifications	Analyzes small oligonucleotides (20–40 mer), tolerates chemical modifications	Analyzes relatively large oligonucleotides (20–40 mer), tolerates chemical modifications	Analyzes >25 mer oligonucleotides, impacted by chemical modifications
Sample pretreatment	Extraction needed	Method specific	Method specific	Nucleic acid extraction needed
Sensitivity (LLOQ*)	Acceptable; study specific (LLOQ: ~5–10 ng/mL)	Medium; study specific (LLOQ: ~1–5 ng/mL)	High (LLOQ: ~0.1–1 ng/mL)	Highest (LLOQ: pg-fg/mL)
Specificity	Highly specific	Endogenous interference potential metabolites	Endogenous interference	Endogenous interference, potential contamination
Reagent	Common reagents and consumables	Probe customization	Probe customization	Primer and Probe customization, two to three enzymes
Throughput	96-well plate	96-well plate	96-well plate	384-well plate
Metabolites identification/ profiling	Detects full-length oligonucleotides and all metabolite	Detects full-length oligonucleotides and truncated metabolites	Not able to detect truncated metabolites	Not able to detect truncated metabolites
Time spend for MD and SA	Long	Long	Long	Short
Regulatory guidance on method validation	Chromatographic assay	Ligand-binding assays	Chromatographic assay	Not available

References

1 MacNeill, R., Hutchinson, T., Acharya, V. et al. (2019). An oligonucleotide bioanalytical LC-SRM methodology entirely liberated from ion-pairing. *Bioanalysis* 11 (12): 1157–1169.

2 Pellestor, F. and Paulasova, P. (2004). The peptide nucleic acids (PNAs), powerful tools for molecular genetics and cytogenetics. *European Journal of Human Genetics* 12: 694–700.

3 Tian, Q., Rogness, J., Meng, M., and Li, Z. (2017). Quantitative determination of a siRNA (AD00370) in rat plasma using peptide nucleic acid probe and HPLC with fluorescence detection. *Bioanalysis* 9 (11): 861–872.

4 Zhang, X., Goel, V., and Robbie, G.J. (2020). Pharmacokinetics of Patisiran, the first approved RNA interference therapy in patients with hereditary transthyretin-mediated amyloidosis. *Journal of Clinical Pharmacology* 60 (5): 573–585.

5 Ji, Y., Liu, Y., Xia, W. et al. (2019). Importance of probe design for bioanalysis of oligonucleotides using hybridization-based LC-fluorescence assays. *Bioanalysis* 11 (21): 1917–1925.

6 Kotapati, S., Deshpande, M., Jashnani, A. et al. (2021). The role of ligand-binding assay and LC-MS in the bioanalysis of complex protein and oligonucleotide therapeutics. *Bioanalysis* 13 (11): 931–954.

7 Karkare, S. and Bhatnagar, D. (2006). Promising nucleic acid analogs and mimics: characteristic features and applications of PNA, LNA, and morpholino. *Applied Microbiology and Biotechnology* 71: 575–586.

8 Thayer, M.B., Lade, J.M., Doherty, D. et al. (2019). Application of locked nucleic acid oligonucleotides for siRNA preclinical bioanalytics. *Scientific Reports* 9 (1): 3566.

9 https://www.microsynth.com/files/Inhalte/PDFs/Oligosynthesis/Flyer_Oligo_LNA.pdf

10 Raymond, C.K., Roberts, B.S., Garrett-Engele, P. et al. (2005). Simple, quantitative primer-extension PCR assay for direct monitoring of microRNAs and short-interfering RNAs. *RNA* 11: 1737–1744. https://doi.org/10.1261/rna.2148705.

11 Liu, W.L., Stevenson, M., Seymour, L.W. et al. (2009). Quantification of siRNA using competitive qPCR. *Nucleic Acids Research* 37: e4. https://doi.org/10.1093/nar/gkn903.

12 Zhang, J.Y., Li, Z.P., Wang, H. et al. (2011). Ultrasensitive quantification of mature microRNAs by real-time PCR based on ligation of a ribonucleotide-modified DNA probe. *Chemical Communications (Cambridge, England)* 47: 9465–9467. https://doi.org/10.1039/c1 cc13466c.

13 Chen, C., Ridzon, D.A., Broomer, A.J. et al. (2005). Real-time quantification of microRNAs by stem–loop RT–PCR. *Nucleic Acids Research* 33 (20): e179. https://doi.org/10.1093/nar/gni178.

14 Chen, C., Ridzon, D.A., Broomer, A.J. et al. (2005). Real-time quantification of microRNAs by stem-loop RT-PCR. *Nucleic Acids Research* 33 (20): e179.

15 Castellanos-Rizaldos, E., Brown, C.R., Dennin, S. et al. (2020). RT-qPCR methods to support pharmacokinetics and drug mechanism of action to advance development of RNAi therapeutics. *Nucleic Acid Therapeutics* 30 (3): 133–142. https://doi.org/10.1089/nat.2019.0840. Epub 2020 Mar 23. PMID: 32202961.

5.4 Oligonucleotide Drugs: Strategies for Metabolism and Metabolite Profiling and Identification

Gengyao Qin, Weiqun Cao

5.4.1 Introduction

In recent years, oligonucleotides have achieved rapid development in research and development (R&D), production, and commercialization due to their unique advantages in gene expression regulation. A global data search revealed that between 1998 and 2023, 19 oligonucleotide drugs were approved, and hundreds of phase II or III clinical trials are underway. These oligonucleotides target 66 genes and have been approved or are being investigated for 102 different indications across 14 therapeutic areas [1]. Similar to traditional small-molecule drugs, it is crucial to conduct *in vitro* and *in vivo* metabolism studies during the preclinical research stage of oligonucleotide drugs. These studies aid in understanding whether oligonucleotide drugs are capable of reaching therapeutic targets and provide valuable information about metabolic pathways and metabolites. *In vitro* and *in vivo* metabolism studies are an essential part of filing materials for oligonucleotide drugs, providing insights into metabolic behavior and clearance pathways. This chapter focuses on the enzymes involved in the metabolism of oligonucleotide drugs, as well as strategies for conducting *in vitro* and *in vivo* metabolism research. Additionally, this review covers methods for metabolite profiling and identification, aiming to serve as a reference for the development of new oligonucleotide drugs.

5.4.2 Metabolism Studies, Metabolite Profiling, and Identification of Oligonucleotide Drugs

5.4.2.1 Metabolic Enzymes and Metabolic Pathways

Oligonucleotide drugs are metabolized and degraded into shortmers by exonucleases or endonucleases. First-generation ASOs with PS backbone modifications are metabolized into shortmers mainly by cleavage from the 3′ end by exonucleases in the blood. In addition, these ASOs are mainly cleaved from both ends from the outside to the inside by 3′ and 5′ exonucleases in tissues. Second-generation ASOs usually feature a PS backbone in the middle and a nonphosphorothioate backbone containing a 2′-O-modified gapmer at both ends. These compounds are cleaved in the middle by endonucleases in tissues and then further cleaved into small fragments by exonucleases.

The metabolism of the SS and AS of GalNAc-conjugated siRNA differs. The AS is metabolized into shortmers by exonucleases, whereas the SS is primarily metabolized by detachment from GalNAc and the linker, followed by metabolism into shortmers by exonucleases (Figure 5.26) [2].

5.4.2.2 Strategies and Methods for the Metabolism Study of Oligonucleotides

Over the past two decades, the study of the metabolism of approved and investigational oligonucleotide drugs has undergone significant changes. Initially, the focus was on *in vivo* assays conducted on target tissues, plasma, and urine. However, more recent approaches have shifted toward *in vitro* incubation assays using cells, microsomes, S9, plasma, or serum, as well as the analysis of plasma, urine, feces, and target tissues following *in vivo* administration. Scientists continue to explore scientific, rational, and comprehensive approaches to metabolism studies. Extensive research has demonstrated that the metabolic stability of oligonucleotides in the

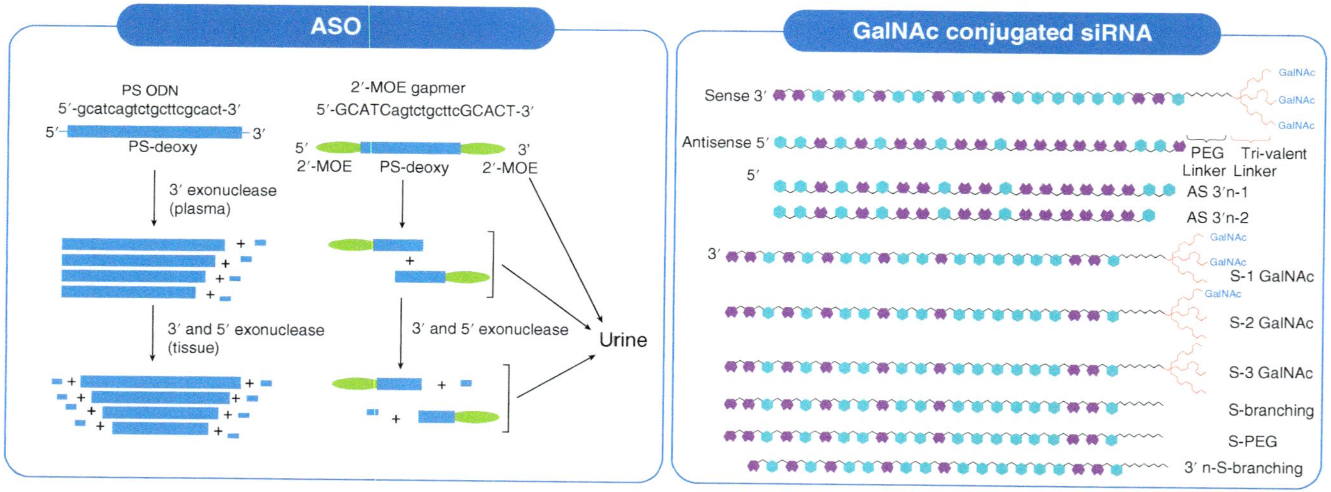

Figure 5.26 Schema of the metabolic pathways of ASOs and GalNAc–siRNAs. *Source:* A Bouchard 2010 (page 6) [3] and Adapted from Basiri 2020 (Figure 1) [2].

presence of hydrolytic enzymes in the blood can be effectively investigated through *in vitro* incubation in plasma or serum. Similarly, the metabolic behavior of oligonucleotides in target tissues *in vivo* can be assessed by *in vitro* incubation of tissue S9 fractions. Furthermore, the use of human *in vitro* models, such as tissue homogenates and subcellular fractions (e.g., S9), has facilitated the development of methods for qualitative and quantitative analysis of clinical *in vivo* metabolites [4].

5.4.2.3 Strategies for *In Vitro* and *In Vivo* Metabolism Studies of Oligonucleotides

In vitro MetID of oligonucleotides in different species can provide insights into species-specific metabolic differences, which can then be compared to *in vivo* animal metabolic data. These comparisons help to establish correlations between *in vitro* and *in vivo* metabolism, enabling the identification of animal species with metabolic behaviors closely resembling those of humans. Common *in vitro* models for studying oligonucleotide metabolism include plasma, S9, and tissue homogenates. For instance, liver S9 can be employed for assessing metabolic stability and identifying metabolites in liver-targeted oligonucleotides. Preclinical *in vivo* metabolism studies primarily focus on MetID and quantitative bioanalysis of plasma, urine, feces, and target tissues (such as liver or kidney) in animals (Table 5.3).

We comprehensively investigated oligonucleotide drug research methods published in the past two decades and verified and evaluated the research systems for oligonucleotide drugs by continuously optimizing various experimental conditions and then established the metabolism research platform of oligonucleotide.

5.4.2.4 Analytical Methods for Oligonucleotide Metabolite Identification Studies

At present, the main bioanalytical methods for identifying oligonucleotides include liquid chromatography-ultraviolet (LC-UV), LC-tandem MS (LC-MS/MS), immunocapture hybridization mesoscale discovery (MSD), ELISA, LC-FL, and quantitative reverse transcription-polymerase chain reaction (qRT-PCR). Among these methods, LC-UV, LC-MS/MS, and LC-FL can be utilized for metabolite profiling. HRMS, which is based on high resolution, sensitivity, specificity, and mass accuracy, is an effective method for the qualitative or quantitative analysis of unknown metabolites using full mass scan and multiple ion scan modes.

The combination of HRMS and LC-UV offers a notable advantage in the analysis and detection of oligonucleotides and their metabolites. In our study on oligonucleotide metabolism, we utilized a dedicated UPLC-UV-HRMS instrument for precise analysis and detection. By following the study

Table 5.3 Strategies for *in vitro* and *in vivo* metabolism studies of oligonucleotides.

	Assays	Biological matrices	Target analytes
In vitro	Metabolic stability	Plasma, S9, and tissue homogenate	Oligonucleotide drugs
	Metabolite identification	Plasma, S9, and tissue homogenate	Oligonucleotide drugs and metabolites
In vivo	Metabolite identification	Plasma, urine, feces, and target tissue	Oligonucleotide drugs and metabolites
	PK/TK	Plasma, urine, feces, and target tissue	Oligonucleotide drugs and high proportion metabolites

PK: pharmacokinetics; TK: toxicokinetics.

Table 5.4 Comparison of oligonucleotide metabolite identification.

WuXi AppTec DMPK		Literature	
Oligo 1	**Oligo 3**	**Oligo 1**	**Oligo 3**
Full-length	Full-length	/	Full-length
3′ *n*−1	3′ *n*−1	3′ *n*−1	3′ *n*−1
3′ *n*−2	3′ *n*−2	3′ *n*−2	3′ *n*−2
3′ *n*−3	3′ *n*−3	3′ *n*−3	3′ *n*−3
3′ *n*−4	3′ *n*−4	3′ *n*−4	3′ *n*−4
3′ *n*−5	3′ *n*−10	3′ *n*−5	
3′ *n*−6	3′ *n*−15	3′ *n*−6	
3′ *n*−7	5′ *n*−1	3′ *n*−7	
3′ *n*−8	5′ *n*−2	3′ *n*−8	
3′ *n*−9	5′ *n*−3	3′ *n*−9	
3′ *n*−10			
3′ *n*−11			
3′ *n*−12			
3′ *n*−13			

conditions outlined in the literature and comparing our results with published findings [5] (Table 5.4), we were able to identify additional low-concentration metabolites. This was made possible through the utilization of our highly sensitive HRMS and refined analytical techniques, surpassing the metabolites reported in the literature.

5.4.3 Challenges and Solutions of Oligonucleotide Analysis

5.4.3.1 Sample Pretreatment

The conventional sample processing methods include protein precipitation, LLE, and SPE. Because of their unique structural characteristics, oligonucleotide drugs have a high protein binding rate. Although the protein precipitation method is simple, its recovery is low with obvious endogenous matrix interference; thus, it is not recommended. LLE, SPE, or a combination of both are generally used to increase recovery and reduce the effect of matrix interference.

- LLE method: Phenol–chloroform–isoamyl alcohol is commonly employed as an extractant to remove endogenous organic substances such as proteins and phospholipids. According to the compound properties, the pH is adjusted to improve oligonucleotide solubility in the aqueous phase, such as by adding ammonia.
- SPE method: This is the most common method for oligonucleotide drug extraction from biological matrices. After the sample is loaded to a fully activated SPE column or plate, the interfering substances, such as proteins, are removed by washing, and the oligonucleotide-related components are subsequently eluted from the eluted solution and collected for LC-MS/MS analysis (Figure 5.27) [6].

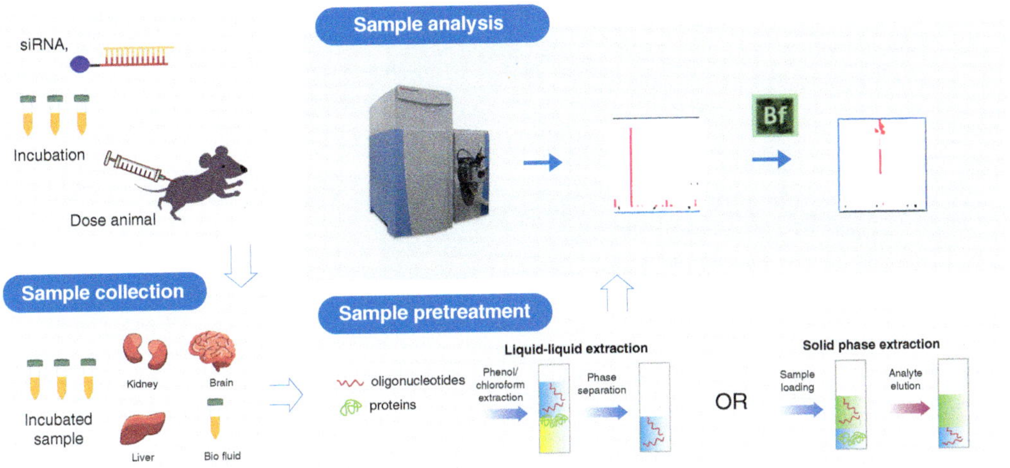

Figure 5.27 Flowchart of the bioanalysis and metabolite identification of oligonucleotides. *Source:* Adapted from reference [6] / ELSEVIER.

In addition, oligonucleotides usually bind nonspecifically to polypropylene tubes/plates or glass containers. Therefore, it is vital to select appropriate consumables such as low-binding pipette tips, centrifuge tubes, and sample plates during sample processing to reduce the inaccuracy of the study results.

5.4.3.2 LC-MS/MS Sample Analysis

Oligonucleotides are highly polar compounds, making it difficult to obtain good peak shapes using conventional reversed-phase chromatography. Therefore, ion-pair reversed-phase chromatography, hydrophilic interaction chromatography (HILIC), and ion-exchange chromatography are commonly used, among which the most common method is ion-pair reversed-phase chromatography. It mainly uses amines such as TEA, *N,N*-diisopropylethylamine, triethylammonium acetate, and HFIP to form ion-pair reagents.

5.4.4 LC-MS/MS Data Analysis and Metabolite Identification

HRMS full MS scans and MS^2 scans (full MS/data-dependent MS^2 [ddMS^2]) can provide valuable information about oligonucleotides, their metabolites, and other impurities. To perform metabolite profiling and identify and confirm the chemical modification positions and sequences of oligonucleotides, specialized software such as BioPharma Finder and ProMass can be used. By analyzing the HRMS data of a published ASO (Oligo 2 [5]) and its metabolite 3′ $n-1$ with BioPharma Finder, we obtained the following structural identification results (Figures 5.28 and 5.29).

This chapter presents a case study involving *in vitro* and *in vivo* metabolite profiling and identification of oligonucleotides.

For this purpose, UPLC-UV-HRMS was employed for metabolite profiling and identification of Oligo A. First, MetID in *in vitro* incubation with liver S9 was conducted, followed by *in vivo* MetID of monkey liver samples after four weeks of repeated administration to Cynomolgus monkeys. The results demonstrated a strong correlation between the major metabolites observed in liver S9 *in vitro* and those found in liver samples *in vivo* (Figure 5.30 and Table 5.5). The identification of

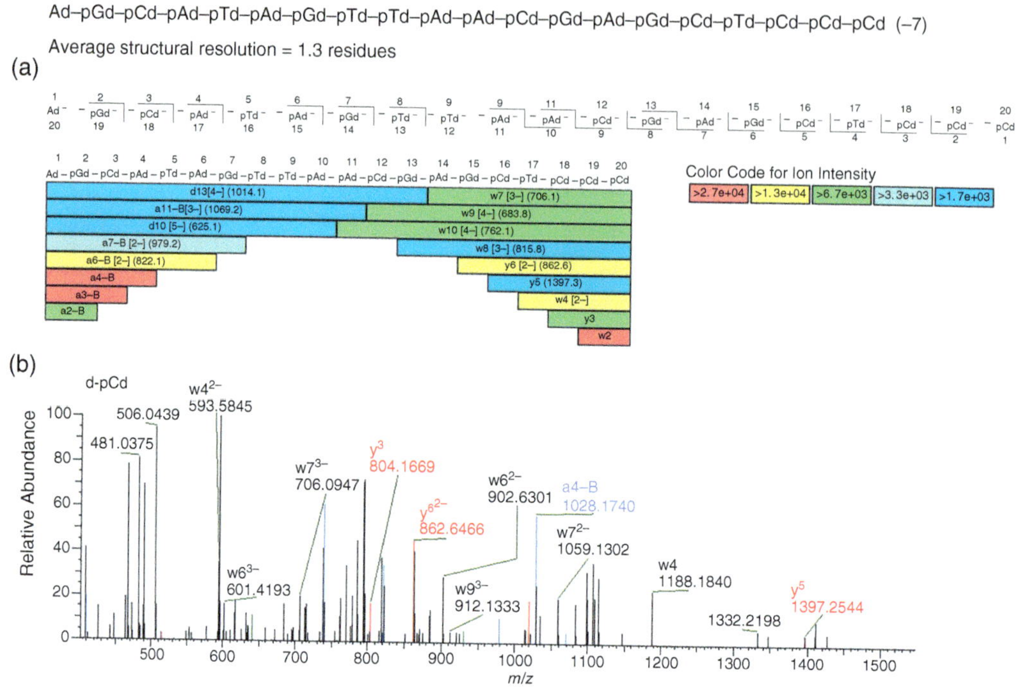

Figure 5.28 MS[2] spectrum (b) of ASO and its structural fragment coverage map (a).

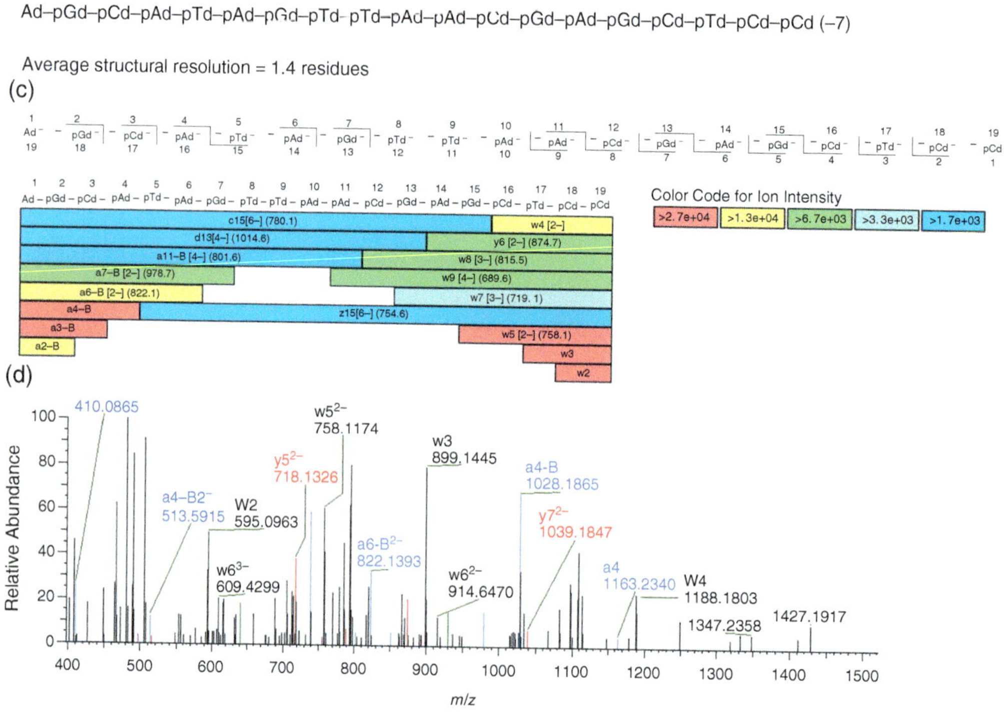

Figure 5.29 MS[2] spectrum (d) of the ASO metabolite 3′n−1 and its structural fragment coverage map (c).

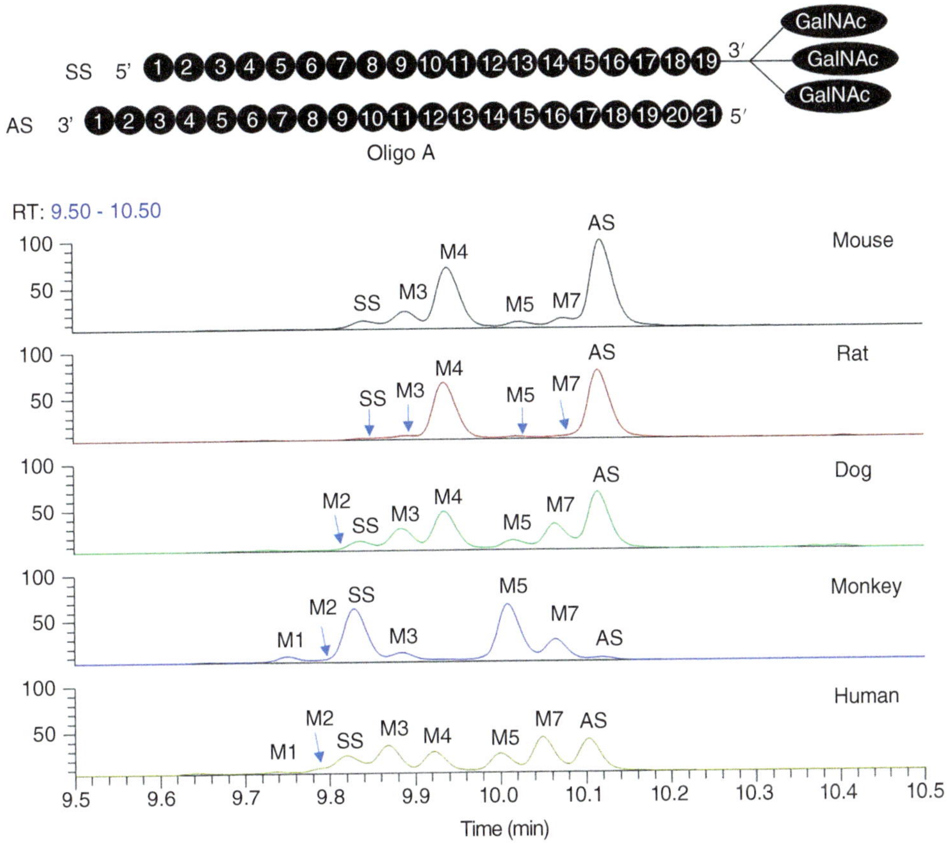

Figure 5.30 LC-UV metabolite profile of Oligo A in liver S9 fractions after incubation for 48 hours.

Table 5.5 *In vitro* and *in vivo* identification of Oligo A metabolites.

		In vitro: liver S9 fractions					*In vivo*: liver
Code	**Metabolic change**	**Mouse**	**Rat**	**Dog**	**Monkey**	**Human**	**Monkey**
M1	SS-(3GalNAc_linker)	ND	ND	ND	★	★	★
M2	SS-GalNAc & 5′ $n-1$)	ND	ND	★	★	★	ND
SS	SS	★	★	★	★	★	★
M3	SS-GalNAc	★	★	★	★	★	ND
M4	SS-2GalNAc	★	★	★	ND	★	ND
M5	SS-3GalNAc	★	★	★	★	★	★
M6	AS_3′ $n-2$	ND	ND	ND	ND	ND	★
M7	AS_3′ $n-1$	★	★	★	★	★	★
AS	AS	★	★	★	★	★	★

★: Detected; ND: not detected.

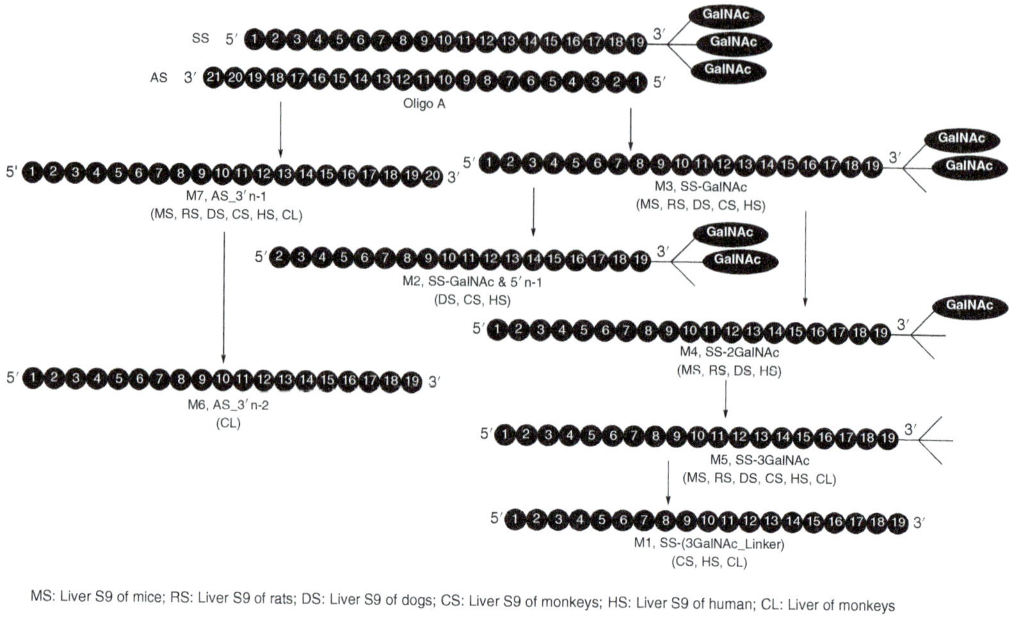

Figure 5.31 Schema of the metabolic pathway of Oligo A in *in vitro* liver S9 after 48 h of incubation and in the liver of Cynomolgus monkeys after 4 weeks of repeated administration.

metabolites in liver S9 provides valuable insights into predicting *in vivo* metabolism, thereby enabling the development of appropriate methods for conducting *in vivo* metabolism studies. Figure 5.31 illustrates the metabolic pathway of Oligo A in liver S9 after 48 hours of incubation and in the liver of Cynomolgus monkeys after 4 weeks of repeated administration.

We have established a comprehensive platform for conducting metabolism studies and MetID of oligonucleotides, including ASOs and siRNAs. This platform utilizes HRMS and professional software and encompasses both *in vitro* incubation systems (such as plasma, liver S9, and liver homogenates) and *in vivo* biological samples (such as plasma, urine, feces, bile, CSF, brain, liver, and kidney).

With the aid of this platform, we can perform metabolite profiling and identification of oligonucleotides, as well as analyze and confirm the chemical structure sequence of oligonucleotides. For instance, Figure 5.32 illustrates the analysis of the chemical structure sequence of Patisiran, including both the SS and the AS, which allowed us to rule out any mismatched structures based on the assay results.

5.4.5 Summary

Metabolite profiling and identification play a crucial role in evaluating the clinical safety and PKs of oligonucleotides in clinical trials. To address this need, we developed a robust platform for metabolite profiling and identification of oligonucleotides, leveraging UPLC-UV-HRMS technology. This platform ensures the delivery of high-quality results, enabling the efficient progress of oligonucleotide drug development worldwide.

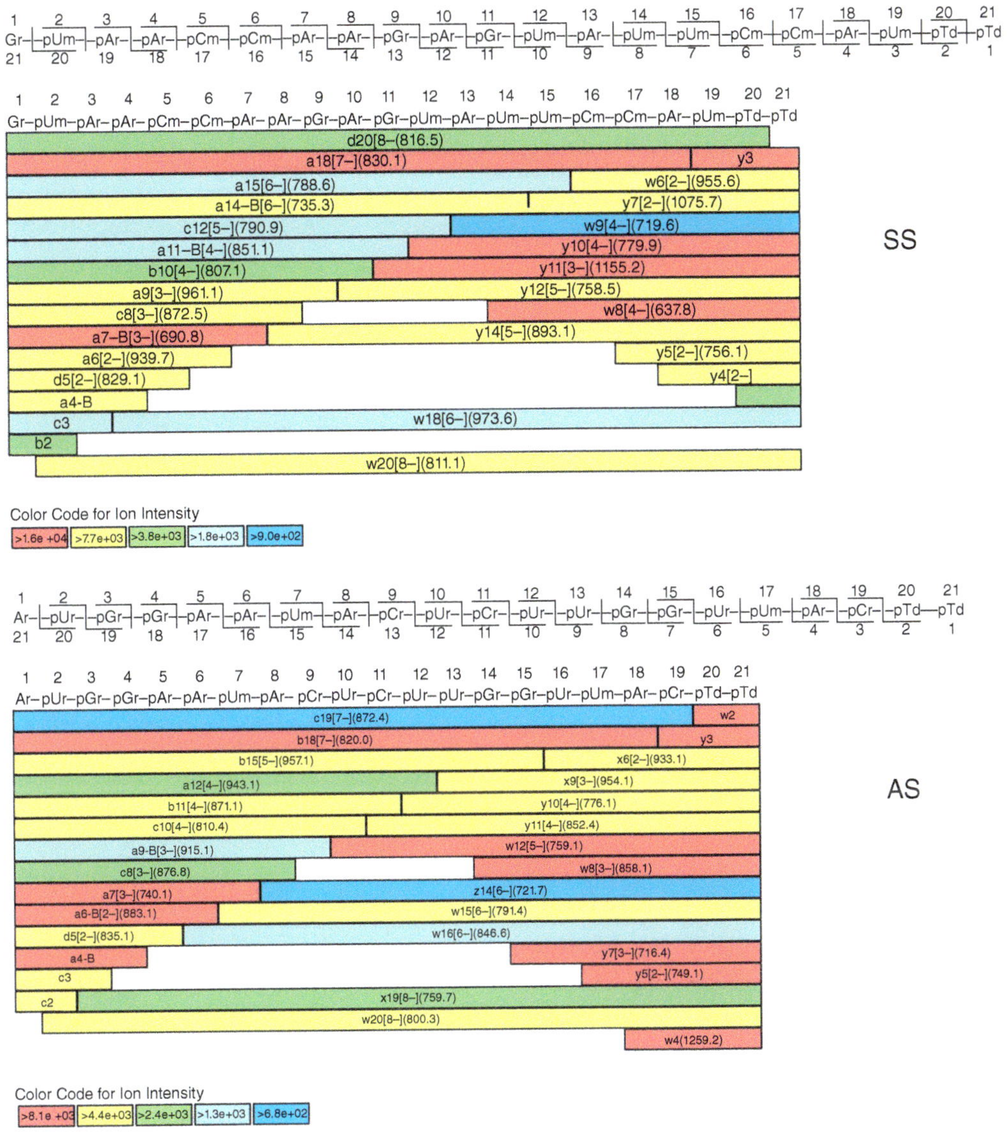

Figure 5.32 Fragment coverage mapping of oligonucleotides.

References

1 Moumné, L. et al. (2022). Oligonucleotide therapeutics: from discovery and development to patentability. *Pharmaceutics* 14: 260. https://doi.org/10.3390/pharmaceutics14020260.

2 Basiri, B. et al. (2020). Introducing an in vitro liver stability assay capable of predicting the in vivo pharmacodynamic efficacy of siRNAs for IVIVC. *Molecular Therapy - Nucleic Acids* 21: 725–736.

3 https://www.toxicology.org/groups/rc/NorCal/docs/2010Spring/2010_3ToxConsider_OligonucleotideTherap.pdf

4 Kilanpwska, A. et al. (2020). In vivo and in vitro studies of antisense oligonucleotides – a review. *RSC Advances* 10: 34501–34516.

5 Kim, J. et al. (2020). In vitro metabolism of 2′-ribose unmodified and modified phosphorothioate oligonucleotide therapeutics using liquid chromatography-mass spectrometry. *Biomedical Chromatography* 34: e4839. https://doi.org/10.1002/bmc.4839.

6 Nuckowski, Ł. et al. (2018). Review on sample preparation methods for oligonucleotides analysis by liquid chromatography. *Journal of Chromatography B* 1090: 90–100.

5.5 ADME Characteristics of Approved siRNA and ASO Drugs

Yan Pan, Jie Wang, Xiangling Wang, Jiaqi Yuan, Genfu Chen, Jing Jin

5.5.1 Introduction

This chapter presents a comparative analysis of the ADME characteristics of approved siRNA (part A) and ASO (part B) drugs. We summarize the study designs, data analysis, and regulatory findings from preclinical PK studies of these drugs to provide insights into their ADME profiles. Understanding the ADME characteristics of these drugs is crucial for optimizing their efficacy and safety profiles.

5.5.2 Overview of siRNA Drugs (Part A)

This part provides a comprehensive overview of approved siRNA drugs, focusing on their ADME, and DDIs. The study details the design and data analysis of five marketed siRNA drugs until July 2023, namely, Patisiran, Givosiran, Lumasiran, Inclisiran, and Vutrisiran.

All five marketed siRNA drugs are designed to target the liver, with a focus on treating liver-related diseases. The pioneer drug, Patisiran, received approval in 2018. Patisiran utilizes an LNP formulation containing double-stranded siRNAs and four lipid excipients. Encapsulation within LNPs safeguards siRNAs from endogenous nucleases and facilitates targeted delivery to the liver. Subsequent siRNA drugs introduced between 2019 and 2022 employ a modified delivery system incorporating N-acetyl-D-galactosamine (GalNAc). Intravenous (IV) infusion is generally used for LNP-targeted siRNAs, while GalNAc-modified delivery systems are administered via subcutaneous (SC) injection (Table 5.6).

The majority of approved siRNAs and those in late-stage clinical trials feature PS backbone modifications along with multiple 2′-OMe and 2′-F modifications (Table 5.7). These alterations significantly influence the *in vivo* and *in vitro* PK properties of these compounds.

After administration *in vivo*, siRNA drugs initially enter the blood circulation and subsequently undergo metabolism or excretion in the liver or kidneys (Figure 5.33). Next, we analyzed the PK study data, including ADME and DDI, included in the regulatory approval applications of the aforementioned five siRNA drugs.

Table 5.6 Introduction of EMA/FDA-approved siRNA drugs.

First approval year	Drug name	Company	Drug type	Target	Indication	Route of administration
2018	Patisiran	Alnylam	siRNA	TTR	Hereditary TTR amyloidosis	IV
2019	Givosiran	Alnylam	siRNA	ALAS1	Acute hepatic porphyria	SC
2020	Lumasiran	Alnylam	siRNA	HAO1	Rare genetic liver disease: primary hyperoxaluria type 1	SC
2020	Inclisiran	Novartis	siRNA	PCSK9	Familial hypercholesterolemia or cardiovascular disease	SC
2022	Vutrisiran	Alnylam	siRNA	TTR	Hereditary TTR amyloidosis	SC

Table 5.7 EMA/FDA-approved siRNA drugs and their structural modifications[a].

		Chemical modifications		
		Iminosugar rings		
Drugs	Skeleton	2′-OMe	2′-F	Delivery systems
Patisiran	NA	√	—	LNP
Givosiran	PS(6)	√(28)	√(16)	GalNAc
Lumasiran	PS(6)	√(34)	√(10)	GalNAc
Inclisiran	PS(6)	√(32)	√(11)	GalNAc
Vutrisiran	PS(6)	√(35)	√(9)	GalNAc

[a] The numbers in parentheses within the table indicate the total number of chemical modifications in each siRNA.

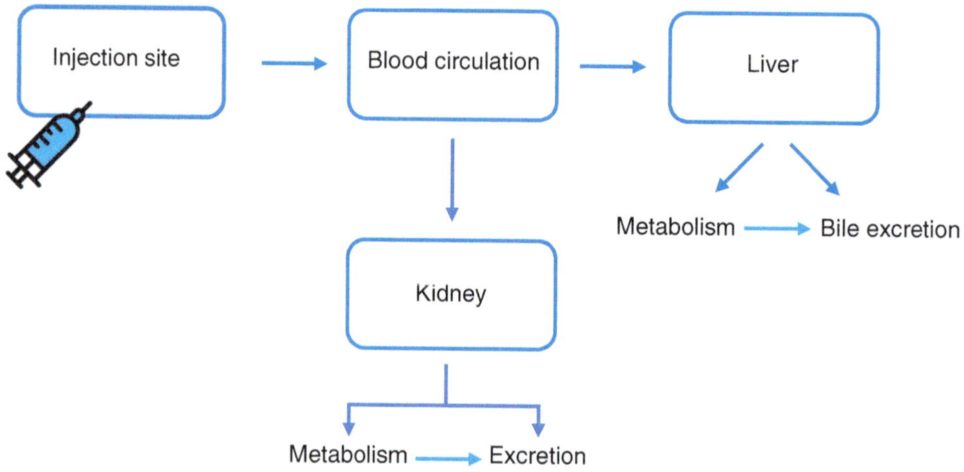

Figure 5.33 *In vivo* disposition of siRNA after administration. Adapted from reference [1].

5.5.2.1 Absorption of Approved siRNA Drugs

How do we measure the preclinical absorption of marketed siRNA drugs?

Absorption refers to the process by which a drug enters the body from the administration site and is distributed throughout the body via the bloodstream. For Patisiran, male and female Sprague–Dawley (SD) rats and male and female cynomolgus monkeys received intravenous administration at low, medium, and high doses. Plasma concentrations were then analyzed to assess the PK profiles of the siRNAs and two novel lipid components. Four GalNAc-modified siRNAs, namely, Givosiran, Lumasiran, Inclisiran, and Vutrisiran, were administered intravenously at a single dose and repeatedly subcutaneously at multiple doses to male and female SD rats and cynomolgus monkeys. This allowed for the evaluation of the PK profiles of the siRNAs after intravenous and subcutaneous administration.

Rationale for the route of administration of Patisiran and the other four GalNAc-modified siRNAs

A significant challenge in developing oligonucleotide drugs lies in achieving adequate exposure to target tissues. Unlike most small-molecule drugs, oligonucleotide drugs are not readily absorbed in the gastrointestinal tract, necessitating alternative routes of administration, such as IV and SC routes, to achieve sufficient exposure to the target tissues.

Compared with subcutaneous SC injection, Patisiran encapsulated in LNPs with a diameter of approximately 60–100 nm achieves notably greater exposure in the liver after IV injection [2]. Thus, IV administration was selected for preclinical and clinical testing. Following IV administration, Patisiran is transported to the liver with the aid of plasma ApoE, facilitating liver-targeted drug delivery.

Givosiran, Lumasiran, Inclisiran, and Vutrisiran utilize a delivery system modified with GalNAc. These siRNAs are designed to target hepatocytes, taking advantage of the high expression of the ASGPR on the hepatocyte membrane. GalNAc serves as a ligand for ASGPR, facilitating siRNA entry into the target tissue through ASGPR-mediated cellular endocytosis. Hepatocytes have abundant ASGPRs, with approximately 500,000 receptors per cell [3], leading to saturation effects. IV administration results in high drug concentrations in the bloodstream, potentially causing ASGPR saturation and reducing the number of drug molecules that can enter hepatocytes. Conversely, SC administration offers a sustained release effect, allowing drug molecules to enter hepatocytes at a slower pace, thereby achieving sufficient exposure to the target tissue.

Summary and analysis of PK data on the absorption of EMA/FDA-approved siRNA drugs

Representative human plasma PK data for Patisiran revealed a characteristic two-compartment model (Figure 5.34). Following a single IV administration, the blood concentration–time curve shows a rapid decrease in concentration, followed by a secondary peak and linear elimination. The initial decrease can be attributed to the rapid distribution of the drug in tissues after it enters the circulation, with a substantial proportion of the Patisiran molecules entering the liver from the bloodstream. Subsequently, some Patisiran molecules re-enter the circulation from the liver, resulting in a secondary increase in plasma concentration. The terminal elimination half-life ($t_{1/2}$) of Patisiran is approximately 2–3 days [4], with a relatively small apparent distribution volume and plasma clearance rate (Table 5.8) and a significantly greater area under the concentration–time curve (AUC) in humans than in rats and monkeys.

The PK profiles of Givosiran, Lumasiran, Inclisiran, and Vutrisiran following single-dose IV administration in rats and monkeys were comparable, **with $t_{1/2}$ values of approximately 0.2–0.6 hours and relatively small apparent volumes of distribution and plasma clearance** (Table 5.8). Upon single-dose SC administration in rats, monkeys, and humans, these siRNAs exhibited plasma T_{max} values ranging from 0.4 to 6 hours, with a slightly delayed peak in humans. Moreover, the observed $t_{1/2}$ ranged from 1 to 9.58 hours, showing a longer half-life in humans than in rats and monkeys. In rats, the bioavailability was between 5% and 35.1%, while in monkeys, between 8% and 25% (Table 5.8). Following SC absorption, the plasma PK profile conformed to a two-compartment model [5–9]. Systemic exposure, measured by plasma C_{max} and AUC_{0-t} values, increased roughly proportionally with dose, and no significant sex differences were observed between male and female animals [3–5, 7, 10].

5.5.2.2 Distribution of EMA/FDA-Approved siRNA Drugs

How have these marketed siRNA drugs been evaluated for drug distribution in preclinical studies?

***In vitro* studies** play a crucial role in understanding compound distribution. For example, permeability assays are employed to characterize the ability of a compound to enter cells, while plasma protein binding (PPB) assays quantify the degree of binding to plasma proteins. Higher PPB rates can potentially limit therapeutic efficacy by reducing the concentration of free drugs available to interact with the intended target. The Patisiran PPB studies primarily utilized fast protein liquid

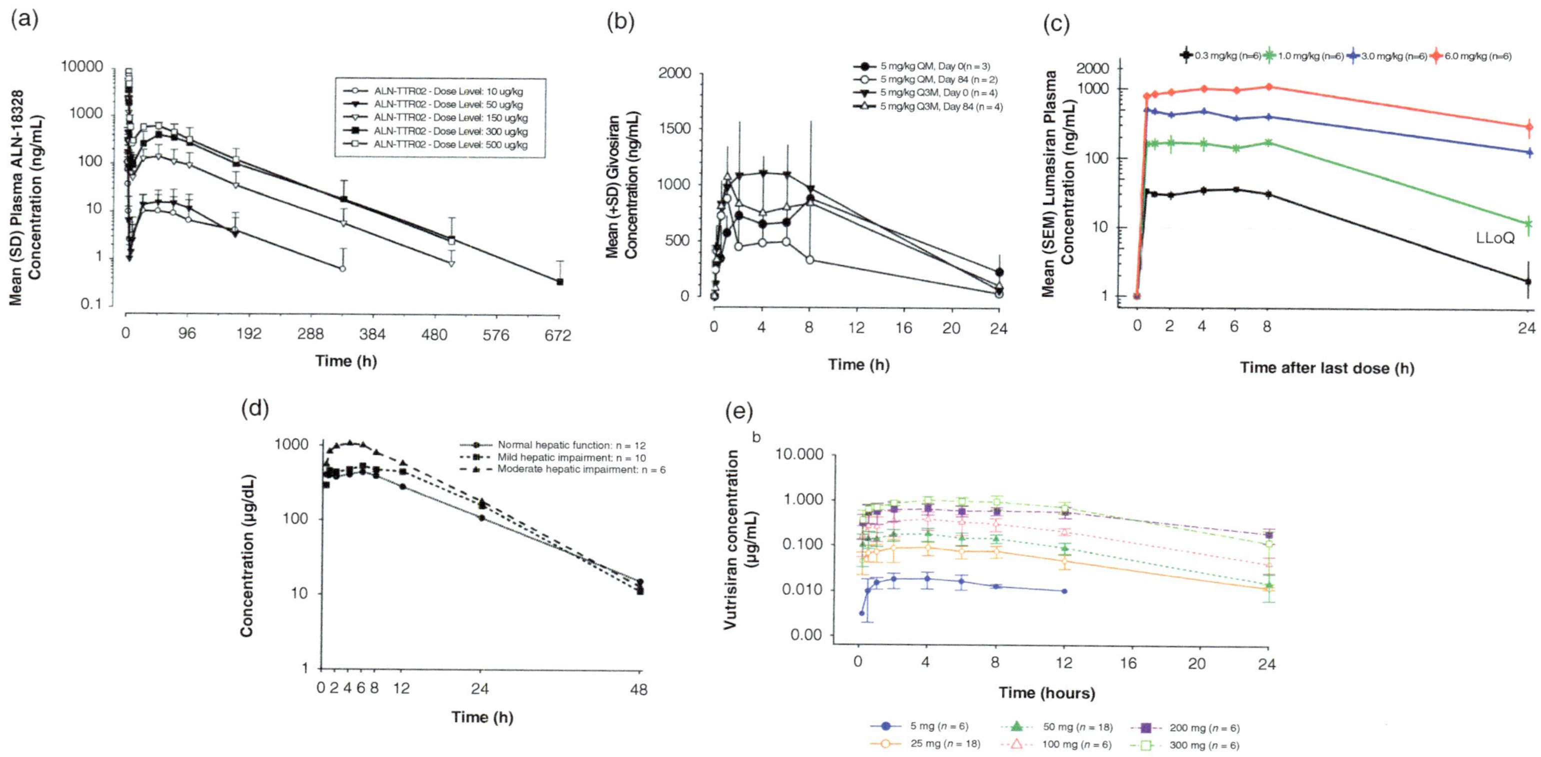

Figure 5.34 PK profiles of approved siRNA drugs after human administration. (a) Patisiran, (b) Givosiran, (c) Lumasiran, (d) Inclisiran, and (e) Vutrisiran. *Source:* [5]. NickKozauer, 2016; [6]. Alnylam. GivosiranNDA MULTI-DISCIPLINE REVIEW; [7]. European Medicines Agency, 2020; [8]. Kallend et al., 2022; [9]. Habtemariam et al., 2021.

Table 5.8 Main PK parameters of EMA/FDA-approved siRNA drugs.

	Patisiran	Givosiran		Lumasiran		Inclisiran		Vutrisiran	
	IV bolus (R, C) or infusion (H)	**IV bolus**	**SC**	**IV bolus**	**SC**	**IV bolus**	**SC**	**IV bolus**	**SC**
Dose (mg/kg)	0.3	10	5	5 (R); 10 (C);	5 (R); 10 (C); 3 (H)	5 (R); 6 (C);	5 (R); 3 (C); 300 mg (H)	5 (R); 10 (C);	3 (R); 3 (C); 300 mg (H)
T_{max} (h)	—	—	0.4 (R); 2.4 (C); 4 (H)	—	1 (R); 2.4 (C); 3.17 (H)	—	1 (R); 1.8 (C); 6 (H)	—	0.5–1 (R); 1.9–2.8 (C); 5 (H)
C_{max} (μg/mL)	7.07 (R); 3.50 (C); 5.49 (H)	—	0.43 (R); 0.83 (C); 0.975 (H)	—	0.581 (R); 2.71 (C); 0.533 (H)	—	0.787 (R); 0.526 (C); 0.509 (H)	—	0.492 (R); 0.552 (C); 1.09 (H)
AUC (μg·h/mL)	3.40 (R); 6.68 (C); 61.5 (H)	11.8 (R); 30.5 (C); – (H)	1.29 (R); 4.65 (C); 7.83 (H)	7.95 (R); 58.9 (C); – (H)	1.22 (R); 13.5 (C); 7.21 (H)	5.36 (R); 26.4 (C); – (H)	1.88 (R); 3.06 (C); 7.98 (H)	13.3 (R); 77.6 (C); – (H)	0.889 (R); 3.26 (C); 15.9 (H)
%F	—	—	24 (R)	—	30 (R), 25 (C)	—	35.1 (R), 24.7 (C)	—	5–18 (R), 8–22.3 (C)
$t_{1/2}$	– (R); – (C); 2.18 d (H)	0.2 h (R); 0.4 h (C); – (H)	2.7 h (R); 3.5 h (C); 9.29 (H)	0.4 h (R); 0.6 h (C); – (H)	1 h (R); 2.6 h (C); 5.98 (H)	0.3 h (R); 0.6 h (C); – (H)	1.7 h (R); 2.4 h (C); 9.58 (H)	– (R); – (C); – (H)	– (R); – (C); 5.53(H)
Vdss (mL/kg)	27.5 (R); 447 (C); 552 (H)	181 (R); 104 (C); – (H)	—	332 (R); 163 (C); – (H)	—	141 (R); 42.6 (C); – (H)	—	— (R); – (C); – (H)	—
CL (mL/h/kg)	88.2 (R); 46.3 (C); 9.35 (H)	870 (R); 340 (C); — (H)	—	646 (R); 189 (C); — (H)	—	950 (R); 229 (C); – (H)	—	376 (R); 129 (C); – (H)	—

SC = subcutaneous, R = rat, C = cynomolgus monkey, H = human; M = mouse. The abbreviations apply to all subsequent tables.

chromatography followed by sodium dodecyl sulfate–polyacrylamide gel electrophoresis to assess PPB. Other siRNA drug studies have employed electrophoretic mobility shift assays to study PPB. These findings provide valuable insights into the distribution behavior of siRNA drugs during preclinical investigations.

The *in vivo* distribution of a drug after administration involves its reversible transfer between different locations within the body. To track changes in drug concentration over time in various tissues and organs, researchers typically employ tissue collection or *in vivo* ADME studies, such as QWBA, utilizing radiolabeled drugs. In the case of these approved siRNA drugs, *in vivo* evaluations of their distribution have been conducted through major tissue collection after the administration of unlabeled compounds or via QWBA assays, providing valuable insights into their *in vivo* distribution patterns.

Summary and analysis of approved siRNA drug distribution data

In vitro protein binding data revealed that LNPs encapsulated in Patisiran displayed a relatively low binding rate to plasma proteins, including albumin and alpha-1-acid glycoprotein. In contrast, the lipid component PEG2000-C-DMG alone exhibited relatively high PPB. Givosiran, Lumasiran, Inclisiran, and Vutrisiran demonstrated an inverse concentration-dependent effect on PPB, where lower concentrations correlated with greater protein binding, and higher concentrations correlated with lower protein binding (Table 5.9).

The available data on *in vivo* tissue concentrations of the marketed siRNA drugs consistently demonstrate strong liver targeting, with significantly greater exposure in liver tissues (measured by AUC or C_{max}) than in the plasma after administration. Notably, the liver exposure of GalNAc-modified siRNAs was consistently much greater than that of Patisiran (Table 5.10).

Radiolabeled QWBA assays revealed that these approved siRNA drugs are predominantly distributed in the liver. Additionally, other blood flow-rich tissues, such as kidneys and lymph nodes in immune organs, also exhibited greater drug distribution (Table 5.11).

5.5.2.3 Metabolism of Approved siRNAs

Oligonucleotides are metabolized primarily by endonucleases and exonucleases rather than phase I and II drug-metabolizing enzymes in the liver [11]. Exonucleases remove one nucleotide at a time from the end of a single strand of siRNA. Endonucleases cleave the strand from the middle of the strand, leaving fragments of different lengths. Following administration, the GalNAc moieties of GalNAc–siRNA are rapidly cleaved by β-N-acetylglucosaminidase during endo/lysosomal trafficking,

Table 5.9 *In vitro* PPB of EMA/FDA-approved siRNA drugs.

Species	Patisiran	Givosiran	Lumasiran	Inclisiran	Vutrisiran
Human	97% (PEG2000-C-DMG), 0.46% (serum albumin), 2.07% (alpha-1-acid glycoprotein)	21–92% (1–50 ug/mL)	20–85% (0.5–50 ug/mL)	86–93% (0.5–1 ug/mL)	19–82% (0.5–50 ug/mL)
Monkey	–	26–90% (1–50 ug/mL)	37–86% (0.5–50 ug/mL)	86–93% (0.5–1 ug/mL)	16–87% (0.5–50 ug/mL)
Rat	0.89% (serum)	28–93% (1–50 ug/mL)	35–96% (0.5–50 ug/mL)	86–93% (0.5–1 ug/mL)	26–84% (0.5–50 ug/mL)
Mouse	—	10–91% (1–50 ug/mL)	–	86–93% (0.5–1 ug/mL)	12–75% (0.5–50 ug/mL)

Table 5.10 Target tissue concentration data of marketed siRNA drugs.

	Patisiran		Givosiran		Lumasiran		Inclisiran		Vutrisiran
Species	**Rat IV**	**Monkey IV**	**Rat SC**	**Rat IV**	**Rat SC**	**Monkey SC**	**Rat SC**	**Monkey SC**	**Rat SC**
Dose (mg/kg)	0.3		10	10	1, 5, 10	1, 5, 10	1, 5, 25	1, 3, 6	3
Liver T_{max} (h)	24–72 h (Hepatocyte uptake Patisiran-LNP)	2 h (Hepatocyte uptake Patisiran-LNP)	1.1	2.1	4, 6, 8	16, 16, 48	—	—	—
Liver C_{max} (μg/g)	1.53	0.065	197	102	11.1, 65.7, 120	31.0, 90.4, 169	—	—	—
Liver AUC (μg·h/g)	5.99	—	10,494	5390	1210, 5820, 10,700	13,485, 41,600, 76,700	—	—	3890
Liver $t_{1/2}$ (h)	—	—	120	55	191, 105, 138	292, 371, 409	—	—	—
Plasma C_{max} (μg/mL)	7.07	3.5	1.07	148	0.581 (5 mg/kg)	2.71 (10 mg/kg)	—	—	—
Plasma AUC (μg·h/mL)	3.4	6.68	2.79	11.8	1.22 (5 mg/kg)	13.5 (10 mg/kg)	—	—	33.2

Table 5.11 QWBA assay results for radiolabeled marketed siRNA drugs.

	Patisiran (^{14}C-DLin-MC3-DMA)	[^{3}H] Givosiran	[^{14}C] Lumasiran	[^{14}C] Inclisiran	[^{3}H] Vutrisiran
Species	Rat	Rat	Rat	Monkey	Rat
Dose (mg/kg)	0.3	10	10	20	3
Tissue uptake	Liver > spleen > kidney > lung > heart > adrenals	Liver > kidney cortex > lymph node > kidney	Dose site > liver > kidney > adrenal > CNS	Liver > kidney > lymph node	Liver > lymph node > kidney > thoracic duct

Table 5.12 *In vitro* stability data of EMA/FDA-approved siRNAs[a].

	Liver S9		Serum	
	AS	SS	AS	SS
Patisiran	42–77%[b]		62–71% (M, R)[c],42–44% (C)[c], 100% (H)[c]	
Givosiran	103% (M), 68% (C), 49% (H), 36%(R)	102% (M), 88% (C), 65% (H), 64%(R)	75% (M), 59%(R), 63% (C), 89% (H)	95% (M), 95% (R), 100% (C), 95% (H)
Lumasiran	Stable in many species[d]	80–85%[d]	>90%[d]	
Inclisiran	>41%[d]	>64%[d]	85% (M), 82%(R), 72% (C), 80% (H)	95% (M), 91% (R), 91% (C), 87% (H)
Vutrisiran	83% (M), 70% (C), 90% (H), 86%(R)	72% (M), 59% (C), 89% (H), 89%(R)	95–99%[d]	98–101%[d]

[a] Data in the table show the percentages of AS or SS remaining after 24 h of incubation with siRNA in the serum or liver S9 matrix.
[b] Amount of free siRNA after 6 or 24 h of incubation in serum or the S9 matrix.
[c] Amount of LNP-encapsulated siRNA after 24 h of incubation in the serum.
[d] The value contains data from rats, mice, monkeys, and humans.

and the linker is further metabolized by amidase in hepatocytes [12]. The marketed siRNAs are structurally modified and are less likely to be metabolized in systemic circulation. Typically, the vast majority of siRNAs are rapidly taken up by the liver, with small proportions distributed in other tissues. **Then, siRNAs are metabolized by nucleases in the liver and other tissues.**

How is the metabolism of approved siRNAs evaluated in preclinical studies?

The *in vivo* and *in vitro* metabolism studies results for the approved siRNAs are summarized in Tables 5.12 and 5.13. For *in vitro* metabolism studies, **liver S9 and serum/plasma stability assays are typically performed**. For *in vivo* metabolism studies, the **plasma and liver tissues generated during PK studies are usually collected for metabolite profiling**. For GalNAc–siRNA, metabolite profiling was conducted for both the AS and the SS. For Patisiran, the LNP-encapsulated siRNAs, in addition to studying the metabolism of the double strands, there was also an additional investigation into the *in vitro* and *in vivo* metabolism of two novel lipid molecules.

Summary and analysis of metabolism data for approved siRNAs

The metabolic data of the marketed siRNAs showed that **the *in vitro* and *in vivo* metabolic profiles of the siRNAs were similar, and these siRNAs shared similar metabolic pathways**

Table 5.13 *In vivo* metabolism data of EMA/FDA-approved siRNAs[a].

	Patisiran	Givosiran	Lumasiran	Inclisiran	Vutrisiran
Rat/ monkey	1) **Plasma (R)**: mainly ALN-18238 (siRNA) 2) **Plasma (R)**: mainly ALN-18238	1) **Plasma (R, 0–48 h)** 54% Givosiran, 13% AS(*N*–1)3′ Givosiran 2) **The remaining of SS strand in the liver** 10% (2 h, R), 0% (8 h, R), 1% (8 h, C)	**Plasma and liver (R, C)** AS: AS (*N*–1)3′ Lumasiran[b]SS: GalNAc moieties were removed from the 3′ end of SS (small amount)	1) **Plasma (R, C)** AS: AS(*N*–1)5′ Inclisiran, AS(*N*–7)3′ Inclisiran (C) SS: GalNAc moieties were removed from 3′ end of SS 2) **Liver (R, C)** AS: AS(*N*–1)3′ Inclisiran and other short strands SS: GalNAc moieties were removed from the 3′ end of SS (82–100%)	**Plasma and liver (R, C)**[c] AS: AS(*N*–1)3′ Vutrisiran and AS(*N*–3)3′ Vutrisiran SS: GalNAc moieties were removed from the 3′ end of SS
Human	**Plasma**: ALN-18328 (less than 5%).[d]	**Plasma**: AS (*N*–1)3′ Givosiran[e]	**Plasma**: AS (*N*–1)3′ Lumasiran[f]	Approximately 82.5% of Inclisiran was taken up by the liver	No metabolites were detected in human plasma
Major metabolites	Related short strands	**AS:** AS(*N*–1)3′ Givosiran **SS:** 1 to 3 GalNAc moieties were removed from the 3′ end of SS	**AS:** AS (*N*–1)3′ Lumasiran **SS:** GalNAc moieties and terminal nucleosides were removed from the 3′ end of SS	**AS:** AS(*N*–1)3′ Inclisiran, AS(*N*–1)5′ Inclisiran **SS:** GalNAc moieties were removed from the 3′ end of SS.	**AS:** AS(*N*–1)3′ Vutrisiran **SS:** GalNAc moieties were removed from the 3′ end of SS

[a] Data in the table were summarized from FDA/EMA information for each marketed drug.
[b] In plasma, the percentage of AS (*N*–1)3′ Lumasiran was no more than 10%. The 3′ deamino metabolite was detected in the monkey's liver.
[c] Fifteen days after administration, the major drug-related substance in the liver of rats and monkeys was still Vutrisiran.
[d] Fewer than 5% of the free double-stranded siRNA was detected in human plasma, indicating that the siRNA was still encapsulated by the LNPs.
[e] In human plasma, the concentration of AS(*N*–1)3′ Givosiran was approximately half that of Givosiran.
[f] The amount of AS (*N*–1)3′ Lumasiran in plasma was less than 10% of that of Lumasiran.

and metabolites across species, including humans (Tables 5.12 and 5.13). **The main metabolic pathway of antisense RNA** is the loss of one nucleotide at the 3′ end to form the major metabolite, AS(N-1)3′. In addition, one to three GalNAc moieties are lost in the SS. Usually, AS(N-1)3′ is the main active metabolite. More comprehensive PK evaluations of active metabolites have been performed by researchers in preclinical and clinical studies when the percentage of active metabolites exceeds 10% (e.g., Givosiran).

Plasma/serum stability data for AS and SS strands of GalNAc–siRNA
As shown in Tables 5.12 and 5.13, the stability of GalNAc–siRNA across the tested species was similar between serum *in vitro* and plasma *in vivo*. The SS strand was more stable than the AS strand in the plasma. This may be due to the presence of the GalNAc group attached to the 3′ end of the SS strand, which protects the 3′ end of the SS strand from exonuclease-mediated cleavage to a certain extent[13]. Moreover, the presence of the GalNAc motif in the SS strand also protects the 5′ end of the AS strand by causing steric hindrance, which may explain why GalNAc–siRNA prefers to form AS(*N*–1)3′ rather than AS(*N*–1)5′.

5.5.2.4 Excretion of Approved siRNAs

Oligonucleotides are primarily excreted via the urine due to their polarity, hydrophilicity, and negative charges. During systemic circulation, the primary elimination pathway of both GalNAc–siRNA and LNP–siRNA is rapid liver uptake followed by slow metabolism in the liver. The secondary elimination pathway occurs through the kidney.

Summary and analysis of the excretion data of the approved siRNAs
As shown in Tables 5.14 and 5.15, the excretion of the five approved siRNAs was evaluated mainly through **mass balance studies of radiolabeled and unlabeled test articles.** For Patisiran, the excretion of two novel lipid molecules was evaluated. Based on the data from the radiolabeled mass balance study in intact rats (Table 5.14), the radioactivity recovered in the urine of intact rats ranged from 19.5% to 57.8%, and that in feces ranged from 14.3% to 33.9%. The radioactivity recovered in the bile ranged from 20% to 28.3% in bile duct-cannulated rats. The radioactivity was mainly excreted through urine and bile. The mass balance data for the cynomolgus monkeys showed that the amount of radioactivity excreted through urine ranged from 32% to 50% and that through feces

Table 5.14 The mass balance data of EMA/FDA-approved siRNAs (hot)[a].

	Rat			Monkey	
	Urine	Feces	Bile	Urine	Feces
Patisiran	49%	24%	NA	50%	10%
Givosiran	30.4%[b]/57.8%	1.42%[b]/14.3%	27.4	NA	NA
Lumasiran	15.3%[b]/19.5%	0.666%[b]/33.9%	28.3%	38%	8%
Inclisiran	29%	23%	Approximately 20%	32%	1.60%
Vutrisiran	25.8%[b]/52.5%	1.50%[b]/15.2%	24.0%	NA	NA

[a] Hot refers to radiolabeled compounds. The data in this table were derived from studies of the excretion of hot compounds. The data in the table were obtained from FDA/EMA information on each marketed drug. "NA" indicates no relevant data.
[b] Mass balance data from bile duct-cannulated rats.

Table 5.15 The mass balance data of EMA/FDA-approved siRNAs (cold)[a].

	Rat			Monkey		Human	
	Urine	Feces	Bile	Urine	Feces	Urine	Feces
Patisiran	—	—	—	—	—	<1%	NA
Givosiran	10.40%	0.09%	NA	15.90%	—	5–14%	NA
Lumasiran	<1%	—	NA	20%	NA	17.4–25.8%	NA
Inclisiran	NA	NA	NA	NA	NA	15.90%	NA
Vutrisiran	9%	<1%	NA	11–24%	0.01%	15.4–25.4%	NA

[a] Cold refers to compounds that are not radiolabeled. The data in this table were derived from excretion studies with cold compounds. The data in the table were obtained from FDA/EMA information on each marketed drug. "—" indicates not detected, and "NA" indicates no relevant data.

ranged from 1.6% to 10%. The results were consistent with those from rats, suggesting that siRNAs were primarily excreted via the urine.

Preclinical data have indicated that **the metabolic pathway of siRNAs is well defined and that siRNAs have long elimination half-lives in the liver.** For safety reasons, no radiolabeled mass balance study has been conducted in human subjects for the approved siRNAs. As shown in Table 5.15, 14–25.8% of the intact siRNA was excreted through the urine in humans, most of which was excreted within 24 hours after administration, indicating that most of the drug was rapidly taken up by the liver. In rats and monkeys, less than 24% of intact siRNA was excreted via the urine. The clinical data were relatively consistent with the preclinical data. The main elimination route of siRNAs in humans is hepatic uptake, and the secondary route is renal excretion.

Special considerations for the use of approved siRNAs in individuals with renal impairment

Of the five siRNA drugs on the market, only Inclisiran has been studied in a dedicated clinical PK study in patients with renal impairment. The other four siRNA drugs have only been evaluated for the impact of mild and/or moderate renal impairment on drug concentration and efficacy in population PK studies [13]. Although the C_{max} and AUC increased in patients with renal impairment, **they did not affect the efficacy of siRNA drugs.** No dose adjustments were made in patients with mild or moderate renal impairment.

5.5.2.5 DDI of Approved siRNAs

Although oligonucleotides are not expected to act as substrates of CYP450 enzymes or transporters, theoretically, oligonucleotides could affect the expression of enzymes and transporters by influencing regulatory pathways. For example, some mRNAs that can be targeted by oligonucleotides may be involved in transcription and translation processes related to CYP enzyme formation, thus affecting the expression of the CYP enzyme.

Table 5.16 summarizes the results and types of DDI assays for each marketed siRNA drug. Except for Givosiran, none of the marketed siRNA drugs **inhibited CYP enzymes or were substrates or inhibitors of transporters.** Givosiran has been shown to have some inhibitory effect on P-gp [14], but this inhibition is unlikely to occur at clinically relevant dosages. In addition, Givosiran is known to reduce the activity of CYP enzymes *via* its pharmacological activity [15]. For this reason, a dedicated clinical DDI study was conducted on Givosiran.

Table 5.16 Summary of DDI evaluation for EMA/FDA-approved siRNAs[a].

		CYP inhibitor	CYP inducer	Transporter substrate	Transporter inhibitor	
Oligo	**CYP substrate**	**Direct**	**Time dependent**			
Patisiran	Not evaluated	No	Not evaluated	No	No	No
Givosirna	Not evaluated	No	No	No	No	P-gp
Lumasiran	No	No	No	Not evaluated	Not evaluated	Not evaluated
Inclisiran	Not evaluated	No	No	No	No	No
Vutrisiran	No	No	No	Not evaluated	Not evaluated	Not evaluated

[a] No: no relevant activity; not evaluated: no relevant assay was performed. The data in the table were obtained from FDA/EMA information on each marketed drug.

In general, there is a lack of DDI guidelines related to oligonucleotide drugs, and it is recommended that DDI potential be evaluated in as many DDI assays as possible.

5.5.3 Overview of ASO Drugs (Part B)

In early 2022, researchers from the University of Connecticut published a review article titled "Absorption, Distribution, Metabolism, and Excretion of US Food and Drug Administration-Approved Antisense Oligonucleotide Drugs" in *Drug Metabolism and Disposition*, a leading journal in the field of drug metabolism [16]. They selected 10 ASOs (Fomivirsen, Pegaptanib, Mipomersen, Nusinersen, Inotersen, Defibrotide, Eteplirsen, Golodirsen, Viltolarsen, and Casimersen) that have been approved by the US FDA, summarizing the main ADME characteristics of ASOs and discussing current knowledge gaps and future research perspectives. In this part, we will interpret some of the highlights from their findings and share our insights.

5.5.3.1 Introduction to ASO Drugs

The ADME is an essential biological process that determines PK parameters, directly impacting efficacy and adverse drug reactions. Chemical structure, dosage form, administration site, and route of administration are the main determinants affecting ADME characteristics, which in turn affect efficacy and adverse reactions. In recent years, significant progress has been made in the development of ASOs. ASOs are synthetic single-stranded ASOs with immense potential for disease treatment. They have unique ADME characteristics that are substantially different from those of small-molecule drugs. Understanding their ADME characteristics is fundamental to the development of new ASOs with improved efficacy and safety.

The development of ASO drugs can be traced back to 1978 when an oligonucleotide was first discovered to inhibit the translation of Rous sarcoma virus RNA. Nearly 20 years later, the first ASO drug, Fomivirsen, was approved by the FDA in 1998 to treat cytomegalovirus (CMV) retinitis (Table 5.17). After years of slow approval of new ASOs, the FDA has consecutively approved eight ASOs since 2013. In 2022, the discovery and development platform for ASOs was further improved, and more than 100 ASOs are being evaluated in various phases of clinical trials.

Table 5.17 Ten FDA-approved ASO drugs.

Date	Drug name	Brand name	Company	Disease
26 August 1998	Fomivirsen	Vitravene	Ionis	CMV
17 December 2004	Pegaptanib	Macugen	Eyetech	Macular degeneration
29 January 2013	Mipomersen	Kynamro	Genzyme	FH
30 Mar 2016	Defibrotide	Defitelio	Jazz	VOD
19 September 2016	Eteplirsen	Exondys 51	Sarepta	DMD
23 December 2016	Nusinersen	Spinraza	Biogen	SMA
05 October 2018	Inotersen	Tegsedi	Ionis	PNA
12 December 2019	Golodirsen	Vyondys 53	Sarepta	DMD
12 August 2020	Viltolarsen	Viltepso	Nippon Shinyaku	DMD
25 February 2021	Casimersen	Amondys 45	Sarepta	DMD

Source: Adapted from reference [16].

According to their mechanisms of action, the 10 ASOs approved by the FDA can be divided into five groups.

A) Inducing RNase H-mediated mRNA degradation: Fomivirsen, Mipomersen, and Inotersen.
B) Inducing exon skipping: Eteplirsen, Golodirsen, Viltolarsen, and Casimersen.
C) Nusinersen plays a role in exon restoration.
D) Pegaptanib is a nucleic acid aptamer that interacts with vascular endothelial growth factor receptors.
E) Defibrotide is a polydisperse mixture of various single-stranded oligonucleotides that target multiple sites to treat veno-occlusive disorders through antagonistic mechanisms.

These drugs can be divided into four groups according to the route of administration (Table 5.18).

A) Intravitreal injection of fomivirsen and pegaptanib for the treatment of CMV retinitis in the eye and neovascular macular degeneration in the back of the eye, respectively.
B) Intrathecal injection: Nusinersen for the treatment of spinal muscular atrophy.
C) Subcutaneous injection (SC): Mipomersen and Inotersen for the treatment of familial hypercholesterolemia and polyneuropathy of amyloidosis, respectively, administered once weekly.
D) Intravenous infusion (IV): Defibrotide, Eteplirsen, Golodirsen, Viltolarsen, and Casimersen. Defibrotide is used to treat hepatic veno-occlusive disease by intravenous infusion for two hours every six hours. The other four drugs are used to treat Duchenne muscular dystrophy with weekly infusions.

The ADME of ASO drugs differs from that of small-molecule drugs because the ADME of small-molecule drugs is mediated by phases I and II drug-metabolizing enzymes, uptake and efflux transporters, transcription factors, and nuclear receptors, whereas the ADME of ASO drugs is determined by completely different factors. ASOs enter target cells mainly by endocytosis, which packages them into early endosomes. They then move intracellularly via subcellular trafficking. The metabolism of ASOs is primarily catalyzed by exonucleases and endonucleases, which produce shortened-chain oligonucleotides, which are then released into the urinary tract via membrane leakage, vesicles, or exocytosis. These characteristics determine the half-life and dosing frequency of ASOs in the circulatory system.

Table 5.18 Administration routes and dosage forms of ASOs.

Route of administration	ASOs	Dosage form	Administration
Intravitreal injection	Fomivirsen	0.33 mg drug dissolved in 50 μL solution	Loading dose: administered twice every other week Maintenance dose: every 4 weeks
	Pegaptanib	0.30 mg drug dissolved in 90 μL solution	Dosing: once every 6 weeks
Intrathecal injection	Nusinersen	12 mg of drug dissolved in 5 mL solution	Loading dose: first three doses with 14-day interval, the 4th dose 30 days after the 3rd dose Maintenance dose: administered every 4 months
Subcutaneous injection	Mipomersen	200 mg of drug dissolved in 1 mL solution	Dosing: once a week
	Inotersen	284 mg of drug dissolved in 1 mL solution	Dosing: once a week
Intravenous infusion	Defibrotide	6.25 mg/kg	Dosing: 2 h of infusion every 6 h for 21 days
	Eteplirsen	30 mg/kg	Dosing: 35–60 min infusion, once a week
	Golodirsen	30 mg/kg	Dosing: 35–60 min infusion, once a week
	Casimersen	30 mg/kg	Dosing: 35–60 min infusion, once a week
	Viltolarsen	80 mg/kg	Dosing: 60 min infusion, once a week

Source: Adapted from reference [16].

5.5.3.2 ADME Characteristics of FDA-Approved ASOs

Absorption: The 10 approved ASO drugs exhibit multiple absorption patterns that correlate with their indications and administration routes (Table 5.19).

Fomivirsen and pegaptanib are used to treat ocular diseases and are administered by local injection into the eye. In rabbits, Fomivirsen is detectable in the retina within hours after injection, with concentrations continuing to increase after 3–5 days of dosing. An intravitreal dose is administered at a low dose and is slowly distributed in the eye, resulting in limited systemic exposure. Similarly, pegaptanib is present at the highest concentrations in rabbit eyes, where it is slowly absorbed from the eye into the systemic circulation and is ultimately excreted via the kidneys. Nusinersen is administered via intrathecal injection, allowing it to bypass the blood–brain barrier to reach the central nervous system. Its concentration in the blood is lower than that in the cerebrospinal fluid (CSF). The other ASO drugs are absorbed systemically, affecting a range of organs throughout the body.

Comments: Unlike small-molecule drugs, which are mainly absorbed after oral administration, most ASOs require SC or IV administration. After IV administration, ASO drugs are mainly distributed in tissues such as the liver and kidneys, with a relatively low distribution in other tissues. Effective delivery of ASOs to specific tissues and organs is a major challenge for future R&D. Various techniques are being explored, including conjugating ASOs with fatty acids, peptides, GalNAc, or cell-penetrating peptides. Some of these strategies have been validated clinically [17].

Distribution: PPB affects *in vivo* drug distribution. The PPB rates and distributions of these 10 ASOs vary considerably (Table 5.20). Inotersen, Nusinersen, Defibrotide, and Mipomersen exhibit high PPB (>90%), while Viltolarsen, Golodirsen, Casimersen, and Eteplirsen display low PPB (<40%). Fomivirsen exhibited lower protein binding values in vitreous samples from rabbit and

Table 5.19 Absorption of ASO drugs.

Drug	The site and route of administration	Local/systemic absorption	Blood–brain barrier penetration ability	Tissue with the highest drug concentration	Peak plasma concentration-time (h)
Fomivirsen	Intravitreal injection	Local, eye	No	Retina, iris; low systemic exposure (below the lower limit of quantification)	Extremely low
Pegaptanib	Intravitreal injection	Local, eye	No	Vitreous, retina, aqueous humor, and kidneys	Slow absorption into the systemic circulation
Nusinersen	Intrathecal injection	Central nervous system, systemic	Yes	The central nervous system, plasma, peripheral tissues (skeletal muscle, liver, and kidneys), and fat, bone, bone marrow, and spleen	1.7–6
Mipomersen	Subcutaneous injection	Systematic	No	Liver, kidneys, bone marrow, adipose tissue, and lymph nodes	3–4
Inotersen	Subcutaneous injection	Systematic	No	Liver, kidneys, and other tissues	2–4
Defibrotide	Intravenous infusion	Systematic	No	Epithelial cells lining blood vessels in the liver	2
Eteplirsen	Intravenous infusion	Systematic	No	Kidneys and skeletal muscle	1–2
Golodirsen	Intravenous infusion	Systematic	No	Kidneys and all tissues except central the nervous system	1
Viltolarsen	Intravenous infusion	Systematic	No	Kidneys and skeletal muscle	1
Casimersen	Intravenous infusion	Systematic	No	Kidneys and skeletal muscle	1

Source: Adapted from reference [16].

monkey eyes (<40%). The *in vivo* steady-state volume distribution of a drug represents its distribution range *in vivo*. The steady-state distribution volumes of these ASOs in descending order are as follows: Golodirsen > Eteplirsen > Casimersen, Viltolarsen, Inotersen, and Nusinersen. Notably, the AUC for defibrotide increased by 50–60% in patients with renal impairment and end-stage renal disease compared with patients without renal impairment. There was also a 35–37% increase in the peak concentrations in patients with renal impairment who received treatment.

Comments: ASOs modified with PS (Fomivirsen, Mipomersen, Nusinersen, and Inotersen) are highly bound to proteins in plasma (especially albumin) due to their negatively charged backbones. PPB in plasma was not reported in the Fomivirsen application data. The results from our in-house laboratory test showed that PPB is highly abundant (>99%) in mouse, rat, monkey, and human plasma at a concentration of 1 μM. ASOs with structures modified by the morpholino

Table 5.20 Biodistribution of ASOs.

Drugs	Biodistribution			
	Protein binding	**Bioavailability**	**AUC**	**Distribution volume**
Fomivirsen	40% (vitreous samples from treated rabbits and monkeys)	No specific information	No specific information	No specific information
Pegaptanib	No specific information is available	No specific information	25 μg/h/mL (3 mg, monocular dose)	No specific information
Nusinersen	Cerebrospinal fluid: <25% Plasma: >94%	100% (intrathecal administration)	No specific information	Cerebrospinal fluid: 0.4 L Plasma: 29 L
Mipomersen	≥90% clinical therapeutic concentration (1–8 μg/mL)	~54–78%	No specific information	No specific information
Inotersen	>94% (dose independent)	No specific information	90 μg/h/mL	293 L
Defibrotide	Average of 93%	100% (IV)	26.9–48.1 g/mL × h	8.1–9.1 L
Eteplirsen	6–17%	100% (IV)	No specific information	600 mL/kg
Golodirsen	33–39% (dose independent)	100% (IV)	34–44%[a]	668 mL/kg (at a dose of 30 mg/kg)
Viltolarsen	~40%	100% (IV)	16–27%[a]	300 mL/kg
Casimersen	8–32%	100% (IV)	16–34%[a]	367 mL/kg

[a] Intersubject variability.
Source: Adapted from reference [16].

subunit (PMO), such as Viltolarsen, Golodirsen, Casimersen, and Eteplirsen, all have neutral backbones and exhibit low PPB. Several PMO-ASOs have common features and a high Vd, indicating that they can achieve high distributions in surrounding tissues such as muscles.

Metabolism: These 10 ASOs are metabolized in both blood and target cells. However, they are not metabolized by the liver or by hepatic microsomes *in vitro*. Therefore, most ASOs (Inotersen, Golodirsen, Viltolarsen, Casimersen, Nusinersen, and Eteplirsen) do not interact with conventional drugs or other transporters (Golodirsen, Viltolarsen, and Casimersen). ASOs are primarily metabolized by endonucleases and exonucleases in the blood and in target organs. Most of these drugs are first metabolized by endonucleases, which cleave the oligonucleotides into fragments, and then, the exonucleases can induce hydrolysis reactions at the 3′ and 5′ ends of the fragments. These drugs are chemically modified to better resist nuclease degradation. There are PS bonds in Fomivirsen in its backbone, so it is considered to be the first generation of therapeutic oligonucleotides. Mipomersen and Inotersen contain 2'-O-methoxyethyl ribose modifications and are considered second-generation ASOs with improved resistance to nucleases. Their metabolites have not been found to be pharmacologically active or to pose safety risks to cellular processes.

Comments: ASOs are not expected to be metabolized by CYP450 enzymes due to their high molecular weights and hydrophilicity. In addition, ASOs enter cells primarily through endocytosis. Intracellular free ASOs are thus less likely to be exposed to CYP enzymes in the endoplasmic reticulum [18, 19].

Excretion: These ASO drugs are primarily excreted through the urinary system. Fomivirsen, Eteplirsen, Golodirsen, Viltolarsen, and Casimersen are mainly excreted through the kidneys in unchanged form. Nusinersen, Inotersen, and Mipomersen are mainly excreted as metabolites via the kidneys. These ASO drugs also have different half-lives *in vivo*. Some have longer half-lives, including Fomivirsen, which has a half-life of approximately 55 h in humans and 78 h in monkeys. Nusinersen has a half-life of 135–177 days in CSF and 63–87 days in plasma. Inotersen was administered at a dose of 284 mg/1.5 mL for 1 week, and the total clearance and half-life were 3.18 L/h and 32.3 days, respectively. The half-life of mipomersen is 1–2 months. Pegaptanib has a half-life of approximately 10 days in humans after intravitreal injection, which is 10 times greater than the clinical dose. Viltolarsen, Eteplirsen, Golodirsen, and Casimersen have half-lives of 2.5–3.5 h, and defibrotide has a half-life of several hours. All five ASOs with a short half-life need to be administered by intravenous infusion.

Comments: ASOs with higher PPB have reduced glomerular filtration and decreased renal clearance and thus exhibit longer plasma half-lives. According to the FDA Pharmacology Review for Mipomersen, its half-life is shorter and its clearance is greater in mice than in humans, which may be associated with the relatively lower PPB in mice [20, 21]. ASOs with morpholine (PMO) backbones, such as Eteplirsen and Viltolarsen, are very stable to nucleases and do not produce any metabolites after incubation with plasma or hepatic subcellular fractions *in vitro*, so they are excreted almost entirely in their original form.

5.5.3.3 Current Knowledge Gaps and Future Prospects

Based on the regulatory approval for these 10 ASO drugs and research data in related fields, a review published by researchers from the University of Connecticut revealed that there are gaps in the knowledge regarding the cellular uptake, subcellular trafficking, and cellular efflux of ASOs.

Cellular uptake: It is generally accepted that there are two competing uptake pathways: (i) a "productive uptake pathway" that brings the drug to its target and (ii) a "nonproductive pathway" that brings the drug to saturable sinks. To date, the key determinants governing the proportions of drugs entering these two pathways are unknown.

Subcellular trafficking and distribution: Small-molecule drugs rapidly reach their respective subcellular regions through uptake by transporters or by diffusion across membranes. ASOs enter cells through endocytic uptake, resulting in their encapsulation in endosomes and subsequent transport to their corresponding targets. Endosomal release and intracellular transport are essential factors for the efficacy of ASOs. Thus, the corresponding mechanisms need to be clarified in the future. It is also important to identify the intracellular locations associated with ASO-mediated cellular damage and ways to prevent ASO accumulation in these locations.

Cellular efflux: The mechanisms by which the remaining parent and shortened-chain products are excreted from cells after ASOs are broken down by nucleases are unknown. These mechanisms may affect the plasma concentrations of ASOs and their targets.

Comments: ASOs belong to a new class of molecular entities. Not only are there gaps in scientific understanding, but there are also many uncertainties in the preclinical development of ASOs, such as whether investigators need to carry out relevant research similar to that on small-molecule drugs and include them in regulatory filing packages and how to carry out corresponding studies, since regulatory authorities have not yet issued formal research guidelines in DMPK studies for this class of drugs.

References

1 Chris Maclauchlin (2020). The ADME of siRNA GalNAc conjugates 2020. *16th Oligonucleotide Ther Soc.*

2 Chen, S., Tam, Y.Y.C., Lin, P.J.C. et al. (2014). Development of lipid nanoparticle formulations of siRNA for hepatocyte gene silencing following subcutaneous administration. *Journal of Controlled Release [Internet]* 196: 106–112. Available from: https://doi.org/10.1016/j.jconrel.2014.09.025.

3 Witzigmann, D., Quagliata, L., Schenk, S.H. et al. (2016). Variable asialoglycoprotein receptor 1 expression in liver disease: Implications for therapeutic intervention. *Hepatology Research [Internet]* 46 (7): 686–696. Available from: https://onlinelibrary.wiley.com/doi/abs/10.1111/hepr.12599.

4 European Medicines Agency (2018). European Public Assessment Report: Onpattro (Patisiran). 44 (July). Available from: www.ema.europa.eu/contact

5 Nick Kozauer, MD (Alnylam Pharmaceuticals I. Cross-Discipline Team Leader Review. Patisiran NDA.

6 Alnylam. Givosiran NDA Multi-Discipline Review.

7 European Medicines Agency (EMA)/Committee for Medicinal Products for Human Use (CHMP) 2020 Assessment report for Oxlumo. Vol. 31.

8 Kallend, D., Stoekenbroek, R., He, Y.L. et al. (2022). Pharmacokinetics and pharmacodynamics of Inclisiran, a small interfering RNA therapy, in patients with hepatic impairment. *Journal of Clinical Lipidology [Internet]* 16 (2): 208–219. Available from: https://doi.org/10.1016/j.jacl.2022.01.001.

9 Habtemariam, B.A., Karsten, V., Attarwala, H. et al. (2021). Single-dose pharmacokinetics and pharmacodynamics of transthyretin targeting N-acetylgalactosamine–small interfering ribonucleic acid conjugate, Vutrisiran, in Healthy Subjects. *Clinical Pharmacology and Therapeutics* 109 (2): 372–382.

10 Springer, A.D. and Dowdy, S.F. (2018). GalNAc-siRNA conjugates: leading the way for delivery of RNAi therapeutics. *Nucleic Acid Therapeutics* 28 (3): 109–118.

11 Andersson, P. and Den Besten, C. (2019). CHAPTER 20: preclinical and clinical drug-metabolism, pharmacokinetics and safety of therapeutic oligonucleotides. *RSC Drug Discovery Series* 474–531.

12 McDougall, R., Ramsden, D., Agarwal, S. et al. (2021). The nonclinical disposition and PK/PD properties of GalNAc-conjugated siRNA are highly predictable and build confidence in translation to man. *Drug Metabolism and Disposition* 50 (6): 781–797. DMD-MR-2021-000428.

13 Jing, X., Arya, V., Reynolds, K.S., and Rogers, H. (2022). Clinical pharmacology of RNAi-based therapeutics: a summary based on FDA-approved small-interfering RNAs. *Drug Metabolism and Disposition* DMD-MR-2022-001107. 51(2): 193–198

14 Li, J., Liu, J., Zhang, X. et al. (2021). Nonclinical pharmacokinetics and absorption, distribution, metabolism, and excretion of Givosiran, the first approved n-acetylgalactosamine-conjugated rna interference therapeutic. *Drug Metabolism and Disposition* 49 (7): 572–580.

15 Humphreys, S.C., Davis, J.A., Iqbal, S. et al. (2022). Considerations and recommendations for assessment of plasma protein binding and drug-drug interactions for siRNA therapeutics. *Nucleic Acids Research* 50 (11): 6020–6037.

16 Migliorati, J.M., Liu, S., Liu, A. et al. (2022). Absorption, distribution, metabolism, and excretion of US food and drug administration-approved antisense oligonucleotide drugs. *Drug Metabolism and Disposition* 50 (6): 888–897.

17 Zanardi, T.A., Korbmacher, B., Boone, L. et al. (2021). Safety, pharmacokinetic, and pharmacodynamic evaluation of a 2′-(2-methoxyethyl)-D-ribose antisense oligonucleotide-triantenarry N-Acetyl-galactosamine conjugate that targets the human transmembrane protease serine 6. *The Journal of Pharmacology and Experimental Therapeutics* 377 (1): 51–63.

18 Andersson, S., Antonsson, M., Elebring, M. et al. (2018). Drug metabolism and pharmacokinetic strategies for oligonucleotide- and mRNA-based drug development. *Drug Discovery Today* 23 (10): 1733–1745.

19 Kazmi, F., Yerino, P., McCoy, C. et al. (2018). An assessment of the *in vitro* inhibition of cytochrome P450 enzymes, UDP-glucuronosyltransferases, and transporters by phosphodiester- or phosphorothioate-linked oligonucleotides. *Drug Metabolism and Disposition* 46 (8): 1066–1074.

20 Geary, R.S., Baker, B.F., and Crooke, S.T. (2015). Clinical and preclinical pharmacokinetics and pharmacodynamics of mipomersen (Kynamro®): a second generation antisense oligonucleotide inhibitor of apolipoprotein B. *Clinical Pharmacokinetics* 54 (2): 133–146.

21 Yu, R.Z., Kim, T.W., Hong, A. et al. (2007). Cross-species pharmacokinetic comparison from mouse to man of a second-generation antisense oligonucleotide, ISIS 301012, targeting human apolipoprotein B-100. *Drug Metabolism and Disposition* 35 (3): 460–468.

6

mRNA-Based Vaccines and Therapeutics

6.1 What Are mRNA Vaccines and Their Future

Lian Guo, Lingling Zhang, Weiqun Cao

6.1.1 Introduction

Since the emergence of the pandemic in late 2019, more than one billion people worldwide have been infected by severe acute respiratory syndrome coronavirus 2 (SARS-CoV-2). In February 2020, the disease caused by SARS-CoV-2 was officially named coronavirus disease (COVID-19). With the persistent spread of SARS-CoV-2, mutated virus strains have continuously emerged, ranging from the Alpha and the Delta variants to the Omicron variants that spread globally. These viral variants have contributed to recurrent COVID-19 outbreaks on a global scale. COVID-19 vaccines have garnered significant public attention and sparked discussions as a crucial tool in combating the pandemic. An overview of the progress made in the development of vaccines and the applications of mRNA vaccines during the COVID-19 pandemic are given in this chapter. Furthermore, this chapter discusses the methods used to conduct PK studies on mRNA vaccines, based on the interpretation of relevant guidelines.

6.1.2 The Overview of Vaccine Development Throughout the COVID-19 Pandemic

6.1.2.1 Pathogenesis of COVID-19

The mode of SARS-CoV-2 infection is mainly through the respiratory tract, and the most severe illness it can potentially induce in humans is pneumonia. The virus can infect humans through the mucous membranes of the mouth, nose, and eyes and migrate along the upper respiratory trachea to reach the lower respiratory tract, eventually reaching the lungs. Once the virus reaches the alveoli, the spike (S) protein within the virus binds to the ACE2 receptor located on type II alveolar cells, allowing the virus to enter the cells through a process called endocytosis. The virus then releases RNA to seize control of the function of the original cell and achieve massive replication. After fully utilizing the cell's replicative capacity, the virus causes the cell to rupture, releasing viral particles that infect surrounding type II alveolar cells (Figure 6.1) [1]. After sufficient exposure to the virus, the adaptive immune system of the body becomes activated. While it aims to attack infected tissues, it may also kill healthy alveolar cells, resulting in impaired gas exchange and reduced respiratory function. In severe cases, this can progress to acute respiratory distress syndrome and even death.

Drug Metabolism and Pharmacokinetics: Frontiers, Strategies, and Applications, First Edition.
Edited by Liang Shen and WuXi AppTec DMPK.

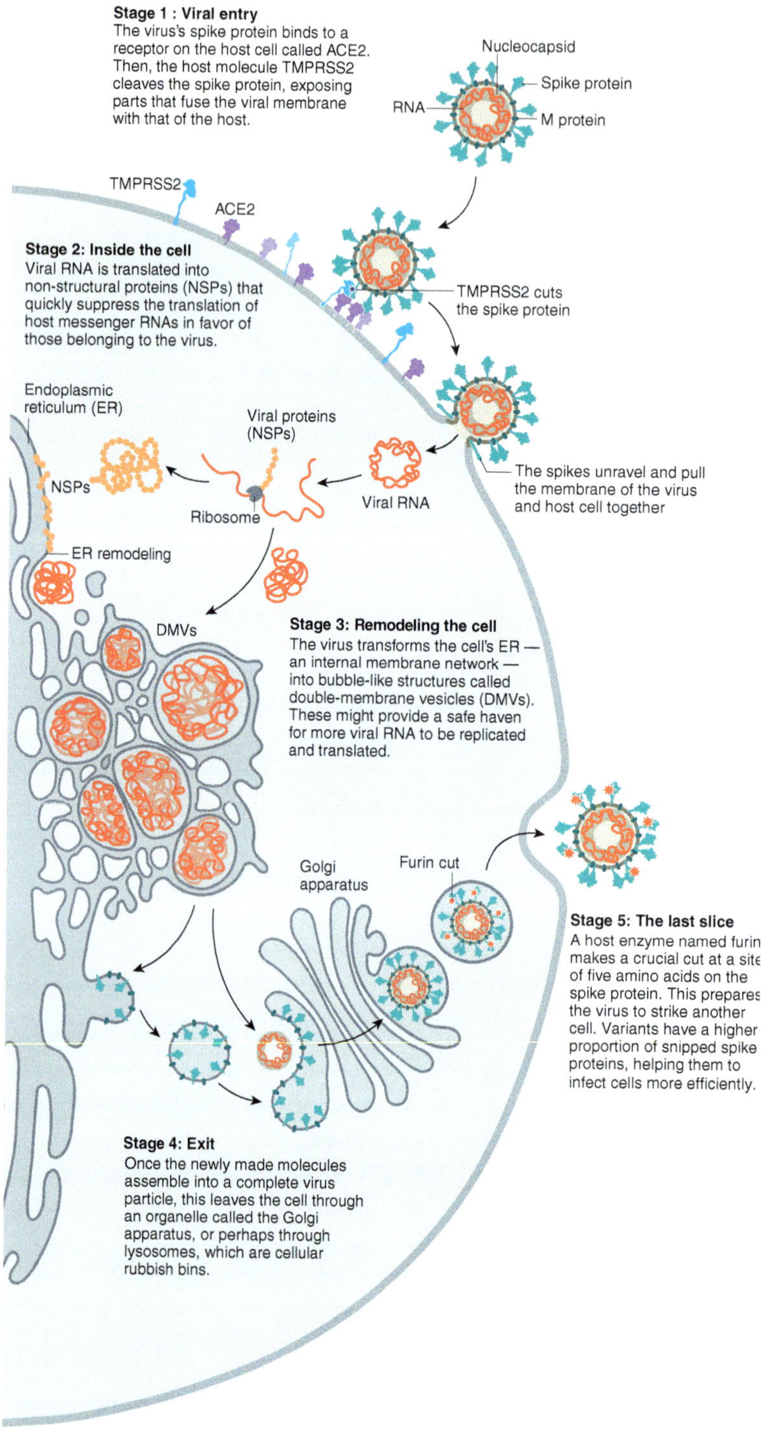

Figure 6.1 Schematic diagram of cellular entry and exit of SARS-CoV-2. *Source:* [1] / with permission of Springer Nature.

6.1.2.2 Mechanism of Action of Vaccines

Vaccines utilize pathogenic components to stimulate the immune system, triggering the production of specific memory B and T cells. This establishes acquired immunity, enabling the body to mount a rapid response when encountering the same pathogen in the future. By swiftly eliminating the pathogen and infected cells, widespread infection can be prevented. In the development of COVID-19 vaccines, the S protein, which plays a crucial role in the pathogenesis of the disease, has been employed as an antigen.

6.1.2.3 Analysis of the Advantages and Disadvantages of Different Types of Vaccines

The types of vaccines designed and developed to ameliorate SARS-CoV-2 pathogenesis include inactivated virus vaccines, recombinant protein subunit vaccines, adenovirus vector vaccines, and mRNA vaccines.

Inactivated vaccines are developed by cultivating the pathogen in a special culture medium and subsequently rendering the virus non-replicative through physical or chemical methods. This process ensures the preservation of the virus's antigenicity. These vaccines offer the advantage of being relatively simple and quick to prepare, as well as being considered safe. They are commonly utilized for addressing acute disease transmission. However, their disadvantages include the requirement for a single immunization route, the necessity for large-scale culture of the virus under the conditions of the P3 laboratory, and the potential for antibody-dependent enhancement, which can exacerbate viral infections.

Recombinant protein subunit vaccines involve the insertion of gene segments encoding "effective antigenic components" from the pathogen's genome into the genome of host cells. These host cells are then able to express the desired antigenic components *in vitro*. This type of vaccine does not involve live pathogens in the production process, thus eliminating the risk of pathogen leakage. Additionally, host cells can be rapidly produced on a large scale through industrial fermentation and similar techniques, which offers significant cost and efficiency advantages to this method.

Adenovirus vector vaccines are produced by introducing genes that encode specific antigens of a virus into the genome of a non-replicable adenovirus through recombination. This enables the expression of desired antigens in human cells. This type of vaccine is associated with minimal adverse reactions and can be easily produced on a large scale. However, since many individuals have been exposed to various types of adenoviruses, they may possess antibodies that can neutralize the adenoviral vector. This can potentially lead to the reduced efficacy of the vaccine.

mRNA vaccines involve the direct injection of modified mRNA encoding pathogen-specific antigens into the human body following certain processing steps. This allows the body's protein synthesis system to produce the desired antigens. This type of vaccine does not involve live pathogens in either the production or vaccination processes, thereby eliminating the risk of virus transmission and infection. Additionally, mRNA vaccines can specifically target the expression of desired antigens, avoiding any adverse effects that other components of pathogens may have on the human body. Furthermore, mRNA vaccines involve the "infection" of cells, which can stimulate both humoral and cellular immunity. However, because of the shorter development timeline of this vaccine technology compared to other methods, extensive validation is necessary for its application.

6.1.2.4 Approved and Emergency Authorized COVID-19 Vaccines

There are multiple COVID-19 vaccines following different technological pathways approved and granted emergency authorization for global use, following different technological pathways. By the end of 2022, thirteen COVID-19 vaccines had been approved or authorized for emergency use in China. These vaccines have demonstrated varying levels of protective effects during the clinical trials. Based on phase III trial data publicly available, several inactivated virus and adenovirus vector vaccines have demonstrated protective efficacies ranging from 50% to 85%. In contrast, mRNA vaccines developed by both Pfizer/BioNtech and Moderna have shown protection rates exceeding 90% [2, 3]. These mRNA vaccines play an important role in controlling the global outbreak of COVID-19. As SARS-CoV-2 continues to mutate, the Pfizer/BioNTech mRNA vaccine has also been studied for protection against mutant strains. The reported efficacy of two doses of the Pfizer/BioNTech mRNA vaccine against the Delta strain is 88% in England [4], 79% in Scotland [5], and 87% in Canada [6]. For the Omicron variant, the reported efficacy of protection is 75.5% in England [7]. It is suggested that this protection may have been derived from cellular immunity. Studies have shown that a majority of T-cell responses are generated by invariant epitopes between the prototype B-lineage virus and the variant of concern [8].

The ongoing COVID-19 pandemic has highlighted the importance of vaccines and the urgent global demand for them. mRNA vaccines have garnered global attention, and it is crucial to recognize their potential for application in other fields and domains.

6.1.3 Application Prospects for mRNA Vaccines

Theoretically, mRNA can express any protein, offering a highly promising technology platform for the treatment of diverse diseases. From 2003 to 2020, over 140 mRNA vaccines have been evaluated in clinical trials, encompassing a range of disease areas such as cancer and infectious diseases (Figure 6.2). This demonstrates the increasing application and attention being devoted to mRNA vaccines in various fields [9].

6.1.3.1 mRNA Vaccines Against Infectious Diseases

Developing preventive vaccines targeting infectious pathogens is the most effective approach to control and prevent epidemics. The commercial vaccine development and approval process for traditional vaccines is slow, posing challenges in addressing rapidly emerging acute viral diseases. Consequently, mRNA vaccines have garnered increased attention owing to their short development timelines and streamlined production processes.

6.1.3.2 mRNA Vaccines Against Cancer

Cancer vaccines and other immunotherapies offer novel strategies for treating malignant tumors. These vaccines utilize tumor antigens to specifically stimulate the body's immune response against tumor cells. mRNA vaccines activate cellular immunity, which involves T cells that efficiently target and kill tumor cells. The mRNA sequence can be precisely edited to control the production of specific antigen types and sequences. By selecting specific antigen protein markers that are exclusively secreted in tumor cells, potential damage to normal cells can be avoided, allowing for more precise targeting of specific immune targets. Over 50% of clinical trials in mRNA vaccine development are dedicated to the treatment of melanoma, prostate cancer, and brain cancer (Figure 6.3). The majority of these trials are still in the early stages, specifically phases I and II [9].

In summary, with the growing number of clinical trials and expanding applications of mRNA vaccines, coupled with the approval of mRNA COVID-19 vaccines, the mRNA vaccine technology is rapidly advancing and displaying promising prospects for further development.

6.1.4 Insight into the DMPK Studies for mRNA Vaccine Development

mRNA vaccines belong to the class of nucleic acid vaccines. In this type of vaccine, a synthetic mRNA fragment derived from a virus is produced *in vitro* and injected into humans. This mRNA

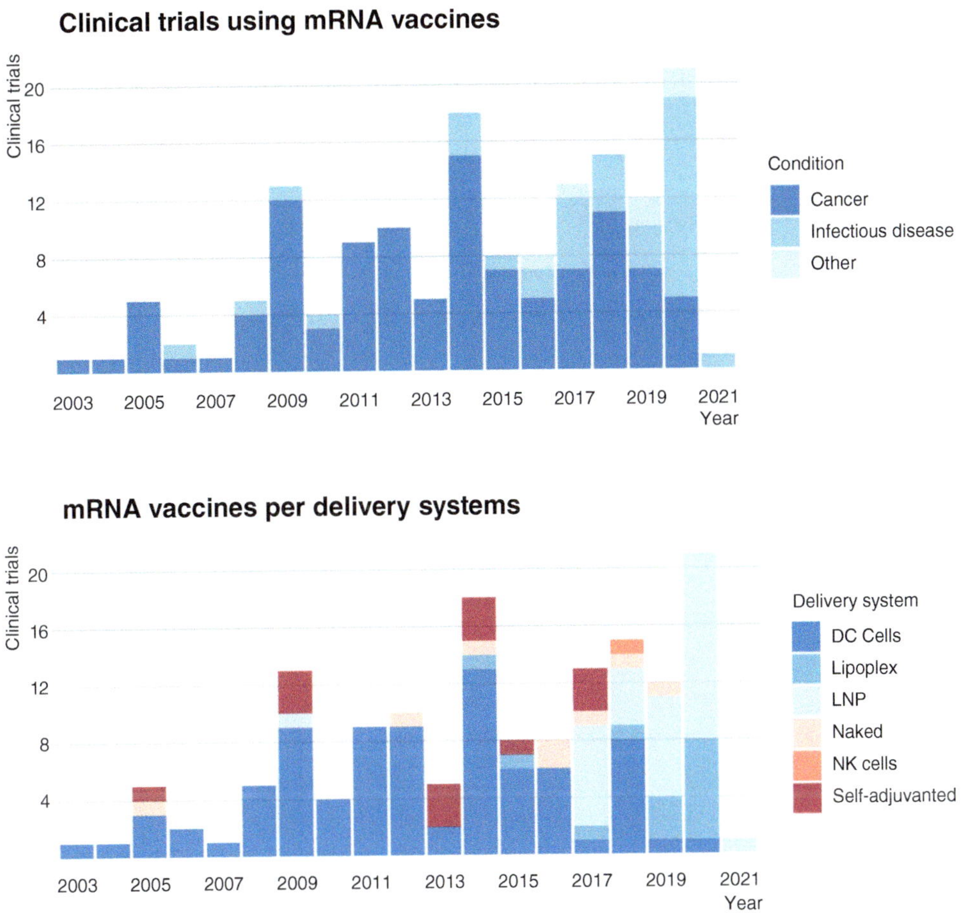

Figure 6.2 Number of clinical trials using mRNA vaccines summarized by disease type (top) and delivery system (bottom). *Source:* [9] / ELSEVIER.

fragment instructs the cells to produce the corresponding antigen, which stimulates the specific immune response in the body and leads to the development of immune memory (Figure 6.4). The entry of intact mRNA into the cytoplasm is pivotal to vaccine function. However, the single-stranded structure of mRNA renders it highly unstable and susceptible to degradation. mRNA is a large molecule carrying a negative charge, posing a challenge in its delivery across the cell membrane composed of anionic lipids. Therefore, special modifications or encapsulation delivery systems are required to achieve intracellular expression of mRNA.

LNPs are widely employed as a delivery system for mRNA. This approach has been utilized in both approved mRNA vaccines for SARS-CoV-2. LNPs refer to a class of lipid particles, including lipid gels and liposomes. In April 2018, the FDA released guidance concerning the utilization of liposomal drugs [11]. This guidance emphasized the necessity of studying their PK and mass balance, as well as metabolism information, such as the major metabolites associated with the therapeutic and toxic effects. In August 2021, the NMPA published the Technical Guidelines for Non-Clinical PK Studies of Nanomedicines (Trial) [12]. This guidance underscored the importance of investigating the *in vivo* release rate and distribution of particles, as well as the PK characteristics of new excipients or carrier materials.

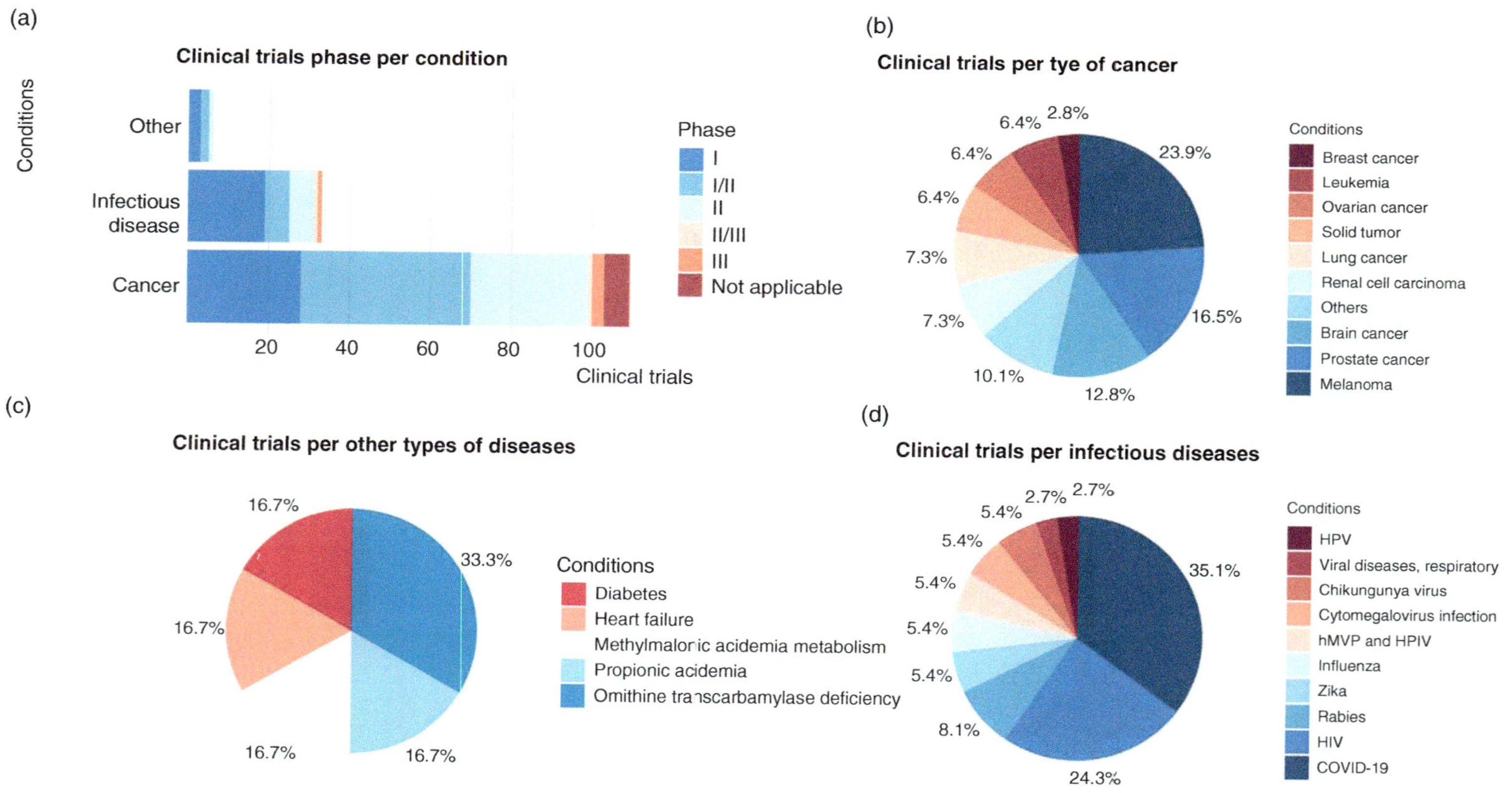

Figure 6.3 Number of clinical trials and application areas for mRNA vaccines. *Source:* [9] / ELSEVIER.

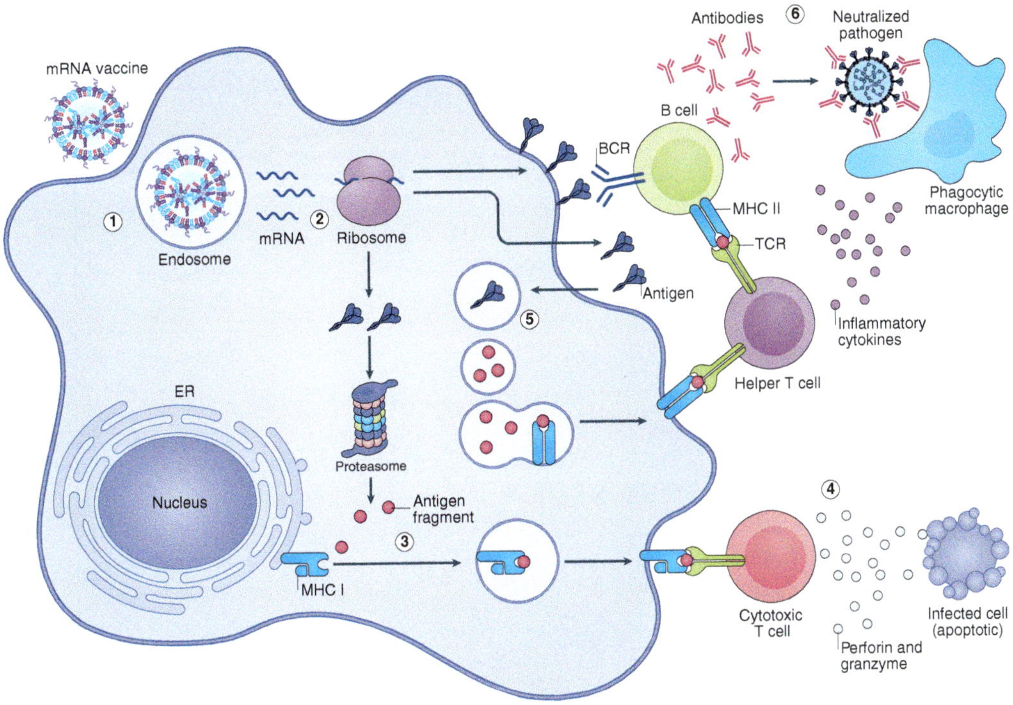

Figure 6.4 Mechanism of action of mRNA vaccines. *Source:* [10] / with permission of Springer Nature.

The liposomal-siRNA drug, Onpattro, which was approved in 2018 underwent adequate pre-clinical DMPK studies, which involved studies on siRNA (ALN-18328) and components of the LNP, including DLin-MC3-DMA and PEG2000-C-DMG. This included investigating the distribution of 14C-labeled DLin-MC3-DMA in rats using the QWBA technique, as well as studying the metabolism and excretion of Onpattro and the LNP component (DLin-MC3-DMA) using^{14}C-labeled DLin-MC3-DMA [13].

Hence, prior to the utilization of mRNA vaccine formulations in healthy populations, it is crucial to carry out comprehensive risk assessments. These assessments should encompass evaluations of both the mRNA itself and the delivery system employed. The duration of mRNA present in the body, as the agent responsible for generating antigenic proteins, can influence the efficacy of antibodies. Therefore, it is essential to conduct comprehensive research on the absorption, metabolism, and excretion properties of LNPs, as exogenous agents in delivery systems, within the human body. Radiolabeling techniques can be employed to investigate the distribution, metabolism, and excretion of mRNA vaccines and their LNP components.

6.1.5 Summary

mRNA vaccines offer several advantages, including safety, efficacy, and high production efficiency. In recent years, there has been an upward trend in the number of clinical trials focused on mRNA vaccines, spanning diverse therapeutical areas such as infectious diseases and oncology. The approval of mRNA vaccines for COVID-19 has validated the applicability of the mRNA technology platform in the field of vaccines. As a result, there has been an increasing emphasis on mRNA vaccines. Although the development of mRNA vaccines encounters various challenges, the continuous emergence of new delivery methods holds the potential for significant breakthroughs in mRNA vaccine research. The development of cancer vaccines offers new possibilities for the treatment of malignant tumors, instilling hope in a larger population of patients.

References

1 Megan, S. (2021). How the coronavirus infects cells — and why Delta is so dangerous. *Nature* 595: 29.
2 COMIRNATY® (COVID-19 Vaccine, mRNA) suspension for injection, for intramuscular use Initial U.S. Approval. (2021).
3 MODERNAT X, Inc. Safety and Immunogenicity of a 50 μg Booster Dose of mRNA-1273 (Moderna COVID-19 Vaccine).(2021)
4 Jamie Lopez Bernal, F.F.P.H., Ph.D., Nick Andrews, Ph.D., et al. Effectiveness of Covid-19 Vaccines against the B.1.617.2 (Delta) Variant. N Engl J Med 2021;385:585–594
5 Aziz Sheikha· Jim McMenaminb · Bob Taylorb, et al. SARS-CoV-2 Delta VOC in Scotland: demographics, risk of hospital admission, and vaccine effectiveness. CorrespondenceVolume 397, Issue 10293p2461-2462June 26, 2021
6 Sharifa Nasreen, Hannah Chung, Siyi He, Kevin A. Brown, et al. Effectiveness of COVID-19 vaccines against symptomatic SARS-CoV-2 infection and severe outcomes with variants of concern in Ontario. Nature Microbiology volume 7, pages379–385 (2022)
7 Nick Andrews, Ph.D., Julia Stowe, Ph.D., Freja Kirsebom, Ph.D., Samuel Toffa, Ph.D., , et al. (2022). Covid-19 Vaccine Effectiveness against the Omicron (B.1.1.529) Variant, N Engl J Med 2022;386:1532–1546
8 Pinja, J., Pekka, K., Hanni, K.H. et al. (2021). COVID-19 mRNA vaccine-induced antibody responses against three SARS-CoV-2 variants. *Nature Communications* 12: 3991.
9 Sara, S.R., Duarte, M.F.P., Ana, M. et al. (2021). mRNA vaccines manufacturing: challenges and bottlenecks. *Vaccine* 39: 2190–2200.
10 Namit, C., Drew, W., and Kathryn, A. (2021). Whitehead. mRNA vaccines for infectious diseases: principles, delivery, and clinical translation. *Nature Reviews Drug Discovery* 20: 817–838.
11 CDER, FDA (2018). *Guidance for Industry: Liposome Drug Products: Chemistry, Manufacturing, and Controls; Human Pharmacokinetics and Bioavailability; and Labeling Documentation*. Silver Spring, MD.
12 NMPA. (2021). Technical guidelines for non-clinical pharmacokinetic research of nanodrugs trials.
13 EMA. (2018). Onpattro-epar-public-assessment-report.

6.2 Strategies for Preclinical PK Evaluation of mRNA-Based Therapeutics: mRNA Vaccines, mRNA Therapeutic Drugs, and Novel Excipients

Liping Ma, Huan Liu, Jing Jin, Maotian Zhou

6.2.1 Introduction

mRNA is a single-stranded ribonucleic acid transcribed from DNA, which carries the genetic information for the synthesis of proteins and can be translated into various types of functional proteins and peptides. The earliest experiments that demonstrate the successful expression of *in vitro* transcribed (IVT) mRNAs, which encoded related proteins in murine muscle cells, have laid the foundation for the field of mRNA therapy [1]. mRNA therapeutics can theoretically be translated into any protein or peptide by the ribosome expression system. Compared to DNA-based gene therapy, mRNA drugs have relatively high transfection efficiency and lower toxicity, as their mode of action does not require nuclear entry. Critically, mRNA drugs circumvent the risk of insertional mutagenesis and viral infection, as well as the nontarget or off-target toxicity due to modifications to the mRNA sequence. In contrast to traditional protein therapies, mRNA therapeutics can be continuously translated into encoded proteins, thereby offering better efficacy and therapeutical potential.

6.2.2 Development History and Mechanism of Action of mRNA Drugs

The development of mRNA drugs can be divided into three stages, the first stage is from 1961 to 1990, and the key discoveries include the development of mRNA, the mechanisms of mRNA capping and tailing, *in vitro* synthesis of mRNA, elucidation of mRNA translation process, and the construction of early nucleic acid delivery systems. This stage contributes to a deeper understanding of mRNA's structure and function and lays the groundwork for the development of early protamine- or liposome-based delivery systems. The second stage, which spanned from 1990 to 2019, was characterized by a growing comprehension of mRNA mechanisms, the establishment of businesses specializing in mRNA therapeutics, and the emergence of various therapeutic techniques, such as experimental mRNA-based cancer immunotherapies, activation of anti-tumor T cells, targeted vaccines for influenza and experimental vaccines for respiratory syncytial virus (RSV), as well as personalized cancer vaccines. The third stage, from 2019 to the present, has been marked by the commercialization of mRNA-1273 (Moderna) and BNT-162b (Pfizer-BioNTech) vaccines against COVID-19. In this period, mRNA drugs have entered a new stage of flourishing, giving rise to a variety of therapeutic modalities, carrier technologies, mRNA modifications, and coding strategies. The development history and critical findings of mRNA drugs are shown in Figure 6.5.

Broadly speaking, mRNA drugs can be categorized into mRNA vaccines and mRNA therapeutics drugs. mRNA vaccines mainly function by inducing immune cells to generate corresponding antibodies by expressing pathogen-associated proteins (usually antigens that are easily recognized by immune cells, such as viral capsid protein). Specifically, this process involves three steps: first, mRNA is delivered into antigen-presenting cells (APCs); second, antigenic proteins are translated by ribosomes after endocytosis, and subsequently transported into the cell surface and presented by MHCI to CD8-positive cytotoxic T cells. The presentation ultimately triggers B cells to differentiate into plasma cells and generate the corresponding antibodies. Third, additional antigenic proteins released into the extracellular environment are taken up by MHCII and broken down into polypeptide antigenic fragments and presented to CD4-positive helper T cells. (Figure 6.6) In

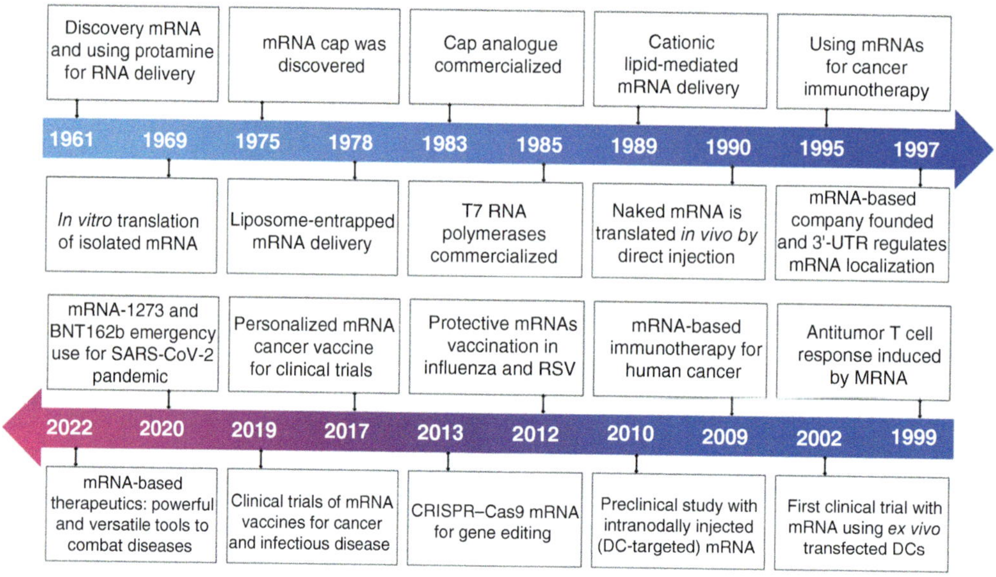

Figure 6.5 Development history and key findings of mRNA drugs. *Source:* [2] / Springer Nature / CC BY 4.0.

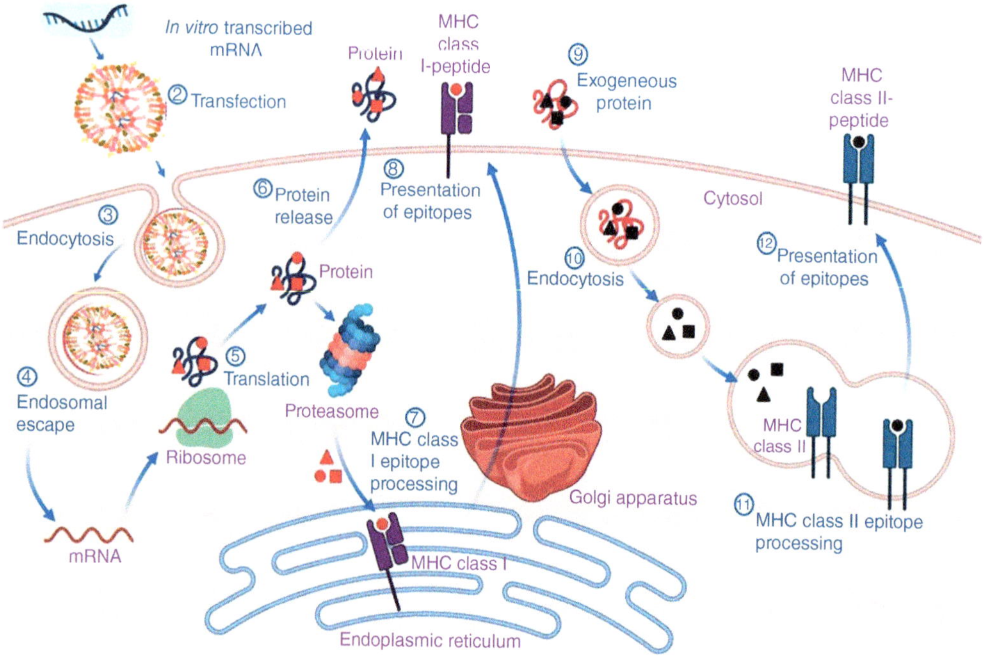

Figure 6.6 Schematic diagram of the onset of action of mRNA vaccines, including liposome-encapsulated mRNA entering cells and activating humoral and cellular immunity, respectively, by acting as endogenous and exogenous antigens after expression. *Source:* [3] / MDPI. / CC BY 4.0.

contrast to traditional inactivated vaccines or antigen protein vaccines, mRNA vaccines have the potential to activate both cell-mediated immunity and humoral immunity.

The potential of mRNA vaccines has been highlighted by the great success of COVID-19 mRNA vaccines, with an increasing array of novel mRNA vaccines continuing to emerge. According to the Next Pharma database, more than 500 mRNA vaccines are under development, including mRNA vaccines targeting infectious diseases such as influenza (Influenza), RSV, human papillomavirus (HPV), Epstein–Barr virus (EBV) as well as human immunodeficiency virus (HIV). There have been numerous endeavors to develop mRNA cancer vaccines due to their ability to stimulate both humoral and cellular immunity. Additionally, mRNA vaccines can effectively cope with the difficulties of tumor individual heterogeneity and tumor microenvironment suppression due to their multiple coding capacity, which allows for the simultaneous encoding of hundreds of tumor antigens and regulatory proteins, such as cytokines. In comparison to conventional cancer vaccines, mRNA vaccines are tailored with protein or peptide antigens and have significant advantages in personalized medicine owing to their design and synthesis flexibility, fast research and development cycle, and mature delivery methods.

mRNA therapeutic drugs, which essentially function as expressed functional effector proteins for medical purposes, are often known as protein replacement therapies. This approach has been explored across a range of disease research areas, including cancer, cardiovascular, pulmonary, hematological, metabolic, neurological disease, and muscular dystrophies. However, to date, only a limited number of mRNA therapeutic drugs have entered the clinical stage, with the majority remaining in preclinical development [4]. Examples of such drugs include vascular endothelial growth factor A (VEGF-A) for heart failure, human fibrosis transmembrane regulatory protein (cystic fibrosis transmembrane conductance regulator [CFTR]) to cure pulmonary fibrosis, as well as methylmalonyl-CoA mutase (MUT) in the treatment of methylmalonic acidemia, etc.

With the extraordinary clinical success of two mRNA vaccines, mRNA-based therapeutics are experiencing a global development boom. mRNA holds immense potential in the fields of vaccines and therapeutic drugs. Theoretically, mRNA can encode any protein, making it possible for mRNA-based therapeutics to potentially replace all protein-based therapeutics. Moreover, mRNA can express therapeutic proteins within target cells, overcoming the challenge associated with delivering proteins to intracellular targets. Compared to protein-based drugs, mRNA-based therapeutics offer considerable advantages, such as cost-effectiveness and a shorter development cycle. Given the promising nature of mRNA-based therapeutics, it is crucial to understand how to conduct the PK evaluation for this new modality.

6.2.3 Types and Properties of mRNA-Based Therapeutics: mRNA Vaccines and mRNA Therapeutic Drugs

mRNA-based therapeutics can be classified into two categories based on the encoded protein type. The first category is known as mRNA vaccines, which encode antigenic proteins. Depending on the type of disease, they can be further divided into infectious disease vaccines and tumor vaccines. The second category refers to mRNA therapeutic drugs, which encode nonantigenic proteins. This category can be subdivided into mRNA-encoded antibodies, enzymes, cytokines, growth factors, and so on (Figure 6.7). There are several drugs in each of these categories that have entered the clinical stage [5, 6].

There are several differences between mRNA vaccines and therapeutic drugs. First, mRNA vaccines encode antigenic proteins and generally require only a low level of protein expression to exert their effects. The induction of antigenicity in the body reflects their efficacy. In contrast, mRNA therapeutic drugs generally require a high level of protein expression. It is desirable to minimize

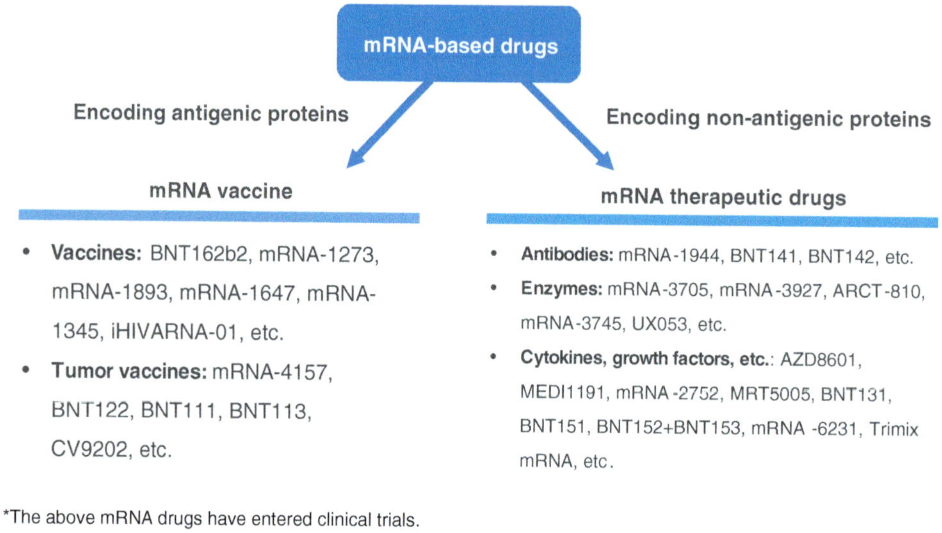

Figure 6.7 Classification of mRNA-based therapeutics.

the stimulation of immunogenicity as it may reduce the concentration of the therapeutic protein. Second, vaccines do not require tissue-specific delivery, whereas mRNA therapeutic drugs must be delivered to specific tissues for translation. Third, most vaccines are injected into muscles because muscle tissue has abundant blood flow and high levels of infiltrating immune cells, whereas mRNA therapeutic drugs are mostly administered systemically or topically to target tissues according to their therapeutic purpose. Overall, the technology for mRNA vaccine development has been well established, whereas the development of mRNA therapeutic drugs has encountered several challenges. These challenges include enhancing protein expression levels, reducing immunogenicity, improving tissue targeting, achieving long-term effects, and ensuring long-term safety for repeated use.

6.2.4 Insights into PK Studies of mRNA-Based Therapeutics

According to relevant technical guidelines for vaccines issued by the FDA and the NMPA, **PK studies are typically not required for vaccines. However, several specific vaccines, including mRNA vaccines, need biodistribution studies.** Based on the mechanism of action of mRNA, mRNA vaccines encode antigen proteins and utilize the immune system to amplify their effects. The duration of efficacy is not directly related to the expression level and duration of the antigen. Moreover, mRNA therapeutic drugs encode functional proteins that directly exert their effects. The duration of their efficacy is related to the level and duration of protein expression, indicating a dose–effect relationship. As a result, mRNA vaccines typically do not require PK studies, while PK studies of mRNA therapeutic drugs are conducive to understanding the dose–response relationship of the drugs.

The structure of mRNA-based therapeutics also affects PK studies. A complete mRNA structure contains five components, including a 5′ cap structure, 5′ and 3′ untranslated regions (UTRs), a coding region, and a 3′ poly(A) tail region. Several strategies have been developed to optimize mRNA structures as below (Figure 6.8). From the perspective of PK studies, the administration dose of amplified mRNA is relatively low, which requires higher detection sensitivity. PCR

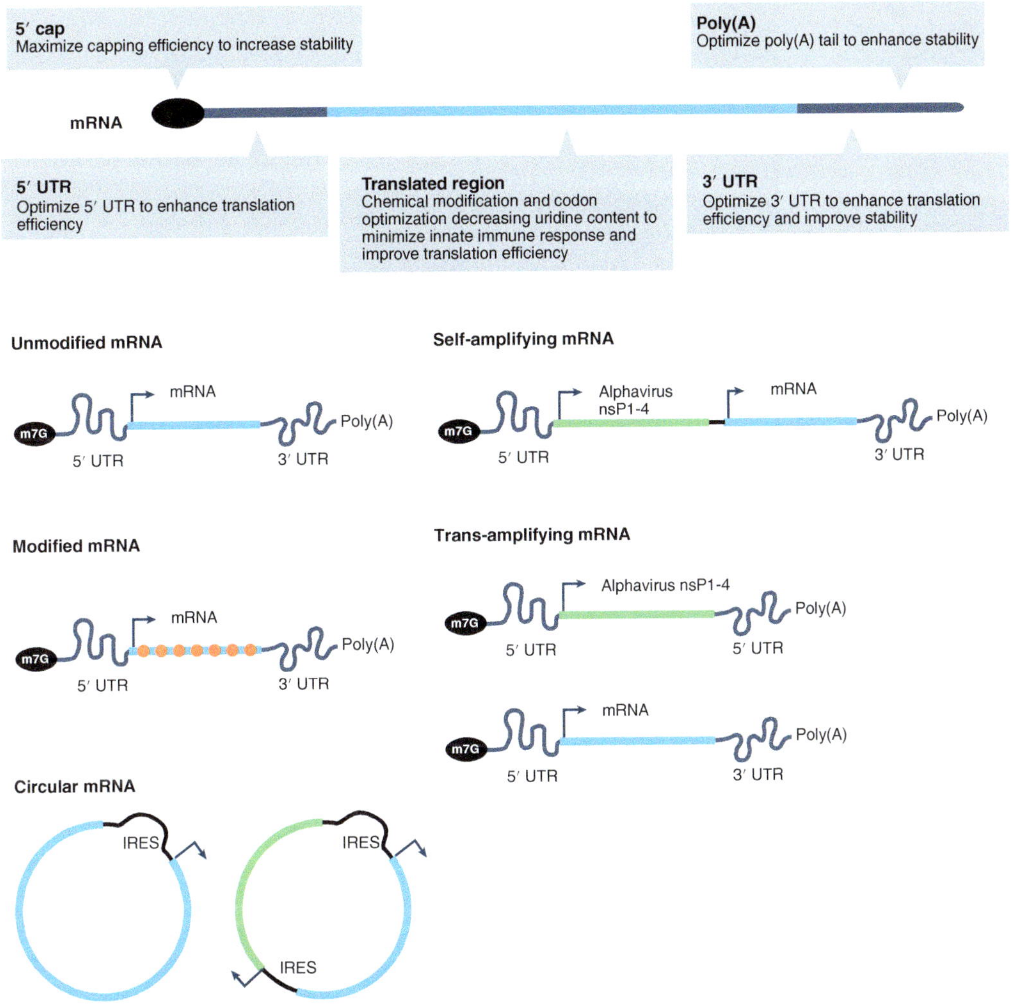

Figure 6.8 Structures of mRNA-based therapeutics and common optimization strategies. *Source:* [6] / with permission of Springer Nature.

amplification of circular mRNA needs to comprise the Junction site to demonstrate its circularization feature.

1) 5′ cap and 3′ tail modifications improve mRNA stability.
2) 5′ UTR and 3′ UTR optimizations improve translation efficiency.
3) Chemical modification of the coding region improves translation efficiency and reduces innate immunity.
4) Self-amplifying mRNA and trans-amplifying mRNA.
5) Circular mRNA.

The development of mRNA delivery systems is also one of the hot spots in research and development. Single-stranded mRNA is easily degraded by nucleases and its highly electronegative property makes it difficult to cross the cell membrane. Thus, specific carriers are required for delivering mRNA into target cells. The most commonly used delivery system is LNPs, and extracellular

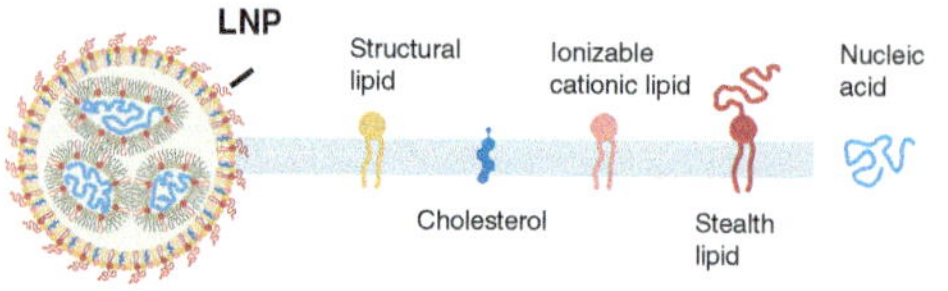

Figure 6.9 Illustrated composition of the lipid nanoparticle (LNP) delivery system. *Source:* [6] / with permission of Springer Nature.

vesicles such as exosomes, bionic vesicles, and even cells are also employed for mRNA delivery. PK studies have primarily focused on the LNP delivery system because one of its principal constituents, ionizable cationic lipids, predominantly contributes to cytotoxic effects, whereas PEGylated lipids tend to cause immunogenicity. If the cationic lipids or other LNP components are novel pharmaceutical excipients, it is necessary to follow the Guidelines "Nonclinical Studies for the Safety Evaluation of Pharmaceutical Excipients," which mandate that "It is recommended that the absorption, distribution, metabolism, and excretion of the excipient be studied following administration by the clinically relevant routes to the same species that are used in the nonclinical safety studies" (Figure 6.9).

6.2.5 PK Studies and Case Analysis of mRNA-Based Therapeutics

PK studies typically include assessments of ADME, and DDIs, as shown in Table 6.1. For mRNA vaccines, PK studies are generally not required, with only biodistribution studies typically being conducted. In the case of mRNA therapeutic drugs, studies of absorption and distribution are necessary. For novel pharmaceutical excipients, a comprehensive ADME study is essential. The potential risk of DDI with pharmaceutical excipients should be evaluated on a case-by-case basis.

Case Study 1: Biodistribution Study of mRNA Vaccines

The left panel of Figure 6.10 illustrates the biodistribution of the BNT162b2 vaccine, which was studied using an LNP-encapsulated luciferase mRNA substitution. Theoretically, the role of LNP determines the tissue specificity of delivery, and loading different mRNAs does not affect the delivery result. Therefore, the tissue-specific expression of luciferase mRNA can reflect the biodistribution of mRNA delivered by the same LNP. The detection method used in this study was *in vivo* bioluminescence imaging. The right panel of Figure 6.10 shows the biodistribution results presented in the mRNA-1273 application files. These results were also generated using an

Table 6.1 PK studies of mRNA drugs.

Pharmacokinetic study content	mRNA vaccines	mRNA therapeutic drugs	New pharmaceutical excipients
Absorption	—	√	√
Distribution	√	√	√
Metabolism	—	—	√
Excretion	—	—	√
Drug interactions	—	—	Optional
Common bioanalytical methods	qPCR, bDNA, ddPCR, *in vivo* imaging, LBA	qPCR, bDNA, ddPCR, *in vivo* imaging, LBA	LC-MS/MS, Radioactivity analysis

qPCR, quantitative polymerase chain reaction; bDNA, branched DNA; ddPCR, droplet digital polymerase chain reaction; LBA, ligand binding assay; LC-MS/MS, liquid chromatography–tandem mass spectrometry.

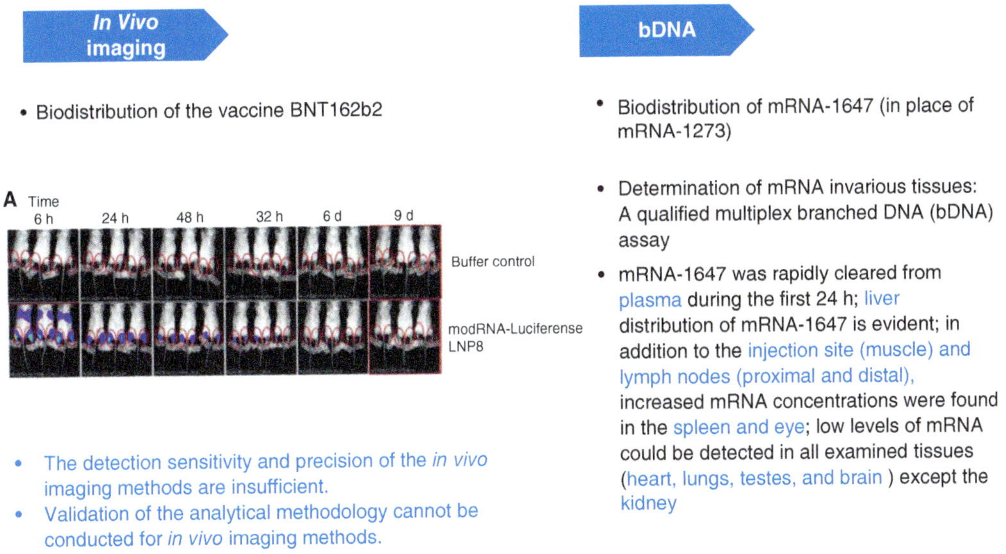

Figure 6.10 Biodistribution studies of approved mRNA vaccines. *Source:* Adapted from [7] / Public Health and Medical Professionals for Transparency.

LNP-encapsulated alternative mRNA with a bDNA-based detection method. The comparison of the biodistribution results of these two approved mRNA vaccines shows that *in vivo* imaging may lack sensitivity and precision in the study of tissue distribution analysis. Furthermore, *in vivo* imaging differs from conventional bioanalytical methods in that validation of the analytical methodology cannot be conducted.

Case Study 2: Biodistribution Comparison of mRNA and Protein

It is crucial to know whether the distribution patterns of the mRNA and protein are identical. Figure 6.11 demonstrates the distribution of the mRNA and the protein it encodes in different tissues using LNPs composed of three different cationic lipids for mRNA delivery. The mRNA distribution and protein distribution were not identical. The liver exhibited the highest mRNA distribution, while the lung displayed the highest protein expression level. This suggests that the

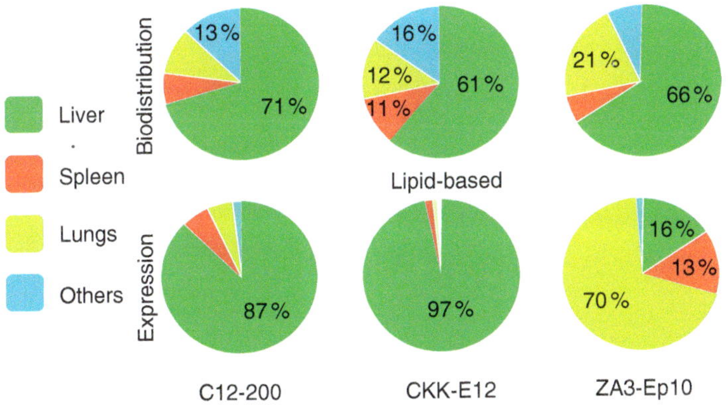

Figure 6.11 The distribution of mRNA and its encoded protein in different tissues. *Source:* [8] / with permission of Royal Society of Chemistry.

translation efficiency from mRNA to protein expression may vary across different tissues. It is crucial to integrate the biodistribution results of both mRNA and protein when analyzing biodistribution studies of mRNA therapeutic drugs. Specifically, for mRNA therapeutic drugs, the expression level and duration of the encoded protein in the target tissue affect the generation and duration of drug efficacy. Alternative approaches used for biodistribution studies of mRNA vaccines may not apply to mRNA therapeutic drugs.

6.2.6 Strategies for Preclinical PK Assessment of mRNA Vaccines, mRNA Therapeutic Drugs, and Novel Excipients

Preclinical PK studies of mRNA-based therapeutics can be categorized into three types, which are focused on mRNA vaccines, mRNA therapeutic drugs, and novel excipients (Figure 6.12).

1) It is necessary for mRNA vaccines to conduct biodistribution studies using tissue harvesting methods, *in vivo* imaging methods, or LNP-encapsulated alternative mRNAs.
2) PK and biodistribution studies are necessary for mRNA therapeutic drugs. The mRNA, encoded protein, immunogenicity, and disease-related biomarkers can be detected in plasma PK studies, and the selection of analytes is based on the mechanism of drug action, the purpose of the assay, and the difficulty of detection. Biodistribution studies can first examine mRNA distribution and then investigate the concentrations of the encoded proteins or biomarkers in key tissues based on the mRNA distribution results.
3) Studies of novel excipients include *in vitro*, *in vivo*, and drug interaction studies. *In vitro* studies include metabolic stability studies and metabolite identification in various systems. *In vivo* studies include ADME studies in animals to characterize the *in vivo* processes. The examination of drug interactions for novel excipients depends on the level of risk involved. In general, the

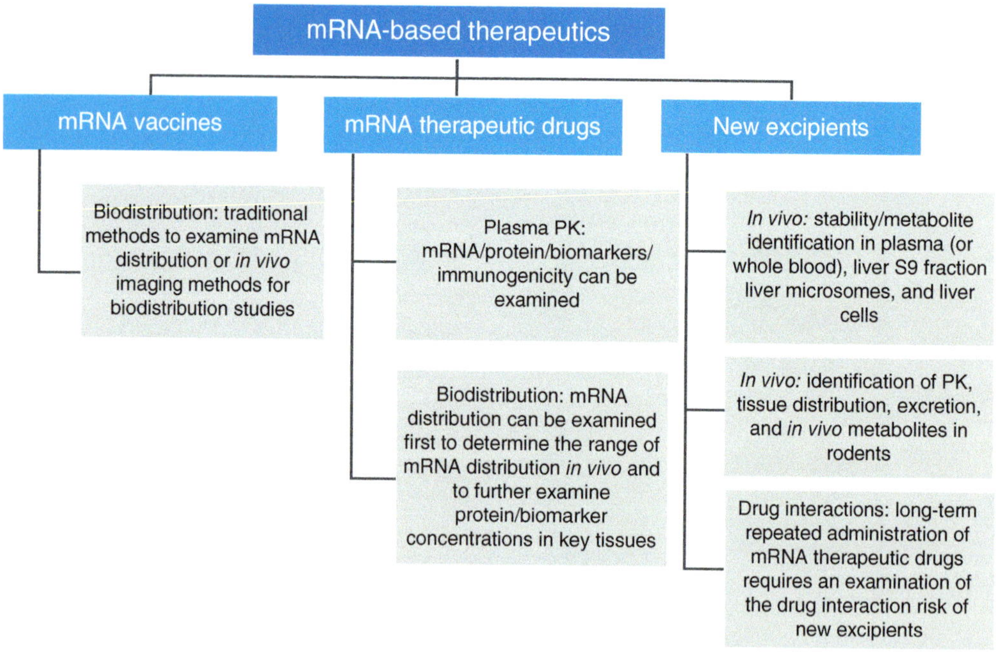

Figure 6.12 Preclinical PK evaluation strategies for mRNA-based therapeutics.

mRNA vaccine has a low-dose frequency, with long intervals, resulting in a low-drug interaction risk for novel excipients. However, it is recommended to assess the potential drug interactions with novel excipients.

References

1 Sahin, U., Karikó, K., and Türeci, Ö. (2014). mRNA-based therapeutics—developing a new class of drugs. *Nature Reviews Drug Discovery* 13 (10): 759–780.

2 Qin, S., Tang, X., Chen, Y. et al. (2022). mRNA-based therapeutics: powerful and versatile tools to combat diseases. Signal Transduction and Targeted. *Therapy* 7 (1): 166.

3 Gote, V., Bolla, P.K., Kommineni, N. et al. (2023). A comprehensive review of mRNA vaccines. *International Journal of Molecular Sciences* 24 (3): 2700.

4 Vavilis, T., Stamoula, E., Ainatzoglou, A. et al. (2023). mRNA in the context of protein replacement therapy. *Pharmaceutics* 15 (1): 166.

5 Barbier, A.J., Jiang, A.Y., Zhang, P. et al. (2022). The clinical progress of mRNA vaccines and immunotherapies. *Nature Biotechnology* 40 (6): 840–854.

6 Rohner, E., Yang, R., Foo, K.S. et al. (2022). Unlocking the promise of mRNA therapeutics. *Nature Biotechnology* 40 (11): 1586–1600.

7 Module, B. (2022). 2.4. Nonclinical Overview.

8 Zadory, M., Lopez, E., Babity, S. et al. (2022). Current knowledge on the tissue distribution of mRNA nanocarriers for therapeutic protein expression. *Biomaterial Sciences* 10 (21): 6077–6115.

6.3 FDA-Approved mRNA Vaccines: Interpretation of Preclinical Pharmacokinetic (PK) Data

Liping Ma, Huan Liu, Jing Jin

6.3.1 Introduction

The mechanism of mRNA vaccines is to deliver modified mRNA that expresses virus antigens into cells. This process activates the immune system and enhances the ability to resist viruses. Compared to traditional vaccines, mRNA vaccines offer various advantages, such as a faster research and development cycle, a simpler production process, and lower research costs. Furthermore, mRNA vaccines are considered safer than DNA vaccines. This is because mRNA is expressed in ribosomes without entering the cell nucleus, eliminating the risk of mutations that may occur when exogenous nucleic acid sequences integrate into the host genome. However, the accelerated development of mRNA vaccines and therapeutics still faces numerous challenges. For instance, there is a lack of comprehensive guidance on the DMPK research of mRNA vaccines, and the PK profiles of mRNA vaccines are not well understood. This chapter offers valuable insights into the preclinical DMPK data of the two FDA-approved mRNA vaccines, serving as a valuable reference for the DMPK research of mRNA vaccines.

The preclinical PK studies of BNT162b2 (COMIRNATY®) and mRNA-1273 (SPIKEVAX®) can be categorized into two main areas: (i) research on the biodistribution of the mRNA and (ii) studies on the *in vivo* PK, tissue distribution, excretion, and metabolism of the novel lipid components used in the vaccine. Studies on mRNA are primarily based on the FDA's vaccine development and licensing guidelines, which specify that "Biodistribution studies in an animal species should be considered if the vaccine construct is novel in nature and there are no existing biodistribution data from the platform technology" [1]. As for the research on LNP delivery systems, we can refer to the "Guidelines for Pharmacological Research on Prophylactic mRNA Vaccines to Prevent SARS-CoV-2." For the delivery substances that have not been used globally, research should be conducted following the "Guidelines for the Non-clinical Safety Evaluation of New Pharmaceutical Excipients." According to the guidelines, it is recommended to investigate the ADME of excipients in the same animal species used for the nonclinical safety evaluation. The dose routes should correspond to those used in clinical trials [2, 3].

6.3.2 Interpretation of Preclinical PK Data for BNT162b2 (COMIRNATY)

Based on the available information, preclinical PK studies of BNT162b2 (COMIRNATY) are summarized in Table 6.2.

6.3.2.1 Biodistribution Study of BNT162b2 mRNA

The biodistribution of BNT162b2 mRNA was examined through *in vivo* fluorescence imaging. The luciferase-encoding modRNA, formulated in LNP, was used as a surrogate mRNA. It was administered to BALB/c mice via IM injection. At six hours post-dose, luciferase protein was detected both at the injection site and in the liver. The luciferase levels in the liver were relatively low and were undetectable at 48 hours post-dose. In contrast, the fluorescence at the injection site persisted for a longer duration. The total systemic fluorescence levels in the mice returned to baseline levels by day 9 after administration. However, owing to the limitations of sensitivity and resolution of fluorescence imaging, accurate quantification of the luciferase protein (indicating the mRNA distribution) in each tissue was challenging [4].

Table 6.2 Summary of preclinical PK studies on BNT162b2 (COMIRNATY).

Study content	Animal species	Study design	Analysis content
In vivo mRNA distribution study	Male BALB/c mice	Intramuscular injection of modRNA encoding luciferase formulated in LNP comparable to BNT162b2 with RNA dosage at 2 μg per animal	Fluorescence imaging
In vivo PK and excretion studies of ALC-0315 and ALC-0159 in rats	Male Wistar–Han rats	Intravenous injection of modRNA encoding luciferase formulated in LNP comparable to BNT162b2 with RNA dosage at 1 mg/kg	Concentrations of ALC-0315 and ALC-0159 in plasma, liver, urine, and feces
In vivo distribution study of the LNPs	Male and female Wistar–Han rats	Intramuscular injection of modRNA encoding luciferase formulated in LNP comparable to BNT162b2 with trace amounts of [^{3}H]-CHE as a nondiffusible label. The RNA dosage is 50 μg per animal	Radioactivity of ^{3}H-CHE in whole blood, plasma, and tissues
In vivo metabolite identification of ALC-0315 and ALC-0159	Male Wistar–Han rats	Intravenous injection of modRNA encoding luciferase formulated in LNP comparable to BNT162b2 with mRNA dosage at 1 mg/kg	Metabolite identification in plasma, liver, urine, and feces
In vitro metabolic stability and metabolite identification of ALC-0315 and ALC-0159	CD-1 mouse, Wistar Han rat, cynomolgus monkey, and human	Incubation in whole blood, the liver S9 fraction, and hepatocyte systems of different species for up to 120 min	Remaining% of ALC-0315 and ALC-0159 in different incubation systems and metabolite identification

LNP, lipid nanoparticle; [^{3}H]-CHE, [cholesteryl-1,2-^{3}H(N)]-cholesteryl hexadecyl ether.

6.3.2.2 *In Vivo* Pharmacokinetic (PK) Study of ALC-0315 and ALC-0159 in Rats

ALC-0315 and ALC-0159 are two novel lipids used in the LNP delivery system of BNT162b2. ALC-0315 is a cationic lipid that plays a role in regulating the endosomal escape process of mRNA after entry into the cell. ALC-0159 is a PEG-modified lipid that forms a hydrophilic layer to encapsulate mRNA, thus facilitating the stability of RNA and reducing specific binding to proteins (Figure 6.13).

The *in vivo* PK of ALC-0315 and ALC-0159 was investigated in rats using modRNA encoding luciferase as surrogate mRNA. The corresponding dosage of ALC-0315 and ALC-0159 was 15.3 mg/kg and 1.96 mg/kg, respectively, with dosing modRNA at 1 mg/kg in rats. After intravenous injection, both ALC-0315 and ALC-0159 were rapidly distributed from the blood to the liver. The PK curves

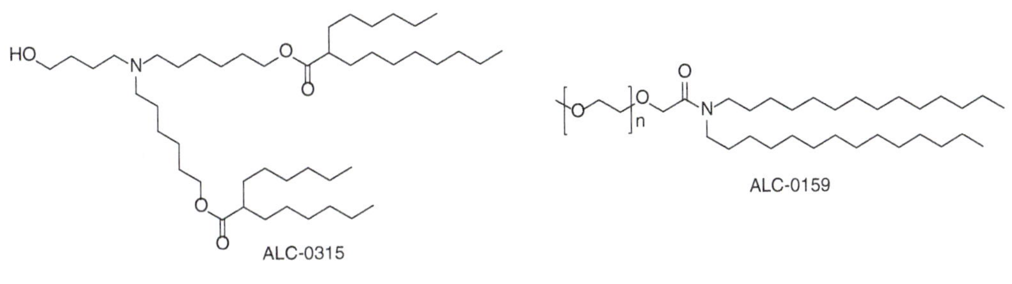

Figure 6.13 The chemical structure of ALC-0315 and ALC-0159.

of the two lipids were similar. However, the exposure of ALC-0315 in the liver was significantly higher than that of ALC-0159, indicating a prolonged distribution of ALC-0315 in the liver. Urinary and fecal excretion of both lipids were also investigated in this study. ALC-0315 and ALC-0159 were not detected in the urine. However, the excretion rate in feces was approximately 1% for ALC-0315 and approximately 50% for ALC-0159. The limited recovery of both lipids suggests that they undergo metabolic clearance to some degree *in vivo* [4].

6.3.2.3 *In Vivo* Distribution Study of BNT162b2 LNP

The *in vivo* distribution of LNP was assessed in rats by monitoring [cholesteryl-1,2-^{3}H(N)]-cholesteryl hexadecyl ether ([^{3}H]-CHE). A trace amount of [^{3}H]-CHE was spiked into the LNP delivery system as a non-exchangeable and non-metabolizable lipid marker. After intramuscular injection in male and female Wistar Han rats, the highest levels of radioactivity were consistently observed at the injection site. In addition to the injection site, radioactivity was also detected in the plasma and most tissues with the greatest levels in plasma observed during one to four hours post-dose. Within 48 hours after administration, the LNP was mainly distributed in the liver, adrenal glands, spleen, and ovaries, with the T_{max} ranging from 8 to 48 hours post-dose. There was no significant difference in the distribution of LNP between male and female rats. Apart from the injection site, the highest total recovery rate (% of the administered dose) of the LNPs in male and female rats was observed in the liver (18%), followed by the spleen (≤1.0%), adrenal gland (≤0.11%), and ovaries (≤0.095%). These findings indicate the hepatic targeting characteristics of the LNP delivery system [4].

6.3.2.4 *In Vitro* and *In Vivo* Metabolic Studies of ALC-0315 and ALC-0159

Four metabolites of ALC-0315 generated from ester bond hydrolysis were detected in both *in vivo* and *in vitro* incubation systems, and the hydrolyzed products were conjugated subsequently with glucuronide *in vivo*. Based on the results of metabolite identification *in vivo* and *in vitro*, the primary metabolic pathways of ALC-0315 were hydrolysis and glucuronidation. The main metabolic pathway of ALC-0159 was the hydrolysis of amide bond to form N,N-ditetradecylamine. The hydrolytic metabolism of both lipids was observed in all four species (human, rat, mouse, and monkey) [4].

6.3.3 Interpretation of Preclinical PK Data for mRNA Vaccine: mRNA-1273

6.3.3.1 Biodistribution of mRNA-1273 mRNA

Moderna submitted the biodistribution data of mRNA-1647 to facilitate the approval of the vaccine mRNA-1273 because the two vaccines employ the same LNP delivery system. However, there are two significant differences between the two vaccines. First, mRNA-1273 contains one mRNA encoding the full-length SARS-CoV-2 S-2P protein, whereas mRNA-1647 contains six mRNAs that encode the full-length cytomegalovirus gB and gH/gL/UL128/UL130/UL131A pentameric glycoproteins, respectively. Second, although the LNP delivery systems used in both vaccines have the same lipid composition, the LNP size of mRNA-1273 is slightly smaller than that of mRNA-1647. Currently, it remains uncertain whether the nanoparticle size substantially affects the tissue distribution of the mRNA. However, Moderna believes that the diameters of both nanoparticles are smaller than the aperture on the sinusoidal endothelial cells in rat liver (161 nm). This suggests that the difference in nanoparticle size has minimal impact on the distribution of the LNP delivery system in the liver [5].

Tissue distribution studies of mRNA-1647 were conducted in male SD rats. Following intramuscular injection of 100 μg of mRNA-1647 in rats, mRNA was detected in most tissues two hours after administration, including the brain, heart, lung, eye, and testis. The C_{max} was achieved at 2–24 hours post-dose, and all tissue exposure levels were higher than the plasma exposure level. Similar to BNT162b2, the remarkable liver distribution of mRNA-1647 confirmed the liver-targeting characteristic of the LNP delivery system. Furthermore, mRNA-1647 was found to be distributed predominantly at the injection site, in the lymph nodes, spleen, and eye, with a concentration higher than plasma concentration [5].

mRNA-1647 exhibited rapid elimination from plasma, with a half-life ranging from 2.7 to 3.8 hours. In contrast, the elimination rate in tissues was significantly lower. The average $T_{1/2}$ of the six mRNAs at the injection site, proximal popliteal lymph nodes, distal axillary lymph nodes, and spleen were 14.9, 34.8, 31.1, and 63.0 hours, respectively. In other tissues, the mRNA levels were below the minimum detection limit three days after administration, and the T_{last} of the mRNAs in brain tissue was approximately 25 hours after administration [5].

In addition to the above tissue distribution data for mRNA-1647 in rats, the mRNA concentrations in tissues were also monitored in a toxicology test for mRNA-1273. mRNA was detected in all tissues, including the liver, heart, lung, testis, and brain, albeit at low concentrations, except for the kidney. The mRNA concentration in the brain was found to be approximately 2%–4% of the plasma concentration. This suggests that LNP-delivered mRNA can cross the blood–brain barrier to distribute in the brain to a certain extent [5].

6.3.3.2 PK, Distribution, Metabolism, and Excretion of Lipids

LNP used in the mRNA-1273 vaccine contains two novel lipid molecules, SM-102 and PEG2000-DMG. Their structures are illustrated in Figure 6.14.

The researcher did not specifically investigate the distribution, metabolism, or PK of SM-102. However, the *in vivo* distribution, metabolism, and excretion characteristics were discussed based on data of its structural analog (SM-86). SM-86 was rapidly metabolized *in vivo* through ester bond hydrolysis, and the fatty acid metabolites can be rapidly excreted via bile and kidney. In addition, QWBA revealed that both SM-86 and its metabolites were distributed in all tissues for a period longer than 168 hours. In summary, these results suggest SM-102 does not tend to accumulate *in vivo* with repeated administration and therefore a low likelihood of toxicity [5].

6.3.3.2.1 Discussion of the PK Characteristics of mRNA Vaccines

The biodistribution of mRNA will be dependent on the LNP delivery system. Based on the application materials for FDA-approved vaccines, it is acceptable to use surrogate mRNA formulated in the same LNP as the test article for evaluating tissue distribution.

After intramuscular injection of an mRNA vaccine, mRNA will have higher exposure in tissues compared to plasma. It also shows a prolonged distribution at the injection site. Meanwhile, mRNA

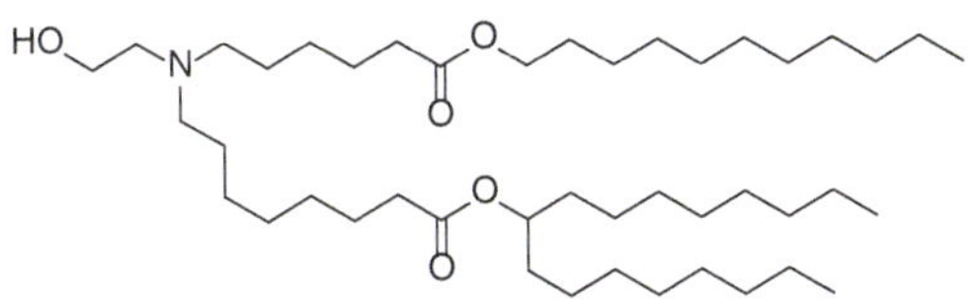
SM-102

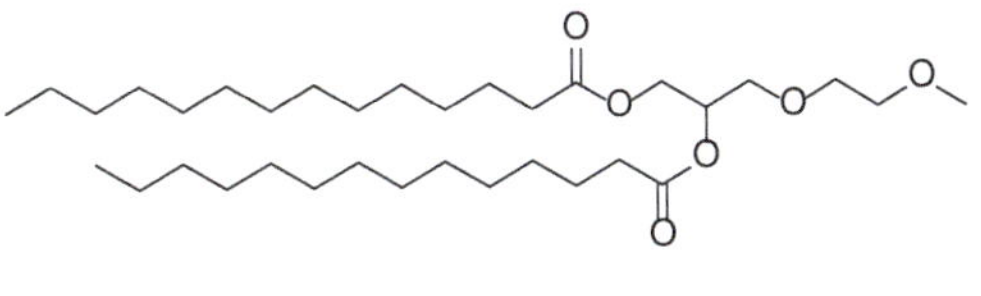
PEG2000-DMG

Figure 6.14 The chemical structure of SM-102 and PEG2000-DMG.

tends to largely distribute in the liver due to the liver-targeting property of LNP and showed certain exposure in the lymph nodes and spleen (immune system organs) which should be paid attention to. In general, mRNA undergoes rapid distribution and elimination in most tissues, with T_{max} within 24 hours and T_{last} within 72 hours post-dose. Notably, mRNA may exhibit persistent distribution at the injection site, lymph nodes, and spleen for a period of two to three weeks, with a slow elimination rate.

Among the components of the LNP delivery system, phospholipids and cholesterol are similar to natural endogenous lipids, and their PK profile does not require specific evaluation. However, ionizable cationic lipids and PEG lipids, as exogenous components, should be assessed for their ADME processes, considering the unknown *in vivo* disposition, toxicity, and immunogenicity of these components.

Based on the structural features and study results of several lipid molecules in the BNT162b2 and mRNA-1273 delivery systems, it has been observed that both lipid molecules undergo varying degrees of hydrolytic metabolism *in vivo*. Excretion studies indicated that the recovery rate of the parent lipid molecules is generally insufficient in urine, feces, and bile. The tissue distribution of lipids is relatively consistent with that of mRNA, exhibiting the most abundant distribution at the injection site, followed by the liver (LNP target tissue) and immune tissues, such as the spleen and lymph nodes.

References

1 U.S. FDA. (2023). Development and licensure of vaccines to prevent COVID-19. https://www.fda.gov/regulatory-information/search-fda-guidance-documents/development-and-licensure-vaccines-prevent-covid-19.

2 Guidelines for pharmacological research on prophylactic mRNA vaccines to prevent SARS-CoV-2. https://www.cde.org.cn/zdyz/opinioninfopage?zdyzIdCODE=9efa70b504dc5c66965001181a49c30b&rddt=1

3 Guidelines for the non-clinical safety evaluation of new pharmaceutical excipients. https://www.cde.org.cn/zdyz/domesticinfopage?zdyzIdCODE=9320dbe8ae0bdb1f798ad9971bcd0dc1

4 U.S. FDA. (2021). Public assessment report authorisation for temporary supply COVID-19 mRNA vaccine BNT162. https://www.fda.gov/vaccines-blood-biologics/comirnaty.

5 EMA Science Medicines Health. (2021). EMA public assessment-report of COVID-19 vaccine Moderna. https://www.ema.europa.eu/en/documents/assessment-report/comirnaty-epar-public-assessment-report_en.pdf.

6.4 Application of Integrative Bioanalysis in mRNA Drugs DMPK Research

Maotian Zhou, Huan Yan, Jinlian Lu, Miaomiao Song, Huan Liu, Lili Xing

6.4.1 Introduction

DMPK studies of mRNA drugs can be broadly divided into two categories, DMPK studies of mRNA vaccines and mRNA therapeutic drugs. mRNA vaccine research mainly focuses on the tissue distribution of mRNA and novel carrier excipients within the systemic circulation. Apart from this, the research on mRNA therapeutics must encompass the proteins expressed post-administration, any associated immunogenicity, and PK/PD studies. To offer the necessary bioanalytical support for these diverse DMPK research areas and topics, we have developed a set of integrated bioanalytical platforms specifically for mRNA medicines. Here mRNA drugs are encapsulated with representative LNPs, including analysis of the mRNA itself, analysis of expressed proteins, novel carrier excipients (cationic lipid or PEG lipid), immunogenicity, and biomarkers (Figure 6.15). Below we will detail the specific bioanalytical content of each type of component, as well as the corresponding internal validation experiments.

6.4.2 Contents and Platforms of mRNA Integration Bioanalysis

6.4.2.1 mRNA Bioanalytical Platform

The bioanalysis of mRNAs is mainly performed by two types of platforms, one based on polymerase chain reaction (PCR) and the other based on the signal amplification reaction (mimic ligand binding assay), respectively. See Figure 6.16 for the basic principle, mRNA is reverse transcribed into cDNA and then is further subjected to amplification by polymerase. The mRNA increment in the RT-qPCR reaction is shown by the fluorescence of SYBR green or probe (TaqMan method).

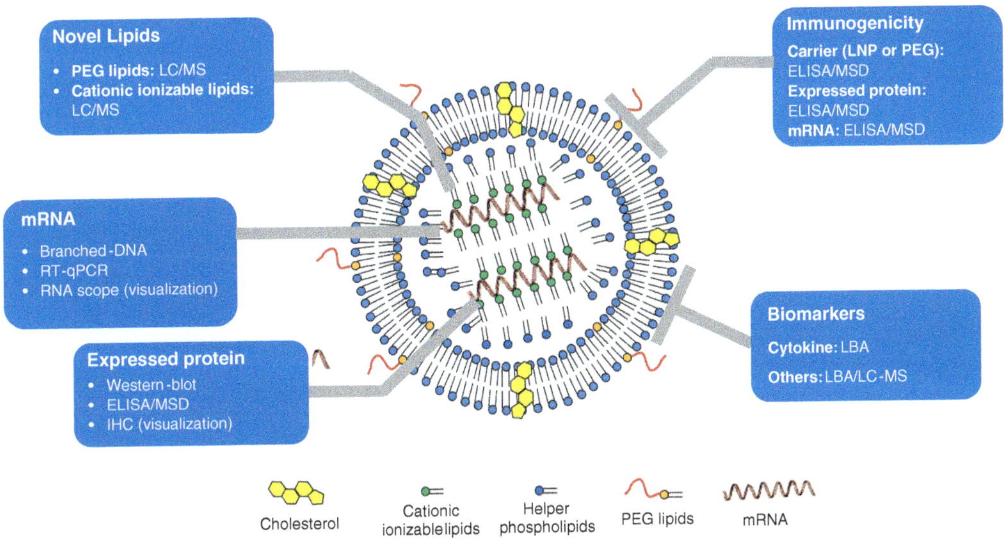

Figure 6.15 Integrative bioanalysis content and corresponding bioanalysis platform of mRNA drugs (once again with LNP-mRNA drugs).

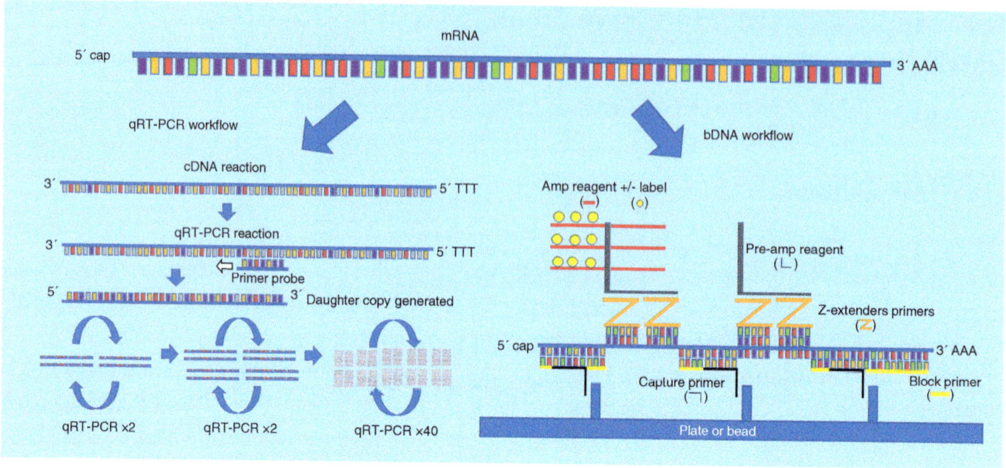

Figure 6.16 Schematic diagram of basic principles of RT-qPCR and b-DNA for mRNA analysis. *Source:* [1] / with permission of Taylor & Francis Group.

The absolute amount of mRNA is indicated by Ct values, which are defined by the number of cycles required to a specific threshold (a certain fluorescence intensity). As for the signal amplification reaction, branched-DNA (b-DNA) relies on nucleic acid probes of complementary sequences to capture mRNA on a solid phase carrier, then multiplex probes are incorporated to precisely amplify the signal in a step-by-step manner. Eventually, the linked enzymes on the probes react with the specific substrates, thus the signal is proportional to the absolute amount of mRNA.

Notably, RNA-scope microimaging analysis is another popular platform, in addition to the two absolute quantification-based analytic systems mentioned above. Despite it being a semi-quantitation technique, its excellent visualization for delivering finer space of mRNA and other information for cellular structures still attracted many early mRNA researchers. For example, Moderna employed it to investigate the tissue distribution of the mRNA-LNP vaccine. The RNA-scope technology is similar to b-DNA in principle since it is likewise based on nucleic acid probe hybridization and signal amplification.

The RT-qPCR platform is primarily used in the bioanalysis of mRNA drugs, especially for IND approval. However, the regulatory guidelines are still lagging. Thus, we developed our own RT-qPCR method and acceptance criteria for verification based on industry white papers as well as the Global CRO Council (GCC) [2]. Method validation includes but is not limited to the standard curve, sensitivity, inter/intra accuracy and precision assessment, selectivity, stability, extraction recovery, etc.

We used simulated mRNA drugs (EGFP mRNA) to validate RT-qPCR analysis platforms. Considering that the liver is the main targeted and metabolic tissue typically for LNP-mRNA drugs, then we selected SD rat liver tissue to fully validate the qPCR method. We designed our in-house primers and probes for EGFP and theoretically validated its specificity by NCBI Blast. Through experimentation, these primers and probes exhibit good specificity and amplification efficiency. RT-qPCR linear range is between 1 pg /mL and 1000 ng/mL. The standard curves for whole blood and liver are listed in Figure 6.17.

We determined the within-batch/between-batch precision and accuracy, selectivity, and stability for the RT-qPCR method validation. Three independent runs were evaluated spanning three days by two analysts, showing the accuracy (%Bias) and precision (%CV) of within-batch/ between-batch.

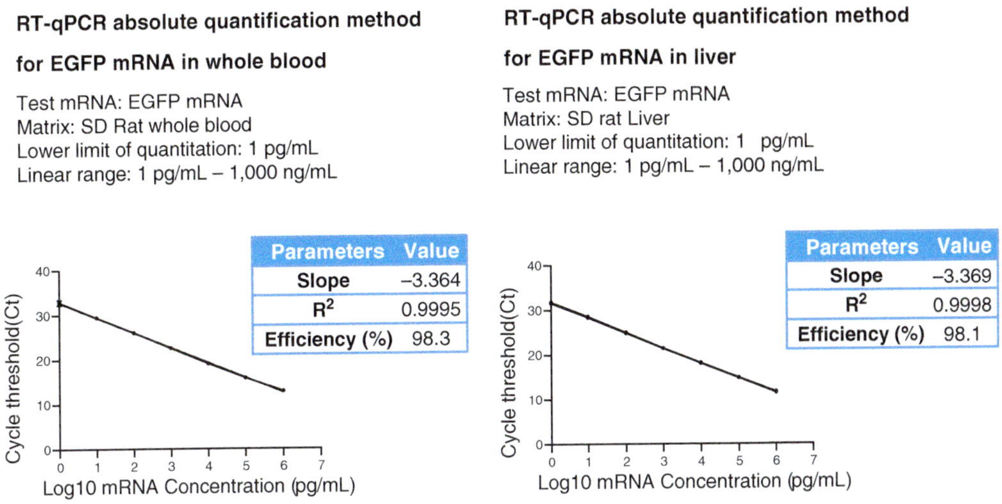

Figure 6.17 Schematic diagram of absolute quantification of EGFP mRNA in the liver with cycle number threshold (Ct) and mRNA concentration after log10 (pg/mL) in the ordinate.

The data demonstrated three batches were robust and reproducible as the %CV and %Bias all met the acceptance criteria (Tables 6.3 and 6.4). Three levels of spiked quality control samples (upper limit of quantitation [ULOQ], High QC [HQC], and LLOQ) were prepared in ten individuals to check the selectivity parameter of the method. Exceeding 80% of spiked QC samples (17/18) satisfied the acceptance criteria, suggesting the method exerted good selectivity. Meanwhile, the short-term stability study concluded that both high- and low-spiked quality control samples (HQC and LQC) were stable during freeze-thaw cycles and bench-top storage.

6.4.2.2 Novel Lipid Bioanalysis

LNPs typically consist of four classes of components: cholesterol, neutral auxiliary phospholipids, ionizable cationic lipids, and PEG lipids. Cholesterol and neutral auxiliary phospholipids are commonly found in other similar carriers and acquired FDA approval. Then, more kinetic research and related biological studies are frequently not needed for these two lipids. The ionizable cationic lipids are versatile parts and optimization units for different LNPs, and they determine the targeting features as well as the kinetic characteristics. These novel lipid components are not applied in the marketed drugs, thus related bioanalysis and the *in vitro* and *in vivo* ADME assay must be carried out.

The cationic lipid determines the delivery and transfection efficiency of the LNP delivery system. It consists of three crucial structural regions including the polar head, the connecting chain, and the hydrophobic tail. Among them, the polar head affects the lipid charge situation and lysosomal escape process. The connecting chain impacts the chemical stability and biodegradability of cationic liposomes. The hydrophobic tail affects the lipid behavior, not only providing enough mobile phase for the lipid bilayer but also facilitating the lipid fusion of cationic liposomes *in vivo*.

The LC-MS/MS platform is the main platform for novel lipid components of LNPs. However, the methods are tough because of the intricate structure, large number of isomers, and susceptibility to the interference of endogenous substances *in vivo*. During the analysis, special attention should be paid to the problems of nonspecific adsorption, strong binding to the chromatographic column, and poor extraction recovery due to its special structure.

Table 6.3 Within-batch/between-batch precision accuracy assessment of standard curve samples.

Inter/intra accuracy and precision evaluation of standard samples (liver)								
		Nominal concentration (ng/mL)						
Run identification		1×10^3	1×10^2	1×10^1	1×10^0	1×10^{-1}	1×10^{-2}	1×10^{-3}
Run1	Ct_CV (%)	1.1	0.2	0.4	0.2	0.5	0.5	0.5
	Mean determined concentration (ng/mL)	9.29×10^2	9.98×10^1	1.05×10^1	1.11×10^0	1.03×10^{-1}	9.10×10^{-3}	9.98×10^{-4}
	Intra-assay precision (%)	8.8	2.5	4.8	2.8	8.4	10.5	10.8
	Bias (%)	−7.1	−0.2	5.1	10.8	2.9	−9.0	−0.2
	n	2	2	2	2	2	2	2
Run2	Ct_CV (%)	0.7	0.9	0.7	0.6	0.9	0.3	0.8
	Mean determined concentration (ng/mL)	9.53×10^2	1.02×10^2	1.02×10^1	1.04×10^0	1.04×10^{-1}	9.41×10^{-3}	1.01×10^{-3}
	Intra-assay precision (%)	5.2	8.9	8.6	8.9	14.6	5.7	17.0
	Bias (%)	−4.7	2.3	2.2	3.7	4.1	−5.9	0.8
	n	2	2	2	2	2	2	2
Run3	Ct_CV (%)	0.6	0.7	0.2	0.4	0.3	0.3	0.3
	Mean determined concentration (ng/mL)	9.27×10^2	1.03×10^2	1.07×10^1	1.09×10^0	9.96×10^{-2}	8.32×10^{-3}	1.09×10^{-3}
	Intra-assay precision (%)	4.8	7.3	2.9	5.8	4.4	6.3	5.8
	Bias (%)	−7.3	2.8	7.0	8.9	−0.4	−16.8	9.3
	n	2	2	2	2	2	2	2
Overall	Mean determined concentration (ng/mL)	9.36×10^2	1.02×10^2	1.05×10^1	1.08×10^0	1.02×10^{-1}	8.94×10^{-3}	1.03×10^{-3}
	Inter-assay precision (%)	5.2	5.5	5.0	5.7	8.1	8.3	10.2
	Bias (%)	−6.4	1.6	4.8	7.8	2.2	−10.6	3.3
	n	6	6	6	6	6	6	6

Table 6.4 Within-batch/between-batch precision accuracy assessment of quality control samples.

Inter/intra accuracy and precision evaluation of QC samples (liver)						
		Nominal concentration (ng/mL)				
Run identification		1×10^{3}	5×10^{2}	5×10^{0}	5×10^{-2}	1×10^{-3}
Run1	Mean determined concentration (ng/mL)	1.01×10^{3}	5.52×10^{2}	4.87×10^{0}	4.87×10^{-2}	9.11×10^{-4}
	Intra-assay precision (%)	19.7	12.3	6.2	6.7	12.9
	Bias (%)	0.8	10.4	−2.7	−2.5	−8.9
	n	6	6	6	6	6
Run2	Mean determined concentration (ng/mL)	8.37×10^{2}	4.94×10^{2}	4.73×10^{0}	4.12×10^{-2}	8.08×10^{-4}
	Intra-assay precision (%)	17.5	14.0	7.3	9.4	32.8
	Bias (%)	−16.3	−1.2	−5.4	−17.7	−19.2
	n	6	6	6	6	6
Run3	Mean determined concentration (ng/mL)	9.91×10^{2}	5.64×10^{2}	4.84×10^{0}	6.25×10^{-2}	7.13×10^{-4}
	Intra-assay precision (%)	27.1	10.6	10.8	45.7	19.3
	Bias (%)	−0.9	12.7	−3.1	25.1	−28.7
	n	6	6	6	6	6
Overall	Mean determined concentration (ng/mL)	9.45×10^{2}	5.37×10^{2}	4.81×10^{0}	5.08×10^{-2}	8.11×10^{-4}
	Inter-assay precision (%)	0.2	0.1	0.1	0.4	0.2
	Bias (%)	−5.5	7.3	−3.7	1.6	−18.9
	n	18	18	18	18	18

Overall, the RT-qPCR platform, which is frequently employed for mRNA analysis, offers excellent sensitivity, specificity, and robustness for the quantitation of mRNA therapeutics. In addition to liver tissue, we established and optimized sample pretreatment methods for other matrix types such as serum/plasma and other challenging-to-homogenize tissues such as muscle and brain. All these efforts may help to meet the research needs of mRNA drug PK and tissue distribution.

6.4.2.3 Bioanalysis of Expressed Proteins

The analysis of expressed proteins can be performed using conventional semi-quantitation Western blot, LBA-based absolute quantitation platforms such as ELISA or Meso Scale Discovery (MSD), or microscopic imaging-based IHC, depending on the goal and stage of the investigation. Table 6.5 lists the advantages and disadvantages of the corresponding platform. At present, LBA-based platforms can be developed to quantify expressed proteins in various types of body fluids and tissues. A sandwich- or competition-ELISA is commonly used based on the availability of key reagents.

6.4.2.4 Immunogenicity Analysis

Due to the relative complexity of mRNA drug components, distinct components may generate distinct immunogenicity. Therefore, it is important to take account of the varying immunogenicity

Table 6.5 Different analysis platforms for advantages and disadvantages of mRNA expression protein analysis.

Platform	Advantages	Disadvantages
Ligand binding assay (LBA)	• Absolute quantitation • Good specificity • High sensitivity	• High dependence on critical reagents • No spatial information
Western blot (WB)	• The method is simple and easy • Less critical reagent dependency	• Susceptible to endogenous interference • Semi-quantitative • No spatial information
Immunohistochemistry (IHC)	• Spatial distribution information • Multi-staining (co-localization study)	• Semi-quantitative or qualitative • Susceptible to endogenous interference

of distinct components. This is the case with LNP-carried mRNA drugs. The immunogenicity investigation usually includes the following three parts: immunogenicity analysis of the expressed protein, the mRNA itself, and the carrier (especially PEG).

- It is widely acknowledged that the immunogenicity of mRNA itself is very low, and it is difficult to trigger the production of ADAs. However, its sequence has the potential to activate specific Toll-like receptors (TLRs) as well as immune pathways. Generally, the samples can be retained for testing at the preclinical stage, and there is no regulatory requirement for investigation.
- The immunogenicity of the carrier varies greatly depending on the carrier, and LNPs are generally considered to be relatively lower immunogenic vectors. LNP ADA (positive control) preparation is difficult and generally not investigated. However, it is worth noting that PEG-modified ingredients are potentially immunogenic substances, and the body may have pre-existing anti-PEG antibodies that also cause accelerated drug clearance. Therefore, it will be recommended to investigate the immunogenicity of PEG components for the LNP carrier's mRNA drugs.
- The protein expressed by mRNA drugs represents another significant potential source of immunogenicity. It may have potential risks related to the differences in protein sequence and folding mode from endogenous proteins. It may also increase the risk associated with the quantity and duration of mRNA protein expression.

The immunogenicity of mRNAs as well as deliveries usually employ the direct ADA method, while immunogenicity assays of expressed proteins often use the ADA bridging method. A schematic of the test format is shown in Figure 6.18.

6.4.2.5 Analysis of mRNA Drug Biomarkers

Preclinical biomarker analysis of mRNA drugs mainly includes immune response analysis (including analysis of immunophenotyping, cytokine/immune factor, and immunoglobulin subtype), effector protein-related pathway biomarker analysis, etc. The analysis of biomarkers for mRNA therapeutics or vaccines is different, generally centered around assessing the safety and efficacy of mRNA drugs, choosing appropriate biomarkers, and validating analytical methods by the fit-for-purpose design (Figure 6.19).

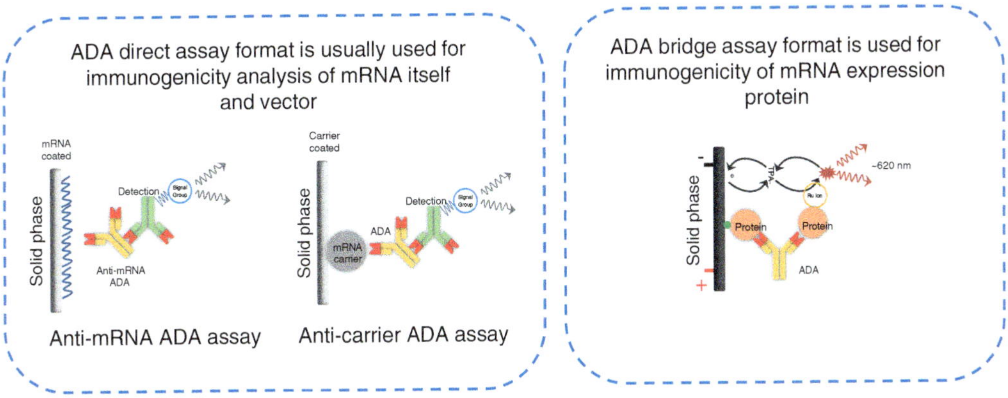

Figure 6.18 Schematic diagram of mRNA immunogenicity assay test format. ADA direct assay format is usually used for immunogenicity analysis of mRNA itself and vector, while ADA bridge assay format is used for immunogenicity of mRNA expression protein.

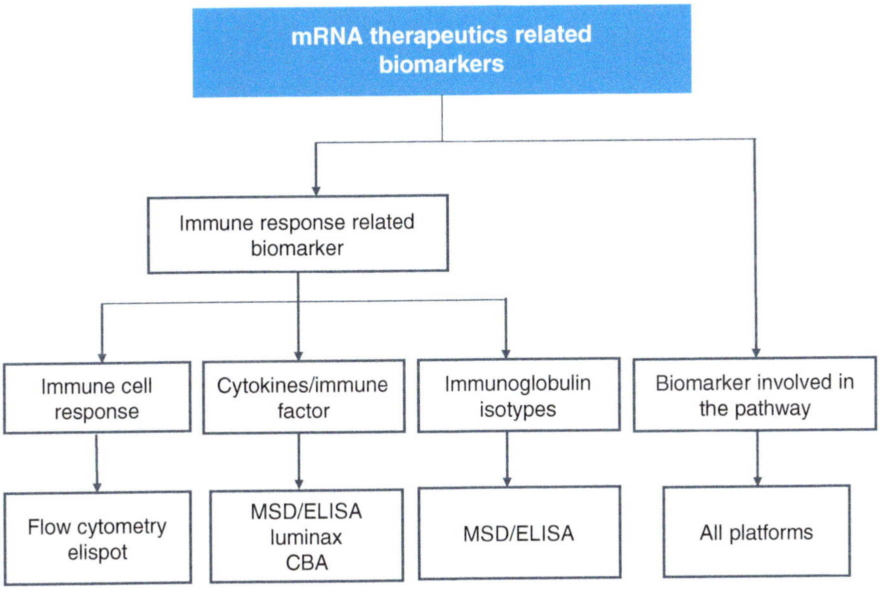

Figure 6.19 Schematic diagram of mRNA drug-related biomarker analysis, including immune response analysis and effector protein downstream pathway biomarker analysis.

6.4.3 Conclusion

mRNA can be translated into any protein or peptide by the ribosome expression system, opening new avenues for the development of mRNA therapeutics, vaccinations, as well as personalized treatments. With the huge success of COVID-19 mRNA vaccines, enthusiasm for the research and development of mRNA vaccines or therapeutic drugs is also rising. Relative to traditional macromolecular drugs, mRNA drugs have a more intricate structure. Thus, more bioanalytical platforms and strategies are involved in the study of the ADME, efficacy, and safety characteristics. Given the differences in mRNA drug designs, different mRNA drugs also need tailored bioanalytical methods.

References

1 Henderson, N. and Wilson, A. (2019). Measurement of mRNA therapeutics: method development and validation challenges. *Bioanalysis* 11 (21): 2003–2010.

2 Wissel, M., Poirier, M., Satterwhite, C. et al. (2022). Recommendations on qPCR/ddPCR assay validation by GCC. *Bioanalysis* 14 (12): 853–863.

Section II

Pharmacokinetic and Metabolism Studies of Therapeutics for Disease Areas

7

Ophthalmic Drugs

7.1 Ophthalmic Drugs: Preclinical Pharmacokinetic (PK) Profiles and Research Strategies

Binbin Tian, Guang Yang, Jing Jin

7.1.1 Introduction

Ocular disease and visual impairments are very common and affect a large number of patients all over the world. According to the "World Report on Vision" published by the World Health Organization (WHO), there are at least 2.2 billion people worldwide with visual impairments, of which at least 1 billion could have been prevented or have the potential for treatment. Common ocular diseases include refractive errors, cataracts, age-related macular degeneration (AMD), glaucoma, diabetic retinopathy (DR), corneal opacity, trachoma, dry eye syndrome, and conjunctivitis. Table 7.1 illustrates the causes, clinical manifestations, and treatments of common ophthalmic diseases. Uncorrected refractive errors (e.g., myopia or hyperopia), cataracts, and AMD are the main causes of moderate to severe visual impairments or blindness, which substantially affect the patients' quality of life. This chapter introduces the eye's physiological structure, ocular drug delivery routes, drug delivery barriers, and preclinical research strategies, providing references for PK studies in ophthalmology drugs.

7.1.2 Physiological Structure of the Eye

The eye consists of two major parts: the eyeball wall and the content of the eye. The eyeball wall is composed of three layers, including the outer, the middle, and the inner layers from the outside to the inside. The outer layer is a fibrous membrane composed of the cornea and sclera. The middle layer is the uvea composed of the iris, ciliary body, and choroid, and it is also known as the vascular layer or uveal layer. The inner layer is the retina.

Anatomically, the eye is divided into the anterior and posterior segments with the lens as the boundary (Figure 7.1). The structures in front of the lens are the anterior segment of the eye, which accounts for approximately one-third of the eye. The anterior segment includes the conjunctiva, cornea, iris, ciliary body, lens, and aqueous humor, playing an important role in studying ocular surface diseases and treatments. The structures behind the lens are the posterior segment of the eye, which accounts for approximately two-thirds of the eye. The posterior segment comprises the sclera, choroid, retina, vitreous body, and optic nerve. In the

Drug Metabolism and Pharmacokinetics: Frontiers, Strategies, and Applications, First Edition.
Edited by Liang Shen and WuXi AppTec DMPK.

Table 7.1 Common ocular disease.

Ocular disease	Disease description	Treatment methods
Refractive error	Due to the abnormal eyeball shape or length, light cannot be focused on the retina, resulting in blurry vision, including hyperopia, myopia, astigmatism, etc.	Refractive surgery, prescription glasses, corneal reshaping lenses, drug treatment (low-dose atropine), etc.
Cataract	The cloudy eye lens leads to progressively blurry vision. The risk of cataracts increases with age.	Cataract surgery.
Age-related macular degeneration (AMD)	Damage to the central part of retina causes the appearance of dark patches, shadows, or distorted central vision. The risk of AMD increases with age.	Laser therapy, surgery, drug treatment (antioxidant and anti-vascular endothelial growth factor [VEGF]), etc.
Glaucoma	Progressive damage to the optic nerve; classified as closed-angle and open-angle glaucoma. Open-angle glaucoma is the most common type, which begins with peripheral vision loss and may worsen over time, leading to severe visual impairment.	Laser therapy, surgery, drug treatment (prostaglandins, β-blockers, α-adrenergic receptor agonists, carbonic anhydrase inhibitors, Rho-kinase inhibitors, miotics, or cholinergic drugs), etc.
Diabetic retinopathy	Complications of diabetes cause damage to the retinal blood vessels, resulting in leakage or blockage. The most common cause of vision loss is swelling of the central part of the retina.	Laser therapy, surgery, drug treatment (anti-VEGF), etc.
Corneal opacity	A type of condition that causes corneal scarring or clouding. The most common causes of corneal opacity are injury, infection, or vitamin A deficiency in children.	Laser therapy, surgery, drug treatment, corneal transplantation, etc.
Trachoma	The cause is a bacterial infection. After years of repeated infections, the eyelashes may turn inward (known as trichiasis), leading to corneal scarring and even blindness.	Drug treatment (antibiotic treatment), surgery, etc.
Dry eye syndrome	Insufficient tear production may cause irritation and blurry vision.	Artificial tears, etc.
Conjunctivitis	Inflammation of the conjunctiva is usually caused by allergies or infections.	Drug treatment, etc.

treatment of posterior segment diseases, ocular drug delivery methods, such as intravitreal injections, are commonly used to facilitate drug diffusion to the target areas and exert therapeutic effects. The distinction between the anterior and posterior segments of the eye during drug development plays a crucial role in studying the delivery and distribution of ophthalmic drugs.

7.1.3 Routes for Ocular Drug Delivery

Routes for ocular drug delivery can be classified into systemic and topical administrations. Systemic administration includes the traditional routes of administration, such as oral administration, intravenous injection, or intramuscular injection. Due to factors such as the blood–eye barriers, the

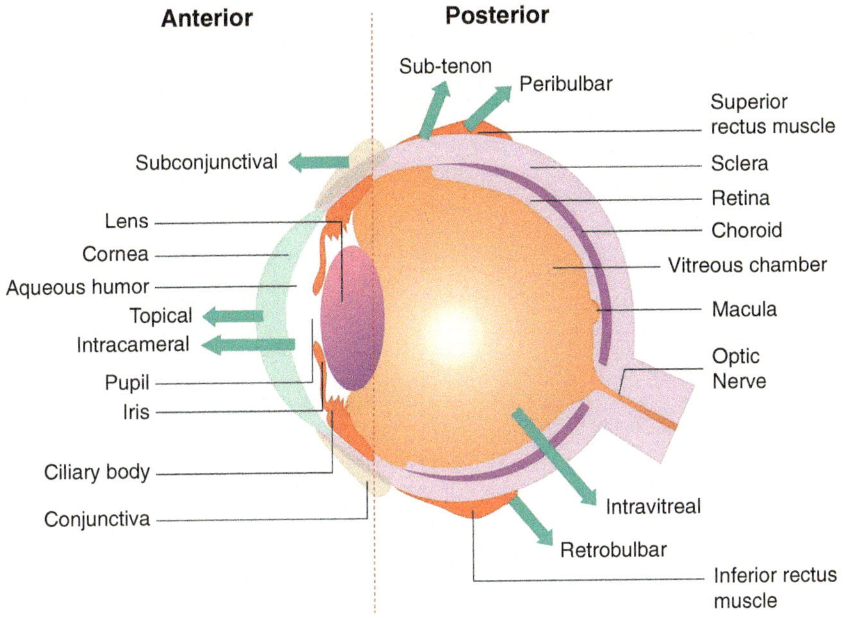

Figure 7.1 Anatomy of the eye and routes of drug administration. *Source:* [1] / MDPI. / CC BY 4.0.

amount of drug reaching the target tissues by systemic administration is limited, which leads to inadequate therapeutic effects. Therefore, for ophthalmic drugs, topical drug delivery is primarily used in clinical settings.

Topical ocular drug delivery is classified into three main categories: (i) topical administration, such as eye drops, eye ointments, and eye gels; (ii) intraocular administration, such as intravitreal injections, intracameral injections, and intravitreal implants; and (iii) periocular administration, such as subconjunctival injections, retrobulbar injections, peribulbar injections, and sub-tenon administrations. Ophthalmic drugs used to treat anterior segment diseases are often administered through ocular surface administration, while drugs used to treat posterior segment diseases are often administered through intraocular or periocular routes.

1) For topical administration, tear drainage, blinking, and entering the systemic circulation through the nasolacrimal pathway result in only 5–10% of the drug entering the corneal barrier [2]. After entering the aqueous humor through the cornea, the drug is quickly distributed into the iris and ciliary body, but it is difficult to enter the posterior segment of the eye. In addition, the eyelids have rich blood vessels, which may allow drugs to enter the bloodstream. Therefore, it is recommended that blood samples should be collected during the early stages of drug screening to confirm whether or not the drug enters the bloodstream and to provide support for later-stage studies on the effect of potential systemic exposure.
2) For intravitreal injections, the drugs can diffuse to both the anterior segment tissues and the retina. It is worth noting that the drug distribution to the choroid occurs at a relatively slow rate due to the presence of tight junctions on the retinal pigment epithelium (RPE).
3) For periocular injections, the drugs can penetrate the sclera to reach the retina and vitreous. This administration route is commonly used for drugs whose targets are in the posterior

Table 7.2 Routes for ocular drug delivery and ocular pharmacokinetic (PK) profiles.

	Topical administration	Intraocular or implant administration	Periocular administration
Absorption	Corneal barrier; blood-aqueous barrier; conjunctival blood/lymph circulation; drainage and turnover of tears and aqueous humor; and efflux pumps expressed by corneal endothelial cells	Blood–retinal inner barrier; choroidal blood/lymph circulation; and efflux pumps expressed by retinal endothelial cells	Blood–retinal outer barrier; choroidal blood/lymph circulation; conjunctival barrier; and efflux pumps expressed by retinal endothelial cells
Distribution	Cornea, conjunctiva, iris, ciliary body, aqueous humor, vitreous, and lens; melanin binding; and plasma protein binding	Aqueous humor, lens, vitreous, iris, conjunctiva, retina, and choroid; and melanin binding	Sclera, choroid, retina, and vitreous; melanin binding; and plasma protein binding
Metabolism [3, 4]	CYP450 enzymes and esterases in the cornea and ciliary body	CYP450 enzymes and esterases in the retina	CYP450 enzymes and esterases in the retina

segment. The absorption, distribution, and metabolism of drugs through topical administration, intraocular or implant, and periocular administration are summarized in Table 7.2.

7.1.4 Barriers to Ocular Drug Delivery

The biggest challenge in ocular drug delivery research is overcoming the static and dynamic biological barriers to effectively deliver the drug to the target ocular tissues. These barriers can be classified based on their anatomical location and functional characteristics, which are generally categorized as anterior segment barriers and posterior segment barriers [5]. Anterior segment barriers include static barriers (such as the cornea, conjunctiva, blood-aqueous barrier, and efflux pumps) and dynamic barriers (such as tear drainage, conjunctival lymph, blood flow, and the aqueous humor) [6]. Posterior segment barriers include static barriers (such as the sclera, Bruch's membrane, blood–retinal barrier, and efflux pumps) and dynamic barriers (such as the choroidal blood and lymph circulation) (Figure 7.2) [7].

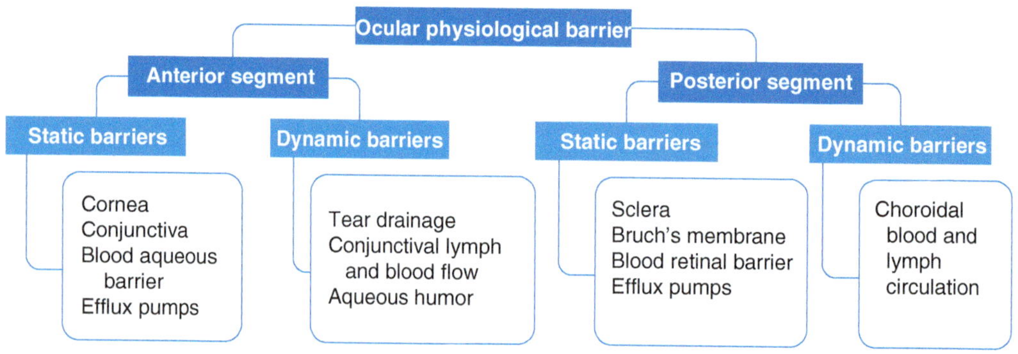

Figure 7.2 Ocular biological barriers.

7.1.5 Strategies for Ophthalmic Preclinical Pharmacokinetics Studies

7.1.5.1 Selection of Animal Species in Ophthalmic Preclinical Pharmacokinetics Studies

Rabbits and monkeys are the most commonly used animal species in preclinical ophthalmic drug research due to their comparability to human eyes [8, 9]. Considering the research costs and experimental operations, for small-molecule drugs, rabbits are often used for preclinical PK studies, whereas species relevant to drug efficacy are often adopted for the preclinical studies of large-molecule drugs.

The RPE plays a crucial role in drug distribution in the ocular tissues. It is also known as the uvea and consists of three continuous parts: the iris in the anterior segment, the ciliary body in the middle segment, and the choroid in the posterior segment. The pigment epithelial cells contain melanin, which can affect ophthalmic drug distribution and release. Drugs may also bind to melanin, leading to slow drug release or accumulation, thus affecting efficacy or toxicity evaluation [10]. Albino experimental animals (e.g., CD-1 mice, Sprague–Dawley rats, and New Zealand white rabbits) lack melanin in the choroid. Therefore, it is important to select appropriate animal models based on the drug properties for ophthalmic drug research in preclinical studies.

It is also necessary to consider the specificity of the ocular anatomical structures of different animal species. For example, only humans and nonhuman primates have well-developed macular structures, making monkeys the preferred experimental animals for pharmacodynamic (PD) studies of macular degeneration-like diseases [9]. In general, a single animal model could not be used to fully simulate the drug efficacy in the human body. It is recommended to use a combination of multiple animal models for preclinical studies to comprehensively evaluate drug properties.

7.1.5.2 Pharmacokinetics Models in Ophthalmic Preclinical Pharmacokinetics Studies

The distribution and elimination of drugs vary in the different ocular tissues due to the influence of barriers, such as tears, the cornea, the blood-retinal barriers, and blood-aqueous barriers. The different ocular tissues can be recognized as separate compartments separated by barriers. The classic compartmental models could be applied to study the PK profiles for ophthalmic drugs.

Using eyedrops as an example, the tears can be modeled as a single-compartment model to study the PK of ophthalmic drugs in the tear. If a portion of the drug enters the body through the corneal barrier, the single-compartment model is not appropriate. In this case, it is recommended to use the anterior cornea, such as the tears, as one compartment and the cornea as the second compartment to perform data analysis using a two-compartment model. Moreover, in the study of aqueous humor and posterior segment tissues, based on the drug delivery process, a three-compartment model analysis can be performed with the anterior cornea as one compartment, the cornea as the second compartment, and the aqueous humor, vitreous humor, or retina as the third compartment (Figure 7.3) [11].

Appropriate multi-compartment models are conducive to topical ocular PK studies. However, data modeling of this approach is based on different assumptions and is relatively complex and mathematically intricate. Improper use of this method can compromise the accuracy of the data. In preclinical studies, noncompartmental models are often preferred for calculating ophthalmic PK parameters.

7.1.5.3 Strategies for Preclinical Pharmacokinetics Studies

Currently, there is no specific guidance from regulatory agencies for nonclinical PK research of ophthalmic drugs. The following points should be considered based on our experiences:

- For drugs using topical drug delivery (e.g., intravitreal injection), both topical and systemic exposure should be assessed.

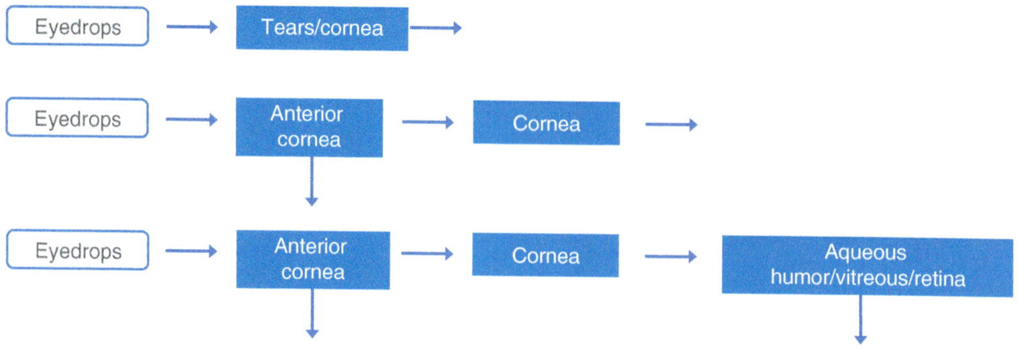

Figure 7.3 Ophthalmic compartment models. *Source:* Adapted from reference [11].

- For drugs with systemic exposure or those administered through systemic administration, conventional systemic PK studies can be applied to assess systemic exposure and PK profiles such as drug distribution, metabolism, and elimination.
- For drugs distributed mainly in the ocular tissues and exerting drug efficacy within the eye, topical ocular PK evaluation should be performed, including the distribution, metabolism, and elimination processes of the drug in ocular tissues such as the aqueous and vitreous humor. Additionally, the distribution and target binding of drugs should be studied, with a focus on systemic exposure and drug accumulation in nontarget organs after topical drug delivery, or drug distribution and target binding within eyes after systemic administration.
- The challenges of DMPK preclinical studies in ophthalmology include: (i) the high demand for high-level animal experimental skills due to the delicate eye structures and the need for professional veterinary and anatomical teams to perform drug administration and ocular tissue collection; (ii) the difficulty in ophthalmic sample bioanalysis due to the various ophthalmic matrices, the small volume of samples and high requirements of sample handling.

In general, comprehensive PK and PD evaluations are recommended at the preclinical stage. Regardless of systemic or ocular topical administration, tailored DMPK studies need to be carried out according to the actual distribution of the drug.

7.1.6 Case Study

The research involves various animal species, like rodents, rabbits, dogs, and monkeys. Multiple ocular drug administrations (e.g., topical administration as eye drops, ointments or gels, anterior chamber injections, and intravitreal injections) can be offered, allowing for precise anatomical dissection and collection of ocular tissues, followed by sample processing and bioanalysis.

Taking intravitreal injection of commercialized vancomycin in New Zealand rabbits as an example, after a single intravitreal injection of vancomycin at 1.00 mg/eye, aqueous humor, vitreous humor, sclera, iris, cornea, conjunctiva, retina, lens, and plasma were collected to evaluate the distribution of the drug in ocular tissues.

Various homogenization ratios and homogenization devices were used to produce high-quality samples for bioanalysis. For example, cornea and sclera tissues are tough and not easily homogenized. If the homogenization ratio is inappropriate, the samples may become gelatinous and difficult to handle. After optimizing the homogenization ratio, a multifunctional biological sample homogenizer with three-dimensional high-speed vibration capabilities is used for sample processing, providing uniformly textured and well-flowing tissue homogenates.

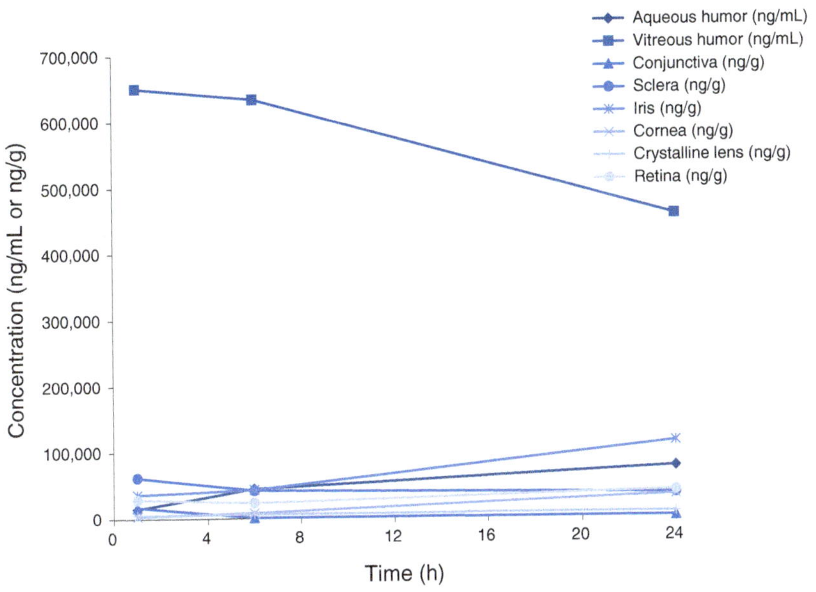

Figure 7.4 Average tissue drug concentrations after a single intravitreal injection of commercialized vancomycin in New Zealand rabbits.

Bioanalysis: Due to various physiological mechanisms and barriers, extremely low drug concentrations are suspected in some ocular tissues, which places a high demand for the sensitivity of bioanalysis. Through the development and validation of analytical methods for various tissues, we have achieved a lower limit of quantification (LLOQ) of 0.1 ng/mL, enabling effective analysis of tissue samples.

The results show that the drug concentration in plasma was extremely low (C_{max} at 60.2 ng/mL). The drug is mainly distributed in the vitreous, which is consistent with the PK characteristics of conventional intravitreal injection formulations. The study provided data support for the use of intravitreal injection as a route of administration for this drug (Figure 7.4).

7.1.7 Summary

Ocular diseases are usually chronic diseases, and their prevalence increases dramatically with age. It is expected that the population of ophthalmic patients would continue to grow due to factors such as population aging, excessive use of electronic devices, and air pollution. The rising prevalence of ophthalmic diseases and increased public health awareness are driving the growth of the ophthalmic drug market. We believe that ophthalmic drug research will become more comprehensive and patients with ophthalmic diseases will receive better treatment.

References

1 Koppa Raghu, P., Bansal, K.K., Thakor, P. et al. (2020). Evolution of nanotechnology in delivering drugs to eyes, skin, and wounds via topical route[J]. *Pharmaceuticals* 13 (8): 167.

2 Dubald, M., Bourgeois, S., Andrieu, V. et al. (2018). Ophthalmic drug delivery systems for antibiotherapy – a review[J]. *Pharmaceutics* 10 (1): 10.

3 Duvvuri, S., Majumdar, S., and Mitra, A.K. (2004). Role of metabolism in ocular drug delivery[J]. *Current Drug Metabolism* 5 (6): 507–515.
4 Nakano, M., Lockhart, C.M., Kelly, E.J. et al. (2014). Ocular cytochrome P450s and transporters: roles in disease and endobiotic and xenobiotic disposition[J]. *Drug Metabolism Reviews* 46 (3): 247–260.
5 Cholkar, K., Dasari, S.R., Pal, D. et al. (2013). Eye: anatomy, physiology, and barriers to drug delivery. In: *Ocular Transporters and Receptors*, 1–36. Woodhead Publishing.
6 Bachu, R.D., Chowdhury, P., Al-Saedi, Z.H.F. et al. (2018). Ocular drug delivery barriers – the role of nanocarriers in the treatment of anterior segment ocular diseases[J]. *Pharmaceutics* 10 (1): 26.
7 Burhan, A.M., Klahan, B., Cummins, W. et al. (2021). Posterior segment ophthalmic drug delivery: Role of muco-adhesion with a special focus on chitosan[J]. *Pharmaceutics* 13 (10): 1685.
8 Zernii, E.Y., Baksheeva, V.E., Iomdina, E.N. et al. (2016). Rabbit models of ocular diseases: new relcvance for classical approaches[J]. *CNS & Neurological Disorders-Drug Targets (Formerly Current Drug Targets-CNS & Neurological Disorders)* 15 (3): 267–291.
9 Vézina, M. (2012). *Comparative ocular anatomy in commonly used laboratory animals[M]// Assessing ocular toxicology in laboratory animals*, 1–21. Totowa, NJ: Humana Press.
10 Rimpelä, A.K., Reinisalo, M., Hellinen, L. et al. (2018). Implications of melanin binding in ocular drug delivery[J]. *Advanced Drug Delivery Reviews* 126: 23–43.
11 Agrahari, V., Mandal, A., Agrahari, V. et al. (2016). A comprehensive insight on ocular pharmacokinetics[J]. *Drug Delivery and Translational Research* 6 (6): 735–754.

7.2 *In Vitro* Methods and Significance of Drug–Melanin Binding

Chunhong Lu, Xue Tang, Jie Wang, Xiangling Wang, Genfu Chen

7.2.1 Introduction

Melanin is widely distributed in various tissues of the body, including the eyes and skin. Drugs that bind to melanin can accumulate in these tissues, affecting the distribution and elimination of the ophthalmic drugs [1]. The binding will lead to changes in ocular PK and PD, potentially affecting normal retinal function. Furthermore, prolonged accumulation of drugs that strongly bind to melanin within tissues can result in adverse effects, such as increased ocular phototoxicity and inner ear toxicity. To address these concerns, the 2015 ICH *S10 Photosafety Evaluation of Pharmaceuticals: Guidance for Industry* recommended assessing the *in vitro* binding of drugs to melanin. This chapter discusses the *in vitro* research methods for drug binding to melanin and highlights the significance and impact of drug–melanin binding, aiming to provide valuable insights to facilitate the development of the drugs.

7.2.2 Definition and Function of Melanin

Melanin, an anionic polymer derived from tyrosine through enzymatic and spontaneous reactions, is synthesized within melanosomes which are the organelles covered by a lipid membrane. Melanin is widely distributed in the eyes, skin, hair, brain, and inner ears. It is classified into two main types (eumelanin and pheomelanin), which differ in color and molecular structure. Notably, melanin exhibits exceptional antioxidant activity and reduces the generation of reactive oxygen species through the chelation of metal ions. These characteristics play a crucial role in safeguarding pigment cells against the adverse impacts of natural toxins [3].

7.2.3 Effects of Drug Binding to Melanin

It is reported that the majority of drugs binding to melanin are small molecules. Although some reports show that proteins and oligonucleotides can bind to melanin, large molecules encounter challenges in crossing the lipid membrane and melanosomal membrane due to their poor permeability, making it difficult for them to reach the melanin *in vivo*. In comparison to the *in vivo* methods for studying melanin binding, *in vitro* melanin binding methods offer two distinct advantages:

1) *In vitro* studies have the advantages of low cost and simple operation.
2) *In vitro* melanin binding serves as a crucial reference for *in vivo* preclinical studies.

7.2.3.1 Pharmacological Effects of Drug Binding to Melanin

The binding of drugs to melanin has a significant impact on ocular PK and PD. When a drug binds to melanin, it could increase the drug's retention in the eyes, leading to a prolonged half-life. Moreover, melanin-bound drugs can be slowly released, allowing the free drug to reach the target site and further enhance drug efficacy (Figure 7.5). Therefore, it is crucial to assess melanin binding during drug development to evaluate its role in ocular drug retention.

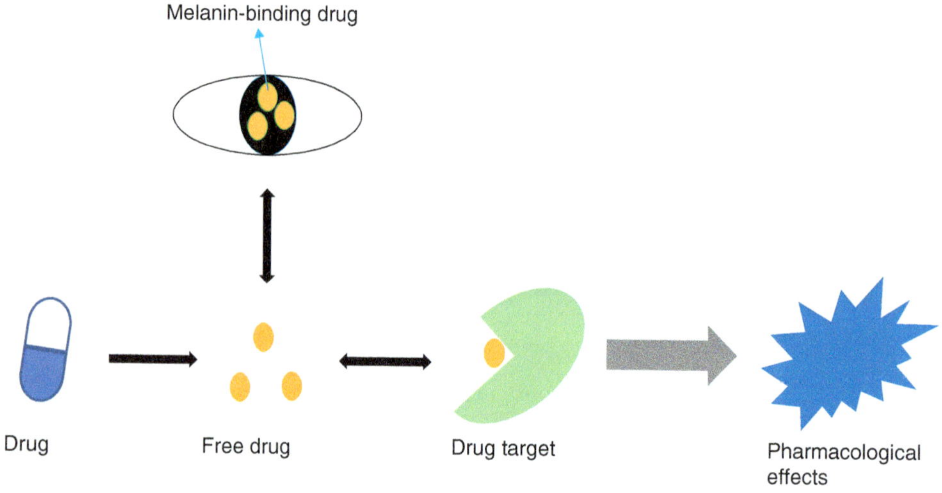

Figure 7.5 Principles of pharmacological effects resulting from the binding of drugs to ocular melanin.

7.2.3.1.1 Prolonged Drug Half-Life

Since the 1970s, the differences in the pharmacological effects of certain ophthalmic drugs in pigmented and albino animals have been investigated. In pigmented animals, drug–melanin binding could result in a small reservoir that gradually releases the drug into tissues, affecting the PK, efficacy, and safety of the drug. For instance, in comparison to albino animals, pigmented animals exhibited significantly longer atropine retention and action times [2](with half-lives of 96 hours and 43 hours, respectively). The accumulation of drugs in melanin tissues extends the drug efficacy and may also lead to potential toxicity. Therefore, research on drug binding to melanin is very necessary.

7.2.3.1.2 Enhanced Drug Targeting

Using the characteristics of drug–melanin binding in drug delivery systems for the treatment of retinal diseases to modify compounds to enhance their basicity and lipophilicity, can increase the binding of compounds to melanin and targeting of drugs for the treatment of retinal diseases. For example, the anti-vascular endothelial growth factor (anti-VEGF) drug pazopanib [4] forms a drug reservoir in the choroid uvea after a single oral dosing, allowing for a slow release that effectively hinders angiogenesis in a choroidal neovascularization (CNV) model. These indicate that the drug binding to melanin influences the extent and process of the drug's pharmacological effects, offering a promising approach for treating CNV lesions.

7.2.3.2 Potential Adverse Effects of Drug–Melanin Binding

Due to the binding of drugs to melanin, melanosomes in pigmented ocular cells can serve as reservoirs for drug accumulation compared to those in nonpigmented tissues. Many drugs, including nonocular drugs administered systemically (such as chloroquine and chlorpromazine), may accumulate in the melanin-rich tissues and cause adverse reactions, such as interference with retinal function, unexpected visual consequences, and phototoxicity and ototoxicity. Therefore, studying the binding of drugs to melanin should be prioritized.

7.2.3.2.1 Ocular Phototoxicity

Certain drugs (such as sparfloxacin) can act as photosensitizers, inducing phototoxic effects in the eyes [5]. When these drugs absorb sunlight, they trigger a phototoxic reaction and generate reactive oxygen species and oxidative stress within the cells, leading to phototoxic reactions and cellular damage [5, 6]. Among the melanin-binding drugs, quinolone antibiotics [7, 8] and chlorpromazine [9] exhibit phototoxicity. For instance, substantial cellular damage has been observed in the RPE of mice and rabbits after the administration of sparfloxacin and ofloxacin, respectively [7, 10]. Shimoda and Kato [11] also observed a phototoxic response in the retina of albino animals after quinolone treatment followed by exposure to ultraviolet light.

7.2.3.2.2 Inner Ear Toxicity

Although aminoglycoside antibiotics are extensively used as bacteriostatic drugs, they have the potential to induce severe nephrotoxicity and ototoxicity. These antibiotics display notably higher inner ear toxicity in pigmented guinea pigs than in albino guinea pigs, suggesting a possible correlation between their ototoxic effects and melanin [12]. This could lead to the accumulation of antibiotics in the pigmented tissues of the inner ear and promote the toxic effects on surrounding cells.

Summary: Various adverse reactions may arise because of the long-term accumulation of drugs in ocular tissues. Therefore, it is necessary to develop research methods for drug binding to melanin *in vitro* and corroborate them with *in vivo* assays to elucidate the specific mechanisms of drug–melanin binding. This will help reduce the toxicity arising from the long-term accumulation of drugs in the eye or other tissues.

7.2.4 Research Methods for Drug Binding to Melanin *In Vitro*

The currently available *in vitro* methods for studying drug binding to melanin include equilibrium dialysis, rapid equilibrium dialysis, and centrifugation. WuXi AppTec DMPK has selected equilibrium dialysis and centrifugation for validation and established an *in vitro* research method for evaluating drug binding to melanin. Melanin is derived from various sources, with the most prevalent natural melanin being derived from the eyes of cattle, pigs, and squid ink.

7.2.4.1 Equilibrium Dialysis Method

The equilibrium dialysis method, which is very similar to the plasma equilibrium dialysis, was employed to determine the binding rate of compounds to melanin *in vitro*. The bound and free drugs are separated through equilibrium dialysis after a specific incubation period (Figure 7.6). The unbound, bound, and recovery rates of the compounds with melanin are subsequently calculated to determine the distribution of the compounds in the melanin. As test compounds, nine

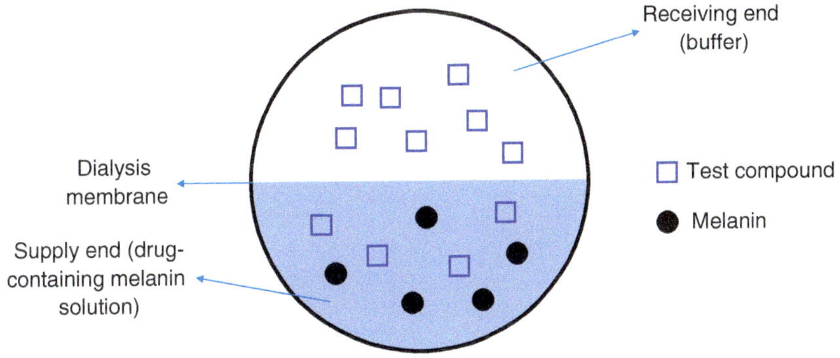

Figure 7.6 Equilibrium dialysis method.

approved commercial drugs, such as atenolol, chloroquine, propranolol, diclofenac, nadolol, quinidine, fluconazole, timolol maleate, and chlorpromazine, were selected. Their binding to squid melanin was confirmed through equilibrium dialysis (bound%). With a goodness of fit R^2 value exceeding 0.9, the study's finding showed a strong correlation with the values in the literature (Figure 7.7).

7.2.4.2 Centrifugation Method

Centrifugation is employed to assess the binding of various compounds to melanin *in vitro*. Under specific conditions of speed, time, etc., the bound and free drugs are separated through centrifugation (Figure 7.8). Liquid chromatography–mass spectrometry (LC–MS)/MS analysis is used to detect the F (free sample), T (total sample), and T_0 (sample at time zero) of the compounds. These data are then used to calculate the unbound, bound, and recovery rates of the compounds with melanin, which is followed by assessing the distribution of the compounds in the melanin.

To verify the feasibility of the method, the nine commercially available standard compounds mentioned above were selected to verify the binding of these compounds to melanin using the centrifugation method, obtaining the binding rate (Bound%) in cuttlefish melanin. The study results strongly correlated with the values in the literature, with a goodness of fit R^2 value exceeding 0.9 (Figure 7.9).

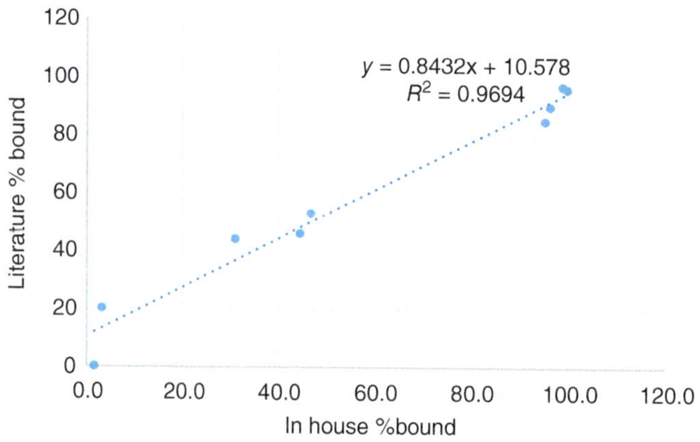

Figure 7.7 Correlation between drug binding to melanin as determined using the equilibrium dialysis method (in-house %bound) and the values in the literature (literature %bound).

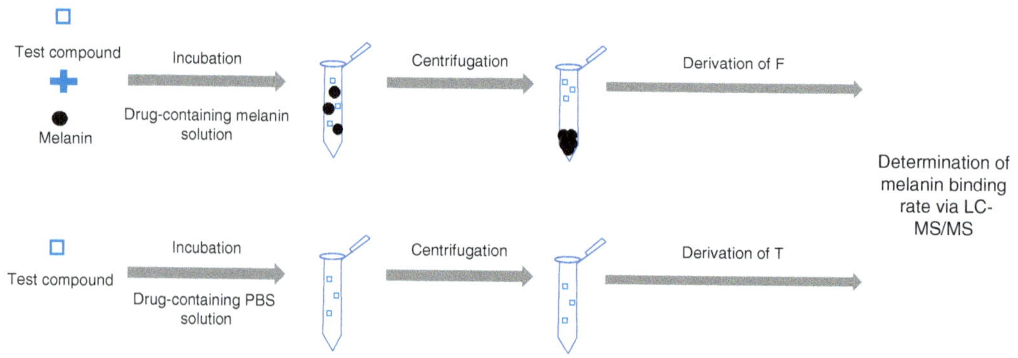

Figure 7.8 Centrifugation method.

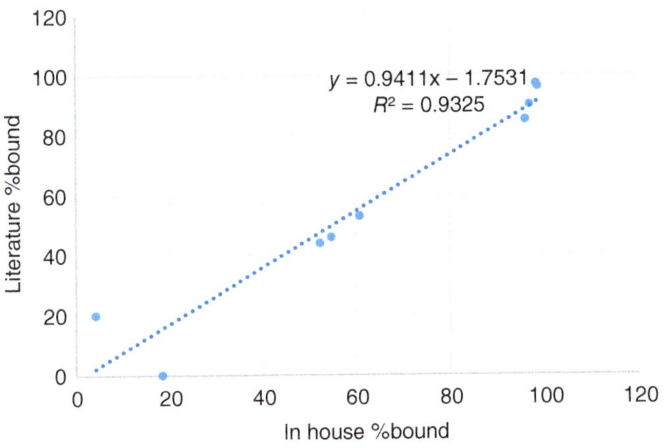

Figure 7.9 The correlation between drug binding to melanin was determined using the centrifugation method (in-house %bound) and the values in the literature (literature %bound).

7.2.5 Integrated Application of *In Vitro* Melanin–Drug Binding Models in Drug Research and Development

Despite numerous examples demonstrating the biological relevance of melanin binding in drug development, the PK transformation from *in vitro* to *in vivo* is not fully elucidated [13]. Studies have shown that alkaline, lipophilic, and aromatic molecules exhibit the highest degree of binding to melanin when assessed using melanin that was isolated using the *ex vivo* method. Although melanosome binding degree can be quantified using the dissociation constant (K_d), the accurate measurement of K_d at various concentrations and time points is expensive and prone to error. Our *in vitro* assay generally determined the free fraction (f_u) value for drug–melanin binding and obtained results with low error and cost, providing a reference for *in vivo* preclinical studies. Therefore, employing this *in vitro* model and comprehending the mechanistic interactions in drug–melanin binding with the *in vivo* metabolic processes may aid in applying *in vitro* assay data to predict the *in vivo* binding of drugs to melanin. This approach offers guidance for the rational design of new drug candidates with desirable physicochemical properties and absorption, distribution, metabolism, and excretion (ADME) characteristics.

7.2.6 Summary

We have successfully established a fast, efficient, and economical high-throughput *in vitro* model for drug–melanin binding studies. This model offers a more precise understanding of the accumulation and distribution of test compounds in melanin, providing a valuable reference for relevant PK and pharmacological studies.

References

1 Robbie, S.J., von Leithner, P.L., Ju, M. et al. (2013). Assessing a novel depot delivery strategy for noninvasive administration of VEGF/PDGF RTK inhibitors for ocular neovascular disease. *Investigative Ophthalmology & Visual Science* 54 (2): 1490–1500.

2 Rimpelä, A.-K., Reinisalo, M., Hellinen, L. et al. (2018). Implications of melanin binding in ocular drug delivery. *Advanced Drug Delivery Reviews* 126: 23–43.

3 Buszman, E., Wrześniok, D., Surażyński, A. et al. (2006). Effect of melanin on netilmicin-induced inhibition of collagen biosynthesis in human skin fibroblasts. *Bioorganic & Medicinal Chemistry* 14 (24): 8155–8161.

4 Jakubiak, P., Cantrill, C., Urtti, A., and Alvarez-Sánchez, R. (2019). Establishment of an *In Vitro–In vivo* Correlation for Melanin Binding and the Extension of the Ocular Half-Life of Small-Molecule Drugs. *Molecular Pharmaceutics* 16 (12): 4890–4901.

5 Onoue, S., Kawamura, K., Igarashi, N. et al. (2008). Reactive oxygen species assay-based risk assessment of drug-induced phototoxicity: Classification criteria and application to drug candidates. *Journal of Pharmaceutical and Biomedical Analysis* 47: 967–972.

6 Shimoda, K., Nomura, M., and Kato, M. (1996). Effect of antioxidants, anti-inflammatory drugs, and histamine antagonists on sparfloxacin-induced phototoxicity in mice. *Fundamental and Applied Toxicology* 31: 133–140.

7 Rampal, S., Kaur, R., Sethi, R. et al. (2008). Ofloxacin-associated retinopathy in rabbits: role of oxidative stress. *Human & Experimental Toxicology* 27: 409–415.

8 Beberok, A., Otręba, M., Wrześniok, D., and Buszman, E. (2013). Cytotoxic effect of lomefloxacin in culture of human epidermal melanocytes. *Pharmacological Reports* 65: 689–699.

9 Persad, S., Menon, I.A., Basu, P.K., and Carre, F. (1988). Phototoxicity of chlorpromazine on retinal pigment epithelial cells. *Current Eye Research* 7: 1–9.

10 Shimoda, K., Yoshida, M., Wagai, N. et al. (1993). Phototoxic lesions induced by quinolone antibacterial agents in auricular skin and retina of albino mice. *Toxicologic Pathology* 21: 554–561.

11 Shimoda, K. and Kato, M. (1999). Apoptotic photoreceptor cell death induced by quinolone phototoxicity in mice. *Toxicology Letters* 105: 9–15.

12 Wästerström SA. Accumulation of drugs on inner ear melanin. Therapeutic and ototoxic mechanisms. *Scandinavian Audiology. Supplementum* 1984;23 : 1–40.

13 Jakubiak, P., Cantrill, C., Urtti, A., and Alvarez-Sánchez, R. (2019). Establishment of an *In Vitro – In vivo* Correlation for Melanin Binding and the Extension of the Ocular Half-Life of Small-Molecule Drugs. *Molecular Pharmaceutics* 16: 4890–4901.

7.3 Preclinical Pharmacokinetic Research of Ophthalmic Drugs: *In Vitro* and *In Vivo* Research Methods

Chao Zhang, Jie Li, Chen Ning, Zhihai Li, Xuan Dong, Hongfeng Qian, Furong Jiao, Cheng Tang, Jacob Zhi Chen, Shoutao Liu

7.3.1 Introduction

Preclinical PK studies of medications used in the eyes are important because they help us better understand how medicines are absorbed, distributed, metabolized, and excreted in the eye. The study also provides insights into the concentration variations and the steady state of the drug within the ocular tissues, as well as the potential interactions among the concurrently dosed ophthalmic drugs. The result is essential to determine the appropriate dosage regimens, dose levels, and frequency to ensure effective optimal drug concentrations and desired drug duration in the target ocular tissues.

The scientist can screen the target compounds and optimize the candidates to enhance their efficacy and safety by assessing PK properties specific to the ocular drugs. Moreover, the assessment of PK properties of ocular drugs provides valuable support in the design and interpretation of clinical trials. Considering the specific characteristics like small eye volume and precise tissue structure, a high level of technical expertise is required in ophthalmic PK research which needs to ensure accurate drug administration, precise sample collection, and tissue isolation.

The physiological barriers that exist in the eye limit drug absorption and distribution. Therefore, these barriers must be considered while selecting administration routes and drug delivery methods. Additionally, ophthalmic examinations are essential in the experiments to assess the condition and health of the eyes before and after administration, determine the baseline state, and monitor potential pathologic changes or adverse symptoms.

Due to anatomical and physiological differences in the eye structures among different animal species, it is crucial to ensure the reliability and comparability of experimental results by selecting the appropriate animal species. In addition, the selection and design of ophthalmic drug formulations can also impact the bioavailability of the ophthalmic drugs. This chapter will provide detailed exploration and application research methods in the preclinical PK studies of ophthalmic drugs from *in vitro* assays, and *in vivo* assays including ophthalmic administration, sample collection and processing, ophthalmic examinations, and the formulation of ophthalmic drugs.

7.3.2 *In Vitro* Assays

7.3.2.1 *In Vitro* Ophthalmic Permeability Study

1) Introduction of *in vitro* ophthalmic permeability study

Eye drops are noninvasive and have excellent patient compliance. Therefore, the application amount of eye drops in ophthalmic therapy accounts for more than 70%. The principal absorption site of eye drops is the cornea; otherwise, most drugs have poor permeating into the eyes by breaking through the entire corneal layer because of the particular physiological structure of lipid solubility-water solubility-lipid solubility in the cornea. *In vitro* ophthalmic permeability study can directly evaluate permeability using cornea and sclera, which excludes the interference of other biological factors, meanwhile, facilitating preclinical *in vivo* study design and development of non-corneal penetration pathways of ophthalmic drugs.

2) **Introduction of *in vitro* cornea and sclera permeability study**

In vitro cornea and sclera permeability studies are processed in the Franze cell diffusion cell system, using the cornea and sclera tissue of New Zealand white rabbits (see Figure 7.10). The studies stimulate the drug delivery process of eye drops and subconjunctival injection (see Figure 7.11).

We comprehensively analyze the literature and WuXi AppTec DMPK *in vitro* ophthalmic permeability platform, and we can conclude that the quality of *in vitro* ocular tissue is the key factor for the experimental results [1]. The reliability and accuracy of the experimental data can be improved by optimizing screening methods of *in vitro* eye tissue samples.

3) ***In vitro* ocular tissue screening methods**

Fluorescein sodium staining (before the experiment).

Evaluation of corneal integrity. After staining with fluorescein sodium, detection is performed using a slit lamp cobalt blue filter.

Determination of corneal hydration level (after the experiment).

The corneal hydration level (HL) is calculated by weighing method (weighing the wet cornea W_b at the endpoint of the experiment and the dry cornea W_a dried at 70°C for 12 hours).

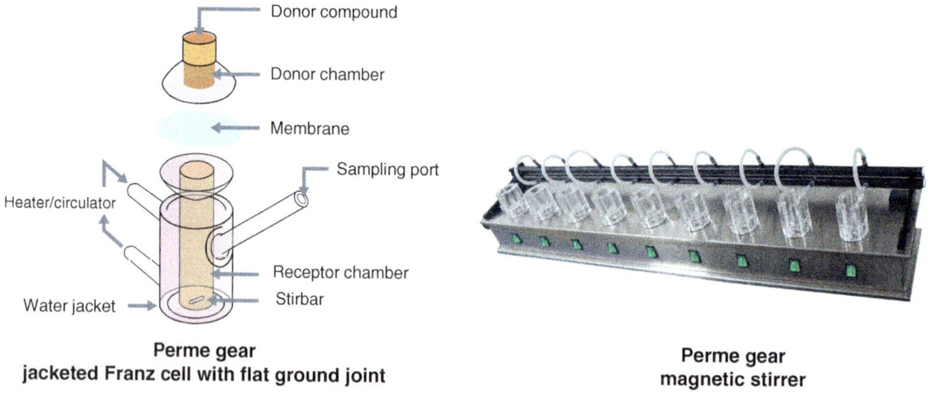

Figure 7.10 Schematic diagram of the diffusion cell. *Source:* [2] / with permission of John Wiley & Sons.

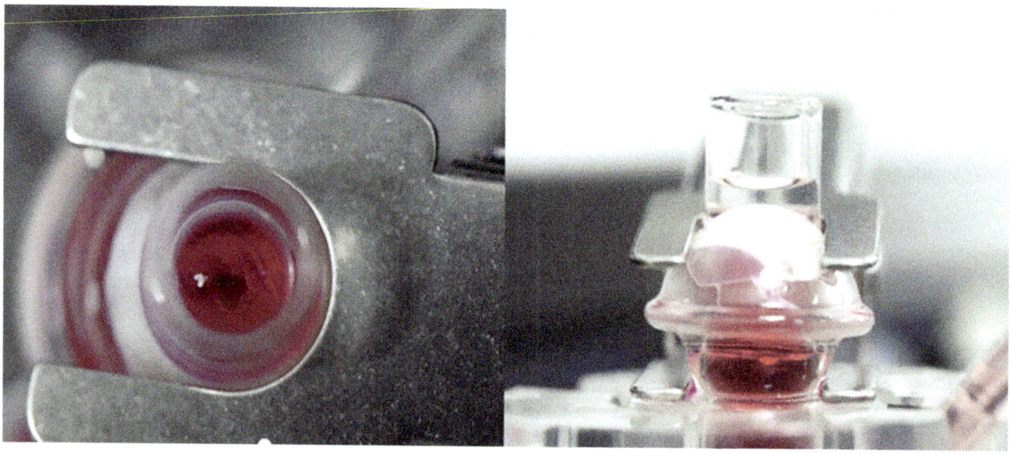

Figure 7.11 Experimental endpoint cornea (left) and sclera (right).

4) **Validation experiment: Permeability study of different dosage forms of the drug in cornea and sclera tissue**

The experimental results show (Figure 7.12) that the cumulative permeation amount (Qn) and apparent permeability parameter (Papp) of 0.3% gatifloxacin in the cornea and sclera eye drops group are both higher than the ophthalmic gel group, and Qn and Papp of 0.3% gatifloxacin in sclera are both higher than in cornea.

The data show that adding excipients can improve the retention time of the drug on the ocular surface. At the same time, it affects the penetration rate of the drug in the ocular surface tissue. Therefore, using excipients requires a balance between retention time and permeation rate. Compared to the complex physiological barrier of the cornea, the sclera has a simple structure, which is composed of a large amount of collagen and elastic fiber. The results of the permeability test show that most drugs have unique advantages in developing noncorneal penetration pathways [3].

5) **Advantages of *in vitro* ophthalmic permeability study**

In vitro ophthalmic permeability study has unique advantages as a preliminary screening experiment (see Figure 7.13). WuXi AppTec DMPK has established the ophthalmic permeability study

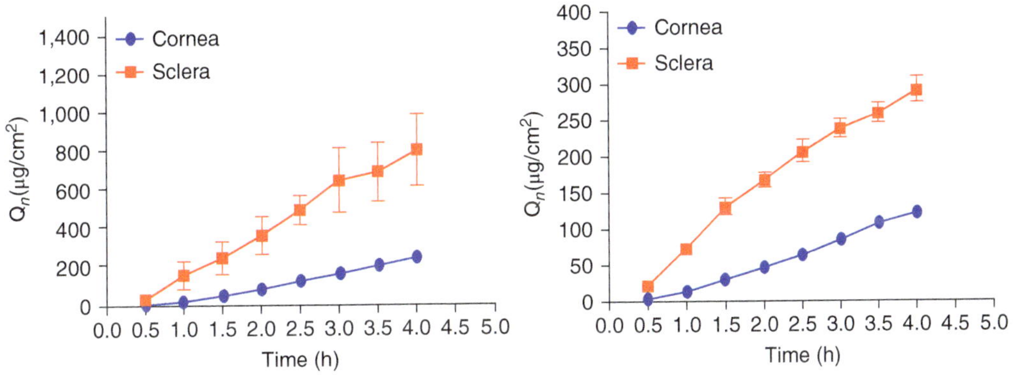

Figure 7.12 The cumulative permeability (Qn) of 0.3% gatifloxacin eye drops (left) and gels (right) in the cornea or sclera of rabbits.

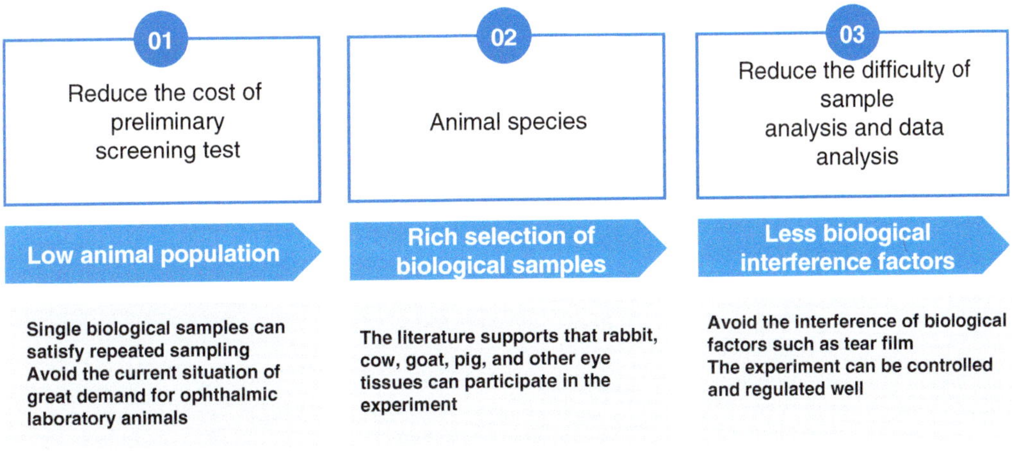

Figure 7.13 Advantages of *in vitro* ophthalmic permeability study.

platform and completed corneoscleral permeability studies of various compounds and commercial drugs.

7.3.3 *In Vivo* Assays

Animal species selection is crucial for *in vivo* ophthalmic drug research. For the selection of animal species in ophthalmic preclinical PK studies, refer to Section 7.1, that is, Ophthalmic Drugs: Preclinical Pharmacokinetic (PK) Profiles and Research Strategies.

7.3.3.1 Ocular Administration Routes and Their Advantages and Disadvantages

The eyeball can be anatomically divided into two segments: the anterior segment and the posterior segment. The anterior segment is situated in front of the lens and comprises approximately one-third of the eyeball, consisting of the cornea, conjunctiva, iris, ciliary body, lens, and aqueous humor. The posterior segment is situated behind the lens and comprises approximately two-thirds of the eyeball, including the sclera, choroid, retina, and vitreous body [4]. Precisely identifying the location of the anterior and posterior segments in ocular drug administration is critical to facilitate drug diffusion to the target locations. Generally, topical administration on the ocular surface, including intracameral injection, is commonly used for the anterior segment, while intraocular or periocular administration routes are often adopted in the posterior segment.

Ocular drug administration routes primarily involve topical administration through eye drops and ointments/gels, as well as subconjunctival injections, intracameral injections, intravitreal injections, and retrobulbar injections. Sedation or anesthesia is generally required for these injection methods, except for the eye drops and ointments/gels. It is important to process with caution during intraocular injections in the eyeball to avoid blood vessels, the posterior capsule of the lens, and the retina. Different administration routes have their advantages and disadvantages, which are outlined in Table 7.3 and Figure 7.14.

Table 7.3 Different ocular drug delivery methods.

Drug delivery	Advantages	Disadvantages
Topical administration	Noninvasive and high patient compliance	Frequent drug administration, nontarget tissue drug exposure, and significant drug loss
Intracameral injection	Reduced systemic and corneal adverse effects, and high drug concentration in the anterior chamber	Invasive, unable to achieve effective drug concentration in the ocular posterior segment
Intravitreal injection	Low frequency of administration, long-lasting effect, topical pharmacology, and high bioavailability	Invasive, superior ocular complications, and retinal toxicity
Periocular injection	No change in intraocular pressure, high dosage, and long duration of drug action	Invasive, frequent administration, and risk of retrobulbar hemorrhage
Sub-tenon injection	Relatively less invasive, fewer complications, and maintenance of high drug concentrations	Risk of conjunctival edema or subconjunctival hemorrhage and need to cross the retinal pigment epithelial barrier

Source: [5]

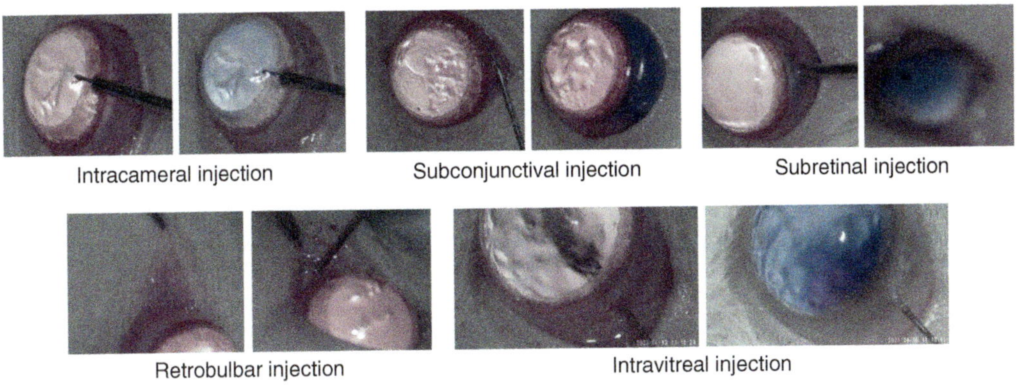

Figure 7.14 Multiple injection routes of drug administration in the eye.

7.3.3.2 Ocular Tissue Collection and Sample Processing

Ocular tissues can be collected by anatomical separation, including the lacrimal gland, conjunctiva, eyelids, cornea, iris, ciliary body, lens, retina, choroid, sclera, and optic nerve. The aqueous humor and vitreous humor can be collected by puncture sampling techniques. The tear can be collected by capillary or filter paper strip method. The types of ocular biological samples that can be collected from different animal species are shown in Table 7.4.

7.3.3.3 Ophthalmic Sample Processing

Ocular tissues vary in size, structure, and characteristics and require different tissue processing methods. For tissue processing, various devices, such as high-shear homogenizers, ball mills, and multifunctional biological sample homogenizers, can be used.

High-shear homogenizers (Figure 7.15, left) use electric motor-driven blades to shear tissues into fine particles. They have the advantage of high rotational power, making them suitable for homogenizing voluminous tissues such as lacrimal gland and lens. However, the drawback of the device is that the samples need to be processed individually, resulting in low efficiency and potential risks of cross-contamination during sample handling.

The ball mill (Figure 7.15, middle) utilizes the collision and friction of grinding beads to pulverize samples. The advantages include no cross-contamination between samples and the ability to process multiple samples simultaneously making them suitable for processing smaller and fragile tissue samples such as the iris, ciliary body, retina, choroid, and optic nerve. However, it is less effective for tougher tissues such as cornea, sclera, and conjunctiva.

The multifunctional biological sample homogenizer (referred to as a homogenizer, Figure 7.15, right) possesses three-dimensional high-speed vibration capabilities, enhancing its homogenization efficiency and providing enhanced homogenization capabilities of tissues such as the cornea, sclera, and conjunctiva (Figure 7.16). Moreover, it can maintain a low temperature for sample homogenization with the help of the BR-Cryo cooling system.

7.3.3.4 Ophthalmic Examination

Following ocular drug administration, in addition to sample collection for PK analysis, close observation of the potential abnormalities is essential. Using ophthalmic devices, experienced veterinarians will conduct detailed ocular examinations and evaluations.

The experienced ocular study team should possess extensive experience in ocular drug administration, be knowledgeable in ocular veterinary practices, and showcase skillful use of diverse ophthalmic

Table 7.4 Types of ocular biological samples that can be collected from different animal species.

Type of sample	Conjunctiva	Cornea	Iris	Retina	Choroid	Sclera	Crystalline lens	Ciliary body	Vitreous humor	Lacrimal gland	Lower eyelid	Optic nerve	Aqueous humor	Tears
Mouse	✔	✔	✔	✔	✔	✔	✔	✔	✔	✔	✔	✔	✔	✔
Rat	✔	✔	✔	✔	✔	✔	✔	✔	✔	✔	✔	✔	✔	✔
Rabbit	✔	✔	✔	✔	✔	✔	✔	✔	✔	✔	✔	✔	✔	✔
Dog	✔	✔	✔	✔	✔	✔	✔	✔	✔	✔	✔	✔	✔	✔
Monkey	✔	✔	✔	✔	✔	✔	✔	✔	✔	✔	✔	✔	✔	✔

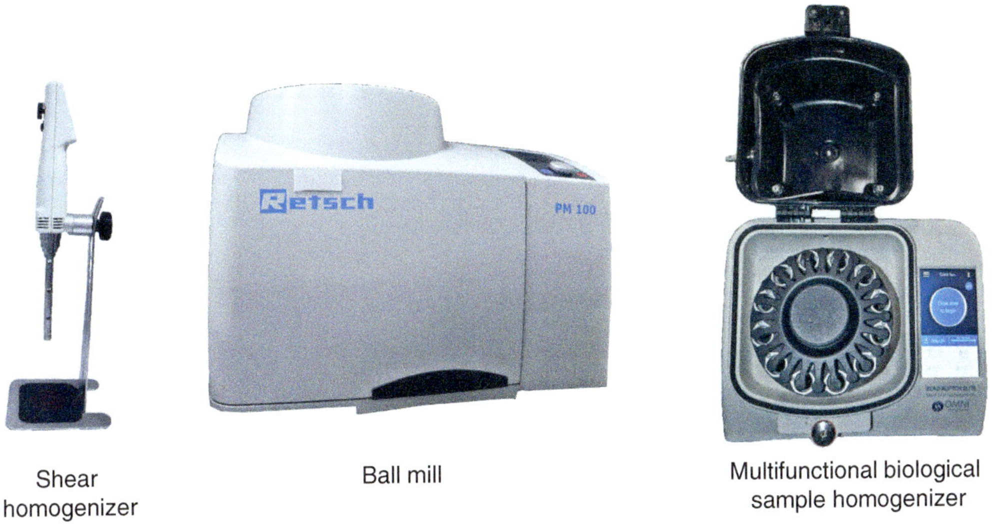

Figure 7.15 Ophthalmic sample processing apparatuses.

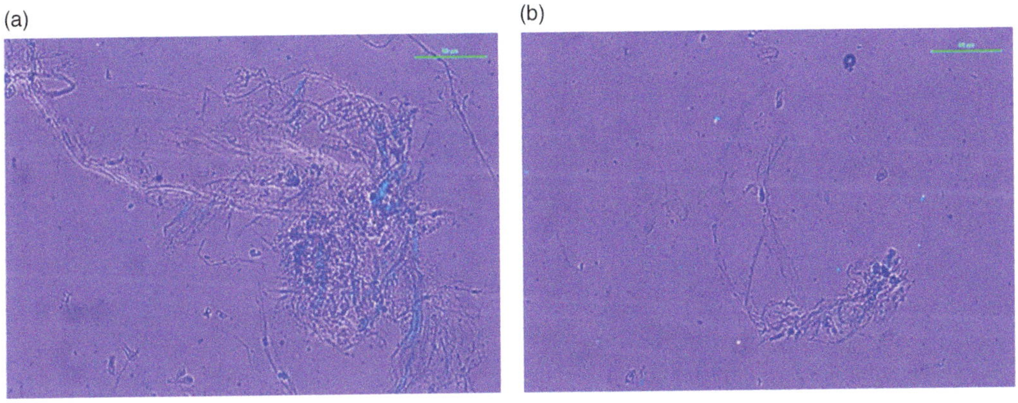

Figure 7.16 Comparison of the homogenizing effects of a high-shear homogenizer and a homogenizer (200×). a: The cornea was homogenized for 2 minutes at 30,000 rpm using a high-shear homogenizer. b: The cornea was homogenized for 2 minutes at 6 m/s using a biological sample homogenizer.

examination devices (Figure 7.17). The team should also have comprehensive ocular scoring criteria (Draize eye irritation score and McDonald–Shadduck scoring system) for ophthalmic examinations.

7.3.3.4.1 Ophthalmologic Examination Devices and Items

Tonometer. Intraocular pressure testing is commonly performed using the tonometer. Intraocular pressure measurement before and after the delivery of invasive drugs can effectively assess whether or not the dosage and operation affect normal aqueous circulation.

Slit lamp. The slit lamp is generally used to examine the transparent tissues of the eye, such as the cornea, anterior chamber, lens, and vitreous humor, to determine the location, nature, size, and depth of the lesions.

Direct/indirect ophthalmoscope. Under dim lighting conditions, the eyes are observed using a binocular indirect ophthalmoscope combined with appropriate refractive lenses to assess the

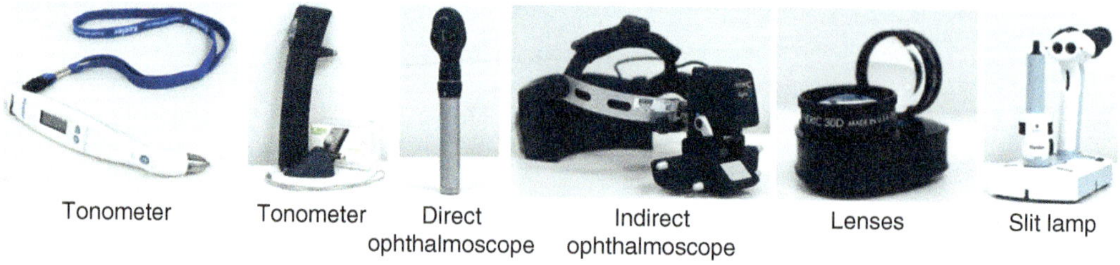

Figure 7.17 Ophthalmic instruments.

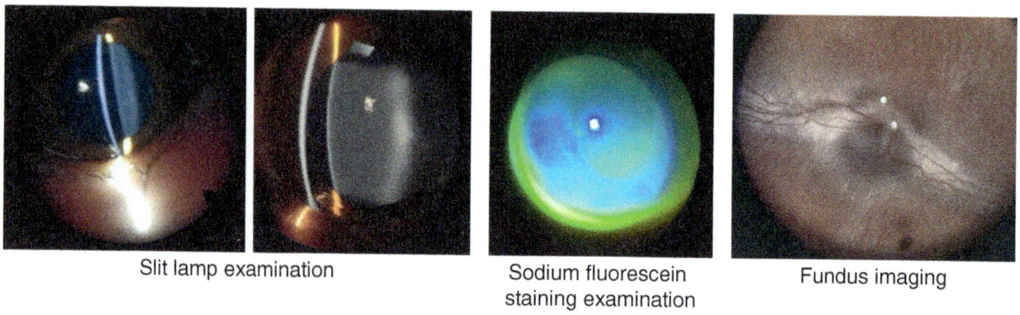

Figure 7.18 Examples of ophthalmic examinations.

overall eye condition and eye structure. In some cases, a direct ophthalmoscope can be used for further examination of retinal lesions. Fundus examination represents the condition of the fundus and can also monitor changes in the fundus. A fundus examination is strictly performed before and after the posterior segment invasive process.

Ophthalmic microscope. The ophthalmic microscope is used to assist in drug administration and confirm the tissue integrity of the ocular posterior segment before and after drug administration. The leading evaluation indicators include fundus vessels and omental status, which can be an essential evaluation factor for ocular barrier integrity and ocular status.

Examples of ophthalmic examinations are shown in Figure 7.18.

7.3.3.4.2 Ocular Scoring

Draize ocular irritation score

The drug that potentially irritates the eyes can be assessed using the Draize ocular irritation score. The scoring system is based on a comprehensive assessment of corneal opacity, including the area of opacity, the iris condition, conjunctival congestion and swelling, and secretion levels, to determine the ocular irritation score of the study animals [6].

McDonald–Shadduck scoring system

This scoring system is used to determine the condition of ocular tissues by evaluating parameters, such as conjunctival congestion, conjunctival edema, conjunctival secretions, aqueous flashes, iris involvement, corneal clouding, corneal neovascularization, fluorescein staining results, and the lens condition [7].

7.3.3.5 Preclinical Ophthalmic Formulations

To enhance the bioavailability and safety of ophthalmic drugs, different administration routes require different ophthalmic formulations. Topical formulations for ocular surface administration need to achieve high corneal residence time and proper drug permeability. Ophthalmic injection formulations need to consider the factors of isotonicity and sterility.

In the design of ophthalmic injection formulations, isotonicity is crucial to ensure that the formulation does not adversely affect ocular tissues/cells during intraocular administration. Additionally, sterility is an essential factor in ophthalmic drugs, as it minimizes the risk of infection.

1) **Ocular Topical Delivery Formulations**

Ocular topical administration formulations are ideal for the patient, as they are noninvasive, and have fast onset of action for the ocular surface diseases. Common ocular topical administration formulations comprise solutions, suspensions, emulsions, and gels.

Solutions

Solutions are common formulations for eye drops, providing immediate effects. Drugs administered through eye drops could rapidly distribute in the corneal and conjunctival tissues, thus achieving a high drug concentration at the target sites. To optimize solution formulations, the following strategies are often adopted:

- Cyclodextrin complexation is adopted to improve the solubility of poorly water-soluble compounds.
- Increasing the viscosity of eye drop solutions to prolong the residence time of the drug in the cornea.
- Using penetration-enhancing reagents (polyvinylpyrrolidone, sodium dodecyl sulfate, and sodium decanoate) to enhance drug permeability.

Suspensions

Suspensions are homogeneously dispersed formulations containing insoluble drug particles and typically including dispersants and solubilizers. The size of the drug particles determines the time of the drug molecule absorbed by the corneal tissue. Also, the particles will stimulate the tear secretion and ultimately affect the proportion of drugs cleared via tears. Methods for particle size optimization in suspensions include ball milling and ultrasonication.

Emulsions

Emulsion is a homogeneous formulation consisting of aqueous phase (denoted by W), oil phase (denoted by O), and emulsifier. The advantages of emulsions include improving lipophilic drug solubility, facilitating corneal penetration, and prolonging the residence time in the anterior segment of the cornea.

2) **Ophthalmic injectable formulations**

Injectables are commonly used for posterior segment ocular drug delivery, such as intravitreal administration. Both traditional small molecules and novel molecular entities can be delivered to the target tissues by injection. The quality of the formulation is a critical factor influencing the injectability of the ophthalmic drug. Common quality parameters include:

- pH: the eye's tolerance range is pH 6–8.
- Osmolality: isotonic.
- Aqueous solution: sterile and free from bacteriostatic and antioxidants.

Vehicle plays an important role in the formulation development. With the rapid development of scientific technology and material science, a large number of ophthalmic formulation vehicles are available (Table 7.5). Additionally, lots of drug delivery technologies have been developed, such as the cutting-edge Punctum Pug Delivery Systems (PPDS), which enhance the delivery of the drug to the target site.

Table 7.5 Common excipients used in ophthalmic solutions.

Types	Representative agents	
pH adjustment	Phosphate buffer, borate buffer	
Aqueous isotonic solution	Sodium chloride, boric acid, and glucose	
Viscosity modifier	Methylcellulose, sodium carboxymethyl cellulose, hydroxypropyl methyl cellulose, hydroxyethyl cellulose, hyaluronic acid, carbomer	
Solubilizer	Co-solvents	Ethanol, polyethylene glycol 200–400, gropylene glycol, glycerol,
	Surfactants	Tween, solutol HS 15, poloxamer, cremophor, sodium dodecyl sulfate
	Complexation	Hydroxypropyl-α-cyclodextrin, hydroxypropyl-β-cyclodextrin, sulfobutylether β-cyclodextrin
	Oil	Liquid paraffin, olive oil, corn oil

7.3.4 Others

Efflux transporters are located on the cornea and retina and play a vital role in the topical treatment of ocular tissues. Efflux transporters protect tissues from foreign substances by efflux excretion, and it is a barrier within the permeation and absorption of drugs; in contrast, uptake transporters can effectively promote the absorption and distribution of drugs. However, the relevant mechanistic studies on ocular tissue transporters still need to be clarified, and more in-depth studies and validation are to be discovered [8].

References

1 Yu, N., Xun, Y. et al. (2010). Effect of sperminated pullulans on drug permeation through isolated rabbit cornea and determination of ocular irritation. *The Journal of International Medical Research [J]*. 38 (2): 526–535.

2 Finnin, B., Walters, K.A., and Franz, T.J. (2012). In Vitro Skin Permeation Methodology. In: *Transdermal and Topical Drug Delivery: Principles and Practice*, 85–108. https://doi.org/10.1002/9781118140505.ch5.

3 Zhao, X., Lu, H. et al. (2018). Study on the permeability of cornea and sclera *in vitro* by cyclodextrin-coated eye drops[J]. *Liaoning Chemical Industry.* 47 (07): 636–638.

4 Koppa Raghu, P., Bansal, K.K., Thakor, P. et al. (2020). Evolution of nanotechnology in delivering drugs to eyes, skin and wounds via topical route[J]. *Pharmaceuticals* 13 (8): 167.

5 Song, S., Wang, N.N. et al. (2021). Pharmacokinetic research strategy of ophthalmic drugs [J]. Chinese Journal of New. *Drugs* 30 (18): 1668–1674.

6 Wang, Q.L. and Peng, J. (2005). Evaluation of Draize eye irritation test [J]. *Traditional Chinese Drug Research and Clinical Pharmacology* 16 (04): 301–304.

7 GB/T 28538–2012, Ophthalmic optics. Contact lenses and contact lenses care products. Determination of biocompatibility by ocular study with rabbit eyes [S].

8 Mandal, A., Agrahari, V., Khurana, V. et al. (2017). Transporter effects on cell permeability in drug delivery. *Expert Opinion on Drug Delivery* 14 (3): 385–401.

7.4 *In Vivo* Bioanalysis in Ophthalmology: Challenges and Strategies

Tingting Li, Xinfa Fu, Yanfu Ren, Lili Xing

7.4.1 Introduction

The intricate structure of the eye poses several difficulties for ophthalmic bioanalysis, including limited sample volumes, low analyte concentrations, and challenges in acquiring an authentic blank matrix. This chapter focuses on these challenges and delineates strategies for *in vivo* bioanalysis in ophthalmic.

7.4.2 Challenges and Strategies of Ophthalmic Bioanalysis After Ocular Drug Administration

7.4.2.1 Low Exposure of Ocular Drugs in Plasma

Regardless of the routes of administration (ocular or systemic), crit is vital to measure drug concentrations in plasma to evaluate the potential systemic exposure-related drug toxicity [1]. The "blood-eye barrier" makes it difficult for endogenous drugs to enter the eye after systemic administration. Therefore, topical ocular drug delivery is commonly used for ocular disease treatment. However, due to the small dosage of topical ocular drug administration and various barriers, the systemic drug concentration usually remains at a low level after local administration, which significantly challenges the determination of plasma drug concentration.

WuXi AppTec DMPK provides highly efficient strategies using LC–MS, primarily based on the optimization of sample pretreatment, LC, and MS conditions (Figure 7.19). This ensures the detection sensitivity of developed LC–MS analytical methods can reach the pg/mL level, effectively dealing with this challenge.

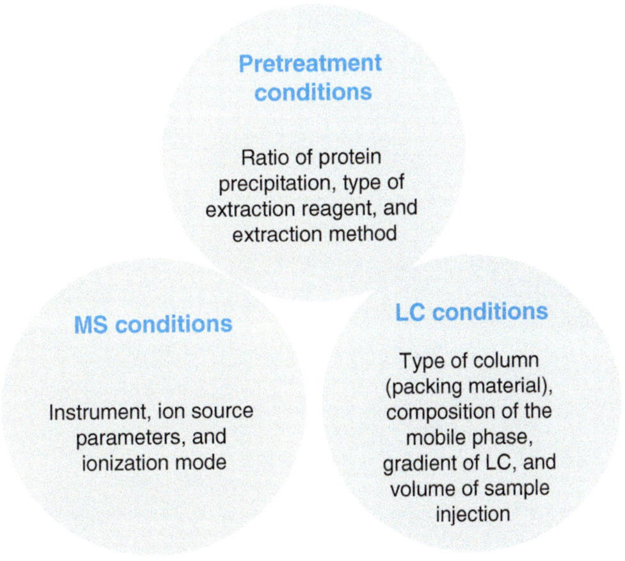

Figure 7.19 Optimization strategies for LC–MS plasma methods after ocular drug administration.

7.4.2.2 Nonspecific Binding Issues of Ocular Drugs

Ophthalmic fluid samples (aqueous humor, vitreous humor, and tears) present potential nonspecific issue binding, which is mainly due to their unique composition. The main component of aqueous humor is water, but it also contains small amounts of chloride, vitamin C, urea, inorganic salts, and proteins, of which, the protein content accounts for approximately 0.5% of that in the plasma. Additionally, vitreous humor is mainly composed of water, making up 99% of vitreous humor, but it also contains small amounts of inorganic salts, sugars, and less than 1% protein. Tears consist of a variety of electrolytes, lipids, and small-molecule metabolites, with proteins accounting for approximately 10–15% of that in the plasma. As a result of the low protein content in these three matrix samples, drugs are prone to have nonspecific adsorption issues.

7.4.2.3 Difficulty in Acquiring an Authentic Blank Matrix

Matrix types are complex and diverse during ophthalmic PK research. Therefore, it is not only vital to investigate the ADME of ophthalmic drugs in the entire body, but also to focus on their distribution in ocular tissues. Consequently, meticulous separation of each tissue or fluid is necessary for PK studies. However, when it comes to the bioanalysis of ocular tissue samples, obtaining an authentic blank matrix is also a significant challenge due to the limited volume of each ocular tissue and the impossibility of using a large number of animals to procure sufficient blank matrices.

To address this challenge, two main aspects require attention. First, it is crucial to select an appropriate surrogate matrix, such as one that closely resembles ocular tissues in physical character and chemical composition. If such similar matrices are unavailable, alternatively biological matrices or appropriate reagents can be used as the surrogate matrix. Second, the procurement of commercialized matrices can be adopted, such as artificial tears. The findings of the present study show that artificial tears can be used as a surrogate matrix for analyzing the concentration of Natamycin in rabbit tears, as shown in Figure 7.20.

7.4.3 Challenges and Strategies for Tear Sample Bioanalysis of Ocular Drugs

The challenges of tear sample bioanalysis are mainly related to its special collection methods, which mainly include two methods: the capillary tube method and the filter paper strip method.

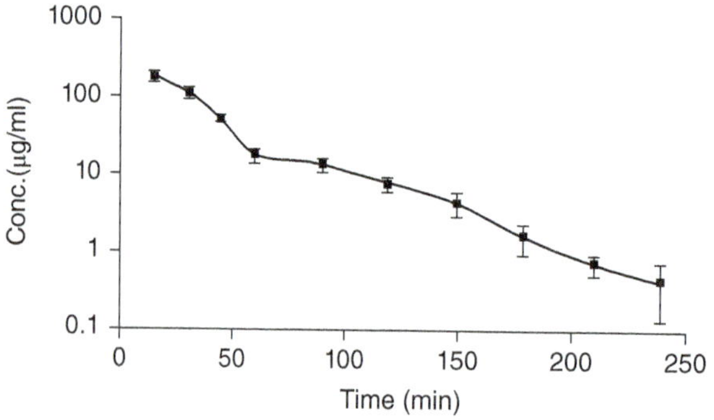

Figure 7.20 Time–drug concentration curve of Natamycin in rabbit tear fluid after topical administration of Natamycin eye drops (5%, w/v). *Source:* [2] / ELSEVIER.

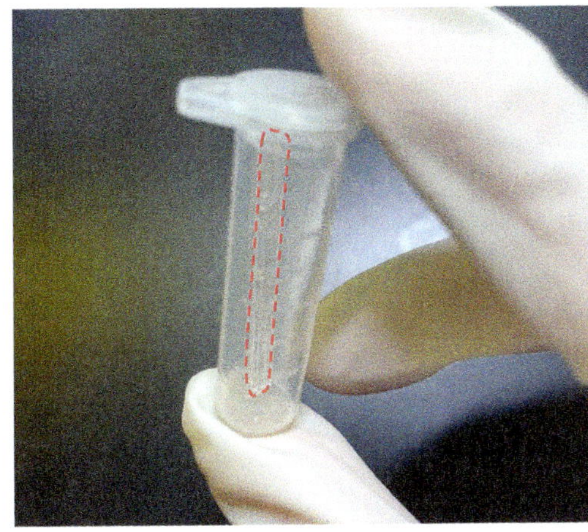

Figure 7.21 The collection of tear samples using the capillary tube method.

7.4.3.1 Capillary Tube Method

During sample collection of tear samples, gently touch the tear meniscus at the lower eyelid margin with the capillary tube, allowing the tear fluid to enter the tube through capillary action. For capillary tube sampling, the single collection volume is about 10 μL. As shown in Figure 7.21, tear fluid samples have small volumes and tend to remain within the capillary tube, making it difficult to pipette the sample accurately. Accurately aspirating the sample and obtaining a sufficient blank matrix for sample analysis pose challenges to the capillary tube collection method. To address this question, organic reagents are generally used to dilute the tear samples and wet the samples well in the sample tubes to facilitate accurate sampling. Additionally, organic reagents are also used as surrogate matrices to prepare calibration standard curves and quality control samples for the quantification of tear samples.

7.4.3.2 Filter Paper Strip Method

During tear sampling using the filter paper strip method, one end of the filter paper strip is inserted under the conjunctival sac of the animal's eyes, and retained in place for a certain period to allow the tear infiltrating to the paper strip. As shown in Figure 7.22, the filter paper strip serves as a solid sample, making it impossible to obtain a suitable liquid for sampling even after homogenization. The challenges of the filter paper strip method involve how to extract the drugs from the filter paper strip sample to obtain a liquid state for sampling, and how to use the filter paper strip as a blank matrix for preparing standard curves and quality control samples. To deal with these two challenges, we designed a specific process. First, the filter paper strip was soaked in an organic solvent for a period to extract the drug from the strip, which transformed the filter paper strip from a solid to a liquid form for sampling. Second, for the preparation of standard calibration samples and quality control samples, there was an alternative method that directly added the working solution of the standard curve and quality control samples to the supernatant solution containing the blank filter paper strip for subsequent analysis, and the method has been verified: during the qualification process, the calibration curve samples were obtained using the alternative approach, and the quality control samples were added to the filter paper strip for processing and extracting for data validation. As shown in Figure 7.22, the qualified results indicated that the assay accuracy achieved using the alternative approach meets the analytical requirements.

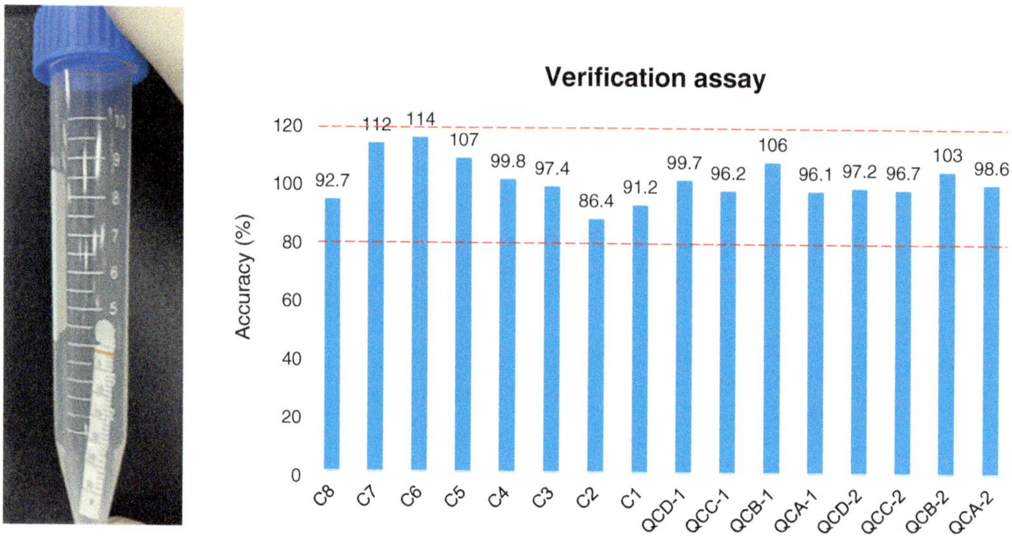

Figure 7.22 The results of the validation assay for the tear samples were collected using the filter paper strip method.

7.4.4 Challenges and Strategies in Drug–Melanin Binding

In pigmented species, ocular melanin is located in the anterior portion of the ciliary body and iris, the posterior portion of the choroid, and within the RPE [3]. The binding of drugs to melanin could form a drug reservoir, which leads to the slow drug release into the surrounding tissues, and affects the PK, efficacy, and safety of drugs [1, 4]. The binding of drugs to melanin brings a potential risk of detecting inaccurate concentrations and encountering unstable false positives, posing a significant challenge in bioanalysis.

The solutions should focus on:

1) Adding high concentrations of salts, such as NaCl and $MgCl_2$. This approach utilizes metal cations binding to the melanin, competing with the drug for the binding site [3], thus releasing the drug.
2) Adjusting the pH value is significant. For weak basic drugs with pKa values exceeding 9, lowering the pH below 7 does not affect the ionization of the drug, but it alters the ionization of melanin, thereby reducing the drug binding to melanin [3].
3) Adding a protease disrupts the hydration of melanin's surface protein, consequently destroying its structure. This process does not affect the structure of the small molecule drug but can reduce the binding rate and facilitate the release of the drug.

7.4.5 Summary

Due to the various barriers encountered in ocular drug delivery, drugs are difficult to reach target tissues to exert their efficacy, with short drug action duration. However, with the development of advanced drug delivery systems, such as liposomes and nanoparticles, the duration of drug action has been prolonged. And novel types of ophthalmic drugs, such as oligonucleotides and exosomes, are continually being developed, bringing new hope to patients with eye diseases. At the same time, ophthalmic bioanalysis will face additional challenges. We are committed to closely monitor the advancements in the industry, improving the service capacity and supporting the research and development of more ophthalmic drugs.

References

1 Velagaleti, P.R. and Buonarati, M.H. (2013). Challenges and Strategies in Drug Residue Measurement (Bioanalysis) of Ocular Tissues. In: *Ocular Pharmacology and Toxicology. Methods in Pharmacology and Toxicology* (ed. B. Gilger). Totowa, NJ: Humana Press.

2 Bhatta, R.S., Chandasana, H., Rathi, C. et al. (2011). Bioanalytical method development and validation of natamycin in rabbit tears and its application to ocular pharmacokinetic studies. *Journal of Pharmaceutical and Biomedical Analysis* 54 (5): 1096–1100.

3 Rimpelä, A.K., Reinisalo, M., Hellinen, L. et al. (2018). Implications of melanin binding in ocular drug delivery. *Advanced Drug Delivery Reviews* 126: 23–43.

4 Novack, G.D. and Moyer, E.D. (2016). How Much Nonclinical Safety Data Are Required for a Clinical Study in Ophthalmology? *Journal of Ocular Pharmacology and Therapeutics* 32 (1): 5–10.

8

Respiratory Drugs

8.1 Inhaled Medications: Challenges and Strategies for Preclinical *In Vivo* Pharmacokinetic Studies

Yingying Li, Lijuan Hou, Xuan Dong, Jing Jin, Cheng Tang

8.1.1 Introduction

Owing to the advancement of medical and pharmaceutical sciences, as well as our understanding of pulmonary function and diseases such as asthma and chronic obstructive pulmonary disease (COPD), inhaled delivery has been considered the simplest and most effective administration route of respiratory diseases [1]. According to market statistics, the sales of inhaled drugs, such as aerosols, dry powders, and sprays, have consistently risen over the past decade. In 2022, the market sales were approximately 31.5 billion USD and are expected to grow at a compound annual growth rate (CAGR) of 6.6% from 2023 to 2030 [2]. As such, inhaled drug delivery is garnering increasing attention in the pharmaceutical industry.

To understand the inhaled delivery system, a comprehensive knowledge of the intricate structure of the respiratory system is necessary. The system consists of the respiratory tract (the nose, pharynx, larynx, trachea, and bronchi) and alveoli. Lesions from respiratory diseases primarily occur in the trachea, bronchi, lungs, and chest cavity. Therefore, inhaled drugs can directly target the disease areas to treat respiratory system diseases. The processes of drug absorption, distribution, metabolism, and excretion within the body after inhalation are illustrated in Figure 8.1 [3].

As shown in Figure 8.1, inhaled medications directly target tissues in the lung, provide rapid pharmaceutical effects, and reduce the required drug dose and systemic exposure, thereby lowering the risk of potential systemic toxicity. Additionally, the alveolar cells are thin and there are abundant blood vessels on the alveolar surface, which facilitates the absorption of drugs intended for systemic therapeutic effects through inhalation. This approach also bypasses first-pass metabolism in the liver, increasing bioavailability in the lung. Moreover, the treatment of respiratory diseases often requires long-term medication, and inhaled delivery is noninvasive and relies on simple devices, enhancing patient compliance. Because of these pronounced advantages, inhalation drug delivery has attracted considerable attention in the pharmaceutical industry. However, there are still numerous challenges in the development of inhaled drugs. This chapter discusses the challenges and strategies for the development of inhaled drugs.

Drug Metabolism and Pharmacokinetics: Frontiers, Strategies, and Applications, First Edition.
Edited by Liang Shen and WuXi AppTec DMPK.

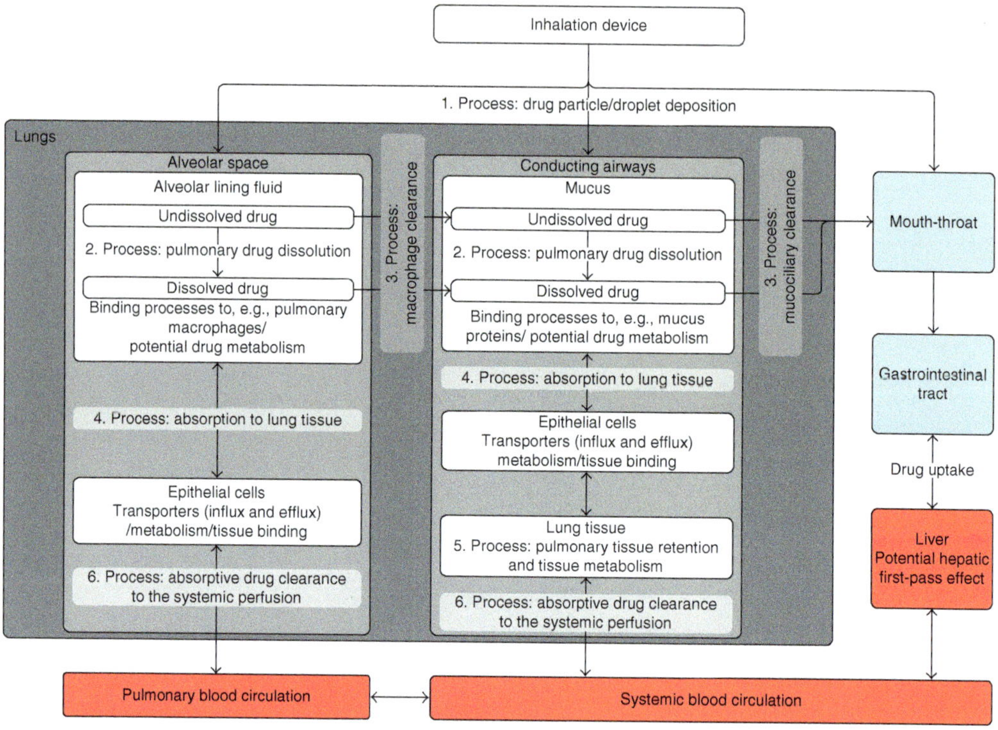

Figure 8.1 *In vivo* pharmacokinetic processes after inhaled delivery. *Source:* [3] / John Wiley & Sons / CC BY 4.0.

8.1.2 Challenges in the Development of Inhaled Drugs

After inhalation, not all of the drugs are deposited in the lung, with 40–90% of drugs getting deposited in the oropharynx instead [4]. One major challenge faced by the development of inhaled drugs is the improvement of pulmonary bioavailability while decreasing systemic exposure.

Increasing lung deposition rates and prolonging lung retention are the two ways to improve pulmonary bioavailability.

8.1.2.1 Increasing Lung Deposition Rates

There are three forms of movement in which drug particles enter the respiratory tract after inhalation: impaction (inertial), sedimentation (gravity), and Brownian diffusion (Figure 8.2). In the upper respiratory tract, larger particles are more likely to deposit owing to inertial impaction against the tracheal wall. In the middle to end regions of the bronchi, drug particles deposit at different locations under gravitational influence due to different particle sizes. However, in the alveoli, the drug particles may be exhaled during expiration if the size of the drug particles is too small. According to the literature [3], larger particles tend to deposit more in the oropharynx. On the other hand, extremely small drug particles tend to have low total deposition in the respiratory tract. Therefore, lung deposition is closely related to the size of the drug particles. The general distribution pattern based on the particle size is summarized as follows (Figure 8.3):

- Particles with aerodynamic diameters of 1–5 μm show the best lung deposition rates.
- Particles >5 μm tend to deposit in the oral cavity and trachea.
- Particles <1 μm will be quickly exhaled from the lungs with the airflow.

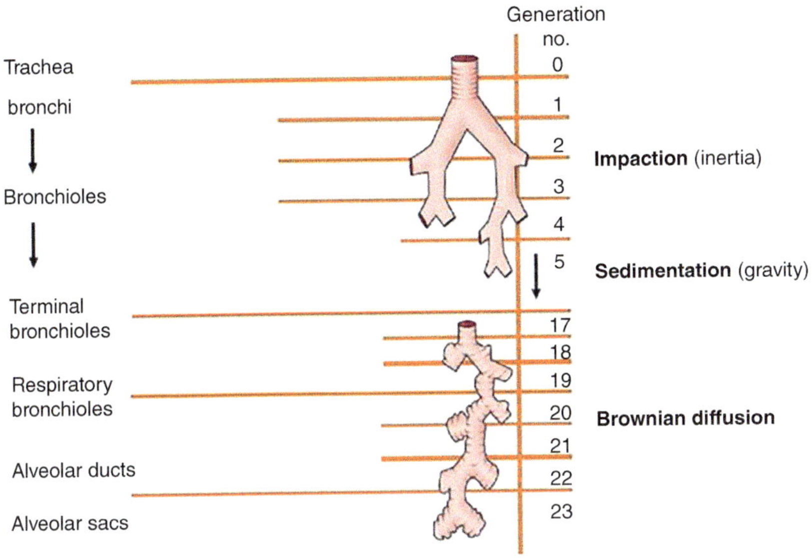

Figure 8.2 Deposition of drug particles in different parts of the respiratory system. *Source:* [5] / with permission of Springer Nature.

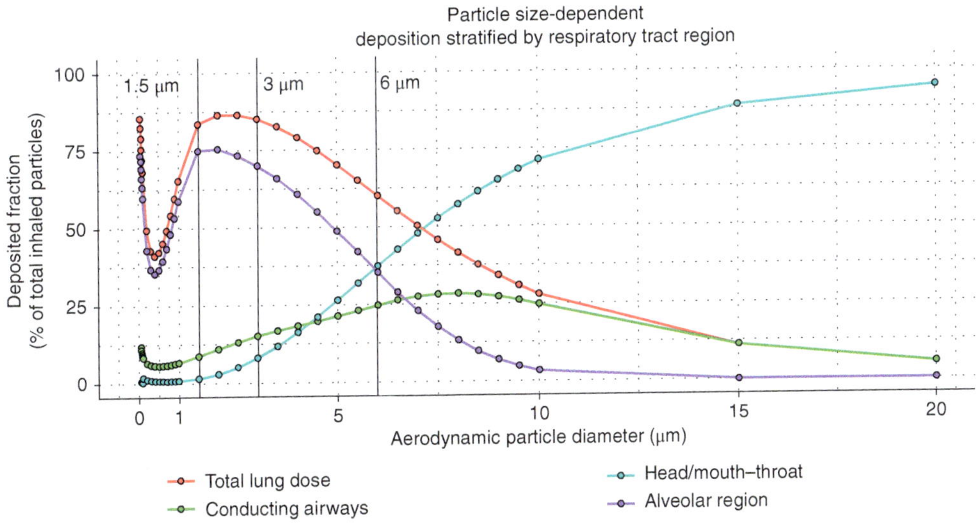

Figure 8.3 Relationship between location of deposition and drug particle size. *Source:* [3] / John Wiley & Sons / CC BY 4.0.

8.1.2.2 Prolonging Lung Retention

The rate at which a drug dissolves in the epithelial lining fluid (ELF) determines its absorption rate. Rapid dissolution results in rapid systemic absorption, whereas slower dissolution promotes longer retention in the lungs. Consequently, prolonging the dissolution time increases the retention time of drugs in the lungs.

The intrinsic solubility of a compound is also a major factor affecting the dissolution rate. Structural modifications could be used to reduce a compound's solubility or alter its crystal structure. For example, fluticasone used in the treatment of asthma and COPD has low solubility (approximately 100 ng/mL). It takes more than eight hours for fluticasone microcrystals to completely dissolve in human bronchial fluid, resulting in a relatively long average absorption time in the lungs (5–7 hours).

New formulation technologies can be employed to regulate the rate of drug release in the lungs. For instance, liposomes or microspheres can be used to achieve optimal drug release rates, and drug coatings, such as biodegradable polymer nanofilm coatings or trehalose-based coatings, can ensure slow drug release in the lungs.

In addition to enhancing the distribution and absorption of drugs in the lungs, the rate of drug–target dissociation can be lowered by improving target binding kinetics and drug structure. Fluticasone, as an example of this strategy, is a long-acting receptor antagonist and a new-generation inhaled corticosteroid. Compared to dexamethasone, fluticasone exhibits a high binding rate and percentage to the glucocorticoid receptor (binding constant k_l is 23.9×10^5 L/mol·min for fluticasone and 12.5×10^5 L/mol·min for dexamethasone). Moreover, the fluticasone–receptor complex has a slower dissociation rate (dissociation constant k_{-1} is 13.2×10^{-4}/min for fluticasone and 11.7×10^{-3}/min for dexamethasone) [6].

To reduce systemic exposure to inhaled drugs, it is necessary to consider three aspects: absorption, distribution, and metabolism, listed below:

- Absorption: Reduce oral bioavailability by lowering permeability, decreasing solubility, and increasing the molecular weight.
- Distribution: Decrease the free concentration of drugs in the plasma by increasing plasma protein binding.
- Metabolism: Enable rapid metabolism of drugs into inactive metabolites in the liver.

In addition to optimizing the drug's physicochemical properties, formulations, and target binding, selecting an appropriate inhalation device can further optimize drug delivery and deposition in the lungs.

8.1.3 Choice of Inhalation Device

There are three common types of inhalation devices for inhaled delivery in clinical trials: pressurized metered-dose inhalers (pMDIs), dry powder inhalers (DPIs), and nebulizers (Figure 8.4).

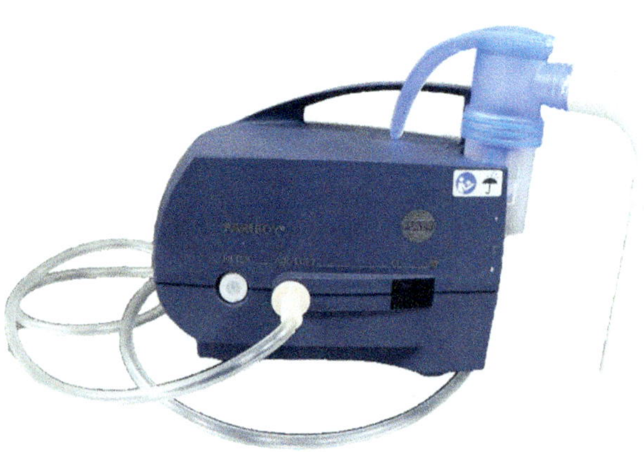

Figure 8.4 Three types of inhalation devices in clinical use.

Moreover, four common types of inhalation devices are commonly used in preclinical studies: aerosol exposure chambers, intranasal delivery devices, pulmonary delivery nebulizers, and nose-only exposure inhalation towers.

8.1.4 Inhaled Delivery Techniques and Study Design for Preclinical *In Vivo* PK Studies

8.1.4.1 Selection of Inhalation Devices for Preclinical PK Studies

The choice of drug delivery device should be based on various factors, including the drug development stage, formulation type, and quantity of the compound. In the early stages of drug development, when only limited quantities of the compound are available, intranasal or intratracheal administration can be chosen to investigate the pulmonary PK properties using a small amount of the compound. If the compound amount is sufficient, a nose-only exposure inhalation tower can be used for drug research. The strategies for inhalation device selection are shown in Figure 8.5.

8.1.4.2 Preclinical Study Design for Inhaled Delivery

In preclinical PK studies, different inhalation devices are used for different study designs. The dosage designs are summarized in Table 8.1.

8.1.4.3 Sampling Design

In PK studies on respiratory diseases, in addition to routine plasma collection, special attention should be paid to the distribution of the drug in the respiratory tract and drug concentration in the bronchoalveolar lavage fluid (BAL) (Table 8.2).

8.1.4.4 Data Analysis

Taking propranolol, a small-molecule drug, as an example, the bioavailability of conventional oral administration (PO), intratracheal nebulization administration (IT), and nose-only exposure inhalation administration was investigated.

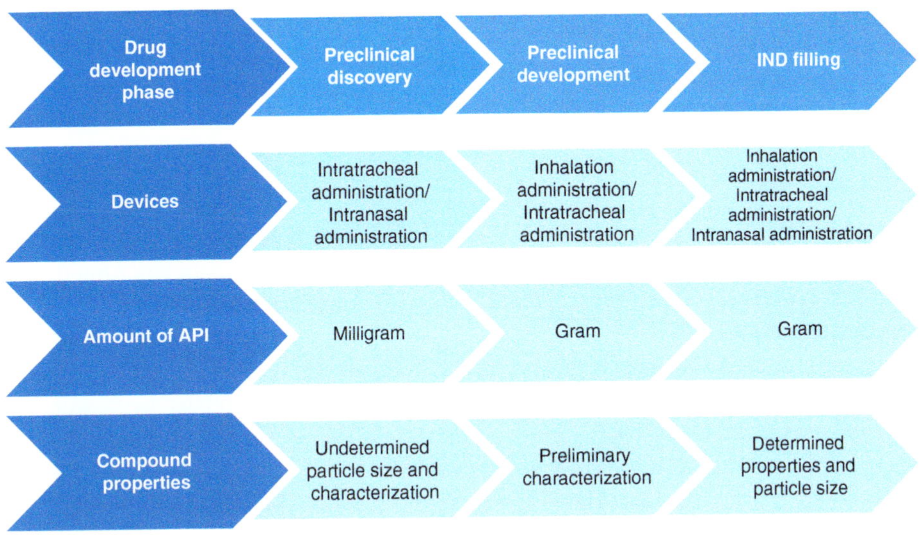

Figure 8.5 Preclinical study design for inhaled delivery.

Table 8.1 Dosage designs for different administration routes.

Routes of administration	Formulation types	Dosage calculation
Intranasal administration	Solution Dry powder	
Intra-tracheal administration	Solution Dry powder	Solution formulation: dose administered (mg/kg) = formulation concentration (mg/mL) × volume administered (mL/kg)
Nose-only exposure inhalation	Solution Dry powder	Delivered dose[a] (mg/kg) = (aerosol concentration[b] × animal ventilation per min[c] × inhalation duration)/ animal weight [5]

[a] The delivered doses are not fully deposited in the lungs. According to the literature, lung deposition rates are generally 10% of the delivered dose in rodents and 25% in non-rodents [7].
[b] Aerosol concentration is determined by instruments such as filters, photometers, U-tubes, aerodynamic particle size analyzers, and cascades (disc impingers).
[c] Ventilation per min (L/min) = 0.608 × body weight (kg)^0.852.

Table 8.2 Sampling designs of inhaled delivery in airway tissues.

Sampling site	Sample type
Nasal sampling	Nasal contents
	Nasal lavage
Lung sampling	BAL (calculating ELF concentration[a])
	Trachea
	Lung

[a] The ELF of the lungs is important for understanding the mechanisms of acute lung injury, inflammatory lung disease, and PK studies. Bronchoalveolar lavage is a technique for sampling the ELF from the lung, but it may also result in the dilution of the ELF. To quantify the apparent volume of ELF obtained by bronchoalveolar lavage, urea is used as an endogenous marker. Since urea diffuses readily in the body and the concentration of urea in the plasma/serum and ELF is the same, a simple dilution principle can be used to calculate the ELF volume [8], which can be converted to the drug concentration in the ELF.

Experimental design: Twelve mice were randomly divided into four groups and given propranolol via IV, oral administration, IT administration, and nose-only exposure inhalation. After dosing, whole blood samples were collected continuously and plasma samples were prepared and tested for the drug concentration.

The PK curves after dosing via each route are shown in Figure 8.6. The results showed that the low bioavailability (Table 8.3) of propranolol after oral administration was related to the extensive hepatic first-pass effect, and the bioavailability increased significantly after IT administration and nose-only exposure inhalation. The results of this study provided data to support altering the administration route, reducing the treatment dose, and improving the bioavailability of drugs with low oral bioavailability, especially for pulmonary diseases.

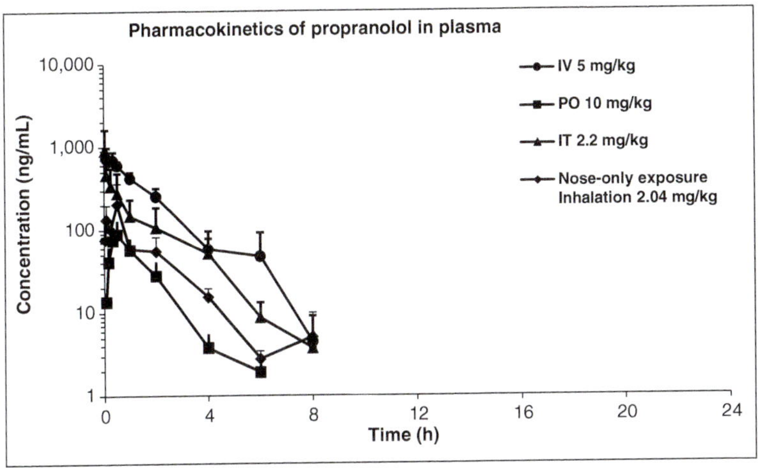

Figure 8.6 Pharmacokinetics of propranolol after IV bolus, oral, intratracheal, and nose-only exposure inhalation administration.

Table 8.3 Plasma exposure after propranolol hydrochloride administration via different routes.

Dose route	Dose level (mg/kg)	C_0/C_{max} (ng/mL)	$AUC_{0\text{-}inf}$ (ng. h/mL)	Bioavailability
IV	5	832	1,341	–
PO	10	90.7	135	5.05%
IT	2.2	907	633	107%
Nose-only exposure inhalation	2.04	210	286	52.2%

C_0: initial plasma concentration.
C_{max}: peak plasma concentration.
$AUC_{0\text{-}inf}$: the area under the plasma concentration–time curve from time zero to infinity.

8.1.5 Preclinical Inhaled Delivery Platform

The inhaled delivery platform of WuXi AppTec DMPK includes intranasal administration, intratracheal administration, and nose-only exposure inhalation systems, which satisfies the needs of inhalation drug research at all preclinical stages (Table 8.4).

Intranasal administration (Figures 8.7 and 8.8).

Intratracheal administration (Figures 8.9 and 8.10).

Inhalation delivery through nose-only exposure tower (Figure 8.11).

8.1.6 Summary

Approximately 300 million people worldwide suffer from chronic pulmonary diseases and endure breathing difficulties. There remains a clinical need for more effective inhaled drugs with fewer side effects. This chapter summarizes the challenges and strategies for preclinical *in vivo* PK studies with inhaled delivery, aiming to facilitate the development of respiratory disease medications and bring hope to patients.

Table 8.4 Preclinical inhaled drug delivery platform.

Drug delivery device	Characteristics	Scope
Intranasal delivery devices	Direct administration, simple operation, precise dosing, and drugs can easily cross the blood–brain barrier	Animal studies with rats, mice, monkeys, dogs, etc.
Intratracheal nebulizer	Direct administration, precise dosing, avoiding deposition in the oral or nasal cavity, rapid drug delivery, and minimal drug loss	Animal studies with rats, mice, monkeys, dogs, etc.
	Performed on animals under mild sedation or anesthesia, suitable for a single dose or multiple-day intervals; multiple doses may cause local irritation	
Nose-only exposure inhalation tower	Spontaneous inhalation of the drug by the animals	Animal studies with rats, mice, etc.
	Administered when the animal is awake, can be used for multiple dosages	
	Deposition of the drug in the lungs depends on the animal's respiration, and deposition will occur in the oropharynx, where drug consumption levels are high	

Figure 8.7 Intranasal instillation.

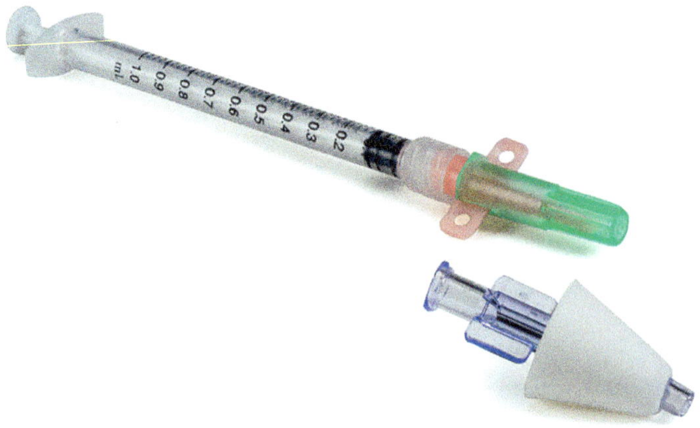

Figure 8.8 Intranasal nebulizer.

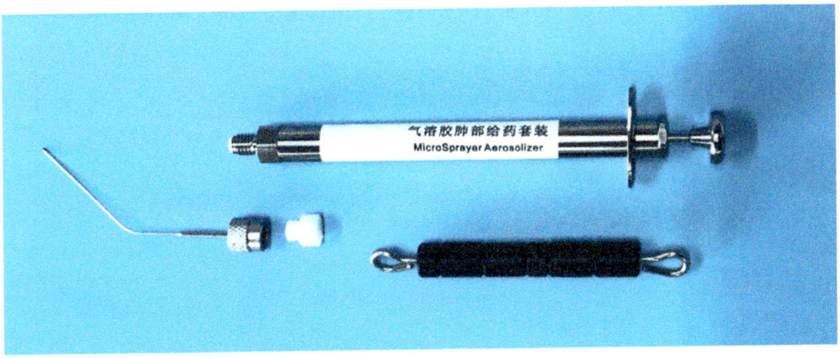

Figure 8.9 Nebulization syringe.

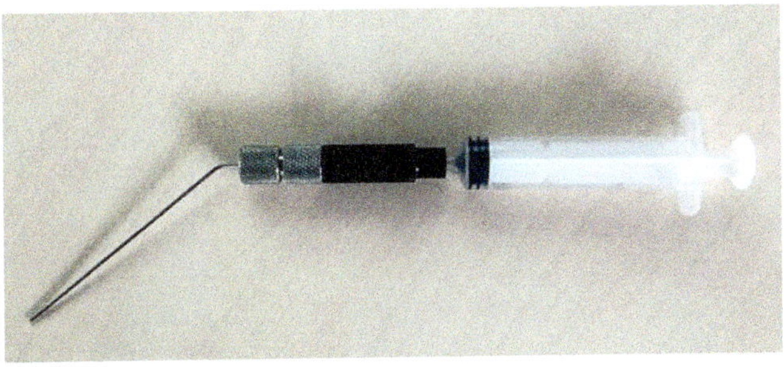

Figure 8.10 DPI-specific syringe.

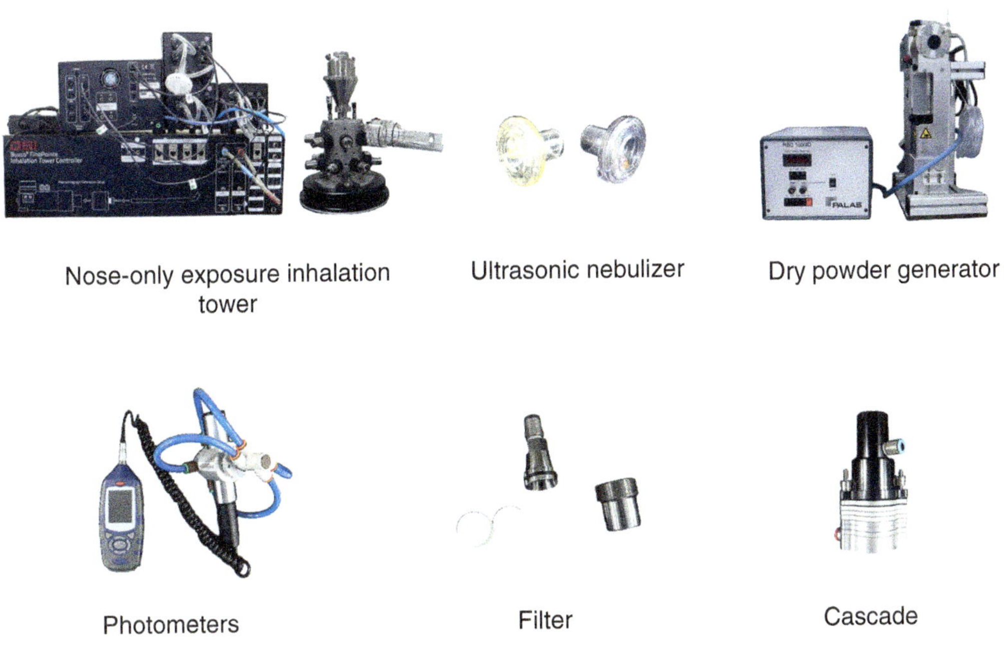

Figure 8.11 Inhalation delivery through nose-only exposure tower.

References

1 Lei, T. and Zhao, R. (2016). Current clinical application status and research progress of inhaled drug delivery[J]. *Clinical Medication Journal* 14 (3).

2 Market Analysis Report. Inhalable Drugs Market Size, Share & Trends Analysis Report By Product (Aerosol, Dry Powder Formulation, Spray), By Application (Respiratory & Non-Respiratory Diseases), And Segment Forecasts, 2023–2030

3 Borghardt JM, Kloft C, and Sharma A.Inhaled therapy in respiratory disease: The complex interplay of pulmonary kinetic processes. *Canadian Respiratory Journal*, Volume 2018, Article ID 2732017, 11 pages

4 Hochhaus, G., Horhota, S., Hendeles, L. et al. (2015). Pharmacokinetics of orally inhaled drug products. *The AAPS Journal* 17 (3).

5 Smyth, H.D. and Hickey, A.J. (ed.) (2011). *Controlled Pulmonary Drug Delivery*. Springer Science & Business Media.

6 Högger, P. and Rohdewald, P. (1994). Binding kinetics of fluticasone propionate to the human glucocorticoid receptor. *Steroids.* 59: 597–602.

7 Alexander, D.J., Collins, C.J., Coombs, D.W. et al. (2008). Association of inhalation toxicologists (AIT) working party recommendation for standard delivered dose calculation and expression in non-clinical aerosol inhalation toxicology studies with pharmaceuticals. *Inhalation Toxicology* 20: 1179–1189.

8 Rennard, S.I., Basset, G., Lecossier, D. et al. (1985). Estimation of volume of epithelial lining fluid recovered by lavage using urea as a marker of dilution. *Journal of Applied Physiology* 60 (2): 532–538.

8.2 Chronic Obstructive Pulmonary Disease (COPD): Analysis of PK Profiles of Approved Inhaled Medications

Qian Li, Lijuan Hou, Lijia Gao, Jing Jin

8.2.1 Introduction

COPD is a chronic condition involving bronchitis and/or emphysema, characterized by airflow obstruction, which can further develop into common chronic diseases such as pulmonary heart disease and respiratory failure. Its incidence rate increases with age. COPD is a partially irreversible chronic disease that affects approximately 384 million individuals globally [1]. According to the World Health Organization's statistics in 2019, approximately 3.23 million people died from COPD, accounting for approximately 6% of global deaths and making it the third leading cause of death worldwide [2].

The causative factors of COPD include smoking, air pollution, kitchen smoke inhalation, airway hyperresponsiveness diseases, and genetic factors. The pathophysiology of COPD is generally associated with oxidative stress, protease–antiprotease imbalances, immune mechanisms, cellular senescence, impaired cell-repair mechanisms, cell necrosis, and autophagy [3].

The diagnosis and treatment of COPD primarily follow the guidelines provided by the Global Initiative for Chronic Obstructive Lung Disease (GOLD). Patients with COPD are classified into groups A, B, C, and D based on the severity of airflow limitation (forced expiratory volume $[FEV]_1$ value) and the nature and severity of their symptoms (mean response time and COPD assessment test) [4]. The commonly used medications for stable COPD include bronchodilators and anti-inflammatory drugs. Bronchodilators primarily encompass β2 receptor agonists, anticholinergic agents, and methylxanthines, whereas anti-inflammatory drugs mainly comprise glucocorticoids, PDE-4 inhibitors, antibiotics, mucolytics, and antioxidants. This chapter illustrates the structural characteristics and PK profiles of the bronchodilator (tiotropium) and the combination drug (olodaterol hydrochloride).

8.2.2 Overview of Medications for Stable Chronic Obstructive Pulmonary Disease (COPD)

For stable COPD treatment, there are nine approved classes of maintenance therapy medications. These include short-acting β2 agonists (SABAs), long-acting β2 agonists (LABAs), short-acting anticholinergic agents (SAMAs), long-acting anticholinergic agents (LAMAs), a combination of a SABA and a LAMA (SABA + SAMA), a combination of a LABA and a LAMA (LABA + LAMA), a combination of a LABA and inhaled corticosteroids (LABA + ICS), triple therapy (LABA + LAMA + ICS), and PDE-4 inhibitors. Over the past two decades, only one COPD medication with a new target (PDE-4), roflumilast, has been approved [5].

According to GOLD guidelines, for GOLD A or B patients in a stable phase with mild symptoms, LAMAs or LABAs may be administered. LABA + LAMA can be used for the treatment of patients with moderate to severe disease. Compared to monotherapy, dual therapy (LABA + LAMA) considerably improves patient outcomes and reduces COPD exacerbation. For patients with moderate to very severe COPD, triple therapy (LABA + LAMA + ICS) may be used.

Tiotropium, the first approved long-acting anticholinergic drug, continues to play an important role as a first-line treatment for COPD and has shown more significant efficacy when used in combination with olodaterol hydrochloride.

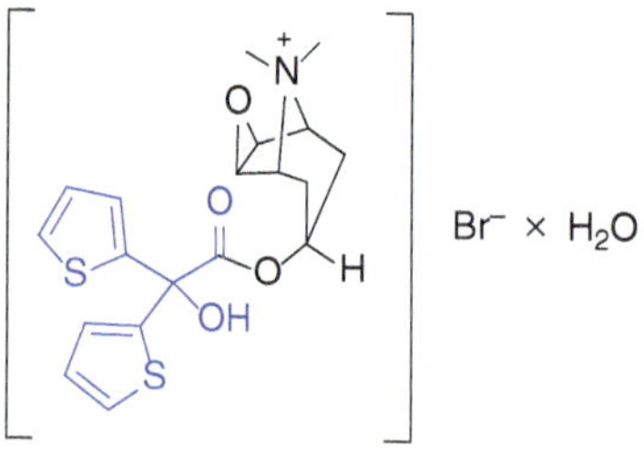

Figure 8.12 Structure of tiotropium.

8.2.3 What Is Tiotropium (LAMA) and Its Structural Characteristics?

According to GOLD guidelines, tiotropium is a long-acting bronchodilator that should be introduced early in the disease trajectory (GOLD II stage). It can also be used consistently in all subsequent stages of COPD (GOLD III and IV stages). Tiotropium is a LAMA that inhibits smooth muscle cell contraction by inhibiting the binding of acetylcholine to M3 muscarinic receptors [6].

The structural features of tiotropium contribute significantly to its efficacy in COPD treatment (Figure 8.12). The presence of two thienyl rings enhances the affinity of tiotropium to muscarinic receptors and controls its dissociation half-life. Its high affinity with M3 receptors and its long dissociation half-life make tiotropium a long-acting anticholinergic medication. Similar to its predecessor, ipratropium or tiotropium also contains a quaternary ammonium group, which partially restricts its gastrointestinal absorption and prevents it from crossing the blood–brain barrier. These properties reduce the risk of systemic anticholinergic side effects and central nervous system toxicity [7].

8.2.4 PK Profile of Tiotropium

8.2.4.1 Absorption of Tiotropium

Inhalation delivery device: Inhalation is a crucial administration route for tiotropium, allowing the drugs to directly reach the target site in the bronchi and lungs. Tiotropium can be administered by inhalation using an aerosol device.

Tiotropium exhibits low oral bioavailability and high lung deposition in various animal species and humans. The absolute bioavailability of tiotropium in healthy humans is only 2–3% of the oral dose, and the oral bioavailability in mice, rats, rabbits, and dogs is 0.02%, 0.5%, <0.1%, and 6.3%, respectively, indicating poor gastrointestinal absorption [8]. After inhalation administration of tiotropium, the absolute bioavailability in healthy humans is 19.5%. Approximately 80% of the inhaled tiotropium dose is swallowed and enters the gastrointestinal tract, while an additional 17% of the inhaled dose reaches the lung (Figure 8.13). As the dose reaching the lung is almost fully bioavailable, and the systemic bioavailability of the total administered dose is low, fewer systemic effects are relatively pharmacologically reflected.

Plasma concentration: After inhalation administration, tiotropium is rapidly absorbed and effectively distributed in the lung, thus providing rapid and effective treatment for COPD's airway obstruction symptoms. After inhalation, a small amount of tiotropium enters the systemic circulation for distribution and metabolism. The maximum plasma concentration of tiotropium can be observed within 5 minutes after inhalation, and the maximum plasma concentration in steady state is 15–19 pg/mL after continuous administration once daily for 2–3 weeks. Based on maximum plasma concentration (C_{max}), the area under the plasma concentration–time curve ($AUC_{0\text{-}t}$), and urine excretion data, multiple doses to steady state can result in 2–3 times accumulation, and continued use of tiotropium will not lead to further accumulation [7] (Figure 8.14).

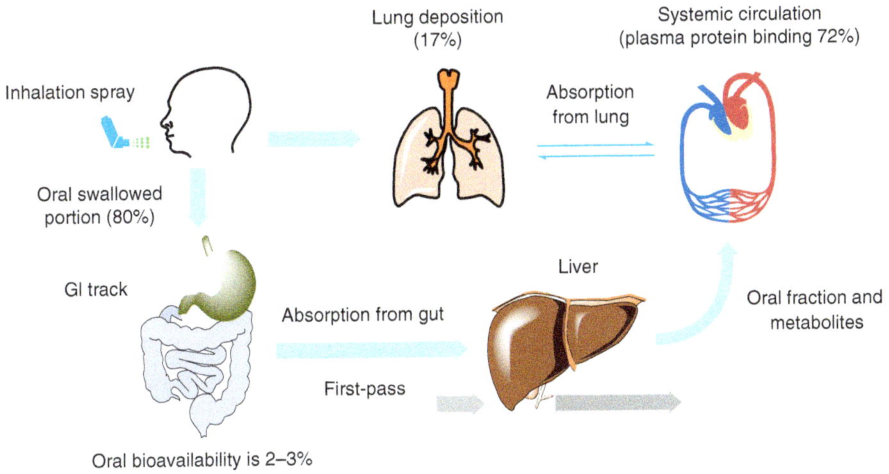

Figure 8.13 Absorption process of tiotropium after inhalation.

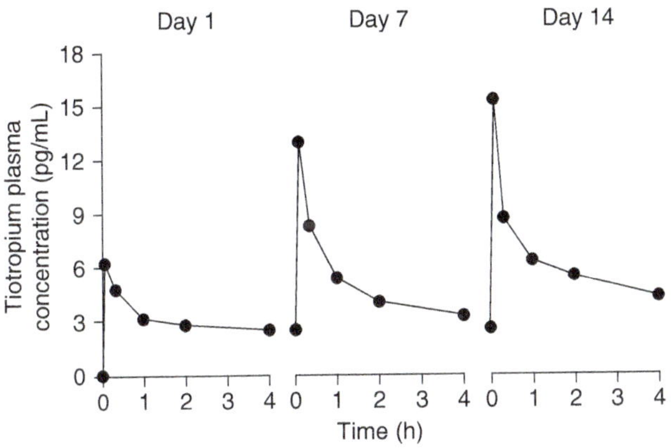

Figure 8.14 Plasma concentration after inhalation administration for 14 consecutive days. *Source:* [7] / with permission of Taylor & Francis Group.

8.2.4.2 Distribution of Tiotropium

Tiotropium has a volume of distribution of 32 L/kg in humans and approximately 26 L/kg in animals, indicating that the drug has extensive tissue distribution and no central nervous system distribution. About 72% of tiotropium is bound to human plasma proteins, which is approximately three times higher than that in animal species (15.5–21.7%). The moderate to high plasma protein binding reduces the concentration of free drugs in the systemic circulation, thereby reducing the risk of adverse systemic effects.

8.2.4.3 Metabolism of Tiotropium

In vitro studies have shown that the first step of tiotropium metabolism is hydrolysis of the ester bond without the need for esterase, producing two metabolites (*N*-methylscopine and dithienylglycolic acid). Both of these metabolites are pharmacologically inert, meaning they neither enhance the efficacy of tiotropium nor alter its safety or tolerability. The oxidized metabolites of *N*-methylscopine are metabolized by CYP2D6 and CYP3A4, which undergo thienyl ring oxidation and subsequent conjugation with glutathione.

8.2.4.4 Clearance (Excretion) of Tiotropium

Drugs that enter the systemic circulation need to be rapidly cleared or excreted to minimize the risk of drug accumulation in the body. After IV, 74% of the dose is excreted through the urine in its original form. After dry powder inhalation, 14% of the dose is excreted through the urine. The remaining dose is not absorbed by the intestines and is excreted in feces. After IV infusion in healthy male subjects, the total clearance of tiotropium is 880 mL/min, with renal clearance at 669 mL/min. The terminal elimination half-life of inhaled tiotropium is 5–6 days.

8.2.4.5 Drug Interactions (DDIs) of Tiotropium

Tiotropium has a low risk of DDI. It does not inhibit cytochrome P450 enzymes 1A1, 1A2, 2B1, 2C9, 2C19, 2D6, 2E1, or 3A4. Additionally, the extremely low systemic exposure further reduces the risk of drug interactions.

8.2.5 Drug Combination: Tiotropium (LAMA)/Olodaterol Hydrochloride (LABA)

Olodaterol hydrochloride is a LABA, and the combination drug of olodaterol hydrochloride (LABA) with tiotropium (LAMA) was approved by the FDA in 2015 [9, 10]. Tiotropium and olodaterol hydrochloride act on the M3 muscarinic receptors and β2 receptors on the airway smooth muscles, respectively, achieving bronchodilatory.

Olodaterol hydrochloride shows a similar PK profile to tiotropium, presenting poor oral absorption, rapid onset of action, low systemic exposure, no central nervous system distribution, and rapid clearance. When the combination of tiotropium and olodaterol hydrochloride is administered by inhalation, the PK parameters of each component are similar to those observed with individual administration. Human PK data for tiotropium and olodaterol hydrochloride are shown in Table 8.5.

As demonstrated in Figure 8.15, the combination therapy of olodaterol hydrochloride and tiotropium has more pronounced therapeutic effects than monotherapy.

8.2.6 Summary

Using the examples of tiotropium inhalation formulation and tiotropium bromide/olodaterol hydrochloride combination inhalation formulation, the characteristics and PK profiles of inhaled COPD medications were summarized as follows:

Table 8.5 Summary of human pharmacokinetic data for tiotropium and olodaterol hydrochloride.

Name	Bioavailability for oral administration	Bioavailability for inhalation administration	[a]C_{max}(pg/mL)	[a]T_{max} (min)	Plasma protein binding (PPB)	V_{ss} (L/kg)	CL (mL/min)
Tiotropium (5 μg)	2–3%	19.5%	15–19	5	72%	32	880
Olodaterol hydrochloride (5 μg)	<1%	30%	2.82–3.35	10–20	60%	18.5	872

[a] Inhalation administration; C_{max}, maximum plasma concentration; T_{max}, time to reach the maximum plasma concentration; V_{ss}, volume of distribution at steady state; CL, total body clearance.

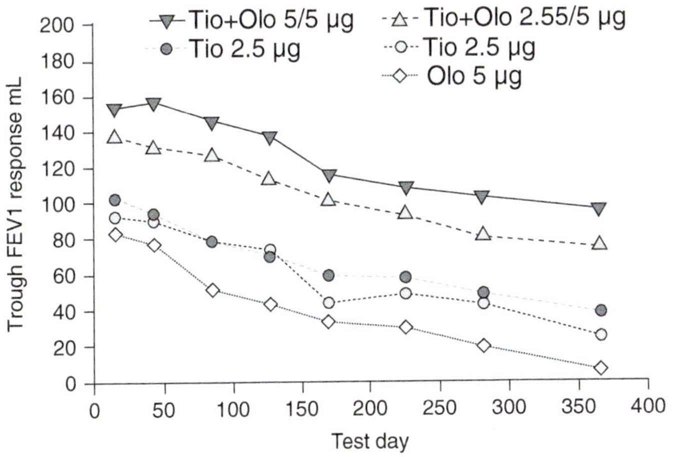

Figure 8.15 Comparison of the therapeutic effects of tiotropium and olodaterol hydrochloride in combination therapy versus monotherapy. *Source:* [11] / with permission of The European Respiratory Society.

- High potency, which meets the requirement for low-dose inhalation administration.
- Low gastrointestinal absorption, which reduces the risk of toxicity owing to systemic exposure.
- Limited ability to cross the blood–brain barrier, which reduces the impact on the central nervous system.
- Rapid clearance from the circulatory system, which decreases drug exposure to nontarget tissues and organs.

Given the widespread prevalence of patients with COPD worldwide, the market prospects for inhaled COPD drugs are promising. Therefore, developing favorable characteristics and PK profiles of COPD drugs requires more effort.

References

1 Research and Markets (2021). Worldwide Chronic Obstructive Pulmonary Disease Industry to 2030 - Market Insights, Epidemiology and Forecasts - http://ResearchAndMarkets.com (http://yahoo.com)

2 The World Health Organisation (2020). The top 10 causes of death. www.who.int.

3 Xu, Y., Liu, H., and Song, L. (2020). Novel drug delivery systems targeting oxidative stress in chronic obstructive pulmonary disease: a review[J]. *Journal of Nanobiotechnology* 18: 1–25.

4 Global Initiative for Chronic Obstructive Lung Disease (GOLD), 2022 Global Strategy for Prevention, Diagnosis and Management of COPD 2022 version.

5 van Haarst, A., McGarvey, L., and Paglialunga, S. (2019). Review of drug development guidance to treat chronic obstructive pulmonary disease: US and EU perspectives[J]. *Clinical Pharmacology & Therapeutics* 106 (6): 1222–1235.

6 Nardini, S., Camiciottoli, G., Locicero, S. et al. (2014). COPD: maximization of bronchodilation[J]. *Multidisciplinary Respiratory Medicine* 9: 1–10.

7 Price, D., Sharma, A., and Cerasoli, F. (2009). Biochemical properties, pharmacokinetics and pharmacological response of tiotropium in chronic obstructive pulmonary disease patients[J]. *Expert Opinion on Drug Metabolism & Toxicology* 5 (4): 417–424.

8 FDA. FDA Drug Approval Package: Pharmacology Review(s). Application Number: 21–395

9 Deeks, E.D. (2015). Olodaterol: a review of its use in chronic obstructive pulmonary disease[J]. *Drugs* 75: 665–673.

10 FDA. FDA Approved Drug Products: Tiotropium and Olodaterol Metered Inhalation Spray

11 Buhl, R., Maltais, F., Abrahams, R. et al. (2015). Tiotropium and olodaterol fixed-dose combination versus mono-components in COPD (GOLD 2–4)[J]. *European Respiratory Journal* 45 (4): 969–979.

9

Transdermal and Topical Drugs

9.1 Preclinical Pharmacokinetic Strategies for Topical and Transdermal Drug Products

Zhangpei Zhu, Jing Jin

9.1.1 Introduction

The skin is a key site for both local and systemic drug delivery. It is a complex, multilayered structure and is generally divided into four layers (see Figure 9.1). The stratum corneum (SC), composed of dead cells, is the very thin (approximately 10–20 μm) and outermost layer of the skin. The viable epidermis, with a thickness of about 0.2 mm, is the region for drug binding, metabolism, and active transport. The dermis, about 3–4 mm thick, is rich in collagen. The hypodermis, which can be up to several centimeters thick, is the innermost layer of the skin and serves as an important storage site for fat [2, 3].

Topical and transdermal drug delivery products refer to those that are administered on the skin surface, which pass through the skin and enter the body to exert local or systemic therapeutic effects. These products offer a noninvasive and highly compliant dosage form. Topical dermatologic products are designed to deliver the active ingredient to dermal topical application or specific layers of skin. Their therapeutic areas include infections caused by bacteria, fungi, and viruses, as well as conditions such as psoriasis, atopic dermatitis, and acne. Transdermal products, are designed to deliver the active ingredient (drug substance) across the skin and into systemic circulation. It is usually prescribed to treat pain, the central nervous system, the cardiovascular system, and other diseases. With the aging of the global population and the increase in the number of patients with chronic diseases, the market demand for such drugs is increasing.

Based on the drug characteristics and therapeutic purposes, topical and transdermal drug delivery products are broadly divided into liquid, semi-solid, and matrix-type delivery systems, such as sprays, gels, emulsions, ointments, and gel patches. New drug delivery systems, such as microneedles and ultrasonic introduction, are under development, [4]. The diversity of formulations also makes the research more complex. This chapter explored preclinical DMPK assessment strategies for topical and transdermal drug delivery products based on literature and our experience (Figure 9.2).

Drug Metabolism and Pharmacokinetics: Frontiers, Strategies, and Applications, First Edition.
Edited by Liang Shen and WuXi AppTec DMPK.

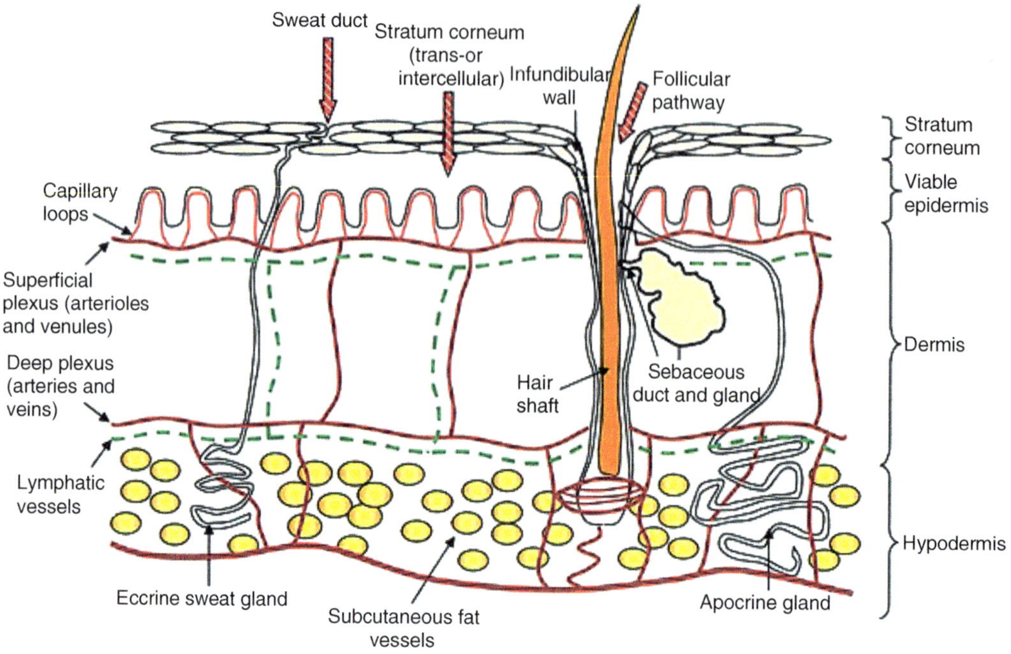

Figure 9.1 Structure of the skin. *Source:* It's from Dove Medical Press under Creative Commons CC BY-NC 3.0 [1].

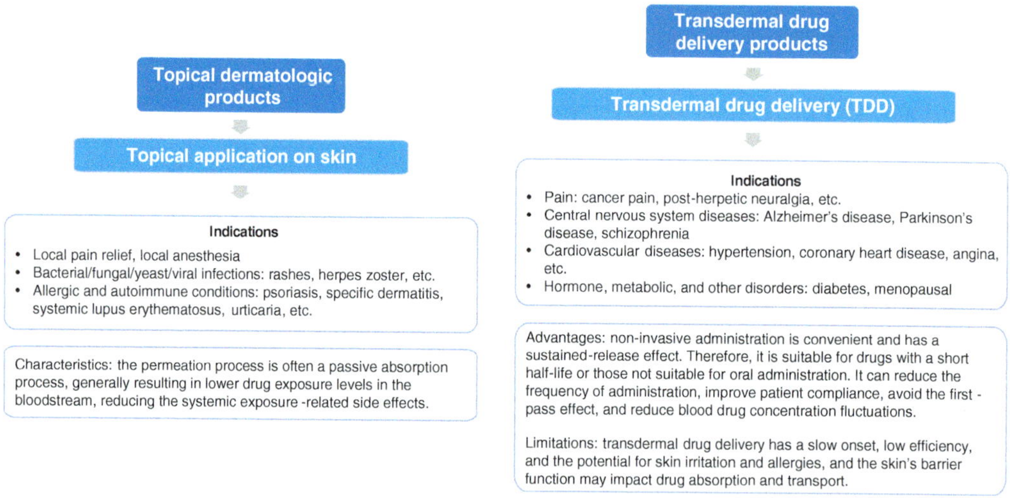

Figure 9.2 Brief introduction to topical and transdermal drug delivery systems.

9.1.2 Drug Permeation and Research Method of Topical and Transdermal Drugs

There are several physiological and physicochemical barriers that affect drug penetration through the skin. Microscopically, from outermost to innermost, skin constitutes three main histological layers: epidermal (including SC and active epidermal layer), dermis, and subcutaneous layer. It also includes skin appendages such as hair follicles, sweat glands, and sebaceous glands. There are mainly three routes of drug penetration into the systemic bloodstream [5]:

1) The transcellular pathway: This pathway passes through corneocytes and into the skin. Corneocytes, which contain highly hydrated keratins, provide a hydrophilic environment to facilitate the passage of hydrophilic drugs.
2) The intercellular pathway: This route involves diffusion through intercellular lipid matrix. This path is generally considered to be most favored by lipophilic small molecules and uncharged drugs.
3) Trans skin appendage pathway: This route of drug penetration permeates through skin appendages such as hair follicles, sweat glands, and sebaceous glands.

Topical dermatologic products diffuse to the epidermis or dermis, where they may be adsorbed or bound by the relevant cells and components of the skin, acting on the target site to produce local efficacy. Transdermal products need to further penetrate the dermal vascular-lymphatic system and eventually enter the bloodstream (Figure 9.3).

The absorption of drugs mainly depends on the first two routes, which require drugs to pass through the SC layer, the outermost layer of the skin, before they penetrate deeper layers. Among all the skin layers, the SC barrier is the most effective barrier. The SC cells and the brick-wall structure composed of intercellular lipids block the absorption of drugs [6], directly or indirectly leading to the lower bioavailability of drugs (Figure 9.4).

The effectiveness and acceptability of skin formulations are directly related to the properties of the carrier/vehicle used. Faster drug release commonly indicates more effective drug penetration. Therefore, it is necessary to pay attention to the permeability of the formulation through the various layers of the skin in the early phases of formulation development. Both *in vitro* models and *in vivo* animal studies are used in the evaluation of delivery systems. Franz diffusion cell is generally used to assess skin permeability and evaluation of transdermal formulations [7], which is regarded as the industry gold standard. This method is normally used with excised human or animal skin. However, when biological skin is not available, synthetic membranes are employed. The artificial model membranes enable the testing and evaluation of various drugs and formulations as an initial screening approach to narrow the selection of the formulations to be evaluated with a more biologically intact model, such as

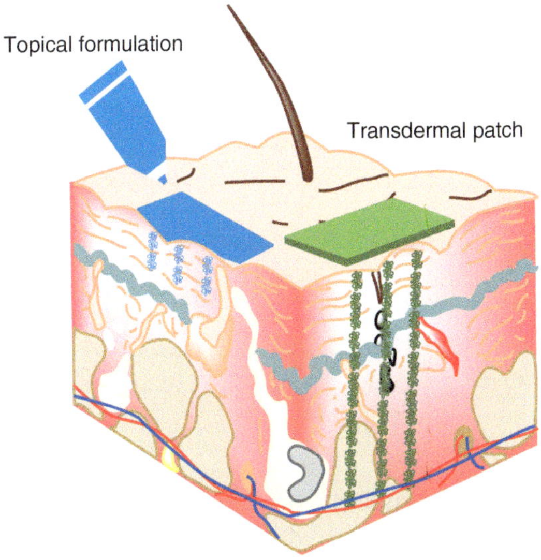

Figure 9.3 Schematic representation of drug permeation for topical and transdermal drug delivery products.

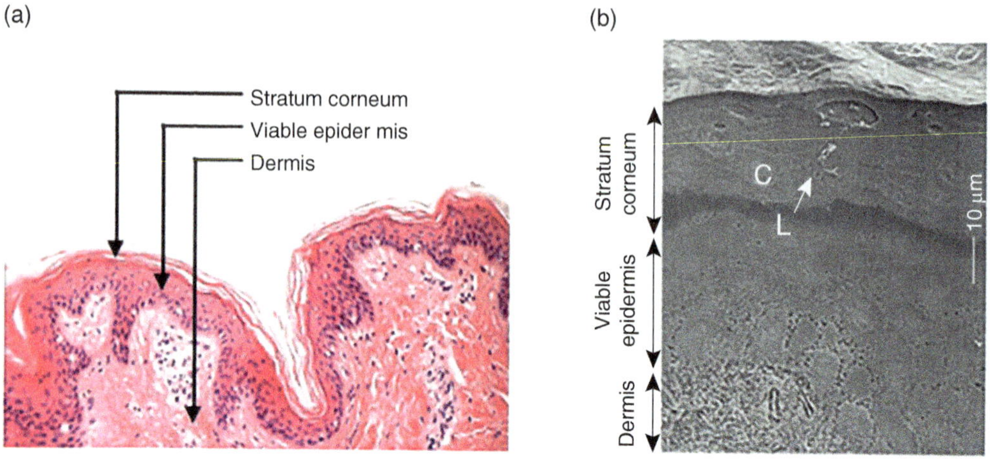

Figure 9.4 Histological structure of mammalian skin: (a) skin structure (image of H&E stained porcine skin); (b) SC structure (cryoscan electron microscopy image) lipids organized in bilayer structures (L), corneocytes (C). *Source:* [4]/Springer Nature.

silicone, parallel artificial membrane permeability assay (PAMPA), and phospholipid vesicle permeation assay (PVPA) models [8]. *In vivo* studies have seminal techniques such as skin biopsy, skin curettage, negative pressure suction bubble technique and microdialysis, and nontraumatic techniques such as, QWBA, laser scanning confocal microscopy, and Roman Confocal microscope [5].

9.1.3 DMPK Research Strategies in the Development of Topical and Transdermal Drug Products

To provide regulatory support for the application and marketing of topical and transdermal products, each regulatory authority gave continuously updated policy guidance, mainly focusing on drug product quality control and generic drug evaluation, as shown in Table 9.1. A number of recommendations for the development of individual drugs have also been issued and further expatiated here.

Preclinical development of topical and transdermal delivery products should follow the general principles of preclinical DMPK of pharmaceuticals. At the same time, compared to drugs administrated by other routes, they have particularities in formulation composition and dosage form. Therefore, evaluation strategies should be designed purposefully (Figure 9.5).

9.1.3.1 General Considerations

1) **The delivery system should be selected to be representative of the clinical trial**

For the test articles for preclinical PK study, the delivery system should be at pilot scale or above that represents the samples for clinical trial, and the physical and chemical properties of the formulation, such as volatility, stability, and solubility, are necessary to provide a reasonable basis for clinical trial design.

2) **Selection of preclinical species**

When assessing the local skin kinetics or systemic absorption of such drugs, experimental animal species for preclinical PK should be close to human skin and have sufficient background research data. For the selection of rodents, the draft guidance for preclinical safety evaluation of dermal administration published by the OECD [8] recommends using rats or guinea pigs for

Table 9.1 FDA and CDE guidance for topical and transdermal drug products.

Regulatory authority	Guidance	Date issued	Version
FDA	*In Vitro* Permeation Test Studies for Topical Drug Products Submitted in ANDAs	2022/10/21	DRAFT
	In Vitro Release Test Studies for Topical Drug Products Submitted in ANDAs	2022/10/21	DRAFT
	Physicochemical and Structural (Q3) Characterization of Topical Drug Products Submitted in ANDAs	202 210/21	DRAFT
	Transdermal and Topical Delivery Systems – Product Development and Quality Considerations	21 Nov 2019	DRAFT
	Assessing Adhesion with Transdermal Delivery Systems and Topical Patches for ANDAs: Draft Guidance for Industry	April 2023	DRAFT
	Assessing the Irritation and Sensitization Potential of Transdermal and Topical Delivery Systems for ANDAs: Draft Guidance for Industry	April 2023	DRAFT
	Residual Drug in Transdermal and Related Drug Delivery Systems: Guidance for Industry	17 Aug 2011	FINAL
CDE	Technical Guidance for Clinical Trials of Locally Applied Locally Products	30 May 2022	Issued
	Technical Guidance for Chemical Generic Drugs for Topical Dermatology (Trial)	16 March 2021	Trial
	Technical Guidance for Pharmaceutical Study of Chemical Generic Transdermal Patch (Trial)	25 December 2020	Trial
	Technical Guidance for Irritation, Sensitization, and Hemolysis Studies of Drugs	13 May 2014	Issued

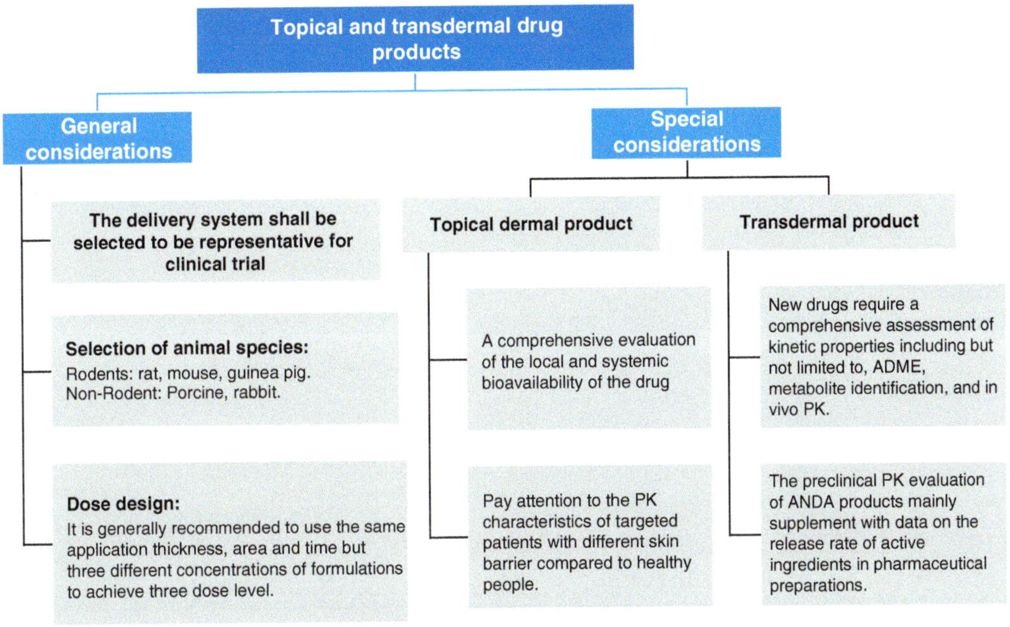

Figure 9.5 Preclinical pharmacokinetics assessment strategy for topical and transdermal drugs.

safety evaluation of systemic exposure, and this recommendation is also applicable to PK studies. When it comes to non-rodent selection, the skin of minipigs has good similarity with that of humans in terms of anatomical structure and physiological status, with similar skin penetration data of compounds [9]. Moreover, the pig skin is light in color, has few hair follicles and offers good feasibility.

3) **Considerations of the dose**

The considerations of administration dose levels in the *in vivo* PK experiment are the same as conventional routes, and routinely choose three dose levels. It is generally recommended to use three concentrations of formulation to achieve three dose levels with the same application thickness, area, and application time. In the preclinical safety assessment, the administration dose can also be increased by expanding the administration area and increasing the frequency of administration. OECD Draft Guidance *Repeated Dose Dermal Toxicity: 21/28-day Study* [8] suggested the maximum administration area should not exceed 10% of the body surface area.

9.1.3.2 Special Considerations for Topical Dermatological Products

1) **The bioavailability of the drug in the skin and circulation system should be comprehensively evaluated**

The local kinetic trail in the skin can help in understanding whether the drug reaches the target site, and the speed and quantity of the drug reaching the target site. The study of local kinetics in humans is not easy to conduct due to the skin sampling techniques and ethics. The data of preclinical local kinetics can support the subsequent human clinical trial design. In addition, systemic bioavailability is important for the evaluation of absorption and *in vivo* loading toxicity. Therefore, topical skin preparations should be evaluated comprehensively for their local and systemic bioavailability.

2) **Attention should be paid to the impact of skin disease types, their barriers, and their PK profile**

For some diseases, the patient's skin is damaged, and its skin absorption may be quite different from that of intact skin. It should be studied for its PK behavior before clinical use. Table 9.2 lists several disease models and changes in the skin barrier.

The topical animal model can be used for preclinical assessment. Different animal models can be obtained by different methods. For example, the model of SC loss can be obtained by tape stripping. It is necessary to determine the frequency of stripping according to the skin healing

Table 9.2 Skin barrier system in different skin diseases.

Disease	Target tissue	Skin barrier
Fungal infection	Epidermis	Not damaged
Herpes simplex	Superficial epidermis	Severely damaged (trauma to skin)
Melanoma	Deep epidermis (basal layer)	Not damaged
Psoriasis (plaque type)	Epidermis and dermis	Not damaged
Atopic dermatitis	Epidermis and dermis	Partly damaged
Acne	Sebaceous glands and dermis	Partly damaged (face)
Muscle pain	Subcutaneous and intramuscular	Not damaged

characteristics and the administration modes in the study protocol, ensuring the consistency of drug absorption.

3) **Case**

VTAMA® (Tapinarof) is a novel small molecule for mild, moderate, and severe plaque psoriasis in adults [10], which was approved by the FDA in May 2022. Tapinarof is applied as a thin layer of cream to the lesion once a day, which binds and activates the aryl hydrocarbon receptor (AhR), thereby modulating the imbalance of proinflammatory cytokine levels in the body. The preclinical PK studies conducted are shown in Table 9.3.

9.1.3.3 Special Considerations for Transdermal Products

1) **The transdermal delivery system requires an assessment of absorption characteristics and systemic exposure, separately**

Transdermal preparations are systemically active dermal preparations, and their systemic exposure and ADME properties of drugs can be assessed by referring to drugs dosed by conventional systems. The assessment of systemic exposure requires the selection of one rodent and one non-rodent species that are metabolically similar to humans, while transdermal absorption assessment often selects minipigs.

2) **Preclinical PK assessment of the transdermal system of the modified formulation**

Most of the transdermal products approved by the FDA belong to the NADA products. The application mainly relies on the available literature data and the real data of the marketed product to provide evidence of safety and effectiveness, while the preclinical test plays a supplementary role. Preclinical PK primarily assesses factors such as the release rate of active ingredients in the delivery system. Relevant animal species can be used for *in vivo* PK studies.

3) **Case**

EXELON® PATCH (rivastigmine transdermal patch) is a novel formulation modified from the original oral capsule formulation [11] for the treatment of Alzheimer's disease with mild, moderate, and severe cognitive impairment. The results of plasma protein binding and *in vitro* and *in vivo* metabolism were based on the existing data of oral preparations, supplemented with the PK data of pigs (isotope-labeled) under the transdermal route of administration, patch drug residue, and *in vitro* human skin metabolism studies. When comparing the transdermal product with a the oral product, the pig's bioavailability was increased by 23.8-fold with a longer plasma concentrations curve over time. It also reduces the gastrointestinal side effects of rivastigmine and is preferred by Alzheimer's patients.

9.1.4 Summary

Topical and transdermal drug delivery products are relatively complex as there is no universal assessment system and regulators have different degrees of acceptance. Therefore, it is advisable to select appropriate research methods based on drug physicochemical properties, disease targets, or to combine multiple methods, to provide a more comprehensive understanding of drugs' absorption, distribution, and metabolism at the target site. We established a universal set of drug research systems to help the whole process research from formulation screening at an early stage to new drug applications, including *in vitro* release test (IVRT) and *in vitro* permeation test (IVPT) assay platform, *in vivo* dermal pharmacokinetics, and visualized QWBA technology.

Table 9.3 Preclinical pharmacokinetics of tapinarof.

Experiment type	Test item	Main test results
Absorption	An *in vitro* investigation of the passive membrane Permeability of [^{14}C] tapinarof in MDCKII-MDRl cells	[^{14}C]-tapinarof was determined to have a high passive membrane permeability of 670 nm/s at pH 7.4 and 500 nm/s at pH 5.5 in Madin-Darby canine kidney (MDCKII-MDR1) cells in the presence of the potent P-glycoprotein (P-gp) inhibitor
	Multiple-dose topical dermal toxicokinetics in Göttingen minipigs	T_{max}: 6 h AUC_{0-t}: 5.09 ng·h/mL at 3% C_{max}: 0.686 ng/mL at 3% Accumulation: No Dose proportionality: Exposure generally increased dose proportionally (0.5, 1.0, and 3%) on week 39
Distribution	Plasma protein binding in pooled rat, minipig, and human plasma using equilibrium dialysis	The plasma protein binding of tapinarof was very high, ranging from 99.1% to 99.4%, with no apparent concentration dependence in the three species tested (rat, minipig, and human) over the concentration range from 20 to 1000 ng/mL.
	In vitro protein binding in human and minipig and human skin homogenate	*In vitro* protein binding study of tapinarof showed that the skin protein binding values for the minipig and human homogenates were 90.7% and 92.7%, respectively. The binding appeared to be concentration-dependent at high concentrations.
	Quantitative whole-body autoradiography tissue Distribution study in rat after single subcutaneous dosing	All tissues arrived at maximum concentrations at approximately 1 h after dosing. It was mainly distributed in the treated skin site, urinary bladder wall, adrenal cortex, small intestinal wall, liver, fat, and kidney cortex. The exposure in tissues was higher than in blood
Metabolism	An *in vitro* investigation of the hepatic metabolite profiling of WBI-1001 in human liver microsomes and cryopreserved hepatocytes from rat, rabbit, Gottingen minipig, and human Metabolism of [^{14}C] GSK2894512 in the human Crl: WI (Han) and Crl: ICR (CD-1) mouse	The major *in vitro* biotransformation pathways of the test article in all of the species were attributed to the phase II enzyme activities, for example, sulfate and glucuronic acid conjugations on the phenolic group, alone, or in combination with oxidation. All metabolites noted in human hepatocyte incubations were observed in rat or minipig hepatocyte incubations. Only incubation of the test article with rabbit hepatocytes produced a phase I metabolite, which was formed likely due to an aromatic hydroxylation reaction
	Metabolism in male rats following a single subcutaneous administration of [^{14}C] tapinarof at a target dose level of 2 mg freebase/kg	The major routes of metabolism *in vivo* in mice and rats were oxidation, glucuronidation, and combinations thereof
Excretion	The excretion of radioactivity following a single subcutaneous administration of [^{14}C] free base to the CD-1 mouse and rat (2 mg/kg)	In rats and mice, the major route of elimination was in the feces. For bile duct-cannulated rats, the major route of elimination was in the bile. Elimination of drug-related material was rapid following SC administration, with 80–88% of the dose recovered by 24 h post-dose. Blood-to-plasma ratios ranged from 0.66 to 1.2 in the rat and 0.60 to 0.77 in the mouse, and liver-to-plasma ratios ranged from 7.1 to 14.0 in the rat and 4.0 to 14.1 in the mouse, respectively
	Excretion of radioactivity following dermal administration of [^{14}C]GSK2894512 free base to the minipig (10 mg/kg)	In the minipig study, the majority of the radioactivity (82.5% of the dose) was recovered in the interim dressing removed from the dose site 10 h after dosing. Less than 1% of the administered radiolabeled material was recovered in excreta (urine and feces), suggesting minimal absorption of the radioactive test material. Blood and plasma concentrations of radioactivity were below the limit of detection

Table 9.4 Pharmacokinetics data of EXELON® PATCH in three administration routes to minipig.

Pharmacokinetics parameters	Intravenous administration (0.1 mg/kg)	Oral administration (1.0 mg/kg)	Transdermal drug delivery (18 mg)	Transdermal drug delivery (54 mg)
C_{max} (ng/mL)	27.04	0.67	1.03	4.06
T_{max} (h)	0.083	0.5	18	24
AUC_{0-96} (ng. h/mL)	14.6	0.7	43.0	140.8
F%	—	0.5	11.9	13.0

References

1 Abd, E., Yousef, S.A., Pastore, M.N. et al. (2016). Skin models for the testing of transdermal drugs. *Clinical Pharmacology: Advances and Applications* 8: 163–176.

2 Patel, R., Patel, A., Prajapati, B. et al. Transdermal drug delivery systems: a mini review. *International Journal of Advanced Research* 6 (5): 891–900.

3 Rastogi, V. and Yadavf, P. (2012). Transdermal drug delivery system: an overview. *Asian Journal of Pharmaceutics* 6 (3): 161–170.

4 Prausnitz, M.R. and Langer, R. (2008). Transdermal drug delivery. *Nature Biotechnology* 26 (11): 1261–1268.

5 Supe, S. and Takudage, P. (2021). Methods for evaluating drug penetration into the skin: a review. *Skin Research and Technology* 27 (3): 299–308.

6 Gorzelanny, C., Mess, C., Schneider, S.W. et al. (2020). Skin barriers in dermal drug delivery: which barriers have to be overcome and how can we measure them? *Pharmaceutics* 12 (7): 684.

7 Santos, L., Swofford, N.J., and Santiago, B.G. (2020). *In vitro* permeation test (IVPT) for pharmacokinetic assessment of topical dermatological formulations. *Current Protocols in Pharmacology* 91 (1): e79.

8 OECD Ilibrary. OECD guidelines for the testing of chemicals, Section 4. https://www.oecd-ilibrary.org/environment/oecd-guidelines-for-the-testing-of-chemicals-section-4-health-effects_20745788.

9 Jung, E.C. and Maibach, H.I. (2014). Animal models for percutaneous absorption. In: *Topical Drug Bioavailability, Bioequivalence, and Penetration*, 21–40. Wiley article. https://pubmed.ncbi.nlm.nih.gov/25345378

10 Min, S., Hong-bin, L., Jia-li, C. et al. (2015). A comparative study of bulleyaconitin Penetration rate on three damaged ways of skin. *Drug Evaluation Research* 38 (5): 491–496.

11 https://www.accessdata.fda.gov/drugsatfda_docs/nda/2007/022083_exelon_toc.cfm

9.2 Application of *In Vitro* Permeation Test (IVPT) for the Development of Transdermal and Topical Drugs

Xuan Dong, Cheng Tang

9.2.1 Introduction

Since 3000 BCE, ancient Egyptians and Babylonians used extracts derived from animals, minerals, or plants on the skin for therapeutic purposes [1]. In 1979, the US FDA approved the first systemic transdermal drug delivery system (TDDS), the scopolamine patch. This was a significant milestone for TDDS and initiated a new phase in its development [2].

TDDS offers notable advantages compared to conventional methods of drug administration, as it can circumvent first-pass metabolism in the liver, enhance patient compliance through noninvasive delivery, and maintain stable drug concentrations in the bloodstream for long-term therapeutic effects [3]. The methods used to evaluate transdermal drug delivery are broadly divided into *in vivo* and *in vitro*. Several experimental studies demonstrated that *in vitro* studies under simulated *in vivo* conditions, specifically IVPT, can provide data equally important to those obtained from conventional *in vivo* studies [4]. In this chapter, we have discussed the applications, relevant guidance requirements, skin models, and commonly used test designs of IVPT and have demonstrated the use of IVPT for formulation screening through a validation study.

9.2.2 Applications of *In Vitro* Permeation Test (IVPT)

Unlike traditional *in vivo* tests, IVPT measures the extent and the rate of a compound across the dermal barrier and facilitates precise measurement of the distribution in various layers of the skin. It helps reveal the differences in skin permeation during the early screening phase as well as the late development and submission phases. IVPT reduces the need for live animals and facilitates repeated testing on the same donor, making it particularly suitable for screening tests to compare different formulations of transdermal administration drugs. IVPT studies also play an important role in assessing bioequivalence (BE) for several topical products. Several recent FDA product-specific guidance listed IVPT as a comparable method to replace clinical trials, with additional support in product quality (Q1) and quantity (Q2). In combination with the IVRT results, IVPT can provide crucial data for microstructural equivalence (Q3) in evaluating the BE of topical formulations.

Typical applications of IVPT experiments include:

- Early screening of compounds.
- Formulation development and optimization of topical dermatological products.
- Assessment of drug BE and bioavailability.
- Safety assessments of pharmaceutical or cosmetic products.

9.2.3 Relevant Guidance Requirements

The pertinent points of current global guidelines are listed in Table 9.5 [5–8].

Skin type: In the FDA and European Medicines Agency (EMA) guidelines, only human-isolated skin is specified as a test model for IVPT, whereas the Japanese Pharmaceuticals and Medical Devices Agency (PMDA) allows the use of animal-isolated skin for the test.

Table 9.5 Comparison between global guidelines.

Organization	Experimental design	End points	Acceptance criteria
FDA	Franz cell method Human-isolated skin Skin temperature: 32 ± 1°C Test duration: until a complete permeation curve is obtained to determine the maximum permeation rate and subsequent decrease in permeation rate at multiple time points Sampling frequency: at least eight nonzero sampling time points recommended throughout the study period (e.g., 48 h) Skin thickness: 500 ± 250 μm Sink condition: saturation solubility greater than three times the maximum drug concentration in the receiving solution, ideally by an order of magnitude	Maximum flux (J_{max}) Cumulative permeation Recovery	When the within-reference standard deviation (S_{WR}) is <0.294, the $(1-2\alpha) \times 100\%$ confidence interval (CI) must be contained in [0.8, 1.25] When $S_{WR} \geq 0.294$, the 95% upper confidence bound for ($\mu T - \mu R$)2–$\theta\sigma 2WR$ must be no larger than zero, and the point estimated should fall in the prespecified [0.8, 1.25] range
EMA	Franz cell method Human-isolated skin Skin temperature: 32 ± 1°C Test time: no more than 24 h or until a complete permeation rate curve is available Sampling frequency: at least six-time points Minimum of 24 tests for control, test, and negative control formulations Number of donors ≥12, at least two sets of parallel tests per donor Sink condition: maximum concentration of drug reached in receiving solution <30% of maximum solubility (3–10 times)	Maximum flux (J_{max}) Total permeation (A) Recovery	The 90% CI for the geometric ratio of means of the test and reference products should be contained within the acceptance interval of 80.00–125.00% Wider 90% CI limits, to a maximum of 69.84–143.19%, are accepted in the case of high variability observed with low strength and limited diffusion drug products, and if clinically justified. Recovery: 90–110%
PMDA	Franz cell method Animal-isolated skin Skin temperature: 32 ± 0.5°C Test time: not exceeding 24 h/permeation rate reaches a certain value for over 6 h/permeation curves of standard formulations reach a steady state Sampling frequency: more than five sampling points in total including the end point More than six trials per formulation Select the amount of receiving fluid media that corresponds to the device	The ratio of the average permeation rate of the test and reference formulations at a specific test time or one-time point after the permeation rate has stabilized and at the same time point at which half the rate of permeation has been achieved	The ratio of average permeation rates of the test and reference formulations should range from 0.7 to 1.3 The permeation rate changes of the test formulation must be equal to or less than those of the reference formulation.

FDA: US Food and Drug Administration; EMA: European Medicines Agency; PMDA: Pharmaceuticals and Medical Devices Agency.

Duration of the test: In both the EMA and PMDA guidelines, the test duration is recommended to be no longer than 24 hours, whereas the FDA recommends that the sampling frequency should be selected to provide a suitable resolution for the flux profile, and a minimum of eight nonzero sampling time points is recommended across the study duration (e.g., 48 hours).

Sampling frequency: The FDA recommends a minimum of eight nonzero sampling time points; the EMA recommends a minimum of six sampling time points; and the PMDA recommends more than five sampling time points, including the end point.

The guidelines share consensus on the parameters, such as test device selection and skin temperature control.

While designing each study to be compliant with various guidelines, the actual objectives and conditions of the product should be considered comprehensively.

9.2.4 How to Choose and Build Suitable Skin Models

The choice of species is the first consideration when preparing skin materials. Although human-isolated skin is the gold standard for IVPT experiments, animal skin is often used as a substitute due to ethical and cultural concerns and availability challenges.

Pig skin bears the closest resemblance to human skin, including similar SC, epidermal thickness, and hair follicle density [9, 10]. The skin characteristics of different animal species are compared in Table 9.6.

Rodent skin is one of the most readily available types of skin material. However, owing to structural differences such as fat composition and hair follicle density, the drug permeation parameters of rodent-isolated skin tend to be higher than those of human-isolated skin, which can lead to overestimations of drug absorption [11]. Therefore, the preferred skin for IVPT experiments is the isolated skin of pigs, which is similar to human skin. Table 9.7 shows the skin models commonly used in IVPT experiments.

9.2.5 Commonly Used Apparatus and Parameter Configuration

Figure 9.6 shows devices commonly used for IVPT, which include static cells (SCs) and continuous flow cells (CFCs) [12]. SCs have a fixed volume receiving chamber, and the receiving fluid is continuously agitated during the test. SCs are the most commonly used type of diffusion cell. In CFCs, the receiving medium flows continuously to simulate blood flow in the body. However, due to the introduction of new parameters, additional preliminary tests are needed to determine the relevant test parameters.

Table 9.6 Skin characteristics of different animal species.

Species (anatomical site)	Follicle density (pcs/cm^2)	Stratum corneum (μm)	Epidermis (μm)	Intact skin (μm)
Human	29±8	16.8	46.9	2970
Pig	22±3	26.4	65.8	3430
Pig (ear)	NA	10	50	1300
Rat	1598±179	18	32	2090
Mouse (back)	5045±921	5	13	800

Source: [9]. Todo, 2017; [10]. Susanne Mangelsdorf, 2014.

Table 9.7 Common IVPT skin models.

Model		Characteristics
Simulated skin models	Strat-M®	Strat-M® membranes are designed to mimic the layered structure and lipid chemistry of human skin Relatively low cost Can be used to test and optimize drug formulations, has good reproducibility, and is suitable for early drug/formulation development
Skin models	Rodents	Small size and relatively low cost Often overestimates drug permeation properties
	Bama pig	Skin histology is similar to human skin High acquisition costs
	Pig ear skin	Skin histology is similar to human skin Often taken from food industry waste Difficult to control the age and strain of the animals

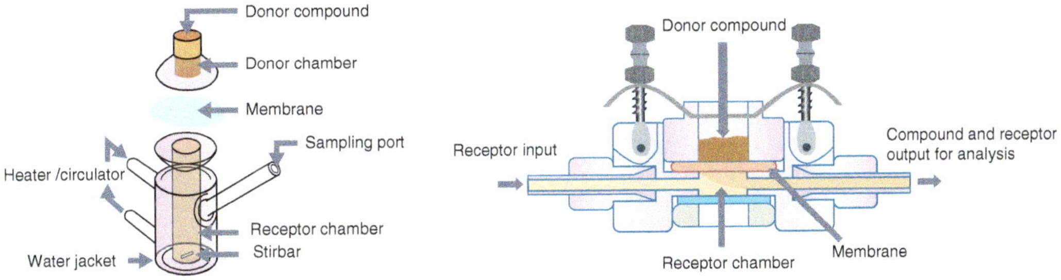

Figure 9.6 Schematic diagram of commonly used diffusion cells (left, SC; right, CFC). Adapted from references. *Source:* [12]/John Wiley & Sons.

When conducting experiments with diffusion cells, it is necessary to configure the parameters appropriately. Figure 9.7 describes a typical set of parameters for IVPT in a static cell [13]. This design is generally consistent with the requirements outlined in the "Draft Guidance on Acyclovir" provided by the FDA, for example, more than four donors per group, more than four replicates per donor, at least eight sampling time points during a test, and single-dose administration with a dosage range of 5–15 mg/cm^2. When selecting parameters for BE testing, it is imperative to perform method development and validation before critical tests are conducted.

9.2.6 IVPT Validation Study

Compound A is a widely used nonsteroidal anti-inflammatory drug with rapid absorption and clearance and a short half-life when administered systemically. Using compound A as a model drug, the *in vitro* permeation of six different formulations was evaluated using the dorsal skin of Bama pigs mounted on vertical Franz diffusion cells (PermeGear, Inc., Hellertown, PA, USA). In the comparison, Group 1 was a clarified solution, Group 2 was a commercialized gel, and Groups 3–6 were independently screened Poloxamer formulations.

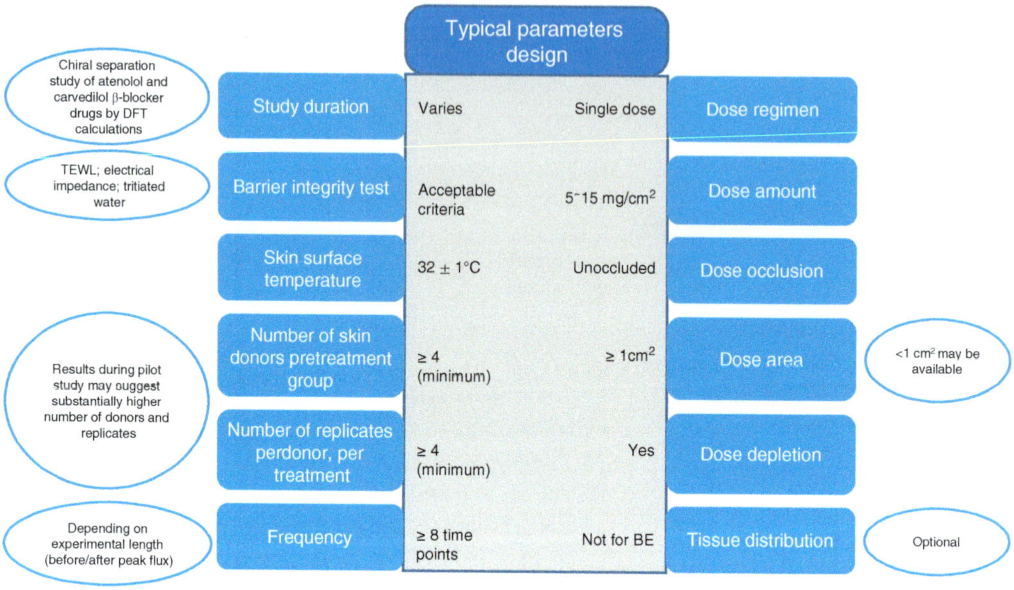

Figure 9.7 Configuration of parameters for IVPT. *Source:* Adapted from reference [13].

The dorsal skin of Bama pigs at 750 um was used as the test system. The prepared skin was placed between the donor and receptor chambers with the SC facing up, and only skin samples with qualified transepidermal water loss (TEWL) values were used in the actual comparison. A single dose of a commercial formulation of compound A (1.16%, w/w) was applied to the skin in the donor chamber. Normal skin temperature conditions were simulated during the test, with the skin temperature maintained at 32 ± 1°C. Receiver fluid (200 μL) was collected at predetermined time points. After the test, the SC was cleaned and removed. The epidermis and dermis were subsequently separated and collected to determine the distribution of the compound in the skin.

The vertical Franz diffusion cell clearly distinguished the formulations in terms of their skin permeation characteristics, as shown in Figure 9.8. Compared to the clarified solution formulations in Group 1, the commercialized gel formulations in Group 2 had a significantly lower permeation rate and cumulative drug permeation rate. For the four Poloxamer formulations, the solvent ratio affected the transdermal permeability of the drug. These results pointed out the direction for subsequent formulation screening and optimization processes. Compound A's epidermal and intradermal retention was found significantly lower in the Poloxamer-containing formulations (Groups 3–6), as shown in Figure 9.9.

9.2.7 Summary

TDDS aroused widespread attention among researchers and pharmaceutical companies, while the development of high-quality transdermal drug formulations remains challenging. Percutaneous absorption of drugs is a multifactorial and multistep process. Various factors, including the type of animal, skin type, skin pretreatment steps, and physicochemical properties of the tested compound and delivery system, all influence the IVPT outcome. IVPT not only serves as a highly valuable research tool for Q3 characterization during generic evaluations of topical formulations but also provides useful information for compound screening and formulation optimization. We have

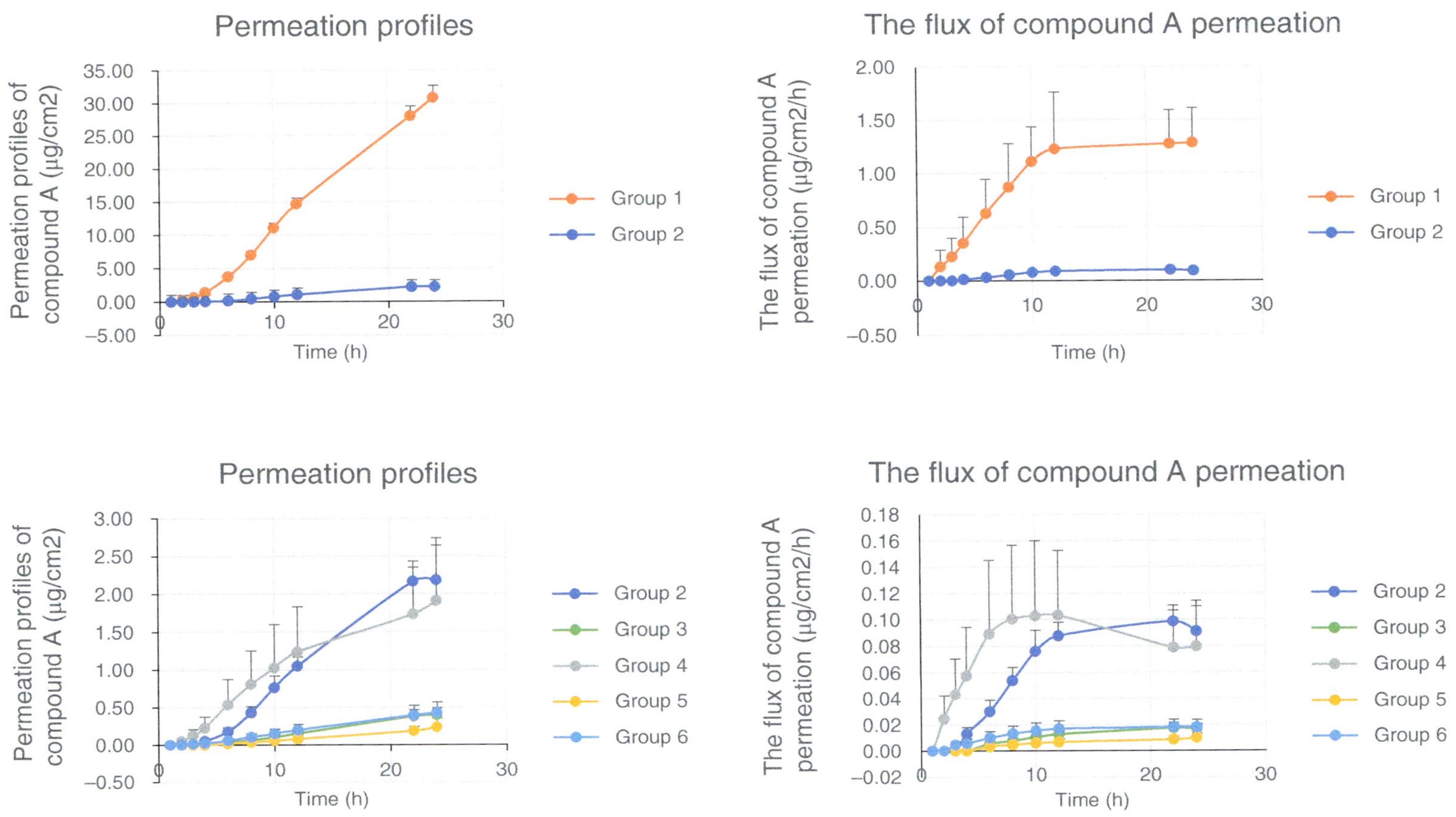

Figure 9.8 Cumulative permeation (left) and permeation rate (right) of each group of formulations over time.

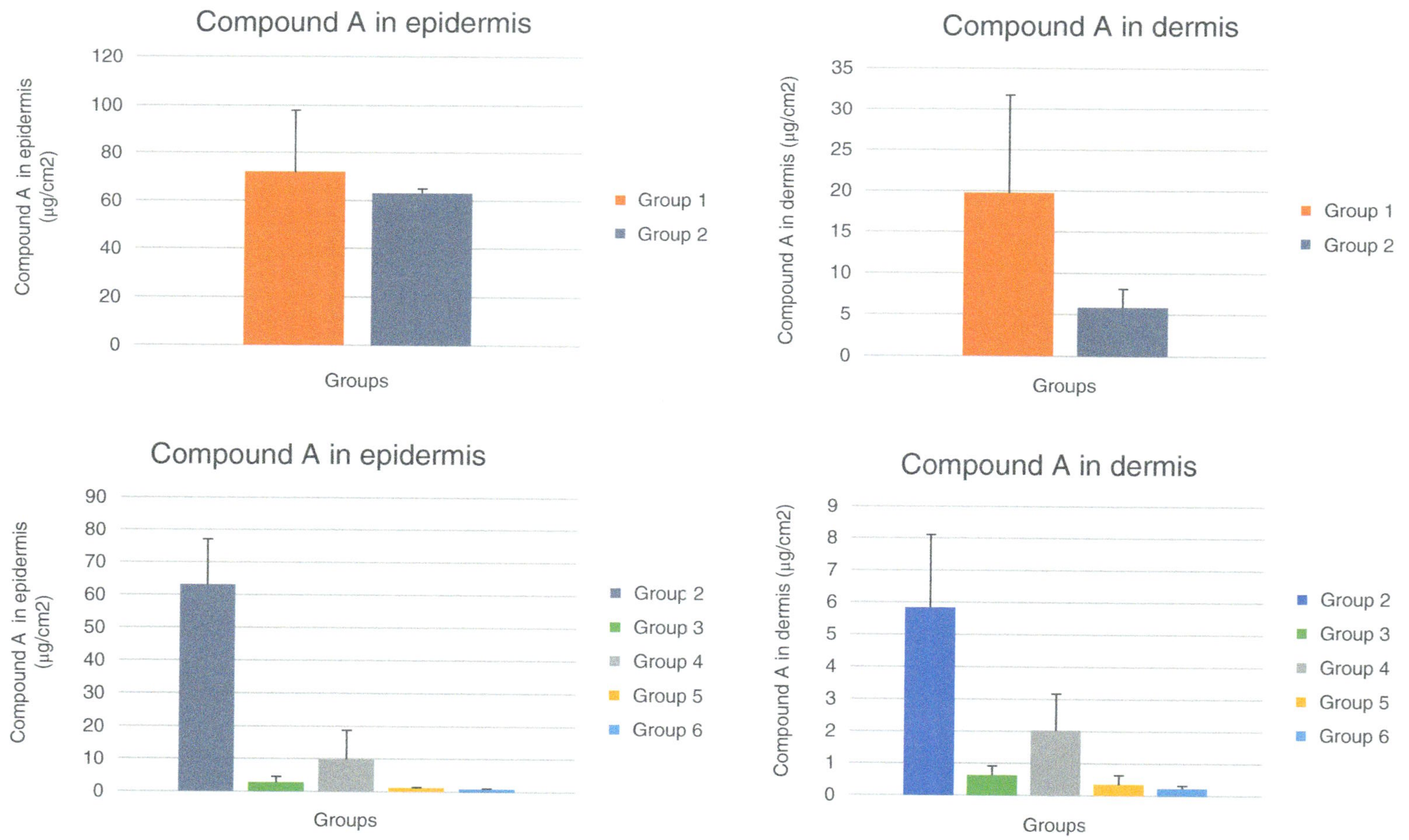

Figure 9.9 Retention of compound A in the epidermis (left) and dermis (right) 24 h after drug administration.

established an IVPT testing platform and this valuable tool will facilitate the rapid development and optimization of complex formulations for topical and transdermal administration along with the advancement of permeation models.

References

1 Benson, H.A.E. et al. (2019). Topical and transdermal drug delivery: from simple potions to smart technologies. *Current Drug Delivery* 16 (5): 444–460.

2 Prausnitz, M.R. and Langer, R. (2008). Transdermal drug delivery. *Nature Biotechnology* 26 (11): 1261–1268.

3 Lionel, T. (2017). *Dermal Drug Selection and Development*. Berlin: Springer.

4 Organization for Economic Cooperation and Development (2004). 428-Guideline for the testing of chemicals-skin absorption: *in vitro* method.

5 Food and Drug Administration 2016. Draft guidance on acyclovir.

6 European Medicines Agency (2018). CHMP: draft guideline on quality and equivalence of topical products.

7 Pharmaceuticals and Medical Devices Agency (2010). Bioequivalence testing guidelines for formulation changes for locally applied dermatological products (semi-solid formulations and patches).

8 US Food and Drug Administration (2022). *In vitro* permeation test studies for topical drug products submitted in ANDAs, guidance for industry.

9 Todo, H. (2017). Transdermal permeation of drugs in various animal species. *Pharmaceutics* 9 (3): 33.

10 Mangelsdorf, S., Vergou, T., Sterry, W. et al. (2014). Comparative study of hair follicle morphology in eight mammalian species and humans. *Skin Research and Technology* 20 (2): 147–154.

11 Bronaugh, R.L., Stewart, R.F., and Congdon, E.R. (1982). Methods for *in vitro* percutaneous absorption studies II. Animal models for human skin. *Toxicology and Applied Pharmacology* 62 (3): 481–488.

12 Finnin, Barrie, et al. (2011) *In vitro* skin permeation methodology. 85–108.

13 Santos, L., Swofford, N.J., and Santiago, B.G. (2020). *In vitro* permeation test (IVPT) for pharmacokinetic assessment of topical dermatological formulations. *Current Protocols in Pharmacology* 91 (1): e79.

9.3 Evaluation of Transdermal Drugs *in In Vivo* Pharmacokinetic Studies

Linlin Pei, Huan He, Chao Zhang, Zhihai Li, Shoutao Liu

9.3.1 Introduction

Transdermal drug delivery is more convenient and flexible compared with oral or injection administration. It has the potential to greatly improve safety and patient compliance in certain populations. As such, the development of TDDS has become an important field within the pharmaceutical industry.

In vivo absorption of transdermal drugs is a crucial factor in evaluating the quality of transdermal agents. Evaluating skin permeability and PK parameters of drugs using both *in vivo* and *in vitro* models can guide research and optimize the development of transdermal agents. *In vivo* PK studies play a vital role in the optimization and evaluation of transdermal products [1].

In vivo experiments involve collecting specific body fluids (such as blood and urine) or tissues (such as dermis, subcutaneous fat, and muscle tissue) at predetermined time points after administering the drug to various animal species through the skin. By measuring drug concentrations in these samples, the effectiveness and bioavailability of a particular drug dosage form can be directly determined.

This chapter provides a comprehensive overview of *in vivo* PK studies related to TDDS. It discusses the strategies employed in the design of these studies, including the selection of animal species. A comparison of damaged and normal skin models, along with the methods for skin sample collection and processing, is also elaborately discussed.

9.3.2 Species Selection and Evaluation

In the early stages of transdermal formulation development, swine (porcine) skin is the most commonly used due to its histological and biological similarities to human skin [2, 3]. In Table 9.8, some of the main skin parameters are compared across different animal species [3].

We conducted a comprehensive evaluation of Bama mini pig skin histology using gross examination, light microscopy, and immunohistochemical staining. Figure 9.10 presents the skin sections from the Bama pig used in our laboratory, as demonstrated by hematoxylin–eosin (HE) staining. We also measured the thickness of each skin layer; comparative data related to human skin features are provided in Table 9.9 [4]. Our findings indicate that Bama mini pigs have features comparable to human skin, making them suitable for evaluating transdermal formulations.

9.3.3 Comparison of Administration Methods Across Different Skin Models

The natural behaviors of Bama mini pigs, such as rooting and rubbing their skin, pose multiple challenges in terms of animal restraint and drug administration. After years of technical exploration and continuous optimization, we have developed a comprehensive operating procedure and evaluation method with the aid of various practical, homegrown experimental tools. These

Table 9.8 Comparative analysis of skin across five different species.

Parameters	Human	Pig	Guinea pig	Rat	Mouse
Skin attachment	Firmly attached	Firmly attached	Loose attached	Loose	Loose
Hair coat	Sparse	Sparse	Sparse or dense	Dense (except for some breeds)	Dense (except for some species)
Epidermis	Thick	Thick	Thick	Thin	Thin
Dermis	Thick	Thick	Thick	Thin	Thin
Panniculus carnosus	Absent	Absent	Present	Present	Present
Healing mechanism	Re-epithelialization	Re-epithelialization	Contraction	Contraction	Contraction

Source: [3] / ELSEVIER.

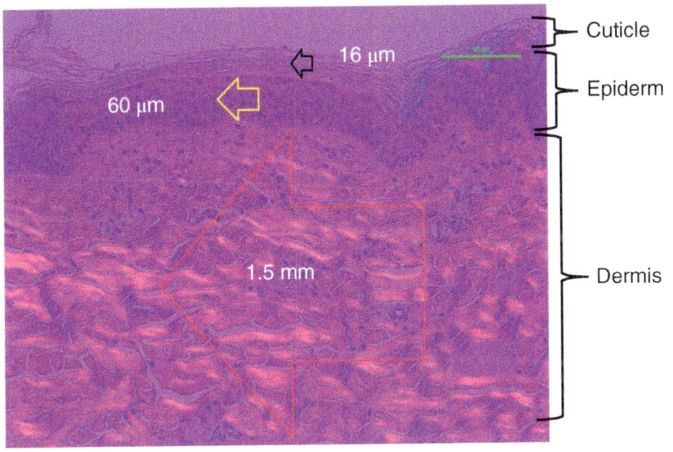

Figure 9.10 HE staining of Bama mini pig skin.

Table 9.9 Comparison of the thickness of each skin layer between Bama Mini Pigs and humans [4].

Type of skin	Stratum corneum (μm)	Epidermis (μm)	Dermis (mm)
4-Month-old Bama pig	14.900 ± 1.370	84.800 ± 3.360	1.271 ± 0.068
6-Month-old Bama pig	15.600 ± 1.506	97.100 ± 4.532	1.933 ± 0.066
Adult human	15.100 ± 1.663	86.200 ± 6.579	1.315 ± 0.069

methods meet the need for applying different types of formulations, ensuring consistency in drug application, and ultimately, the reliability of experimental data.

Depending on the needs of each experimental design, the skin condition is evaluated accordingly. For experiments involving normal and intact skin, we evaluate the skin from an animal using

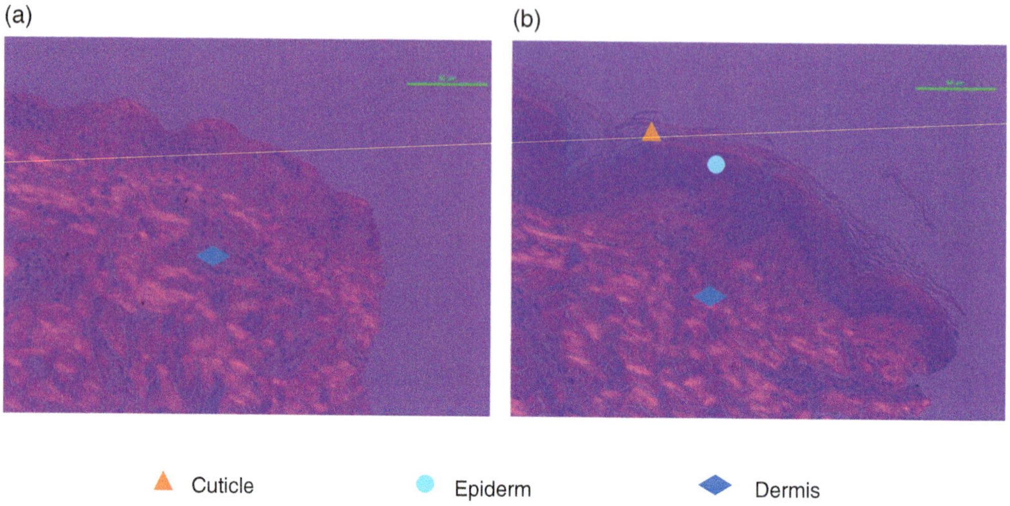

Figure 9.11 HE staining results of Bama mini pig skin. ((a): damaged skin; (b): normal/intact skin).

our established standards. For damaged or diseased skin models, we employ specific methods that comply with animal welfare. The degree of skin damage is assessed according to established standards to ensure reliability and reproducibility in each experiment. For instance, Figure 9.11 displays a representative HE-staining of the skin of a Bama mini pig. The left panel shows skin with the SC and epidermis removed, compared to the normal skin in the right panel. The concentrations of the experimental drug "A" in the dermis and plasma are significantly higher in the damaged skin group than in the intact group (Figure 9.12), indicating that a compromised skin barrier in the damaged skin model increases percutaneous absorption of drug "A."

9.3.4 Collection and Processing of Skin Samples

The design of the experimental sample collection and processing is tailored to meet the needs of the study. In the case of topical formulations with only local effects, the collection and stratification of skin samples are more crucial than blood collection. For instance, we use a skin punch to collect a series of skin samples at different time points from a single animal, ensuring sample uniformity while minimizing damage to the animal. We then use microtomes to stratify skin layers according to the varying thicknesses of each layer. Depending on the study design, SC, epidermis, dermis, and subcutaneous adipose tissue are separated. Each separated skin layer is subsequently homogenized using a bead-milling method at low temperatures to ensure sample homogeneity and uniformity. Traditional homogenization methods are unsuitable for handling small volumes, are prone to cross-contamination, and produce homogenates with large particle sizes. In contrast, the bead-milling method can handle small-volume samples, provide a low-temperature environment, process samples with high throughput, and produce homogenates with smaller particle sizes (Figure 9.13).

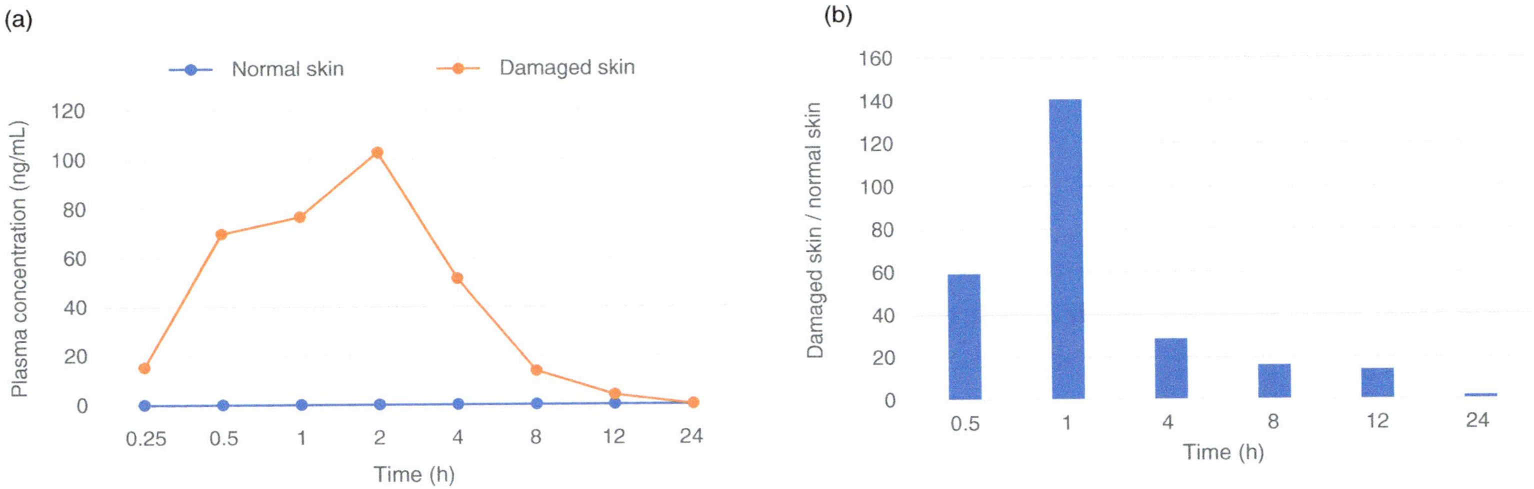

Figure 9.12 Analysis of drug "A" concentrations in plasma and dermis across different skin models.
(a) The average drug concentrations in plasma from damaged and intact skin groups.
(b) The ratios of drug concentration in the dermis from damaged and normal skin groups.

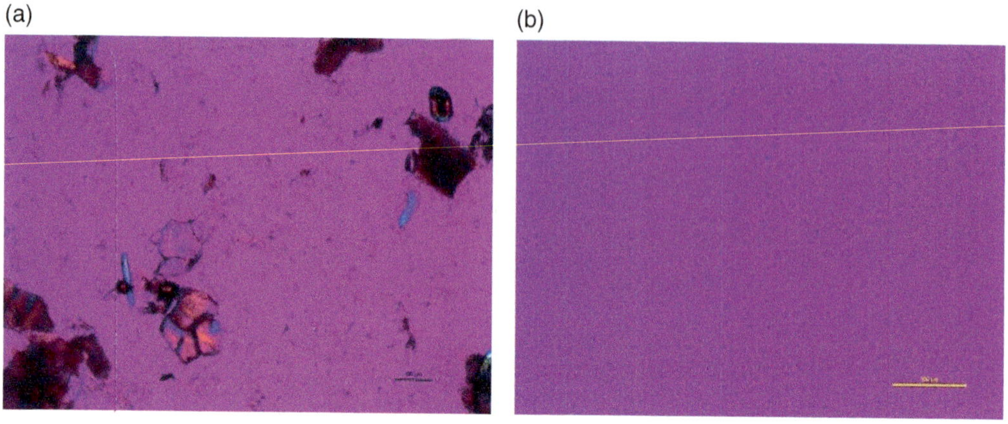

Figure 9.13 Tissue homogenized ((a) using the traditional method; (b) by bead grinding method).

References

1 Jiang, X., Ma, X., Liu, W. et al. (2023). Research progress on permeability of transdermal patches. *Chinese Pharmaceutical Affairs* 3: 312–320.

2 Jacobi, U., Kaiser, M., Toll, R. et al. (2010). Porcine ear skin: an in vitro model for human skin. *Skin Research and Technology* 13 (1): 19–24.

3 Summerfield, A., Meurens, F., and Ricklin, M.E. (2015). The immunology of the porcine skin and its value as a model for human skin. *Molecular Immunology* 66 (1): 14–21.

4 Chen, J., Hu, J., and Wei, H. (2006). Comparative study on skin between BAMA miniature pig and human. *Chinese Journal of Comparative Medicine* 16 (5): 288–290. 319.

9.4 Challenges and Bioanalytical Strategies for Quantitative Detection of *In Vivo* Biological Samples from Skin Administration

Li Gao, Yanfu Ren, Lili Xing

9.4.1 Introduction

The skin is the largest but most easily overlooked organ of the human body. It serves as a major physical, immunological, and sensory barrier to our environment. While it has long been used as a portal for drug delivery, it is a formidable barrier that requires appropriate technology for successful delivery [1]. However, the quantitative analysis of *in vivo* biological samples of skin administration is challenging. The bioanalytical method of plasma should have high sensitivity to achieve a better LLOQ. Moreover, for different skin tissues, corresponding bioanalytical methods need to be developed separately. This chapter mainly focused on the quantitative determination of *in vivo* biological samples for dermal drug delivery.

Minipig skin, apart from its light color, is similar to human skin in many aspects, making it suitable for *in vivo* tests. Due to the multilayered structure of the skin, the matrices of quantitative detection in dermal PK could include plasma, epidermis, dermis, subcutaneous tissue, and whole skin, as well as the SC. Based on analytical characteristics, these matrices can be divided into three categories: plasma, skin tissues, and SC. The analytical difficulties and precautions of these three types of matrices will be introduced.

9.4.2 Strategies for Quantitative Detection of Biological Samples

9.4.2.1 Plasma Sample Analysis

The plasma sample is the most common type of matrix. For topical delivery, extremely low systemic exposure is expected, and the high sensitivity is critical in the plasma concentration detection method, which is one of the greatest challenges in plasma sample analysis.

To improve the signal response of test compounds, LC, MS, and sample pretreatment are considered comprehensively. For example, after optimizing the LC–MS/MS method, the LLOQ for the marketed compound dexmedetomidine could achieve 3 pg/mL in the plasma sample, with a good linear regression ($r = 0.9998$). Moreover, in the analysis of plasma samples, it is necessary to pay attention to matrix stability and NSB of compounds, which are used to assess the method of sample collection, storage, pretreatment, and integrity in the supernatant to ensure that the measured concentrations are reliable.

9.4.2.2 Skin Tissue Sample Analysis

1) Significance of skin tissue sample analysis

The bioanalysis of skin tissue samples is essential at all stages, from discovery and development to post-marketing therapeutic monitoring of new drugs. This analysis is crucial to confirm the combination between drug candidates and their targets, to understand their safety, efficacy, and distribution within the body, and to establish a PK–PD relationship [2]. The tissue uptake of the drug can be assessed by the quantitative analysis of tissue samples. This is particularly important for topical drug delivery, which often exhibits nearly undetectable systemic exposure. Therefore, the exposure of the drug in the target tissue, namely, skin tissue, becomes even more critical.

2) **Homogenization of skin tissue samples**

Skin layers will be prepared with a cryosectioning technique after collection. A homogenizer is often preferred to convert the solid samples into liquid form for LC–MS/MS analysis, which is a key step. First, skin tissue samples contain collagen and elastin, making them more challenging to homogenize compared to other tissue samples, especially dermal samples. Second, factors such as compound solubility and NSB also need to be considered. Therefore, the choice of homogenate medium and the ratio of tissue weight to medium volume is essential. The homogenate medium for skin samples is organic and PBS mixed solution generally, and a large homogenate ratio can usually improve the homogeneity of tissue samples. It is worth noting that the larger sampling amount during sample pretreatment can somehow counteract the effect of poor homogenate dispersion to improve the reproducibility of the analytical result.

3) **Analysis of skin tissue samples**

The analysis of specific tissue samples using LC-MS/MS is generally optimized or adjusted based on the analysis method of plasma samples. For skin tissue samples, two points need attention. First, compared with plasma samples, skin tissue homogenate samples contain a variety of cells, different types of phospholipids, and other endogenous compounds [3]. The matrix effect can not be neglected, otherwise the measured concentration will be less reliable due to the co-elution on LC of interfering components and target analytes. Second, there are generally many tissue matrices to be analyzed while with a limited sample number and tissue amount, especially for the epidermis. Based on this, appropriate surrogate matrices should be selected to prepare calibration standards and quality controls (QCs) and shorten the analysis duration.

4) **Other platforms for skin tissue sample analysis**

For the analysis of skin tissue samples, in addition to the LC–MS/MS method, which is the general procedure as shown in Figure 9.14, QWBA is an alternative technology that allows visualization and adequately shows tissue distribution results.

9.4.2.3 Analysis of Stratum Corneum (SC) Samples

The SC is thin and requires a special sample collection method. The tape-stripping method which is also recommended by the FDA is widely accepted and used. This method is a noninvasive technique by repeating stripping of the appropriate tape to collect SC samples. This unconventional collection

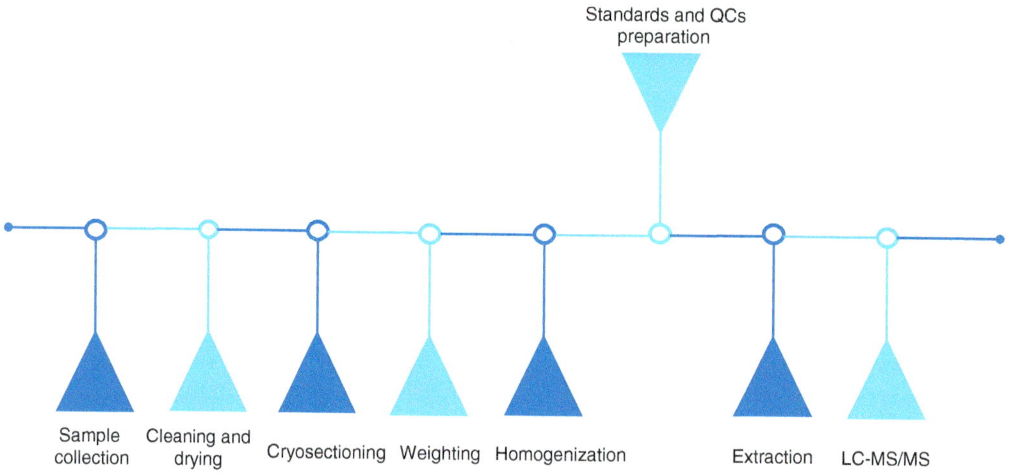

Figure 9.14 LC–MS/MS general procedure for analysis of skin tissue samples.

Figure 9.15 Before sample pretreatment.

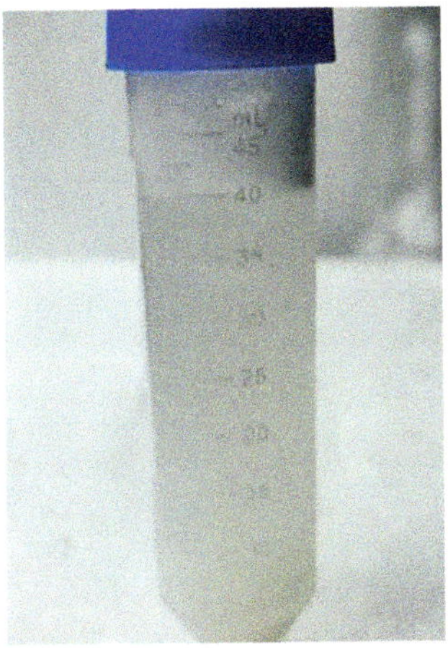

Figure 9.16 After sample pretreatment.

method inevitably brings some challenges for sample analysis. The solid samples striped by tapes (see Figure 9.15) must convert to a suitable liquid state for injection (see Figure 9.16), but they cannot be homogenized like the skin tissues. Moreover, the calibration standards and QCs cannot be easily prepared by the conventional method due to the particularity of the blank matrix.

1) **Pretreatment of stratum corneum (SC) samples**

Considering the two problems mentioned above, we designed an assay as shown in Figure 9.17. On the one hand, for unknown samples and a blank tape, an appropriate solvent was developed to "soak" the analyte as completely as possible. The tape was cut into small pieces using a scissor, the appropriate solvent was added to a constant volume, ultrasound was applied, and the sample was left overnight to ensure maximum extraction of analyte. On the other hand, for standards and QC samples, theoretically, the pretreatment process should exactly follow that of the unknown

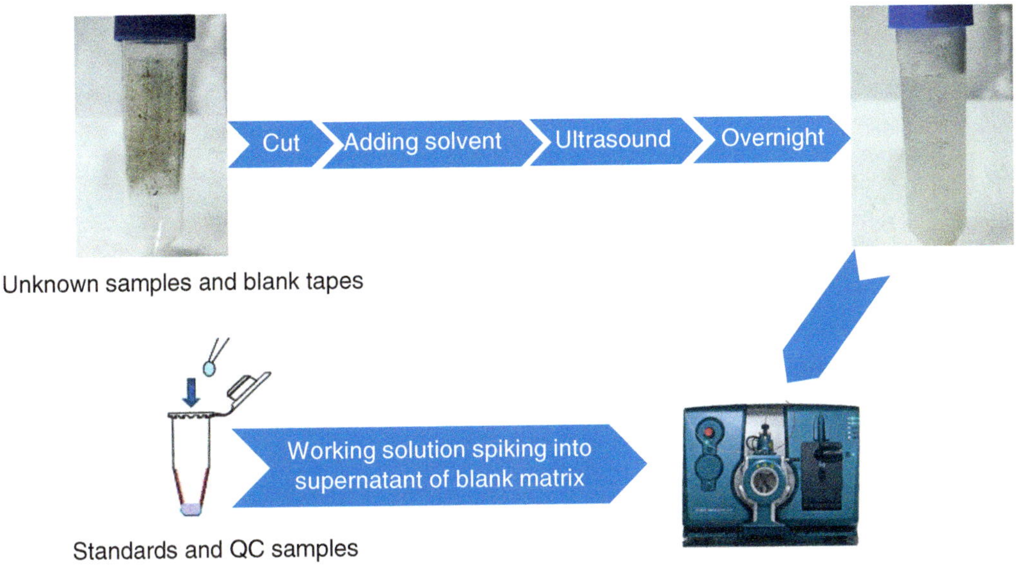

Figure 9.17 Workflow of stratum corneum analysis.

Table 9.10 Validation assay results.

Calibration standards (ng/mL)	Theoretical concentration	0.1	0.2	1	5	10	50	100	300
	Measured concentration	0.100	0.198	1.01	4.96	10.6	48.4	95.6	307
% Accuracy		100	99.0	101	99.2	106	96.8	95.6	102
Quality controls (ng/mL)	Theoretical concentration	0.3	4	80	240	0.3	4	80	240
	Measured concentration	0.343	4.08	87.9	246	0.379	3.76	77.0	239
% Accuracy		114	102	110	103	126	94.0	96.2	99.6

samples. However, a convenient alternative approach, in which working solutions of the standards and QC samples are added directly to the supernatant of the blank tape containing blank SC matrix, was designed due to the tedious sample pretreatment one by one.

2) **Validation**

Due to the inconsistency pretreatment between the calibration standards and incurred samples, it is necessary to validate the reliability of the process. The QC samples were used to simulate unknown samples and followed the same processing as mentioned in Section 9.4.2.3, while calibration standards still used a convenient alternative approach as mentioned. Validation results showed that the accuracy met the analytical criterion (Table 9.10). Because the collection method of SC is special, more attention should be paid to the tape brand, scissors-cutting process, internal standard addition, and other operations.

9.4.3 Summary

As the population ages, the number of patients with chronic diseases increases, and consumers spend more on health care or beauty, the market for skin administration is increasing. Due to the advantages of skin administration, more and more scientists are committed to research in this field, and the possibility of skin drug delivery has been expanded.

For example, marketed oral drugs can change their route of administration to transdermal drug delivery to achieve new applications. Additionally, other active ingredients for skin drug delivery are continuously being developed, and the research and development of various formulation technologies, such as flexible liposomes, make the delivery of macromolecules possible [3]. We have developed a comprehensive protocol for the optimization of formulations, administration, sample collection, and quantitative analysis, which can efficiently provide accurate and reliable data and assist the drug research and development process of skin drug delivery.

References

1 Abd, E., Yousef, S.A., Pastore, M.N. et al. (2016). Skin models for the testing of transdermal drugs. *Clinical Pharmacology: Advances and Applications* 8: 163–176.

2 Li, W., Zhang, J., and Xie, L. (2017). *Handbook of LC-MS Bioanalysis: Best Practices, Experimental Protocols and Related Regulations*. Science Press.

3 Wang, W. and Huang, Q. (2021). Research progress of flexible liposomes in drug delivery system. *Chinese Journal of Modern Applied Pharmacy* 38 (4): 495–502.

10

Central Nervous System Drugs

10.1 Overview of Pharmacokinetic Research Strategies for Central Nervous System Drugs

Huihui Li, Tingting Ruan, Qing Liu, Tao Xiong, Xuan Dong, Cheng Tang, Genfu Chen, Jing Jin

10.1.1 Introduction

The nervous system plays a dominant role in physiological activity regulation. It consists of two major parts: the CNS and the peripheral nervous system, among which the CNS is the main part, including the brain and spinal cord. If disorders happen in the CNS, an individual's physiological function or cognitive function could be damaged (Table 10.1).

According to the World Mental Health Report released by the WHO in June 2022, approximately one-eighth of people in the world are suffering from mental disorders [2]. The number of people with dementia (Alzheimer's disease accounts for 60–70%) increased each year (Figure 10.1) and was the fifth leading cause of death worldwide [3]. Due to the uncertainty of the market caused by the COVID-19 pandemic, it is expected that the CNS market will increase to $137 billion by 2025. However, CNS drugs have a low success rate (~ 8%), which is even lower for Alzheimer's disease (~1%) because of various reasons, such as the low regeneration ability of CNS, complex mechanism of action, low reliability of preclinical animal models, and low permeability to the blood–brain barrier (BBB) [4]. Still, CNS drug development remains active. Based on the number of pipelines under development worldwide, neurological drugs (including drugs for CNS disease and mental health) have become the second largest therapeutic area after oncology [5]. For CNS drug development, opportunities always come along with challenges. High risk also means high benefit.

Improving permeability to BBB via compound structure modification is one of the effective methods to improve the development success rate. In this chapter, we will mainly focus on the introduction of BBB permeability evaluation assays and parameters for small molecules from the DMPK perspective.

10.1.2 Brain Barrier

The brain is a carefully protected organ in our bodies. In addition to a 7-mm skull, which resists external mechanical damage, there is a special brain barrier to protect the brain from harmful substances in the systemic circulation. The brain barrier includes BBB, blood–cerebrospinal fluid barrier (BCSFB), and cerebrospinal fluid–brain barrier (CSF–brain barrier) (Figure 10.2), among which the BBB has the lowest permeability.

Drug Metabolism and Pharmacokinetics: Frontiers, Strategies, and Applications, First Edition.
Edited by Liang Shen and WuXi AppTec DMPK.

Table 10.1 Classification of central nervous system disorders.

Disease Type	Disease Examples
Vascular disorders	Stroke, transient ischemic attack (TIA), etc.
Infections	Meningitis, encephalitis, etc.
Structural disorders	Brain or spinal cord injury, Bell's palsy, etc.
Functional disorders	Headache, epilepsy, dizziness, etc.
Degeneration	Parkinson's disease, Alzheimer's disease (AD), etc.

Source: Adapted from reference [1].

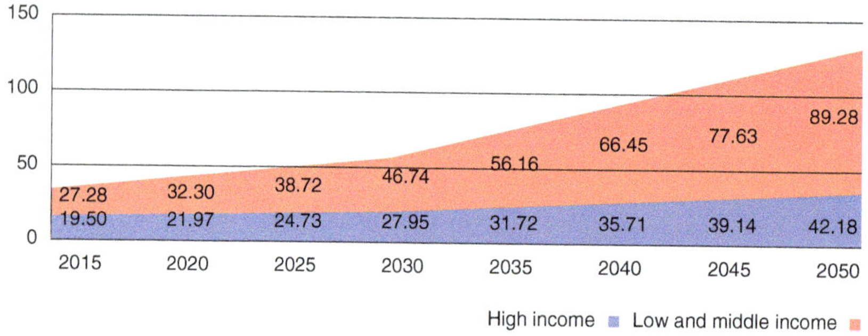

Figure 10.1 The growth in the number of people with dementia (millions) in high-income and low- and middle-income countries. *Source:* [3].

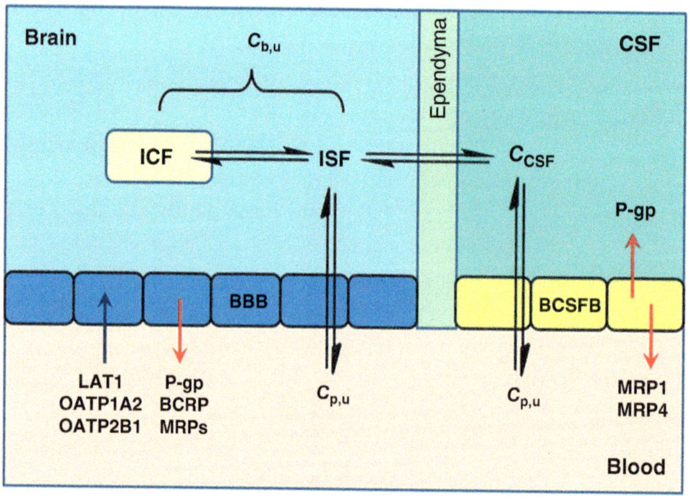

Figure 10.2 A three-compartment model illustrates unbound drug distribution equilibrium between plasma, brain (ISF and ICF), and CSF. *Source:* [4] / with permission of American Chemical Society.

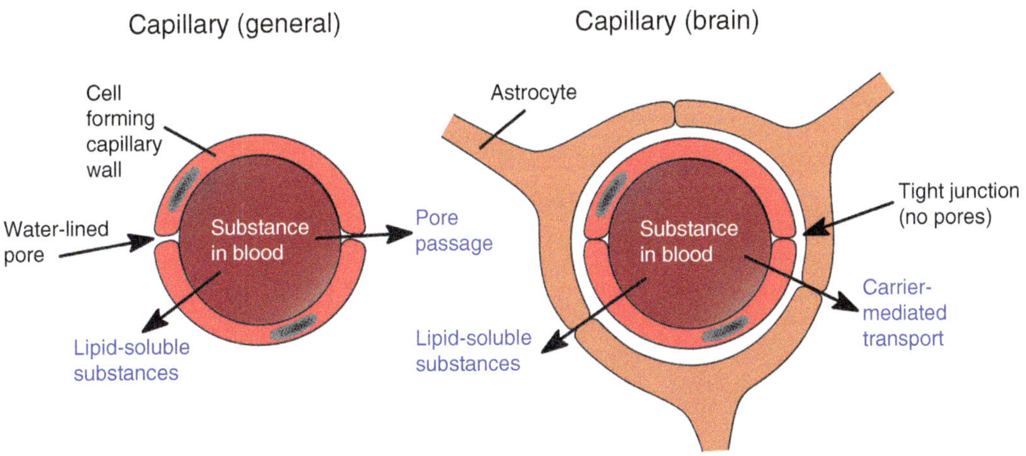

Figure 10.3 Comparison of BBB to common capillaries. *Source:* [6] / Frontiers Media S.A. / CC BY 4.0.

The poor permeability of BBB is mainly determined by its structure. There are continuous capillaries and tight junctions between endothelial cells in BBB, which make it more difficult to get through compared with the general capillaries (Figure 10.3).

10.1.3 The Way Substances Enter the Blood–Brain Barrier

Although the BBB structure is tight, our brain is constantly exchanging substances with blood, CSF, etc. There are six pathways for substances to get in and out of the BBB (Figure 10.4). Most of the small hydrophobic molecules enter the brain by passive diffusion. A few water-soluble small molecules pass through the paracellular transport pathway. Endogenous small molecules and

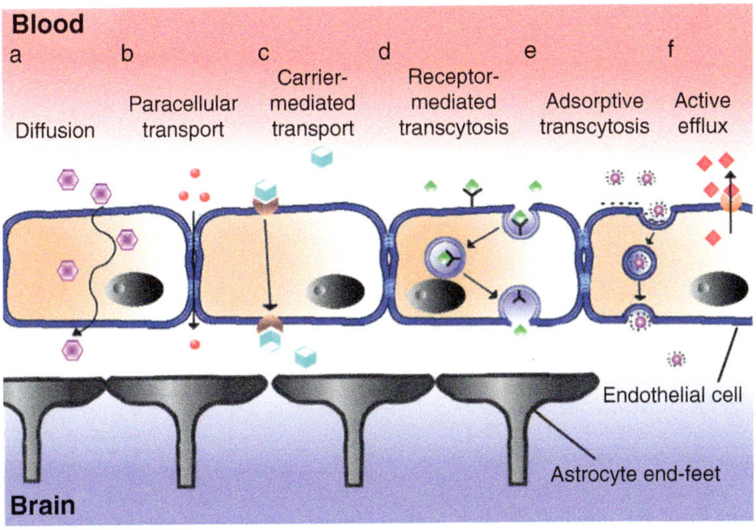

Figure 10.4 The way substances enter and leave the blood–brain barrier. *Source:* [7] / MDPI. / CC BY 4.0.

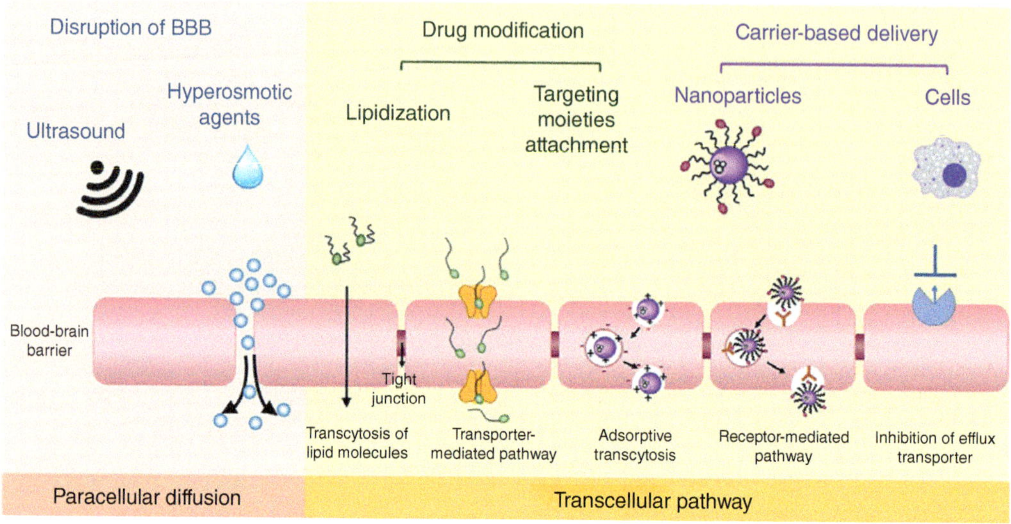

Figure 10.5 Transport routes across the BBB and the strategies for delivery of CNS-active therapeutics across the BBB. *Source:* [8] / ELSEVIER.

analogs are mainly via transporter-mediated uptake, which has been used to deliver biological drugs. Substances that enter the brain via various ways may be excreted by efflux transporters.

10.1.3.1 How to Improve Permeability of Small Molecules to the Blood–Brain Barrier

Structure modification strategies have been used to increase the compound's ability to cross the BBB, including (Figure 10.5):

- Changing physicochemical properties, such as molecular weight, H-bond acceptor number, H-bond donor number, and lipid solubility.
- Modifying structure to make use of transporter/receptor/carrier-mediated transcytosis.
- Using delivery systems, such as lipid nanoparticles, viral vectors, and cell vectors.
- Transiently disrupting tight junctions using sonication or hyperosmotic reagents (mannitol, arabinose, urea, etc.). This method should be used with caution, as it has little selectivity on the substances entering the brain, potentially causing neurotoxicity or brain damage.

For small molecules, improving physicochemical properties through structure modification is usually the most efficient way to increase brain exposure. The rules listed in Figure 10.6 are believed to be effective [9]. Generally, these compounds will go through *in vitro* assays to evaluate permeability, stability, solubility, etc., and the resulting promising candidates will then get into *in vivo* pharmacokinetics (PK) study evaluations to check the free concentrations in plasma and brain. The *in vitro* and *in vivo* evaluation assays for small molecules will be introduced in the coming sections.

10.1.4 Brain Permeability Evaluation Assays

10.1.4.1 *In Vitro* Permeability and Transporter Evaluation Assays

Brain permeability is a combined effect of passive diffusion, receptor/transporter-mediated efflux, and uptake. CNS drugs are suggested to have high passive diffusion capacity and are not substrates of efflux transporters. The ATP-binding cassette (ABC) transporters are the main efflux transporters

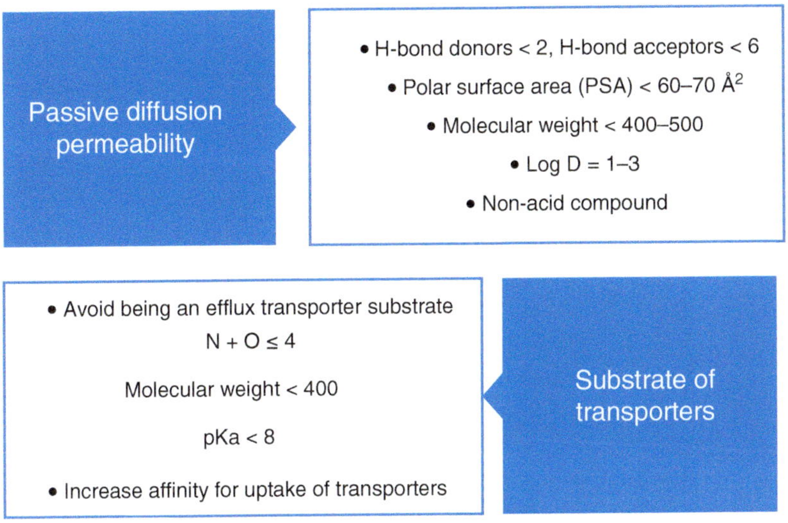

Figure 10.6 Rules for CNS drug design and modification.

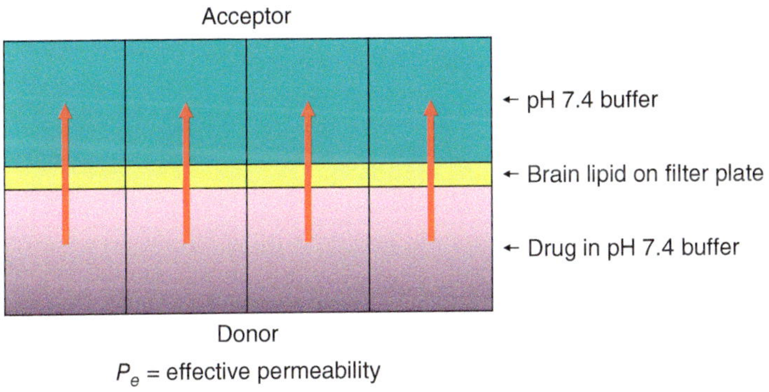

Figure 10.7 Schematic diagram of BBB–PAMPA model. *Source:* [9] / ELSEVIER.

in BBB, including P-glycoprotein (P-gp) and breast cancer resistance protein (BCRP) efflux transporters, among which P-gp plays the main role [10]. Therefore, P-gp substrate assessment is recommended during the early screening period. In the following part of this section, we will introduce the *in vitro* BBB passive diffusion evaluation assay BBB–PAMPA, the efflux transporter evaluation assay (MDR1-MDCK I cell model) and a screening strategy for BBB permeability evaluation.

PAMPA is a popular *in vitro* model used to evaluate the passive transcellular permeation of drugs. The model is established by coating an organic solution of lecithin on a polyvinylidene fluoride (PVDF) membrane, where the lecithin forms a film in the membrane pores, mimicking the lipid bilayer of the cell membrane. In BBB–PAMPA, lecithin is replaced with porcine brain lipid extract (Figure 10.7). Because there are no transporters involved, BBB–PAMPA evaluates the passive permeation of drugs. BBB–PAMPA was reported to have a good correlation with the in situ brain perfusion data in rats [11]. It is a high-throughput, low-cost method to assess the permeability of compounds to BBB.

MDR1-MDCK I cells were derived from the National Institutes of Health (NIH), and the P-gp expression level is higher than that in MDR1-MDCK II cells, which has been commonly used in permeability experiments, resulting in higher efflux ratio (ER). The ER values obtained from the

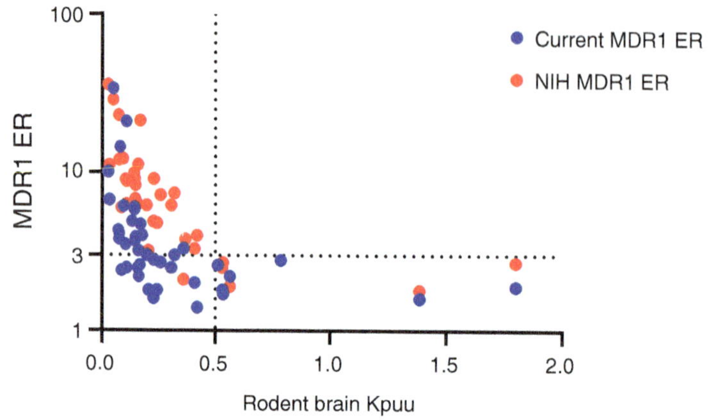

Figure 10.8 Correlation between rat brain-to-plasma $K_{p,uu}$ and efflux ratios from MDR1-MDCK II and NIH MDR1-MDCK I cell lines. *Source:* [12] / John Wiley & Sons.

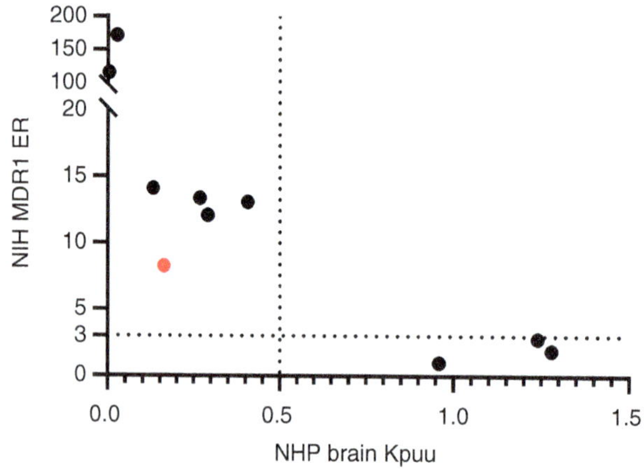

Figure 10.9 Correlation between nonhuman primate (NHP) $K_{p,uu}$ and efflux ratios from MDR1-MDCK I cell line. *Source:* [12] / with permission of John Wiley & Sons.

MDR1-MDCK I cells have been reported to correlate better with the $K_{p,uu}$ in rats (Figure 10.8) and nonhuman primates (Figure 10.9). Moreover, for compounds with ER < 3 in MDR1-MDCK II cells, the MDR1-MDCK I cell line demonstrated a substantially greater dynamic range, with ERs ranging from 1 to ~30, which suggests that MDR1-MDCK I cell line is more sensitive to differentiate compounds [12] (Figure 10.10).

To verify the accuracy of BBB–PAMPA and MDR1-MDCK I in predicting brain penetration, 24 commercial compounds, including 10 BBB permeable (CNS+) compounds and 14 BBB impermeable compounds (CNS-) [13–17], were selected to run the two assays. $Pe = 1.5 * 10^{-6}$ cm/s (Log Pe = – 5.85) in BBB–PAMPA and ER = 4 in MDR1-MDCK I were used as cut-off values to group these compounds. As a result, the accuracy was 83.3% (20/24) if either of the two assays was used as the only criteria to group compounds (Figures 10.11 and 10.12). However, if BBB–PAMPA was used as the first run to exclude compounds with poor passive permeability, and MDR1-MDCK I was then used to exclude efflux transporters substrates, the 10 CNS+ compounds could be selected accurately (Figure 10.13). Based on these tests, along with others (not reported), we propose to use the two assays (a "Funnel" model) for selecting BBB permeable compounds at the early screening stage (Figure 10.14).

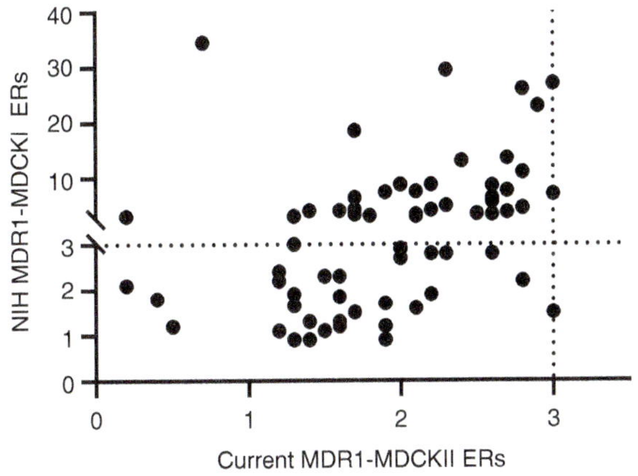

Figure 10.10 Comparison of efflux ratios between MDR1-MDCK II and MDR1-MDCK I cell line. *Source:* [12] / with permission of John Wiley & Sons.

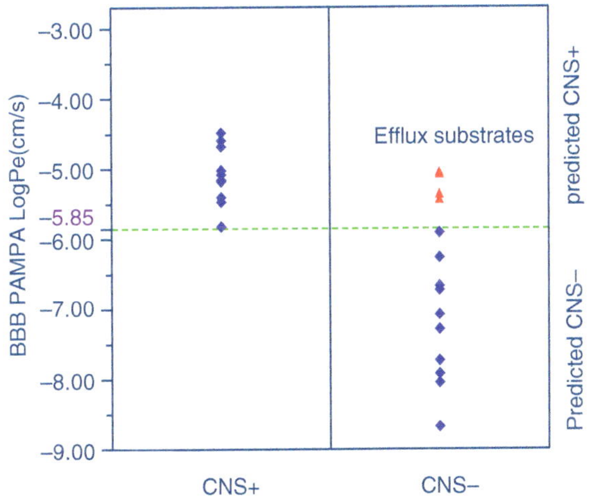

Figure 10.11 Accuracy of BBB-PAMPA in predicting BBB permeability. Three compounds in the red triangle were false positive errors, as they are reported to be efflux substrates and impermeable to BBB.

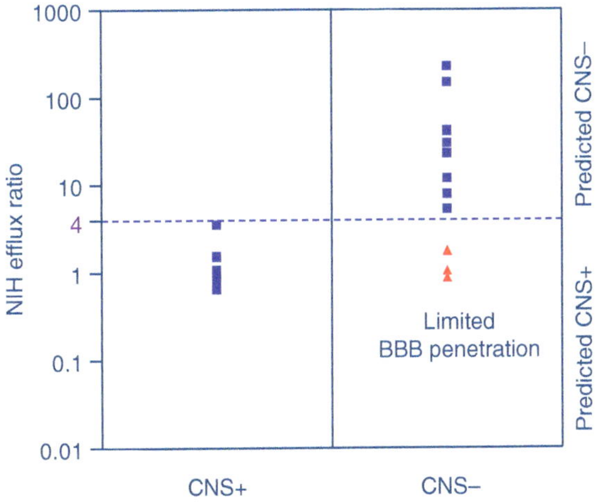

Figure 10.12 Accuracy of MDR1-MDCK I in predicting BBB permeability. Three compounds in the red triangle were false positive errors. They are reported to be impermeable to BBB even though ER <4. As shown in BBB-PAMPA data, they had limited passive permeability to BBB.

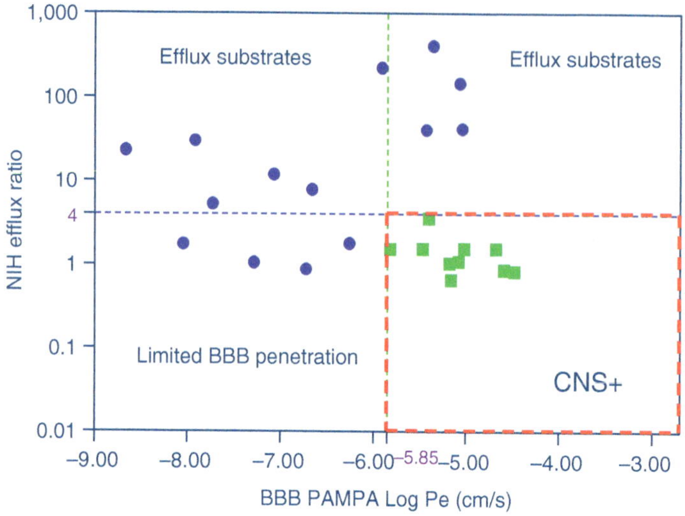

Figure 10.13 Combining BBB–PAMAPA and MDR1-MDCK I assay, the 10 CNS+ compounds were selected accurately (green square).

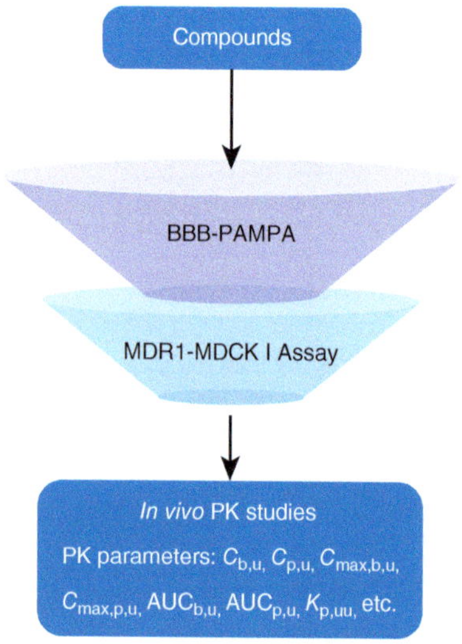

Figure 10.14 "Funnel" model to evaluate and select BBB permeable compounds.

10.1.4.2 *In Vivo* Brain Penetration Evaluation Models

10.1.4.2.1 In Situ Brain Perfusion Model

The *in situ* brain perfusion model is a sensitive method to evaluate BBB permeability. The perfusate-containing testing compound is perfused into the brain via the carotid artery (Figure 10.15). At the end of perfusion, the cerebral vessels are flushed with blank perfusate, and the brain is removed and homogenized to check compound concentration. There are many advantages of this model: (i) it provides a direct and accurate way to reflect the amount of compound that penetrates the brain; (ii) the effect of drugs remaining in cerebral vessels can be excluded; (iii) efflux can also be evaluated if co-dosing with transporter inhibitors or using transporter knockout animals.

However, this model is not extensively used because of some obvious limitations, such as technically demanding operation, low throughput, and high animal cost.

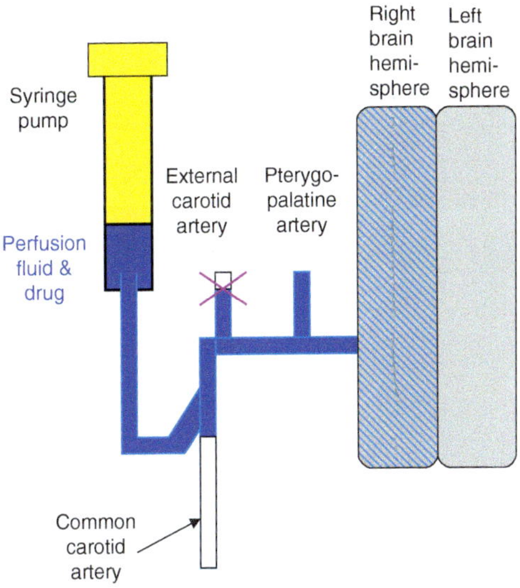

Figure 10.15 *In situ* brain perfusion assay of the rat. *Source:* [9] / ELSEVIER.

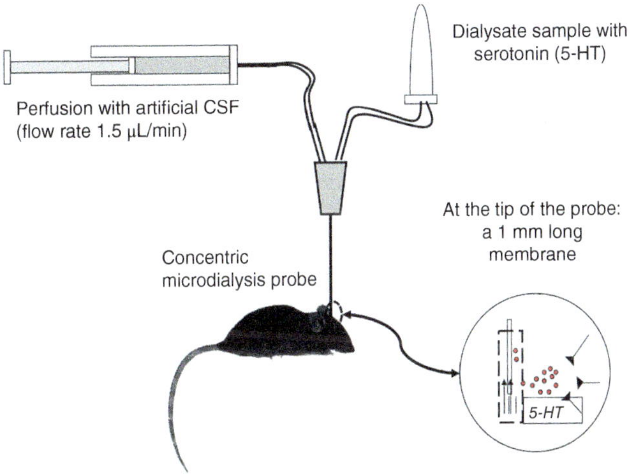

Figure 10.16 Diagram of brain microdialysis in awake, freely moving mice. *Source:* [18] / Frontiers Media S.A. / CC BY 4.0.

10.1.4.2.2 *Brain Microdialysis Model*

The brain microdialysis (MD) model is used to directly obtain free drug concentrations in the brain. An MD probe containing a dialysis membrane is implanted into the brain, which is continuously perfused with artificial extracellular fluid (Figure 10.16). Due to the concentration gradient on both sides of the dialysis membrane, the substances in the brain are extracted through exchange. This model allows continuous sampling in conscious animals, and the free drug concentrations in the brain can be monitored in a real-time manner. In addition, if dialysis of the brain and blood is performed simultaneously on the same animal, $K_{p,uu}$ could be obtained directly.

Unfortunately, not all compounds are suitable to evaluate in this model, especially for compounds with high nonspecific binding to dialysis membranes or perfusion tubes. Therefore, *in vitro* preliminary experiments need to be performed to confirm the feasibility of the assay conditions.

10.1.4.2.3 *NeuroPK Model*

NeuroPK studies are widely used to check PK properties *in vivo*. In addition to these commonly used administration routes (such as intravenous, oral, and subcutaneous administration), intrathecal injection (IT), lateral ventricle administration, and intranasal administration have been widely used to increase brain exposure. After administration, plasma, CSF, and brain can be collected from animals at different time points, from which the total concentration or exposure of compounds ($C_{p,t}$, $C_{b,t}$, C_{CSF}, $AUC_{p,t}$, $AUC_{b,t}$, and AUC_{CSF}) could be obtained. Combined with *in vitro* protein binding assays, the free compound concentration and free AUC in the brain and plasma can be obtained ($C_{p,u}$, $C_{b,u}$, $AUC_{p,u}$, and $AUC_{b,u}$) with the following calculation:

$$C_{b,u} = C_{b,t} \times f_{b,u};$$
$$C_{p,u} = C_{p,t} \times f_{p,u};$$
$$\text{AUC}_{b,u} = \text{AUC}_{b,t} \times f_{b,u};$$
$$\text{AUC}_{p,u} = \text{AUC}_{p,t} \times f_{p,u};$$
$$K_{p,uu} = C_{b,u} / C_{p,u} = \text{AUC}_{b,u} / \text{AUC}_{p,u}$$

$K_{p,uu}$ is a critical parameter to evaluate free frug exposure in the brain and assess whether distribution equilibrium has been established between plasma and brain compartments.

- When $K_{p,uu} < 1$, $C_{b,u} < C_{p,u}$, the compound is possibly a substrate of efflux transports, and/or has low passive penetration across the BBB. Structural modification should be considered to improve brain permeability.
- When $K_{p,uu} > 1$, $C_{b,u} > C_{p,u}$, it suggests that active update is involved in brain penetration.
- When $K_{p,uu} \approx 1$, $C_{b,u} \approx C_{p,u}$, the compound is at distribution equilibrium between the plasma and the brain, and the unbound plasma concentration can be used to estimate the unbound brain concentration.

AUC_{CSF}:$AUC_{p,u}$ and $AUC_{b,u}$:AUC_{CSF} ratios can also be calculated to evaluate whether CSF concentration can be used as a surrogate of unbound brain concentration.

10.1.5 Common Misconceptions in Central Nervous System Drugs Evaluations and Development

Misconception No. 1: the higher the B/P ratio, the better the drug is for CNS penetration
The B/P ratio is calculated using the total drug concentrations in the brain and plasma. However, it is the unbound B/P ratio ($K_{p,uu}$ or $C_{b,u}/C_{p,u}$) that should be used to evaluate the penetration to BBB. The total concentration is largely influenced by the nonspecific binding of compounds to proteins and lipids in plasma or brain homogenates.

Misconception No. 2: To increase unbound brain concentration, the brain fraction unbound ($f_{b,u}$) should be increased by structural optimization
The $f_{b,u}$ values are not suitable to guide the structural optimization of compounds, as when $f_{b,u}$ changed, the $C_{b,t}$ may be changed as well [19]. Therefore, *in vitro* plasma protein binding assay and brain tissue homogenate binding assay are not recommended to serve as one of the parameters for compound screening at an early stage.

Misconception No. 3: CSF concentration is used as a surrogate for $C_{b,u}$ regardless of compound property
C_{CSF} is considered to be within a threefold range of $C_{b,u}$ (obtained through MD detection) [20]. However, it's not always accurate to use CSF concentration to represent $C_{b,u}$, especially for compounds that are transporters substrates. P-gp in BCSFB is transporting compounds into the CSF (Figure 10.2), and BSCFB is more permeable than BBB, so C_{CSF} may tend to be higher than $C_{b,u}$. In addition, CSF is not a well-mixed compartment and is refreshed rapidly (once every five hours). CSF concentrations may vary significantly at different sampling sites [21].

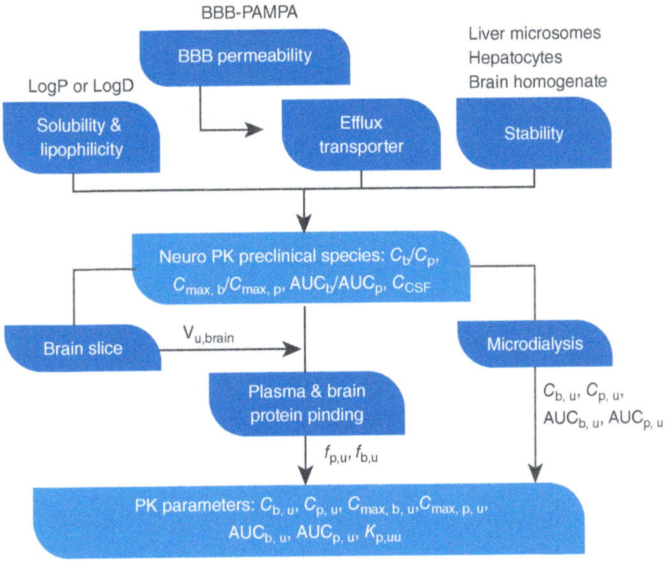

Figure 10.17 Screening paradigm for CNS drugs selection and optimization.

Misconception No. 4: CNS PK should use large animals (e.g., monkeys) to better predict human CNS penetration
For compounds that are not substrates of efflux transporters, $K_{p,uu}$ is similar in different species, rat (or other species) $K_{p,uu}$ can be used to predict free concentration in the human brain with the following conversion [4]:

$$\left[C_{b,u,human} \approx K_{p,uu,rat}{}^{*}C_{p,u,human}\right]$$

For compounds that are transporter substrates, monkeys will be considered a better species for human brain exposure prediction due to species differences in transporters [10].

10.1.6 Summary

CNS drug development is very difficult because of the special BBB structure. Compound selection and optimization based on misleading parameters often lead to unsuccessful discovery programs. In the early drug screening period, the recommended strategy to optimize drug candidates is to reduce efflux transport (by P-gp/BCRP), minimize metabolism and clearance, and increase permeability and solubility. A screening paradigm is proposed to guide CNS drug selection and optimization (Figure 10.17).

References

1 Hopkins Medicine. Overview-of-nervous-system-disorders. https://www.hopkinsmedicine.org/health/conditions-and-diseases/overview-of-nervous-system-disorders

2 World mental health report: Transforming mental health for all (Fig. who.int). https://www.who.int/publications/i/item/9789240049338&#32

3 World Alzheimer Report 2015, The Global Impact of Dementia: An analysis of prevalence, incidence, cost and trends (http://alzint.org)

4 Di L, Rong H, and Feng B. Demystifying brain penetration in central nervous system drug discovery: Miniperspective. *Journal of Medicinal Chemistry* 2013; 56 (1): 2–12. https://doi.org/10.1021/jm301297f. Epub 2012 Nov 6. 23075026.

5 Global Pharmaceutical R & D Panoramic Prospect 2023 - IQVIA

6 Boonstra E, de Kleijn R, Colzato LS, et al., Neurotransmitters as food supplements: the effects of GABA on brain and behavior. *Frontiers in Psychology* 2015; 6:1520. https://doi.org/10.3389/fpsyg.2015.01520. 26500584; PMC4594160.

7 Georgieva, J.V., Hoekstra, D., and Zuhorn, I.S. (2014). Smuggling drugs into the brain: an overview of ligands targeting Transcytosis for drug delivery across the blood – brain barrier. *Pharmaceutics* 6: 557–583. https://doi.org/10.3390/pharmaceutics6040557.

8 Li, G., Shao, K., and Umeshappa, C.S. (2019). Recent progress in blood-brain barrier transport research. In: *Brain Targeted Drug Delivery System* (ed. H. Gao and X. Gao), 33–51. Academic Press. ISBN: 9780128140017 https://doi.org/10.1016/B978-0-12-814001-7.00003-2.

9 Di, L. and Kerns, E.H. (2016). Chapter 28 - Blood-brain barrier. In: *Drug-Like Properties*, Seconde (ed. L. Di and E.H. Kerns), 351–369. Academic Press. ISBN: 9780128010761 https://doi.org/10.1016/B978-0-12-801076-1.00028-9.

10 Di, L. and Kerns, E.H. (2016). Chapter 10 - Blood-brain barrier. In: *Drug-Like Properties*, Seconde (ed. L. Di and E.H. Kerns), 141–159. Academic Press. ISBN: 9780128010761 https://doi.org/10.1016/B978-0-12-801076-1.00010-1.

11 Di L, Kerns EH, Bezar IF, et al. Comparison of blood-brain barrier permeability assays: in situ brain perfusion, MDR1-MDCKII and PAMPA-BBB. *Journal of Pharmaceutical Sciences* 2009; 98 (6): 1980–91. https://doi.org/10.1002/jps. 21580. 18837012.

12 Jiang L, Kumar S, Nuechterlein M, et al. Application of a high-resolution *in vitro* human MDR1-MDCK assay and *in vivo* studies in preclinical species to improve prediction of CNS drug penetration. *Pharmacology Research & Perspectives* 2022 Feb; 10 (1): e00932. https://doi.org/10.1002/prp2.932. 35156331; PMC8929335.

13 Tsinman, O., Tsinman, K., Sun, N., and Avdeef, A. (2011). Physicochemical selectivity of the BBB microenvironment governing passive diffusion-matching with a porcine brain lipid extract artificial membrane permeability model. *Pharmaceutical Research* 28 (2): 337–363.

14 Drion, N., Lemaire, M., Lefauconnier, J.M., and Scherrmann, J.M. (1996). Role of P - glycoprotein in the blood - brain transport of colchicine and vinblastine. *Journal of Neurochemistry* 67 (4): 1688–1693.

15 Karssen, A.M., Meijer, O.C., and de Kloet, E.R. (2005). Chapter 3.4 - Corticosteroids and the blood-brain barrier. In: *Techniques in the Behavioral and Neural Sciences*, Volume 15, Part 1e (ed. T. Steckler, N.H. Kalin, and J.M.H.M. Reul), 329–340. Elsevier.

16 Taskar KS, Mariappan TT, Kurawattimath V, et al. Unmasking the role of uptake transporters for digoxin uptake across the barriers of the central nervous system in rat. *Journal of Central Nervous System Disease* 2017; 9:1179573517693596. https://doi.org/10.1177/1179573517693596. 28469522; PMC5392048.

17 Parkinson, F.E., Friesen, J., Krizanac-Bengez, L., and Janigro, D. (2003). Use of a three-dimensional *in vitro* model of the rat blood-brain barrier to assay nucleoside efflux from the brain. *Brain Research* 980 (2): 233–241.

18 Gardier, A.M. (2013). Antidepressant activity: contribution of brain microdialysis in knock-out mice to the understanding of BDNF/5-HT transporter/5-HT autoreceptor interactions. *Frontiers in Pharmacology* 4: 98.

19 Smith DA, Di L, and Kerns EH. The effect of plasma protein binding on *in vivo* efficacy: misconceptions in drug discovery. *Nature Reviews. Drug Discovery* 2010; 9 (12): 929–39. https://doi.org/10.1038/nrd3287. 21119731.

20 Liu, X., Van Natta, K., Yeo, H. et al. (2009). Unbound drug concentration in brain homogenate and cerebral spinal fluid at steady state as a surrogate for unbound concentration in brain interstitial fluid. *Drug Metabolism and Disposition* 37: 787–793.

21 Shen, D.D., Artru, A.A., and Adkison, K.K. (2004). Principles and applicability of CSF sampling for the assessment of CNS drug delivery and pharmacodynamics. *Advanced Drug Delivery Reviews* 56: 1825–1857.

10.2 Application of Microdialysis in Pharmacokinetic Research

Furong Jiao, Xuan Dong, Cheng Tang

10.2.1 Introduction

10.2.1.1 Background

The concept of " dialysis" was first proposed by Scottish chemist Thomas Graham back in 1854. In the 1970s, the term "microdialysis" was created with a combination of perfusion technology and dialysis [1, 2]. MD has become an international cutting-edge sampling technique to evaluate the concentration changes of endogenous and exogenous substances in the body, and study metabolic, biochemical, and pharmacological events in tissues. To date, MD has been used as a tool for drug research and development in more than 18,000 articles. This chapter introduces the principles, and properties of MD technology and its application in PK research.

10.2.1.2 Introduction of Microdialysis Technology

Blood sampling and tissue homogenate have been commonly used in *in vivo* PK studies. However, this sampling method cannot directly get the free drug concentrations that exert the pharmacological effect in the extracellular fluid, cannot sample without any loss of biological fluids, and cannot dynamically capture transient drug changes *in vivo*. In contrast, MD technology can sample and quantify free endogenous and exogenous target molecules in the extracellular fluid of almost every specific tissue in animals and humans in real time.

A microprobe with a dialysis membrane, continuously perfused with artificial extracellular fluid, is implanted into the target tissue of conscious animals. Due to the concentration gradient on both sides of the dialysis membrane, the free drug enters the probe through the dialysis membrane and is continuously brought out by the perfusion fluid. The accurate concentration of the target substance can then be monitored (Figure 10.18) [3, 4, and 22].

MD technology can be used to detect the whole ADME process of drugs in *in vivo* PK studies. Furthermore, biomarkers can be monitored simultaneously to assess the PK/PD relationship *in vivo* [5–8].

10.2.1.3 Microdialysis System Device

MD system device (Figure 10.19) is mainly composed of a micro-pump, MD probe, collector, connecting tube, and supporting equipment. MD probe is the core of this technology. Until 2024, the commercially available probes are concentric circle probes, linear probes, and ring probes, among which concentric probes are primarily used in the brain and blood; linear probes are mainly used in peripheral tissues, such as skin, eyes, and muscles (Figure 10.20). The most widely used device is the concentric circular probe, whose structure is similar to a capillary. The probe comprises an inlet tube, outlet tube, shaft, and tip vessel implanted into the tissue. The vessel, with a semipermeable (white) tip, is where the free drug in the interstitial space and the probe liquid are exchanged (Figure 10.18).

10.2.1.4 Microdialysis Application

So far, MD technology has been used not only in laboratories but also extensively in preclinical and clinical research to sample from various tissues, with the encouragement of regulatory authorities. For example, in bioequivalence (BE) assessment, MD technology is used to directly sample from target tissues (Figure 10.21) [9].

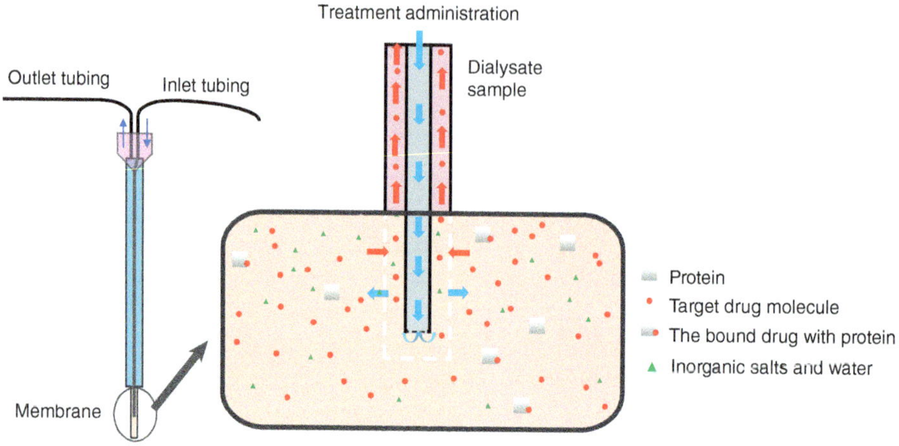

Figure 10.18 Schematic diagram of MD technical principle.

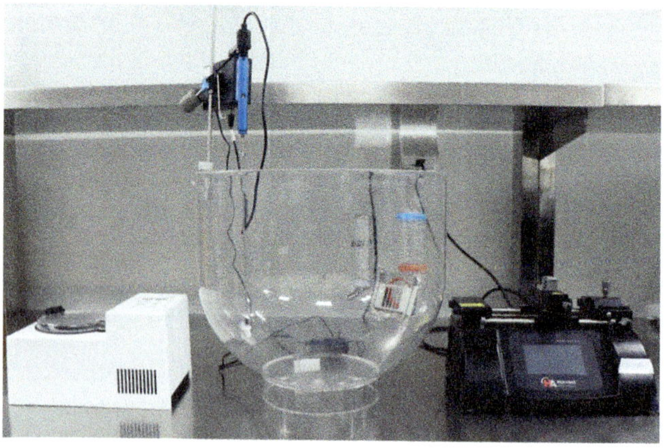

Figure 10.19 Microdialysis device.

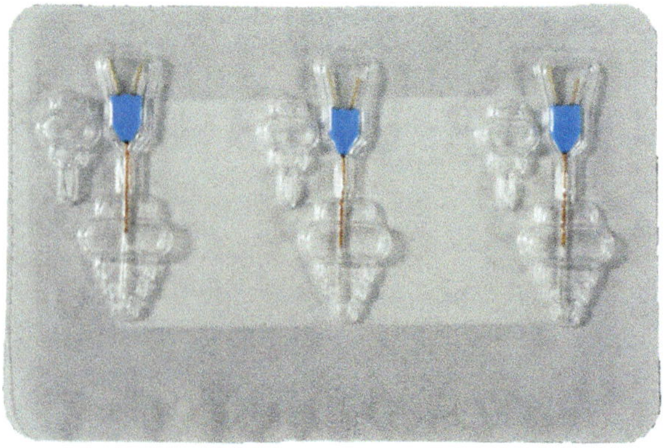

Figure 10.20 Mircrodialysis probes.

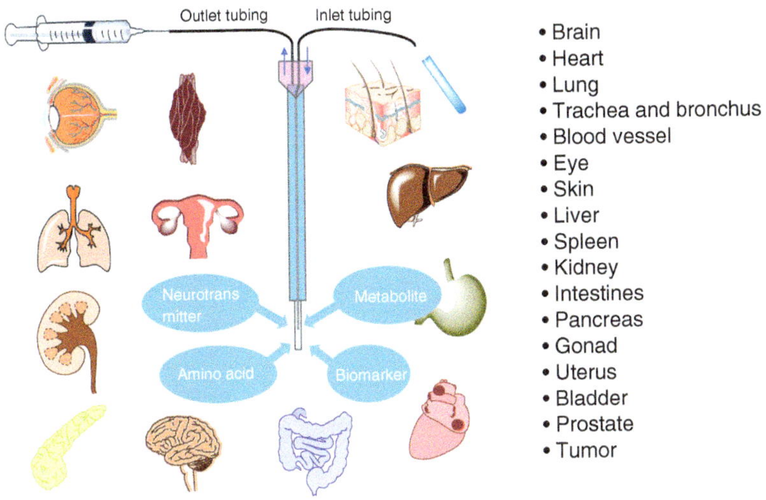

Figure 10.21 Application diagram of MD technology.

WuXi AppTec DMPK has established the ability of MD sampling from the brain, blood vessels, skin, and kidneys, among which the brain and skin MD sampling is performed in conscious animals, while the other tissues are sampled in anesthetized animals.

10.2.2 Microdialysis Case Sharing

10.2.2.1 Brain Microdialysis

Empirically, the drug concentration in CSF is commonly used as the surrogate for the free drug concentration in the brain. However, it's not always accurate, especially for compounds that are transporter substrates. The application of MD technology in CNS drug research, which can simultaneously monitor free drug concentrations in the brain and blood from the same animal, is of great value in elucidating the permeability of the compound into the brain.

As shown in Figure 10.22, the free drug concentrations determined via the MD method match well with the data coming from the serial bleeding method (after correction with plasma fraction unbound obtained from the *in vitro* plasma protein binding assay). The combination of brain MD and intravenous cannulation (blood) MD enables the data sampling from the same animal to get free drug concentrations in the brain and plasma, unbound AUC, and $K_{p,\,uu}$ (Figures 10.23–10.25) [10–12]. Moreover, biomarkers, such as neurotransmitters, can be monitored simultaneously to assess the PK/PD relationship.

10.2.2.2 Skin Microdialysis

PK evaluation of antimicrobial agents is primarily based on the measurement of total plasma concentration, which is often inappropriate and misleading, as the action site of antimicrobial agents is generally on the skin surface or under the skin. Skin MD is to obtain the free drug concentration in the intradermal site (as shown in Figure 10.26), which reflects the drug concentration at the site of action to guide the adjustment of drug dose and dosage form [14–16]. Recent findings from clinical skin MD studies indicate that skin MD is considered the gold standard for improving antimicrobial therapy and dose adjustment.

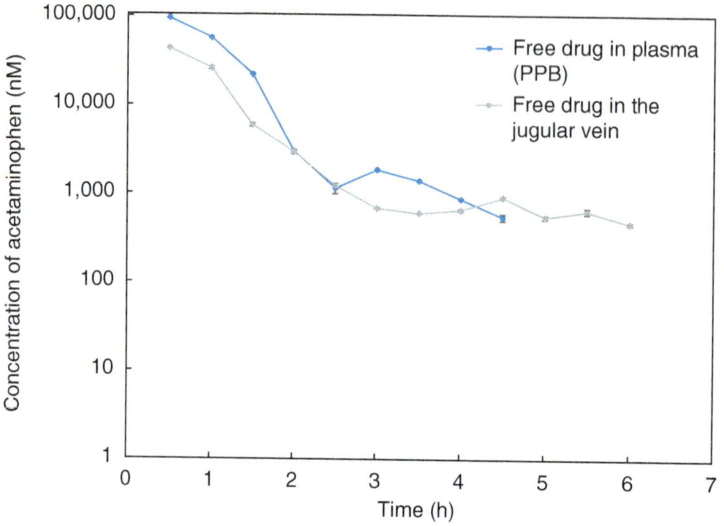

Figure 10.22 Comparison of free drug concentration in plasma obtained via the serial bleeding method (blue line, corrected by $f_{p,u}$ obtained from *in vitro* plasma protein binding assay) and the MD method (gray line, obtained directly by MD probe in the jugular vein). Acetaminophen was purchased from Sigma (St. Louis, MO, USA).

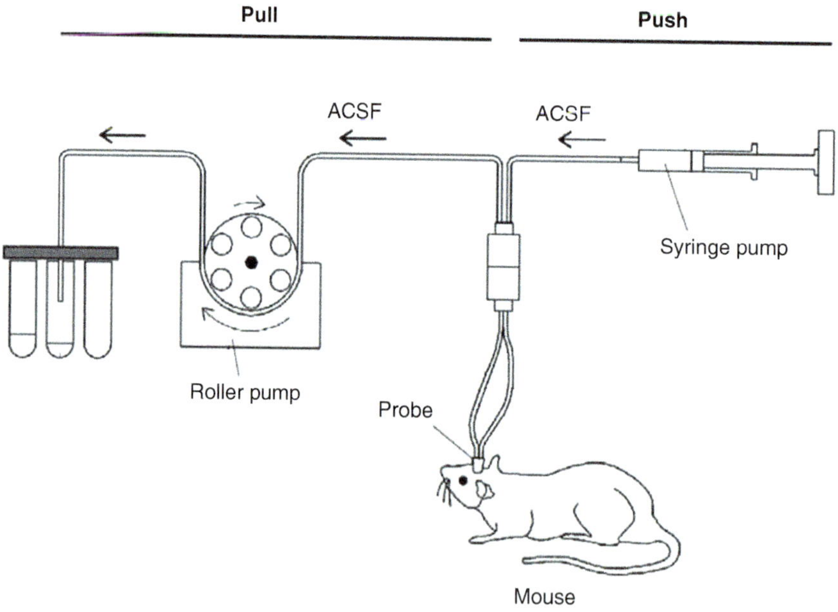

Figure 10.23 Schematic of brain MD. *Source:* [12] / ELSEVIER.

In a verification experiment shown in Figure 10.27, both skin MD and vascular MD were used to obtain the free drug concentrations in plasma and dermis. Combined with the reported minimum inhibitory concentration (MIC), the free concentration in the dermis and plasma will help to evaluate the onset time and effective interval of bacteria inhibition [18].

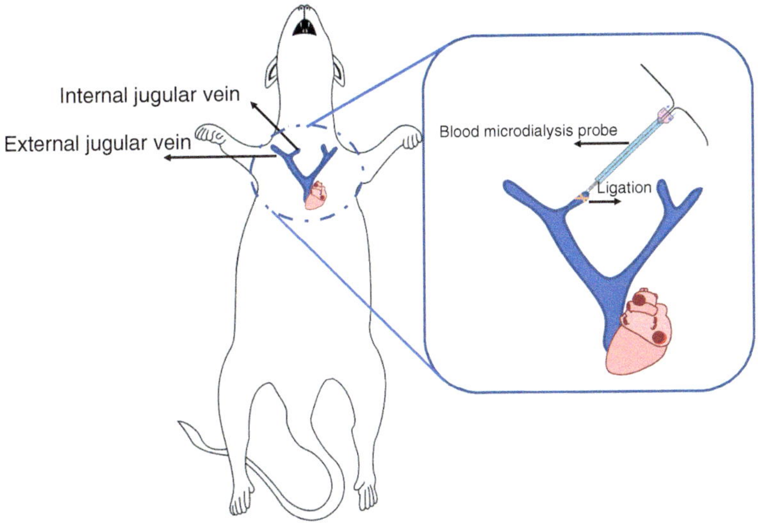

Figure 10.24 Schematic diagram of venous cannula (blood) MD. *Source:* [13] / American Chemical Society.

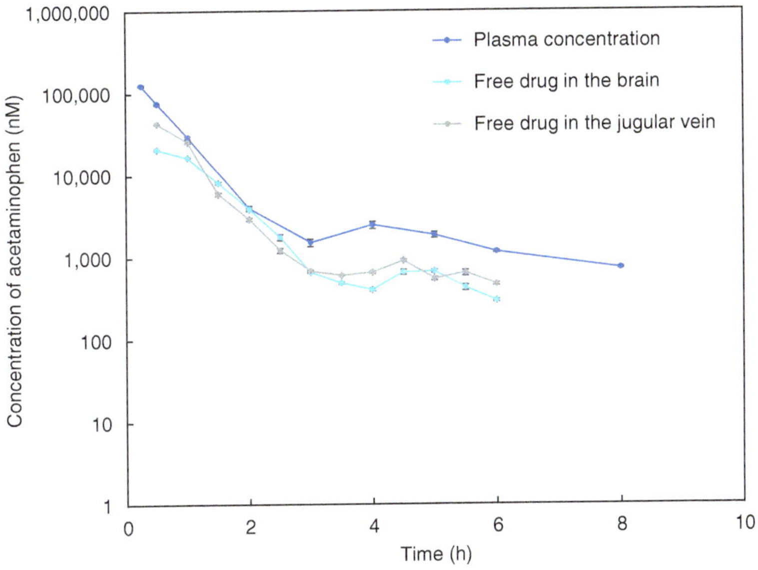

Figure 10.25 Brain MD in combination with venous cannula (blood) MD (dark blue line: total drug concentrations in plasma collected by serial bleeding; light blue line: free-drug concentrations in brain obtained directly by MD probe in the brain; gray line: free-drug concentrations in blood obtained directly by MD probe in the jugular vein). Acetaminophen was purchased from Sigma (St. Louis, MO, USA).

10.2.2.3 Renal Microdialysis

Renal excretion is the primary elimination route for many drugs and drug's excretion is closely related to the efficacy, duration of drug action, and side effects [19–21]. When using kidney homogenization to detect the free drug concentrations, data are usually incorrect because of the drug in urine that remains in the tissue. By contrast, renal MD technology can directly test the drug concentration in the renal cortex, which is more straightforward and accurate (Figure 10.28).

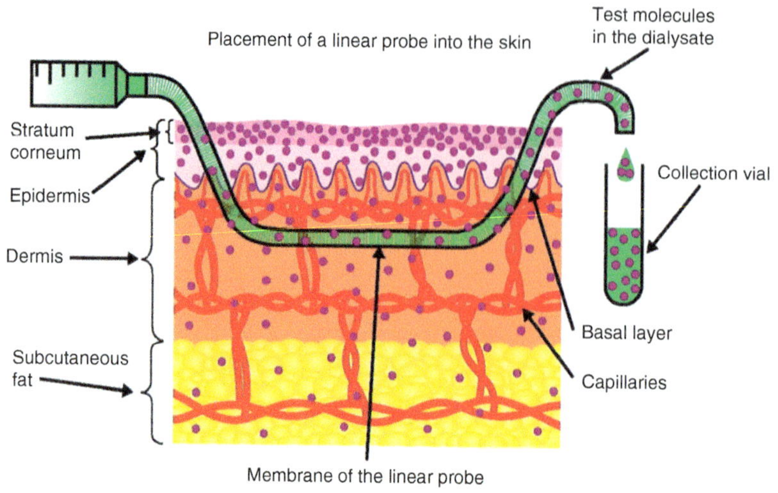

Figure 10.26 Schematic of skin MD. *Source:* [17] / Helen Rea.

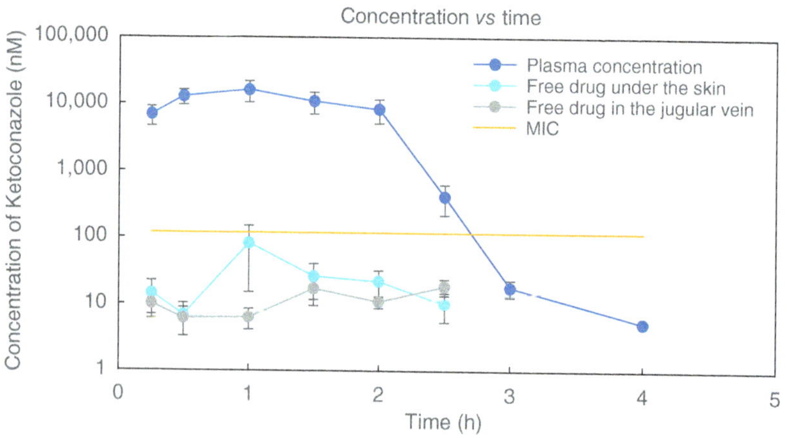

Figure 10.27 Skin MD and *in vitro* bacteriostasis test results (dark blue line: total drug concentrations in plasma collected by serial bleeding; light blue line: free-drug concentrations in skin obtained directly by MD probe in the skin; gray line: free-drug concentrations in blood obtained directly by MD probe in the jugular vein; yellow line: MIC of the drug obtained by *in vitro* bacteriostasis test).

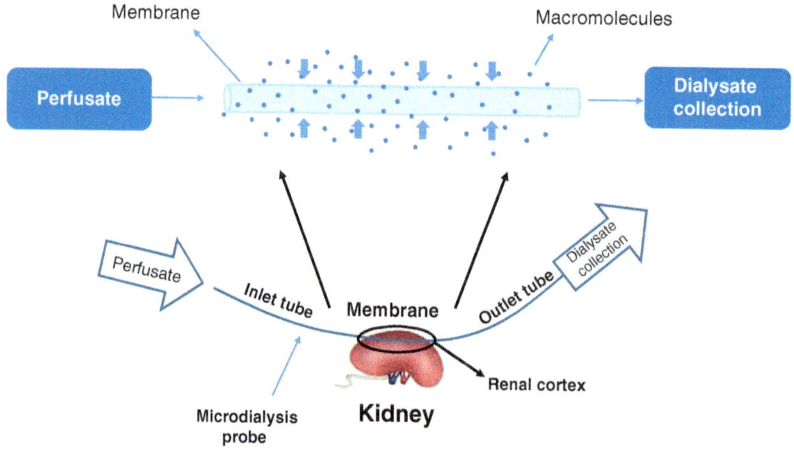

Figure 10.28 Schematic of renal MD.

10.2.3 Summary

The advantage of MD lies in the real-time measurement of free substance concentrations, which can be provoked by spontaneous events or by the topical or systemic administration of provoking agents. However, different experimental strategies should be developed for various target analytes the traditional MD pore size of the membrane is 20 kD, and the test substance less than 3 kD can quickly diffuse. However, for " large" molecules (i.e., drugs with molecular weights greater than 3 kD), although membrane pore sizes with molecular cut-off >100 kD or greater can be used, the technical difficulty increases with the increase in membrane pore size and the lipophilicity of molecules [12]. Therefore, different experimental strategies should be developed to overcome the difficulties caused by factors such as tube adsorption and inappropriate membrane pore size and to improve the accuracy and reliability of the analyte concentration in target tissues.

References

1. Ungerstedt, U. and Pycock, C.H. (1974). Functional correlates of dopamine neurotransmission. *Bulletin der Schweizerischen Akademie der Medizinischen Wissenschaften* 30 (1–3): 44–55.
2. Ungerstedt, U. (1991). Microdialysis—principles and applications for studies in animals and man. *Journal of Internal Medicine* 230 (4): 365–373.
3. Müller, M. (2013). *Microdialysis in Drug Development*. New York: Springer.
4. Chaurasia CS, Müller M, Bashaw ED, et al. (2007). AAPS-FDA workshop white paper: microdialysis principles, application and regulatory perspectives. pp. 1014–1025.
5. Hocht, C., Opezzo, J.A., and Taira, C.A. (2004). Microdialysis in drug discovery. *Current Drug Discovery Technologies* 1 (4): 269–285.
6. Wang, X. and Stenken, J.A. (2006). Microdialysis sampling membrane performance during *in vitro* macromolecule collection. *Analytical Chemistry* 78 (17): 6026–6034.
7. Hocht, C., Opezzo, J.A., Bramuglia, G.F., and Taira, C.A. (2006). Application of microdialysis in clinical pharmacology. *Current Clinical Pharmacology* 1 (2): 163–183.
8. Lionberger, R.A. (2008). FDA critical path initiatives: opportunities for generic drug development. *The AAPS journal.* 10: 103–109.
9. Fridén, M., Winiwarter, S., Jerndal, G. et al. (2009). Structure– brain exposure relationships in rat and human using a novel data set of unbound drug concentrations in brain interstitial and cerebrospinal fluids. *Journal of Medicinal Chemistry* 52 (20): 6233–6243.
10. Di, L. and Kerns, E.H. (2016). Chapter 10 - Blood-brain barrier. In: Drug-Like Properties, Seconde (ed. L.Di and E.H. Kerns), 141–159. Academic Press. ISBN: 9780128010761 https://doi.org/10.1016/B978-0-12-801076-1.00010-1.
11. Stevens, J., Suidgeest, E., van der Graaf, P.H. et al. (2009). A new minimal-stress freely-moving rat model for preclinical studies on intranasal administration of CNS drugs. *Pharmaceutical Research* 26: 1911–1917.
12. Jadhav, S.B., Khaowroongrueng, V., and Derendorf, H. (2016). Microdialysis of large molecules. *Journal of Pharmaceutical Sciences* 105 (11): 3233–3242.
13. Hou, Y., Bai, J., Zhang, Y. et al. (2022). Dynamic continuous blood extraction from rat heart via noninvasive microdialysis technique. *JoVE (Journal of Visualized Experiments).* 187: e64531.
14. Holmgaard, R., Benfeldt, E., Bangsgaard, N. et al. (2011). Probe depth matters in dermal microdialysis sampling of benzoic acid after topical application: an ex vivo study in human skin. *Skin Pharmacology and Physiology* 25 (1): 9–16.

15 Baumann, K., Falkencrone, S., Knudsen, N.P. et al. (2020). The skin reservoir model: a tool for evaluating microdialysis sampling of large biomarkers from human skin. *Acta Dermato-Venereologica* 100 (1): 1–8.

16 EMEA (2000). Points to consider on pharmacokinetics and pharmacodynamics in the development of medicinal antibacterial products. (www.emea.eu.int/pdfs/human/ewp/265599en.pdf)

17 Rea, H. and Kirby, B. (2019). A review of cutaneous microdialysis of inflammatory dermatoses. *Acta Dermato-Venereologica* 99 (11): 945–952.

18 Sun, N., Xie, Y., Sheng, C. et al. (2013). *In vivo* pharmacokinetics and *in vitro* antifungal activity of iodiconazole, a new triazole, determined by microdialysis sampling. *International Journal of Antimicrobial Agents* 41 (3): 229–235.

19 Abdel-Rahman, E.M., Abadir, P.M., and Siragy, H.M. (2008). Regulation of renal 12 (S)-hydroxyeicosatetraenoic acid in diabetes by angiotensin AT1 and AT2 receptors. *American Journal of Physiology - Regulatory, Integrative and Comparative Physiology* 295 (5): R1473–R1478.

20 Awad, A.S., Webb, R.L., Carey, R.M., and Siragy, H.M. (2005). Increased renal production of angiotensin II and thromboxane B2 in conscious diabetic rats. *American Journal of Hypertension* 18 (4): 544–548.

21 Alanazi, W., Fakhruddin, S., and Jackson, K.E. (2015). Microdialysis sampling of renal interstitial fluid in acute studies. *International Journal of Biology.* 8 (1): 69–79.

22 Pierce, C.F., Kwasnicki, A., Lakka, S.S., and Engelhard, H.H. (2021). Cerebral microdialysis as a tool for assessing the delivery of chemotherapy in brain tumor patients. *World Neurosurgery* 145: 187–196.

10.3 Central Nervous System Drug Development Techniques: Exploration and Application in Large Animal Pharmacokinetics

Liping Tang, Jie Li, Zhihai Li, Shoutao Liu, Jacob Zhi Chen

10.3.1 Introduction

CNS drug discovery and development is a highly specialized field that presents unique challenges and characteristics. One of the key factors contributing to its distinctiveness is the presence of the BBB. This barrier poses a significant hurdle in delivering drugs to target sites within the brain, requiring innovative strategies to overcome its limitations. Furthermore, the irreversibility of nerve cell damage in the CNS adds another layer of complexity. Unlike other tissues in the body, nerve cells have limited regenerative capacity, making it crucial to develop therapies that can effectively halt or slow down the progression of neurological disorders [1].

To address these complexities, WuXi AppTec DMPK has successfully enabled steady serial sampling of CSF, a valuable source of information for assessing drug concentrations and efficacy within the CNS. Our consistent study results, using both contralateral ventricle and lumbar puncture administration methods, bypass the BBB and have demonstrated the reliability and effectiveness of delivering test articles (TAs) into the CSF.

10.3.2 Application of Cisterna Magna Cannulation (CMC) Animal Models in IV Administration

Collecting CSF samples from the same animal at multiple time points can be achieved through puncture or surgical intubation [2]. Puncture requires anesthetizing animals during sampling, and obtaining serially multiple CSF samples is difficult due to the risk of blood contamination, making it suitable for low frequency. In contrast, surgical CMC (see Figure 10.29) has the advantages of not requiring anesthesia during sampling, no blood contamination with continuous multiple sampling, and the ability to collect samples serially.

Three beagle dogs, previously subjected to surgical implantation of a CMC Port, were each administered 25 mg of lidocaine intravenously. Blood and CSF samples were then serially collected at the same time points, respectively. The time concentration profiles of Lidocaine in both plasma and CSF,

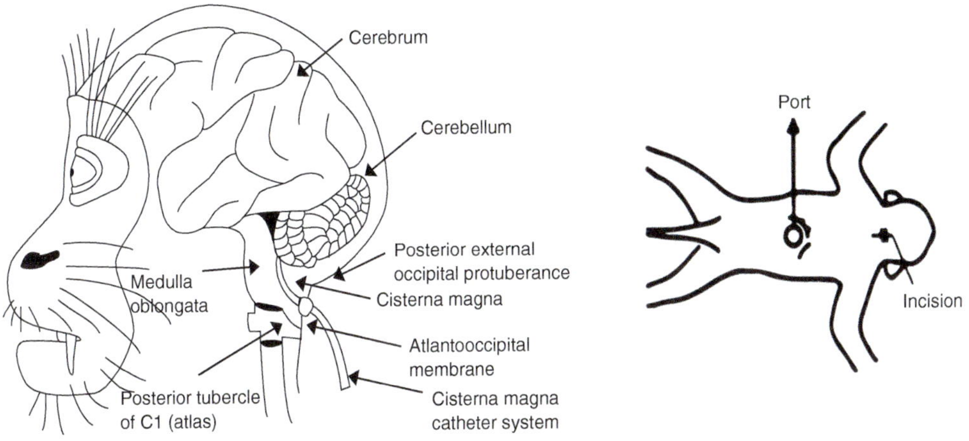

Figure 10.29 Position of the cistern magna cannulation catheter and port.

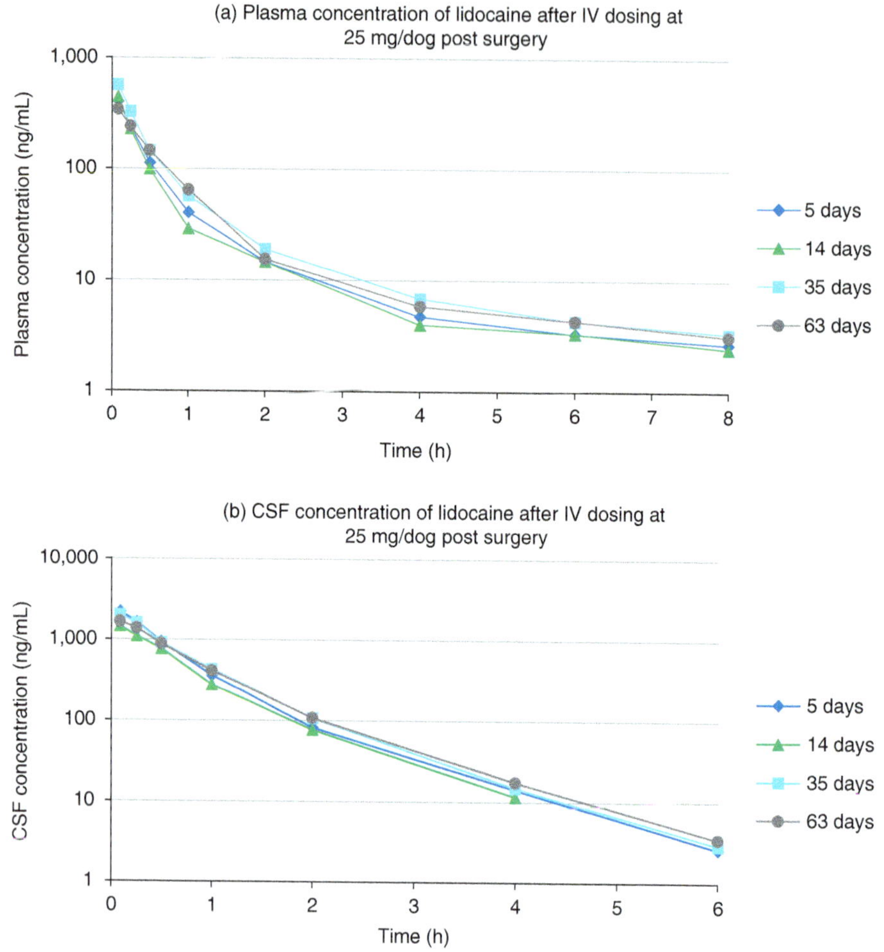

Figure 10.30 Lidocaine concentrations in plasma (a) and CSF (b) following a 25.0 mg/dog intravenous single bolus dose across different days.

taken on day 5, day 14, day 35, and day 63 (as shown in Figure 10.30) exhibited notably similar curves. These concentration profiles demonstrated a consistent pattern across the five days, with minimal variation observed. The close resemblance among the curves suggests that not only did the drug's distribution and elimination remain relatively constant but also that the effectiveness of CSF sampling, one of the major experimental factors, remained consistent throughout the different days. Furthermore, the analysis of PK parameters (as presented in Table 10.2) including AUC, C_{max} (or C_0), T_{max}, and $T_{1/2}$ consistently demonstrated that the exposure of Lidocaine in both plasma and CSF was highly comparable and maintained a consistent pattern across the five different days. The consistent and similar AUC ratios of 0.182 ± 0.0307 between CSF and plasma indicated a highly unlikely scenario of blood contamination entering the CSF on different days. These findings further support the notion that the effectiveness of CSF sampling remained consistent throughout the experiments.

10.3.3 Administration Methods That Bypass the Blood–Brain Barrier: Lateral Ventricle and Intrathecal Injection

The presence of the BBB and other mechanisms limits the entry of drugs into the CNS, with only about 5% of small molecular compounds able to penetrate it, while large molecules can hardly

Table 10.2 PK parameters of lidocaine following a 25.0 mg/dog intravenous single bolus dose across different days.

PK parameter		Days 5	10	14	35	63	Mean	SD
Plasma	C_0 (ng/mL)	2,573	2,255	1,683	2,308	1832	2,130	365
	$T_{1/2}$ (h)	0.818	0.653	0.738	0.767	0.823	0.760	0.0695
	Vd_{ss} (L/kg)	1.29	1.45	1.90	1.71	2.69	1.81	0.546
	Cl (mL/min/kg)	33.0	37.7	45.8	37.5	48.2	40.4	6.33
	AUC_{inf} (ng·h/mL)	1,405	1,262	1,053	1,446	1,330	1,299	155
CSF	C_{max} (ng/mL)	406	488	445	568	340	449	85.8
	T_{max} (h)	0.0830	0.0830	0.0830	0.0830	0.0915	0.0847	0.00380
	$T_{1/2}$ (h)	3.20	3.41	6.36	3.94	4.11	4.20	1.26
	AUC_{inf} (ng·h/mL)	215	221	212	296	249	239	35.3
[a]AUC ratio		0.144	0.164	0.179	0.202	0.222	0.182	0.0307

[a] CSF AUC/plasma AUC.

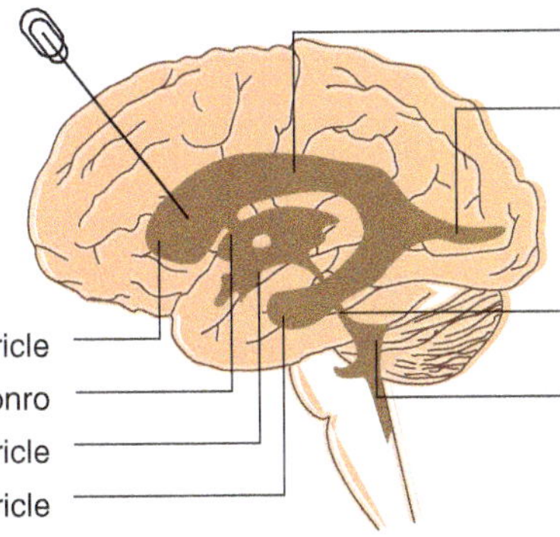

Figure 10.31 Site of lateral ventricle injection.

permeate. To enhance drug exposure to the CNS and mitigate systemic adverse effects from high doses of intravenous or oral medication, TAs can be directly administered into the CSF through intra-cerebroventricular injection (ICV), which includes lateral ventricle and IT. This allows for the direct distribution of drugs into brain tissue and the CNS through CSF circulation, achieving the target concentration in the CNS with a minimal amount of drug administration. [3]

10.3.3.1 Intra-cerebroventricular Injection (ICV)

The injection site, shown in Figure 10.31, is in the anterior corner of the lateral ventricle. After the puncture, CSF presents in the needle, indicating that the needle has reached the lateral ventricle, and the injection can proceed.

Three cynomolgus monkeys, each of which had previously undergone surgical implantation of a CMC Port, were injected into the lateral ventricle with 0.5 mg of dexamethasone sodium phosphate each. Blood and CSF samples were then serially collected at the same time points, respectively (Figure 10.32).

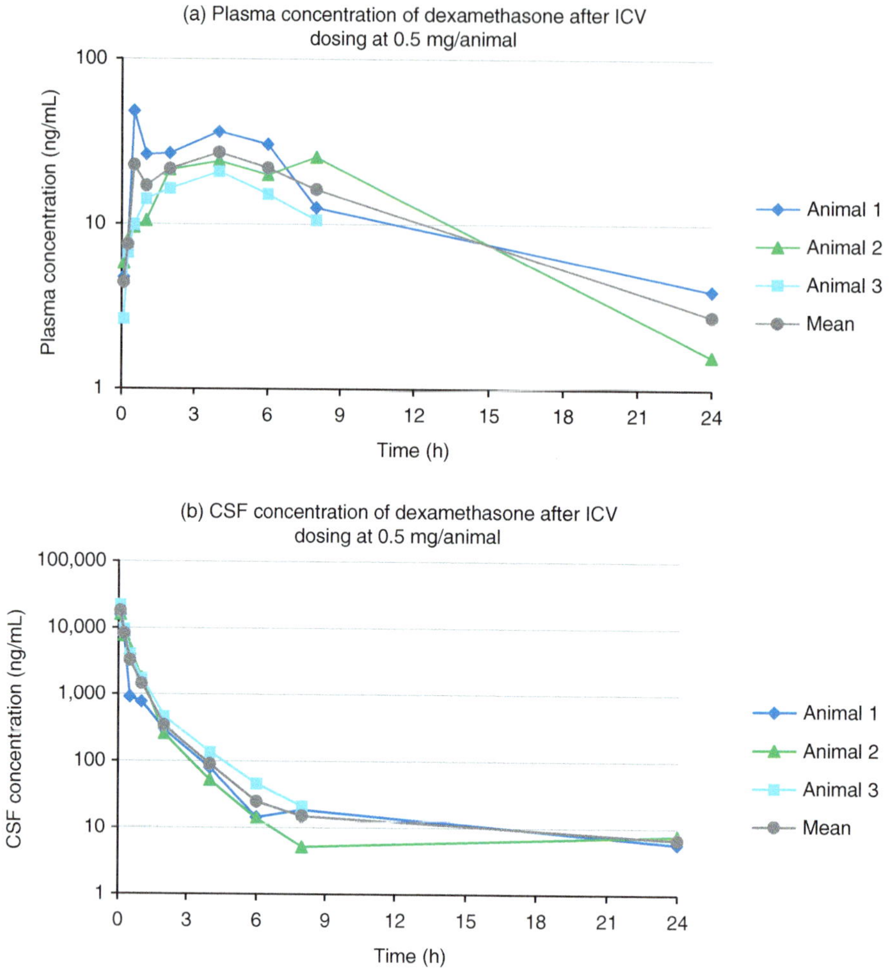

Figure 10.32 Drug concentrations in plasma (a) and CSF (b) after ICV injection.

Based on the experimental results, the average $C_{\max}$ and AUC of dexamethasone in CSF were 76,767 and 68,706, while the average $C_{\max}$ and AUC of dexamethasone in plasma were only 31.5 and 76,767, respectively. The average AUC ratio between CSF and plasma was 334 (Table 10.3). This indicates that the injection was successfully administered into the lateral ventricle. If the injection had not been administered into the lateral ventricle, the plasma exposure of dexamethasone would have been higher than that in CSF due to direct absorption into the bloodstream.

10.3.3.2 Intrathecal Injection (IT)

An IT (a lumbar puncture) is administered at the injection site shown in Figure 10.33, which is typically located between the third and fourth, or between the fourth and fifth lumbar vertebrae.

Table 10.3 PK parameters: dexamethasone ICV dosing at 0.5 mg/animal.

PK Parameters		Animal 1	Animal 2	Animal 3	Mean	SD
Plasma	C_{max} (ng/mL)	48.3	25.4	20.8	31.5	14.7
	T_{max} (h)	0.500	8.00	4.00	4.17	3.75
	$T_{1/2}$(h)	6.57	ND	ND	ND	ND
	$AUC_{0\text{-}inf}$ (ng·h/mL)	379	ND	ND	ND	ND
CSF	C_{max} (ng/mL)	74,400	92,200	63,700	76,767	14,397
	T_{max} (h)	0.0830	0.0830	0.250	0.139	0.0964
	$T_{1/2}$(h)	4.38	4.23	4.58	4.40	0.178
	$AUC_{0\text{-}inf}$ (ng·h/mL)	70,438	64,256	71,424	68,706	3,885
[a]AUC ratio		206	216	581	334	213

[a] CSF AUC/plasma AUC.
ND, not determined (parameters not determined due to inadequately defined terminal elimination phase).

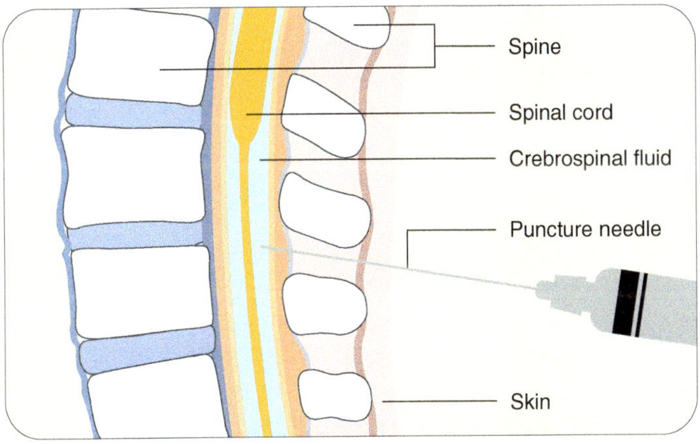

Figure 10.33 Sites of intrathecal injection.

Following intrathecal administration of a TA at 10 mg per cyno, the average C_{max} and AUC of the TA in plasma were 6,000 ng/mL and 26,000 ng·h/mL, while the C_{max} and AUC of TA1 in CSF were 900,000 ng/mL and 1,550,000 ng·h/mL. The AUC ratio between CSF and plasma was 60.0. These results indicate that the injection was successful and intrathecal administration significantly increased the exposure of the drug in the CNS, while the exposure of the drug in the bloodstream was significantly reduced.

The comparable drug $T_{1/2}$ observed in both CSF and plasma following lateral ventricle or intrathecal administration indicates that the drug, after successful injection into the lateral ventricle or the cerebellar bulbar cisterna, is being eliminated from both compartments at a similar rate. This finding suggests a balanced distribution and elimination between CSF and plasma, highlighting the effective distribution of the drug throughout the CNS.

References

1 Xiong, B., Wang, Y., Chen, Y. et al. (2021). Strategies for Structural Modification of Small Molecules to Improve Blood–Brain BarrierPenetration: A Recent Perspective. *Journal of Medicinal Chemistry* 64 (18): 13152–13173.

2 Schematic of rhesus monkey head, posterior view of implant placement, reprinted from David B. Gilberto's paper.

3 Hongxing, L., Xiaoxiao, P., Kai, Z. et al. (2020). Comparison of the efficacy of intrathecal and intravenous dexamethasone injection in aseptic meningitis after microvascular decompression [J]. *Chinese Brain Diseases and Journal of Rehabilitation* 10 (05).

Section III

Research Strategies, Methods, and Applications in Pharmacokinetics and Metabolism

11

In Vitro ADME

11.1 Rapid Determination of Lipophilicity: Establishment and Application of Reversed-Phase High-Performance Liquid Chromatography (RP-HPLC)

Liwei Cheng, Lijin Zheng, Quanli Feng, Cheng Tang

11.1.1 Introduction

Lipophilicity, also known as lipid solubility, refers to the ability of a compound to dissolve in nonpolar solvents. It is typically assessed by observing the partitioning behavior of the compound in a liquid–liquid or liquid–solid two-phase system. The lipophilicity of drugs plays a pivotal role in the fields of pharmacology, PK, and toxicology [1, 2]. Therefore, determining the lipophilicity of test articles is conducive to establishing the structure–activity relationships of drugs. Research into the absorption, distribution, metabolism, excretion, and toxicity (ADMET) properties of compounds is essential for understanding their druggability. The prediction of lipophilicity of test articles contributes to understanding the ADMET properties of these compounds, which assists researchers in identifying druggability issues in the early stage and reducing the failure rate of candidate drugs during the clinical study stage. Given the substantial amount of testing in the early stages of drug screening, there is an urgent need for rapid and accurate determination of lipophilicity to facilitate drug screening research.

This chapter provides an overview of a method for measuring the lipophilicity of compounds, with a focus on the reversed-phase high-performance liquid chromatography (RP-HPLC) that is specifically suitable for highly lipophilic compounds. The differences between the applications of RP-HPLC and the conventional shake-flask method are also discussed.

11.1.2 Characterizing Lipophilicity Based on Log P Values

Meyer and Overton pioneered the use of partition coefficients to explain the biological activity of active substances [3]. In 1964, Hansch et al. established the shake-flask method as an experimental and theoretical standard for determining lipophilicity using an octanol-water system [4]. The lipophilicity of compounds is recognized as a critical parameter in quantitative structure–activity relationship (QSAR) studies. Log P and D are two characteristic parameters for measuring lipophilicity. Log P refers to the logarithm of the partition coefficient between the organic phase and an aqueous phase when the compound exists entirely as non-ionized

Drug Metabolism and Pharmacokinetics: Frontiers, Strategies, and Applications, First Edition.
Edited by Liang Shen and WuXi AppTec DMPK.

molecules in the two phases at a specific pH. Log D refers to the logarithm of the distribution coefficient between the organic phase and an aqueous phase when the compound exists as ionized and non-ionized forms in the two phases at a specific pH [5]. Higher log P and log D_{pHx} values indicate stronger lipophilicity. The magnitude of log P is only related to the basic properties of the compound, such as molecular volume, dipole moment, and hydrogen bond acidity and basicity. Except for these fundamental properties, the magnitude of log D relates to the pH of the environment in which the compound is present. Given that the magnitude of log P provides a more direct indication of the overall trend of lipophilicity, this chapter will next focus primarily on methods for determining log P.

11.1.3 Methods for Determining Log P

Current methods for determining log P include computer simulation, the shake-flask method, and RP-HPLC. Table 11.1 shows a comparison of the three methods. Computer simulation is more cost-effective and faster than the other methods. However, its predictive ability depends on whether the selected software accounts for all substructure contributions of the compound. If not, the results may be less accurate than assay-determined values. The shake-flask method, as the gold standard in lipophilicity measurement, could directly measure log P. It offers the advantages of being accurate and minimal sample requirements. However, there are limitations such as being relatively time consuming, the need for high compound purity, being unsuitable for unstable compounds, and a limited measurement range of $-2 < \log P < 4$.

The proportion of highly lipophilic compounds used in drug development has risen with the steady increase in drug development pipelines [6]. Highly lipophilic compounds ($\log P > 5$) exhibit extremely low solubility in the aqueous phase, posing challenges to obtaining accurate values using instrumental measurements. Therefore, the lipophilicity of such compounds is usually expressed as a range of log P values. Compared with the shake-flask method, RP-HPLC provides several advantages, such as higher speed of measurement, milder operating conditions, small sample volume, low purity requirements, and a broader detection range (applicability may

Table 11.1 Comparison of different log P measurement methods.

Measurement Method	Prediction Range (Log Value)	Interferences	Speed of Measurement	Required Sample Volume	Reproducibility	Predictive Ability
Computer simulation	Broad	—	Rapid	—	—	★
Shake-flask method	−2–4	Not applicable to degradable compounds; demand for the high sample purity	Slow	Small	★★	★★★
Reversed-phase liquid chromatography	0–6	Mild; insensitive to impurities	Rapid	Small	★★★	★★

★ Indicates magnitude.

be expanded to compounds with log P > 6 under certain circumstances). Therefore, RP-HPLC has garnered considerable interest among researchers and is widely utilized in studies.

The "Partition Coefficient (n-octanol/water), High-Performance Liquid Chromatography (HPLC) Method" from the OECD Guidelines for the Testing of Chemicals (2004 edition) indicates the scope of application and requirements for determining log P using HPLC [7] and states that RP-HPLC has a distinct advantage in determining the log P of highly lipophilic compounds. Relevant studies [8] have also reported that this method is suitable for new modalities with high lipophilicity, such as PROTACs.

11.1.4 Establishing RP-HPLC Methods for Lipophilicity Determination

The basic steps for measuring the log P value of a compound using RP-HPLC are as follows:

1) Inject the selected reference compound into the chromatography system to obtain the retention time for the calculation of the capacity factor (k).
2) The logarithms of the capacity factors are calculated in advance and then the logarithms are plotted against their respective known log P values. This equation is referred to as the standard equation.
3) Based on the retention time obtained for the test compound under the same chromatographic conditions, calculate the capacity factor and substitute it into the standard equation to determine the log P value of the test articles.

Establishing the standard equation is the most important step in determining lipophilicity by RP-HPLC.

11.1.4.1 Reversed-Phase High-Performance Liquid Chromatography Lipophilicity Determination Method 1

RP-HPLC (hereinafter referred to as Method 1) was established to determine lipophilicity based on the OECD requirements. Six compounds of various lipophilicities were selected as the reference compounds (Table 11.2). Subsequently, chromatographic parameters such as the mobile phase, gradient, and column were optimized to determine the appropriate chromatographic conditions. Standard Equation (1) was obtained by plotting log k on the x-axis and log P on the y-axis, resulting in the linear correlation coefficient (R^2) of 0.97 (Figure 11.1), which meets the regulatory requirements. Triphenylamine had the highest log P value among the six reference compounds, with a retention time of 27.1 minute. Therefore, this method can be used to detect compounds with log P values below 6 within 30 minutes. The high efficiency of Method 1 is

Table 11.2 List of reference compounds.

Compounds	Log P
4-Acetylpyridine	0.5
Acetophenone	1.7
Chlorobenzene	2.8
Ethylbenzene	3.2
Phenanthrene	4.5
Triphenylamine	5.7

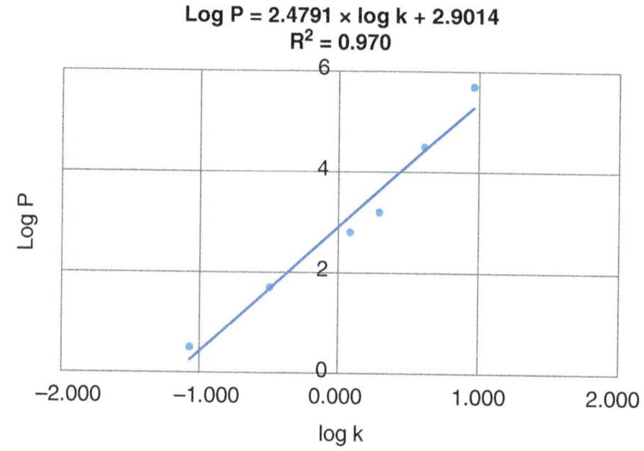

Figure 11.1 Linear fitting of the standard equation using Method 1.

beneficial for drug screening in the early stages. Researchers in this stage need to rapidly analyze a large number of compounds and obtain their log P values to rank the screened compounds.

$$\log \mathrm{P} = \mathrm{a} \times \log \mathrm{k} + \mathrm{b} \tag{11.1}$$

11.1.4.2 Reversed-Phase High-Performance Liquid Chromatography Lipophilicity Determination Method 2

In the late stages of drug development, more accurate log P values are needed to guide subsequent study designs. Therefore, a second method (Method 2) has been established based on Method 1 to satisfy the demand for higher accuracy for log P values.

It has been reported in previous research [9] that water or a buffer with the addition of modifiers such as methanol, acetonitrile, or tetrahydrofuran is used as the mobile phase for RP-HPLC during lipophilicity measurement. Among the various modifiers, methanol does not affect the formation of hydrogen bonds in water and can interact with the stationary phase of the column to form a monolayer, providing hydrogen bonding effects similar to n-octanol. Therefore, methanol can be considered an optimal modifier. It should be noted that organic modifiers can affect the pKa of ionic compounds and the retention time, affecting the accuracy of the log P values. To eliminate this interference from the organic modifiers, RP-HPLC lipophilicity determination Method 2 (hereinafter referred to as Method 2) was established. The same reference compounds as in Method 1 were used in Method 2. In Method 2, log k in Equation (11.1) was replaced by log k_W (capacity factor of the compound in the absence of organic modifiers) to obtain the standard equation (Equation 11.2). For each reference compound, an equation relating log k and methanol content (φ) was established under three mobile phase gradient conditions (Equation 11.3), with the intercept being the log k_W value. As shown in Figure 11.2, the R^2 value for Equation (11.2) is 0.996, indicating that Method 2 has better predictive ability than Method 1. Therefore, more accurate lipophilicity values could be obtained using Method 2 during the late stages of drug development.

$$\log \mathrm{P} = \mathrm{a} \times \log \mathrm{k_w} + \mathrm{b} \tag{11.2}$$

$$\log \mathrm{k} = S\varphi + \log \mathrm{k_w} \tag{11.3}$$

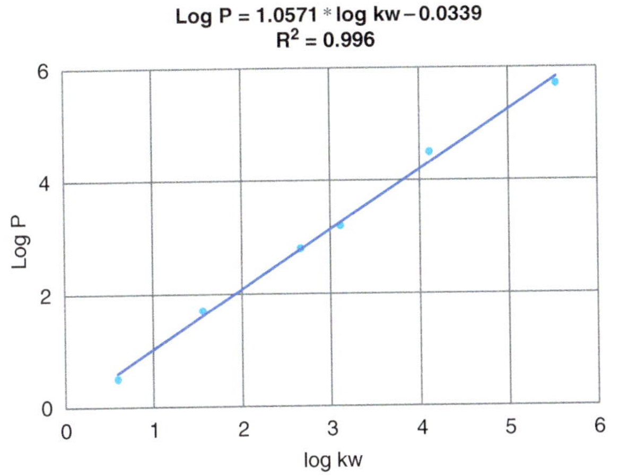

Figure 11.2 Linear fitting of the standard equation for Method 2.

11.1.4.3 Data Validation and Comparison of Methods 1 and 2

Twenty-one compounds with various lipophilicities were selected to measure log P values. For Method 1, the log P values of 85% of the compounds are consistent with those reported in the literature [7, 10], with differences within 0.5 log units. The three compounds with differences exceeding 0.5 log units were retested with Method 2, producing data consistent with the literature values. Therefore, Methods 1 and 2 are suitable for determining compounds' lipophilicity, with Method 2 providing higher accuracy.

The two established RP-HPLC methods offer advantages over the conventional shake-flask method in determining log P values. Method 1 is rapid and cost-effective, while Method 2 possesses higher predictive ability (Table 11.3).

11.1.5 Summary

Lipophilicity, as one of the most crucial physicochemical properties of compounds, is measured during early drug screening to obtain log P values. The classical shake-flask method for determining log P is limited by compound solubility and instrumental detection, affecting the analysis of highly lipophilic compounds. The RP-HPLC for measuring log P enables high-throughput and rapid

Table 11.3 Comparison of Methods 1 and 2 and the shake-flask method

Method	Standard Equation	Correlation Coefficient	Run Time Per Compound	Cost/ Speed	Application Scenario
Method 1	$\log P = a \times \log k + b$	0.970	Within 0.5 hours	Low/fast	Early drug screening of >30 compounds, presence of time constraints
Method 2	$\log P = a \times \log k_w + b$	0.996	2–2.5 hours	High/ slow	Drug screening in late stages of drug development, absence of time constraints
Shake-flask method	NA	NA	4 hours	High/ slow	Absence of time constraints

NA: not applicable.

analysis and can measure the log P of high lipophilicity, thereby effectively overcoming the limitations of the shake-flask method. The two RP-HPLC methods and the shake-flask method enable accurate and efficient log P determination for compounds at different stages of drug development based on project requirements. Method 1 can be selected for rapid log P value determination in the early stages of drug screening when several compounds require screening within a limited time frame to assist researchers in quick decision-making. Method 2, or the shake-flask method, can be adopted at a later stage of drug development, based on different project requirements.

References

1 Smith, D.A., Jones, B.C., and Walker, D.K. (1996). Design of drugs involving the concepts and theories of drug metabolism and pharmacokinetics. *Medicinal Research Reviews* 16 (3): 243–266.
2 Waring, M.J. (2010). Lipophilicity in drug discovery. *Expert Opinion on Drug Discovery* 5 (3): 235–248.
3 Leo, A., Hansch, C., and Elkins, D. (1971). Partition coefficients and their uses. *Chemical Reviews (ACS Publications)* 71: 525.
4 Gordon, L. (1980). Flynn, Substituent Constants for Correlation Analysis in Chemistry and Biology. *Journal of Pharmaceutical Sciences* 69 (9): 1109.
5 Avdeef, A. (2003). *Absorption and drug development: solubility, permeability, and charge state*, 42–66. John Wiley & Sons.
6 Leeson, P.D., Bento, A.P., Gaulton, A. et al. (2021). Target-Based Evaluation of "Drug-Like" Properties and Ligand Efficiencies. *Journal of Medicinal Chemistry* 64 (11): 7210–7230.
7 OECD. (2004) OECD Test No. 117: Partition Coefficient (n-octanol/water), HPLC Method.
8 Pike, A., Williamson, B., Harlfinger, S. et al. (2020). Optimising proteolysis-targeting chimeras (PROTACs) for oral drug delivery: a drug metabolism and pharmacokinetics perspective. *Drug Discovery Today* 25 (10): 1793–1800.
9 Giaginis, C. and Tsantili-Kakoulidou, A. (2008). Current state of the art in HPLC methodology for lipophilicity assessment of basic drugs. A review. *Journal of Liquid Chromatography and Related Technologies* 31: 79–96.
10 Bharate, S.S., Kumar, V., and Vishwakarma, R.A. (2016). Determining partition poefficient (Log P), distribution coefficient (Log D), and ionization constant (pKa) in early drug discovery. *Combinatorial Chemistry & High Throughput Screening* 19 (6): 461–469.

11.2 Plasma Protein Binding Measurement of High Protein-Binding Drugs: Principles and Advantages of the Flux Dialysis Method

Jie Wang, Ziqian Sun, Xiangling Wang, Genfu Chen

11.2.1 Introduction

With the rapid development of new modalities, there is an increasing number of molecules in drug chemistry design that surpass Lipinski's rule of five principles. For instance, PROTACs often exhibit common characteristics such as high molecular weight, strong lipophilicity, poor solubility, and significant nonspecific binding, which exhibit high PPB. As a result, accurately determining PPB is challenging [1]. This chapter introduces flux dialysis, which is a method suitable for determining PPB in compounds with high protein binding. We discuss the assay principle and mathematical model, present our validation data, and explain in detail the derivation of key mathematical formulas.

11.2.2 How to Measure Plasma Protein Binding?

PPB is one of the most controversial topics in the field of drug discovery and development. It is generally not recommended to change the PPB of a drug candidate by optimizing the structure, as there is no significant correlation between PPB and *in vivo* drug efficacy and PPB has a marginal effect on free drug concentrations [2–4]. A compound with high PPB does not necessarily mean that it is a good or bad compound. However, the fraction unbound (f_u) measured by PPB assays is a key parameter that needs to be accurately determined, because it could play a significant role in predicting DDI, estimating therapeutic indexes (TI), and exploiting PK/PD relationships [5].

There are several methods for determining the fraction unbound (f_u) of compounds in biological matrices, including equilibrium dialysis, ultracentrifugation, and ultrafiltration, with equilibrium dialysis being the most commonly used method. The f_u obtained by equilibrium dialysis is reliable if the system is in true equilibrium, the compound is not metabolized or degraded in the biological matrix, and the concentration of the free sample in the buffer meets the requirements of analytical sensitivity.

For highly protein-bound compounds, reaching equilibrium takes a long time. In some cases, it is not able to be achieved, making it difficult to accurately determine f_u. Due to the difficulty of accurately determining f_u, the lower detection limit of the PPB f_u is specified in the DDI guidance of the FDA [6] and EMA [7]. f_u is estimated as 0.01 when the plasma f_u is less than 0.01, which may lead to an overestimation of the DDI effect for highly protein-bound compounds. In recent years, several articles have been published for optimizing PPB determination to improve the accuracy of f_u estimation [5]. Among them, the flux dialysis method was developed based on the principle of equilibrium dialysis.

11.2.3 Principles and Methods of Flux Dialysis

Flux dialysis, which was first developed in the 1950s, can overcome some of the limitations of equilibrium dialysis to determine a low f_u. The basic method is to dialyze a plasma sample with the added compound against a blank plasma. With the known permeability coefficient of the compound in the dialysis membrane, the f_u of the compound in the matrix can be calculated using the initial flux rate of the compound penetrating the receiving chamber.

The basic principle is that the initial flux rate of a compound through a dialysis membrane is proportional to the product of the initial concentration, f_u, and the membrane permeability coefficient. Instead of determining the equilibrium concentrations at equilibrium, f_u can be calculated from the initial slope of the receiver/donor concentration ratio–time curve. This approach bypasses the requirement for equilibrium and has great application for the PPB determination of highly protein-bound compounds.

However, due to various reasons such as the requirement to determine membrane permeability, multiple sampling time points, complex study procedures, and the need for higher radio-purity compounds, this method has not been widely used in practice. With the popularization of high-throughput dialysis devices and the establishment and development of PK models, this method has been improved and has been used in many studies of WuXi AppTec DMPK.

Researchers from the AbbVie research team in the United States provided a detailed mathematical proof of the theoretical model of this novel flux dialysis method [8]. This chapter briefly introduces some key aspects of their findings.

11.2.4 Kinetic Modeling of the Flux Dialysis Methods

The theoretical basis of the method for calculating f_u is demonstrated below. The concentration variation of compounds in the donor and receiver compartments can be described by the two-compartment model (Figure 11.3).

When nonspecific binding and metabolism in the matrix are not taken into account, it can be deduced from the model that $f_{u,\text{ the donor}}$ is proportional to R_{slope} ($R = C_{receiver}/C_{donor}$, R_{slope} is the derivative of the concentration ratio concerning time, that is, dR/dt, at the initial moment, t = 0) and $f_{u,\text{ the donor}}$ can be calculated once the value of R_{slope} is obtained.

$$f_{u,\text{ donor}} = \frac{V_{receiver}}{P_{mem} \times A} \times R_{slope} \tag{11.4}$$

(refer to the end of the article for specific derivation)

However, since it is difficult to directly obtain an accurate R_{slope} at t = 0 through assays, the R values at different time points are used to plot the R-t curve according to Equation (11.5). Then, the initial slope of the curve can be calculated, and the R_{slope} value can be obtained.

$$R = R_{eq} \times \left(1 - e^{-k \times t}\right) \tag{11.5}$$

(refer to the end of the article for specific derivation)

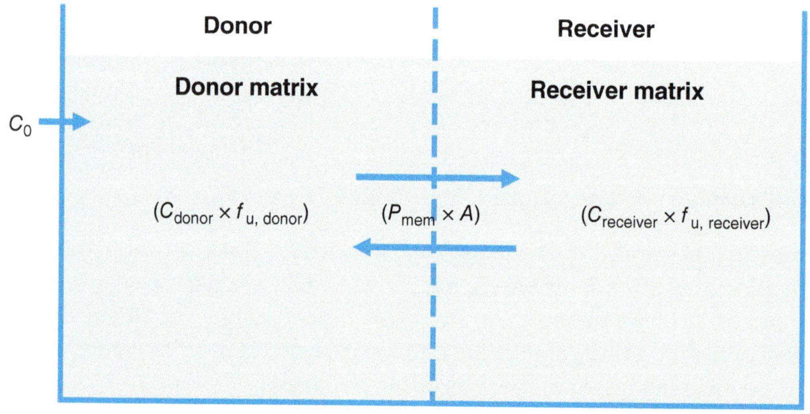

Figure 11.3 Kinetic model of dialysis. P_{mem}: permeability coefficient of the dialysis membrane; *A*: surface area of the membrane; *C*: concentration. *Source:* Adapted from reference [8].

After fitting the complete R–t curve based on the study data at multiple time points, the R_{slope} can be accurately calculated even if there are missing data at earlier time points.

When metabolism in the matrix cannot be ignored, it is further assumed that the rate of matrix degradation is the same on both sides and that the degradation process conforms to the first-order kinetics. The theoretical formula for R–t can be found by mathematical reasoning as Equation ((11.6)), which is independent of K_{loss}. Therefore, the flux method is also applicable to compounds that are unstable in the matrix.

$$R = \frac{f_{\text{u, donor}}}{f_{\text{u, receiver}}} \times \frac{\left(1 - e^{-k_{eq} \times t}\right)}{1 + \beta e^{-k_{eq} \times t}} \tag{11.6}$$

(the specific derivation process is omitted)

11.2.5 Assay Validation Results

Based on the theory of the flux dialysis method, the compound does not need to reach complete equilibrium in the dialysis device to determine its f_u accurately. We selected several commercial compounds for the experimental validation of the flux dialysis method—among which 18 compounds were tested in human plasma to obtain the fraction unbound in human plasma (f_{up}), and 9 compounds were tested in rat brain tissue homogenate to obtain the fraction unbound in brain homogenate (f_{ub}). The study results show a good correlation, with the coefficients of determination (R^2) all greater than 0.98 (Figure 11.4 and Figure 11.5). The receiver/donor ratio–time (R–t) curves of some representative compounds are presented in Figure 11.6.

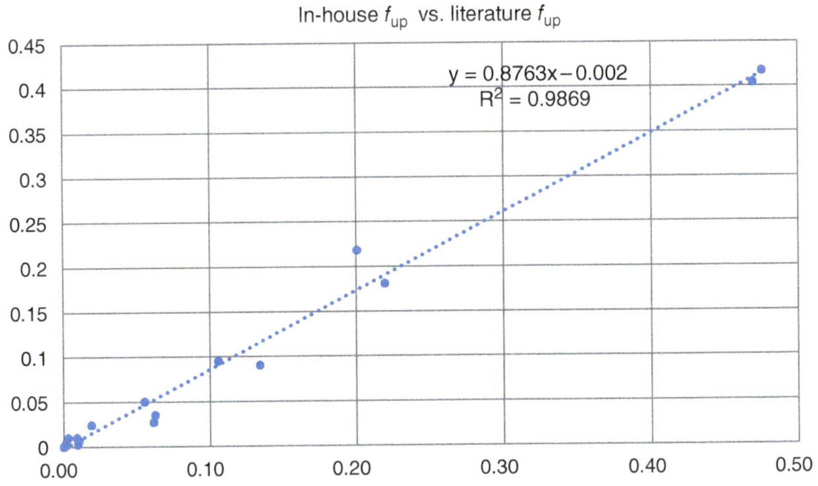

Figure 11.4 The fraction unbound in plasma (f_{up}) of 18 commercial drugs was determined by the flux dialysis method, and the results were compared with literature reports. The x-axis represents the in-house f_{up}, and the y-axis represents the literature f_{up}. The compounds for the validation studies include afatinib, crizotinib, ibrutinib, osimertinib, aspirin, enalapril, imipramine, indomethacin, itraconazole, lapatinib, quinidine sertraline, amiodarone, montelukast, mifepristone, rosiglitazone, ritonavir, and nicardipine.

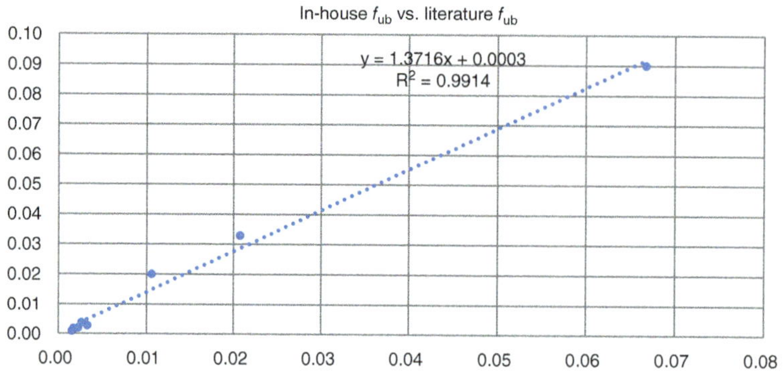

Figure 11.5 f_{ub} of nine commercial drugs in rat brain tissue homogenates as determined by the flux dialysis method compared to the literature reports. The x-axis represents the in-house f_{ub}, and the y-axis represents the Literature f_{ub}. The compounds for the validation studies comprise verapamil hydrochloride, propranolol, saquinavir, nicardipine, paclitaxel, fluoxetine, carbamazepine, chlorpromazine, and sertraline hydrochloride.

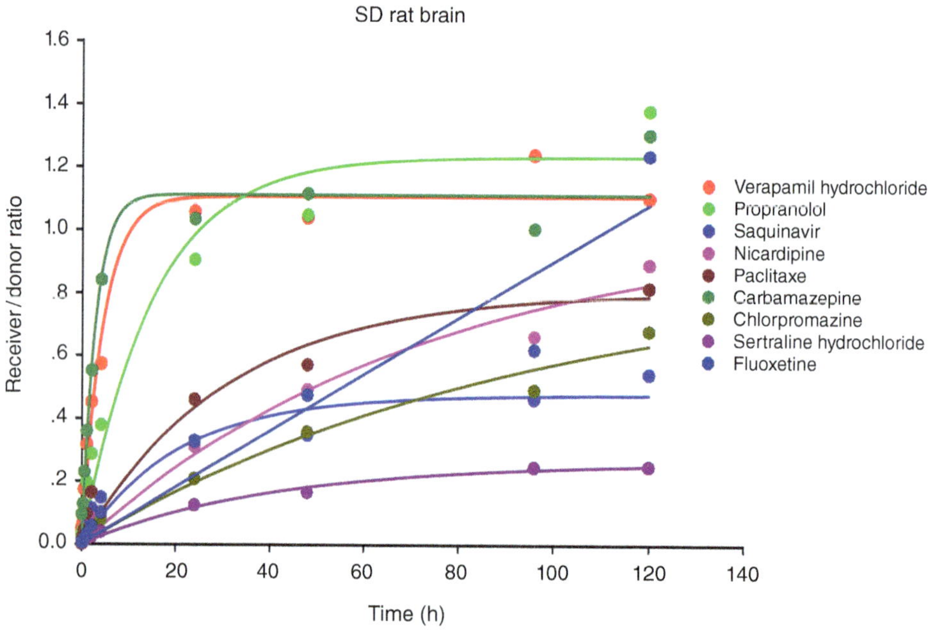

Figure 11.6 R–t curves of some representative compounds.

11.2.6 Advantages of the Flux Dialysis Method

We have established and validated flux dialysis based on the current literature. Compared with the conventional study design of equilibrium dialysis, flux dialysis offers the following advantages:

1) The accurate determination of f_u requires a short theoretical dialysis time. It is difficult for some compounds to reach equilibrium using the conventional equilibrium dialysis method.
2) This method is not affected by nonspecific binding and the stability in the matrix.
3) The sensitivity requirement for the sample analysis is not high.

This method can be used for protein binding determination of highly protein-binding compounds and to categorize and rank the binding rates of the same series of compounds by R–t curves.

11.2.7 Derivation Process of Several Formulas

Derivation process of equation 11.4:
Based on the above model, if nonspecific binding and matrix stability are ignored, the amount of drugs in the donor and receiver compartments can be, respectively, calculated as Equation (11.7) and Equation (11.8):

$$V_{\text{donor}} \times \frac{dC_{\text{donor}}}{dt} = \left(P_{\text{mem}} \times A\right)\left(f_{\text{u, reveiver}} \times C_{\text{reveiver}} - f_{\text{u, donor}} \times C_{\text{donor}}\right) \tag{11.7}$$

$$V_{\text{receiver}} \times \frac{dC_{\text{receiver}}}{dt} = \left(P_{\text{mem}} \times A\right)\left(f_{\text{u, donor}} \times C_{\text{donor}} - f_{\text{u, reveiver}} \times C_{\text{reveiver}}\right) \tag{11.8}$$

At time 0, the drug concentration in the donor compartment is considered to be approximately equal to the initial concentration Equation (11.9):

$$V_{\text{donor}} \times \frac{dC_{\text{donor}}}{dt} = 0;\ C_{\text{donor}} = \frac{X_0}{V_{\text{donor}}} \tag{11.9}$$

flux (the rate of flux of a compound penetrating the receiver end) serves as Equation (11.10):

$$flux = V_{\text{receiver}} \times \frac{dC_{\text{receiver}}}{dt} = f_{\text{u, donor}} \times C_{\text{donor}} \times P_{\text{mem}} \times A \tag{11.10}$$

When the system reaches the steady state (t→∞), the concentration of free drug on both the dosing chamber and the receiver side is the same, subsequently Equation (11.11):

$$f_{\text{u, receiver}} \times C_{\text{receiver}} = f_{\text{u, donor}} \times C_{\text{donor}} \tag{11.11}$$

Define $R = \frac{C_{\text{receiver}}}{C_{\text{donor}}}$, the R of equilibrium, that is, R_{eq} is Equation (11.12):

$$R_{\text{eq}} = \frac{f_{\text{u, donor}}}{f_{\text{u, receiver}}} \tag{11.12}$$

Since C_{donor} can be approximated as a constant when t = 0, it can be obtained by equations (11.10), (11.13) and (11.14):

$$\left(\frac{dR}{dt}\right)_{t=0} = \frac{f_{\text{u, donor}} \times P_{\text{mem}} \times A}{V_{\text{receiver}}} \tag{11.13}$$

Define

$$R_{\text{slope}} = \left(\frac{dR}{dt}\right)_{t=0} = \frac{f_{\text{u, donor}} \times P_{\text{mem}} \times A}{V_{\text{receiver}}} \tag{11.14}$$

It can be seen that $f_{u,\text{ the donor}}$ is proportional to R_{slope}. P, A, and V are constant parameters related to the experimental setup and can be obtained based on the study data of compounds with known f_u values. The R-value under the initial conditions can be obtained through the assays. The slope under the initial conditions can then be fitted to calculate f_u.

Derivation process of Equation 11.5:

The algebraic expression of the R(t) function needs to be obtained by solving the system of differential equations of equations (11.7) and (11.8). The original literature did not provide a detailed derivation process, but directly presented the result: the analytical expression of the R(t) function.

To make this process easier to understand, this chapter provides two approaches to solve:

1) By using Laplace transformation, the Laplace transforms of the concentration functions concerning time in the dosing chamber and the receiver chamber ($C_{\text{donor(t)}}$ and $C_{\text{receiver(t)}}$) are obtained as their respective Laplace domain functions F(s). After calculation, the expression of the original function can be obtained by inverse transformation (see formulas 11.15 and 11.16). Then the ratio of $C_{\text{receiver}}(t)$ and $C_{\text{donor}}(t)$ gives the analytic expression of the R(t) function.

$$C_{\text{donor}}(t) = \frac{X_0}{V_{\text{donor}}} \times \frac{1}{1+\beta}\left(1+\beta e^{-k_{eq} \times t}\right) \tag{11.15}$$

$$C_{\text{receiver}}(t) = \frac{X_0}{V_{\text{receiver}}} \times \frac{\beta}{1+\beta}\left(1-e^{-k_{eq} \times t}\right) \tag{11.16}$$

where $\beta = \dfrac{f_{\text{u, donor}}}{f_{\text{u, receiver}}} \times \dfrac{V_{\text{receiver}}}{V_{\text{donor}}}$

2) Arithmetic rules of differentiation Equation (11.17):

$$d\left(\frac{u}{v}\right) = \frac{v \times du - u \times dv}{v^2} \tag{11.17}$$

where dR/dt is the differentiation of the quotient of $C_{\text{receiver}}(t)$ and $C_{\text{donor}}(t)$ can adopt the above equation, while the differential equations of $C_{\text{receiver}}(t)$ and $C_{\text{donor}}(t)$ are Equations (11.8) and (11.7). The following equation can be obtained after arithmetic simplification Equation (11.18):

$$\frac{dR}{dt} = R_{\text{slope}} - kR \tag{11.18}$$

Further transforming both sides of the equation via the Laplace transformation method yields Equation (11.19):

$$\mathrm{L}\left(\frac{dR}{dt}\right) = L\left(R_{\text{slope}}\right) - L\left(kR\right) \tag{11.19}$$

Further solving this equation yields Equation (11.20):

$$R = \frac{R_{\text{slope}}}{k}\left(1-e^{-kt}\right)$$

which is equal to Equation (11.5):

$$R = R_{\text{eq}}\left(1-e^{-kt}\right) \tag{11.20}$$

References

1 Pike, A., Williamson, B., Harlfinger, S. et al. (2020). Optimising proteolysis-targeting chimeras (PROTACs) for oral drug delivery: a drug metabolism and pharmacokinetics perspective. *Drug Discovery Today* 25 (10): 1793–1800.

2 Smith, D.A., Di, L., and Kerns, E.H. (2010). The effect of plasma protein binding on *in vivo* efficacy: misconceptions in drug discovery. *Nature Reviews Drug Discovery* 9 (12): 929–939.

3 Liu, X., Wright, M., and Hop, C.E.C.A. (2014). Rational use of plasma protein and tissue binding data in drug design. *Journal of Medicinal Chemistry* 57 (20): 8238–8248.

4 Benet, L.Z. and Hoener, B.-A. (2002). Changes in plasma protein binding have little clinical relevance. *Clinical Pharmacology and Therapeutics (St. Louis)* 71 (3): 115–121.

5 Di, L. (2021). An update on the importance of plasma protein binding in drug discovery and development. *Expert Opinion on Drug Discovery* 16 (12): 1453–1465.

6 FDA. *In vitro* drug interaction studies - cytochrome p450 enzyme- and transporter-mediated drug interactions guidance for industry JANUARY 2020. Available from: https://www.fda.gov/regulatory-information/search-fda-guidance-documents/in-vitro-drug-interaction-studies-cytochrome-p450-enzyme-and-transporter-mediated-drug-interactions[cited on 2023 Jan 24].

7 European medicines agency (EMA). Guideline on the investigation of drug interactions. [cited on 2023 Jan 24]. Available from: https://www.ema.europa.eu/en/documents/scientific-guideline/guideline-investigation-drug-interactions-revision-1_en.pdf

8 Kalvass, J.C., Phipps, C., Jenkins Gary, J. et al. (2018). Mathematical and experimental validation of flux dialysis method: an improved approach to measure unbound fraction for compounds with high protein binding and other challenging properties. *Drug Metabolism and Disposition* 46 (4): 458–469.

11.3 How to Measure the Metabolic Clearance of Slowly Metabolized Compounds

Haijuan Liu, Xiangling Wang, Genfu Chen

11.3.1 Introduction

In vitro metabolic stability studies have been widely used in drug discovery to guide structural modifications, understand drug structure–activity relationships, predict *in vivo* clearance, establish IVIVCs, and predict appropriate dosages for human dosing. With the implementation of effective drug design strategies, medicinal chemists can successfully optimize the structure to efficiently block the metabolic site, resulting in an increasing number of compounds that exhibit low clearance [1–3]. These molecules are referred to as slowly metabolized compounds (also known as low clearance compounds) which need prolonged incubation periods and robust enzyme activity systems in *in vitro* studies, making *in vitro* research extremely challenging. To address this challenge, the relevant *in vitro* study models are introduced, and the appropriate model selection methods are provided according to the characteristics of the compounds or the purpose of the study.

11.3.2 *In Vitro* Metabolic Stability Study Models

Liver microsomes and suspension hepatocytes are the most commonly used models for studying metabolic stability *in vitro*. The advantages of liver microsomes are the abundance of metabolic enzymes (especially cytochrome P450 (CYP450) enzymes), easy sample operation, and relatively low cost. Suspension hepatocytes are also frequently employed in drug discovery and development due to their more comprehensive drug-metabolizing enzyme system and cofactors. In addition to CYP450 enzymes, hepatocytes contain many non CYP enzymes, including UGT and AO.

Since drug-metabolizing enzyme activity decreases as the *in vitro* incubation time increases, liver microsomes are usually incubated for up to 1 h. Suspension hepatocytes are incubated for up to 4 h. Such incubation time is not sufficient to detect metabolic reactions of slowly metabolized compounds, thus traditional *in vitro* metabolic stability models have limitations in predicting the clearance of these compounds and their metabolites [4]. It is necessary to establish methods for extending the incubation time while maintaining the enzyme activities.

Current primary research methods for slowly metabolized compounds [3–7] (Figure 11.7):

- **Plated hepatocyte incubation method:** The incubation time can be extended for up to 48 hours, and even up to 1 week with normal enzyme activity by adding the HepExtend™ culture factor.
- **Suspension hepatocyte relay method:** The incubation time can be extended for up to 20 hours or more using suspension hepatocytes.
- **Hepatocyte and stromal cell co-incubation system:** The incubation time can be extended for up to 72 hours based on two well-established and commercial models: HμREL and HepatoPac.
- **HepaRG Incubation System:** Using human hepatocellular carcinoma (HCC) cell lines, the incubation time can be extended to 24 hours.

Among those four methods, the plated hepatocyte method offers the advantage of intact transporter expression, while the suspension hepatocyte relay method provides flexibility in species selection. Both methods exhibit good IVIVC and are easily accessible. This chapter provides detailed descriptions of these two approaches.

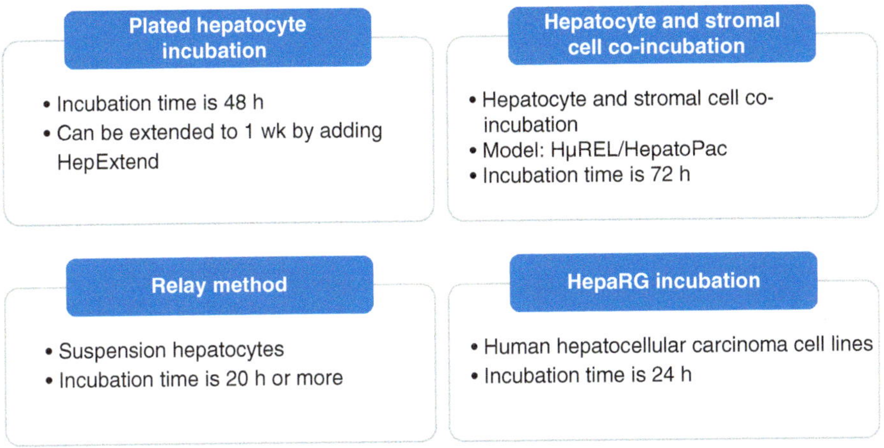

Figure 11.7 Primary methods for slowly metabolized compounds.

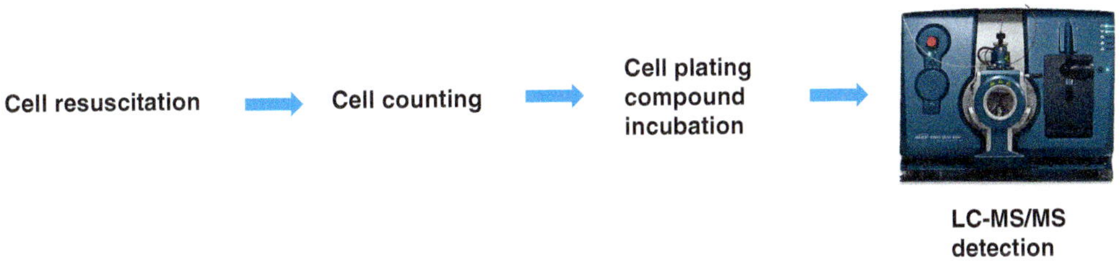

Figure 11.8 Suspension hepatocyte metabolic stability process.

First, we introduce the study procedures of hepatocyte metabolic stability **(Figure 11.8)**.

- **Cell resuscitation:** Cryopreserved hepatocytes were removed from liquid nitrogen and thawed in a 37°C water bath. After gentle centrifugation, pelleted cells were collected.
- **Cell counting:** Add the cells to the incubation medium and dilute them to the appropriate concentration, which is followed by cell counting.
- **Cell plating:** Diluted hepatocyte suspension was added to a corresponding cell incubation plate.
- **Incubation:** After adding the compound working solution, incubation was initiated at different time points. After the incubation was completed, a stop solution was added to terminate the reaction, followed by sample treatment for LC–MS/MS analysis.

11.3.3 The Plated Hepatocyte Method

11.3.3.1 Validation of the Plated Hepatocyte Method

1) **Hepatocyte culture methods (Figure 11.9)**
   a) **Monolayer culture**: Hepatocytes were seeded into collagen-coated cell plates, forming a collagen–hepatocyte–medium monolayer model.
   b) **"Sandwich" culture [8]**: Hepatocytes were seeded into collagen-coated plates. After adhesion, a layer of matrigel was added to form a three-layer model of collagen–hepatocyte–matrigel, with hepatocytes in the middle layer. This model is referred to as the "sandwich" model.

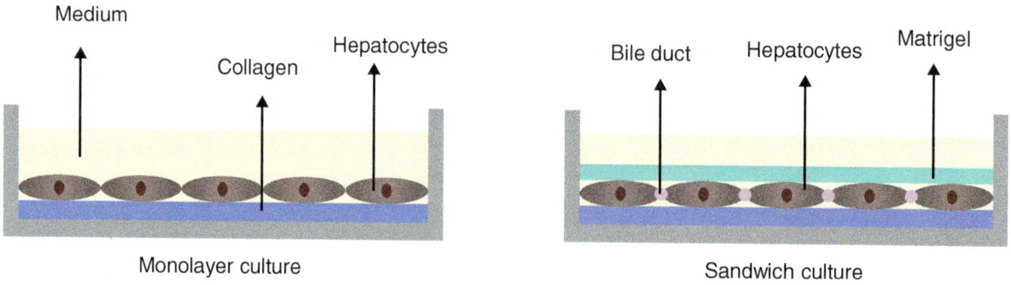

Figure 11.9 Hepatocyte culture methods.

2) **Validation of an incubation system using plated hepatocytes [4, 8–9]**
   a) Selection of slowly metabolized compounds
   The following five slowly metabolized compounds were selected for validation. Their metabolic enzymes include CYP450, UGT, and monoamine oxidase (MAO). Among them, Atorvastatin is the substrate of the transporter OATP1B1 (Table 11.4).
   b) Results
   Accurate clearances of three slowly metabolizing compounds (Timolol, Mephenytoin, and Atorvastatin) were obtained, while those of the other two compounds (Zolmitriptan and Theophylline) were still unable to be obtained due to even lower clearance (Table 11.5 and Figure 11.10).

Table 11.4 Selection of slowly metabolized compounds and their metabolic enzymes.

Compound Name	Metabolic Enzyme
Atorvastatin	CYP3A, UGT
Mephenytoin	CYP2C19
Timolol	CYP2D6
Zolmitriptan	CYP1A2, MAO
Theophylline	CYP1A2

Table 11.5 Intrinsic hepatic clearance of slowly metabolized compounds.

Compound Name	$CL_{int\ (liver)}$ (mL/min/kg)
Atorvastatin	4.5
Mephenytoin	3.8
Timolol	1.1
Zolmitriptan	<0.4
Theophylline	<0.4

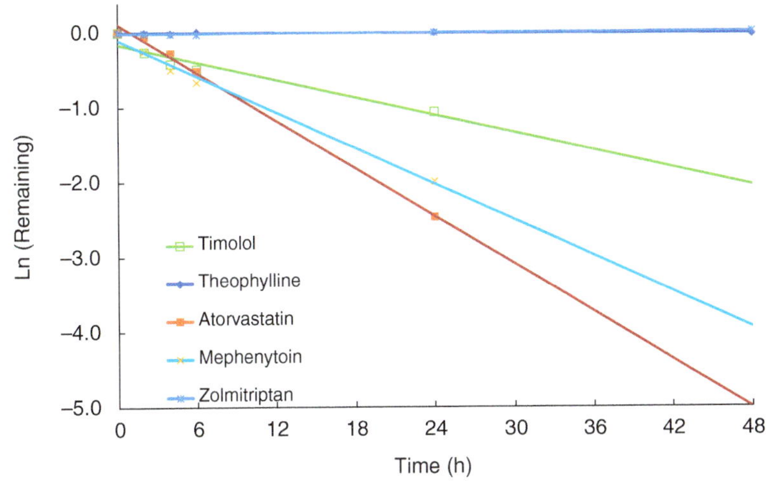

Figure 11.10 Natural logarithm of remaining data–time curves of the compounds in hepatocyte.

The natural logarithm of the remaining data–time curves of the compounds in hepatocytes are shown in Figure 11.10.

The plated hepatocyte incubation method can achieve accurate clearance data for slowly metabolized compounds. However, due to the limitations of their enzyme activity, the maximum incubation time is 48 hours. For compounds with even slower metabolism and lower clearance rates, accurate clearance data cannot be obtained.

11.3.4 Suspension Hepatocyte Relay Method

11.3.4.1 Assay Principle and Protocol

The enzyme activity of suspended hepatocytes can be maintained for 4 hours [3]. The incubation mixture is centrifuged at the end of the 4 hours incubation. The supernatant is then collected and frozen until the next incubation. The supernatant is thawed at the beginning of the next incubation, and freshly resuscitated cryopreserved hepatocytes are added continuously. This method facilitates the incubation of slowly metabolized compounds for up to 20 hours. Because fresh hepatocyte suspension is used every 4 hours, the cells maintain their metabolic capacity throughout the entire study process (Figure 11.11).

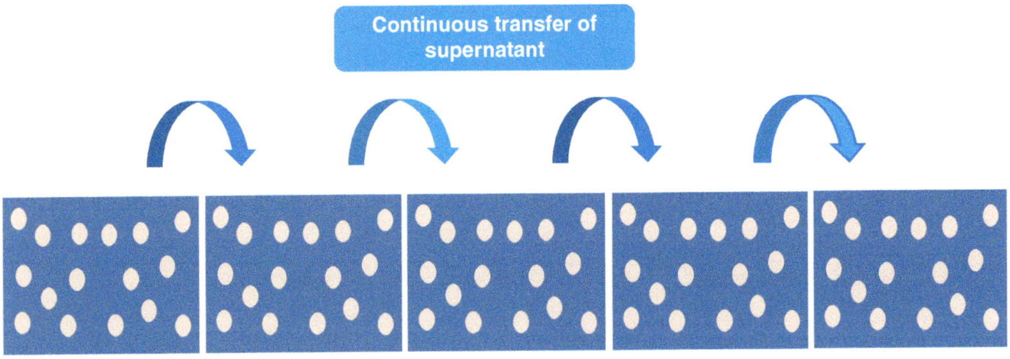

Figure 11.11 Transferring process of the suspension hepatocyte relay method.

After resuscitation of cryopreserved suspended hepatocytes, the cell number and cell viability are measured by trypan blue staining. The diluted hepatocyte suspension is then added to a cell culture plate at each time point, and the incubation is performed by adding the diluted compound working solution.

11.3.4.2 Results

1) Clearance

Accurate clearance values for all five slowly metabolized compounds were obtained (Table 11.6 and Figure 11.12), with mean clearance values of Timolol, Zolmitriptan, and Theophylline being consistent with the literature (less than 1.5-fold difference).

2) The natural logarithm of the remaining data–time curves of the compounds in hepatocytes is shown in Figure 11.12.

Accurate clearance values of five slowly metabolized compounds were obtained using this suspension hepatocyte relay method. In addition, the choice of hepatocyte species in the suspended hepatocyte can be flexible.

Table 11.6 Intrinsic hepatic clearance of slowly metabolized compounds.

	$CL_{int\ (liver)}$ (mL/min/kg)	
Compound Name	**WuXi AppTec DMPK Value**	**Literature Value**
Atorvastatin	**14.7**	—
Mephenytoin	**12.3**	—
Timolol	**9.7**	**14.0**
Zolmitriptan	**2.6**	**3.5**
Theophylline	**2.7**	**2.8**

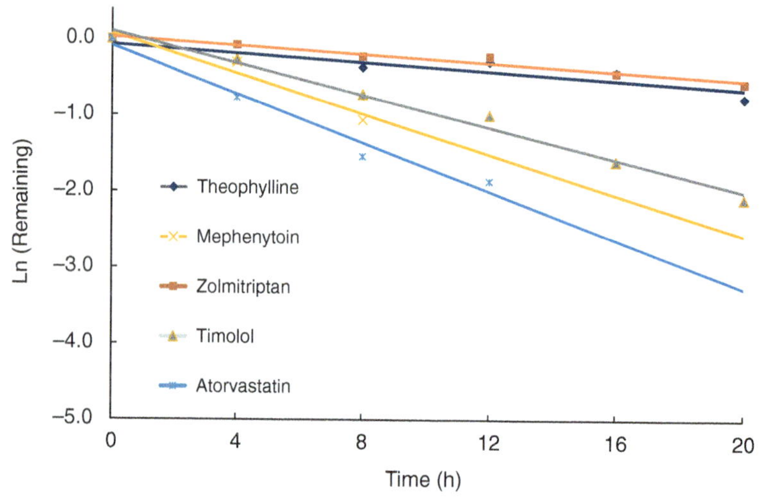

Figure 11.12 Natural logarithm of remaining data–time curves of the compounds in hepatocytes.

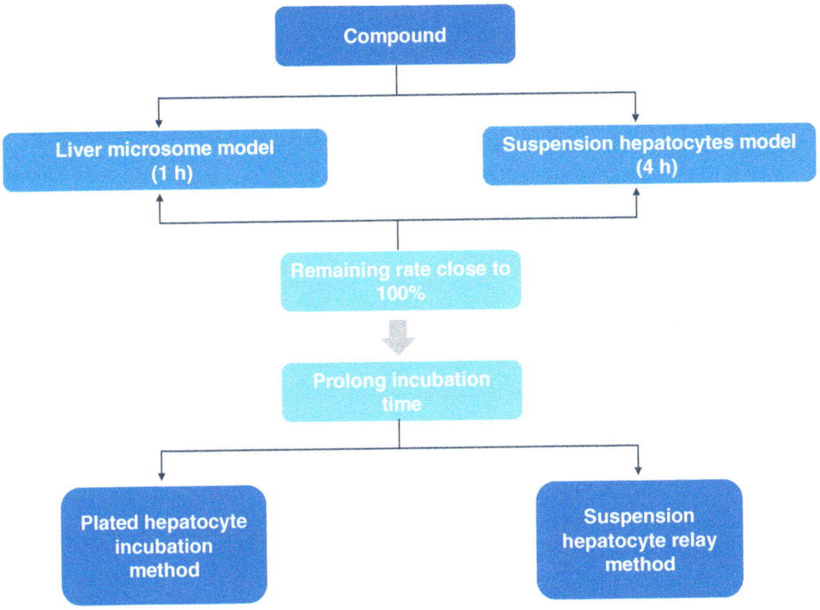

Figure 11.13 Application of the metabolic stability models for slowly metabolized compounds.

By comparing the clearance rates of five slowly metabolized compounds, it can be observed that the metabolism of the five slowly metabolized compounds in plated hepatocyte cultures is slower than in suspended hepatocytes. Possible reasons are: (i) compared to plated hepatocytes, the expression of efflux transporters is incomplete in suspension hepatocytes. (ii) The rotational speed of suspension hepatocytes is higher during incubation, resulting in an increased contact area between the compounds and the cells, allowing the compounds to enter the cells more quickly.

11.3.5 Practical Application of Metabolic Stability Models

If a compound does not show significant metabolism in stability studies with liver microsomes or suspension hepatocytes, the remaining value will be close to 100% at the longest incubation time point. To accurately calculate the clearance of such compounds, two metabolic stability models, the plated hepatocytes culture method and the suspension hepatocyte relay method, can be employed (Figure 11.13). An appropriate model can be selected according to the characteristics of these two methods: (i) to assess the effect of transporters on clearance, plated hepatocyte culture should be given priority because it exhibits relatively complete transporter expression. (ii) To assess metabolic differences between different species, the suspended hepatocyte relay method is the preferred method because a variety of suspended hepatocytes are available.

11.3.6 Summary

By applying the two *in vitro* metabolic stability study models, more accurate *in vitro* clearance data of slowly metabolized compounds can be obtained.

References

1 Obach, R.S. et al. (2011). Predicting clearance in humans from *in vitro* data. *Current Topics in Medicinal Chemistry* 11: 334–339.

2 Blanchard, N., Alexandre, E., Abadie, C. et al. (2005). Comparison of clearance predictions using primary cultures and suspensions of human hepatocytes. *Xenobiotica* 35 (1): 1–15.

3 Di, L., Trapa, P., Obach, R.S. et al. (2012). A novel relay method for determining low-clearance values. *Drug Metabolism and Disposition* 40: 1860–1865.

4 Matthew Hutzler, J., Ring, B.J., and Anderson, S.R. (2015). Low-Turnover Drug Molecules: A Current Challenge for Drug Metabolism Scientists. *Drug Metabolism and Disposition* 43: 1917–1928.

5 Bonn, B., Svanberg, P., Janefeldt, A. et al. (2016). Determination of human hepatocyte intrinsic clearance for slowly metabolized compounds: comparison of a primary hepatocyte/stromal cell co-culture with plated primary hepatocytes and hepaRG. *Drug Metabolism and Disposition* 44: 527–533.

6 Lancett, P., Williamson, B., Barton, P., and Riley, R.J. (2018). Development and Characterization of a Human Hepatocyte Low Intrinsic Clearance Assay for Use in Drug Discovery. *Drug Metabolism and Disposition* 46: 1169–1178.

7 Phil Butler, et al. Hurel corporation and Cyprotex. 2019 ISSX Poster

8 Lau, Y., Sapidou, E., Cui, X. et al. (2002). Development of a novel *in vitro* model to predict hepatic clearance using fresh, cryopreserved, and sandwich-cultured hepatocytes. *Drug Metabolism and Disposition* 30: 1446–1454.

9 Smith, C.M., Nolan, C.K., Edwards, M.A. et al. (2012). A comprehensive evaluation of metabolic activity and intrinsic clearance in suspensions and monolayer cultures of cryopreserved primary human hepatocytes. *Journal of Pharmaceutical Sciences* 101: 3989–4002.

11.4 Non-P450 Enzyme-Mediated Metabolism and *In Vitro* Assessment Method

Haijuan Liu, Xiangling Wang, Xinyun Zhou, Lingling Zhang, Weiqun Cao, Genfu Chen

11.4.1 Introduction

Drug metabolism refers to the process of chemical structure transformation of drug molecules under the action of enzymes after they are absorbed by the body. Drug metabolism has an important impact on drug efficacy, toxicity, and drug interactions during clinical concomitant medication. Drug-metabolizing enzymes are important catalytic substances, which are mainly found in the liver, intestine, and blood and can convert lipophilic drugs into more hydrophilic drug metabolites to promote their excretion from the body. Drug-metabolizing enzymes are divided into phase I and phase II enzymes. Phase I enzymes are those involved in oxidation, reduction, and hydrolysis reactions. Phase II enzymes are those mainly involved in conjugation reactions [1].

Tables 11.7 and 11.8 summarize the metabolic pathways, names, and tissue or subcellular locations of some drug-metabolizing enzymes [1].

Table 11.7 Some phase I metabolic enzymes and their tissue/subcellular locations.

Metabolic Pathway	Phase I-Metabolizing Enzyme	Tissue/Subcellular Location
Oxidation	Cytochrome P450	Endoplasmic reticulum (ER)
	Aldehyde oxidase	Cytosol
	Flavin monoamine oxidase	ER
	Monoamine oxidase	Mitochondria
	Xanthine oxidase/xanthine dehydrogenase	Cytosol
	Alcohol dehydrogenase	Cytosol, blood vessels
	Aldehyde dehydrogenase	Mitochondria, cytoplasmic
	Diamine oxidase	Cytosol
	Prostaglandin H synthetase	ER
Hydrolysis	Carboxylesterase	ER, cytosol, lysosomal
	Epoxide hydrolase	ER, cytosol
	Beta-glucuronidase	Lysosomes, ER, blood
	Arylesterase/paraoxonase	Cytosol
	Pseudocholinesterase/ butyrylcholinesterase	Cytosol
	Peptidase	Blood, lysosomal
Reduction	Aldo-ketoreductase	Cytosol
	Azoreductase and nitroreductase	Mitochondria, ER, and cytosol
	Carbonyl reductase	Cytosol, blood, ER
	Quinone reductase (NADPH)	Cytosol, ER
	Sulfoxide reductase	Cytosol

Table 11.8 Some phase II metabolic enzymes and their tissue/subcellular locations.

Phase II-Metabolizing Enzyme	Tissue/Subcellular Location
Uridine diphosphate glucuronosyltransferase	ER
Sulfotransferase	Cytosol
Glutathione-S-transferase	Cytosol, ER
N-Acetyltransferase	Cytosol
Methyltransferase	Cytosol, ER, blood

As the most important drug-metabolizing enzyme in the body, CYP450 enzyme is involved in about 75% of drug metabolic reactions. Therefore, this enzyme and related oxidative metabolic reactions have always been the focus of the research and development of new drugs, and corresponding research techniques and evaluation strategies have also matured over the years. In recent years, with the development of combinatorial chemistry, high-throughput screening, and other technologies, more and more drugs with new structures have been developed, and metabolic pathways catalyzed by non-P450 enzymes have gradually been concerned. However, a growing number of cases have shown that a lack of understanding of the specific metabolic pathways catalyzed by non-P450 enzymes has led to poor selection of preclinical animal species, ultimately leading to the termination of drug development in clinical trials. Therefore, this section mainly introduces the metabolism of non-P450 enzymes and *in vitro* evaluation strategies to provide references for the development of new drugs.

11.4.2 The Importance of Non-P450 Enzymes

Drug-metabolizing enzymes that mediate drug metabolism or biotransformation play a crucial role in determining the absorption, distribution, metabolism, and excretion characteristics of drugs. The overall role of drug-metabolizing enzymes of the P450 family in this regard cannot be underestimated, but research in the field of non-P450 enzyme-mediated metabolism has gained increasing attention in recent years [2]. This may be due to the fact that medicinal chemistry developers have introduced more P450-insensitive chemical groups during drug design to reduce P450 metabolism, thereby increasing the proportion of metabolism by non-P450 enzymes. In addition, the effect of non-P450 enzyme metabolism on the site of action of drug therapy has also received increasing attention.

Table 11.9 lists the various types of intravenous and oral small-molecule drugs approved by the FDA from 2006 to 2015, in which the main metabolites are mediated by P450 or Non-p450 or a mix of both, and there are no main metabolites [3, 4].

Table 11.9 Fractional assignment of FDA-approved intravenous and orally administered small-molecule drugs (2006–2015).

Items	Percentage
No major metabolites	17.6%
P450	44.8%
Non-P450	20.8%
Mixed	16.8%

Of the 125 FDA-approved drugs, 20.8% were metabolized by non-P450 enzymes in the main metabolic process, and 16.8% were metabolized by both P450 and non-P450 enzymes. These data indicate that non-P450 enzymes are involved in drug metabolism and play an increasingly important role in new drug research and development.

Non-P450 enzymes include hydrolases, oxidases, and conjugated enzymes. This chapter mainly discusses some of the main non-P450 enzymes in phase I metabolism, namely, AO, XO, aldo–keto reductase (AKR), flavin monooxygenase (FMO), and MAO. These five enzymes account for about 5% of the metabolic participation rate among the top 200 FDA-approved drugs in sales from 2005 to 2016. They have also been paid more and more attention by chemists and become one of the important contents of drug metabolism research [3, 4]. If these enzyme-mediated metabolisms are not discovered in the preclinical stage, there may be unforeseen consequences, leading to the termination of the development.

11.4.3 Aldehyde Oxidase (AO) and Xanthine Oxidase (XO)

Both AO and XO have high homology, and they are very similar in amino acid sequence. The AO and XO structures are consistent and are homodimers composed of two identical subunits of approximately 150 kDa. Each subunit is subdivided into three distinct domains: a 20 kDa N-terminal domain that binds to two iron-containing clusters, a 40 kDa domain that contains the flavin adenine dinucleotide (FAD) binding site, and a C-terminal domain that accommodates the cofactor molybdenum (Mo) atom. Molybdenum and FADs together act as essential components for enzyme catalysis [5]. There are significant species differences in AO, and there is only one active AOX1 subtype in humans. Two subtypes of AOX1 and AOX3 are present in the liver of rats and mice, while only one subtype of AOX1 is present in cynomolgus monkeys and guinea pigs. There was no AO gene expression in dogs.

a) Reaction characteristics

AO and XO are molybdenum-containing, NADPH-independent cytosolic enzymes that catalyze the oxidation and reduction of *N*-heterocyclic compounds [1], the detailed description is listed below.

Items	Description
Subcellular distribution	Liver cytosol
Organ distribution	Liver> lung> kidney> small intestine> brain
Cofactor	No cofactor is required
Typical substrates of AO	Carbazeran, phthalazine
Typical inhibitors of AO	Raloxifene, estradiol
Typical substrates of XO	Pterin, xanthine
Typical inhibitors of XO	Folic acid, allopurinol, alloxanthine

An example of a reaction is shown in Figure 11.14.

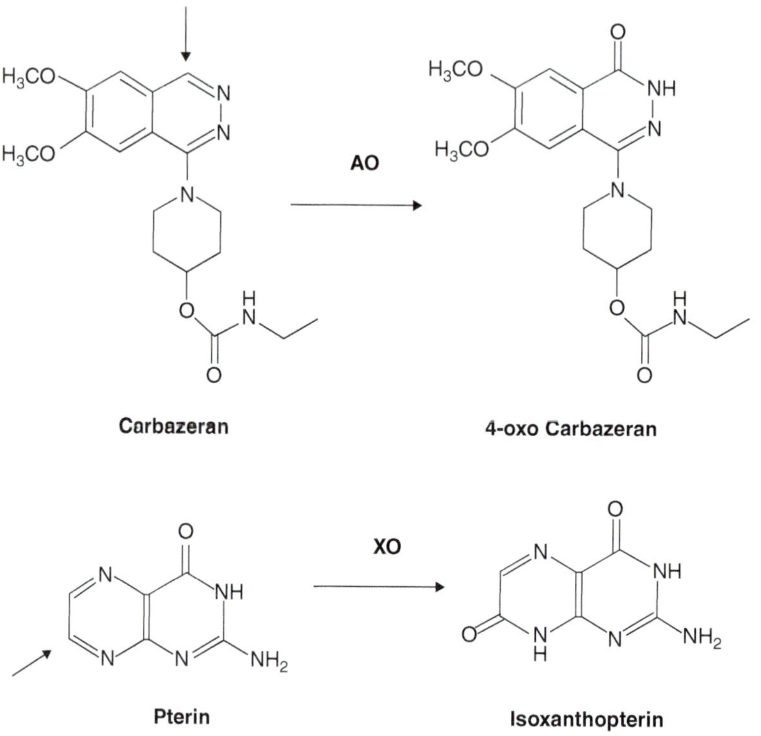

Figure 11.14 AO and XO reactions.

b) Evaluation method
We evaluated AO and XO metabolism by comparing the metabolic differences of compounds in the presence and absence of inhibitors using liver cytosol as the incubation matrix, and carbazeran/pterin and raloxifene/allopurinol as substrates and inhibitors, respectively. The compounds used for validation in this article are all commercially available compounds.

c) Results
The in-house data of carbazeran is consistent with those reported in the literature in human liver cytosol with and without the AO inhibitor raloxifene.

The in-house data of pterin is consistent with those reported in the literature in human liver cytosol with and without the XO inhibitor allopurinol.

The data are shown in Table 11.10.

Table 11.10 Validation data of the substrates metabolized by AO and XO.

		Clearance With and Without Inhibitor (μL/min/mg protein)			
Enzyme	**Substrate**	**(−) Inhibitor**	**(+) Allopurinol**	**(+) Raloxifene**	**Source**
AO	Carbazeran	117.6	105.7	< 3.2	In-house
		98.8/125	71.6	1.83	Literature
XO	Pterin	2.6	1.6	—	In-house
		1.83/1.10	0.7	—	Literature

In summary, inhibitors of AO or XO can be added to the liver cytosol to assess whether the compounds are substrates of AO or XO.

11.4.4 Aldo–Keto Reductase (AKR)

AKR is a structurally related group of proteins that share a common origin. These proteins form (β/α) 8 or triphosphate isomerase-motif, a compact and adaptable structure, and this structural variation can meet the need for diverse binding on the chemical structure of carbonyl substrates. The active site of AKR is located at the C-terminus of the structure and can interact with pyridine nucleotides with high affinity in the absence of a typical Rossman fold [6].

a) Subtype

Up to now, 13 AKR proteins have been identified in humans, such as AKR1A1, AKR1B1 and AKR1B10, AKR1C1, AKR1C2, AKR1C3 and AKR1C4, AKR1D1, AKR6A3, AKR6A5 and AKR6A9, and AKR7A2 and AKR7A3. Among these proteins, AKR1C3 is a hot topic of research. It is mentioned [7] that PR-104 is a phosphate ester of the nitrogen mustard prodrug PR-104A, which showed efficacy in adult leukemia clinical trials. PR-104A was originally designed to target hypoxic cells and can be activated by AKR1C3. It is reported [8, 9] that OBI-3424 is a highly selective prodrug that can be converted by AKR1C3 into a potent DNA alkylating agent. OBI-3424 was granted Orphan Drug Designation by the US FDA for the treatment of HCC, a potential new treatment for acute lymphoblastic leukemia [8].

b) Catalytic reaction characteristics

AKR is a cofactor-dependent superfamily that catalyzes the reduction of ketones and aldehydes to their corresponding alcohols (Figure 11.15) [1]. The detailed description is listed below.

Items	Description
Subcellular location	Cytosol
Organ distribution	liver, mammary gland, and brain
Cofactors	NAD + and NADP +
Typical substrates	Glyceraldehyde, p-nitrobenzaldehyde, and progesterone

c) Method

In a system to evaluate metabolism by AKR, liver cytosol was selected as the incubation substrate with progesterone as the positive control, and its metabolism in five major species was verified through incubations at different time points.

Figure 11.15 Structure of progesterone, one of the typical substrates of AKR.

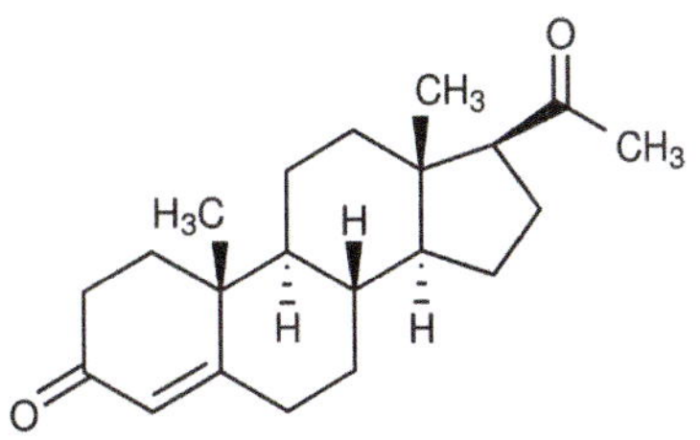

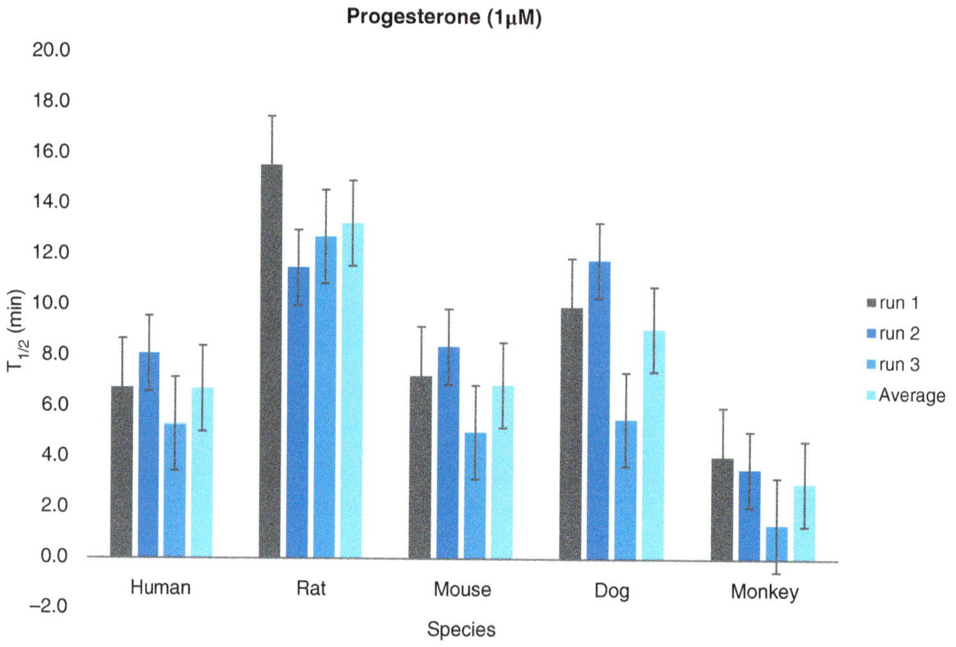

Figure 11.16 Half-life of substrate metabolized by AKR in various species.

d) Results

The metabolism of progesterone in human cytosol was consistent with that in the literature for its human isoforms AKR1C3 and AKR1C1 recombinases (Figure 11.16).

In summary, for unknown compounds, if there is an obvious aldehyde or ketone structure, this system can be used to verify whether it is metabolized by AKR.

11.4.5 Flavin Monooxygenase (FMO)

FMO is a major non-P450 oxidase present in the liver and plays a role second only to P450 enzymes in the biotransformation of therapeutic drugs or xenobiotics by the body. FMO belongs to the class B flavoprotein monooxygenase family, that is, monooxygenases that are dependent on NADPH and molecular oxygen and do not contain heme. FMO proteins have a molecular mass of approximately 60 kDa and contain a highly conserved FAD and NADPH-binding domain in their 532 to 558 amino acid residues [10–13].

a) Subtype

There are obvious tissue, age, sex as well as species differences in the expression levels of FMO [1, 12].

Tissue specificity: FMO3 and FMO5 were predominant in the liver, FMO1 in the kidney and intestine, and FMO2 in the lung (Figure 11.17).

Developmental factors: FMO1 is predominant in the human fetal liver. After birth, the expression of FMO1 is inhibited and gradually dominated by FMO3.

Gender difference: FMO1 was predominant in the liver of male mice, while FMO3 was predominant in the liver of female mice. There was no significant gender difference in human FMO.

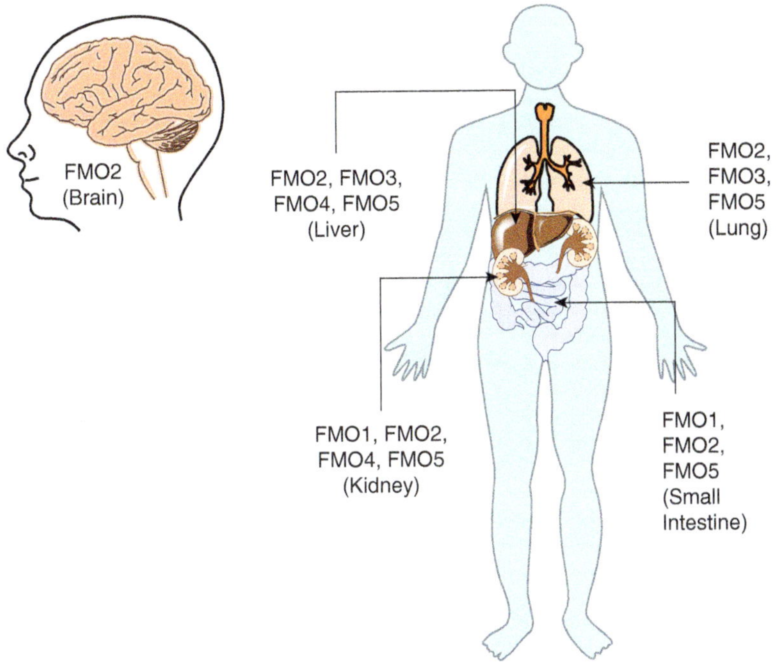

Figure 11.17 Distribution of FMO isoforms in adult tissues.

Species differences: Unlike the low expression of FMO1 in the adult male or female liver, FMO1 is mainly expressed in rat, mouse, and dog liver, but FMO3 is mainly expressed in female mice.

b) Catalytic reaction profile [1] / with permission of Springer Nature.

Items	Description
Subcellular structure	Endoplasmic reticulum
Organ distribution	Liver, kidney, intestine, lung, brain, skin, pancreas, and secretory tissue
Cofactor	NADPH
Substrates	Imipramine (FMO1), cyclobenzaprine (FMO3), benzydamine (FMO1 and FMO3)
Inhibitors	Methimazole, thiourea

c) Method
Liver microsomes were used as the reaction matrix, and benzydamine was used as a positive control to compare the metabolic differences of compounds under two conditions, with and without the FMO inhibitor methimazole.

The major recombinases (FMO1, FMO3, and FMO5) were used to assess the FMO isoforms contributing to metabolizing compounds.

d) Results

There was a significant difference in the metabolism of benzydamine in liver microsomes with and without an FMO inhibitor (Figure 11.18).

The metabolism of benzydamine by recombinant FMO3 was obvious (Figure 11.19).

In summary, liver microsomes can be used to determine whether the unknown compound is metabolized by FMO in the systems with and without the inhibitor methimazole. Recombinant enzymes can be used for phenotypic confirmation.

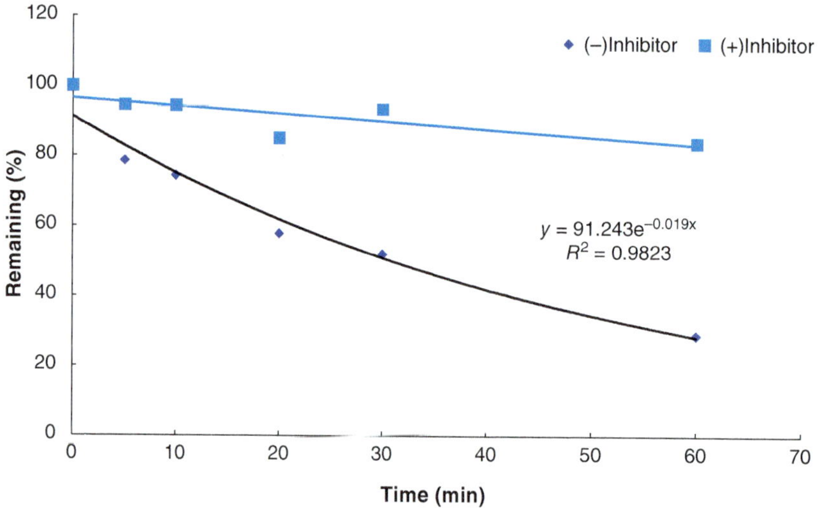

Figure 11.18 There was a significant difference in the metabolism of benzydamine under the two conditions.

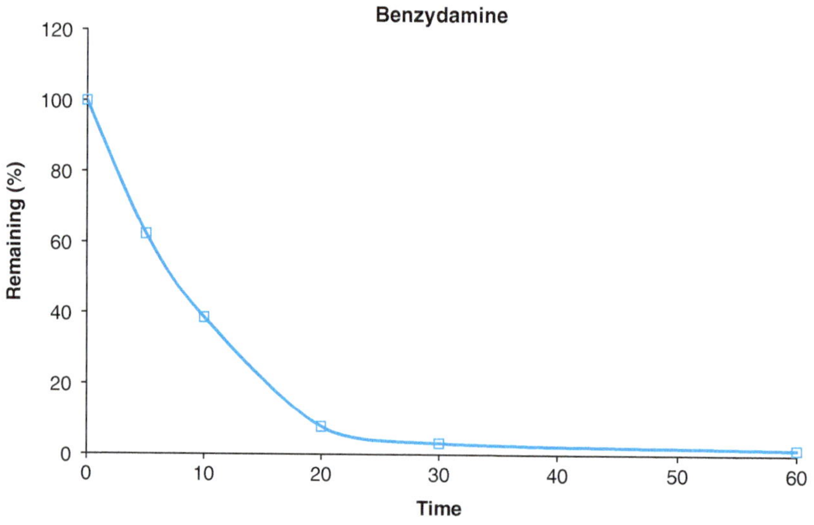

Figure 11.19 The metabolism of benzydamine by recombinant FMO3.

11.4.6 Monoamine Oxidase (MAO)

a) MAO belongs to flavoproteases, which are present in the outer mitochondrial membrane of most cells in the body [1, 14–16]. The main reactions are as follows:

RCH_2NH_2 (substrate) + O_2--RCH = NH + H_2O_2

RCH = NH + H_2O--RCH = O + NH_3

Subtypes: MAO-A and MAO-B, which share 70% homology.

Substrates: 5-Hydroxytryptamine and norepinephrine are substrates of MAO-A; 2-phenylethylamine and benzylamine are substrates of MAO-B.

Inhibitors: clorgyline (MAO-A), selegiline ((R) -deprenyl) (MAO-B), and pargyline (MAO-A and MAO-B).

b) Method

Human liver mitochondria were used as the reaction matrix and 5-hydroxytryptamine as the substrate to study the metabolism of 5-hydroxytryptamine through a series of different time points of incubations.

c) Results

The metabolism of 5-hydroxytryptamine is shown in Figure 11.20.

In summary, liver mitochondria can be used as the matrix, and 5-hydroxytryptamine as the positive control to determine whether the unknown compound may be metabolized by MAO.

11.4.7 The Decision Tree of Non-P450 Oxidase-Mediated Drug Metabolism

a) Substrate type

Since the structure of substrates for different non-P450 oxidases is different, the preliminary judgment could be made according to the structure of compounds in the metabolism study of non-P450 oxidases. FMOs have a wide range of substrate specificity and are able to catalyze the oxidation of nitrogenous, sulfur, selenium, and phosphor compounds, and the oxygen insertion reaction (Baeyer–Villiger) of ketones or aldehydes to generate lactones. The AOs mainly catalyze the formation of

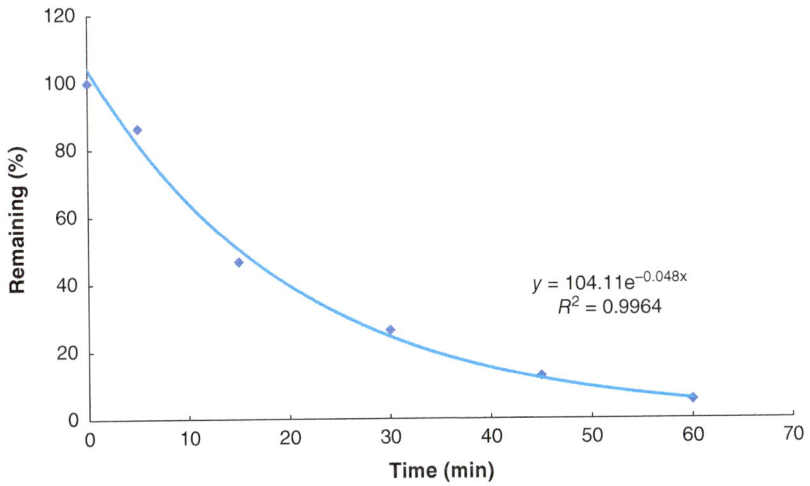

Figure 11.20 Metabolism of 5-hydroxytryptamine by MAO.

carboxylic acids from aldehydes and participate in the oxidation of *O-/N*-heterocyclic ortho-carbon atoms. The substrates of other non-P450 oxidases also have their structural characteristics. Additional attention needs to be paid when relevant characteristic structures are involved in the structure of a compound.

b) Differentiation of compound metabolites in different systems
In addition to the structural characteristics of compounds, the inconsistent results of metabolites in different systems are noteworthy. If the stability results in hepatocytes and liver microsomes differ, as well as different associated oxidation metabolites, this is a good indication that non-P450 oxidase may be involved in drug metabolism. Because specific *in vitro* experiments (such as liver microsomes) do not fully mimic the *in vivo* environment, when metabolite identification results from *in vitro* and *in vivo* studies are inconsistent, it may be caused by non-P450 enzymes.

c) Decision tree of non-P450 oxidase phenotypic experiment
If the compound is assumed to be a non-substrate of P450 oxidase, an enzyme phenotypic study is required. According to the characteristics of each enzyme, reaction type, and metabolite oxidation site, it is possible to determine preliminarily which enzyme is involved. Different methods can be used to identify the responsible enzyme, including the use of different experimental incubation systems and coenzymes (or cofactors), the addition of inhibitors, co-incubation with recombinant enzyme, etc. (Figure 11.21).

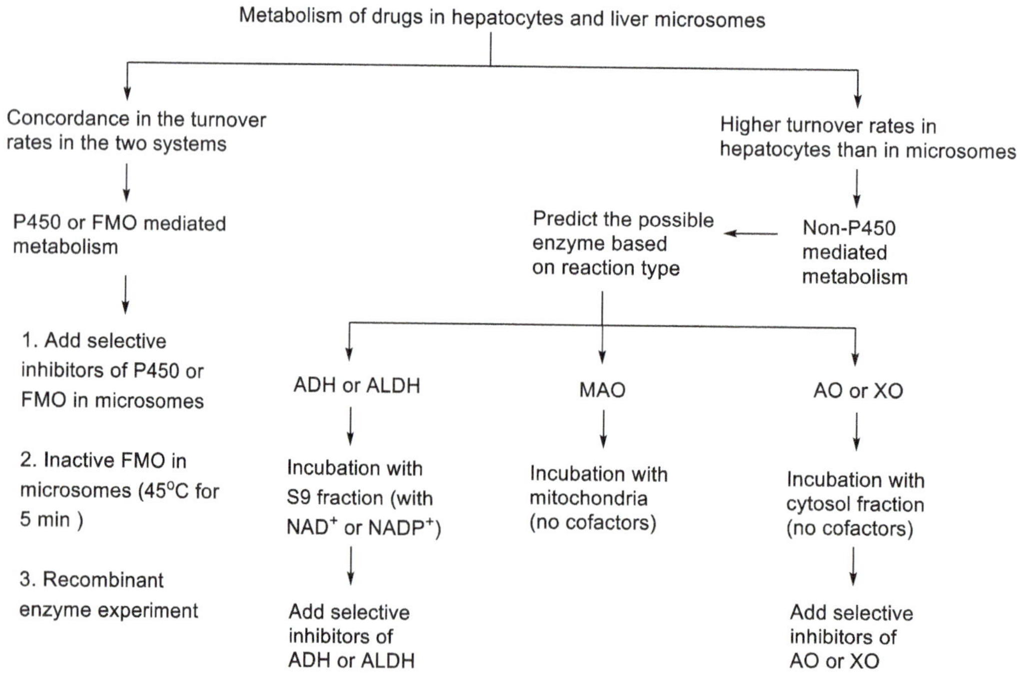

Figure 11.21 Decision tree for identification and validation of non-P450 oxidase-mediated drug metabolism. *Source:* Adapted from reference [17].

11.4.8 Conclusion

In preclinical investigations, it is important to know whether the drug candidate is catalyzed by a non-P450 enzyme. If non-P450 oxidases have a special contribution to drug metabolism, they will influence decision-making in subsequent studies and can provide critical information for chemists to optimize compound structures. This chapter describes the non-P450 enzymes (mainly AO/XO/FMO/MAO/AKR)-mediated metabolic pathways and corresponding *in vitro* study strategies, which will help to identify the presence of non-P450 enzymes involved in metabolism, provide the basis for toxicological species selection, and reduce the risks in clinical trials.

References

1 Khojasteh, S.C., Wong, H., and Hop, C.E. (2011). *Drug Metabolism and Pharmacokinetics Quick Guide*, 1–214. Springer Science & Business Media.

2 Cerny, M.A. (2016). Prevalence of Non - Cytochrome P450 - Mediated Metabolism in Food and Drug Administration - Approved Oral and Intravenous Drugs: 2006 - 2015. *Drug Metabolism and Disposition* 44: 1246–1252.

3 Foti, R.S. and Dalvie, D.K. (2016). Cytochrome P450 and Non - Cytochrome P450 Oxidative Metabolism: Contributions to the Pharmacokinetics, Safety, and Efficacy of Xenobiotics. *Drug Metabolism and Disposition* 44: 1229–1245.

4 Bohnert, T. and Patel, A. (2016). Evaluation of a New Molecular Entity as a Victim of Metabolic Drug-Drug Interactions — an Industry Perspective. *Drug Metabolism and Disposition* 44: 1399–1423.

5 Barr, J.T., Choughule, K.V., Nepal, S. et al. (2014). Why Do Most Human Liver Cytosol Preparations Lack Xanthine Oxidase Activity? *Drug Metabolism and Disposition* 42: 695–699.

6 Barski, O.A., Tipparaju, S.M., and Bhatnagar, A. (2008). The Aldo-keto Reductase Superfamily and Its Role in Drug Metabolism and Detoxification. *Drug Metabolism Reviews* 40: 553–624.

7 Manesh, D.M., El-Hoss, J., Evans, K. et al. (2015). AKR1C3 is a biomarker of sensitivity to PR-104 in preclinical models of T-cell acute lymphoblastic leukemia. *Blood.The Journal of the American Society of Hematology* 126 (10): 1193–1202.

8 Evans, K., Duan, J., Pritchard, T. et al. (2019). OBI-3424, a Novel AKR1C3-Activated Prodrug, Exhibits Potent Efficacy against Preclinical Models of T-All. *Clinical Cancer Research* 25 (14): 4493–4503. *American Association for Cancer Research*.

9 Duan, J. et al. (2016). *Broad In Vitro and In vivo Antitumor Activity of TH-3424: Preclinical Rationale for a Highly Selective AKR1C3 Prodrug for Treating Cancer*, 1369. *AACR Annual Meeting*. Abstract.

10 Lang, D.H. and Rettie, A.E. *In vitro* evaluation of potential *in vivo* probes for human-avin containing monooxygenase (FMO): metabolism of benzydamine and caffeine isoforms by FMO and P450. *Journal of Clinical Pharmacology* 50: 311–314.

11 Lang, D.H., Yeung, C.K., Peter, R.M. et al. (1998). Isoform specificity of trimethylamine N-oxygenation by human flavin-containing monooxygenase (FMO) and P450 enzymes: selective catalysis by FMO3. *Biochemical Pharmacology* 56: 1005–1012.

12 HAINING, R.L. et al. (1997). Baculovirus-mediated expression and purification of human FMO3. *Drug Metabolism and Disposition* 25: 790–797.

13 Fisher, M.B. et al. (2002). Flavin-containing monooxygenase activity in hepatocytes and microsomes: *in vitro* characterization and *in vivo* scaling of benzydamine clearance. *Drug Metabolism and Disposition* 30: 1087–1093.

14 Zhang, W., Guo, H. et al. (2013). Establishment of *in vitro* screening model for monoamine oxidase inhibitory drugs. *Journal of Chongqing Medical University* 38, No. 6.
15 Liu, Z. and Liu, M. (2016). Establishment of *in vitro* screening model for monoamine oxidase inhibitors. *Chinese Journal of Clinical Pharmacy* 25 (1).
16 Saravanakumar, A. et al. (2019). *Clinical Pharmacokinetics* 58: 1281–1294.
17 Pryde D C, Dalvie D, Hu Q, et al (2010). Aldehyde oxidase: an enzyme of emerging importance in drug discovery. Journal of Medicinal Chemistry, 53:8441–8460.

11.5 *In Vitro* Evaluation of CYP450 Time-Dependent Inhibition (TDI): A Novel Area Under the Curve Shift (AUC Shift) Approach

Lifang Jiang, Genfu Chen

11.5.1 Introduction

Many drugs undergo metabolism as a major mechanism of bioactivation or clearance from the body. Drug metabolism occurs primarily through the CYP450 family of enzymes. Therefore, it is important to assess the metabolism-mediated drug interactions on CYP450 enzyme activity. The inhibition of CYP450 enzymes may result in clinically relevant DDIs. Depending on the mechanism of inhibition, the inhibitory effect of a drug on CYP450 enzymes can be classified as reversible inhibition or irreversible inhibition (Figure 11.22).

In general, TDI results from irreversible covalent binding or quasi-irreversible non-covalent tight binding of a chemically reactive intermediate to the enzyme that catalyzes its formation, resulting in a loss of enzyme function. Metabolism-mediated DDIs resulting from TDI can display a delayed onset due to the time dependence on inhibition and can persist even after the inhibitor has been eliminated because enzymatic activity is only restored by de novo protein synthesis. Therefore, TDI has gained increasing attention from regulatory agencies and drug developers over the years. Industrial white papers or guidelines have been issued by organizations, such as Pharmaceutical Research and Manufacturers of America (PhRMA), the FDA, the EMA, and the NMPA. These guidelines state that TDI studies on the seven major drug-metabolizing CYP450 enzymes should be conducted as standard *in vitro* ADME studies. This chapter introduces a novel *in vitro* research approach, the AUC shift approach for assessing TDI. This approach could overcome the limitations of the IC_{50} shift method in evaluating the TDI potential of the compound of interest, aiming to provide more accurate data.

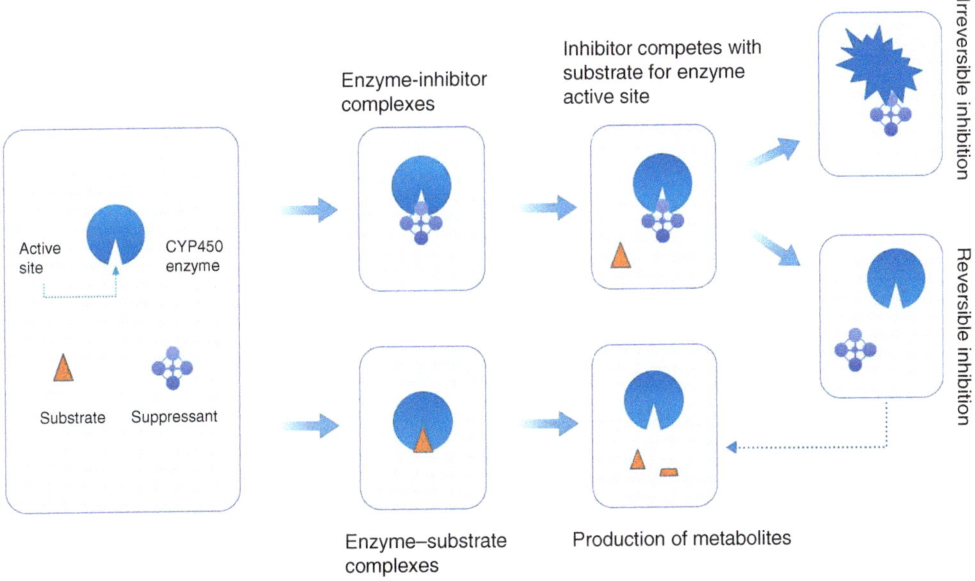

Figure 11.22 Types of enzyme-related drug-drug interactions.

11.5.2 Limitations of the Traditional Approach for Evaluating Time-Dependent CYP Inhibition

Typically, the IC_{50} shift fold of a compound against CYP450 enzymes after preincubation with and without NADPH is used to assess the TDI. A compound is considered to exhibit TDI (i.e., mechanism-based inhibition) when the IC_{50} shift fold is ≥ 1.50. However, it is not possible to calculate the IC_{50} shift fold in some cases, making it difficult to assess a compound's TDI potential accurately.

Our team has identified three scenarios in which it is difficult to obtain the IC_{50} shift fold:

1) The IC_{50} shift fold of the test article falls within a range of > 0 to 1.50 (Figure 11.23a).
2) The solubility of the compound is limited (the ideal maximum test concentration is 50.0 μM), making it difficult to obtain IC_{50} values (Figure 11.23b).
3) The compound is a weak mechanism-based inhibitor, and it is challenging to obtain IC_{50} values under the conditions tested (Figure 11.23c).

The accurate assessment of the TDI potentials of compounds is crucial in drug discovery. If the compounds cannot be ranked in order or compared based on *in vitro* TDI data, it will not help in decision-making. Therefore, it is necessary to find a novel and quantifiable method for assessing

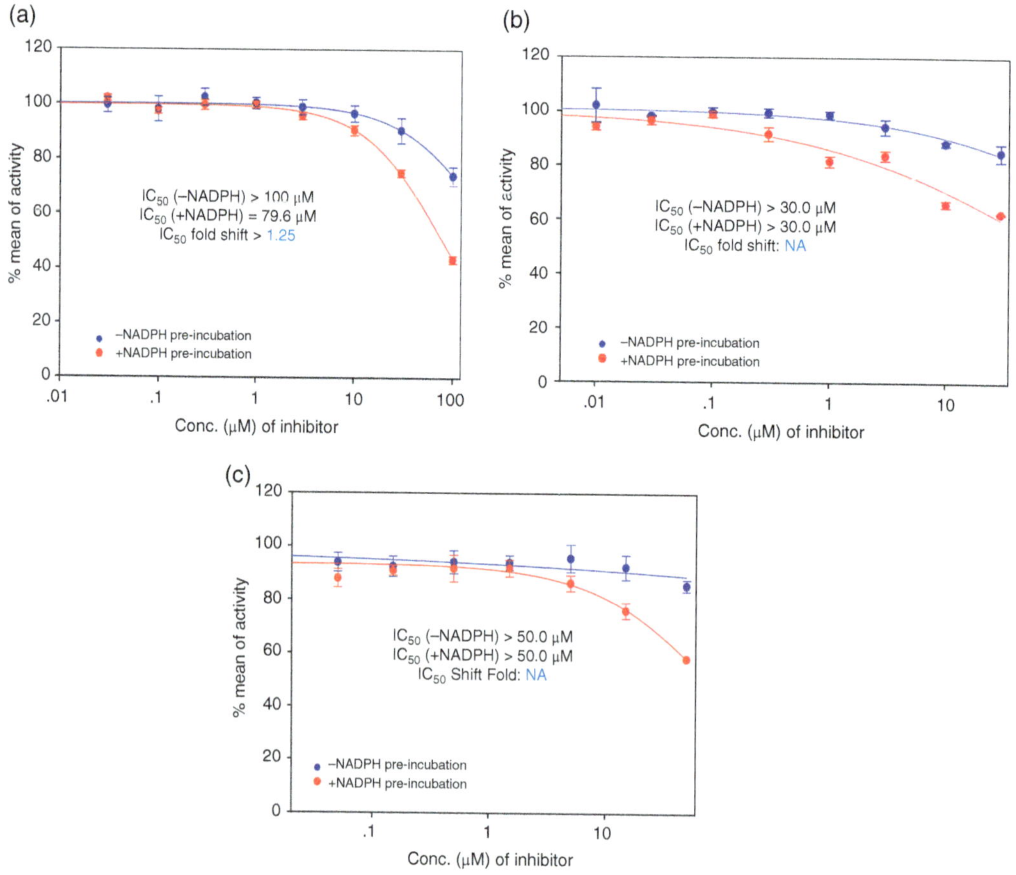

Figure 11.23 Three scenarios where an accurate IC_{50} shift fold cannot be obtained.

the TDI potentials of compounds during the drug development process. The AUC shift approach assesses the TDI potentials of compounds by determining the AUC shift of the inhibition of CYP450 enzymes during preincubation, with and without the presence of NADPH, thereby enhancing the *in vitro* TDI research on CYP450 enzymes and providing a basis for predicting clinical DDI risks.

11.5.3 Establishment of the AUC Shift Approach for Evaluating CYP450 TDI

The IC_{50} shift and AUC shift approaches share the same study design experimental conditions. However, their subsequent approach to data processing is different. Thus, determining the cut-off value that could assess the TDI potentials of compounds is crucial for establishing the AUC shift approach, which can be divided into two steps:

First, 27 commercially available compounds, which have been known as strong, moderate, weak, or no TDI inhibitors for each of the seven CYP450 enzyme subtypes, are selected based on the literature data.

Second, the IC_{50} shift fold and AUC shift value of those selected compounds are calculated separately. Additionally, the cut-off value of the AUC shift approach needs to be confirmed according to whether the compound exhibits TDI potential. The corresponding IC_{50} shift fold value serves as evidence of establishing a stable, reliable, and reproducible AUC shift platform.

The results of the AUC shift approach using commercially available compounds with TDI potentials

We determined the AUC shift value of 27 commercially available compounds (strong, moderate, weak, and no TDI potentials) on human liver microsomal CYP450 enzymes (CYP1A2, CYP2B6, CYP2C8, CYP2C9, CYP2C19, CYP2D6, and CYP3A), and compared the results with the literature values. Based on whether the compounds exhibited TDI potentials and the calculated AUC shift values, the AUC shift value of 15.0% was identified as the cut-off. The following results of CYP3A (with midazolam as the substrate), CYP2C9, and CYP2C19 are presented as examples (Tables 11.11, 11.12, and 11.13).

Case 1: Ketoconazole and troleandomycin are recommended by the FDA and NMPA as the positive controls for determining the reversible and TDI of CYP3A, respectively. The IC_{50} shift fold and AUC shift (%) values of these compounds were consistent with the data in the literature, and the results aligned with the recommendations of the FDA and NMPA (Table 11.11).

Case 2: Sulfaphenazole and tienilic acid are recommended by the FDA and NMPA as the positive controls for assessing the reversible and TDI of CYP2C9, respectively. The IC_{50} shift fold and AUC shift (%) values of these compounds were close to the values reported in the literature, and the results were consistent with FDA and NMPA recommendations (Table 11.12).

Table 11.11 TDI of ketoconazole and troleandomycin by IC_{50} shift fold and AUC shift methods.

Enzyme	Inhibitor	IC_{50} Shift Fold and AUC Shift (%)	AUC Shift (%) vs. Literature Data [1]
CYP3A	Ketoconazole	< 1.50 and < 15.0	
	Troleandomycin	> 1.50 and > 15.0	

Note: The blue dots and blue lines represent the average values and standard deviations of WuXi AppTec DMPK's measurements, respectively (n = 3), while the orange dots and orange lines represent the average values and standard deviations reported in the literature, respectively (n = 10).

Table 11.12 TDI of sulfaphenazole and tienilic acid using IC_{50} shift fold and AUC shift methods.

Enzyme	Inhibitor	IC_{50} Shift Fold and AUC Shift (%)	AUC Shift (%) Versus Literature Data[1]
CYP2C9	Sulfaphenazole	< 1.50 and < 15.0	
	Tienilic acid	> 1.50 and > 15.0	

Note: The blue dots and blue lines represent the average values and standard deviations of WuXi AppTec DMPK's measurements, respectively (n = 3), whereas the orange dots and orange lines represent the average values and standard deviations published in the literature, respectively (n = 10).

Table 11.13 TDI of (+)-N-3-benzylnirvanol and ticlopidine using IC_{50} shift fold and AUC shift methods.

Enzyme	Inhibitor	IC_{50} Shift Fold and AUC Shift (%)	AUC Shift (%) Versus Literature Data[1]
CYP2C19	(S)-(+)-N-3-Benzylnirvanol	< 1.50 and < 15.0	
	Ticlopidine	> 1.50 and > 15.0	

Note: The blue dots and blue lines represent the average values and standard deviations of WuXi AppTec DMPK's measurements, respectively (n = 3), while the orange dots and orange lines represent the average values and standard deviations published in the literature, respectively (n = 10).

Case 3: (+)-N-3-Benzylnirvanol and ticlopidine are recommended by the FDA and NMPA as the positive controls for assessing reversible and TDI of CYP2C19, respectively. The IC_{50} shift fold and AUC shift (%) values of these compounds were close to the values reported in the literature, and the results were consistent with FDA and NMPA recommendations (Table 11.13).

The measured values for our standards were close to the values in the literature (within twofold). We also calculated the IC_{50} shift fold of the inhibitor data as above and compared them with the AUC shift (%) value. The results from both approaches were identical. This demonstrates that we have established a new reliable approach (the AUC shift approach) to study TDI against CYP450 *in vitro*.

11.5.4 Applications of the AUC Shift Approach for Evaluating TDI Potentials

Case 1: For compound A, an IC_{50} shift fold of > 1.33 was obtained, making it difficult to accurately determine whether the compound exhibits TDI potentials. However, a specific value could be obtained by the AUC shift approach, which enables a more accurate assessment of the TDI potentials (Figure 11.24).

Case 2: For compound B, it was difficult to obtain the IC_{50} value due to the limited solubility with the ideal maximum test concentration of 100 μM. Therefore, the IC_{50} shift fold could not be calculated. However, the AUC shift approach could provide specific values and accurately assess the TDI potentials (Figure 11.25). This aligns with the actual characteristics of the compound (lower activity percentage of the vehicle control at the highest concentration tested in the +NADPH group; significant deviation of the two curves in the −NADPH and +NADPH groups).

Case 3: For compound C (weak time-dependent inhibitor), the IC_{50} shift fold was unable to be calculated. However, a specific value can be obtained from the AUC shift approach and assessing the compound's TDI potentials (see Figure 11.26). This also aligns with the compound's actual

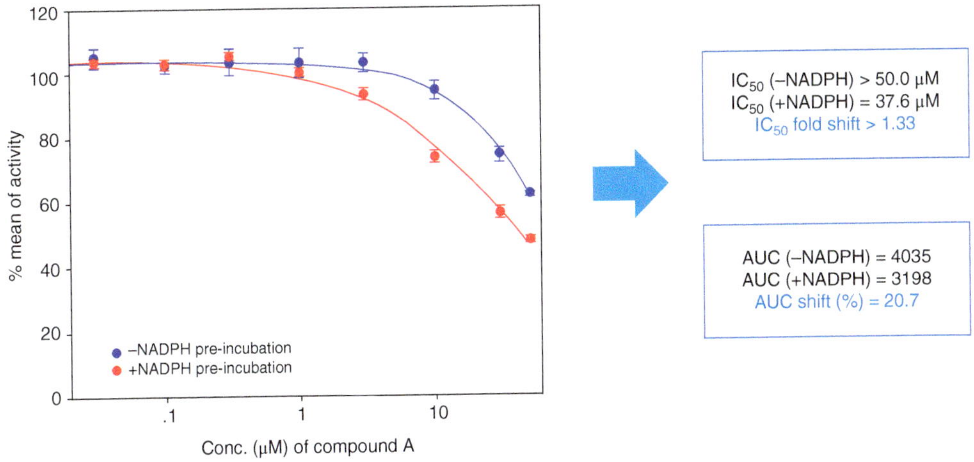

Figure 11.24 Evaluation of a compound with IC_{50} shift fold reported as a range using the AUC shift method.

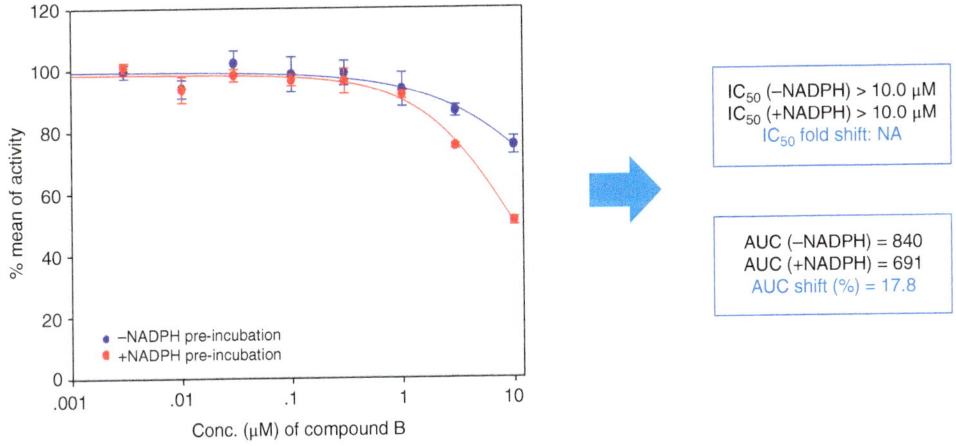

Figure 11.25 Evaluation of a solubility-limited compound using the AUC shift approach.

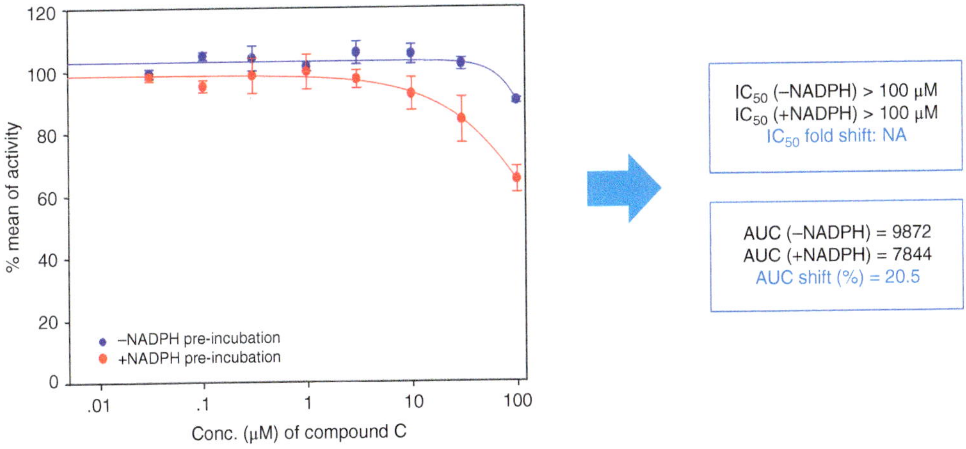

Figure 11.26 Evaluation of weak TDI compounds using the AUC shift approach.

characteristics (lower activity percentage of the vehicle control at the highest concentration tested in the +NADPH group; significant deviation of the two curves in the –NADPH and +NADPH groups).

The AUC shift approach can address the issues that cannot be solved using the IC_{50} shift approach, allowing for a more accurate determination of the TDI potentials of compounds. This provides a basis for the clinical study design of test articles.

11.5.5 Summary

The AUC shift approach has proven to be a novel quantitative method for the *in vitro* evaluation of TDI. The IC_{50} shift and AUC shift approach could be combined to determine the TDI potentials of compounds, providing crucial evidence for the structural optimization of lead compounds and the clinical study design of candidate compounds.

Reference

1 Mukadam, S., Tay, S., Tran, D. et al. (2012). Evaluation of time-dependent cytochrome P450 inhibition in a high-throughput, automated assay: introducing a novel area under the curve shift approach. *Drug Metabolism Letters* 6 (1): 43–53.

11.6 Cytochrome P450 Reaction Phenotyping for Four Additional CYPs: Why and How

Huan Liu, Lifang Jiang, Genfu Chen

11.6.1 Introduction

Metabolism is the major clearance mechanism listed for approximately three-quarters of the top 200 prescribed drugs in the United States in 2002 [1]. Successful identification of the major enzyme(s) of a New Chemical Entity (NCE) is critical for the drug discovery program. CYP enzymes are listed as contributors to clearance for 48% of drugs cleared by metabolism, with CYP3A4 being the major isoform among all CYPs. Besides, it is suggested that additional enzymes, such as CYP2A6, CYP2J2, CYP4F2, and CYP2E1, also play a major role in drug metabolism [2–4]. Therefore, the objective of this chapter is to establish these additional human CYP reaction phenotyping assays using recombinant human CYPs. This chapter will illustrate the significance, methods, and results of the additional CYP reaction phenotyping assays and provide case studies on its applications.

11.6.2 Why Do We Need to Establish Additional Human CYP Reaction Phenotyping Assays?

If a drug is metabolized by a relatively singular enzyme and metabolism is the major route of elimination of the drug from the body, the risk of metabolism-based DDI with coadministered drugs may be very high. Reaction phenotyping is a study to identify and characterize the major enzymes involved in the metabolism of a drug. Accurate identification helps to provide an early risk assessment of their potential to exhibit significant metabolism-based DDI.

11.6.2.1 The requirements of CYP enzyme research scope in FDA Guidance

In the draft FDA Guidance (2006), CYP1A2, CYP2C8, CYP2C9, CYP2C19, CYP2D6, and CYP3A were listed as enzymes that should be evaluated in *in vitro* phenotyping experiments routinely to determine which enzymes are involved in the metabolism of the investigated drug. By 2012, CYP2B6, together with CYP1A2, CYP2C8, CYP2C9, CYP2C19, CYP2D6, and CYP3A, was included as the routine CYPs. Additional enzymes were mentioned for the first time. If a drug is not metabolized by the major CYPs (listed above), the drug's likelihood of being a substrate of the other CYP enzymes (e.g., CYP2A6, CYP2J2, CYP4F2, and CYP2E1) or non-CYP phase I enzymes should be considered. In the final FDA Guidance (2020), CYP1A2, CYP2B6, CYP2C8, CYP2C9, CYP2C19, CYP2D6, and CYP3A were listed as enzymes that should be evaluated in *in vitro* phenotyping experiments routinely to determine the enzymes metabolizing the investigational drug. If the investigational drug does not undergo significant *in vitro* metabolism by these major CYP enzymes, the sponsor should then determine what additional enzymes contribute to the metabolism of the investigational drug. These additional enzymes include but are not limited to:

- CYP enzymes including CYP2A6, CYP2J2, CYP4F2, and CYP2E1
- Other phase I enzymes
- Phase II enzymes

11.6.2.2 Additional enzymes play a major role in the metabolism of a drug

According to the final FDA Guidance (2020), the contribution of a specific metabolizing enzyme to an investigational drug's clearance is considered significant if the enzyme is responsible for > 25% of the drug's elimination based on the *in vitro* phenotyping studies and human PK data. Under these circumstances, the sponsor should conduct clinical DDI studies using strong index inhibitors and/or inducers of the enzyme. Thus, conducting reaction phenotyping assay has become a standard practice in identifying and characterizing the major enzymes in drug discovery and development.

As the number of new molecular entities (NMEs) is increasing, additional enzymes (e.g., CYP2A6, CYP2J2, CYP4F2, and CYP2E1) have played a major role in the metabolism of the investigational drugs. For example, CYP4F2 is the major enzyme in the metabolism of Fingolimod Hydrochloride (trade name: Gilenya) and the prodrug of Gemcitabine, respectively.

Considering (i) the requirements of FDA Guidance, (ii) the research and development trend of new investigational drugs, (iii) the abundance of individual CYPs in native human liver, and (iv) the proportion of CYPs involved in the metabolism of drugs, the scope of investigation for the additional CYP enzymes in this section will be defined as follows: CYP2A6, CYP2J2, CYP4F2, and CYP2E1 using recombinant human CYPs as the experimental system.

11.6.3 When Should We Conduct the Additional Human CYP Reaction Phenotyping Assays?

There are three methods for identifying the individual CYP enzymes responsible for a drug's metabolism: (i) chemicals, drugs, or antibodies as specific enzyme inhibitors in human liver microsomes; (ii) individual human recombinant CYP enzymes; (iii) correlation analysis which involves measuring the reaction velocity of the biotransformation pathway of interest in several individual human liver samples and correlating the reaction rates with the activity of the individual CYP enzymes in the same bank of microsomal samples. The two approaches, including CYP isozyme-selective chemical inhibition and metabolism study using recombinant human CYP enzymes, are widely used in reaction phenotyping assays.

1) CYP enzyme-selective chemical inhibition method: the investigational drug will be incubated with human liver microsome in the presence and absence of a selective chemical inhibitor to each CYP enzyme. The percentage inhibition of metabolism of the investigational drug by the CYP-selective inhibitor will be calculated. The CYP enzyme, which showed significant inhibition in the presence of its selective chemical inhibitor, is considered to be the metabolic enzyme of the investigational drug.
2) Metabolism study using individual human recombinant CYP enzymes: the investigational drug will be incubated with each human recombinant CYP enzyme to measure the reaction velocity. Scaled reaction velocity in microsomes will be calculated by multiplying the reaction velocity by the specific CYP with its expression abundance in native human liver microsomes. %Contribution by individual CYP in human liver microsomes will be calculated.

Generally, both the chemical inhibition study and the recombinant CYP metabolism study can provide consistent results. However, in some cases, the results of the chemical inhibition study and recombinant CYP metabolism study are inconsistent. Thus, the investigational drug is not found to undergo significant metabolism by these major CYP enzymes. Under such circumstances, additional human CYP450 reaction phenotyping assays should be conducted.

11.6.4 How Do We Establish These Additional Cytochrome P450 Reaction Phenotyping Assays?

Unlike the common CYP isozymes, the role of additional human CYP could not be confirmed in both the selective chemical inhibition method and the metabolism study using individual human recombinant CYP enzymes. Thus, the objective of this study is to obtain consistent values of the enzyme kinetic parameters (e.g., K_m and V_{max}) or intrinsic clearance (CL_{int}) using substrates for individual CYPs, as reported in the literature, based on the characteristics of drug metabolic enzymes. Taking CYP2J2 as an example, the methodology is described below.

Example 1: CYP2J2

CYP2J2 is the only member of the human CYP2J subfamily known for metabolizing arachidonic acid to physiologically important epoxides. The mRNA level of CYP2J2 is highest in the small intestine and heart, while it is relatively lower in the liver. There is an overlay of the two active sites with the I helix included for orientation between CYP2J2 and CYP3A4 [10]. The homology model suggests that the active site volume for CYP2J2 is similar to that of CYP3A4. Moreover, Ketoconazole, a CYP3A inhibitor commonly used in the chemical inhibition study in human liver microsomes, can inhibit the activity of CYP2J2 under certain conditions. Therefore, an apparent metabolism of a compound of interest by CYP3A in a chemical inhibitor approach could be mediated by CYP2J2. Since no selective chemical inhibitors are available for CYP2J2, CYP2J2 phenotyping could not be identified using the CYP enzyme-selective chemical inhibition method in human liver microsome. Therefore, a metabolism study using recombinant CYP2J2 was established by measuring the intrinsic clearance (CL_{int}) of Astemizole (CYP2J2 probe substrate).

11.6.5 Results of the Additional Human CYP Reaction Phenotyping Assays

The enzyme kinetic parameters (e.g., K_m and V_{max}) of metabolites from probe substrates for CYP2A6, CYP4F2, and CYP2E1 were determined. The results are shown below.

Under the optimized linear conditions, the enzyme kinetic parameters (e.g., K_m and V_{max}) or intrinsic clearance ($\mathrm{CL_{int}}$) using probe substrates for individual CYPs (CYP2A6, CYP4F2, CYP2E1, and CYP2J2) were determined and showed in Tables 11.14 and 11.15. The values for enzyme kinetic parameters (e.g., K_m and V_{max}) or intrinsic clearance ($\mathrm{CL_{int}}$) are consistent with the values from literature (within two fold). Therefore, additional human CYP reaction phenotyping assays (including CYP2A6, CYP4F2, CYP2E1, and CYP2J2) were established successfully.

The CYP2J2 metabolism study was performed according to the method of Caroline A Lee (2010) [10]. The results indicated that Astemizole was highly metabolized in recombinant human CYP2J2 at 100 pmol/mL. The %remaining of Astemizole was less than 10 at a 5-minute time point, suggesting the value of $\mathrm{CL_{int}}$ might be underestimated at 100 pmol/mL. The in-house $\mathrm{CL_{int}}$ value

Table 11.14 K_m Values between in-house and literature.

Enzyme	K_m (μM) In-house versus Literature [5–9]
CYP2E1	
CYP2A6	
CYP4F2	

Note: In-house (blue) and literature (red).

Table 11.15 $\mathrm{CL_{int}}$ Values between in-house and literature.

Enzyme	$\mathrm{CL_{int}}$ (μL/min/pmol) In-house versus Literature [10]
CYP2J2	

Note: In-house (blue) and literature (red).

for CYP2J2 was higher than the value from the literature using the optimized recombinant enzyme concentration (5 pmol/mL or 10 pmol/mL).

11.6.6 Application of the Additional Human CYP Reaction Phenotyping Assays

Case 1: For the initial metabolite M1 of compound A, the results of the chemical inhibition study and recombinant CYP metabolism study were contradictory. The chemical inhibition study showed that CYP3A was the major metabolic enzyme catalyzing the formation of M1. However, the recombinant CYP metabolism study showed that CYP2C9 was the major metabolic enzyme (Figure 11.27). The additional metabolism study using recombinant human CYP isozyme showed that CYP2J2 was the major metabolic enzyme (Figure 11.28). In conclusion, the above results showed that CYP2J2 was the major metabolic enzyme for the formation of M1.

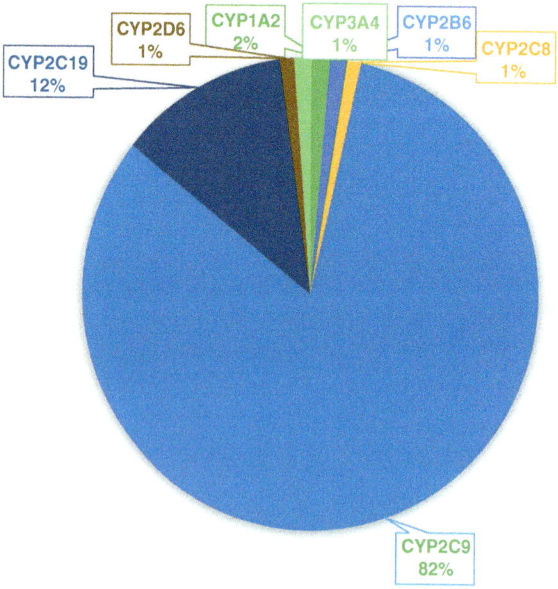

Figure 11.27 %Contribution of individual CYP (7CYPs) to the metabolite M1 from compound A in human liver microsomes.

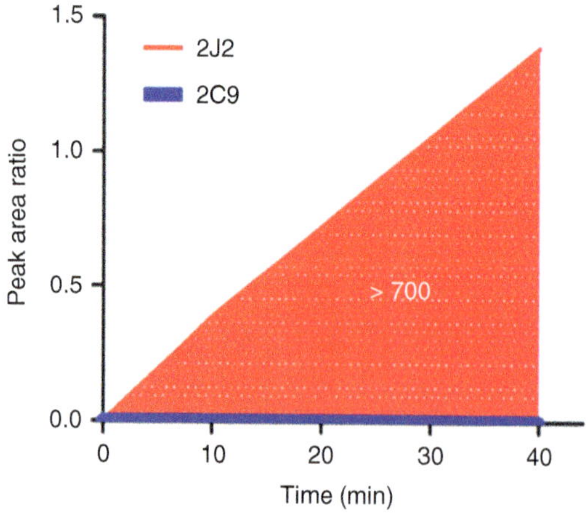

Figure 11.28 The relative formation rate of M1 in recombinant CYP2J2 and CYP2C9.

The cases above indicate that the additional recombinant CYP assays provided more important and reliable information to researchers.

11.6.7 Summary

The *in vitro* studies on identifying enzymes responsible for the metabolism of the drugs provide insights into the PK properties of these drugs. Some additional enzymes, such as CYP2J2, have been identified to play important roles in drug metabolism. Therefore, establishing a reliable system to identify the role of the additional human CYPs involved in drug metabolism is essential.

References

1 Zhang, D. and Surapaneni, S. (2012). *ADME-enabling technologies in drug design and development*, 550–559.

2 Raccor, B.S. and Kaspera, R. (2013). Extra-hepatic isozymes from the CYP1 and CYP2 families as potential chemotherapeutic targets. *Current Topics in Medicinal Chemistry* 13: 1441–1453.

3 Lee, C.A., Jones, J.P. III, and Katayama, J. (2012). Identifying a Selective Substrate and Inhibitor Pair for the Evaluation of CYP2J2 Activity. *Drug Metabolism and Disposition* 40: 943–951.

4 Christensen, H. and Hestad, A.L. (2011). CYP3A5-mediated metabolism of midazolam in recombinant systems is highly sensitive to NADPH cytochrome P450 reductase activity. *Xenobiotica* 41 (1): 1–5.

5 Yamamura, Y., Koyama, N., and Umehara, K. (2015). Comprehensive kinetic analysis and influence of reaction components for chlorzoxazone 6-hydroxylation in human liver microsomes with CYP antibodies. *Xenobiotica* 45 (4): 353–360.

6 Wang, M.Z., Saulter, J.Y., and Usuki, E. (2006, December). CYP4F Enzymes Are the Major Enzymes in Human Liver Microsomes That Catalyze the O-Demethylation of the Antiparasitic Prodrug DB289 [2,5-Bis(4-aminophenyl) furan-bis-O-methylamidoxime]. *Drug Metabolism and Disposition* 34 (12): 1985–1994.

7 Walsky, R.L. and Obach, R.S. (2004). Validated assays for human cytochrome P450 activities. *Drug Metabolism and Disposition* 32: 647–660.

8 Weymarn, L.B.V.O.N. and Murphy, S.E. (2003). CYP2A13-catalysed coumarin metabolism: comparison with CYP2A5 and CYP2A6. *Xenobiotica* 33 (1): 73–81.

9 Zhuo, X., Jun, G., and Zhang, Q.-Y. (1999). Biotransformation of Coumarin by Rodent and Human Cytochromes P-450: Metabolic Basis of Tissue-Selective Toxicity in Olfactory Mucosa of Rats and Mice. *The Journal of Pharmacology and Experimental Therapeutics.* 288 (2): 463–471.

10 Lee, C.A., Neul, D., Clouser-Roche, A., and Dalvie, D. (2010). Identification of Novel Substrates for Human Cytochrome P450 2J2. *Drug Metabolism and Disposition* 38: 347–356.

11.7 Overview of the Application of the Na^+-Taurocholate Cotransporting Polypeptide (NTCP) in Drug Development

Yuxi Wang, Xuan Zhang, Tao Xiong, Genfu Chen

11.7.1 Introduction

In 2012, Dr. Wenhui Li's laboratory discovered that the NTCP is a functional receptor for both hepatitis B virus (HBV) and hepatitis D virus (HDV), pioneering a new era in the drug development for hepatitis B and D. For his outstanding contributions to hepatitis B research and treatment, Dr. Li received the highest honor award in the global hepatitis B research field, the "Baruch S. Blumberg Prize" in November 2020, and the "Life Science Award" of the Future Science Prize on August 21, 2022. In recent years, the development of virus entry inhibitors targeting NTCP has become the mainstream of liver disease drug development. In August 2020, Bulevirtide, the world's first NTCP-targeted drug developed by Gilead for the treatment of hepatitis D, was approved for marketing by the EMA. In November 2021, Gilead also submitted a marketing application to the FDA for the treatment of hepatitis D. In this chapter, the interaction between NTCP and hepatitis virus and the role of NTCP in bile acid enterohepatic circulation were introduced, and the application of NTCP in drug development was described.

11.7.2 Structure of Na^+-Taurocholate Cotransporting Polypeptide

NTCP is the first transporter protein discovered in the solute carrier family 10 (SLC10), also known as SLC10A1 [1], located primarily on the basolateral membrane of hepatocytes. The human NTCP gene spans a genomic region of 21.4 kb, and it is located on chromosomes 6q24 and 14q24, encoding 349 amino acids [2]. NTCP is a multi-transmembrane protein with nine transmembrane domains (Figure 11.29) [3, 4].

11.7.3 Function of Na^+-Taurocholate Cotransporting Polypeptide and Its Application in Drug Development

NTCP is a functional receptor for HBV and HDV, mediating the virus invasion into hepatocytes. It plays a crucial role in regulating bile acids (salts) in the body, primarily transporting bile salts and sulfated compounds, while also mediating the transport of other substrates, such as thyroid hormones, drugs, and toxins [6].

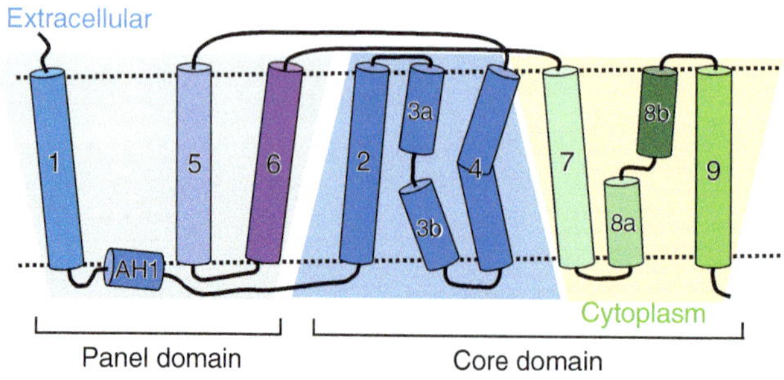

Figure 11.29 NTCP transmembrane topology diagram *Source:* [25] / Springer Nature / CC BY 4.0.

11.7.3.1 NTCP and Hepatitis Viruses

Research and Development of Anti-hepatitis Drugs

According to the WHO, viral hepatitis causes about 1.3 million deaths worldwide each year, mainly due to its progression to chronic liver disease and HCC. Of these, approximately 95% are caused by HBV and HCVs [7]. About 250 million people worldwide are infected with chronic hepatitis viruses [8], mainly HBV, HCV, and HDV, of which about 5% of HBV patients are coinfected with HDV. HDV is a defective virus that must be assisted by HBV or other hepatotropic DNA viruses to replicate and assemble into infectious virus particles. HDV coinfection accelerates the progression of HBV infection, and the current treatment effect for HBV–HDV coinfection is not satisfied [9]. For chronic HBV infection, there are two main treatments: interferon (IFN-α) and nucleoside analogs (Table 11.16), which are used to modulate immunity or inhibit viral replication. Although these treatments can slow the progression of hepatitis to some extent, there is no clinical cure, and in most cases, patients need to take medication for their whole lives [10]. For patients with chronic HBV–HDV coinfection, previous treatment options have been limited to interferon and its pegylated derivatives.

Chronic HBV and HDV infections are the major causes of liver diseases and cancer worldwide, and NTCP, as a functional receptor for HBV and HDV, plays a key role in the entry of viruses into hepatocytes. In August 2020, Bulevirtide (also known as Myrcludex B or MyrB), the world's first NTCP-targeted new drug developed by Gilead for the treatment of hepatitis D, was approved by the EMA, which effectively prevented the spread of the virus in the liver by inhibiting HDV/HBV from invading hepatocytes. In addition, in June 2023, the study of Bulevirtide in the treatment of hepatitis B progressed to the clinical phase III [11]. As a functional receptor for HBV and HDV, the discovery of NTCP provides new targets and strategies for developing therapeutic drugs for HBV/HDV infection [12]. Table 11.17 lists the entry inhibitors approved or under development for the treatment of HBV infection, including a variety of NTCP inhibitors [13]. Recent studies have also shown that NTCP can also regulate HCV infection in hepatocytes by modulating the innate immune response in the liver [14].

The Interaction of NTCP with Hepatitis Viruses

HBV binds to NTCP on the plasma membrane of hepatocytes via the pre-S1 domain of L-hepatitis B surface antigen (L-HbsAg), after which the virus particles are encapsulated, and the nucleocapsid is transported to the nucleus of the infected cell, where part of the double-stranded relaxed circular DNA is released (Figure 11.30). Subsequently, the host's DNA repair mechanism converts the incoming DNA into covalently closed circular DNA (cccDNA), which is packaged by histone and non-histone proteins into chromatin, forming the virus's microchromatin. This microchromatin serves as a transcription template for HBV and a crucial viral genomic intermediate in the HBV lifecycle. It also acts as a reservoir for the viral genome in the nucleus

Table 11.16 The main types of drugs in the clinical treatment of hepatitis B and their advantages and disadvantages.

Type of Drug	Mechanism of Action	Advantages	Disadvantages
Interferons	Immune modulation	Long-lasting effects, good efficacy with the possibility of discontinuation	Low negative conversion rate of hepatitis B surface antigen, poor tolerance
Nucleoside analogs	Inhibition of viral polymerase	Few side effects	Lifelong medication, virus rebound upon discontinuation

Table 11.17 Entry inhibitors for HBV infection that have been approved or are under development.

Substance	Target	Status
Heparin	S, M, L	FDA-approved
Suramin	S, M, L	FDA-approved
Proanthocyanidin	Pre-S1	Preclinical [15]
SALP	HSPGs	Preclinical [13]
Tayrocholic acid, UDCA, TUDCA, GUDCA	NTCP	/
Ezetimib	NTCP/NTC1L1	FDA-approved
Irbesartan	NTCP	FDA-approved
Myrcludex B	NTCP/interferes slightly with bile acid uptake	Phase III [11]
CsA	NTCP/interferes strongly with bile acid uptake	FDA-approved
SCY450	NTCP/no interferes with bile acid uptake	Preclinical [13]
SCY995	NTCP/no interferes with bile acid uptake	Preclinical [13]
Vanitaracin A	NTCP/interferes strongly with bile acid uptake	Preclinical [16]
Robert1-5253	NTCP/antagonist of retinoic acid receptor which regulates NTCP expression	Preclinical [13]
Evans blue	NTCP	FDA-approved
Hepalatide	NTCP	Phase II [17]
hzVSF (IgG4)	NTCP	Phase II [18]
A2342	NTCP	Preclinical [19]
HH-003	NTCP	Phase II [20]
HH-006	NTCP	Phase I [20]

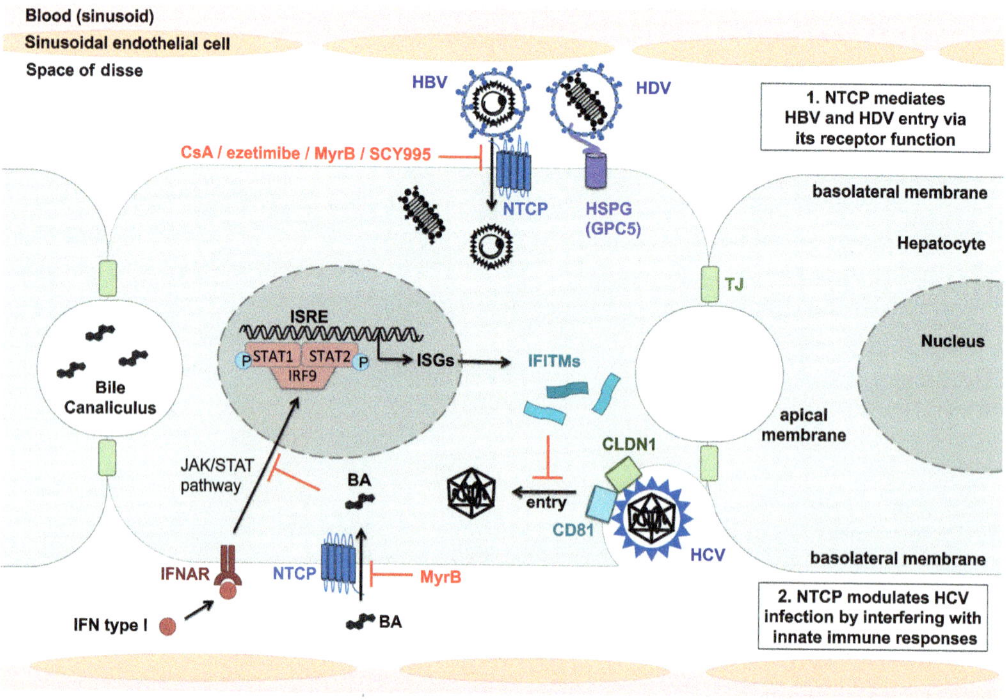

Figure 11.30 Interaction model between NTCP and HBV, HDV, and HCV. *Source:* [20] / with permission of Springer Nature.

of hepatocytes, which is replenished through the intracellular circulation of the nucleocapsid and continuous infection of new hepatocytes. Therefore, NTCP plays a critical role in the invasion of hepatocytes by HBV.

NTCP was identified as the long-sought specific receptor for the pre-S1 domain of HBV, which promoted the studies of HBV hepatotropism and species specificity [21]. Targeting the interaction between the HBV pre-S1 domain and the NTCP receptor required for HBV/HDV entry is a promising strategy to block the invasion of both viruses. Before NTCP was identified as the receptor for HBV/HDV, entry inhibitors targeting HBV pre-S1 had been demonstrated to effectively inhibit HBV infection *in vitro* and *in vivo* [22, 23]. Bulevirtide is a myristoylated pre-S1-derived peptide, which effectively inhibits the spread of HBV *in vivo* and its amplification in the cccDNA pool of infected human hepatocytes [24]. It is the first entry inhibitor targeting NTCP for HBV/HDV and exhibits a high safety profile with only minor interference with bile acid uptake [25, 26].

The identification of NTCP as the entry receptor for HBV/HDV has accelerated the development and discovery of several new potential entry inhibitors. The discovery of NTCP as the only identified receptor for both HBV and HDV is a milestone in the study of the life cycle of these viruses. This discovery has significantly advanced research on HBV/HDV entry into host cells and virus–host interactions. Moreover, based on the discovery, novel infection model systems using cell lines stably expressing NTCP allow for detailed studies on the early processes of the viral life cycle. By examining real infections in cell lines, these model systems could help researchers understand the formation and degradation of HBV cccDNA, a key target for curing HBV. A study evaluating FDA-approved drugs for liver diseases showed that NTCP inhibitors inhibited HDV entry [26], and three compounds, Irbesartan, Ezetimibe, and Ritonavir, inhibited HDV infection *in vitro* [27]. Although the interaction of Ezetimibe with NTCP was previously unknown, its inhibitory effect on HBV infection has been reported. In 2014, Watashi and his colleagues identified Cyclosporin as an HBV entry inhibitor targeting NTCP [28]. In the same year, scientists discovered that Cyclosporin could inhibit the transport of bile salts by NTCP, and it also inhibited HBV entry into liver cells due to interference with the NTCP receptor. Other studies have reported that the binding of HBV to NTCP and the binding of endogenous ligands to NTCP is competitive, thus, identifying inhibitors with strong NTCP inhibitory potency can provide a basis for the development of anti-HBV drugs [12]. Entry inhibitors targeting NTCP for HBV/HDV not only inhibit the transporter function of NTCP but also interfere with the uptake of bile salts by hepatocytes, increasing the risk of adverse reactions like cholestasis. Recently, researchers have proposed the use of the selective inhibition of NTCP to inhibit HBV entry without affecting bile acid uptake. Anna S Lok and others have shown that Cyclosporin derivatives, SCY450 and SCY995, can inhibit HBV/HDV entry without interfering with NTCP transporter activity [29]. Other scientists have also discovered a new class of entry inhibitors, oligomeric proanthocyanidins, and their analogs, targeting the HBV large envelope protein's pre-S1 domain directly, thereby preventing its attachment to NTCP. By directly targeting HBV particles, they impaired HBV's infectivity but did not affect NTCP-mediated bile acid transport activity [30].

In addition to playing a key role in the entry of HBV and HDV into hepatocytes, NTCP also plays an important role in HCV infection [31]. A study found that NTCP was a host-dependent factor for HCV infection, and further emphasized its importance in virus–host interactions. The potential of NTCP-targeted antiviral drugs to enhance innate antiviral response and participate in the host immune system in clearing infections may be a useful characteristic for the treatment of all hepatotropic viruses, including HBV, HCV, and HDV. Animal model systems expressing human NTCP will make studying virus–host interactions and antiviral treatment possible. Furthermore,

NTCP has been identified as an antiviral target, and several molecules targeting NTCP are under clinical development in 2024, to improve future treatment methods.

11.7.3.2 NTCP and Bile Acid (Salt) Circulation

Bile formation is another essential function of the liver. The main components of bile are bile salts, phospholipids, and organic anions [32]. In the small intestine, bile salts are almost quantitatively reabsorbed and transported back to the liver via the portal circulation (Figure 11.31). In the liver, hepatocytes reabsorb bile salts and secrete them back into the bile, facilitating continuous entero-hepatic circulation. Bile salt molecules can act as carriers to remove substances with poor water solubility (such as cholesterol) from the bile. Excessive accumulation of bile salts within hepatocytes can damage mitochondrial function, leading to cytotoxicity [33]. NTCP transports both conjugated and unconjugated bile salts from the blood into hepatocytes in a sodium-dependent manner, cotransporting two sodium ions and one bile salt molecule [34], making it an electrogenic transporter [35]. It's noteworthy that the expression of NTCP can be regulated by phosphorylation/dephosphorylation under various pathophysiological conditions [36]. In the case of impaired bile secretion, NTCP expression is typically downregulated to protect hepatocytes from the accumulation of cytotoxic bile salts. Inhibition of NTCP interferes with the reabsorption of bile salts, which in turn impacts the clearance of bile. This increases the content of bile acids in the blood, leading to toxic side effects. Pfizer's antidiabetic drug Troglitazone and its primary metabolites are inhibitors of the bile salt export pump (BSEP), which is involved in bile secretion and bile salt transport, and can inhibit the excretion of bile salts. At the same time, Troglitazone is also an inhibitor of

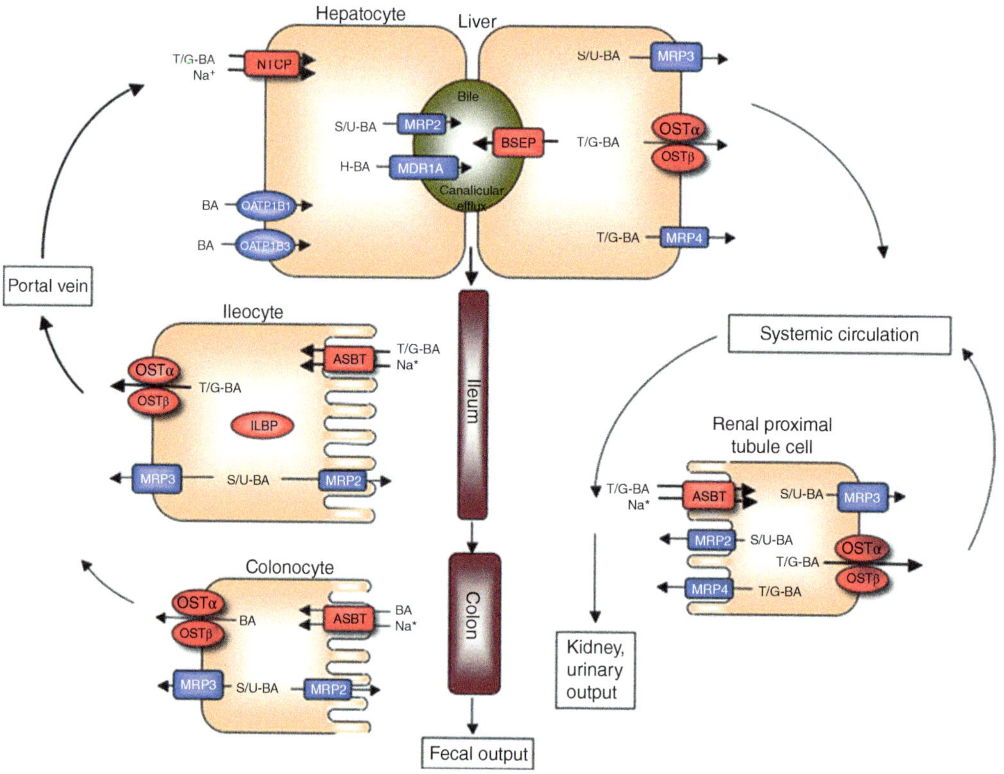

Figure 11.31 Transporter-mediated bile acid cycle. *Source:* [37] / with permission of Springer Nature.

NTCP, reducing the reabsorption of bile salts back into the liver. This simultaneously increases the levels of bile salts in both the liver and blood, leading to cholestatic liver disease (one of the clinical manifestations of drug-induced liver injury). As a result, regulatory agencies like the UK's Medicines and Healthcare Products Regulatory Agency (MHRA) and the US FDA have requested the withdrawal of Troglitazone.

11.7.4 *In Vitro* Na^+-Taurocholate Cotransporting Polypeptide Model and Its Main Applications

Our laboratory has used HEK293 cells transfected with human NTCP to establish a cell-based model for assessing NTCP substrates and inhibitors. The validation data for substrates and inhibitors used in experiments were highly consistent with those reported in the literature [30] (Tables 11.18 and 11.19). Taurocholate is a known NTCP substrate. To reduce endogenous interference and improve experimental sensitivity, Taurocholate salt-D4 was selected as a known substrate for NTCP. The K_m value (the substrate concentration at which half-maximum transport rate was achieved) for Taurocholate-D4 was (12.4 ± 0.82) μmol/L. The IC_{50} values (half-maximal inhibitory concentration, referring to the concentration of a substance that is required for 50% inhibition in a biological process) for inhibitors Cyclosporin and Ritonavir against NTCP were (0.93 ± 0.07) μmol/L and (1.58 ± 0.77) μmol/L, respectively. Cyclosporin was selected as the inhibitor for the NTCP substrate evaluation experiment. The uptake ratio of Taurocholate-D4, which represents the ratio of transport activity of the compound between HEK293-NTCP and HEK293-MOCK cells, was 30.9 ± 6.07, and the Rs/Ri value (the ratio of transport activity of the compound in HEK293-NTCP cells with and without the presence of a positive inhibitor) was 13.9 ± 3.79; both the uptake ratio and Rs/Ri value were greater than 2, indicating that HEK293-NTCP cells exhibited good substrate uptake ability, and the inhibitors can effectively inhibit substrate uptake by NTCP under the test conditions.

Some studies [38, 39] have shown that NTCP binds to substrates such as Taurocholate and this area overlaps with the region that HBV and HDV enter hepatocytes via NTCP (Figure 11.32). Therefore, the invasion of viruses can be prevented by blocking the binding of HBV or HDV to NTCP using inhibitors or substrates of the NTCP transporter, thereby controlling disease progression. Using *in vitro* models to rapidly screen potential inhibitors or substrates of NTCP opens up

Table 11.18 Comparison of in-house validation data with literature data.

Compound Name	Parameter	WuXi AppTec DMPK Data (μmol/L), n = 3	Literature Value (μmol/L)
Taurocholate-D4 (substrate)	K_m	12.4 ± 0.82	10.1 ± 0.60
Cyclosporin (inhibitor)	IC_{50}	0.93 ± 0.07	1.00 ± 0.68
Ritonavir (inhibitor)	IC_{50}	1.58 ± 0.77	2.10 ± 1.30

Table 11.19 Validation data for NTCP uptake of known substrates.

Compound Name	Uptake Ratio, n = 3	Rs/Ri, n = 3
Taurocholate-D4	30.9 ± 6.07	13.9 ± 3.79

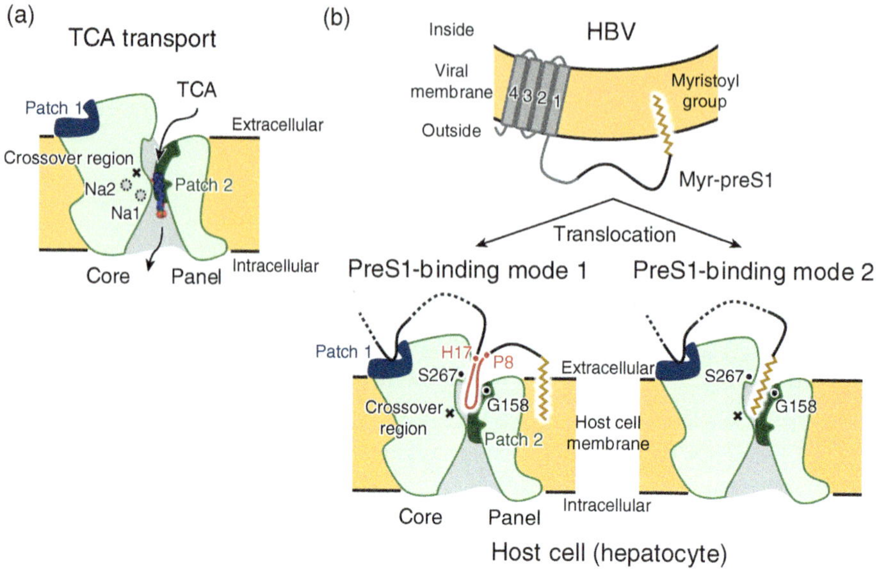

Figure 11.32 Modes of binding to the NTCP transporter. (a) Substrate binding mode. (b) NTCP receptor binding mode. *Source:* [40] / with permission of Springer Nature.

new eras for the development of new drugs to cure HBV/HDV. Moreover, NTCP has a high affinity for conjugated bile salts, and tauroursodeoxycholic acid conjugates can be specifically taken up into hepatocytes by NTCP, becoming a potential HCC targeting treatment method. Thus, cell models stably expressing NTCP can also provide research tools for studying liver-targeted drugs. Furthermore, as a specific uptake transporter expressed in the liver, NTCP plays an important role in the disposition of drugs in the body. Exogenous substances undergo metabolic inactivation or activation after entering the liver; studying the substrate and inhibitory effects of drugs on the NTCP transporter can also predict drug-induced liver injury. Therefore, studying the interactions between drugs and NTCP can provide a basis for clinically rational medication use. With the help of the NTCP cell model, the infection mechanism of HBV and HDV can also be further investigated.

11.7.5 Summary

NTCP is not only a crucial transporter for the absorption of bile salt into the liver but also a functional receptor for HBV and HDV. Therefore, the establishment of the *in vitro* NTCP model provides a novel tool for the rapid screening of potential inhibitors or substrates, research, and development of new drugs for HBV and HDV treatment and is also of great significance for drug safety evaluation.

References

1 Hagenbuch, B. and Dawson, P. (2004). The sodium bile salt cotransport family SLC11. *Pflügers Archiv* 447: 566–570.

2 Hagenbuch, B., Stieger, B., Foguet, M. et al. (1991). Functional expression cloning and characterization of the hepatocyte Naþ/bile acid cotransport system. *Proceedings of the National Academy of Sciences of the United States of America* 88: 10629–10633.

3 Hallen, S., Mareninova, O., Branden, M., and Sachs, G. (2002). Organization of the membrane domain of the human liver sodium/bile acid cotransporter. *Biochemistry* 41: 7253–7266.

4 Mareninova, O., Shin, J.M., Vagin, O. et al. (2005). Topography of the membrane domain of the liver Na+-dependent bile acid transporter. *Biochemistry* 44: 13702–13712.

5 Park, J.-H., Iwamoto, M., and Yun, J.-H. (2022). Structural insights into the HBV receptor and bile acid transporter NTCP. *Nature* 606: 1027–1031.

6 Ni, Y., Lempp, F.A., Mehrle, S. et al. (2014). Hepatitis B and D viruses exploit sodium taurocholate co-transporting polypeptide for species-specific entry into hepatocytes. *Gastroenterology* 146: 1070–1083.

7 World Health Organization (2016). Global health sector strategy on viral hepatitis 2016–2021. Towards ending viral hepatitis. World Health Organization.

8 Schweitzer, A., Horn, J., Mikolajczyk, R.T. et al. (2015). Estimations of worldwide prevalence of chronic hepatitis B virus infection: A systematic review of data published between 1965 and 2013. *Lancet* 386: 1546–1555.

9 World Health Organization (2017). Global hepatitis report 2017. World Health Organization. License: CC BY-NC-SA 3.0 IGO.

10 Urban, S., Bartenschlager, R., Kubitz, R., and Zoulim, F. (2014). Strategies to inhibit entry of HBV and HDV into hepatocytes. *Gastroenterology* 147: 48–64.

11 Wedemeyer, H., Aleman, S., Brunetto, M.R. et al. (2023). A Phase 3, Randomized Trial of Bulevirtide in Chronic Hepatitis D. *The New England Journal of Medicine* 389 (1): 22–32.

12 Yan, H., Zhong, G., Xu, G. et al. (2012). Sodium taurocholate cotransporting polypeptide is a functional receptor for human hepatitis B and D virus. *eLife* 1: e000049.

13 Herrscher, C., Roingeard, P., and Blanchard, E. (2020). Hepatitis B Virus Entry into Cells. *Cells* 9 (6): 1486.

14 Piccini, L.E., Castilla, V., and Damonte, E.B. (2015). Dengue-3 Virus Entry into Vero Cells: Role of Clathrin-Mediated Endocytosis in the Outcome of Infection. *PLoS One* 10: e0140824.

15 Kamisuki S, Shibasaki H, Ashikawa K, et al. (2022). Determining the absolute configuration of vanitaracin A, an anti-hepatitis B virus agent. *The Journal of Antibiotics* (Tokyo).75(2):92–97.

16 Shanghai HEP Pharmaceutical. Hepalatide by Shanghai HEP Pharmaceutical for Hepatitis B: Likelihood of Approval. (2024).

17 Miyakawa, Y., Otsuka, M., Sekiba, K. et al. (2021). Humanized virus-suppressing factor inhibits hepatitis B virus infection by targeting viral cell entry. *Heliyon.* 7 (7): e07586.

18 Britta, B., Ellen, S., Ivana, U. et al. (2022). P15The orally available sodium/taurocholate co-transporting polypeptide inhibitor A2342 blocks hepatitis B and D entry *in vitro*. *BMJ Journals.* 71 (3): A42–A42.

19 Pan, Y., Xia, H., He, Y. et al. (2023). The progress of molecules and strategies for the treatment of HBV infection. *Frontiers in Cellular and Infection Microbiology* 13: 1128807.

20 Eller, C., Heydmann, L., Colpitts, C.C. et al. (2018). The functional role of sodium taurocholate cotransporting polypeptide NTCP in the life cycle of hepatitis B, C, and D viruses. *Cellular and Molecular Life Sciences* 75 (21): 3895–3905.

21 Yoshimori, T., Yamamoto, A., Moriyama, Y. et al. (1991). Bafilomycin A1, a specific inhibitor of vacuolar-type H (+)-ATPase, inhibits acidification and protein degradation in lysosomes of cultured cells. *The Journal of Biological Chemistry* 266: 17707–17712.

22 Chojnacki, J., Anderson, D.A., and Grgacic, E.V.L. (2005). A Hydrophobic Domain in the Large Envelope Protein Is Essential for Fusion of Duck Hepatitis B Virus at the Late Endosome. *Journal of Virology* 79: 14945–14955.

23 Matlin, K.S., Reggio, H., Helenius, A., and Simons, K. (1982). Pathway of vesicular stomatitis virus entry leading to infection. *Journal of Molecular Biology* 156: 609–631.

24 Watashi, K., Sluder, A., Daito, T. et al. (2014). Cyclosporin A and its analogs inhibit hepatitis B virus entry into cultured hepatocytes by targeting a membrane transporter, sodium taurocholate cotransporting polypeptide (NTCP): Watashi et al. *Hepatology* 59: 1726–1737.

25 Macovei, A., Petrareanu, C., Lazar, C. et al. (2013). Regulation of Hepatitis B Virus Infection by Rab5, Rab7, and the Endolysosomal Compartment. *Journal of Virology* 87: 6415–6427.

26 Hong, H.J., Ryu, C.J., Hur, H. et al. (2004). *In vivo* neutralization of hepatitis B virus infection by an anti-preS1 humanized antibody in chimpanzees. *Virology* 318: 134–141.

27 Johannsdottir, H.K., Mancini, R., Kartenbeck, J. et al. (2009). Host Cell Factors and Functions Involved in Vesicular Stomatitis Virus Entry. *Journal of Virology* 83: 440–453.

28 Lok, A.S., Zoulim, F., and Dusheiko, G. (2017). Ghany MG: Hepatitis B cure: from discovery to regulatory approval. *Journal of Hepatology* 67: 847–861.

29 Lok, A.S., Zoulim, F., Dusheiko, G., and Ghany, M.G. (2017). Hepatitis B cure: from discovery to regulatory approval. *Hepatology* 66: 1296–1313.

30 Herrscher, C., Roingeard, P., and Blanchard, E. (2020). Hepatitis B Virus Entry into Cells. *Cells* 9 (6): 1486.

31 Esteller, A. (2008). Physiology of bile secretion. *World Journal of Gastroenterology* 14 (37): 5641–5649.

32 Sokol, R.J., Devereaux, M., Dahl, R., and Gumpricht, E. (2006). "Let there be bile" – understanding hepatic injury in cholestasis. *Journal of Pediatric Gastroenterology and Nutrition* 43 (Suppl 1): S4–S9.

33 Hagenbuch, B. and Meier, P.J. (1996). Sinusoidal (basolateral) bile salt uptake systems of hepatocytes. *Seminars in Liver Disease* 16: 129–136.

34 Weinman, S.A. (1997). Electrogenicity of Na(þ)-coupled bile acid transporters. *The Yale Journal of Biology and Medicine* 70: 331–340.

35 Anwer, M.S. (2004). Cellular regulation of hepatic bile acid transport in health and cholestasis. *Hepatology* 39: 581–590.

36 Kim, R.B., Leake, B., Cvetkovic, M. et al. (1999). Modulation by drugs of human hepatic sodium-dependent bile acid transporter (sodium taurochlate cotranporting polypeptide) activity. *The Journal of Pharmacology and Experimental Therapeutics* 291: 1204–1209.

37 Dawson, P.A. (2011). Role of the intestinal bile acid transporters in bile acid and drug disposition. *Handbook of Experimental Pharmacology* 201: 169–203.

38 McRae, M.P., Lowe, C.M., Tian, X. et al. (2006). Ritonavir, saquinavir, and efavirenz, but nt nevirapin, inhibit bile acid transport in human and rat hepatocytes. *The Journal of Pharmacology and Experimental Therapeutics* 319: 1068–1075.

39 Ignat, M.D., Balta, A.A.S., Barbu, R.E. et al. (2024). Antiviral Therapy of Chronic Hepatitis B Virus between Present and Future. *Journal of Clinical Medicine* 13 (7): 2055.

40 Asami, J., Kimura, K.T., Fujita-Fujiharu, Y. et al. (2022). Structure of the bile acid transporter and HBV receptor NTCP. *Nature* 606 (7916): 1021–1026.

11.8 The Significance and *In Vitro* Methods of Lysosomal Trapping in Drug Development

Yu Zhang, Qing Liu, Tao Xiong, Genfu Chen

11.8.1 Introduction

The lysosome, as a key digestive organelle and a transmission center of metabolic signals, is involved in the regulation of a variety of physiological metabolic processes. It has been reported in recent studies that a variety of marketed drugs exhibit a high affinity for lysosomes. Lipophilic or amphiphilic compounds with basic moieties are likely to be protonated and trapped within lysosomes, which can affect the PK characteristics, efficacy, and safety of these drugs. Long-term administration may result in the accumulation of drugs in tissue lysosomes, leading to drug resistance and some adverse effects. Therefore, it is necessary to determine whether the test compounds can be trapped by lysosomes as early as possible, to better predict the distribution of drugs in tissues and sub-cells, and provide a reference for the prediction of ADME properties and early safety assessments in drug development. This study will describe the mechanism and application of lysosomal trapping based on the structure and function of lysosomes and provide the corresponding *in vitro* research methods.

11.8.2 Overview of Lysosomes

11.8.2.1 Structure of Lysosomes

Lysosomes are membrane-bound, heterogeneous intracellular organelles and are widely distributed in all mammalian cells except mature erythrocytes. There are more than 60 hydrolytic enzymes in lysosomes (Figure 11.33), and the pH value of lysosomal luminal is approximately from 4.5 to 5. Vacuolar H^+- ATP_{ase} (V-type ATP_{ase}), a member of multiple lysosomal membrane proteins, contributes to the acid luminal pH by transporting protons into lysosomes through the energy released from ATP hydrolyses, thereby forming an intracellular pH gradient between the lysosomal lumen and the cytoplasm. The V-type ATPase acts synergistically with other key lysosomal membrane proteins including several ion channels proteins (e.g., Cl^-, Na^+/K^+ channels proteins) to maintain the homeostasis of the acidic luminal environment.

11.8.2.2 Function of Lysosomes

Abundant hydrolytic enzymes such as proteases, lipases, phosphatases, and nucleases are expressed in lysosomes. All of these hydrolytic enzymes have good activities and help lysosomes digest a variety of endogenous or exogenous macromolecules transported by endocytosis, phagocytosis, and autophagy and also digest old or damaged organelles.

In addition to the basic digestive functions, lysosomes also play important roles in the regulation of innate and adaptive immunity, and nutrient sensing. Lysosome-related mechanisms are well reported in the development of diseases including cancer, lysosomal storage disorders, neurodegenerative diseases, and autoimmune disorders [2].

11.8.3 Mechanism of Lysosomal Trapping and Its Effect on Absorption, Distribution, Metabolism, and Excretion (ADME)

Lysosomal trapping is a process whereby lysosomotropic drugs are protonated and trapped within lysosomes. Lysosomotropic drugs have two properties, namely lipophilicity (drugs can passively diffuse across cell membranes) and basicity (basic moieties can be protonated inside lysosomes).

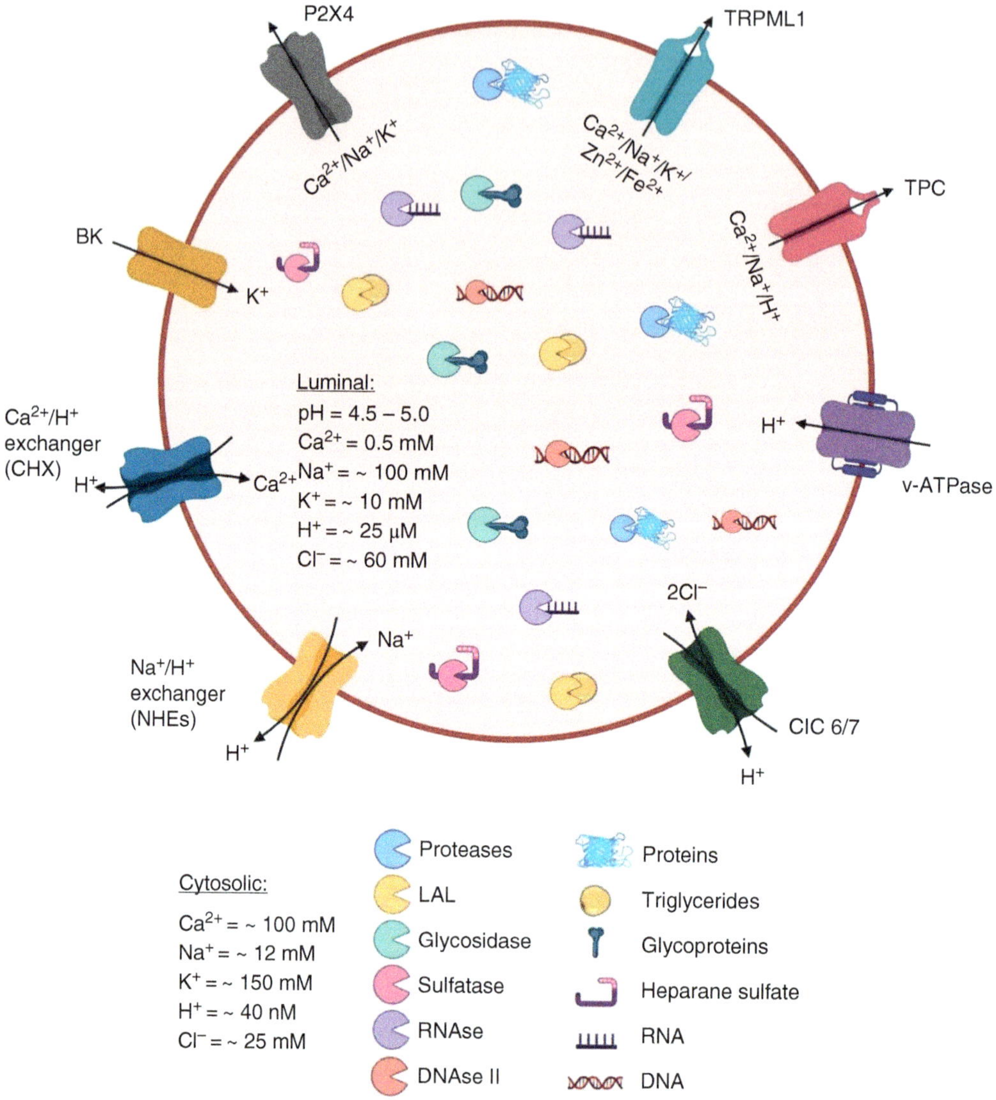

Figure 11.33 Structure of lysosomes. *Source:* [1] / MDPI. / CC BY 4.0.

11.8.3.1 Mechanism of Lysosomal Trapping

Lipophilic weak bases with a Clog P > 2.0 (partition coefficient of the neutral species of a compound between octanol and water) and a pK_a between 6.5 and 11 are probably trapped in lysosomes. Under neutral pH conditions, compounds with both lipophilic or amphiphilic moieties and basic moieties are free to penetrate biological membranes by endocytotic and pinocytotic. Once it reaches an acid compartment, such as lysosomes, the acidic environment promotes protonation and ionization of basic moieties. Ionized compounds have poorer permeability and thus are accumulated in lysosomes, this phenomenon is known as lysosomal trapping (Figure 11.34). Lipophilic weak bases are likely to be protonated in the acidic lumen and then sequestered in the lysosomal compartment.

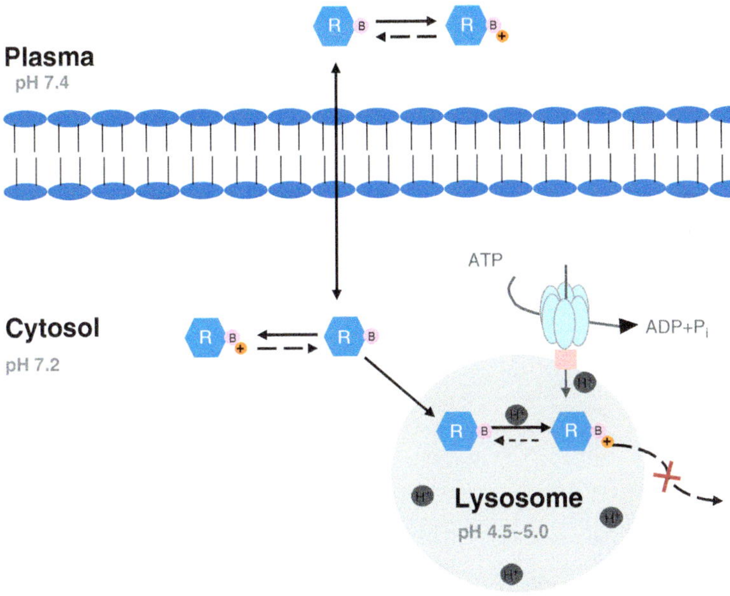

Figure 11.34 Mechanism of lysosomal trapping.

11.8.3.2 Effect of Lysosomal Trapping in Absorption, Distribution, Metabolism, and Excretion (ADME)

The PK parameters of drugs are determined by multiple factors such as administration route (e.g., oral administration, injection administration, transdermal administration, and inhalation administration), physiological factors (e.g., tissue or organ blood flow, humoral pH, and transporter expression), intrinsic properties of drugs (e.g., lipophilic, acidity and alkalinity, and plasma binding), and tissue composition.

Lysosomotropic drugs are widely distributed in fat-rich tissues, such as adipose, liver, and kidney, due to their good lipophilic. Besides, they are likely to be trapped within acidic organelles including lysosomes and Golgi apparatus, thus lysosomotropic drugs show extremely high tissue distribution in lysosomal-rich tissues and organs (such as lung, kidney, liver, spleen, brain, and intestine). The apparent volume of distribution increases, as well as the corresponding elimination half-life and retention time of drugs *in vivo*. In addition, for orally administered drugs, lysosomal trapping results in undesirable drug efficacy by limiting the extent and rate of drug absorption. Dextromethorphan (DEX) is classified as a BCS I compound in the Biopharmaceutical Classification System (BCS) due to its high solubility and permeability throughout the GI tract. However, the plasma levels with T_{max} (time to reach the maximum plasma concentration (C_{max})) of DEX is approximately 2.5 hours after oral administration, which is much slower than other typical BCS I drugs (the T_{max} of pravastatin is 1–1.5 hours) [3]. The delayed absorption is directly correlated with its lysosomotropic properties.

11.8.4 Application of Lysosomal Trapping

11.8.4.1 Lysosomal Membrane Permeabilization-Induced Cell Death

Hydrolytic enzymes are sufficiently sequestered in the lysosomal lumen due to the highly glycosylated lysosomal membrane. Damage of lysosomes often leads to lysosomal membrane permeabilization (LMP). In this case, soluble lysosomal contents will be released into the cytoplasm.

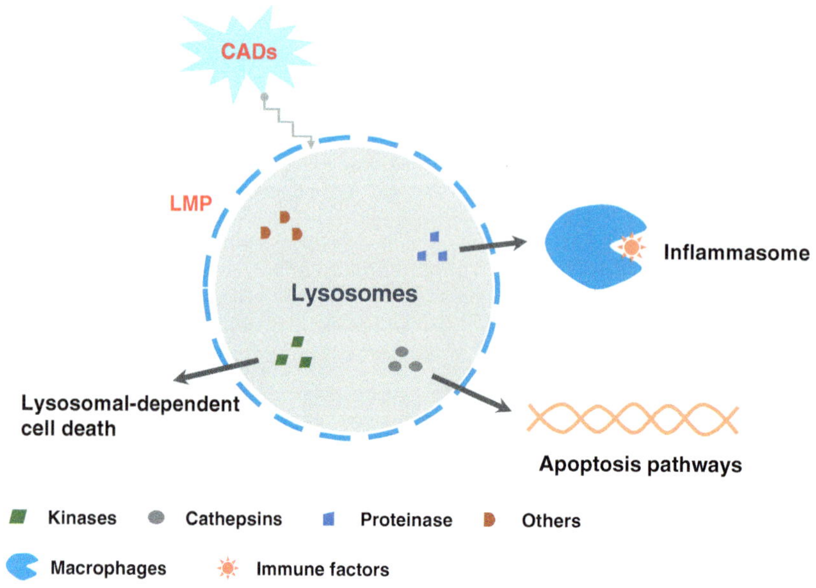

Figure 11.35 Proposed model of LMP-induced cell death for CADs.

The leakage of some lysosomal enzymes, including cathepsin B and cathepsin D, triggers a cascade of hydrolysis of the cytoplasmic contents and/or activates several pathways, thereby inducing cell death (Figure 11.35).

In tumor cells, lysosomes exhibit larger size and increased number, and the levels of lysosomal enzymes also increase. Many cationic amphiphilic drugs (CADs) with antitumor activity are lysosomotropic, they can cross biological membranes, sequestered within the lysosomes, and promote LMP-induced cell death. Accordingly, tumor cells may be more sensitive to the toxicity of CADs than normal cells, and it suggests that LMP-induced cell death may serve as an effective therapeutic strategy for cancer [4]. Long-term treatment with CADs may result in LMP and trigger a cascade of regulated cell death.

11.8.4.2 Drug-Drug Interactions Within Lysosomes

A variety of anti-cancer, anti-malarial, anti-arrhythmic as well as antidepressant drugs are lysosomotropic. If their intracellular target sites are located in the nucleus and/or cytoplasm, lysosomal trapping will reduce drug distribution at the target site and induce chemoresistance. Therefore, the therapeutic effect of lysosomotropic drugs is often limited when used alone and can be improved by the coadministration of similar lysosomotropic drugs or lysosomal pH alkalinization drugs.

Sorafenib (SF), a marketed chemotherapeutic agent for HCC, shows limited drug sensitivity and availability due to lysosomal trapping. Verteporfin (VP), a lysosomotropic autophagy inhibitor and antitumor photosensitizer, can induce lysosomal pH alkalinization and LMP, leading to the dysregulation of autophagy and enhancer tumor-specific proteotoxicity. When SF is in combination with VP, this therapeutic regimen can effectively inhibit tumor proliferation and progression and shows good antitumor effects (Figure 11.36) [5].

SF is passively trapped in the lysosomal compartment, thereby it can reduce drug availability and induce chemoresistance. VP treatment leads to LMP-induced cell death and tumor-specific proteotoxicity. A combination of VP and SF can overcome chemo-insensitivity and exert great antitumor effects.

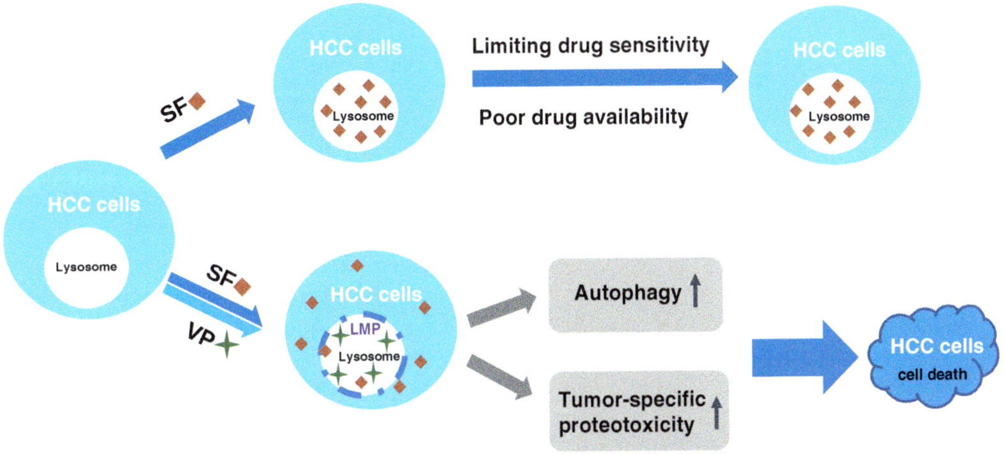

Figure 11.36 Combination of sorafenib and verteporfin in the treatment of liver cancer. *Source:* [5] / Springer Nature / CC BY 4.0.

11.8.4.3 Drug-Induced Phospholipidosis

It is reported that the long-term treatment of CADs contributes to the development of drug-induced phospholipidosis (DIPL) because of an intracellular excessive accumulation of phospholipids. Several potential mechanisms have been proposed that accumulating CADs inhibit lysosomal phospholipid degradation, enhance phospholipid and cholesterol biosynthesis, and induce the formation of nondegradable drug–phospholipid complex.

Tummino et al. [6] reported that the antiviral activity of many CADs including hydroxychloroquine, azithromycin, and amiodarone showed a consistent correlation with phospholipidosis in the treatment of SARS-CoV-2. These results suggested that lysosomal trapping-induced DIPL may play a role in the application of antiviral drugs.

Nevertheless, much research showed that CAD treatment can raise the lysosomal luminal pH, reduce lysosomal hydrolase activities, and alter lysosomal membrane composition. Therefore, accumulating CADs and phospholipids may damage the normal physiological function of lysosomes, resulting in vacuolization of macrophages and a variety of parenchymal cells, which are pathological hallmarks of multiple impaired neural and cardiovascular tissues [7]. Early phospholipid deposition safety risk assessment of drugs in advance can provide a reference for the screening and structural optimization of lead compounds and increase the feasibility of drug candidates to enter clinical trials.

11.8.5 *In Vitro* Methods for Lysosomal Trapping

There are two main methods for lysosomal trapping assessment: (i) direct method based on analysis by liquid chromatography–tandem mass spectrometry (LC–MS/MS); (ii) indirect method using lysosome-specific fluorescent probes (Table 11.20).

11.8.5.1 Direct Method (LC–MS/MS Analysis)

V-type ATPase inhibitors can inhibit lysosomal acidification, thereby reducing the accumulation of drugs within lysosomes. The difference in intracellular or endolysosomal concentrations of lysosomotropic drugs in the absence and presence of V-type ATPase inhibitors reflects whether the compounds have the potential for lysosomal trapping. Rapid, accurate, and high-throughput

Table 11.20 *In vitro* methods for lysosomal trapping.

	Direct Method	Indirect Method
Mechanism	Intracellular and/or endolysosomal drugs decreased with the rising pH values of the lysosomal lumen	Intracellular and/or endolysosomal fluorescence intensity diminished with the accumulation of lysosomotropic drugs
Parameters	Accumulated amount	Fluorescence intensity
Methods	LC–MS/MS method	High-content fluorescence microscopy
Analysis	Accumulation of test compounds in the absence or presence of V-type ATP_{ase} inhibitors	Fluorescence intensity of Lyso Tracker probe in cells treated with different concentrations of the compound

in vitro screening assays have been reported for assessing lysosomal trapping using multiple cellular models including MDCK II cells, rat cardiomyocyte cells H9C2, and human hepatoma cells HepG2 [8–10].

11.8.5.2 Indirect Method (Accumulation of Lyso Tracker Probes)

Lyso Tracker probe is a pH-dependent and lysosome-specific fluorescent probe. Lyso Tracker Red DND-99 (LTR) and Lyso Tracker Green DND-26 (LTG) are commonly used. Accumulating lysosomotropic drugs might competitively decrease the intensity of Lyso Tracker probes in lysosomes. By comparing the fluorescence intensity of Lyso Tracker probes within lysosomes and/or intracellular in the absence and presence of compounds by using high-content fluorescence microscopy, the potential of lysosomal trapping for compounds can be predicted.

11.8.6 Establishment of Lysosomal Trapping Assay

Considering that LC–MS/MS analysis can more directly and accurately reflect the intracellular accumulation level of lysosomotropic drugs, and it requires fewer test concentrations, our laboratory has established two screening models for lysosomal trapping assays using a transwell system and a plated-cell system.

11.8.6.1 Transwell System

MDCK II cells were seeded on the 96-well transwell insert plates and cultured at 37.0°C, 5.0% CO_2, and saturated humidity for a few days before being used. Test compounds were tested in the absence and presence of V-type ATPase inhibitors in apical to the basolateral (A-B) direction. Concentrations of compounds in samples were quantified by LC–MS/MS. Lysosomal trapping could be predicted by comparing the change of solution recovery values of compounds in the A-B direction with and without inhibitors.

Five known lysosomotropic drugs (chlorpromazine, loperamide, imatinib, propranolol, and DEX) and diclofenac (a negative control) were selected as model drugs for validation experiments, compound information is listed in Table 11.21. Under the tested conditions, for lysosomotropic drugs, the %Solution Recovery values were all increased significantly ($p<0.05$) in the presence of inhibitor, while there was no difference in the %Solution Recovery value of diclofenac ($p>0.05$) (Figure 11.37). These results demonstrated that the MDCK II transwell system was considered acceptable.

Table 11.21 Information of model compounds.

Compound Name	Log P	pKa (Strongest Acidic)	pKa (Strongest Basic)	Acid/Alkaline	Lysosomal Trapping
Chlorpromazine	4.6	—	9.2	Alkaline	Yes
Loperamide	5.5	14.0	9.4	Alkaline	Yes
Imatinib	3.5	12.5	8.3	Alkaline	Yes
Propranolol	2.6	—	9.7	Alkaline	Yes
Dextromethorphan	3.5	—	9.9	Alkaline	Yes
Diclofenac	4.3	4.0	—	Acid	No

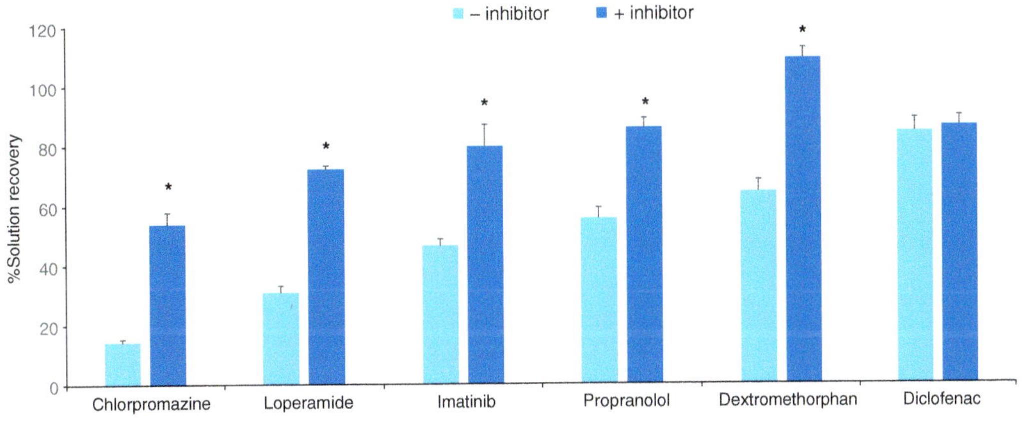

Figure 11.37 Validation of the lysosomal trapping model in MDCK II cell transwell system using model compounds (* means $p<0.05$).

Subsequently, we selected two screening compounds, compounds A and B, to confirm the transwell system. Chlorpromazine was used as a positive control compound, and diclofenac was used as a negative control compound. After incubation, the %Solution Recovery value of compound A was increased significantly, while the %Solution Recovery value of compound B was not changed in the presence of an inhibitor (Figure 11.38). Therefore, compound A is likely to be trapped by lysosomes within MDCK II cells, while compound B is not.

11.8.6.2 Plated-Cell System

MDCK II cells were plated on a PDL-coated 96-well plate and cultured overnight at 37.0°C, 5.0% CO_2, and saturated humidity before being used. MDCK II cells were preincubated with Hank's buffer with or without V-type ATP_{ase} inhibitors, followed by another incubation of test compounds with or without inhibitors. After incubation, the intracellular concentration of test compounds in cell lysate samples was measured by LC–MS/MS method.

In the plated-cell system, under the conditions tested, the treatment of inhibitor reduced the intracellular accumulation of lysosomotropic drugs (chlorpromazine, loperamide, imatinib, propranolol, and DEX), while there was no difference in the intracellular accumulation of the acidic

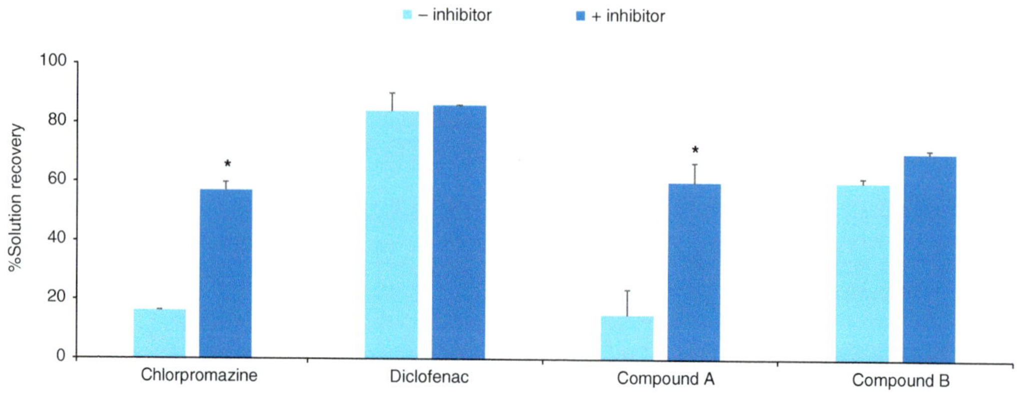

Figure 11.38 Validation of the lysosomal trapping model in MDCK II cell transwell system using two screening compounds (* means $p<0.05$).

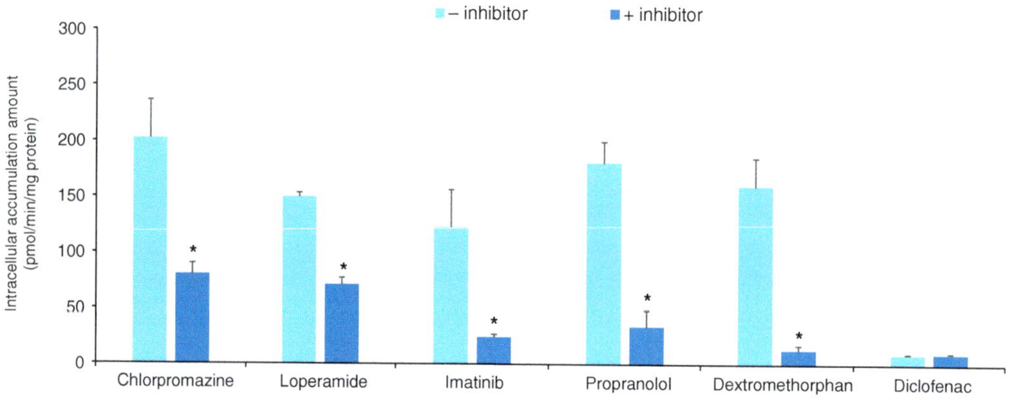

Figure 11.39 Validation of the lysosomal trapping model in the plated-MDCK II cell system using model compounds.

compound diclofenac (Figure 11.39). This result further demonstrated that the plated-MDCK II cell system was considered acceptable.

Similarly, the plated-cell system was utilized to concurrently investigate whether compounds A and B are subject to lysosomal trapping. DEX was used as a positive control compound, and diclofenac was used as a negative control compound. Under the conditions tested, in the presence of an inhibitor, the intracellular accumulation of compound A in MDCK II cells decreased significantly, whereas the intracellular accumulation of compound B showed no significant difference (Figure 11.40). This finding is consistent with the conclusions drawn from the transwell system.

11.8.6.3 Comparison of the Transwell System and the Plated-Cell System

Both the MDCK II cells transwell system and the plated-cell system can be utilized for lysosomal-trapping high-throughput assessment *in vitro*. The MDCK II cells transwell system allows for the evaluation of lysosomal trapping potential for test compounds, while simultaneously predicting the permeability, and exploring the impact of lysosomal trapping effects on the permeability of these test compounds. Compared to the MDCK II cell transwell system, the

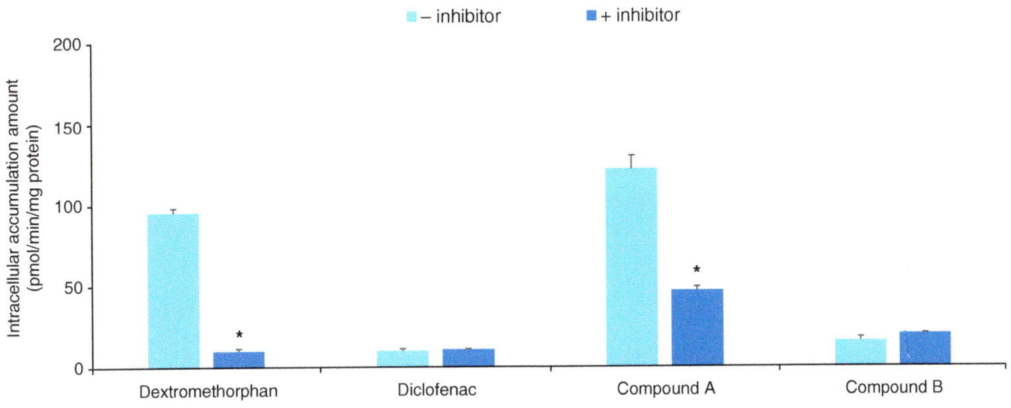

Figure 11.40 Validation of the lysosomal trapping model in plated MDCK II cell system using two screening compounds.

plated-MDCK II cell system has a shorter experimental cycle and lower cost, making it more suitable for high-throughput screening during the early stages of drug development.

11.8.7 Summary

Lysosomal trapping is present in a variety of marketed drugs as well as in the drug candidates under development. Accumulation of drugs in lysosomes can affect the PK characteristics, efficacy, and safety of drugs. Lysosomotropic drugs can also cause DDIs by changing the pH, volume of distribution, and enzyme binding sites in the lysosomal lumen, further affecting efficacy or toxicity. Therefore, early identification of whether the drugs can be trapped by lysosomes can provide more comprehensive guidance for drug development.

References

1 Trivedi, P.C., Bartlett, J.J., and Pulinilkunnil, T. (2020). Lysosomal biology and function: modern view of cellular debris bin. *Cells* 9 (5): 1131–1165.

2 Zhang, Z., Yue, P., Lu, T. et al. (2021). Role of lysosomes in physiological activities, diseases, and therapy. *Journal of Hematology and Oncology* 14 (1): 79–117.

3 Bolger, M.B., Macwan, J.S., Sarfraz, M. et al. (2018). The irrelevance of *in vitro* dissolution in setting product specifications for drugs like dextromethorphan that are subject to lysosomal trapping. *Journal of Pharmaceutical Sciences* 108 (1): 268–278.

4 Serrano-Puebla, A. and Boya, P. (2018). Lysosomal membrane permeabilization as a cell death mechanism in cancer cells. *Biochemical Society Transactions* 46 (2): 207–215.

5 Gavini, J., Dommann, N., Jakob, M.O. et al. (2019). Verteporfin-induced lysosomal compartment dysregulation potentiates the effect of sorafenib in hepatocellular carcinoma. *Cell Death and Disease* 10 (10): 749–765.

6 Tummino, T.A., Rezelj, V.V., Fischer, B. et al. (2021). Drug-induced phospholipidosis confounds drug repurposing for SARS-CoV-2. *Science* 373 (6554): 541–547.

7 Breiden, B. and Sandhoff, K. (2019). Emerging mechanisms of drug-induced phospholipidosis. *Biological Chemistry* 401 (1): 31–46.

8 Bednarczyk, D. and Sanghvi, M.V. (2019). The impact of assay recovery on the apparent permeability, a function of lysosomal trapping. *Xenobiotica* 50 (7): 753–760.

9 Nadanaciva, S., Lu, S., Gebhard, D.F. et al. (2010). A high content screening assay for identifying lysosomotropic compounds. *Toxicology in Vitro* 25 (3): 715–723.

10 Schmitt, M.V., Lienau, P., Fricker, G., and Reichel, A. (2019). Quantitation of lysosomal trapping of basic lipophilic compounds using *in vitro* assays and *in silico* predictions based on the determination of the full pH profile of the endo-/lysosomal system in rat hepatocytes. *Drug Metabolism and Disposition* 47 (1): 49–57.

12

In Vivo Pharmacokinetics

12.1 Enhancing the Bioavailability of Poorly Soluble Compounds: Strategies for Formulation Optimization in Preclinical Studies

Yu Chen, Jacob Zhi Chen, Chao Zhang, Shoutao Liu

12.1.1 Introduction

The increasing prevalence of lead compounds, candidate compounds, and approved drugs featuring high hydrophobicity and low water solubility is becoming a notable trend. This is particularly true for NCEs, where the proportion of poorly soluble compounds appears to be on the rise. The low solubility of these compounds presents a considerable challenge to their absorption in the intestinal tract, resulting in low systemic bioavailability. This creates a hurdle in accurately assessing the PK, efficacy studies, dosage determination, DDI studies, formulation development, safety assessment, etc. for compounds with low water solubility in aqueous system designs.

Given the low probability of these compounds receiving approval, the optimization of formulations in preclinical PK studies necessitates meticulously crafted strategies and designs. Additionally, it is crucial to ensure the safety and tolerance of the excipients in laboratory animals during *in vivo* experiments. This chapter discusses strategies for optimizing formulations in preclinical PK studies of poorly soluble compounds, with a focus on solvent selection and particle size reduction, to expedite the progress of drug development.

12.1.2 Poorly Soluble Compounds and Why Are They Prevalent?

The Biopharmaceutics Classification System (BCS) scientifically categorizes all compounds based on their solubility and permeability [1]. It comprises four classes: high solubility and high permeability (Class I), low solubility and high permeability (Class II), high solubility and low permeability (Class III), and low solubility and low permeability (Class IV) (Figure 12.1). Classes II and IV are poorly soluble compounds.

Poorly soluble Classes II and IV compounds constitute approximately 90% of the NCEs (Figure 12.2) [2]. During preclinical studies, most compounds exhibit insoluble characteristics, making it necessary to optimize preclinical formulations for further PK studies.

Drug Metabolism and Pharmacokinetics: Frontiers, Strategies, and Applications, First Edition.
Edited by Liang Shen and WuXi AppTec DMPK.

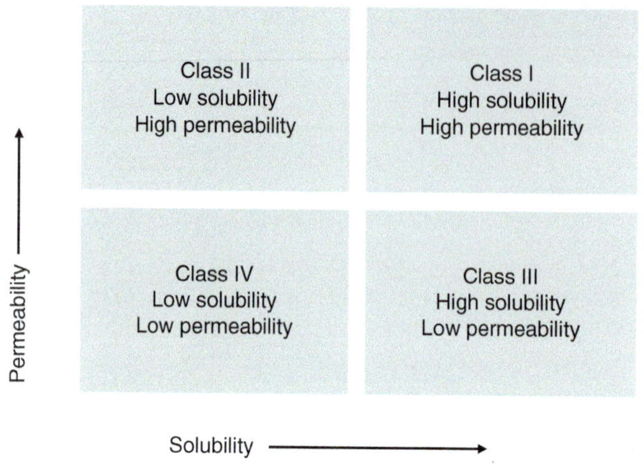

Figure 12.1 Biopharmaceutics classification system.

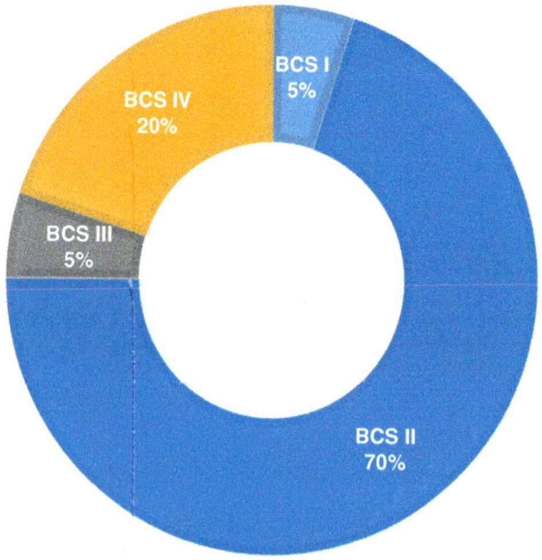

Figure 12.2 The proportion of insoluble compounds of NCEs. Data from reference. *Source:* Adapted from reference [2].

Chemists have proposed various hypotheses for the trend toward poorly soluble compounds. The actual reasons are multifaceted. As chemists pursue high potency in compounds by modifying their structures, considerations such as receptor binding, target and intracellular signaling channels, lipid architecture, and highly lipophilic endogenous ligands come into play. These factors increase the need for highly lipophilic, low water-soluble candidate compounds to access and interact with the target.

12.1.3 How to Improve *In Vivo* Absorption of Poorly Soluble Compounds?

The poor *in vivo* absorption of these compounds not only results in low oral bioavailability within the animal but also poses challenges in establishing a good linear relationship between exposure and dose in dose–range-finding studies. This situation complicates the execution of PK studies and the initiation of IND trials. From a formulation perspective, there are three main approaches to address these challenges:

1) Selection of appropriate solvents.
2) Particle size reduction to increase dissolution rate.
3) Application of novel formulations, such as solid dispersions, liposomes, and the self-emulsifying drug delivery system (SEDDS).

This chapter primarily concentrates on the first two approaches, while the SEDDS will be discussed in Section 12.2.

12.1.4 Selecting Appropriate Solvents

Choosing and using a combination of solvents is crucial for enhancing drug solubility. There are five major categories of solvents used to increase solubility: pH modifiers, cosolvents, inclusion complexes, surfactants, and lipids.

1) pH modifiers

As shown in Figure 12.3, 75% of compounds are basic and 20% are acidic [3]. Most molecules are weak acids or weak bases that can be ionizable; it may be possible to achieve the target concentration through pH modification. Therefore, buffer solutions are widely used, such as citrate buffer, acetic acid buffer, or PBS. Different routes of administration have different pH constraints. Typically, the pH of orally administered formulations should be between 2 and 11, with formulations of pH 4–8 being more common due to lower irritation. The pH range for intravenous administration is narrower, between 3 and 9, to reduce vascular irritation.

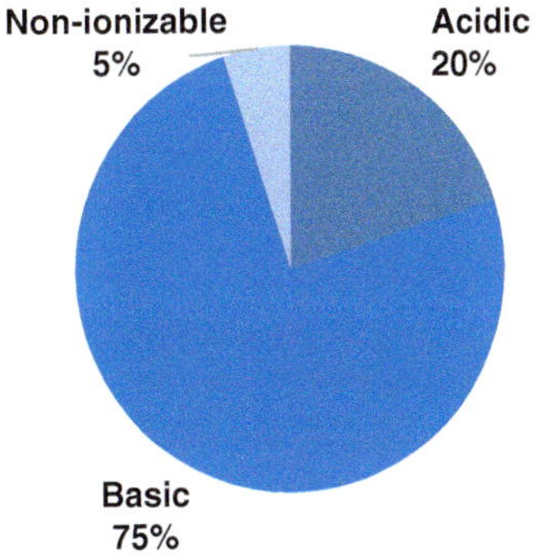

Figure 12.3 Percentage distribution of acidity and basicity of drugs. Data from reference. *Source:* [3].

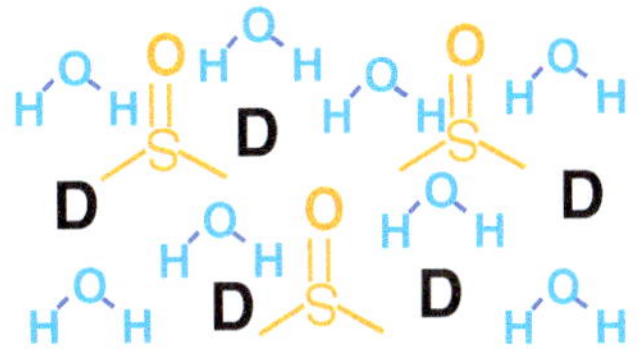

Figure 12.4 Action principle of cosolvents.

2) Cosolvents

Cosolvents are water-miscible organic reagents that are typically used to enhance the solubility of low water-soluble compounds. The use of cosolvents is based on the principle that different solvents have specific affinities for various structural aspects of chemical entities (Figure 12.4). This ensures that the compound with low aqueous solubility achieves maximum solubility at a certain ratio. Common cosolvents range from highly polar reagents such as DMSO, *N*-methyl-2-pyrrolidone (NMP), dimethylacetamide (DMA), and ethanol, to weakly polar ones such as low-molecular-weight polyethylene glycol (PEG) and propylene glycol (PG). Due to the safety profile of highly polar cosolvent reagents, 10% to 15% of US FDA-approved parenteral products incorporate cosolvents. These formulations not only substantially increase the solubility of certain drugs but also reduce the hydrolysis reactions of some drugs in the solution, enhancing their formulation stability. The use of cosolvents for solubility enhancement can meet solubility requirements by increasing the proportion of organic solvents. However, excessive use can lead to adverse reactions, so the quantity needs to be controlled within a certain range.

3) Inclusion complex

As illustrated in Figure 12.5, cyclodextrins (CDs) are widely employed for solubilizing poorly soluble drugs. Their distinctive cage-like structure forms host–guest molecular complexes, wherein nonpolar drug molecules are enclosed within the nonpolar cage structure. The polyhydroxy exterior of CD exhibits a strong affinity for polar water molecules, leading to enhanced solubility. The most commonly used derivatives of β-CD as solubilizers are hydroxypropyl-β-CD (HP-β-CD) and sulfobutyl ether-β-CD (SBE-β-CD). Their advantages include improving drug stability, water solubility, safety, reducing renal toxicity, mitigating drug hemolysis, controlling drug release rates, and masking unpleasant odors. Since the first drug was launched in 1976, the number of drug-CD formulations approved on the market has grown to over 100 products, and this number continues to increase.

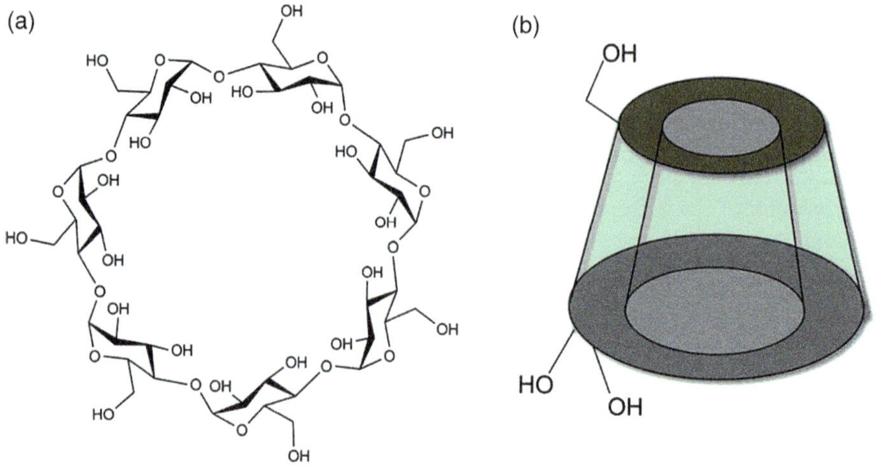

Figure 12.5 The structure of hydroxypropyl-β-cyclodextrin (a) is the molecular structure, and (b) is the three-dimensional structure. *Source:* [1] / with permission of The American Society for Pharmacology and Experimental Therapeutics.

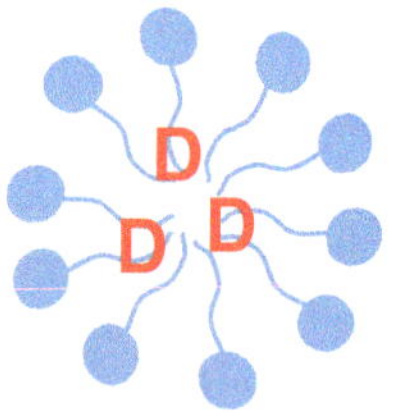

Figure 12.6 The surfactant wraps the drug.

4) Surfactants

Surfactants are commonly used to solubilize poorly water-soluble compounds by incorporating compounds into surfactant micelles, simultaneously increasing the stabilization of suspension formulations (Figure 12.6). In formulations containing proteins, surfactants can also decrease protein degradation. Traditional surfactants include Tween 80, polyoxyethylated castor oil, and so on. Novel surfactants, such as glyceride 44/14 and Polyethylene Glycol 15 Hydroxystearate (Solutol HS-15), exhibit superior solubilization effects, biocompatibility, and safety compared to traditional surfactants. Therefore, Solutol HS-15 is commonly used as a replacement for Tween 80. However, it is worth noting that high surfactant concentrations could potentially cause local or systemic adverse reactions after parenteral administration, such as hypersensitivity reactions.

5) Lipids

Lipids have been proven to be effective in aiding the dissolution of compounds with high lipid solubility, with common options including oils and caprylic/hydroxy decanoate triglycerides.

Lipid-based drug delivery (LBDD) is a method that successfully enhances the solubility of lipophilic drugs in formulations. More importantly, it enables normal absorption and increasing bioavailability in the gastrointestinal tract. Commonly used lipids include Labrafac PG, Maisine® CC, and Transcutol® HP. Lipid excipients have been demonstrated to be applicable for BCS Class II drugs and show potential for Class III and IV drugs. They enhance the drug's solubility in the formulation, ensuring its dissolution in the gastrointestinal tract. Lipid excipients can be digested and absorbed *in vivo*, promoting the absorption of the drug. The collaborative effect of lipid excipients facilitates the absorption of the drug, and the digestion process of lipid excipients helps maintain the drug in a dissolved state (Figures 12.7 and 12.8). Lipid excipients dissolve the drug in the formulation, facilitating the entry of the drug into the body, and crucially, maintaining its dissolved state even after exposure to the aqueous system.

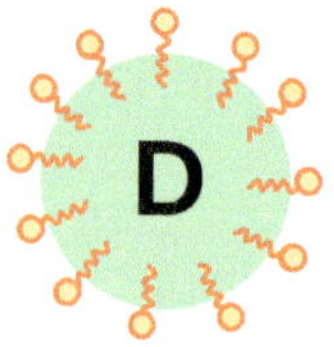

Figure 12.7 The lipids wrap drugs.

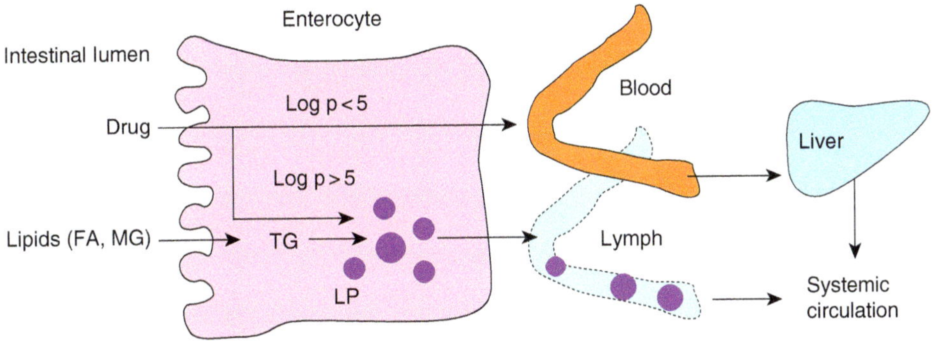

Figure 12.8 Schematic representation of how lipid-based formulations (LBFs) promote the absorption and bioavailability of poorly water-soluble drugs. *Source:* Adapted from reference [1].

For lipid excipient prescription, the enhancement of drug bioavailability involves the use of medium-chain length oils ($C < 12$), which facilitates drug permeation through the intestinal epithelium, directly diffusing into the bloodstream. Unsaturated oils with long chains ($C \geq 12$) stimulate chylomicron secretion, increasing drug absorption through the lymphatic pathway and avoiding the first-pass effect in the liver.

The hot spot formulation SEDDS used for lipophilic compounds involves the combination of different ratios of solubilizers/surfactants/oil esters. Considerations such as the tolerance of different animal species, routes of administration, food effect, and other variables, as well as *in vitro* simulation testing evaluation methods, all have preclinical formulation screening decision trees and successful application cases [4].

The rational utilization of the five major classes of combination solvents mentioned above can effectively enhance drug solubility and avoid the possibility of compound precipitation after administration. Cosolvents, CDs, and surfactants are the most used and can be combined with pH modification for a markedly improved solubilization effect.

12.1.5 Particle Size Reduction to Increase Dissolution Rate

Reducing the particle size of a compound, thereby increasing its specific surface area (the total surface area per unit mass of material), can enhance the dissolution rate of poorly water-soluble compounds (Figure 12.9). The dissolution rate refers to the amount of drug that dissolves into the solution per unit of time at a certain temperature. The dissolution rate is determined by the gravitational

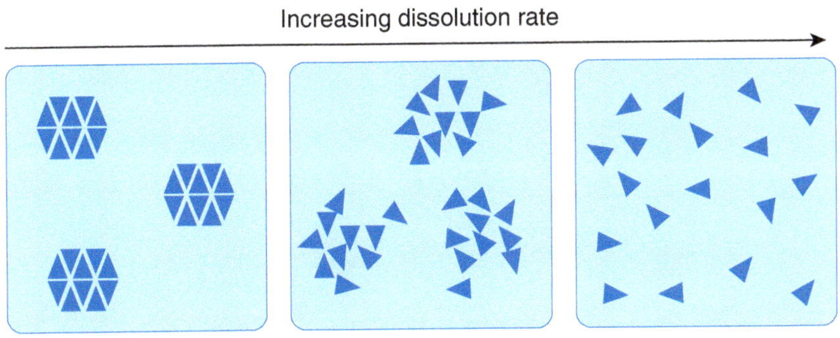

Figure 12.9 The particle size decreases and the specific surface area increases. *Source:* Adapted from reference [1].

forces between the solvent, solute molecules, and the diffusion rate of solute molecules in the solvent. Decreasing the particle size of solid drugs, thereby increasing the specific surface area, enhances the interaction between the drug and the solvent, leading to increased solubility. Therefore, obtaining micronization or nanoscale formulations through size reduction can improve the bioavailability of poorly water-soluble compounds. Commonly used methods for particle size reduction include mortar grinding, ultrasonic fragmentation, ultra-high-speed homogenization, and ball milling.

Traditionally, small particles are obtained through top-down methods, such as mortar grinding, ultrasonic fragmentation, and ultra-high-speed homogenization. These methods yield micrometer-sized particles with diameters of 1–10 μm, a process known as micronization (Figure 12.10). With technological advancements and the widespread use of stabilizers, ball milling can produce even smaller nanoparticles, typically 200–500 nm (Figure 12.10).

Accurate measurement of particle size can be achieved using a polarization microscope, and the particle size distribution of nanoscale formulations can be effectively assessed using a size distribution analyzer, providing valuable insights into the quality of the formulation (Figures 12.11 and 12.12).

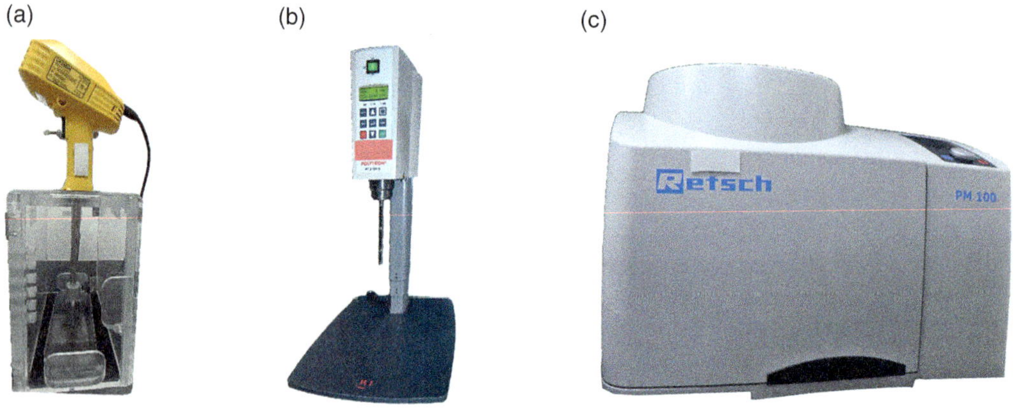

Figure 12.10 Cell disruption (a), high-speed homogenization (b), and ball milling (c).

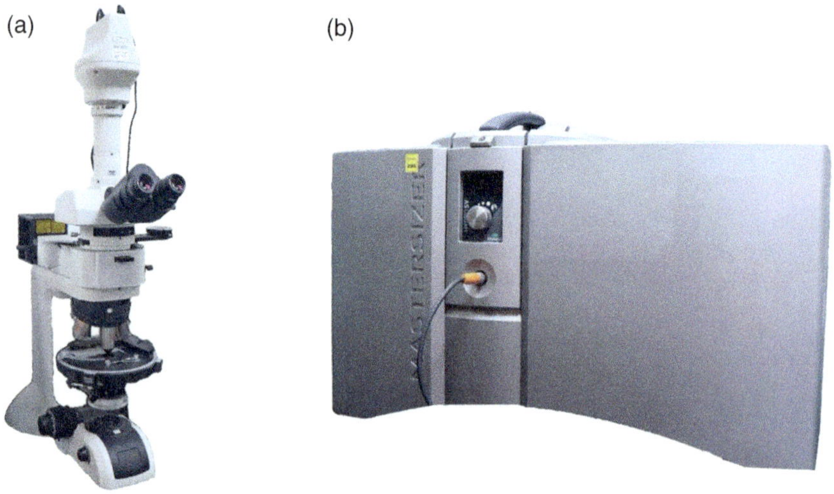

Figure 12.11 (a) Polarizing microscope and (b) particle size analyzer.

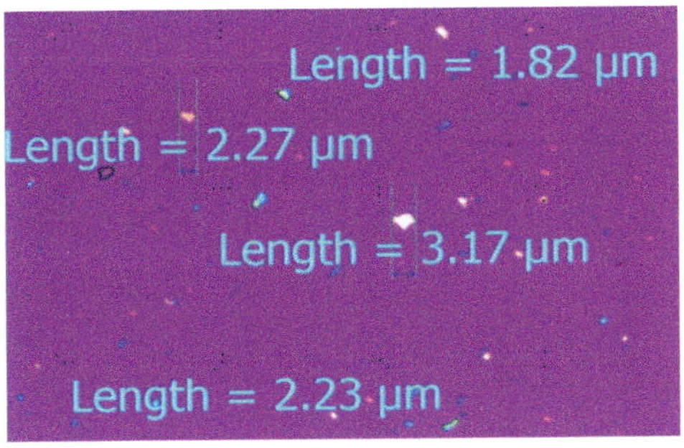

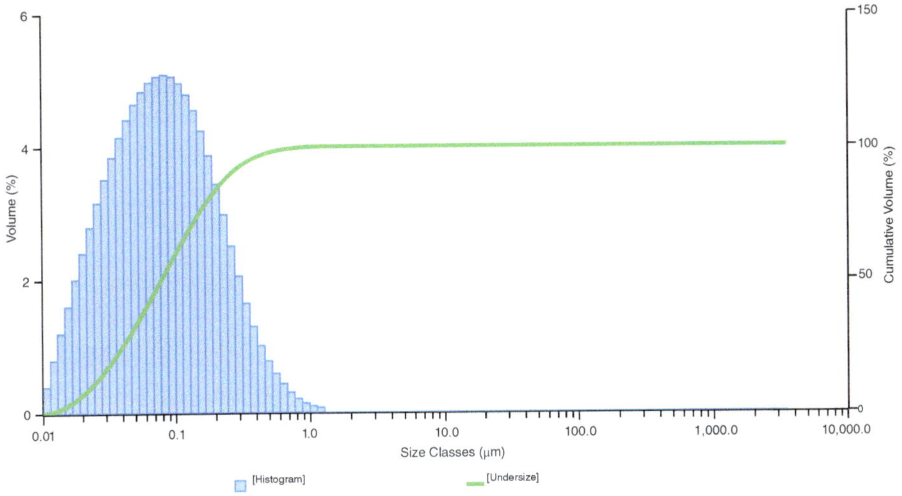

Figure 12.12 Polarizing microscope and particle size distribution detection result.

12.1.6 Summary

Solubility is an important physicochemical property. To improve dissolution through formulation optimization strategies, it is necessary to design a tolerable solvent formulation based on compound properties, animal species, dosing routes, and dosing volume. The *in vitro* simulation experiments, formulation stability, and formulation solubility should also be taken into consideration for optimal solvent combinations for animal experiments.

References

1 Williams, H.D., Trevaskis, N.L., Charman, S.A. et al. (2013). Strategies to address low drug solubility in discovery and development. *Pharmacological Reviews* 321–418.

2 Sherry Ku., M. (2008). Use of the Biopharmaceutical Classification System in Early Drug Development. *The AAPS Journal.* 10 (1): 208–212.

3 Kearns, E.H. and Li, D. Translated by Zhong Dafang(2011). Nature of drug class. In: *Concepts, Structural Design, and Methodology. From ADME to Security Optimization* (ed. et al.). Beijing: Science Press.

4 Leonaviciute, G. and Bernkop-Schnurch, A. (2015). Self-emulsifying drug delivery systems in oral(poly)peptide drug delivery. *Expert Opinion on Drug Delivery* 12 (11): 1703–1716.

12.2 Self-Emulsifying Drug Delivery System (SEDDS): Enhancing the Oral Absorption of Lipophilic Compounds

Xin Li, Jacob Zhi Chen, Chao Zhang, Shoutao Liu

12.2.1 Introduction

Oral formulations have become the preferred choice in the development of most pharmaceutical formulations due to their convenience of administration, painless intake, and lower risk of systemic infection or adverse reactions at the site of administration. Their ease of storage and transportation, cost-effectiveness, and potential for prolonged or controlled drug release further contribute to their appeal in the pharmaceutical industry. However, numerous orally administered drugs with high lipid solubility face challenges such as precipitation and low solubility in the gastrointestinal environment, leading to low bioavailability and hindering their therapeutic effectiveness. Statistics suggest that in the early stages of new drug screening, approximately 90% of drugs are classified as poorly soluble [1]. This low water solubility can result in reduced oral bioavailability and significant interindividual variability. In Section 12.1, we discussed various formulation strategies to enhance the *in vivo* exposure of poorly soluble compounds. This chapter focuses on a novel formulation strategy – the SEDDS – and its application in improving the oral bioavailability of lipophilic compounds.

12.2.2 Components and Advantages of the Self-Emulsifying Drug Delivery System

12.2.2.1 Components of the Self-Emulsifying Drug Delivery System

The SEDDS is a liquid or solid formulation composed of an oil phase, nonionic surfactants, and cosurfactants. When introduced into water or simulated gastric fluid (SGF) and gently stirred under conditions typical of the gastrointestinal tract (usually at body temperature 37°C), the SEDDS spontaneously emulsifies into droplets with a particle size of 100–500 nm.

12.2.2.2 Advantages of the Self-Emulsifying Drug Delivery System

Self-emulsifying technology, considered an advanced formulation approach, with well-established theoretical foundations, has been successfully applied to marketed drugs. This technology enables lipophilic drugs to spontaneously form a stable microparticle system in water, enhancing their dissolution and dispersion capabilities. It simultaneously promotes the passage of the drug through the mucous membrane of intestinal epithelial cells, thereby improving bioavailability. The SEDDS, as a strategy to improve the bioavailability of orally administered drugs, has the following features:

- Increases the intragastric solubility of drugs.
- Inhibits drug precipitation within the gastric solution.
- Prolongs gastrointestinal tract retention.
- Maintains the stability of the drug in the gastrointestinal tract.
- Promotes lymphatic transport.
- Inhibits the efflux transporter P-glycoprotein (P-gp).
- As a drug carrier, the SEDDS can be used for slow-release purposes.
- Facilitates drug storage and transportation since the drug is dissolved in the oil phase.

Gao et al. [2] conducted extensive studies on the *in vivo* absorption mechanisms of the SEDDS and summarized the potential processes of drug absorption (Figure 12.13).

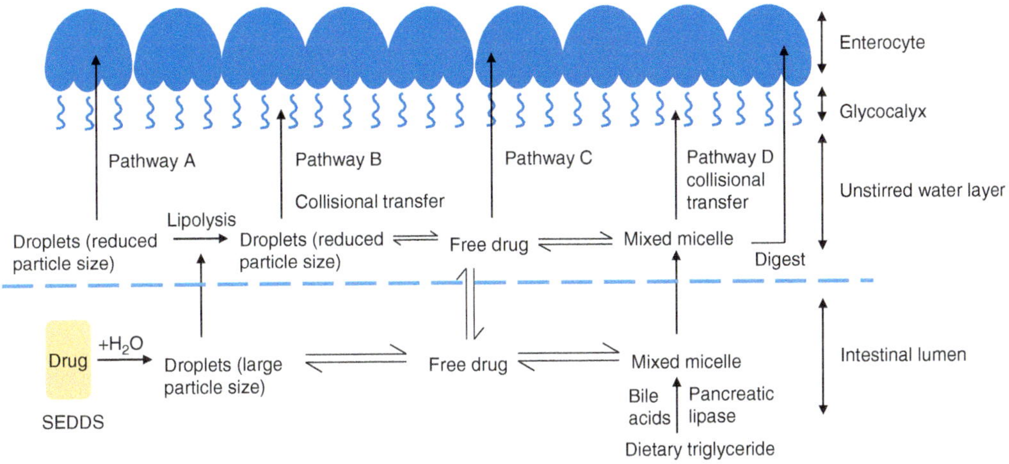

Figure 12.13 Schematic diagram of the intestinal absorption of the SEDDS drugs. *Source:* Adapted from reference [3].

Pathway A: Following oral administration, droplets are initially formed in the gastrointestinal lumen, undergo lipolysis, and are transported to enterocytes.

Pathway B: A small amount of the drug, influenced by the action of freely available surfactants in the gastrointestinal lumen, forms micelles. Droplets that are not transported through Pathway A undergo further lipolysis; both forms of drug carriers undergo collisional transfer to the glycocalyx layer, eventually entering intestinal epithelial cells.

Pathway C: Small amounts of free drugs in the intestine are transported by passive diffusion to enterocytes.

Pathway D: Triglycerides from food form mixed micelles under the action of bile acids and pancreatic lipase. A small amount of freely available drug collides with these mixed micelles and is transported to the glycocalyx layer to enter the enterocyte.

Once the highly lipophilic drugs are transported into the enterocyte, they are digested with food and form chylomicrons, which enter the lymphatic system or, ultimately, the systemic venous circulation. Drugs with moderate lipophilicity, on the other hand, undergo collisional transfer directly into the venous system without forming chylomicrons [3].

12.2.3 Application of Self-Emulsifying Drug Delivery System

On September 9, 2016, Liporaxel, the first oral Paclitaxel product, was approved by Korea's Ministry of Food and Drug Safety (MFDS) for treating gastric cancer. The application of the SEDDS technology substantially enhanced the oral bioavailability of Paclitaxel. Simultaneously, there was a marked reduction in adverse reactions among patients, such as peripheral neuropathy and allergic responses. This innovation avoided effects associated with intravenous injection, such as extravasation and pain, thereby improving medication safety and facilitating home-based treatment for patients.

The SEDDS is applied to deliver NMEs, such as providing a protective effect on peptide drugs. SEDDS has been proven effective in resisting the digestion and degradation of peptide drugs, enhancing the penetration of the intestinal mucosal gel layer and mucosal cell permeability, thereby promoting their absorption. Additionally, SEDDS has been applied in recent research focusing on PROTAC drugs. Due to such drugs' high hydrophilic–lipophilic balance (HLB) coefficients, self-emulsifying formulations can address the issues of low solubility, leading to low *in vivo* exposure.

The components of the SEDDS are all in the nonaqueous phase. Considering the retention time of the drug in the body and its safety, the dosing volume of the SEDDS should not be large; therefore, highly concentrated formulations should be developed. While SEDDS has numerous advantages, its development is a time-consuming process involving compound exploration and lacks a straightforward evaluation method.

Before screening the SEDDS formulations, an assessment is required to identify if compounds are suitable for the SEDDS delivery system. A rapid and concise method aids researchers in making quick determinations and recognizing compounds suitable for SEDDS application [4]. This method relies on the physicochemical parameters of compounds:

- Poor water solubility (<10 μg/mL).
- High lipophilicity (log P > 3).
- Relatively high solubility (>25 mg/mL) for oil or fat solubility.

12.2.4 Formulation Design and Pharmacokinetic (PK) Study Process of Self-Emulsifying Drug Delivery System

12.2.4.1 Developmental Process

The focal point of SEDDS development is to maximize the compound's solubility in the preconcentration solution and assess the emulsification performance after its dilution with water. This is because the size and stability of emulsified droplets can impact compound absorption (Figure 12.14). The developmental process begins by determining the compound's solubility in

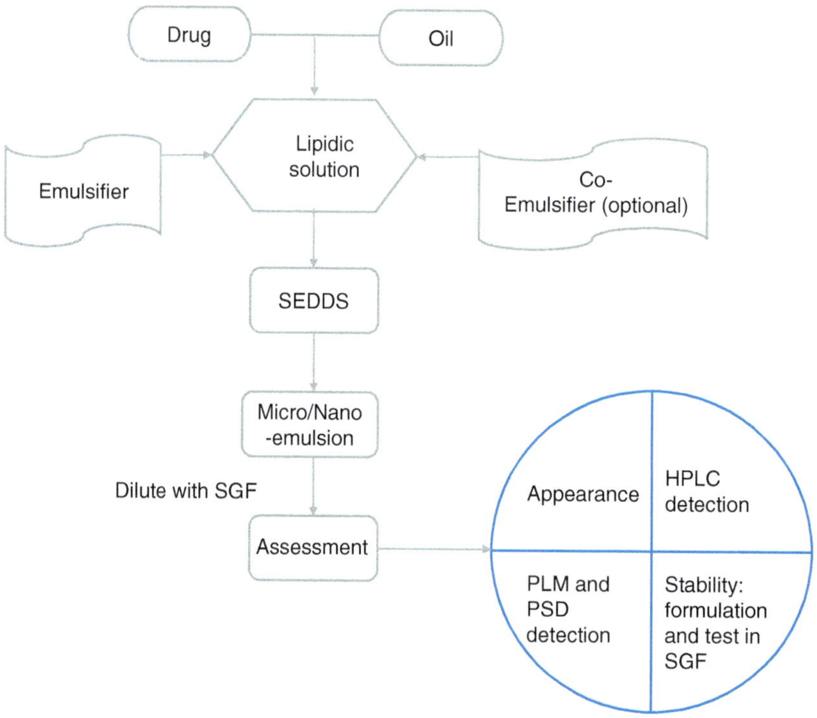

Figure 12.14 SEDDS development flowchart. HPLC, high-performance liquid chromatography; PLM, polarized light microscopy; PSD, particle size distribution; SGF, simulated gastric fluid.

various commonly used excipients. The next step is to select components with higher solubility and mix them in suitable proportions to form a clarified concentrate. The final step includes simulated assessments in *in vitro* assays. Commonly used excipients include medium- and long-chain fatty acids, polyoxyethylene sorbitan esters, and PEG series.

The mechanism by which excipients in SEDDS formulations enhance bioavailability involves emulsification and permeability. The oil phase can alter the biopharmacological properties of drugs and has a complex effect on the bioavailability of orally administered drugs. These effects include lowering gastric emptying time, increasing drug solubility in intestinal fluid, and inducing easier lymphatic transport of highly lipophilic drugs. However, it can affect drug absorption [5], as a single oil phase or a high oil content may result in weak emulsification. For example, soybean oil, due to its slow emulsification in gastrointestinal fluids and the formation of large emulsion droplets, can hinder drug distribution and absorption.

In addition to facilitating emulsification, surfactants aid permeation by altering the fluidity of small intestinal epithelial cell membranes and the tight junctions between cells, thereby increasing drug membrane permeability and improving drug absorption [6]. Many BCS Class II drugs have poor solubility and may be substrates for transporters and metabolizing enzymes, resulting in lower bioavailability. Studies have shown that many nonionic surfactants have inhibitory effects on P-gp [7, 8], which can effectively improve drug absorption. Therefore, SEDDS is often used to improve the absorption of poorly water-soluble or oil-soluble drugs in conventional formulations and has attracted much attention for improving the bioavailability of low-solubility Class II and IV drugs.

12.2.4.2 Formulation Design

For formulation design, SEDDS can be categorized into four types, depending on the components and ratios (Table 12.1). Classes I and II are early-development formulations, which form larger emulsion droplets during evaluation. As SEDDS research has progressed, Class III SEDDS, with the addition of cosolvents, exhibits enhanced solubilization effects and allows for a corresponding reduction in the proportion of surfactants in the system. This not only improves the formulation's flowability and aids dispersion but also makes it safer. Additionally, it facilitates the formation of smaller emulsion droplets, making it more commonly used.

12.2.4.3 Methodology for the *In Vitro* Evaluation of Self-Emulsifying Drug Delivery System Formulations

The *in vitro* dispersion test is a rapid and effective assessment method used to screen SEDDS formulation. The self-emulsification rate indicates the formulation's ability to spontaneously form stable and homogeneous microemulsions.

Table 12.1 Formulation design of different SEDDS.

	Type I	Type II	Type IIIA	Type IIIB
Lipids	100%	40% ~ 80%	40% ~ 80%	<20%
Surfactants	0	20% ~ 60% HLB <12	20% ~ 40%	20% ~ 50%
Cosolvents	0	0	0% ~ 40%	20% ~ 50%

hydrophilic–lipophilic balance (HLB).

Evaluation methods include the following:

a) Visual assessment of emulsification levels [9, 10].
b) The equilibrium time of self-emulsification was determined by analyzing the concentration of the compound using HPLC.
c) Emulsion droplet size is observed under a microscope, and PSD is measured using a Malvern particle size analyzer.

Based on the SEDDS formulation design, 6–10 formulations can be quickly prescreened. If there are similar evaluation results within these formulations (emulsification speed, droplet size), we can also select the optimal recommendation based on the stability results of the formulation in gastric and intestinal fluids.

12.2.4.4 Factors Influencing Animal Experiments and Strategies

Several factors in animal experiments can influence the choice of excipients and the application of formulations in the SEDDS developmental process (Table 12.2).

In response to the challenges posed by the factors mentioned above, several strategies can be employed:

- Scientists have developed S-SEDDS (i.e., a supersaturable SEDDS) to address emulsification challenges in SEDDS and reduce the intake of the total lipid phase. This involves pre-diluting SEDDS with a solution containing inhibitors to achieve supersaturation, facilitating advanced emulsification and dispersion.
- Dosages can be designed based on the tolerance of different species to varying ratios of excipients.
- Formulations can be selected using fasted/fed state simulated gastric fluid (FaSSGF/FeSSGF) or fasted/fed state simulated intestinal fluid (FaSSIF/FeSSIF) emulsification tests, simulating human or canine conditions, to predict potential precipitation in advance.

The impact of food can be considered by studying existing food formulations, defining food based on clinical guidance principles, and designing high-calorie feeds, regular feeds, and low-calorie feeds suitable for different species according to caloric content to meet the design requirements of different experiments.

Table 12.2 Important factors in animal experiments in the SEDDS developmental process.

Factor	Challenge
Different species (mice, rats, dogs, or monkeys)	Differences in the emulsification and digestion processes of the SEDDS.
The high volume of administration	Compared to clinical trials, the administration volume in animal experiments is higher.
Mode of administration	Semisolid SEDDS formulations (capsules/tablets).
High total oil phase intake	Tolerance and safety.
High dose concentrations	Precipitation in the gastrointestinal tract during emulsification.
Food effect	Different feeds from various manufacturers or formulations contain different triglyceride levels.

12.2.5 Summary

Despite facing various challenges during development, particularly as many SEDDS formulations exhibit effective self-emulsification only under specific phase combinations, necessitating extensive screening work, the SEDDS continues to optimize formulation development technology. This includes the production of smaller-sized self-nano-emulsion formulations, more stable and easily producible self-emulsifying solid formulations, and the incorporation of excipients such as hydroxylpropyl methylcellulose (HPMC) to inhibit precipitation, ensuring the safer application of supersaturated self-emulsifiers. As research progresses, self-emulsification technology will be better applied to meet the challenges of drug absorption, ultimately benefiting patients.

References

1 Ku, M.S. (2008). Use of the Biopharmaceutical Classification System in Early Drug Development. *The AAPS Journal.* 10 (1): 208–212.

2 Gao, P. and Morozowieh, W. (2006). Development of supersaturate self-emulsifying drug delivery system formulations for improving the oral absorption of poorly soluble drugs [J]. *Expert Opinion on Drug Delivery* 3 (1): 97–110.

3 Feng, Y., Cheng, Q., and Li, F. (2007). Self-emulsifying drug delivery systems:a recent development. *International Journal of Pharmaceutical Research* 34 (4): 280–281.

4 Chen, X.-Q., Gudmundsson, O.S., and Michael, J. (2012). Hageman Application of lipid-based formulations in drug discovery. *Journal of Medicinal Chemistry* 55: 7945.

5 Gershanik, T. (2000). Benita s self—dispersing lipid formulations for improving oral absorption of lipophilic drugs [J]. *European Journal of Pharmaceutics and Biopharmaceutics* 50 (1): 179–188.

6 Kommuru TR, Gurley B·Khan MA. et a1. Self-emulsifying drug delivery systems (SEDDS) of coenzyme Q10: formulation development and bioavailability assessment [J]. *International Journal of Pharmaceutics*, 2001, 212(2): 233–246.

7 Al-Mohizea, A.M., Zawaneh, F., Alam, M.A. et al. (2014). Effect of pharmaceutical excipients on the permeability of P-glycoprotein substrate [J]. *Journal of Drug Delivery Science and Technology* 24 (5): 491–495.

8 Cornaire, G., Woodley, J., Hermann, P. et al. (2004). Impact of excipients on the absorption of P-glycoprotein substrates *in vitro* and *in vivo* [J]. *International Journal of Pharmaceutics* 278 (1): 119–131.

9 Gershanik, T., Benzeno, S., Benita, S. et al. (1998). Interaction of a self-emulsifying lipid drug delivery system with the everted rat intestinal mucosa as a function of droplet size and surface charge. *Pharmaceutical Research.* 15 (6): 863–869.

10 Griessinger, J., Dunnhaupt, S., Cattoz, B. et al. (2015). Methods to determine the interactions of micro- and nanoparticles with mucus. *European Journal of Pharmaceutics and Biopharmaceutics* 96: 464–476.

12.3 Application of Anesthetics in Rodent Pharmacokinetic Study

Jinyu Yan, Quanli Feng, Xuan Dong, Furong Jiao, Danqing Lin, Qiqi Han, Cheng Tang

12.3.1 Introduction

Preclinical PK studies are a vital part of new drug development. Rodents (rats, mice, guinea pigs, hamsters, etc.) are usually used to conduct preclinical PK studies to evaluate ADME processes of the compounds *in vivo* study. In rodent PK studies, anesthetic drugs are used to keep the animal anesthetized or sedated during the experiments to meet the needs of some special studies (e.g., intrathecal administration, submandibular gland injection, and microdialysis experiments). This chapter describes the background of anesthetics, anesthetic regimens for rodents, and DDI during anesthesia and introduces the application strategies and methods of anesthetics in rodent PK studies.

12.3.2 Introduction to Anesthesia

12.3.2.1 Background of Anesthesia

Anesthesia is a reversible inhibition of the nervous system through the use of drugs or other methods so that the patient temporarily loses general or local sensation to achieve painlessness and provide conditions for surgical treatment or other medical examinations and treatments. The history of anesthesia is divided into ancient, modern, and present-day anesthesia. Ancient anesthesia mainly used Chinese herbal medicines such as ephedra powder and *datura stramonium* to cause paralysis, and some ancient books such as *Classic of the Materia Medica* have related records. On October 16, 1846, American physician William Morton successfully performed a case of ether anesthesia surgery for the first time, which marked the beginning of the modern history of anesthesia. In the 1950s, it was called modern anesthesia time and anesthesia began to enter disciplinary specialization, and the types of anesthetics became diversified, the safety of anesthesia was also greatly improved, and the application range of anesthesia was further expanded [1].

In addition, anesthesia is also closely related to the welfare of laboratory animals, and the application of anesthesia in animal experiments is an effective measure to improve animal welfare. Regulatory guidelines such as the *Guide for the Care and Use of Laboratory Animals: Eighth Edition*, the *3Rs (Replacement, Reduction, and Refinement) Guidelines,* and the *U.S. Public Health Service Policy on Humane Care and Use of Laboratory Animals* clearly state that appropriate anesthetic measures are needed to alleviate pain and distress of animals during studies.

12.3.2.2 Introduction to Rodent Anesthesia Program

1) Classification of anesthetics

As shown in Figure 12.15, anesthetics can be divided into general and local anesthetics according to the action range. General anesthetics can be divided into inhalation anesthetics (e.g., nitrous oxide and isoflurane) and injection anesthetics (e.g., pentobarbital sodium and ketamine) according to the route of administration. Local anesthetics are divided into esters, such as tetracaine, and amides such as lidocaine. Some drugs are not natural anesthetics and are often used in combination with anesthetics to achieve a better anesthetic effect or to meet some other preoperative needs, such as tranquilizers, sedatives, analgesics, and muscle relaxants, which can be collectively referred to as auxiliary anesthesia drugs.

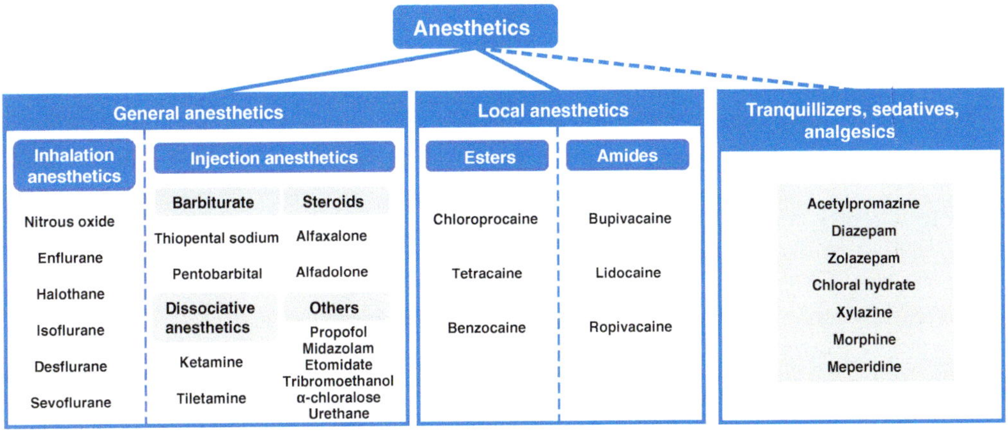

Figure 12.15 Classification of anesthetics. Adapted from reference [1].

2) Selection of anesthetics and establishment of anesthetic regimen

There are a wide variety of anesthetics available. According to the characteristics of various anesthetics and experimental needs, veterinarians and laboratory technicians will comprehensively consider the following factors to select the optimal strategy for the use of anesthetics:

- Easy to use, with a good anesthetic effect.
- Minimal side effects and the corresponding antidote can be used to reverse the anesthetic effect.
- Pharmaceutical-grade drugs with a reasonable price.
- Noncontrolled drugs are preferred with the same anesthetic effect.

Optimal anesthesia regimens are required for different studies or animal strains/stocks. Before the anesthesia regimen is established, first, the veterinarians will search the literature to obtain the reference administration dosage of relevant anesthetics. Second, different dose gradients will be set according to the purpose of dose verification. Third, the veterinarians will observe the anesthetic effect on animals (e.g., checking whether the animals can enter the surgical anesthesia period and whether the anesthesia is overdosed). Finally, the veterinarians will determine the optimal dose regimens. If other abnormalities are found during anesthesia, the regimen will be continuously optimized and updated.

12.3.3 Drug-Drug Interaction (DDI) During Anesthesia

After the surgery, the animals are generally provided a specific postoperative recovery time before the studies. On the day of the study, the anesthetics are entirely eliminated from the animals. In general, no DDI will occur at this time. Most DDI occurs when anesthesia or sedation is used simultaneously during the study, such as submandibular gland injection administration, sublingual administration, intra-articular injection, ophthalmic administration, intrathecal injection, direct injection administration in each intestinal segment, and serial collection of cerebrospinal fluid, among others.

PK study mainly involves the ADME process. When animals are anesthetized, the physical characteristics (e.g., gastrointestinal motility, plasma protein displacement, blood–brain barrier permeability, hemodynamics, hepatic enzyme activities, hepatic blood flow, renal function, and renal blood flow) may be affected, which may impact the ADME process of the test compounds in animals [1]. The following are some relevant cases.

1) Effect of anesthesia on gastrointestinal function

The absorption of drugs occurs primarily in the gastrointestinal tract. The results of literature research have shown that brief anesthesia of animals with isoflurane (only six minutes of exposure) affects gastrointestinal motility, slowing down gastric emptying and intestinal motility. Starting from the pylorus, the marker in the intestine of anesthetized animals moves approximately 50% less than in conscious animals [2]. However, compared with conscious animals, there was no significant difference in gastric emptying and gastric volume change by intraperitoneal injection or subcutaneous injection of another anesthetic, Zoletil 50, indicating that Zoletil 50 anesthesia did not affect gastric function in animals [3]. Therefore, more attention should be paid to the use of isoflurane anesthesia for gastrointestinal administration and related studies under anesthesia.

2) Effect of anesthesia on *in situ* intestinal perfusion study

Animals must be anesthetized throughout the *in situ* intestinal perfusion studies. As shown in Figure 12.16, three test articles absorbed by passive diffusion in the intestine were selected: low permeability (atenolol), medium permeability (metoprolol), and high permeability (antipyrine), and four different anesthetic regimens were set: pentobarbital, ketamine/ midazolam, ketamine/ xylazine, ketamine/xylazine/butorphanol. In addition, intestinal permeability was explored. The study results showed that the highly permeable antipyrine had the weakest intestinal absorption effect under ketamine combined with midazolam anesthesia, while it had the best intestinal absorption effect under ketamine combined with xylazine plus butorphanol anesthesia. As shown in Figure 12.17, the study results also indicated that after adding analgesic butorphanol, it enhanced intestinal absorption as it reduced the painful stress of the animals. The results showed that ketamine/xylazine/butorphanol can be recommended as a standard anesthesia scheme for *in situ* intestinal perfusion studies [4].

3) Effect of anesthesia on blood–brain barrier permeability

Isoflurane and ketamine are two commonly used anesthetics in rodent PK studies. It was reported that isoflurane anesthesia can increase blood–brain barrier permeability and significantly increase

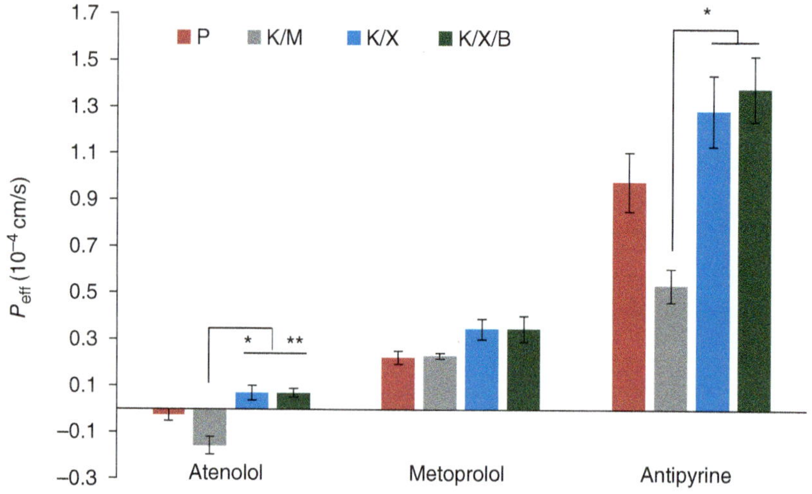

Figure 12.16 Comparison of the permeability of atenolol, metoprolol, and antipyrine under pentobarbital, ketamine/midazolam, ketamine/xylazine, and ketamine/xylazine/butorphanol anesthesia. *Source:* [4] / with permission of Springer Nature.

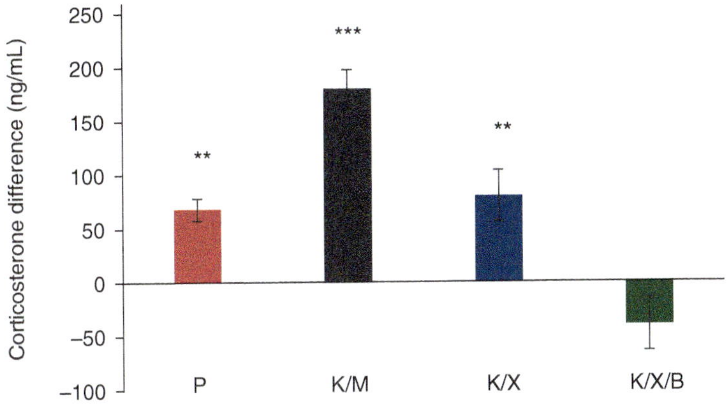

Figure 12.17 Comparison of different corticosterone levels before and after the surgery under different anesthesia regimens. *Source:* [4] / with permission of Springer Nature.

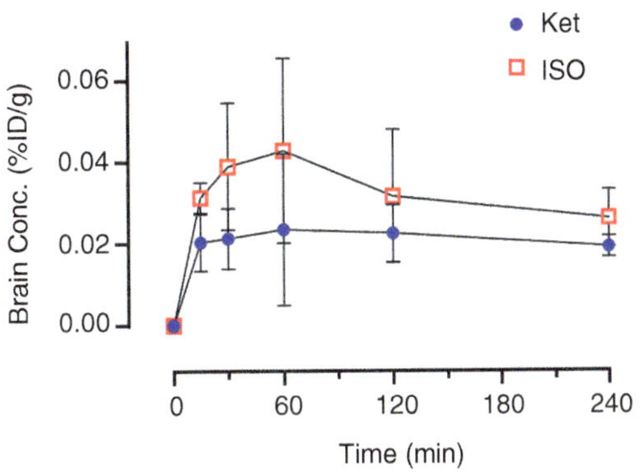

Figure 12.18 Time curve of sucrose in the brain under different anesthesia regimens. Ket is ketamine and ISO is isoflurane. *Source:* [5] / with permission of The American Society for Pharmacology and Experimental Therapeutics.

the exposure of some drugs in the brain. On the contrary, ketamine combined with xylazine anesthesia does not have such an effect (Figures 12.18 and 12.19). The mechanism is that isoflurane affects the permeability of blood–brain barrier cell membranes rather than affects tight junctions between cells [5]. Therefore, caution should be exercised when using isoflurane anesthesia in central nervous system (CNS) drug research.

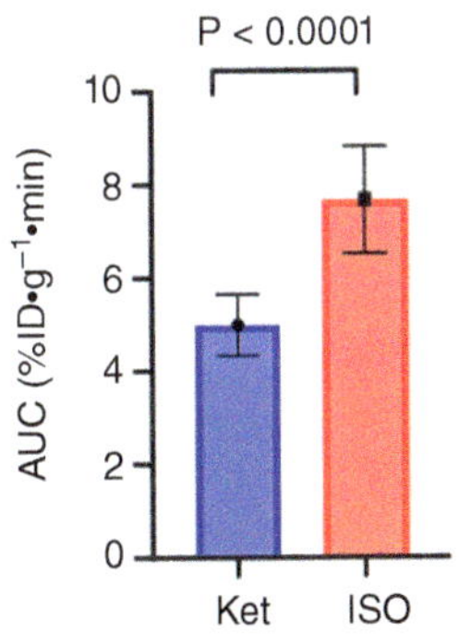

Figure 12.19 Comparison of the AUC of sucrose in the brain during 0–4 hours under different anesthetic regimens. Ket is the ketamine and ISO is the isoflurane. *Source:* [5] / with permission of The American Society for Pharmacology and Experimental Therapeutics.

4) Effect of anesthesia on liver enzyme activity

Drug metabolism primarily occurs via cytochrome enzymes (also known as CYP enzymes, CYP450) in the liver. A literature study found that both pentobarbital sodium and Zoletil 50 increased the activity of the BROD enzyme (7-benzyloxyresorufin-O-debenzylase, a live form of the cytochrome enzyme family) in the liver [6] (Table 12.3). However, isoflurane anesthesia did not affect cytochrome enzymes at either high concentrations for transient exposure or low concentrations for prolonged exposure (less than 0.5%, inhalation for 4 hours per day for 30 weeks) [7]. Therefore, it is necessary to comprehensively consider the factors such as the

Table 12.3 Effect of anesthesia with Telazol, tiletamine, zolazepam, or phenobarbital on total CYP content and BROD activity.

Treatment (dose)	Total CYP content (nmol/mg protein)	BROD activity (pmol/min/mg protein)	EROD activity (pmol/min/mg protein)
Control	1.4±0.1 (1.0)	257±24 (1.0)	409±44 (1.0)
Telazol (120 mg/kg)	1.2±0.0 (0.9)	3,933±634[a] (15.3)	429±70 (1.0)
Tiletamine (60 mg/kg)	1.3±0.1 (0.9)	536±66[b] (2.1)	441±75 (1.1)
Zolazepam (60 mg/kg)	1.2±0.1 (0.9)	2,292±201[a,b] (8.9)	438±70 (1.1)
Control	1.3±0.1 (1.0)	323±22 (1.0)	529±50 (1.0)
PB (60 mg/kg)	1.8±0.1[a] (1.4)	6,974±118[a] (21.6)	1,035±55[a] (2.0)
PB (120 mg/kg)	1.9±0.1[a] (1.5)	7,292±56[a] (22.6)	1,037±92[a] (2.0)

Each value is the mean ± SEM for five or six rats per treatment group. Numbers in parentheses denote the difference relative to the value of the control group.

[a] The mean value of the treatment group was significantly different ($P < 0.05$) from that of the control group.

[b] Mean value of the tiletamine or zolazepam treatment group was significantly different ($P < 0.05$) from that of the Telazol treatment group.

Source: [6] / ELSEVIER.

administration route, the characteristics of the test compounds (such as the main metabolic enzymes), and other factors to select the appropriate anesthesia regimen to minimize the effect of anesthesia in the studies.

5) Effect of anesthesia on renal excretion

The kidney is the primary drug excretion organ. The results of a numerous literature studies have shown that the commonly used isoflurane or ketamine anesthesia has no significant effect on the structure of the kidney, the function of the kidney, the blood flow of the kidney, etc.; therefore, for the animals with normal renal function, the conventional anesthesia methods do not affect the renal excretion function of the test article [8–10].

6) Effect of long-term anesthesia on lymphatic cannulation studies

To date, more than 40% of new active chemicals are lipophilic with poor aqueous solubility. Although these chemicals have good pharmacological activity, they are limited by low oral bioavailability and fail to enter subsequent development. It is well known that oral bioavailability plays an important role in drug discovery. One method of enhancing the oral bioavailability of these chemicals is to improve the drug's intestinal lymphatic system transport capacity. In addition to increasing the overall bioavailability of lipophilic molecules, drug transport through the lymphatic system has other advantages, such as avoiding the first-pass effect in the liver and improving plasma distribution [11]. This type of study generally uses a lymphatic cannulation animal model and calculates the recovery rate of the drugs in the lymphatic system by collecting lymph fluid for analysis at different periods.

There are generally two surgery methods for mesenteric lymphatic cannulation studies: the first one is to perform lymphatic cannulation surgery on animals after short-acting anesthesia and carry out the studies after a period of recovery when animals wake up, which is called a conscious animal model. The second one is short-acting anesthesia followed by lymphatic cannulation surgery and then a continuous injection of anesthetic drugs through the jugular vein cannula to maintain the animal under anesthesia for several hours, which is called the anesthetic animal model. The literature studied the differences between the two models, and the results showed that compared with conscious animals, the lymph flow under continuous anesthesia was significantly reduced (Figure 12.20) [11]. The cumulative recovery of the drugs in the lymph was decreased by about 25% in anesthetized animals compared with conscious animals (Figure 12.21) [11]. In addition, the study also set two control groups. The animals in the control group were not performed with lymphatic cannulation and other experimental operations were consistent. Analysis of jugular blood samples showed that the area under the curve (AUC) of the drugs in blood samples was decreased by about 25% in animals under continuous anesthesia compared with that in the conscious group (Figure 12.22) [11]. This finding is meaningful and indirectly suggests that although anesthesia decreases the drug content in the lymph fluid, it does not affect the drug absorption mechanism; that is, it does not affect the distribution ratio of the drugs through the lymphatic absorption route and other absorption routes. At the same time, the success rate of carrying out such experiments under anesthesia is close to 100%, which can also reduce the number of animals used and is more suitable for some early drug screening studies.

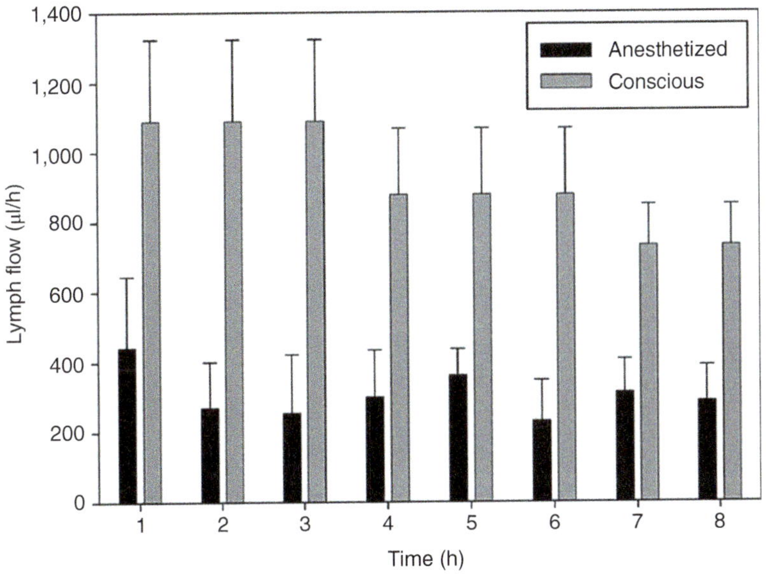

Figure 12.20 Comparison of lymph flow between conscious and anesthetized animals. *Source:* [11] / ELSEVIER.

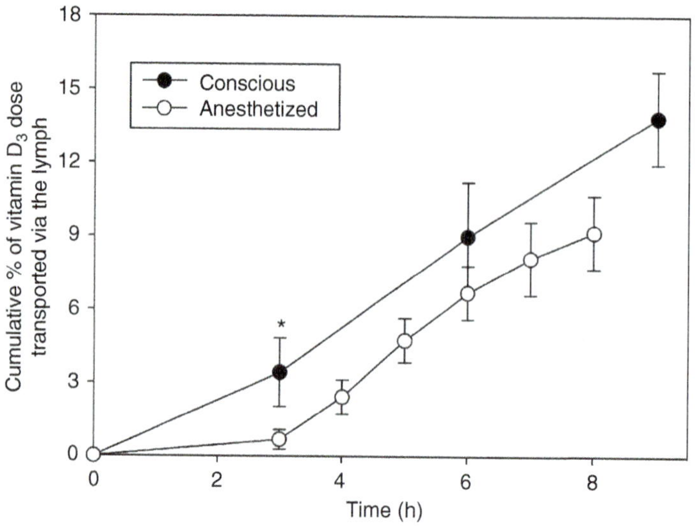

Figure 12.21 Comparison of cumulative recovery of vitamin D3 in lymph between conscious and anesthetized animals. *Source:* [11] / ELSEVIER.

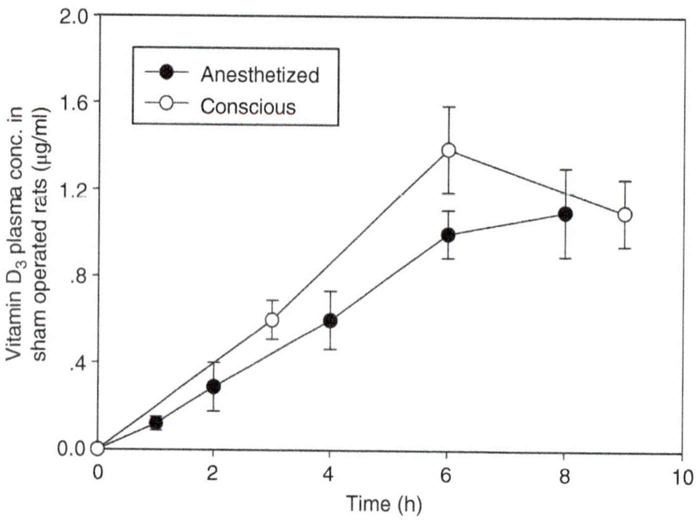

Figure 12.22 Comparison of vitamin D3 concentration in the jugular vein of conscious and anesthetized animals. *Source:* [11] / ELSEVIER.

12.3.4 Summary

Anesthetics are widely used in preclinical PK in rodents, providing a guarantee for surgery and some special studies, as well as ensuring animal welfare. At the same time, because of some pharmacological effects, anesthetics will also have specific effects on the physiological functions of animals, such as affecting gastrointestinal motility, blood–brain barrier permeability, and liver enzyme activity, which will affect the ADME process of the test articles in the body. In the preclinical screening study, the appropriate animal anesthesia method can be selected according to the administration method, absorption method, targeted site, metabolic pathway, and other information from the test articles. It is expected that anesthesiology has a broader application in laboratory animals, providing more options for preclinical *in vivo* studies and facilitating the process of drug discovery.

References

1 Deng, X. et al. (2021). *Modern Anesthesiology*. People's Medical Publishing House.

2 Torjman, M.C., Joseph, J.I., Munsick, C. et al. (2005). Effects of isoflurane on gastrointestinal motility after brief exposure in rats. *International Journal of Pharmaceutics* 294 (1–2): 65–71.

3 Jordi, J., Verrey, F., and Lutz, T.A. (2014). Simultaneous assessment of gastric emptying and secretion in rats by a novel computed tomography-based method. American journal of physiology-gastrointestinal and liver. *Physiology* 306 (3): G173–G182.

4 Saphier S, Yacov G, Wenger A, et al. The effect of anesthetic regimens on intestinal absorption of passively absorbed drugs in rats. *Pharmaceutical Research* 2020;37:1–0.

5 Noorani, B., Chowdhury, E.A., Alqahtani, F. et al. (2023). Effects of volatile anesthetics versus ketamine on blood–brain barrier permeability via lipid-mediated alterations of endothelial cell membranes. *Journal of Pharmacology and Experimental Therapeutics.* 385 (2): 135–145.

6 Wong, A. and Bandiera, S.M. (1998). Induction of hepatic cytochrome P450 2B and P450 3A isozymes in rats by zolazepam, a constituent of Telazol®. *Biochemical Pharmacology.* 55 (2): 201–207.

7 Committee for Medicinal Products for Veterinary Use. European public MRL assessment report (EPMAR) isoflurane (porcine species) (2018).

8 Ruxanda F, Miclăuș V, Rus V, Gal AF, Oana L. (2014) Impact of isoflurane and sevoflurane anesthesia on kidney structure and function in rats. In Bulletin of University of Agricultural Sciences and Veterinary Medicine Cluj-Napoca Veterinary Medicine.

9 Shiga, Y., Minami, K., Uezono, Y. et al. (2003). Effects of the intravenously administered anaesthetics ketamine, propofol, and thiamylal on the cortical renal blood flow in rats. *Pharmacology* 68 (1): 17–23.

10 Bougherara, H. and Bouaziz, O. (2014). Effects of the anaesthetic/tranquillizer treatments (ketamine, ketamine+ Acepromazine, Zoletil) on selected plasma biochemical parameters in laboratory rats. *Central European Journal of Experimental Biology* 3 (2): 1–5.

11 Dahan, A., Mendelman, A., Amsili, S. et al. (2007). The effect of general anesthesia on the intestinal lymphatic transport of lipophilic drugs: comparison between anesthetized and freely moving conscious rat models. *European Journal of Pharmaceutical Sciences* 32 (4–5): 367–374.

12.4 Application of *In Situ* Single-Pass Intestinal Perfusion (SPIP) Model in the Study of Intestinal Absorption of Oral Drugs

Xiang Ren, Furong Jiao, Xuan Dong, Quanli Feng, Cheng Tang

12.4.1 Introduction

With the continuous development of pharmaceutical technology, more and more types of drugs and administration routes have been discovered to treat diseases. Oral drugs dominate the clinical administration routes because of their better compliance, economy, safety, and various dosage forms [1]. The absorption and bioavailability of oral drugs are the key factors that researchers focus on. Intestinal absorption studies of oral drugs are vital to evaluate the bioavailability of drugs, optimize drug dosage forms, and improve therapeutic effects.

The research methods of intestinal absorption in oral drugs centrally include *in vivo*, *in vitro*, and *in situ*. Compared with other methods, the *in situ* method has certain advantages, for instance, it can ensure the integrity of intestinal nerves and the supply of blood and lymph, significantly improve the biological activity of the tissue to be investigated, and more accurately reflect the actual absorption status of the drug in the intestinal segment. It is mainly applied in intestinal permeability study [2]. This chapter introduces the *in situ* SPIP model, which is considered one of the models to study drug absorption by the US FDA, and it centrally includes the SPIP model introduction, the comparison between the SPIP model and *in vitro* and *in vivo* methods, the validation data, and the applications of SPIP, shares the summary and insights of the model in combination with literature reports.

12.4.2 Influencing Factors and Common Research Methods of Intestinal Absorption of Drugs

Solubility and permeability are the basis for the intestinal absorption of oral drugs [3, 4]. Amidon et al. proposed the BCS in 1995, which classified drugs into four categories: Class I (high solubility and high permeability), Class II (low solubility and high permeability), Class III (high solubility and low permeability), and Class IV (low solubility and low permeability) [3]. The solubility of drugs can be improved by some means, such as formulation modification. The permeability of drugs is a property of the drug molecule that is hardly affected by solvent or the process changes in the product. Therefore, permeability studies during intestinal absorption are particularly important for BCS Class I and III drugs. Permeability can be directly measured by the extent of drug absorption in the intestine or by the rate of drug mass transfer across the intestinal membrane, and it can also be indirectly measured by assessing the extent of drug absorption in kinetic studies compared with intravenous doses [5].

Oral bioavailability (F) is a relative terminology that describes the relative ratio of the rate and extent of absorption of an oral drug to that given by a reference route (usually intravenous injection). It is generally acknowledged that oral bioavailability $F = F_a \times F_g \times F_h$, in which F_a is the proportion of drugs absorbed into intestinal epithelial cells after oral administration; F_g is the proportion of drugs that escape from intestinal metabolism; F_h is the proportion of drugs that escape from hepatic metabolism (Figure 12.23).

The methods of intestinal absorption in WuXi AppTec DMPK primarily include *in vivo*, *in vitro*, and *in situ*.

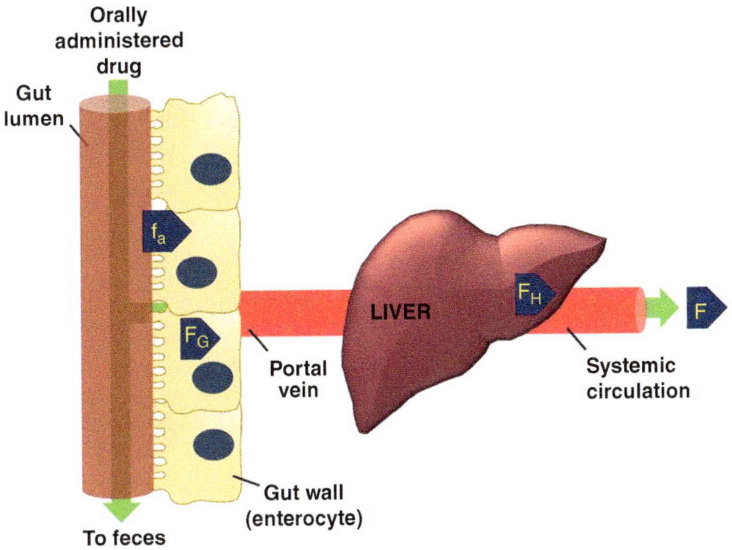

Figure 12.23 Oral bioavailability calculation process simulation. *Source:* [6] / ELSEVIER.

In vivo methods:

The method usually sets intravenous administration (IV) group and oral administration (PO) group, and the complete intestinal absorption ratio of the drug only metabolized by the liver can be obtained by IV clearance rate and theoretical liver blood flow rate:

$$F_a \times F_g = F/F_h = \left[\left(\text{AUC}_{\text{po},0-\infty}/\text{AUC}_{\text{iv},0-\infty}\right)\times\left(\text{Dose}_{\text{iv}}/\text{Dose}_{\text{po}}\right)\right]/\left(1-\text{CL}_b/Q_h\right) \tag{12.1}$$

Through hepatic portal vein and carotid artery catheterization, a more accurate parameter of drug absorption ratio in the intact intestine can be obtained:

$$F_a \times F_g = Q_{\text{por}} \times R_b \times \left(\text{AUC}_{\text{por}} - \text{AUC}_{\text{sys}}\right)/\text{dose} \tag{12.2}$$

Q_{por}, R_b, AUC_{por}, and AUC_{sys} are the portal blood flow, the blood/plasma concentration ratio, AUC in the portal vein, and AUC in the systemic circulation, respectively.

In vitro methods

Caco-2 cell culture and Ussing Chamber are two standard methods. Caco-2 cells are human colon adenocarcinoma cells, which share similar structure and function with differentiated small intestinal epithelial cells and they are usually used to simulate *in vivo* intestinal transport experiments. Therefore, it is one of the most common methods to study intestinal absorption *in vitro*. The Ussing Chamber has two advantages, strong controllability and high throughput. It can detect the activity and integrity of the intestine in real time and is suitable for rapidly determining the intestine permeability of drugs.

In situ methods

The SPIP model is a standard method. SPIP model is operated on an anesthetized rat or mouse, and the target intestinal segments such as both ends of the ileum, are ligated and intubated accordingly. This model involves perfusion at a constant flow rate and subsequent analysis of the absorption of drugs in the intestine by examining the changes between inlet and outlet concentrations at

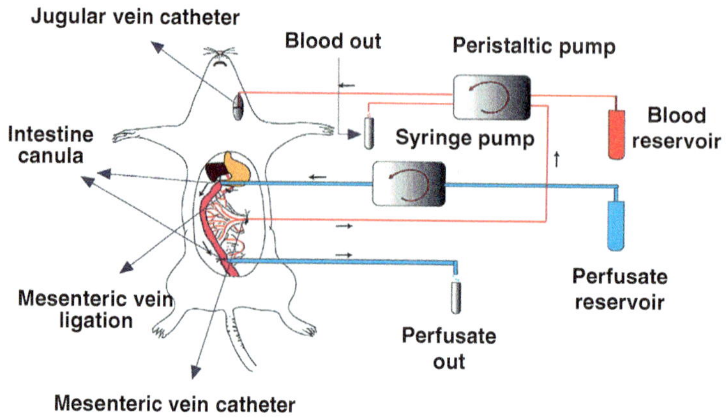

Figure 12.24 SPIP surgical schematic diagram. *Source:* [7] / ELSEVIER.

both ends of the intestinal segment. At the same time, the cannula can be ligated at the mesenteric vein of the target ileal segment for blood sampling to reflect whether the drug is absorbed into the blood, metabolized by the intestine, or adsorbed in the intestinal wall (Figure 12.24).

The SPIP model has the following characteristics:

1) It provides a complete blood supply and a functional intestinal barrier.
2) It retains the physiological characteristics of the small intestine, such as transport routes through paracellular and mediated transportation.
3) It maintains the sink conditions of the drug by blood sampling through the mesenteric vein cannula and by sampling at the outlet of the intestinal lumen, excluding the hepatic first-pass effect, and observing the metabolism of the drug in the intestinal wall.
4) It is mature and controllable and is convenient for controlling the drug concentration, pH, perfusion speed, and flexible for selecting the target intestinal segment in the test and can accurately reflect the absorption status of the drug in the intestinal tract.

Among the three methods above, the *in vitro* method cannot simulate the typical physiological characteristics. Although the *in vivo* model is the most clinically relevant and it includes all the physiological factors that may affect absorption and bioavailability, it is still insufficient to conduct the relevant study on the intestinal absorption mechanism [8]. *In situ* SPIP, as a model recommended by the FDA guidance, can classify permeability characteristics of BCS-based drugs, and this method can be applied for bioequivalence exemption of orally administered drugs, with good prospects for promotion and application.

12.4.3 SPIP Model Establishment and Validation Data

12.4.3.1 SPIP Overview and Model Establishment Method

Since 1950, many researchers have successively reported on intestinal absorption. In 1963, Jacobs et al. designed the SPIP system to study the process of amino acids entering the systemic circulation through intestinal absorption [9]. Since then, the study of SPIP in intestinal absorption has received increasing attention. As shown in Figure 12.24, in the SPIP model method established in this chapter, considering the influence of drug concentration on the experimental results, after a series of verification experiments with different concentration gradients were performed, the results indicate that when it is an unknown drug, the concentration should not be set too high.

In addition, while sampling, multipoint sampling can be performed in the inlet tube to avoid the result deviations caused by the possible nonspecific adsorption between the drug and the tube.

12.4.3.2 Introduction of SPIP Parameters

Similar to the human intestine, the rat's intestine can also secrete or absorb water. The net water flux (NWF) at a steady state is usually used to characterize gut water changes [10]. The actual outlet concentration (C_i) of the drug caused by the change of moisture is obtained by correcting the outlet perfusate with phenol red, and then the absorption fraction Fa value is calculated [11, 12].

$$\mathrm{NWF} = \frac{\left(1 - \mathrm{CPR_{out}} / \mathrm{CPR_{in}}\right) \times Q_{\mathrm{in}}}{l} \tag{12.3}$$

$$C_{\mathrm{i}} = C_{\mathrm{out}} \times \frac{\mathrm{CPR_{in}}}{\mathrm{CPR_{out}}} \tag{12.4}$$

$$F_{\mathrm{a}} = 1 - \frac{C_{\mathrm{i}}}{C_{\mathrm{in}}} \tag{12.5}$$

SPIP model can obtain two most important parameters: the practical permeability coefficient P_{eff} value and the apparent permeability coefficient $\mathrm{P_{app}}$ value.

$$P_{\mathrm{eff}} = \frac{-Q_{\mathrm{in}} \times \ln\left(C_{\mathrm{i}} / C_{\mathrm{in}}\right)}{2\pi r l} \tag{12.6}$$

$$P_{\mathrm{app}} = \frac{\Delta M_b / \Delta t}{2\pi r l \times \langle C \rangle} \qquad \langle C \rangle = \frac{C_{\mathrm{i}} - C_{\mathrm{in}}}{\ln\left(C_{\mathrm{i}} / C_{\mathrm{in}}\right)} \tag{12.7}$$

In the formula, when $\mathrm{NWF} < 0$, it indicates that the intestine absorbs water; when $\mathrm{NWF} > 0$, it indicates that the intestine secretes water; $\mathrm{CPR_{in}}$ and $\mathrm{CPR_{out}}$ are the inlet and outlet concentrations of phenol red in the target perfused intestinal segment, respectively; Q_{in} is the perfusion rate; C_{in} and C_{out} are the import and export test concentrations of drugs, respectively; l is the length of the target intestinal segment; R is the radius of the target perfused intestinal segment; $\Delta M_b / \Delta t$ is the amount of drug permeating in the blood per unit of time under steady-state conditions; $\langle C \rangle$ is the average concentration of the drug in the theoretical cavity.

Generally, it is considered that when $P_{\mathrm{eff}} < 3 \times 10^{-6}$ cm/s, it indicates that the drug is difficult to be absorbed; when $P_{\mathrm{eff}} > 20 \times 10^{-6}$ cm/s, it indicates that the drug is easy to be absorbed [13].

12.4.3.3 Validation Data

The validation experimental results are consistent with the literature reference values by selecting eight compounds, as shown in Table 12.4, for the SPIP study based on the recommended drugs used in BCS to illustrate the applicability of permeability experiments.

At the same time, some data were fitted with human P_{eff} values reported in the literature, and it was found that the P_{eff} values of rats and P_{eff} values of humans had a good correlation (Figure 12.25), and after fitting the P_{eff} values of rats and Caco-2 data, it was also found that there was a correlation (Figure 12.26). These experimental results indicate that we have successfully established the SPIP model platform, and it has good predictability for human permeability.

Table 12.4 Comparison of validation data and literature data.

Group	Drug	Reported values P_{eff} (× 10^{-6} cm/s)	WuXi AppTec Mean P_{eff} (× 10^{-6} cm/s)	WuXi AppTec Mean P_{app} (× 10^{-6} cm/s)
High permeability	Carbamazepine	62.00 ± 6.00	85.60 ± 20.54	34.18 ± 14.47
	Propranolol	68.40 ± 9.81	59.69 ± 7.31	34.47 ± 6.48
	Metoprolol	33.00 ± 15.00	23.71 ± 0.92	1.51 ± 0.03
Moderate permeability	Ranitidine	22.00 ± 10.00	16.84 ± 3.79	13.77 ± 6.64
	Atenolol	32.40 ± 16.00	34.89 ± 8.73	3.15 ± 1.36
Low permeability	Topotecan	0.81 ± 0.30	1.70 ± 2.91	11.49 ± 1.46
	Nadolol	4.30 ± 0.70	3.52 ± 8.89	3.22 ± 2.72

Source: [14]. Dubbelboer et al., 2019; [15]. Zakeri-Milani et al, 2007; [16]. Su et al., 2007; [17]. Castella et al., 2006.

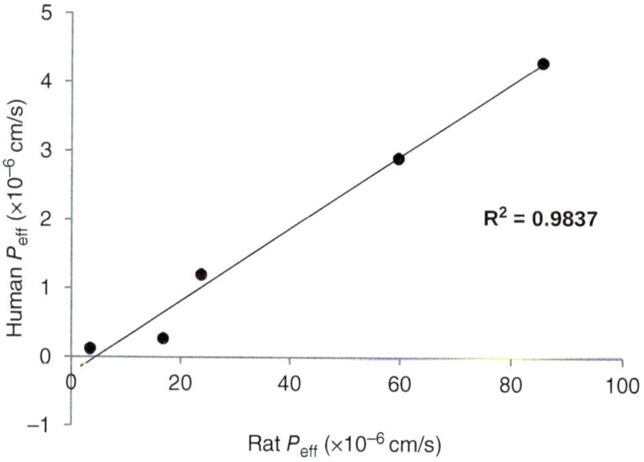

Figure 12.25 Plot of rat P_{eff} on SPIP model versus human P_{eff} on SPIP model.

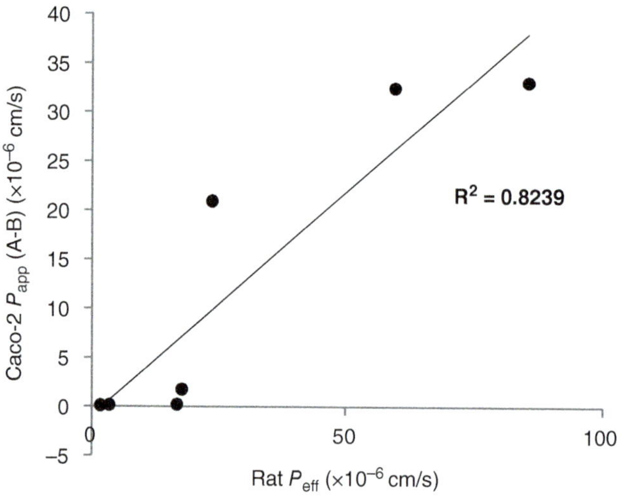

Figure 12.26 Plot of rat P_{eff} on SPIP model versus Caco-2 P_{app} on cell model.

12.4.4 Applications of SPIP

12.4.4.1 Biowaiver and Permeability Classification Identification

In the study of generic drugs, it is necessary to prove the therapeutic equivalence of generic and brand-name drugs. Therapeutic equivalence needs to be illustrated by pharmaceutical equivalence and bioequivalence. SPIP centrally involves the biowaiver in terms of bioequivalence.

In the Guidelines for the Waiver of Bioequivalence Tests in Human Subjects, the definition of the biowaiver is when the oral solid formulation dissolves in the body, gastric emptying occurs rapidly, and the drug has high water solubility. The rate and extent of absorption of the drug do not depend on the dissolution time of the drug or the transit time in the gastrointestinal tract. Therefore, in this case, for BCS Class I and III drugs, other excipients in the formulation do not have a significant effect on the absorption of Active Pharmaceutical Ingredient (API), SPIP, Ussing Chamber, and Caco-2 can verify the permeability consistency of generic drugs and original drugs, eliminating the necessity to run *in vivo* bioequivalence test (Table 12.5).

When the high–medium–low permeability database and system are relatively completed, the permeability of the tested drug can be classified. The permeability reference drug recommended by BCS is used as the control substance, and it is compared with the permeability detection value of the test drug, and the permeability of the test drug could be rated accordingly.

12.4.4.2 Study on Absorption and Mechanism of Different Intestinal Segments and Different Dosage Forms

Compared with *in vitro* and *in vivo* methods, *in vivo* SPIP models can be used to confirm the primary absorption of intestinal segments of drugs, and to research the effects of different dosage forms on the intestinal absorption of drugs. For example, for the same drug, drug absorption may also vary significantly from site to site in different intestinal segments. *In vivo* perfusion studies of different intestinal segments are used to determine the central absorption part of the drug in the intestine by comparing the P_{eff} results of each intestinal segment. At the same time, different dosage forms of the same drug may have different intestinal absorption, which is helpful for formulation screening by comparing the absorption of different dosage forms.

Table 12.5 Commonly used reference drugs with high–medium–low permeability.

High permeability ($F_a \geq 85\%$)	Moderate permeability (F_a = 50%–84%)	Low permeability ($F_a < 50\%$)
Antipyrine	Chlorpheniramine	Famotidine
Caffeine	Creatinine	Nadolol
Ketoprofen	Terbutaline	Sulpiride
Naproxen	Hydrochlorothiazide	Lisinopril
Theophylline	Enalapril	Acyclovir
Metoprolol	Furosemide	Foscarnet
Propranolol	Metformin	Mannitol
Carbamazepine phenytoin	Amiloride	Chlorothiazide
Disopyramide	Atenolol	Enalaprilat
Minoxidil	Ranitidine	

Source: Adapted from reference [4].

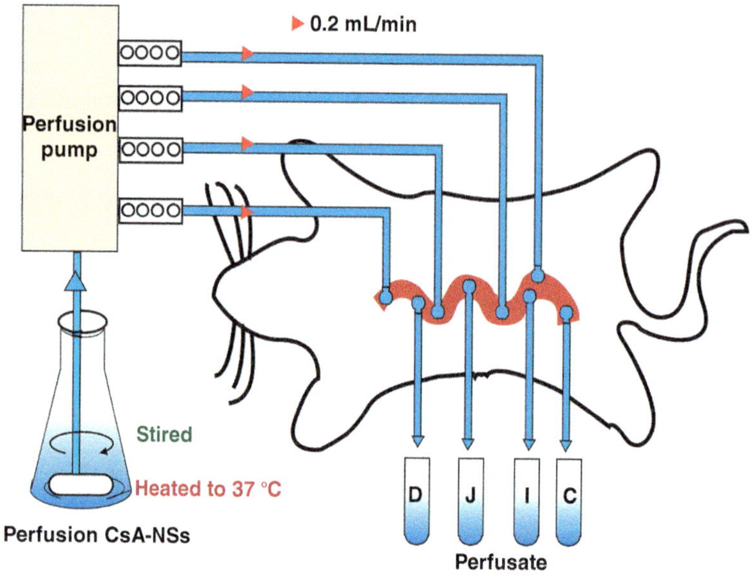

Figure 12.27 Schematic diagram of SPIP in different intestinal segments of the rat. *Source:* It's from Dove Medical Press under Creative Commons CC BY-NC 3.0 [18].

In addition, due to the presence of transporters in intestinal epithelial cells, different types of drugs have different modes of intestinal absorption. Therefore, inhibitors or substrates of transporters such as P-gp or MRP can be added to the perfusate to study the absorption of intestinal transport and DDI (Figure 12.27).

12.4.5 Summary

Intestinal absorption is the central absorption route of most oral drugs. According to BCS, the solubility and permeability of drugs in the gastrointestinal tract are the critical factors in determining the extent of drug absorption. The SPIP model is an effective tool to study the *in vivo* permeability of oral drugs. From the permeability of SPIP, the human permeability will be obtained [5]. SPIP method can evaluate the permeability of drugs and perform permeability classification determination and has been used for absorption evaluation of different intestinal segments and different dosage forms. SPIP model could guide the study of the absorption of drugs in the intestine, is helpful for improving the bioavailability of oral drugs, and eventually accelerates the research and development process of oral drugs.

References

1 Balimane, P.V., Chong, S., and Morrison, R.A. (2000 Jul). Current methodologies used for evaluation of intestinal permeability and absorption. *Journal of Pharmacological and Toxicological Methods* 44 (1): 301–312.

2 Anecdotal, L., Lixiang, R., Xiaohong, L. et al. (2017). Overview of study methods of drug intestinal absorption. *Drug Evaluation Study* 40 (05): 706–710.

3 Amidon, G.L., Lennernäs, H., Shah, V.P. et al. (1995). A theoretical basis for a biopharmaceutic drug classification: the correlation of *in vitro* drug product dissolution and *in vivo* bioavailability. *Pharmaceutical Research* 12 (3): 413–420.

4 FDA (2017). Waiver of *in vivo* bioavailability and bioequivalence studies for immediate-release solid Oral dosage forms based on a biopharmaceutics classification system guidance for industry. *[Guidance for Industry]. U.S. Food and Drug Administration..*

5 Dezani, T.M., Dezani, A.B., Junior, J.B. et al. (2016). Single-pass intestinal perfusion (SPIP) and prediction of fraction absorbed and permeability in humans: a study with antiretroviral drugs. *European Journal of Pharmaceutics and Biopharmaceutics* 104: 131–139.

6 Parodi, A., Buzaeva, P., Nigovora, D. et al. (2021). Nanomedicine for increasing the oral bioavailability of cancer treatments. *Journal of Nanobiotechnology.* 19 (1): 354.

7 Li, H., Dong, L., Liu, Y. et al. (2014). Comparison of two approaches of intestinal absorption by puerarin. *Journal of Pharmacological and Toxicological Methods* 70 (1): 6–11.

8 López-Yerena, A., Pérez, M., Vallverdú-Queralt, A. et al. (2021). Oleacein intestinal permeation and metabolism in rats using an *in situ* perfusion technique. *Pharmaceutics.* 13 (5): 719.

9 Jacobs, F.A., Tarnasky, W.G., and Forks, G. (1963). Primary and secondary transport systems for amino acids in the intact intestine. *Journal of the American Medical Association* 183: 765–767.

10 Nagare, N., Damre, A., Singh, K.S. et al. (2010). Determination of site of absorption of propranolol in rat gut using *in situ* single-pass intestinal perfusion. *Indian Journal of Pharmaceutical Sciences* 72 (5): 625–629.

11 Cummins, C.L., Salphati, L., Reid, M.J. et al. (2003). *In vivo* modulation of intestinal CYP3A metabolism by P-glycoprotein: studies using the rat single-pass intestinal perfusion model. *The Journal of Pharmacology and Experimental Therapeutics* 305 (1): 306–314.

12 Xie YS, Ren XL, Pan GX, et al. The assessment of absorption of periplocin *in situ* via intestinal perfusion of rats by HPLC. *Biomedical Chromatography* 2008; 22 (2): 196–201.

13 Yao, Y., Mi, W., Cao, G. et al. (2020). The absorption characteristics of nonvolatile components in a water extraction from Amomi fructus as determined by *in situ* single-pass intestinal perfusion and high-performance liquid chromatography. *Frontiers in Pharmacology* 11: 711.

14 Dubbelboer, I.R., Dahlgren, D., Sjögren, E. et al. (2019). Rat intestinal drug permeability: a status report and summary of repeated determinations. *European Journal of Pharmaceutics and Biopharmaceutics* 142: 364–376.

15 Zakeri-Milani, P., Valizadeh, H., Tajerzadeh, H. et al. (2007). Predicting human intestinal permeability using single-pass intestinal perfusion in rat. *Journal of Pharmacy and Pharmaceutical Sciences* 10 (3): 368–379.

16 Su, Y., Hu, P., Lee, S.H. et al. (2007). Using novobiocin as a specific inhibitor of breast cancer resistant protein to assess the role of the transporter in the absorption and disposition of topotecan. *Journal of Pharmacy & Pharmaceutical Sciences* 10 (4): 519–536.

17 Castella, M.E., Reist, M., Mayer, J.M. et al. (2006). Development of an *in vitro* rat intestine segmental perfusion model to investigate permeability and predict oral fraction absorbed. *Pharmaceutical Research* 23: 1543–1553.

18 Sun, W., Gao, J., Fan, R. et al. (2022). The effect of particle size on the absorption of Cyclosporin a Nanosuspensions. *International Journal of Nanomedicine* 17: 1741–1755.

12.5 Application of Liver Biopsy in Preclinical Pharmacokinetic Studies

Huayu Li, Hang Zhao, Jie Li, Guang Yang, Shoutao Liu, Jing Jin, Jacob Zhi Chen

12.5.1 Introduction

Liver biopsy, a widely utilized procedure in preclinical PK studies, provides valuable insights into the metabolism, transport, pharmacology, and toxic effects of drugs within the liver. The information is crucial for optimizing drug design, determining appropriate dosage, and enhancing both efficacy and safety. By analyzing liver tissue samples obtained from liver biopsy, scientists can gain a comprehensive understanding of drug metabolism pathways, enzymatic activity, and their impact on drug efficacy and safety. This analysis also provides the concentration information regarding drugs in the liver, allowing for an understanding of their maximum concentrations, duration of efficacy, and interactions with receptors and transporters, which influence the PK and the pharmacological characteristics of drugs. Furthermore, liver biopsy techniques contribute to the evaluation of drug-induced hepatotoxicity and the identification of potential safety concerns by investigating pathological changes in liver tissue. Additionally, it is worth noting that in animal studies, liver biopsy is a cost-effective and animal-saving method for obtaining liver tissue samples.

12.5.2 Introduction of Liver Biopsy

The conventional approach for studying drug exposure in the liver involves obtaining liver tissue through animal necropsy, primarily in rodents, for concentration analysis. However, when it comes to investigating drug exposure in the liver tissue of larger animals such as dogs or monkeys, the anatomical acquisition methods used in rodents are impractical and inefficient. To address this limitation, liver biopsy (Figure 12.28) is an alternative method that enables the extraction of a small amount of liver tissue using a biopsy needle while the animal is under anesthesia. Liver biopsy allows for multiple biopsies in one large animal [1, 2], thereby determining the study of drug exposure in liver tissue.

12.5.3 Risks and Experiences of Liver Biopsy Techniques

Liver biopsy is widely used in clinical practice but still has potential complications, such as infection, abdominal hemorrhage, and intrahepatic and subcapsular hematomas [3]. These complication risks also occur in preclinical trials.

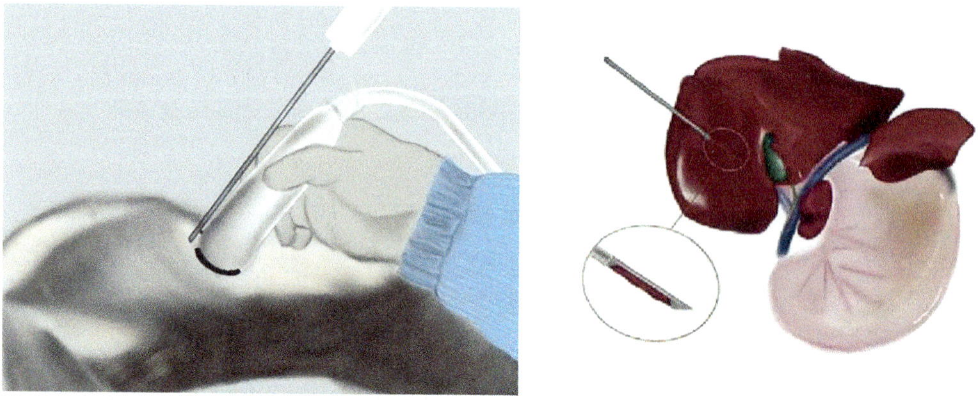

Figure 12.28 Liver biopsy.

Animal welfare is of utmost importance in our experimental operations. Therefore, we require experienced operators who possess a thorough understanding of animal anatomy and B-ultrasound imaging principles. They must also maintain a strong awareness of aseptic techniques. Anesthetics should be selected carefully, taking into account the duration of anesthesia required. The gauge of puncture needles should be selected based on the specific requirements of different study purposes. Additionally, operators should possess the ability to assess the animals' recovery based on their condition, particularly in cases involving multiple liver biopsy operations. Animals without coagulation dysfunction, anemia, abdominal effusion, or any other symptoms can be carefully selected to ensure the smooth execution of the puncture procedure.

After years of technological accumulation and experience summarization, our team has developed a safe, effective, applicable, and feasible operating procedure. The following is the operation procedure: assessing the animal's health, shaving the puncture site, administering anesthetic drugs, disinfecting the puncture site, and preparing related equipment according to the requirements of the experimental program. After confirming that the animal is in deep anesthesia, one scientist utilizes the B-ultrasound to locate the liver puncture site and its operable safety area, avoiding the gallbladder, bile duct, and large blood vessels. Another scientist performs the puncture using a biopsy needle according to the guidance of B-ultrasound (Figure 12.29), takes out the sample, confirms the sample amount, and disinfects the puncture site. The animal's appetite, mental status, and other appearance will be monitored for at least three days post-surgery to monitor the potential postoperative abdominal hemorrhage [4].

12.5.4 Cases and Advantages

Case 1. Reducing Interindividual Errors and Saving Costs

In a preclinical PK study of the treatment of a small molecular drug, concentrations of the drug in both plasma and liver were measured following oral administration in rats. For comparison, another experiment was conducted in three dogs (Table 12.6) where plasma and liver samples were obtained at the same time points. Liver samples were taken from the same animals at certain time points (Figure 12.30). This liver biopsy method reduced interindividual errors compared with rats, conformed to the 3R (Replacement, Reduction, Refinement) principle of humane animal research, and saved costs.

Oligonucleotide drugs are known for their ability to target specific tissues, and the liver is a common target due to its high blood flow and metabolic activity. Two effective delivery systems for oligonucleotide drugs in the liver are: using GalNAc conjugates or targeting the ASGPR. Both

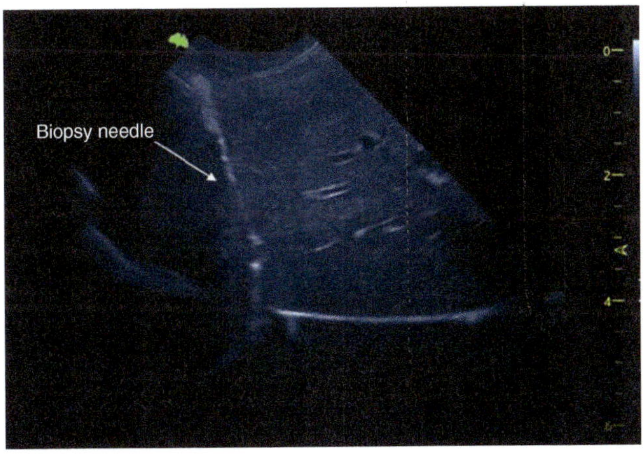

Figure 12.29 Liver biopsy B-ultrasound screen.

Table 12.6 Comparison of animal numbers typically used in pharmacokinetic experiments with dogs and rats.

			Time (h)							
Species	**Group**	**Animal No.**	**0.25**	**0.5**	**1**	**2**	**4**	**6**	**8**	**24**
Rat	1	R01, R02, R03	P	P	P	P	P	P	P	P
	2	R04, R05, R06	—	—	P+L	—	—	—	—	—
	3	R07, R08, R09	—	—	—	P+L	—	—	—	—
	4	R10, R11, R12	—	—	—	—	—	—	P+L	—
Dog	1	D1001, D1002, D1003	P	P	P+L	P+L	P	P	P+L	P

Note: P = plasma, L = liver, R = rat, D = dog.

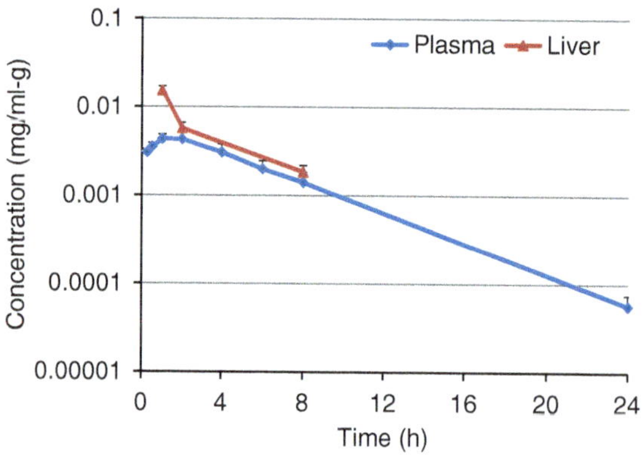

Figure 12.30 Mean (n = 3) liver and plasma concentrations of a small molecular drug after single oral administration in the beagle dogs.

GalNAc and ASGPR have high affinity with liver cells, facilitating the delivery of oligonucleotides into these cells. Nanoliposomes that bind to ApoE and enter hepatocytes via the low-density lipoprotein receptor (LDL-R). This is another strategy to target liver cells and ensure the delivery of therapeutic oligonucleotides. Both approaches aim to increase the specificity and efficiency of oligonucleotide delivery to liver cells, thereby enhancing their therapeutic potential.

The oligonucleotide mainly distributes in the liver after administration, leading to the concentration of oligonucleotide drugs in plasma being relatively low and rapidly decreasing, but the concentration in liver tissue remains higher and lasts longer. Figure 12.31 shows the mean plasma and liver concentrations, and PK parameters of a GalNAC-modified siRNA after a single subcutaneous injection in cynomolgus monkeys. Plasma concentrations dropped below the detection limit of quantification (BLQ=10ng/mL) 16 hours after administration, while liver concentrations remained above the detection limit of quantification until 57 days after administration.

12.5.5 Summary

Liver biopsy plays a significant role in the preclinical drug development phase [5]. Compared with taking the liver after euthanizing the rodents (and large animals), using liver biopsies in large animals can save a substantial number of animals. The study of the metabolism of new molecular drugs in the liver tissue of cynomolgus monkeys shows that the liver biopsy technique is suitable for the development of liver-specific new drugs.

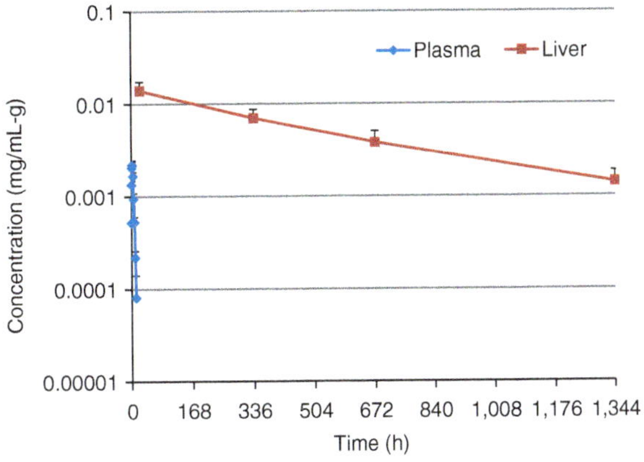

Figure 12.31 Mean (n = 6) plasma and liver concentration of a GalNAC-modified siRNA after a single subcutaneous injection in cynomolgus monkeys.

References

1 Pang, R.Q., Liu, J.K., He, Z.L., and Wang, H.X. (2005). Establishment and evaluation of macaque liver fibrosis model. *Shijie Huaren Xiaohua Zazhi* 13 (16): 1956–1958.

2 He, Y., Wang, J., and Wang, Y. (2001). Advances in methods for diagnosing the degree of chronic liver fibrosis. *World Chinese Journal* 9: 1305–1309.

3 Poterucha, J.J. and Gunneson, T.J. (2010). Liver Biopsy and Paracentesis. *Practical Gastroenterology and Hepatology: Liver and Biliary Disease: Liver and Biliary Disease Wiley* 9: 80–86.

4 Hong, W., Bo, M.A., Qin, S. et al. (2018). Analysis of ultrasound guided percutaneous rhesus monkey liver biopsy[J]. *Technical Acoustics* 37 (1): 62–65.

5 Xuexue He, Hao Qu, Yu Sun, and Shangwei Ji, Research Progress of liver biopsy, Department of Gastroenterology, China-Japan Friendship Hospital, Jilin University, Changchun, 130033, China

13

Metabolite Profiling and Identification

13.1 High-Throughput Identification of Drug Metabolic Soft Spots for Lead Optimization

Wenjun Zhu, Lian Guo, Lingling Zhang, Weiqun Cao

13.1.1 Introduction

In recent years, with the promotion of a series of drug administration reforms and national scientific and technological innovation strategies, the R&D of innovative drugs has been in full swing. Consequently, the biomedical industry has emerged as one of the most rapidly growing sectors. The successful development and marketing of a new drug requires a complex process, including selection of drug targets, screening of lead compounds, determination of clinical candidate compounds, confirmation of clinical efficacy, and drug safety evaluation. When a molecular structure with *in vitro* pharmacodynamic activity is obtained by target screening, the goal is to ensure it has appropriate exposure and action time *in vivo*; however, rapid metabolic clearance of the drug can reduce its exposure. To improve the properties of rapid metabolism, it is important to be able to recognize and protect the metabolic sites of the compound molecules. Chemical groups or individual atoms of compounds prone to *in vivo* or *in vitro* metabolic reactions are termed "soft spots". Early metabolic soft spot identification studies in drug discovery can accurately identify the metabolic soft spots of compound molecules, thereby assisting chemists in purposefully performing lead optimization. Metabolic soft spot identification, which involves the separation and structure elucidation of compounds, is a complex and time-consuming task, while early compound selection stages often require fast and high-throughput techniques to streamline large-scale compound structure optimization. In this chapter, the role of metabolic soft spot identification in compound structure optimization and a software-based method for high-throughput metabolic soft spot identification research will be introduced. These insights aim to assist in the selection of lead compounds and accelerate new drug research and development.

13.1.2 The Importance of Conducting Metabolic Soft Spot Identification Studies

Drugs are eliminated from the body mainly by metabolism and excretion, with metabolism as the major pathway for most drugs. Compounds are metabolized by phase I and phase II enzymes, converting fat-soluble compounds into polar. These water-soluble compounds are then excreted through bile and urine. CYP450 enzymes are the major enzymes of phase I drug metabolism, and

Drug Metabolism and Pharmacokinetics: Frontiers, Strategies, and Applications, First Edition.
Edited by Liang Shen and WuXi AppTec DMPK.

almost three-quarters of marketed drugs have CYP enzymes involved in their elimination process [1]. For many of these drugs, CYP enzyme metabolism has a significant impact on their plasma concentrations and dosing regimens. These enzymes are also related to high inter-individual variability and the risk of drug-drug interactions. When a candidate compound exhibits a high degree of CYP enzyme-dependent clearance, attempts should be made to attenuate the effects of the relevant enzymes and enhance the metabolic stability of the compound [2].

At present, the liver microsomal model is usually selected to investigate the stability of CYP enzyme metabolism. Generally, high-throughput liver microsomal incubation is used to assay the metabolic stability and determine the *in vitro* metabolic half-life of the compounds. Preclinical toxicity and pharmacodynamic properties of compounds were predicted by half-life and *in vivo* clearance rates. However, the liver microsomal clearance test does not provide the metabolic information of compounds in liver microsomes, and only when combined with metabolic soft spot identification studies, the metabolic "soft point" of compounds can be clarified. This information, based on rapid metabolism, can guide the structural optimization of compounds and improve their metabolic stability.

13.1.3 Strategies for Drug Design Based on Metabolism

With the continuous development of structural characterization and computer simulation technology, metabolism-based assisted drug design strategies are also widely used, and a series of structure optimization strategies for lead compounds have been developed at this stage by altering the main metabolic pathways. These strategies mainly include blocking metabolic sites, backbone modifications, bioisosteres, and prodrug modifications [3].

13.1.3.1 Blocking Metabolic Sites

Blocking metabolic sites can block metabolic pathways and improve the metabolic stability of compounds, and fluorine atoms are the most commonly used blocking groups. In drug design, fluorine atom substitution is usually used to block the sites susceptible to oxidative metabolism and selectively prevent the occurrence of oxidative metabolism in small molecule compounds. For example, SCH48461, a lead compound of SCH58235, containing multiple oxidative and demethylated metabolic sites, exhibits a rapid metabolic elimination rate and low bioavailability. SCH58235 exhibits enhanced metabolic stability and improved activity after introducing fluorine atoms to the phenyl ring of the lead compound and blocking the oxidative metabolic site (Figure 13.1) [4].

In addition to fluorine atoms, chlorine atoms, cyanide groups, and deuterium can also be used to seal metabolic sites. Deuteration strategy can be referred to Section 13.5: insights into the R&D strategy of deuterated drugs from the drug metabolism perspective.

13.1.3.2 Backbone Modification

Backbone modifications can also alter metabolic pathways and improve metabolic stability. Backbone modifications include changing the size of the ring, ring modification, length of connecting chain, backbone transition, etc. Through backbone modification, the structure of the compound can be changed, the metabolic position resistance can be increased, or the metabolic soft spot can be reduced. In recent years, with the discovery of small-molecule E3 ligase ligands such as VHL, CRBN, and PROTACs degraders design strategies based on enzyme inhibitors to induce enzymatic degradation has become a research hotspot [5]. Goracci et al. [6] compared the metabolic soft spots and metabolic rates of PROTAC molecules and their constituent ligands; the results are shown in Table 13.1. It can be seen that backbone modification of PROTAC molecules can help enhance metabolic stability and reduce potential off-target toxicity.

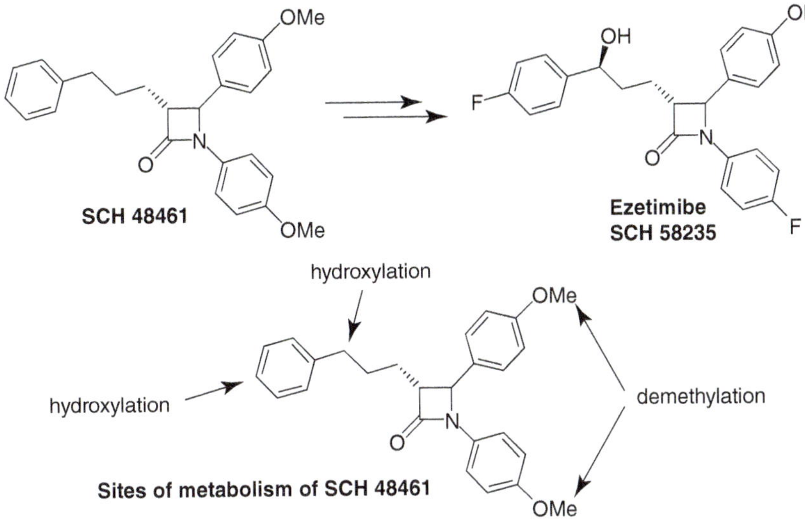

Figure 13.1 Structures of SCH58235 and SCH48461. *Source:* [4] / Royal Society of Chemistry.

Table 13.1 Backbone modification helps enhance metabolic stability.

1 (dBet1) (135.0)	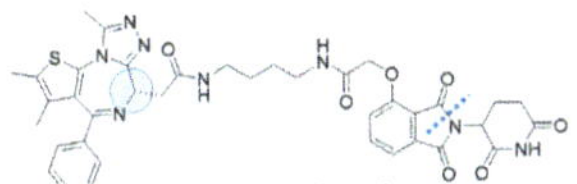	2 (dBet6) (18.2)	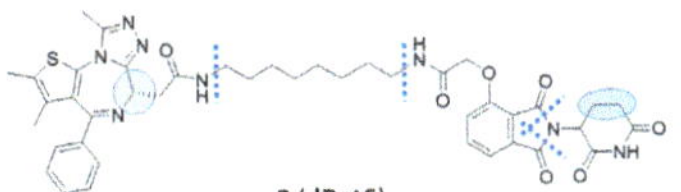	Compared to compound 2, the half-life of compound 1 was reduced by 7.4-fold, possibly because the short linker allows the molecule to have metabolic potential, and high metabolic stability
27	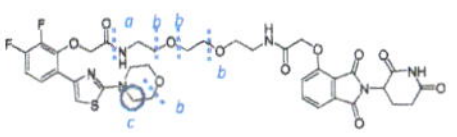	*a. Amide hydrolysis* *b. O-dealkylation (x3)* *c. Aliphatic-hydroxylation*	39.4	Compound 30 is more stable than compound 27 and compound 31, and may be the presence of PEG chains, giving the molecule more soft spots and easier dealkylation
30	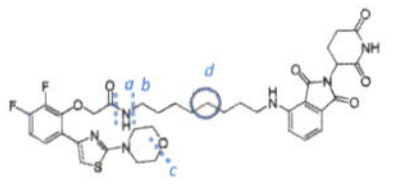	*a. Amide hydrolysis* *b. N-dealkylation* *c. O-dealkylation* *d. Aliphatic-hydroxylation*	47.6	
31	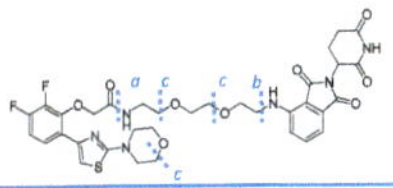	*a. Amide hydrolysis* *b. N-dealkylation* *c. O-dealkylation (x3)*	38.4	
11 (>>240)	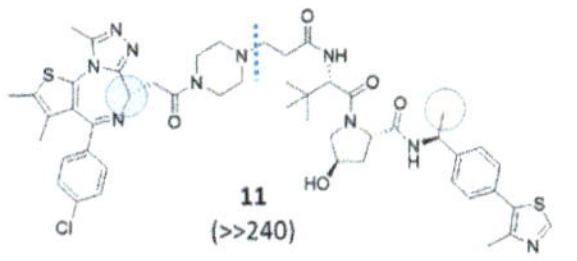	24 (>>240)	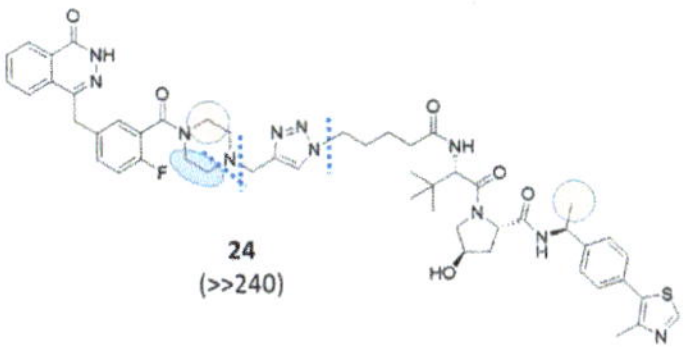	Cyclic linkers such as compound 11 and compound 24 with piperazine and triazole rings have higher metabolic stability than linear linkers

13.1.3.3 "Open ring" and "Snap ring" Strategies

The ring structures prone to metabolism were subjected to "open ring" and "snap ring" to obtain a new class of skeletons that can also block metabolism and improve stability. Chemokine receptor CXCR4 plays an important role in HIV infection, inflammatory/autoimmune diseases, and tumor metastasis, which is considered to be a drug target. Li Zhanhui et al. [7] synthesized a series of aminopyrimidine-based backbones CXCR4 antagonists of good activity, but there was a problem of rapid clearance by liver microsomes, and then scientists found tetrahydroquinoline (THQ) structure may be the culprit for high clearance by extensive structure–metabolism relationship studies.

Wilson et al. [8] have also found in mouse liver microsomes, hydroxylation of THQ fragments resulted in high clearance of the compound (Figure 13.2). Based on this, scientists "opened" the THQ fragment, which was easily hydroxylated and "snapped" it with the *N*-methyl group, resulting in a novel class backbone of CXCR4 antagonists (Figure 13.3), among which compound 46 demonstrated greatly improved metabolic stability in human and rat liver microsomes.

13.1.3.4 Prodrug Modification

The prodrug itself has no biological activity or has very low activity and becomes an active substance after *in vivo* metabolism. The main purpose of prodrug modification is to improve the metabolic stability of the drug, prolong the action time of the drug *in vivo*, increase the bioavailability, and reduce the toxicity and side effects of the drug.

The above describes some metabolism-based strategies to assist in drug design, while the basis of structural modification, first of all, is to study what metabolism occurs and what "soft spots" exist in drugs with poor stability, which will be introduced below.

13.1.4 How to Conduct Metabolic Soft Spot Identification Research with High Throughput

During traditional metabolite identification, long-duration LC–MS analytical methods are used to ensure that each metabolite and biological matrix are chromatographic separated, which requires a lot of manual data analysis work. Therefore, the extended experimental period is challenging to meet the high-throughput requirements during the drug discovery stage. A rapid and high-throughput metabolic soft spot identification method becomes particularly important (Figure 13.4).

Sample Incubation

Sample incubation time can directly affect the degree of metabolism of compounds, and extensive metabolites may not be able to identify metabolic soft spots due to excessive incubation time.

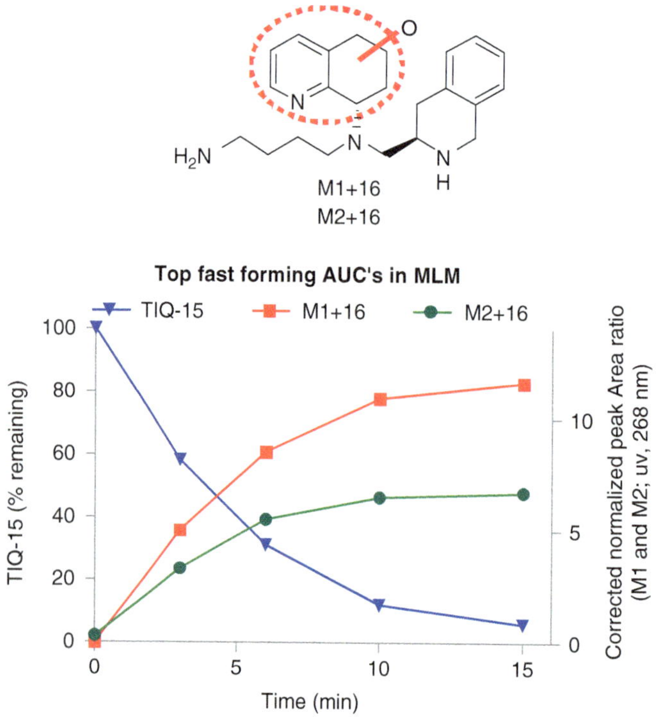

Figure 13.2 Metabolism results of TIQ-15 in mouse liver microsomes. *Source:* [8] / with permission of American Chemical Society.

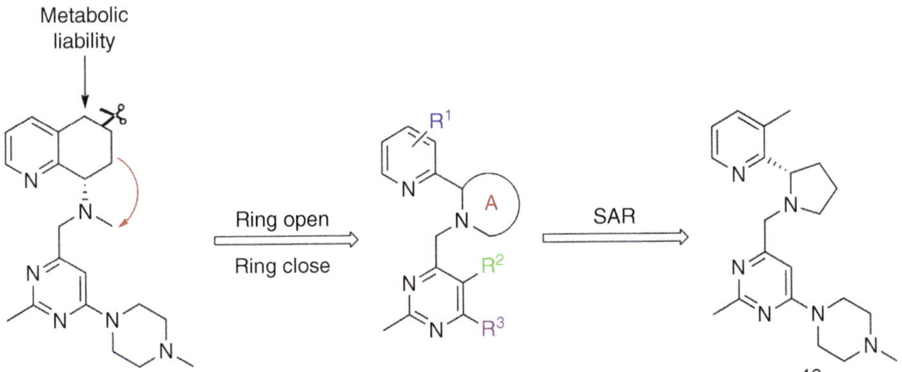

1

Binding IC_{50} = 8.8 ± 1.0 nM
Ca^{2+} flux IC_{50} = 0.021 ± 0.004 nM
hERG: IC_{50} = 8 μM
RLM: Cl_{int} = 1567 mL/min/kg
$t_{1/2}$ = 1.6 min

46

Binding IC_{50} = 79 ± 11 nM
Ca^{2+} flux IC_{50} = 0.25 ± 0.08 nM
hERG: IC_{50} > 30 μM
RLM: Cl_{int} = 123 mL/min/kg
$t_{1/2}$ = 20.2 min

Figure 13.3 "Open Loop" and "Snap Loop" designs. *Source:* [7] / ELSEVIER.

Therefore, it is necessary to customize the incubation time according to the results of compound metabolic stability to reduce the secondary metabolism caused by excessive metabolism and to more clearly expose the primary metabolic soft spot more clearly.

Sample Testing

Different types of compounds have different physical and chemical properties, which present challenges for high-throughput sample detection. In addition to cutting-edge instruments such as ultra-high-performance liquid chromatography (UHPLC) and HRMS, extensive experience in LC–MS method development can shorten the detection time of samples. The advantages of high resolution, high speed, and high sensitivity of UHPLC are used to ensure the retention and separation of compounds and their metabolites in 5–15 minutes. HRMS is used for mass detection, taking advantage of its high resolution, high mass accuracy, wide dynamic range, and other advantages to ensure more accurate mass–charge ratio information. At the same time, a variety of data acquisition modes (full MS/dd-MS2/MSE) and diverse ion filtration functions (mass loss, background subtraction, neutral loss, characteristic product ions, etc.) are used to find metabolites and provide accurate fragment information for the assignment of metabolic soft spots.

Data Interpretation

Massive data analysis work has been one of the key challenges in limiting the efficiency of metabolic soft spot identification. To accelerate the process of the step, many data processing software (Compound Discovery, MetaboLynx, MS Frontier, MetWorks, and Mass-MetaSite) can be used. After adjusting and optimizing various parameters of the software, the spectrum can be automatically interpreted, and then the interpretation results can be effectively combined with manual confirmation to promote efficient and high-quality metabolite structural elucidation.

In addition, combining metabolic soft point detection with metabolic stability experiments can skip the process of repeated incubation, allowing direct detection of the samples in metabolic stability experiments. By comparing and analyzing the results of multiple time points, the dominant metabolic sites at different time points can be identified (Figure 13.5).

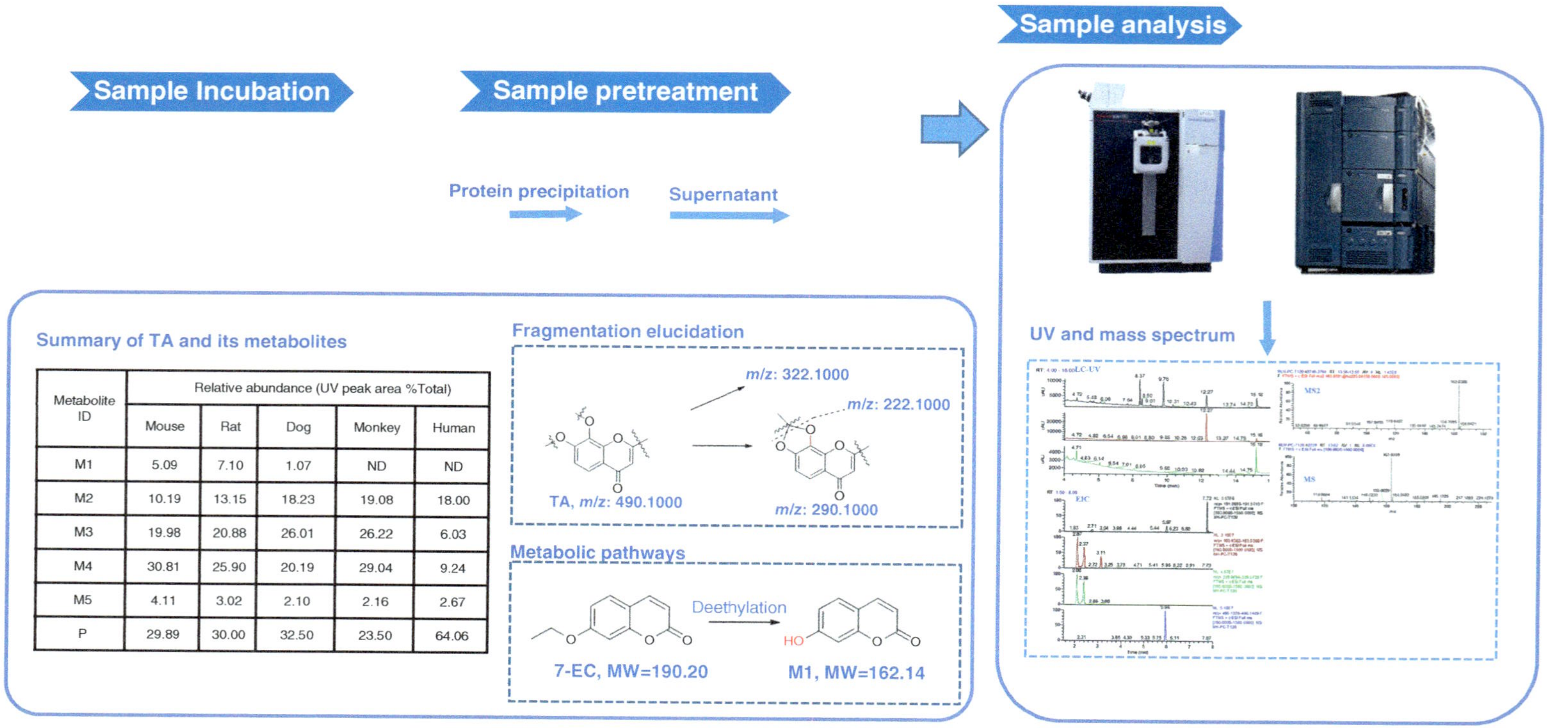

Figure 13.4 High throughput experimental process of metabolic soft spot identification.

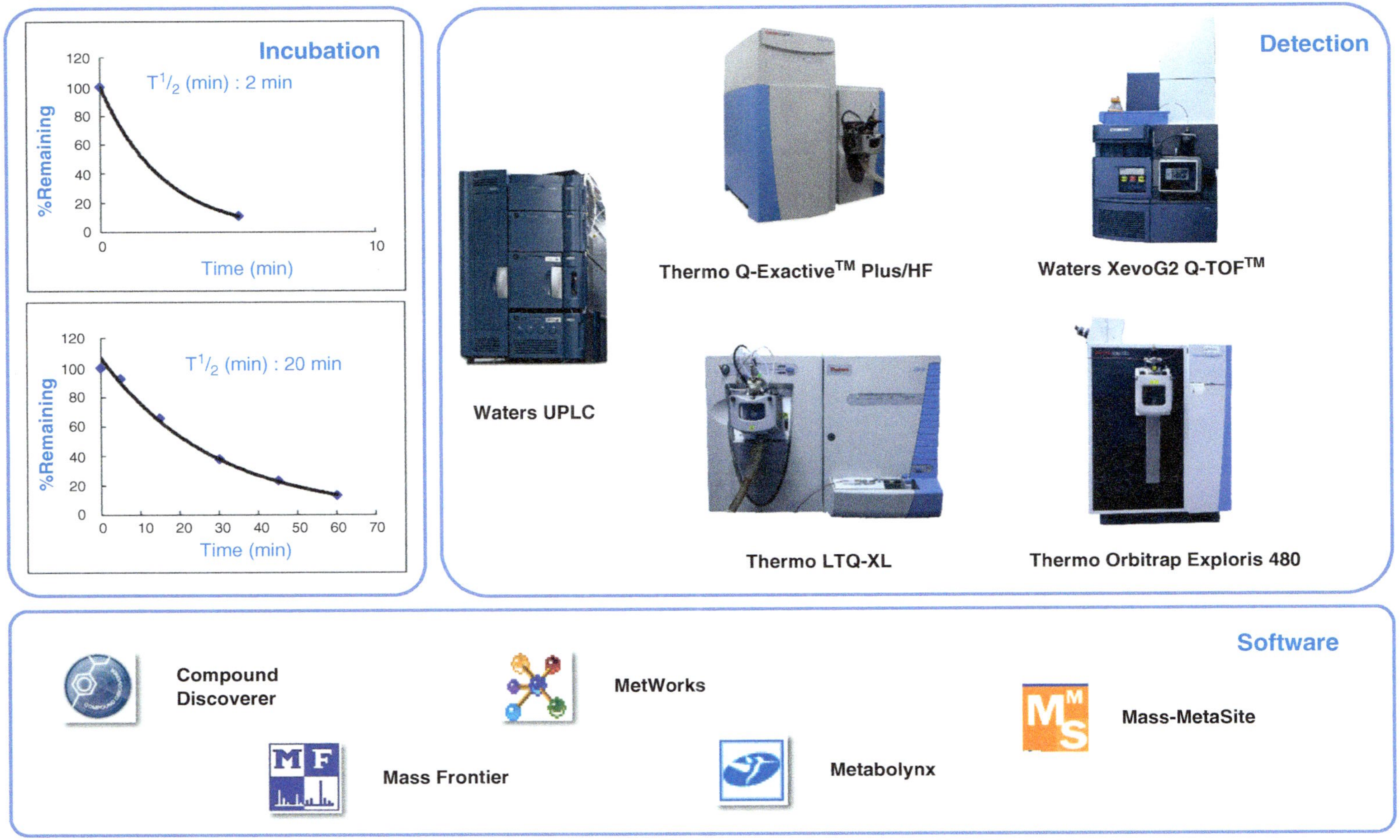

Figure 13.5 Metabolic soft spot identification platform.

13.1.5 Conclusion

Metabolic soft spot identification experiments can not only determine what compounds are metabolized into liver microsomes or hepatocytes but also identify metabolic "soft spots" to help chemists optimize their structure and improve the metabolic stability of compounds by blocking drug metabolic soft spots.

Traditional metabolite identification studies are time-consuming and cannot fully meet the requirements for efficient evaluation and throughput for lead compound optimization in the early stage. In recent years, with the continuous development of technologies such as instruments and data processing software, as well as the multidisciplinary intersection of cheminformatics, stoichiometry, statistics, and computer science with pharmacy, the automated process of metabolite identification has been realized to a certain extent. Although there are some instances of false positive results in the automatic identification, the research process of metabolic soft spots is greatly accelerated by software processing combined with the analysis of senior scientists, which provides better references for compound structure optimization.

References

1 Kennedy, T. (1997). Managing the drug discovery/development interface[J]. *Drug Discovery Today* 2 (10): 436–444.

2 Trunzer, M., Faller, B., and Zimmerlin, A. (2009). Metabolic soft spot identification and compound optimization in early discovery phases using MetaSite and LC-MS/MS validation [J]. *Journal of Medicinal Chemistry* 52 (2): 329.

3 Wang, J. and Liu, H. (2013). Structure optimization strategy of lead compounds (I) - changing metabolic pathways to improve metabolic stability [J]. *Journal of Pharmacy* 48 (10): 11.

4 Purser, S., Moore, P.R., Swallow, S. et al. (2008). Fluorine in medicinal chemistry [J]. *Chemical Society Reviews* 37 (2): 320–330.

5 Gao, H., Sun, X., and Rao, Y. (2020). PROTAC technology: opportunities and challenges[J]. *ACS Medicinal Chemistry Letters* 11 (3): 237–240.

6 Goracci, L., Desantis, J., Valeri, A. et al. (2020). Understanding the metabolism of proteolysis targeting chimeras (PROTACs): the next step toward pharmaceutical applications [J]. *Journal of Medicinal Chemistry* 63 (20): 11615–11638.

7 Zhanhui, L., Wang, X., Lin, Y. et al. (2020). Design, synthesis, and evaluation of pyrrolidine based CXCR4 antagonists with *in vivo* anti-tumor metastatic activity [J]. *European Journal of Medicinal Chemistry* 205: 112537.

8 Wilson, R.J., Jecs, E., Miller, E.J. et al. (2018). Synthesis and SAR of 1, 2, 3, 4-tetrahydroisoquinoline-based CXCR4 antagonists[J]. *ACS Medicinal Chemistry Letters* 9 (1): 17–22.

13.2 Application of Metabolite Biosynthesis Technology in the Precise Identification of Drug Metabolite Structures

Yikuan Lu, Rong Yang, Ruixing Li, Weiqun Cao

13.2.1 Introduction

Drug MetID is one of the main contents of ADME studies, meanwhile, precise metabolite structure identification plays an important role in drug metabolism study, permeating every stage from drug discovery through to IND application. First, during the drug discovery stage, MetID can aid in pinpointing soft spots of the candidates and even lead to the discovery of new active drugs. For instance, SN-38, a metabolite of the anticancer drug irinotecan, exhibits an anticancer activity that is hundreds of times more potent compared with the original drug [1]. Second, during the IND stage, as outlined in the safety testing of drug metabolites guidance for industry from the Food and Drug Administration (FDA) and National Medical Products Administration (NMPA), safety testing of these unique metabolites is required when the exposure of disproportionate drug metabolites exceeds 10% of the total drug-related exposure (calculated by AUC at steady state) in human plasma. If the exposure of metabolites in relevant animal species is insufficient to assess their safety, the metabolites need to be synthesized, and further safety evaluation must be performed by incorporation methods or direct administration [2, 3]. The clarity of metabolite structures directly affects the time, cost, and method employed to obtain their standards and even the IND process. Therefore, obtaining a unique or high proportion of drug metabolites is of paramount importance in drug development. Hence, how to achieve the accurate identification of metabolite structure has become a research hotspot. This article focuses on metabolite biosynthesis technology to provide insights for the precise identification of drug metabolite structures.

13.2.2 Strategy for Precise Identification of Drug Metabolite Structures

At present, the primary method for drug metabolite identification is to use HRMS technology, which can obtain the information of parent ions through full scan and precursor ion (PI) scanning mode, which enables the determination of the metabolite elemental composition and metabolic pathways. With the detailed ion fragment information obtained from multistage mass spectrometry (MS^n), the potential structure of metabolites can be further confirmed. MS^n technology can help to determine or further narrow down the potential metabolic sites. As shown in Figure 13.6(a), the structure of the hydrolysis metabolite of bisacodyl is deduced by the MS^n information and metabolic mechanism. However, in most cases, it is challenging for MS^n to obtain the metabolite's

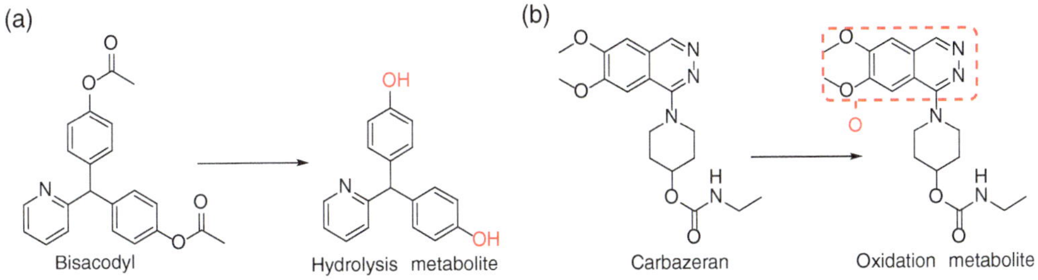

Figure 13.6 Bisacodyl and its metabolite (a) Carbazeran and (b) its metabolite.

precise structure. As shown in Figure 13.6(b), the MS^n data can only localize the oxidative metabolic area to the 6,7-dimethoxyphthalazine part of carbazeran. These results suggest that the MS^n technology has its limitations in structure identification, which may not always confirm the accurate metabolite structures.

Obtaining purified metabolite standards is one of the strategies to achieve precise identification of metabolite structures. High-purity metabolite standards can be analyzed using nuclear magnetic resonance (NMR) to elucidate the metabolite structures. Presently, chemical synthesis and biosynthesis are the mainstream technologies to obtain metabolite standards. If the metabolite structures are clearly confirmed, the chemical method is the preferred option to synthesize the target metabolite. Subsequently, LC–MS/MS is used to match the synthesized compound with real metabolic samples to confirm its identity. However, if the metabolite structure is indeterminate, or the target metabolite has many predicted metabolic sites, it becomes challenging to obtain metabolite standards by chemical synthesis techniques. Under the circumstances, metabolic spots can only be speculated based on the MS^n fragment information of metabolites, drug metabolism principles, and chemical reaction knowledge. In such cases, multiple potential compounds would need to be synthesized to match the target metabolite. Therefore, the feasibility of using the chemical synthesis techniques to obtain metabolite standards is based on the following factors: the number of potential metabolite structures, the complexity of synthesizing potential metabolites, and the timeline of the project cycle. On the other hand, biosynthesis technology offers the advantage of directional synthesis of target metabolites, which obtains target metabolite standards utilizing *in vitro* and *in vivo* metabolism, such as urine, feces, bile or plasma, cellular or subcellular fractions (such as microsomes, cells, cytosol, and S9 fractions), recombinant enzymes (such as CYP enzyme, AO enzyme, and UGT enzyme.), and microorganisms (specially engineered bacteria) in animals.

As shown in Figure 13.7, the application of *in vitro* and *in vivo* biosynthetic technology in the metabolite structure identification has the following advantages: (i) it synthesizes metabolites rapidly and accurately, eliminating the need to match biosynthetic compounds with target metabolites repeatedly against *in vitro* or *in vivo* metabolic samples; (ii) provides an alternative plan when chemical synthesis is difficult; (iii) the process is relatively straightforward and less time-consuming.

13.2.3 Introduction of Metabolite Biosynthesis Technology

Biosynthesis technology, a significant branch of biological research, uses biological systems or organisms to produce substances required by humans. Currently, biosynthesis technologies are widely used in various fields, such as chemicals, biomaterials, bioenergy, and food. In the pharmaceutical field, biosynthesis technology plays a vital role in improving drug yield, enabling cell-free protein synthesis [4], discovering new drugs, and developing novel gene therapies [5], among other applications. For drug metabolite structure identification, biosynthesis technology can be used to obtain metabolite standards by flexibly selecting different biological matrices, such as CYP enzymes or non-CYP enzymes. In the following sections, the application of biosynthesis technology in metabolite structures' precise identification will be detailed (Figure 13.8).

13.2.3.1 *In Vivo* Animal Transformation

If the target metabolite is abundant in a certain matrix in animals, it can potentially be directly separated and purified from this matrix. This approach includes preparing a high-purity i metabolite standards solution and determining its accurate structure using HRMS and NMR.

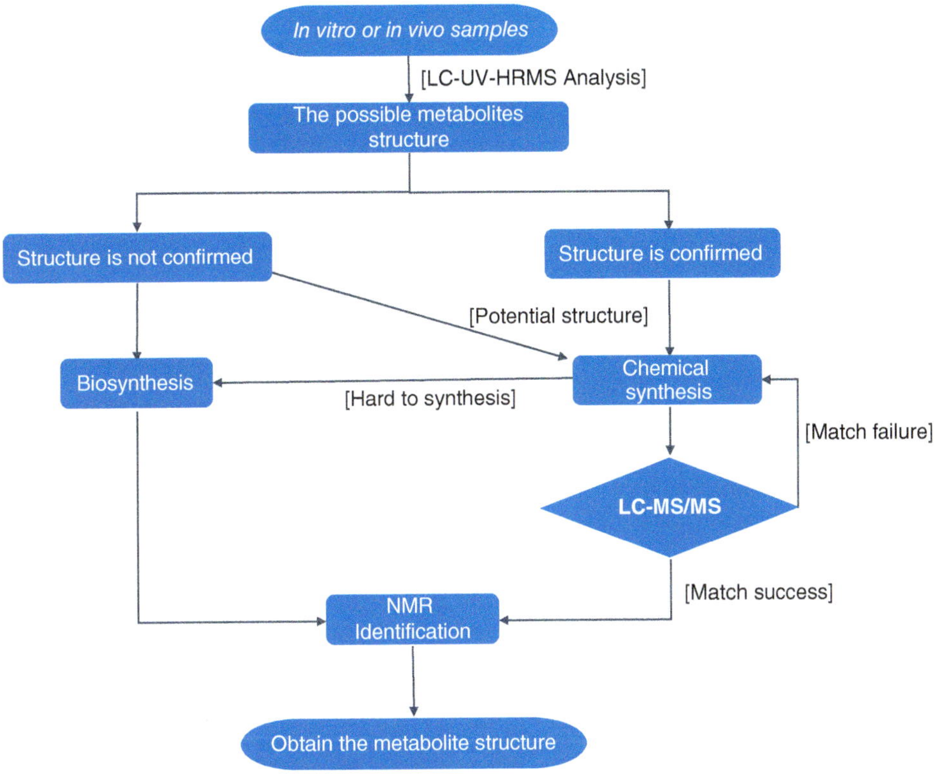

Figure 13.7 Strategy and process of precise identification of metabolite structures.

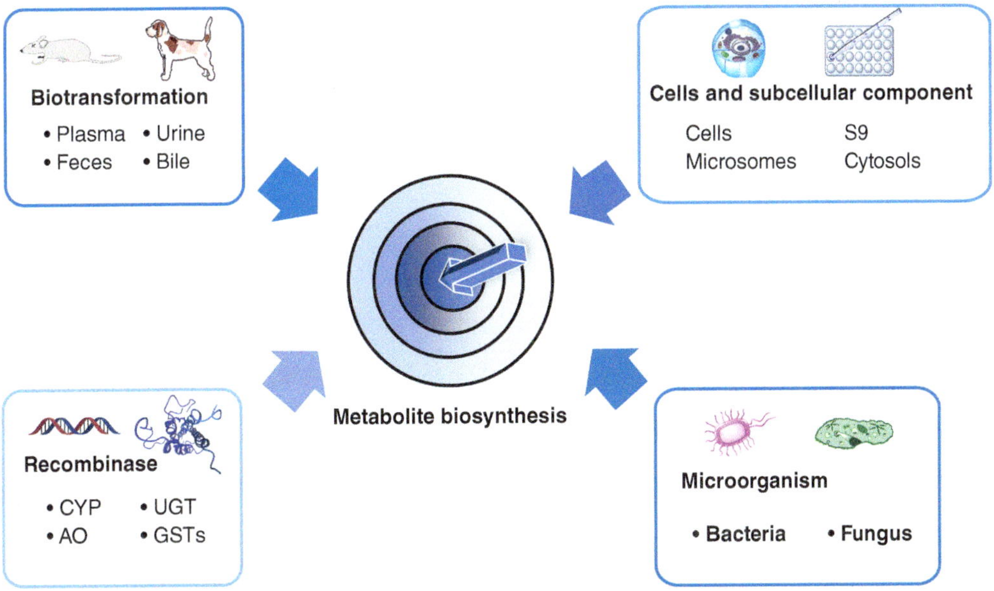

Figure 13.8 Summary of metabolite biosynthesis technologies.

For example, Deng et al. [6] successfully identified the main metabolite of amiodarone as a 4-carboxyl metabolite by isolating and extracting it from patient bile and then confirming its identity using HRMS and NMR.

13.2.3.2 Cellular or Subcellular Components Transformation

If a certain proportion of target metabolites are found to be generated in *in vitro* metabolism experiments, cellular or subcellular components incubation systems (S9, cytosol, microsomes, etc.) can be used for the biosynthesis of target metabolites. The optimal incubation system is selected by optimizing various incubation conditions. Scale-up incubations are performed using the optimized method. After separation and purification, the solution containing the target metabolites is further separated by preparative liquid phase chromatography to obtain highly purified target metabolites. These are then subjected to NMR for structure identification.

13.2.3.3 Recombinase Transformation

If the results of enzyme phenotype identification experiments confirm that the target metabolite is catalyzed by a specific enzyme phenotype, the particular subtype of the enzyme contributing significantly to the target metabolite production could be selected for biosynthesis to obtain a sufficient quantity of the target product for structural identification.

13.2.3.4 Microbial Transformation

When the aforementioned biosynthetic methods are difficult or costly, an alternative approach is to construct *in vivo* and *in vitro* metabolic systems that imitate human or other mammalian species by cultivating microbial-engineered organisms such as bacteria or fungi to achieve the biosynthesis from drugs to metabolites.

13.2.4 Biosynthesis and Structure Identification Process

The procedures for biosynthesis and structure identification of target metabolites mainly include the optimization of the experimental system (including the screening of animal species, matrix species, and experimental conditions), followed by scale-up, isolation, and purification of target metabolites, and finally, HRMS and NMR identification to achieve the precise identification of drug metabolite structures. The specific steps are as follows (Figure 13.9).

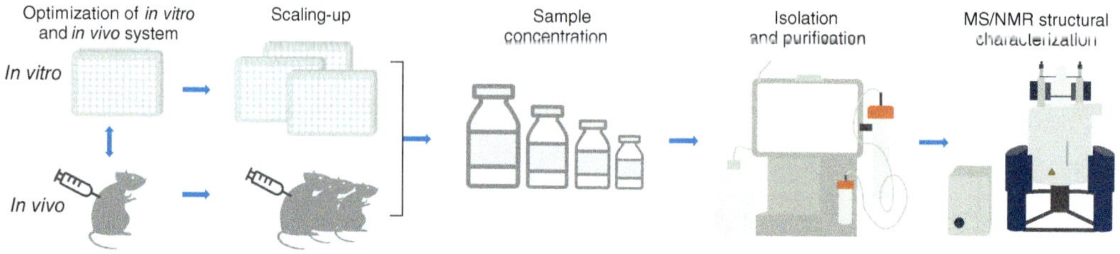

Figure 13.9 Flow diagram for metabolite biosynthesis and structure identification.

13.2.5 Case of Metabolite Biosynthesis and Structure Identification

Carbazeran is an aldehyde oxidase substrate with a nitrogen-containing heterocyclic structure, which undergoes extensive oxidative metabolism in humans, resulting in low levels of drug exposure. In this case, the oxidative metabolites of carbazenran were prepared through the biosynthesis method of *in vitro* transformation. Ultimately, several milligrams of oxidative metabolite were obtained and its structure was further confirmed.

After *in vitro* human hepatocyte incubation, carbazeran predominantly generated a high proportion (~ 60%) of oxidation metabolite (Figure 13.10(a)). Through the comparison of the metabolite profiles of carbazeran in liver microsomes, S9, and hepatocytes, this oxidation metabolite could be presumed to be catalyzed by aldehyde oxidase. In the present case, the effect of different animal species of matrices (liver microsomes, liver S9, and liver cytosol) on target metabolite production has been investigated. Considering the availability of the experimental substrate and the cost, the monkey cytosol, which is rich in aldehyde oxidase, has been selected as the final incubation system. Furthermore, through the control variate approach, the optimization of each experimental condition was confirmed, and approximately 90% of carbazeran was transformed into the target oxidation metabolite (Figure 13.10(b)). Based on the optimal experimental conditions, the scale-up incubation was conducted to transform carbazeran, after extraction using an appropriate method, the solvent containing the target metabolite was obtained. The mixed solution was processed through concentration, separation, purification, and lyophilization, resulting in the highly purified powder of oxidation metabolite. The metabolite powder was then dissolved in deuterated dimethyl sulfoxide solution, and analyzed by NMR to compare the ^{1}H NMR spectrum (Figures 13.11 and 13.12) of carbazeran and its oxidation metabolite. The results showed that the pyridine-4-aminoethyl formate moiety of oxidation metabolite was unchanged, while the signals of pyridazine structure exhibited the addition of one N-H signal (11.98 ppm) and the reduction of one C-H signal (9.11 ppm). Thus, it is hypothesized that the hydrogen at position 7 is replaced by oxygen, forming a carbonyl group, while the tertiary amine at position 7 is converted into a secondary amine and forms a lactam structure. Based on these observations, it can be deduced that this oxidation metabolite is carbazeran 4-oxidation, which is in accordance with the previous research [7].

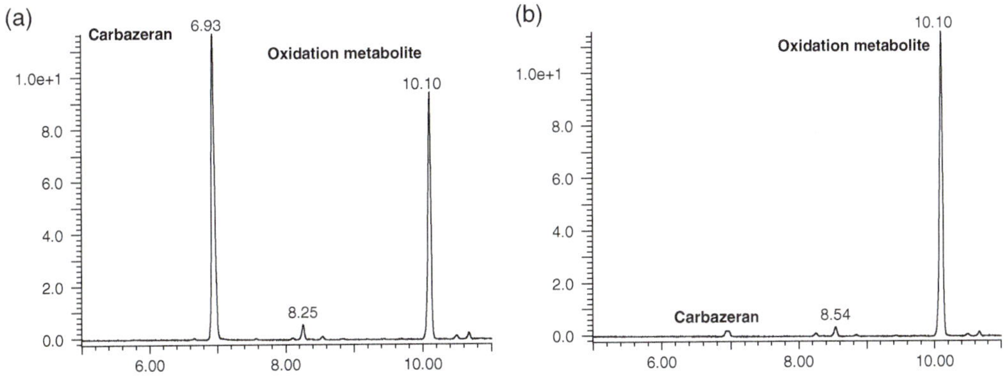

Figure 13.10 LC-UV profiles after 2 hours incubation of Cabazitaxel in human hepatocytes (a) and 4 hours incubation of monkey cytosol (b).

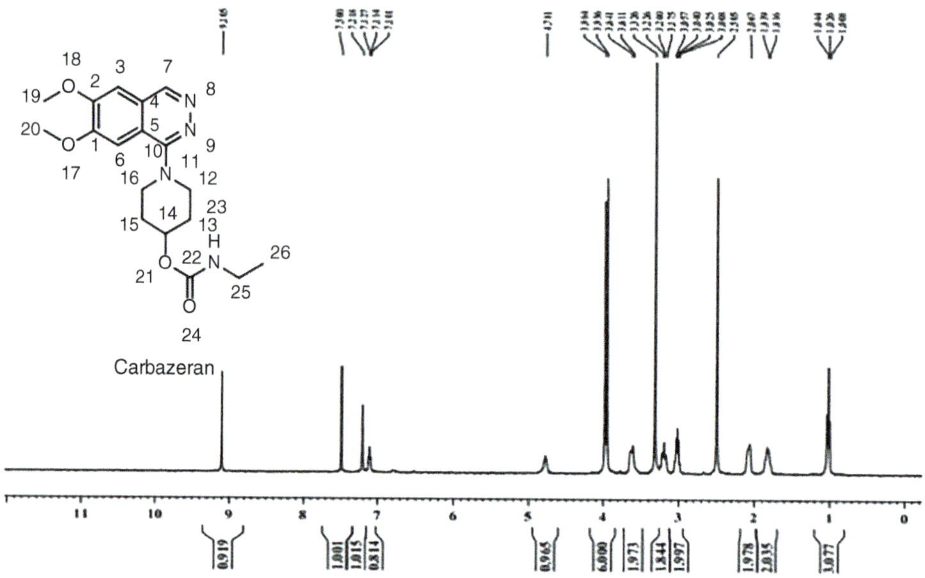

Figure 13.11 Carbazeran ^{1}H NMR spectrum.

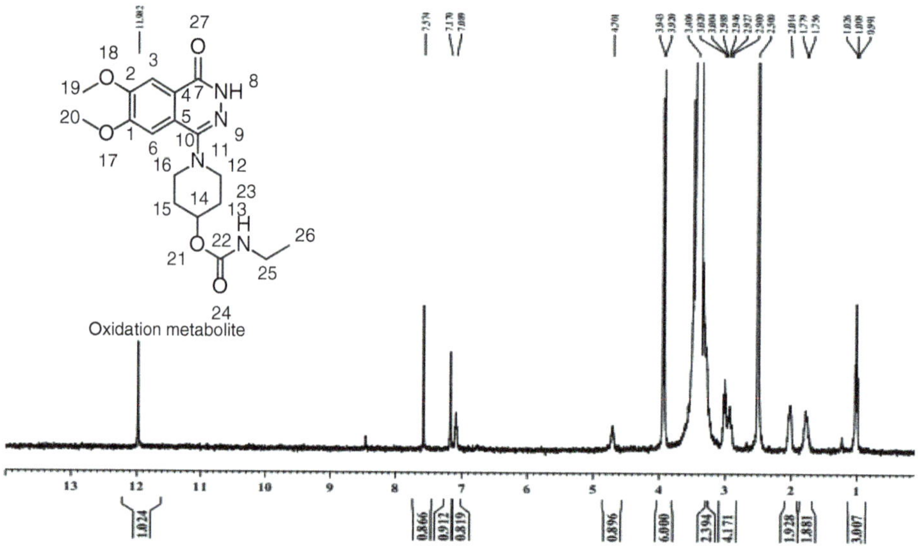

Figure 13.12 Carbazeran 4-oxidation ^{1}H NMR spectrum.

13.2.6 Summary

The directional synthesis and structure identification of key metabolites has an important impact on drug research and development. Both chemical synthesis technology and biosynthesis technology can be employed to obtain metabolite standards, while biosynthesis offers distinct advantages. It allows a more direct and targeted approach to generate metabolites even when their structures are not fully determined. Coupled with HRMS and NMR for structure elucidation, biosynthesis can significantly accelerate the drug research and development process, highlighting its crucial role in this field.

References

1 Ohtsuka, K., Inoue, S., Kameyama, M. et al. (2003). Intracellular conversion of irinotecan to its active form, SN-38, by native carboxylesterase in human non-small cell lung cancer. *Lung Cancer* 41: 187–198.

2 FDA (2020). Safety testing of drug metabolites guidance for industry, Revision 2.; https://www.fda.gov/regulatory-information/search-fda-guidance-documents/safety-testing-drug-metabolites.

3 NMPA (2012). Technical Guidelines for Safety Testing of Drug Metabolites.

4 Tinafar, A., Jaenes, K., and Pardee, K. (2019). Synthetic biology goes cell-free. *BMC Biology* 17: 64.

5 Cubillos-Ruiz, A., Guo, T., Sokolovska, A. et al. (2021). Engineering living therapeutics with synthetic biology. *Nature Reviews Drug Discovery* 20: 941–960.

6 Deng, P., You, T., Chen, X. et al. (2011). Identification of amiodarone metabolites in human bile by ultraperformance liquid chromatography/quadrupole time-of-flight mass spectrometry. *Drug Metabolism and Disposition* 39: 1058–1069.

7 Uehara, S., Yoneda, N., Higuchi, Y. et al. (2020). Human aldehyde oxidase 1 – mediated Carbazeran oxidation in chimeric TK-NOG mice transplanted with human hepatocytes. *Drug Metabolism and Disposition* 48: 580–586.

13.3 Detection of Reactive Metabolites and Its Application in Drug Research and Development

Huijuan Wang, Ruixing Li, Weiqun Cao

13.3.1 Introduction

Drug metabolism usually refers to the biological transformation of drugs under the action of phase I and phase II metabolic enzymes in the body, resulting in increased polarity of metabolites, improved water solubility, weakened or completely inactivated pharmacological activity, serving as a mechanism for drug detoxification. However, in some cases, drug metabolism is also a toxic process. After metabolic activation, it generates some reactive metabolites with electrophilicity, including epoxy, quinone, carbon-positive ions, quinone-imine, etc. The electrophilicity of metabolites allows them to covalently bind to biological macromolecules (including nucleic acids and proteins) in the body, thereby disrupting or altering the normal physiological functions of these macromolecules. This can cause cell damage, hepatotoxicity, immunotoxicity, and cancer (Figure 13.13). Safety issues related to drug metabolism, caused by reactive metabolites, can significantly affect the success or failure of drug development.

In recent years, the safety issue of drug metabolism has received increasing attention, and the production of reactive metabolites by metabolic activation is a crucial aspect of drug safety. This section mainly elaborates on the following aspects; (i) the safety issues of drug metabolism related to reactive metabolites; (ii) types of compounds prone to produce reactive metabolites; (iii) commonly used analytical detection methods for reactive metabolites; and (iv) the application of reactive metabolite screening in the process of drug development.

13.3.2 Drug Safety Concerns Arising from Reactive Metabolites

Initially, people were confused about the biological toxicity exhibited by some chemically relatively inert substances, and this was not solved until the middle of the twentieth century. Covalent modification of proteins by xenobiotics was reported by Elizabeth and James Miller [2] in 1947. They observed irreversible covalent binding of yellow pigment to hepatic proteins in the liver tissue of

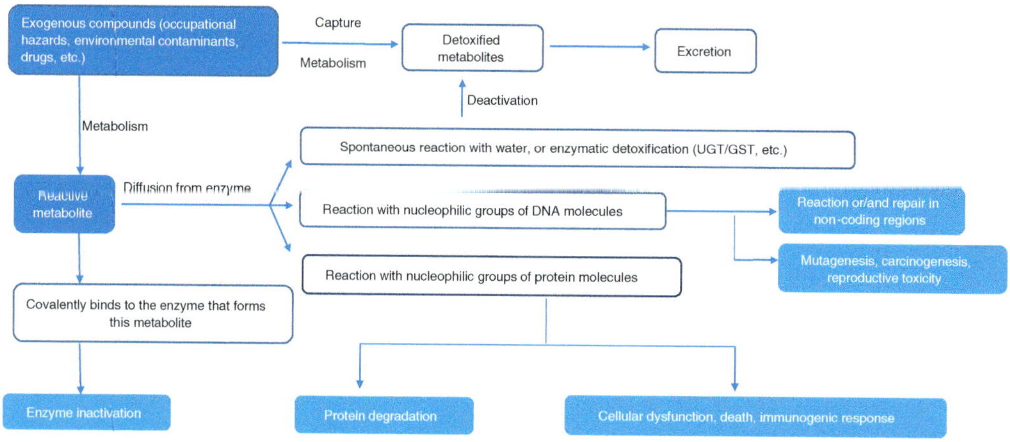

Figure 13.13 Metabolic pathways of xenobiotics. *Source:* Adapted from [1].

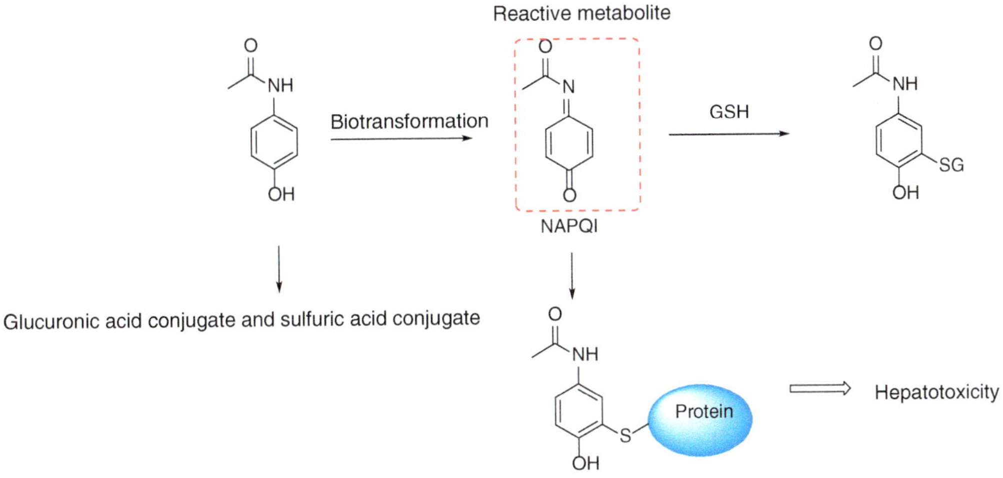

Figure 13.14 Metabolic toxicity of paracetamol. *Source:* [4]. Athersuch et al., 2018; [5]. Jollow et al., 1973.

rats after feeding with methylaminoazobenzene. Subsequently, some researchers characterized the structure and function of DNA and found the conjugate of DNA and carcinogens in animals [3]. Thus, the process of covalent binding of xenobiotics to macromolecules by metabolic activation in organisms is unveiled. The discovery that metabolic activation produces reactive metabolites causing covalent binding opens new space for investigating drug toxicity.

Toxicity studies of paracetamol are classic cases [4–5] (Figure 13.14). Paracetamol is mainly metabolically eliminated by glucuronic acid and sulfuric acid conjugation. Another metabolic pathway is oxidation to *N*-acetyl-p-benzoquinone imine (NAPQI) under the action of CYP450 enzymes. NAPQI rapidly consumes GSH. When the concentration of GSH in the body is low, it covalently binds to the free cysteine residues on proteins, which contain thiol functional groups. At present, it is widely believed that the production of NAPQI by metabolic activation, which then covalently binds to proteins, is the main cause of drug-induced liver injury induced by p-acetophenol. It has been shown that many drugs produce biological toxicity, such as troglitazone leading to liver failure, clozapine causing agranulocytosis, and nefazodone resulting in hepatotoxicity, all of these are related to their metabolism in the body producing reactive metabolites that covalently bind to biological macromolecules [6–8].

In recent years, there have been many drug recalls caused by drug toxicity (Table 13.2). In a study of 200 market-available drugs and 68 drugs that were either recalled or received black box warnings, it was found that at least half of the drug molecule-mediated toxicities were closely associated with reactive metabolites. As one of the direct signals of drug metabolism safety issues, reactive metabolites must be factored into the drug structure design process. Potential reactive metabolites resulting from drug metabolism should be screened as early as possible in drug development to modify or eliminate problematic compounds. By employing rational structural design, the tendency of a drug to produce reactive metabolites can be minimized.

13.3.3 Common Pharmacovigilance Structure

Many drugs that can produce reactive metabolites have been reported in recent years. From the perspective of metabolism, the molecular structure of drugs largely determines whether they are easily metabolically activated and produce toxicity. In addition to the electrophilic groups

Table 13.2 Market drug recall and analysis of reactive metabolites.

Drug name	Structure	Indication	Reason for market withdrawal	Reactive metabolite production (yes/no)
Troglitazone	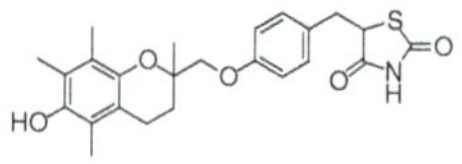	Antidiabetic	Hepatotoxicity	YES
Amineptine	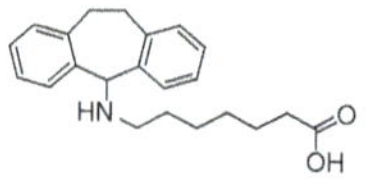	Antidepressant	Hepatotoxicity	YES
Bromfenac	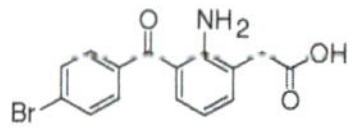	Nonsteroidal anti-inflammatory drug	Hepatotoxicity	YES
Ibufenac	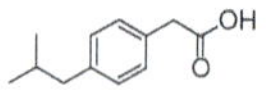	Nonsteroidal anti-inflammatory drug	Hepatotoxicity	YES
Pirprofen	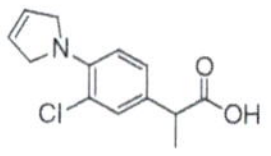	Nonsteroidal anti-inflammatory drug	Hepatotoxicity	YES
Tienilic acid	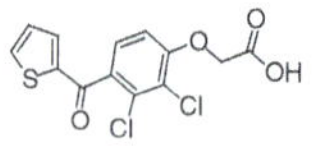	Diuretic	Hepatotoxicity	YES
Aminopyrine	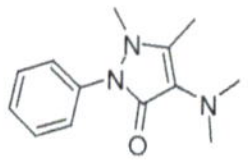	Analgesics	Hematologic Diseases	YES
Tolcapone	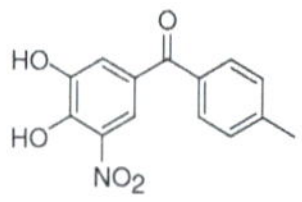	Antiparkinson	Hepatotoxicity	YES

Source: Adapted from reference [7].

(e.g., epoxy, aldehyde, and α, β-unsaturated ketones or esters), Table 13.3 summarizes some of the common pharmacovigilance structures (e.g., aromatic amines, thiophene rings, and furan rings) that may cause toxicity through metabolism leading to reactive metabolites, as well as their possible reactive metabolites. For drug molecules with alert structures, their potential for metabolic activation should be carefully considered.

It should be noted that the knowledge of pharmacovigilance structures is helpful in determining the drug molecules that may produce reactive metabolites, but this is not an absolute predictor. There are some marketed drugs containing alert structures, but the actual application process does not produce toxicity. Whether and to what extent the alert structure is involved in metabolism may be affected by other functional groups within the molecule. The metabolic activation potential of a drug molecule is often assessed with the aid of appropriate analytical detection techniques.

13.3.4 Detection Methods and Technology Platforms for Reactive Metabolites

Metabolic activation of xenobiotics produces reactive metabolites, and the process of metabolic activation is mainly mediated by phase I metabolic enzymes (e.g., CYP450 enzymes). Therefore, *in vitro* incubation experiments are usually performed in the liver microsomal system. If the metabolic

Table 13.3 Pharmacovigilance structures and possible reactive metabolites.

Alert structure	Possible reactive metabolites	Alert structure	Possible reactive metabolites
Aromatic amines	Quinone-imine or nitroso	3-Methyleneindoles	Imine-methide
Hydrazine, hydrazides	Diazene or diazonium ion	Furans	α, β-Unsaturated dicarbonyl
Nitroarenes	Nitroso	Thiophenes	α, β-Unsaturated dicarbonyl
Fused azaheterocycles	Nitrenium ion	Thiazoles	Thioamide, glyoxal
Benzylamines	Nitroso, oxime	Thiazolidinediones	Isocyanate, *S*-oxide
Foramides	Isocyanate	Arenes, bromoarenes	Arene oxide
Sulfonylureas	Isocyanate	Alkynes	Ketenes, oxirene
Thioureas	*S*-Oxides, isocyanate	α, β-Unsaturatedcarbonyl	Intrinsic electrophilicity
Hydroquinones	*p*-Benzoquinone	Aliphatic amines	Iminium ion
o- or *p*-Alkylphenols	*o* or *p*-Quinone methide	Alkylhalides	Acylhalides
Methylenedioxyphenyl	*o*-Quinone	etc.	—

Source: Adapted from reference [9].

activation of drugs involves phase II metabolic enzymes or a mixture of phase I/ II metabolic enzymes, then the experiment shall be conducted by using the liver S9 or hepatocyte system. Reactive metabolites can covalently bind to macromolecules in cells, such as proteins, and their half-life is usually short and difficult to detect directly in *in vitro* and *in vivo* experiments. At present, there are two primary detection methods for reactive metabolites:

1) **Capture with Small Molecule Probes**
Small-molecule capture probes that have been reported in the literature include GSH, cysteine, *N*-acetylcysteine, semicarbazide, methoxyamine, and cyanide anions. Among them, GSH, as a natural trapping agent for chemically reactive metabolites, can capture most reactive metabolites (including epoxides, arene oxides, alkyl halides, Michael acceptors, quinones, etc.) and is widely used in *in vitro* screening experiments for reactive metabolites at the early drug discovery and screening stages. Notably, not all reactive metabolites can be captured by GSH, depending on the nature of the metabolite. By changing or selecting additional capture reagents, reactive metabolites can be purposefully investigated, and experimental data can be supplemented and enriched. For example, reactive intermediates such as cyanide can be used to capture imine ions, or methoxyamine can be used to capture reactive aldehyde substances.

2) **Direct Detection for Covalent Conjugates of Biological Macromolecules**
Most applications of this method require radioisotope labeling of the drug. The labeled drug is incubated with *in vitro* liver microsomal protein, freshly isolated hepatocytes, or liver homogenate for a defined period, or the liver tissue homogenate is prepared post-administration in animals. The level of drug covalent binding and the potential of metabolic activation are evaluated by measuring the radioactive intensity of protein covalent conjugates in the precipitate. Covalent binding experiments with biological macromolecules are considered the gold standard for the quantitative assessment of reactive metabolites.

The workflow of reactive metabolite monitoring and identification is presented in Figure 13.15.

The following is a case analysis and technical validation by WuXi AppTec DMPK based on an *in vitro* reactive metabolite screening study of clozapine:

Clozapine is an atypical antipsychotic with good sedative and hypnotic effects. It is reported in the literature[10] that clozapine undergoes metabolic activation under the action of *in vitro* metabolic enzymes, producing reactive imine ions, which then covalently bind to GSH or protein (Figure 13.16). It is speculated that reactive imine ions covalently bind to target proteins and that clozapine causes adverse reactions such as agranulocytosis and hepatotoxicity in clinical treatment. Proteins that were covalently modified have been detected in neutrophils from patients treated with clozapine, further confirming the above speculation.

In this case, we investigated and validated the reaction metabolites of clozapine under *in vitro* incubation conditions at 37°C for 60 minutes using clozapine as the test substance, GSH as the capture probe, and human liver microsomes as the incubation matrix. By using different data acquisition and processing techniques, reactive metabolite screening was performed on samples incubated with clozapine in the liver microsomal matrix, and the detection results were compared with the literature.

We first used an LC-UV-HRMS system to perform UV scans and full MS/dd-MS^2 mass scans. By comparing the blank sample with the sample after 60 minutes of incubation with clozapine, some

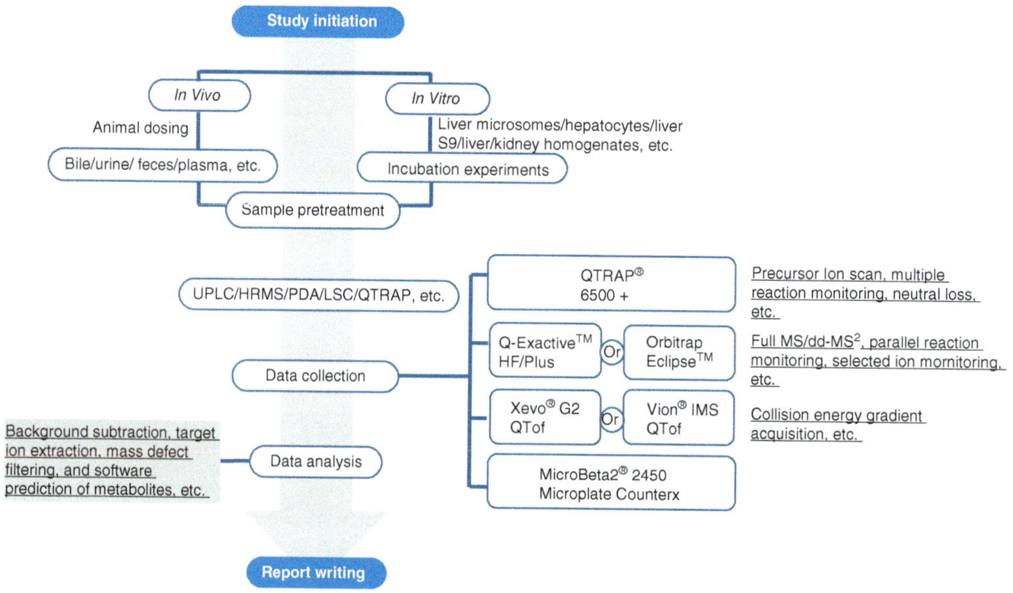

Figure 13.15 Workflow chart of reactive metabolite detection and identification.

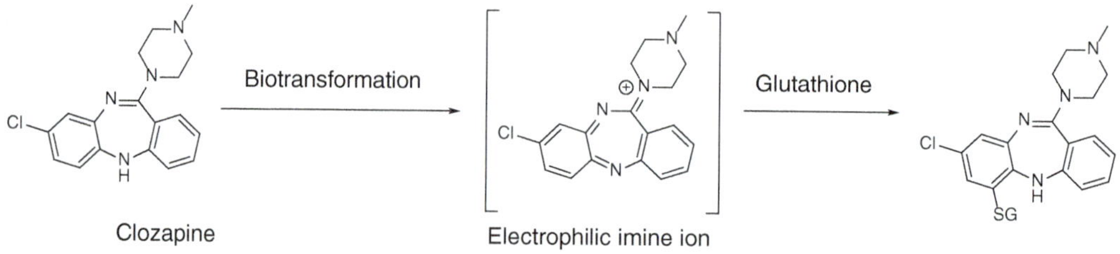

Figure 13.16 Metabolic activation pathway of clozapine.

metabolite peaks with clear UV signals can be observed, such as chromatographic peaks at 8.65 minutes and 9.33 minutes, which are shown in Figure 13.17(a). Not all these observed peaks were GSH conjugation products, requiring structural elucidation by the primary and secondary mass spectra for confirmation. The qualification process for GSH conjugates and non-GSH conjugates necessarily takes up a certain amount of analysis time. To identify GSH conjugates quickly and in a high-throughput manner, selective ion monitoring (SIM) is used to scan the characteristic fragment m/z 272.0888 of GSH conjugates. This detection method can quickly lock GSH conjugates without structural elucidation, and then target to find the primary mass spectrum and secondary mass

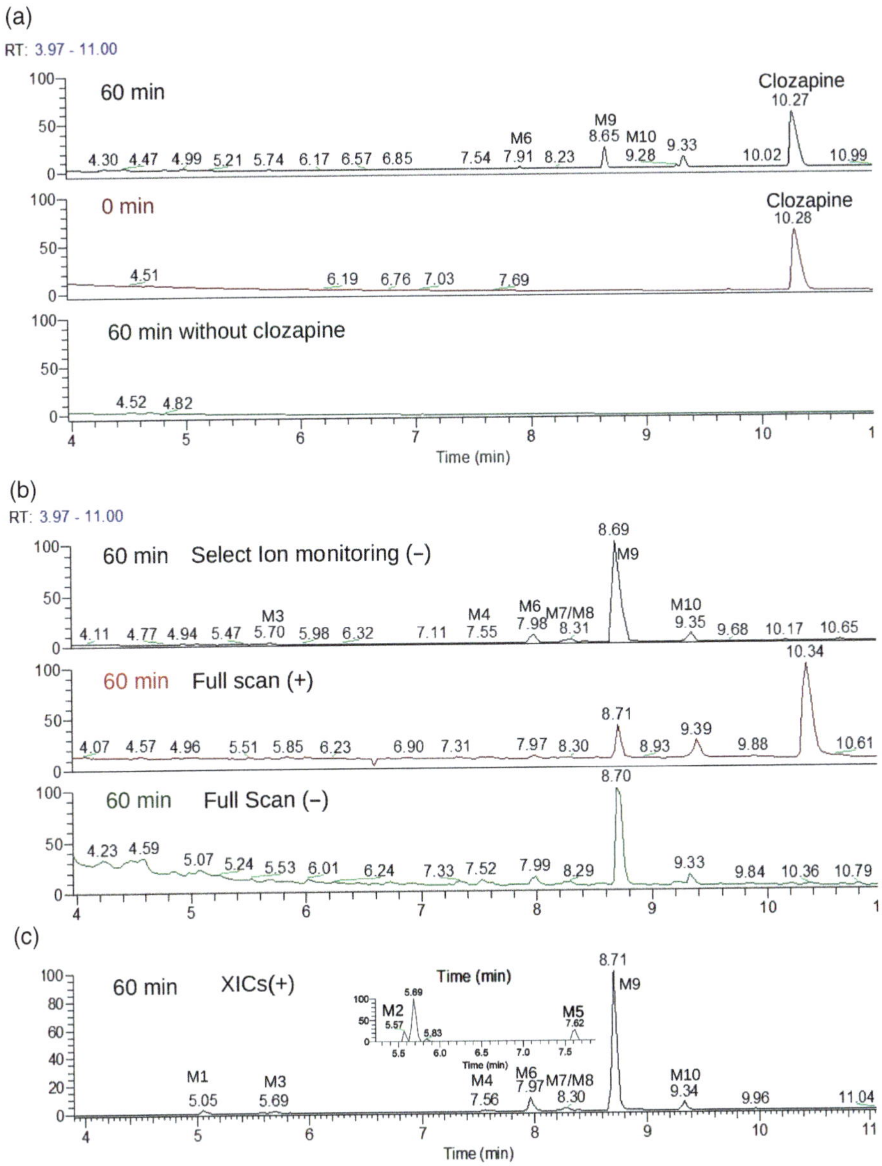

Figure 13.17 LC-UV chromatogram (a), LC–MS/MS chromatogram (b), and extracted ion current chromatogram (c) of clozapine in human liver microsomes after incubation for 0 or 60 minutes in LC-UV-HRMS.

Table 13.4 Reactive metabolites of clozapine on LC-UV-HRMS.

Metabolite code	*m/z*	Retention time (min)	Mass spec area	Relative abundance % (MS peak area)	Metabolic pathways
M1	598.24	5.05	2,948,613	1.99	P − Cl + H + GSH
M2	650.22	5.57	316,556	0.21	P + O + 2H + GSH
M3	650.22	5.69	1,801,373	1.22	P + O + 2H + GSH
M4	632.20	7.56	860,179	0.58	P + GSH
M5	618.19	7.62	457,274	0.31	P − CH_2 + GSH
M6	632.20	7.97	10,342,533	6.98	P + GSH
M7	648.20	8.24	999,968	0.67	P + O + GSH
M8	618.19	8.30	2,487,525	1.68	P − CH_2 + GSH
M9	632.20	8.71	121,505,305	81.99	P + GSH
M10	648.20	9.34	6,467,872	4.36	P + O + GSH

Note: P: parent drug; GSH: $C_{10}H_{15}N_3O_6S$.

spectrum at the corresponding retention time to confirm the metabolites. By this scanning mode, seven GSH conjugates can be detected, as shown in Figure 13.17(b). The metabolite information obtained from SIM scan and UV scan could complement each other, and 10 GSH conjugation products were finally confirmed by target ion extraction, as shown in Figure 13.17(c) and Table 13.4.

For reactive metabolite detection, sample analysis can be performed using a quadrupole tandem ion trap mass spectrometer in addition to the different detection modes using HRMS. Given that GSH conjugates are prone to neutral loss (NL) of 129 Da (pyroglutamic acid) in positive mode, we adopted the NL-IDA-EPI scanning mode. Secondary mass spectral fragmentation information was obtained by automatically triggering product ion scans for parent ions that produced an NL of 129 Da. Meanwhile, to reduce false positive signals, we also employed PI-IDA-EPI scanning for screening. The *m/z* 272 fragment is detected by precursor ion scanning, and most of the matrix interference information is filtered out. Neutral loss scans and PI scans yielded a total of five distinct GSH conjugates (M1, M6, M8, M9, and M10) as confirmed by full MS scan and fragment analysis.

From the above detection results, both quadrupole tandem ion trap MS^n and HRMS can achieve reactive metabolite screening of the test sample, and the main metabolites M6, M9, and M10 can be detected in the above different MS^n acquisition and analysis modes. Among them, the metabolic pathway of M6 and M9 is GSH conjugation, which coincides with the information mentioned above that metabolic activation of clozapine produces reactive imine ions. Naturally, HRMS has certain advantages in terms of instrument resolution, etc., based on the elemental composition to achieve accurate extraction of metabolites, excluding matrix interference signals, so that the detection results have better accuracy and reliability. We compared the detection results of HRMS with those in the literature, and the experimental detection results were consistent with those reported in the literature [11] (Table 13.5).

13.3.5 Application of Reactive Metabolite Studies in Drug Development

As mentioned earlier, the structure optimization of the drug candidate by screening for potential reactive metabolites during the lead compound discovery and optimization stage to minimize the metabolic activation propensity of the drug candidate is also a very important optimization strategy. In

Table 13.5 Comparison of detection results of reactive metabolites of clozapine in liver microsomal incubation system with literature.

Metabolite code	Results of this test	Literature results [11]
M1	Y	Y
M2	Y	Y
M3	Y	Y
M4	Y	ND
M5	Y	Y
M6	Y	Y
M7	Y	Y
M8	Y	Y
M9	Y	Y
M10	Y	Y

Note: Y represents metabolites detected and ND represents metabolites not detected.

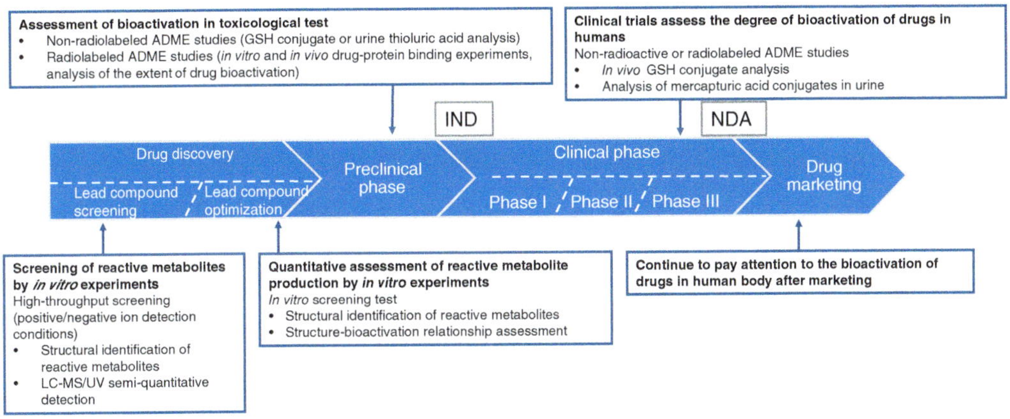

Figure 13.18 Screening strategy for reactive metabolites at various stages of new drug development. Adapted from reference [12].

addition, in all stages of animal toxicological tests and clinical trials, the detection and structural characterization of reactive metabolites and downstream metabolites are conducive to elucidating the mechanism of drug toxicity, helping to find solutions to reduce or eliminate toxicity. The screening strategy for reactive metabolites at various stages of new drug development is depicted in Figure 13.18.

Case Study

Structural modification of felbamate and sudoxicam are typical application cases of reactive metabolite screening in the process of new drug development. Felbamate is an antiepileptic drug marketed in 1993 [7, 13], which is mainly hydrolyzed by esterases and oxidized by aldehyde dehydrogenase in the human body to obtain the main metabolite I (Figure 13.19). The intermediate metabolite, aldehyde carbamate (metabolite II), is very unstable and readily degrades spontaneously to produce phenylacrolein (reactive metabolite). Phenylacrolein consumes GSH in the

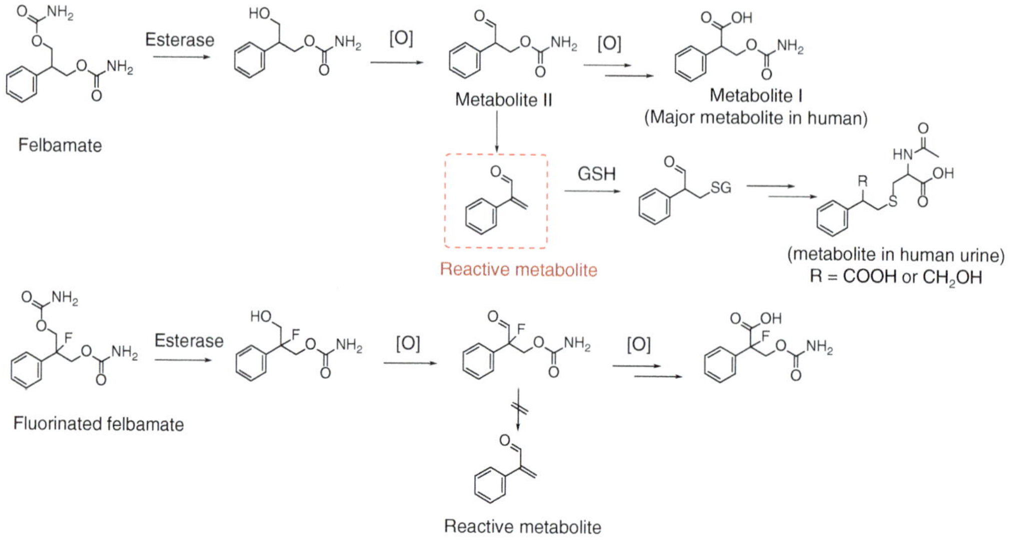

Figure 13.19 Metabolic pathways of felbamate and fluorinated felbamate. *Source:* [7]. Stepan et al., 2011; [13]. Parker et al., 2005.

human body and readily covalently binds to nucleophilic groups on proteins and is considered to be a reactive metabolite closely related to the idiosyncratic toxicity of felbamate. Based on this information, the researchers designed fluorofelbamate to introduce fluorine substituents on the benzyl carbon atoms attached to the benzene ring, and the introduction of fluorine atoms blocked the production of phenylacrolein reactive metabolites and significantly reduced drug hepatotoxicity.

Sudoxicam was developed as an effective nonsteroidal anti-inflammatory drug in the 1970s and later terminated due to severe hepatotoxicity in clinical trials. Its toxicity is thought to be related to the metabolic activation of the thiazole ring to produce an acylthiourea species (reactive metabolite) (Figure 13.20). Subsequently, a methyl substituent was introduced at the 5-position of the thiazole ring to give the same series of drugs, meloxicam. The methyl group acts as a soft spot for the metabolism of the thiazole ring, so meloxicam has a completely different metabolite pathway of the thiazole ring (oxidation of methyl groups to alcohols or further oxidation to carboxylic acids) from sudoxicam, reducing the hepatotoxicity of the drug while maintaining efficacy.

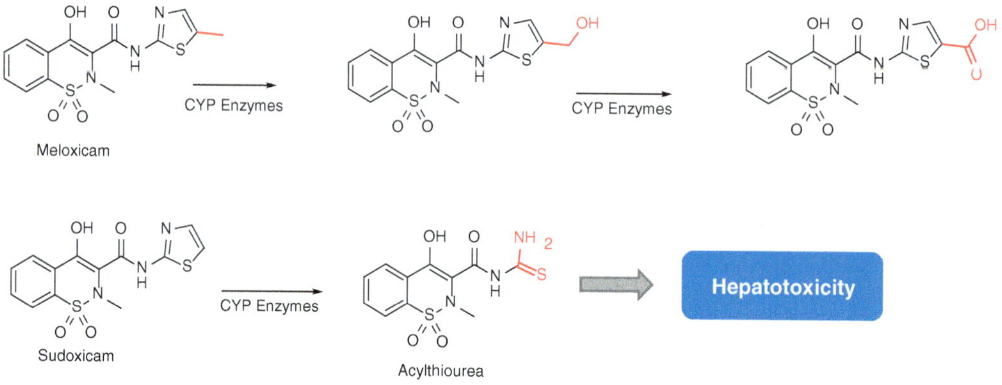

Figure 13.20 Metabolic pathways of sudoxicam and meloxicam. Adapted from reference. *Source:* [14].

13.3.6 Summary

For new drug development, various measures should be taken to reduce or avoid the potential toxicity of the drug. Through reactive metabolite screening, compounds with potential risks of toxicity can be detected and modified as early as possible, which plays a certain predictive role in safety issues. For drug toxicity observed in animals or clinical trials, it is difficult to attribute to a specific metabolite without sufficient experimental data. However, understanding the reactive metabolites that may be generated in this process can provide a direction or propose a hypothesis to increase the success rate of drug development.

In the process of drug development, especially in the early stage of drug discovery, screening for possible reactive metabolites of drug metabolism is helpful in understanding the potential safety issues of a drug, and finding equivalent alternative drug molecules using the method of chemical intervention, which is of great significance for improving drug safety.

References

1 Scott Obach, R. and Kalgutkar, A.S. 1.15-reactive electrophiles and metabolic activation. In: *Comprehensive Toxicology. CA McQueen*.

2 Miller, J.A. (1970). Carcinogenesis by chemicals: an overview – GHA Clowes memorial lecture. *Cancer Research* 30 (3): 559–576.

3 Liebler, D.C. (2008). Protein damage by reactive electrophiles: targets and consequences. *Chemical Research in Toxicology* 21 (1): 117–128.

4 Athersuch, T.J., Antoine, D.J., Boobis, A.R. et al. (2018). Paracetamol metabolism, hepatotoxicity, biomarkers, and therapeutic interventions: a perspective. *Toxicology Research* 7 (3): 347–357.

5 Jollow, D.J., Mitchell, J.R., Potter, W.Z. et al. (1973). Acetaminophen-induced hepatic necrosis. II. Role of covalent binding *in vivo*. *Journal of Pharmacology and Experimental Therapeutics* 187 (1): 195–202.

6 Uetrecht, J.P. (1994). Metabolism of drugs by leukocytes. *Drug Metabolism and Drug Interactions* 11 (4): 259–282.

7 Stepan, A.F., Walker, D.P., Bauman, J. et al. (2011). Structural alert/reactive metabolite concept as applied in medicinal chemistry to mitigate the risk of idiosyncratic drug toxicity: a perspective based on the critical examination of trends in the top 200 drugs marketed in the United States. *Chemical Research in Toxicology* 24 (9): 1345–1410.

8 Craveiro, N.S., Lopes, B.S., Tomás, L., and Almeida, S.F. (2020). Drug withdrawal due to safety: a review of the data supporting withdrawal decision. *Current Drug Safety* 15 (1): 4–12.

9 AS, K. (2005). A comprehensive listing of bioactivation pathways of the organic functional group. *Current Drug Metabolism* 6: 161–225.

10 Kalgutkar, A.S. and Dalvie, D. (2015). Predicting toxicities of reactive metabolite–positive drug candidates. *Annual Review of Pharmacology and Toxicology* 55: 35–54.

11 Wang, Z., Fang, Y., Rock, D., and Ma, J. (2018). Rapid screening and characterization of glutathione-trapped reactive metabolites using a polarity switch-based approach on a high-resolution quadrupole orbitrap mass spectrometer. *Analytical and Bioanalytical Chemistry* 410: 1595–1606.

12 Wen, B. and Fitch, W.L. (2009). Analytical strategies for the screening and evaluation of chemically reactive drug metabolites. *Expert Opinion on Drug Metabolism & Toxicology* 5 (1): 39–55.

13 Parker, R.J., Hartman, N.R., Roecklein, B.A. et al. (2005). Stability and comparative metabolism of selected felbamate metabolites and postulated fluorofelbamate metabolites by postmitochondrial suspensions. *Chemical Research in Toxicology* 18 (12): 1842–1848.

14 Obach, R.S., Kalgutkar, A.S., Ryder, T.F., and Walker, G.S. (2008). *In vitro* metabolism and covalent binding of enol-carboxamide derivatives and anti-inflammatory agents sudoxicam and meloxicam: insights into the hepatotoxicity of sudoxicam. *Chemical Research in Toxicology* 21 (9): 1890–1899.

13.4 UGT Enzymes: Metabolite Identification and Toxicity Assessment Strategies

Yu Xianqing, Weiqun Cao, Lingling Zhang

13.4.1 Introduction

In the process of drug ADME, a crucial point of consideration is the interaction between drugs and metabolic enzymes. Among the common enzymes involved in drug metabolism, CYP enzymes account for the largest proportion, which dominate the phase I metabolism of drugs. Uridine diphosphate glycosyl transferase (UGT), accounting for the second largest portion, is one of the most important enzymes in human phase II metabolism, participating in the metabolic clearance process of many drugs. UGT enzyme uses glucuronic acid as a glycosyl donor and extensively catalyzes the conjugation reaction between endogenous (such as bilirubin, fatty acid, steroidal hormone, compound in food, drug, and environmental pollutant) and exogenous chemical substances. Generally, the compounds after conjugation are more water-soluble and are easier to be excreted from the body, thus serving as a detoxification function.

However, there is a special case for such a binding reaction, namely when the binding site occurs on the carboxyl group, forming potential toxic metabolites known as acyl glucuronide metabolites. In the guidelines for metabolites in safety testing (MIST), it is particularly proposed that additional attention should be paid to the safety of acyl glucuronide metabolites. When there is a carboxyl structure in the drug that can produce acyl glucuronic acid, it is necessary to be vigilant in the potential formation of acyl glucuronide metabolites during metabolism. If acyl glucuronide metabolites are formed, the question arises: will they cause toxicity? This section will introduce the mechanisms by which acyl glucuronide metabolites induce toxicity, how acyl glucuronide metabolites are identified, and how the toxicity of acyl glucuronide metabolites is assessed. This information establishes the foundation for drug development involving acyl glucuronide metabolites.

13.4.2 Mechanism of Toxicity by Acyl Glucuronide Metabolites

The cofactor of the UGT enzyme is uridine-5′-diphosphoglucuronic acid (UDPGA). The reaction process of the UGT enzyme catalyzing the transfer of glucuronic acid structure from UDPGA to the compound is illustrated in Figure 13.21. The UGT enzymes found in the human body fall into four families with a total of 22 subtypes [1] (Table 13.6).

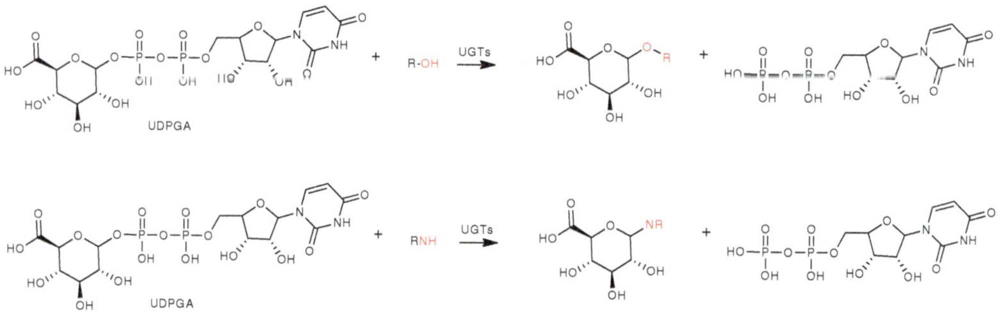

Figure 13.21 Reaction process catalyzed by UGT enzyme.

Table 13.6 UGT enzyme subtype classification.

Family	Subtype
UGT1A	UGT1A1, UGT1A3, UGT1A4, UGT1A5, UGT1A6, UGT1A7, UGT1A8, UGT1A9, UGT1A10
UGT2A/UGT2B	UGT2A1, UGT2A2, UGT2A3 UGT2B4, UGT2B7, UGT2B10, UGT2B11, UGT2B15, UGT2B17, UGT2B28
UGT3A	UGT3A1, UGT3A2
UGT8A	UGT8A1

Under the catalysis of UGT enzymes, compounds bind glucuronide to form metabolites with increased polarity. These conjugates can be partially converted back to unchanged drugs by the action of β-glucuronohydrolase in the gut, excreted from the body, or reabsorbed through the intestinal wall into the systemic circulation. For acyl glucuronide metabolites, on the other hand, some may bind to proteins in hepatocytes and intestinal epithelial cells to form protein adducts, thus inducing toxicity [2].

The specific mechanism by which acyl glucuronide metabolites produce toxicity is shown in Figure 13.22 [3]. The carboxyl group forms 1-O-β-acyl glucuronide through the metabolism of the UGT enzyme. The carbonyl carbon atom in its structure has electrophilic activity, while SH/OH/NH_2 in the amino acid residue of protein acts as a nucleophilic group, attacking the carbon atom to have a transacylation reaction, which results in the detachment of glucuronic acid. It is through this

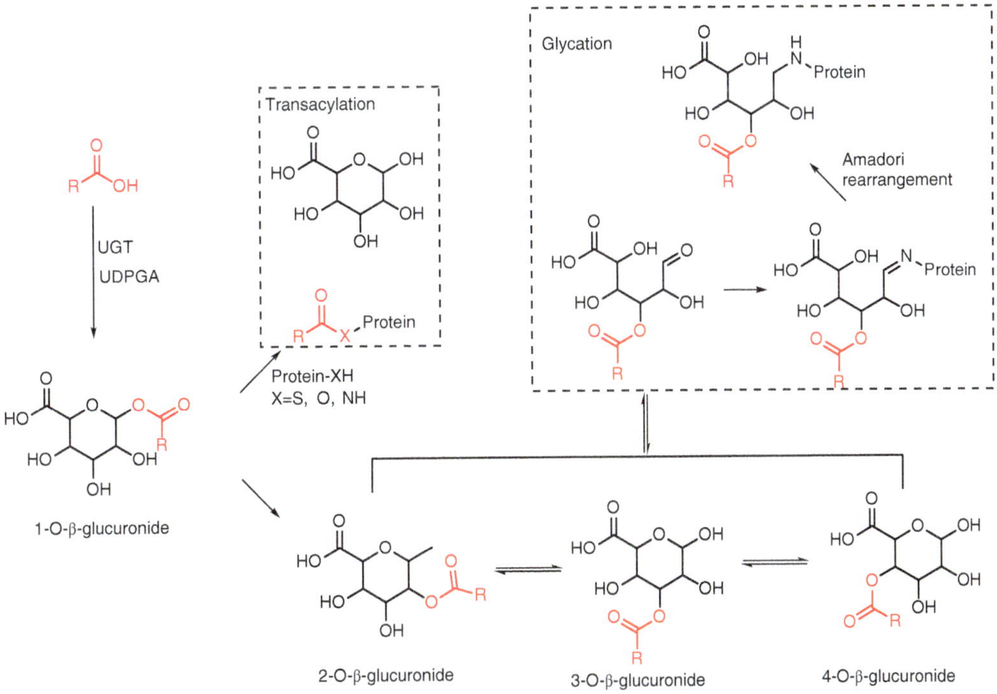

Figure 13.22 Mechanism of toxicity by acylglucuronide metabolites.

"transacylation" that 1-O-β-glucuronide forms a protein-drug adduct, which induces the development of toxic reactions. The 1-O-β-glucuronide is unstable. The reversible intramolecular rearrangement occurs in isomers including 2-O-β-glucuronide, 3-O-β-glucuronide, and 4-O-β-glucuronide. The migrating isomers formed by intramolecular rearrangement form a chain-like aldehyde structure by instantaneous ring opening, and the aldehyde structure formed by ring opening will form an imine structure with the NH_2 structure of amino acid residues of the protein, and induce the occurrence of toxic reactions by rearranging to form stable protein adducts. The process of forming adducts by such means is termed "glycation".

13.4.3 Identification of Acyl Glucuronide Metabolites

If the compound contains sites that may be conjugated with glucuronide such as the carboxyl group, alcohol hydroxyl group, or amino group, it is necessary to determine whether the glucuronic acid conjugation metabolite is an acyl glucuronic acid metabolite. To accurately identify acyl glucuronide metabolites, it is necessary to separate and purify them, and then determine the structure by NMR spectrometer. However, the study period can vary. To meet the need for rapid screening of compounds, the following experimental scheme is provided to help identify acyl glucuronide metabolites:

1) **Capture Experiment [4]**
Based on the characteristics that acyl glucuronide metabolites can bind to proteins, a capture experiment can be designed to identify them: after routine incubation of liver microsomes supplemented with UDPGA, remove microsomal proteins, add peptides in the supernatant for capture experiment, replace the possibly bound proteins with peptides containing lysine NH_2 with known structure, and capture acyl glucuronide products with binding activity.

2) **Mass Spectral Identification [4]**
The formation of acyl glucuronide products can be inversely verified by HRMS searching for binding products linked to specific peptides. At the same time, the adduct captured by the peptide fragment would form *m/z* 470 in the mass spectrum (Figure 13.23). Fragment *m/z* 470 was broken from the carbon base, this fragment is the glucuronide portion that carries away an O in the drug molecular structure. This particular fragment can be directly used to distinguish it from *N*-glucuronide conjugation.

13.4.4 Assessment of the Possible Risks of Acyl Glucuronide Metabolites

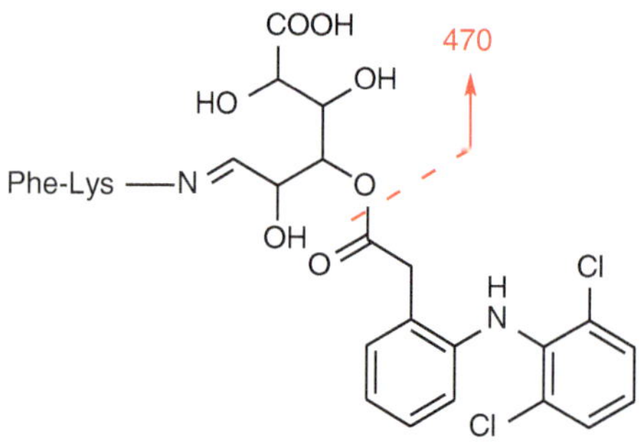

Figure 13.23 Elucidation of mass spectrum of protein adduct in capture experiment.

Not all acyl glucuronide metabolites are toxic, so after determining whether a compound forms an acyl glucuronide metabolite, the next step is to assess whether it is toxic. Assessing the toxicity of acyl glucuronide metabolites as early as possible during drug development can help to develop strategies that circumvent risks or reduce losses.

Among the mechanisms by which acyl glucuronide metabolites cause toxicity, the initially formed 1-O-β-glucuronide is unstable and undergoes intramolecular reversible rearrangement *in vivo* to form migration isomers. A dynamic process, so by identifying the capture experiments

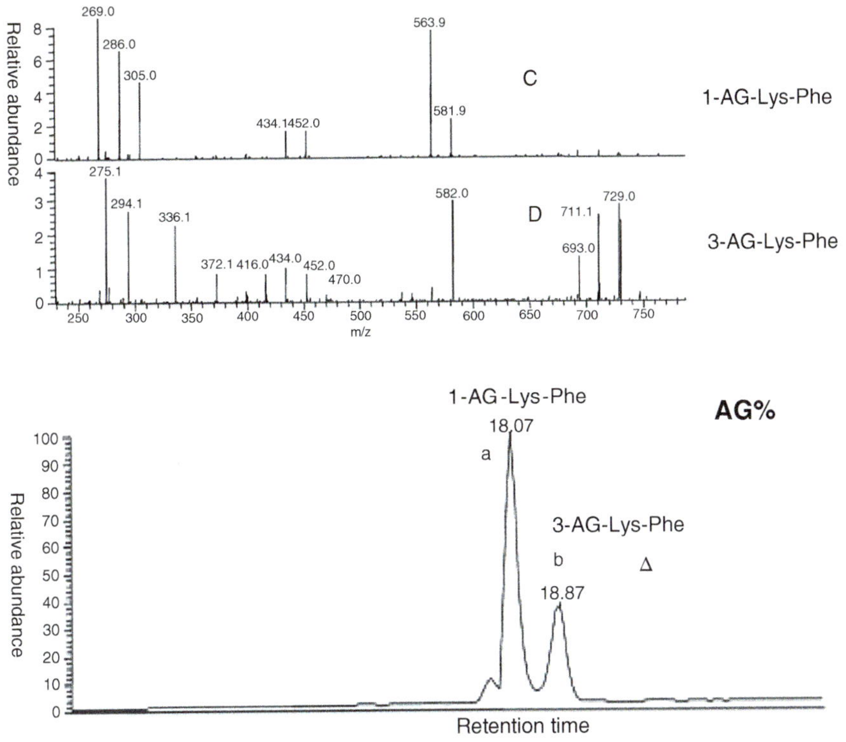

Figure 13.24 Acylglucuronide protein adduct mobility calculation. *Source:* [4] / with permission of American Chemical Society.

mentioned in acyl glucuronide, this dynamic process is fixed, and then the compounds before and after migration are identified by MS^n data, and finally a determined ratio is obtained: migration rate (AG%) = peak area after migration/peak area before migration (Figure 13.24) [4].

The literature [4] summarized the mobility of seven compounds, and the results showed that the two compounds with higher mobility had been withdrawn from the market due to toxic reactions caused by clinical application; the compound with the third mobility was also found to cause hepatocyte injury; the ibuprofen with the lower mobility was a very safe nonsteroidal anti-inflammatory drug in current use. This result suggests that the toxicity of acyl glucuronide metabolites may increase with their mobility.

The literature does not give the exact critical value of the mobility that causes toxicity, but it can still be used as an indicator to assess the safety of acyl glucuronide metabolites. In the process of drug development, the strategy of mobility evaluation provided in the literature can provide the basis for the safety assessment of acyl glucuronide metabolites by comparing the mobility in parallel with the marketed drugs (Table 13.7).

13.4.5 Strategies for Assessing Safety Risks Due to Glucuronide Metabolites

UGT enzyme is an important drug-metabolizing enzyme, which is involved in the clearance and detoxification process of many drugs. Since acyl glucuronide metabolites can bind proteins and have safety risks, additional safety evaluation is required. The research strategy of acyl glucuronide metabolites is shown in Figure 13.25.

Table 13.7 Comparison of mobilities of compounds.

Compound name	Migration rate (AG%)	Clinical studies
Tolmetin	80.0	Withdrawn from the market due to clinical toxicity
Zomepirac	67.5	
Diclofenac	58.4	May cause hepatocellular damage
Ketoprofen	36.4	
Fenoprofen	33.8	
Ibuprofen	15.8	One of the safest nonsteroidal anti-inflammatory drug
Furosemide	2.0	

Source: Data from reference [4].

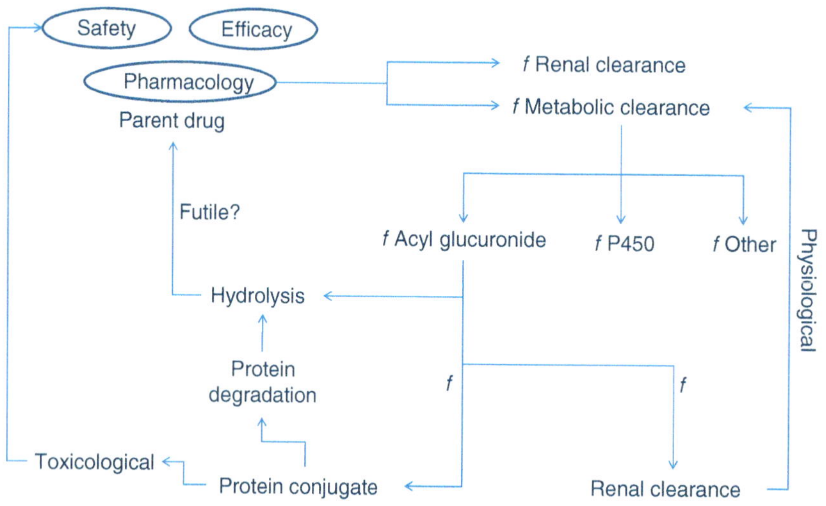

Figure 13.25 Study strategy for acyl glucuronide metabolites. *Source:* Adapted from reference [5].

In the first step, it is necessary to identify whether acyl glucuronide is formed. In addition to the determination by NMR, the active metabolites such as acyl glucuronide can be captured by peptide segment, and the generation of acyl glucuronide products can be determined in combination with MS^n data.

The second step provides the calculation of mobility to help evaluate the potential toxicity of acyl glucuronide metabolites. For compounds with the risk of forming acyl glucuronide metabolites, the ability and mobility of compounds to form acyl glucuronide metabolites as early as possible can help avoid the risks existing in the development process.

13.4.6 Summary

In the process of drug metabolism involving UGT enzymes, scientists often pay more attention to the rapid clearance of drugs, reduced exposure, and enzyme phenotyping caused by UGT enzymes. In addition to detoxification, specific glucuronide conjugation causes safety issues, and acyl glucuronide metabolites that may cause toxicity may lead to accidental or even termination of drug development. Peptide capture, HRMS identification, etc., can help identify acyl glucuronic acid metabolites and assess their potential safety risks, thereby reducing the risk of drug development.

References

1 Fan, Y. (2018). *Metabolism of Exogenous Substances Such as Drugs by Human UGT Enzymes [D]*. Tianjin University.

2 Van Vleet, T.R., Hong, L., Anthony, L. et al. (2017). Acyl glucuronide metabolites: implications for drug safety assessment [J]. *Toxicology Letters* 272: 1–7.

3 Tong, X., Yan, L., Haiping, H. et al. (2009). Progress in glucuronosyltransferases inducing metabolic activation of carboxylic acid drugs [J]. *Journal of Pharmacy* 11: 1193.

4 Jianyao, W., Margaret, D., and Fangbiao, L. (2004). et al, A novel approach for predicting acyl glucuronide reactivity via Schiff base formation: development of rapidly formed peptide adducts for LC/MS/MS measurements [J]. *Chemical Research in Toxicology* 17 (9): 1206–1216.

5 Sophie, L., James, L.M., Thomas, G.H. et al. (2010). Acyl glucuronides: the good, the bad and the ugly [J]. *Biopharmaceutics and Drug Disposition* 31: 367–395.

13.5 Insights into the R&D Strategies of Deuterated Drugs from the Drug Metabolism Perspective

Ya Ding, Lingling Zhang, Weiqun Cao

13.5.1 Introduction

In January 2023, the China National Medical Products Administration approved the oral small molecule drug Deuremidevir Hydrobromide Tablets (Product code: VV116) for marketing in accordance with the special review and approval procedure for drugs. The drug was conditionally approved for the treatment of adult patients with mild to moderate COVID-19 infection, thus making oral small-molecule COVID-19 therapeutics once again a hot topic. It is worth noting that the deuteration strategy was employed in the development process of VV116. This section will discuss the current R&D status of deuterated drugs, and through some case studies, introduce the advantages of deuterated strategies, the uncertainty surrounding deuterated drugs, and outline the R&D strategies of deuterated drugs based on clinical value from a DMPK perspective.

13.5.2 What Deuterated Drugs Are

Deuterium is a stable and safe isotope of hydrogen, one more neutron than ^{1}H, with a natural abundance of approximately 0.01%. Deuterated drugs are drugs obtained by replacing one or more carbon–hydrogen bonds at specific metabolic sites of drug molecules with carbon–deuterium bonds, such as deuterated Tetrabenazine and Donafenib (Figure 13.26). Deuterated tetrabenazine replaces the hydrogen at the methoxy position of tetrabenazine with deuterium. Deuteration reduces the rate of demethylation metabolism, prolongs the half-life, and achieves the expected therapeutic effect at a lower dose. Donafenib is obtained by replacing the hydrogen with deuterium on the methyl group of sorafenib, thereby improving biological activity and slowing down drug metabolism [1].

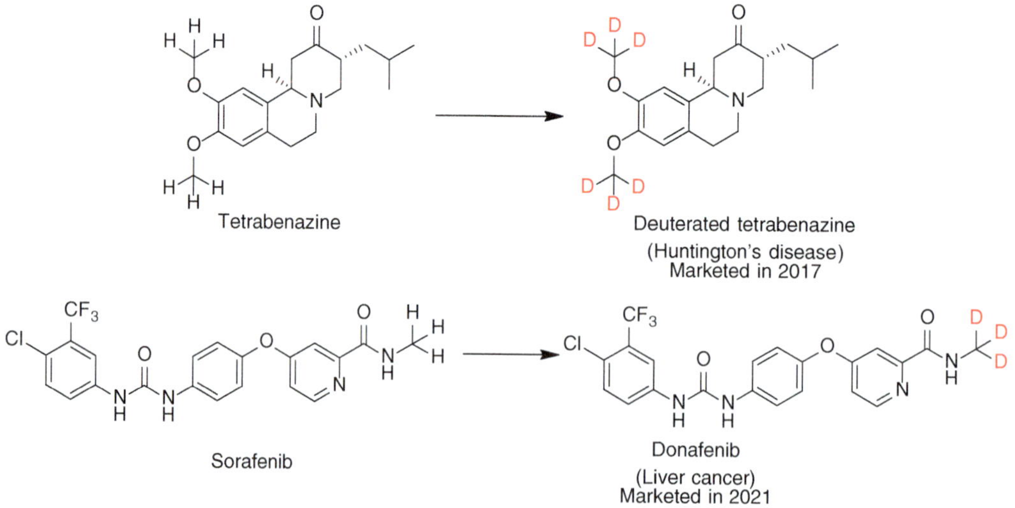

Figure 13.26 Chemical structures of deuterated tetrabenazine, Donafenib, and the original drugs. *Source:* [2] / American Chemical Society.

13.5.3 R&D Status of Deuterated Drugs

In addition to the above two marketed drugs and Deucravacitinib (Psoriasis, marketed by FDA in 2022), many companies worldwide are committing to develop deuterated drugs. Table 13.8 summarizes the clinical trial stages and indications of some deuterated drugs under development.

Deuterated drugs also face great challenges. More and more R&D enterprises are seeking patent protection for the corresponding deuterated derivatives of new drugs. In addition, the cost of raw materials for deuterium is much higher than that of hydrogen, and deuteration requires more advanced production technology, resulting in the production cost of deuterated drugs being higher than their non-deuterated counterparts. Nowadays, it is difficult for simple imitation drugs to meet the clinical value-oriented requirements. Deuterated drugs should have better DMPK properties than non-deuterated drugs to demonstrate their clinical value.

13.5.4 Benefits of Deuteration

The property with small change after deuteration is physical and chemical first. The crystal form and solubility of the compound after deuteration will not change too much, so the preparation can refer to the original drug. Second, in general, the permeability, plasma protein binding rate, and dose–effect relationship of a compound will not be changed due to deuteration, and the appropriate pharmacological model can be selected according to the original drug in terms of drug efficacy. For some safety considerations, such as species selection and immunogenicity, the study content of the original drug can be referred to [9]. Therefore, when the molecular structure of a drug is further optimized,

Table 13.8 Deuterated drugs under study in the clinical trial phase.

Company	Deuterated drugs in the clinic	Original drug	Clinical phase	Indication
Vertex Pharmaceuticals	VX-984	Novel compound	Phase I	Tumor
Concert Pharmaceuticals	CTP-499	Novel compound	Phase III	Diabetes
	CTP-656 (d_9-ivacaftor)	Ivacaftor	Phase II	Rare form of cystic fibrosis
	CTP-543 (d_8-ruxolitinib)	Ruxolitinib	Phase II	Myelofibrosis
	CTP-730 (d_5-apremilast)	Apremilast	Phase I	Psoriasis
Avanir Pharmaceuticals	AVP-786	Dextromethorphan	Phase III	Anxiety
			Phase III	Alzheimer's disease
Revance Therapeutics, Inc.	RT001	Linoleic acid	Phase II	Friedreich's ataxia
Alkeus Pharmaceuticals, Inc.	ALK-001	Vitamin A	Phase II	Fundus xanthomatosis
Suzhou Zelgen Biopharmaceuticals Co., Ltd.	Jacketinib hydrochloride tablets	Ruxolitinib	Phase III	Myelofibrosis
			Phase III	Alopecia areata
	Alkotinib	Osimertinib	Phase II	Non-small-cell lung cancer
Hinova Pharmaceuticals Inc.	Deutenzalutamide (HC-1119 Softgel)	Enzalutamide	Phase III	Prostate cancer

Source: Adapted from reference [2].

the use of a deuteration strategy, compared with other modification methods, such as the substitution of hydrogen atoms with fluorine or methyl, is faster in R&D speed, lower in cost, and less in risk. However, deuterated drugs are not copies of the original drugs, and it is still necessary to pay attention to changes in DMPK properties, mainly changes in metabolic properties. The main reason deuterium affects drug metabolism comes from the kinetic isotope effect caused by deuterium.

1) **Why Deuterium Works**

Compounds are catalyzed by metabolic enzymes and need to bind to proteins in a certain configuration. Deuteration can change the spatial configuration of the molecule and the ability to partially form hydrogen bonds, which in turn affects the binding to proteases. Moreover, the carbon–deuterium bond has higher bond energy than a carbon–hydrogen bond, which will also affect the catalytic rate of metabolic enzymes. Compared with the C–H bond, the breaking of the C–D bond requires a higher activation energy, and the reaction rate is slower. This difference in reaction rate due to the presence of isotope deuterium is called the deuterium kinetic isotope effect [3]. Introducing deuterium into specific metabolic sites of molecules and utilizing the deuterium kinetic isotope effect can improve the DMPK properties of drugs.

2) **Decreased Drug Metabolism**

Remdesivir is the first approved anti-COVID-19 drug in the world. It is an intravenously administered nucleoside prodrug that forms the main active metabolite GS-441524 after conversion in the body, inhibiting the replication of the virus. To further improve the drug efficacy, scientists carried out structural modifications on different positions of GS-441524 and finally found that the compound X1 obtained by introducing deuterium at position 7 of the purine ring moiety could reduce the enzymatic degradation, which was caused by double bond oxidation or triazine ring opening, to slow down the clearance of active metabolites and improve the antiviral activity of the drug. VV116 is then obtained by esterifying the hydroxyl position on the X1 ribose fragment, which can improve its oral bioavailability. Preclinical pharmacodynamic studies show that VV116 exhibits significant anti-COVID-19 effects both *in vivo* and *in vitro* (Figure 13.27) [4].

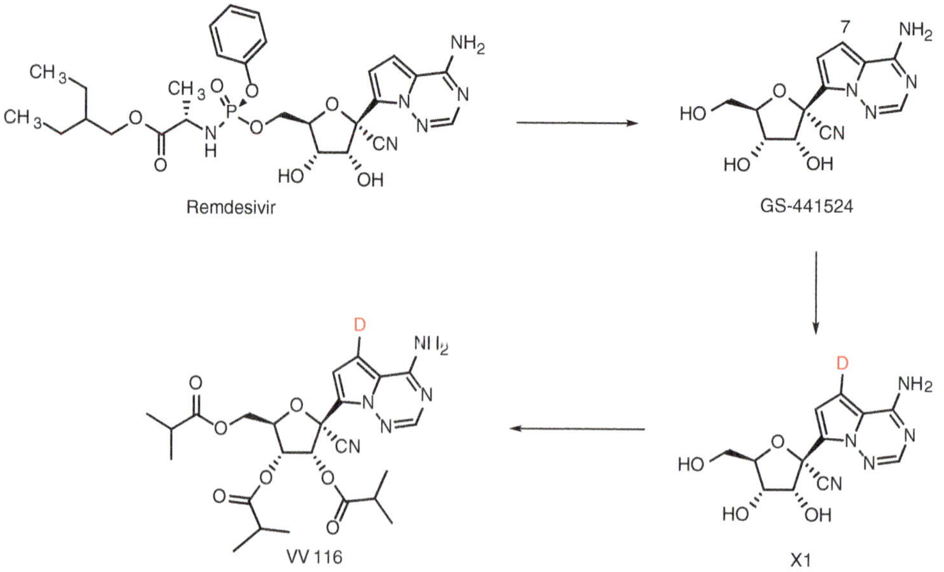

Figure 13.27 Chemical structures of Remdesivir, GS-441524, VV116, and X1. *Source:* [4] / Springer Nature / CC BY 4.0.

3) **Reduced Drug Toxicity**

Deuteration can reduce drug toxicity by reducing the production of toxic metabolites. For example, the original drugs in Table 13.9 all produce toxic metabolites, which have certain toxic side effects in clinical practice. When its specific metabolic site is selectively deuterated, the production of toxic metabolites can be reduced, thereby improving clinical applications.

4) **Improved Bioactivity**

Clopidogrel, an antithrombotic drug, depends on mono-oxidation of the double bonds on its thiophene ring to form an active metabolite for its efficacy, but three methylene groups on the piperidine ring are susceptible to oxidation, which reduces the formation of the active metabolite, so deuteration of the methylene group can increase the formation of the active metabolite (Figure 13.28).

Overall, for a drug to be effective, it requires sufficient exposure to the target tissue. The purpose of deuterated drug R&D is to obtain a longer half-life and higher bioavailability of drug molecules. Both of these goals can be achieved through the characteristics of deuterated effects on biotransformation. Deuteration can selectively inhibit the production of toxic metabolites, improve the conversion of active metabolites, and then achieve the purpose of optimizing the drug's molecular properties (Figure 13.29).

Table 13.9 Examples of drug studies using deuterated structural modifications to reduce toxic metabolites.

Original drug	Toxic active metabolite	Toxic side effects	Deuterated drugs	Post-Deuteration results
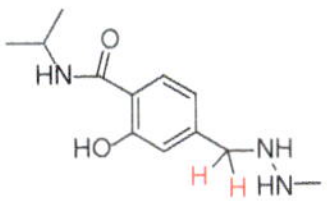	Alkylating agent CH_3^+	Liver toxicity	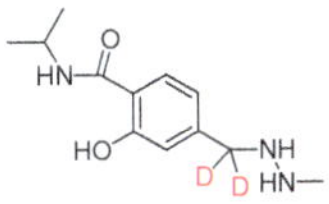	Significantly reduce liver toxicity
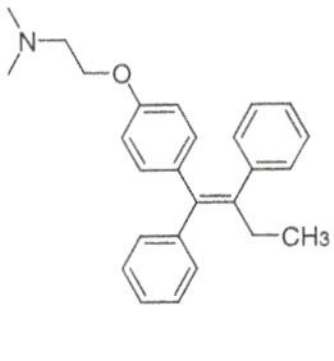	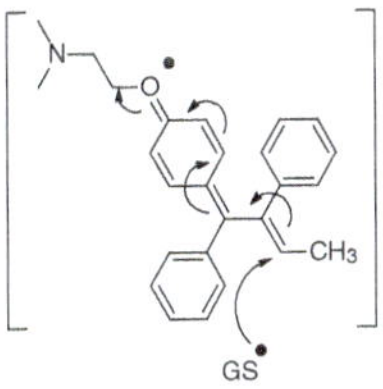	Carcinogenic and toxic to the liver	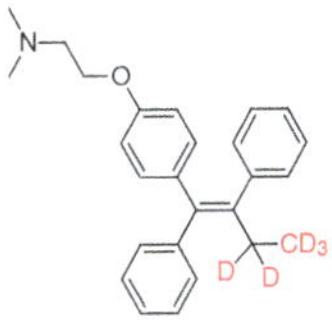	Reduce toxicity by 50%–60%
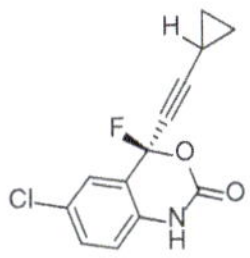	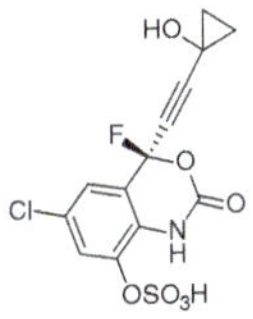	Very toxic to kidneys	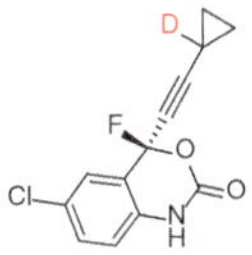	Significantly reduce renal toxicity
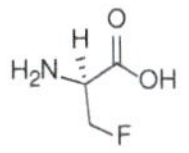	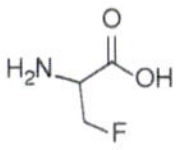	Causes brain vacuolation toxicity	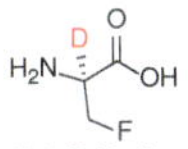 2-D-3-F-Alanine Phase II, Stop	Reduce toxic metabolites but not significantly
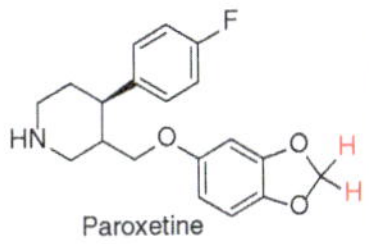 Paroxetine	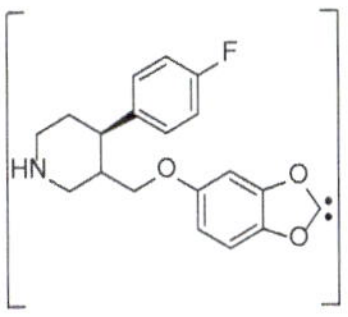	Drug-drug interaction leading to serious adverse reaction	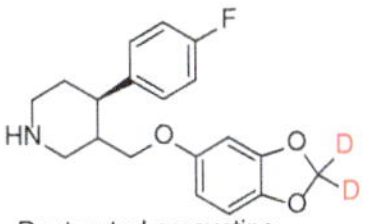 Deuterated paroxetine Phase I	Overcome drug interactions with other drugs

Source: [5].

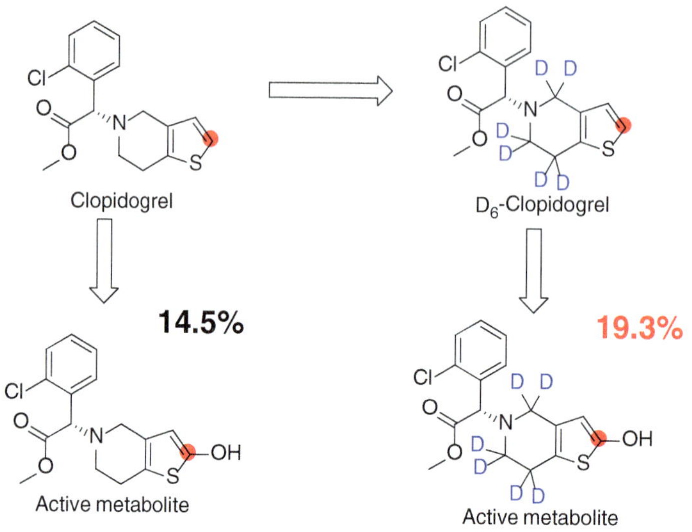

Figure 13.28 Chemical structures of Clopidogrel and its deuterated derivatives. *Source:* [2] / American Chemical Society.

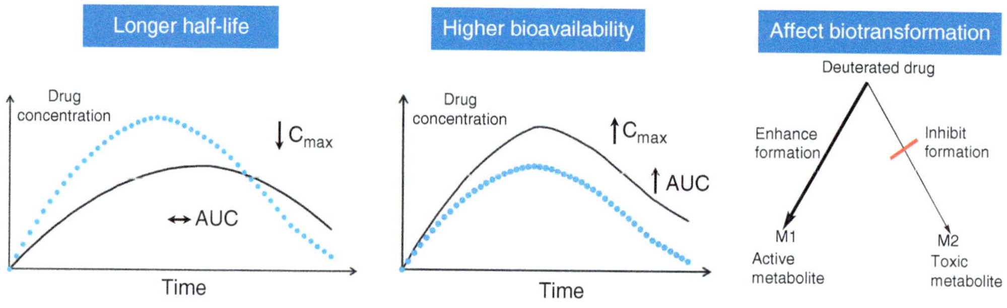

Figure 13.29 Effect of deuteration on drug pharmacology. *Source:* Adapted from reference [6].

13.5.5 Uncertainty of Deuteration Strategy

Deuteration does not necessarily enhance the original properties of the original drug, and it also introduces some uncertainties. The first is its ability to induce metabolic turnover. For example, the anti-HIV drug Maraviroc is prone to mono-oxidation on the methyl group of its triazole. If only this position is deuterated, the half-life of the compound in human liver microsomes is shortened. This is because deuteration inhibits the mono-oxidation of the methyl group, but other sites such as the methylene group at the α-position of the bridged ring N will also be metabolized. Therefore, after further deuteration of the methylene group at the *α*-position of the bridged ring N, the half-life is significantly extended (Figure 13.30). When a drug has two or more metabolic pathways, the introduction of deuterium at the main metabolic site may cause its metabolic switch, that is, one metabolic pathway is weakened, and the other metabolic pathway is enhanced.

Deuteration strategies are sometimes difficult to predict. For example, Dapagliflozin has two major metabolic sites, the methyl group, and the methylene group between the two phenyl rings, where metabolism occurs to form toxic metabolites. When the methyl group is deuterated, its half-life in plasma is improved, but when the methylene group is further deuterated, the half-life is decreased (Figure 13.31). It can be seen that for molecules with multiple oxidative metabolic sites, the deuteration of more sites is not necessarily better.

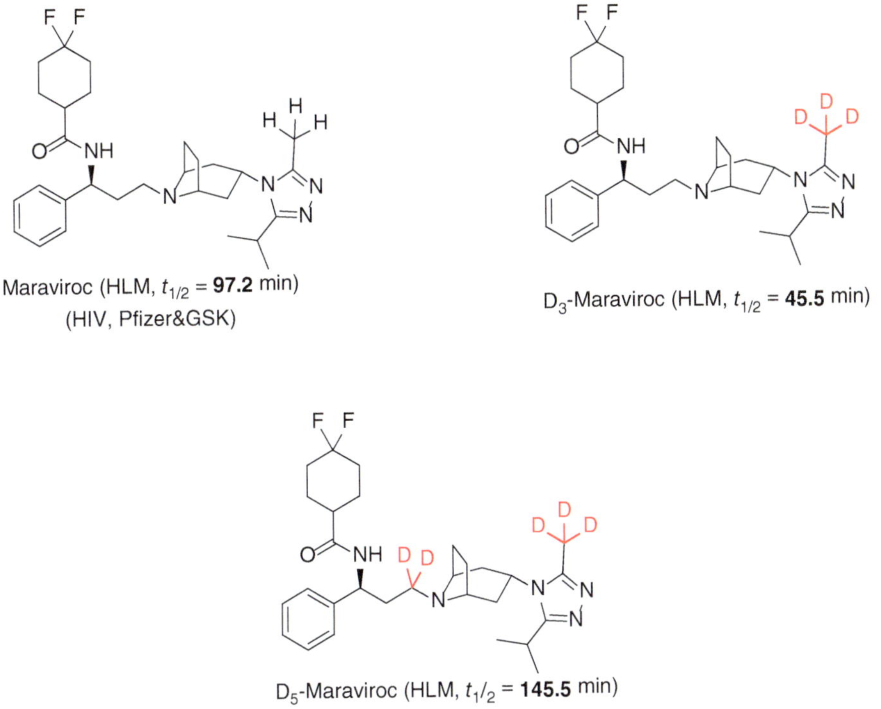

Figure 13.30 Chemical structures of Maraviroc and its deuterated derivatives. *Source:* [2] / American Chemical Society.

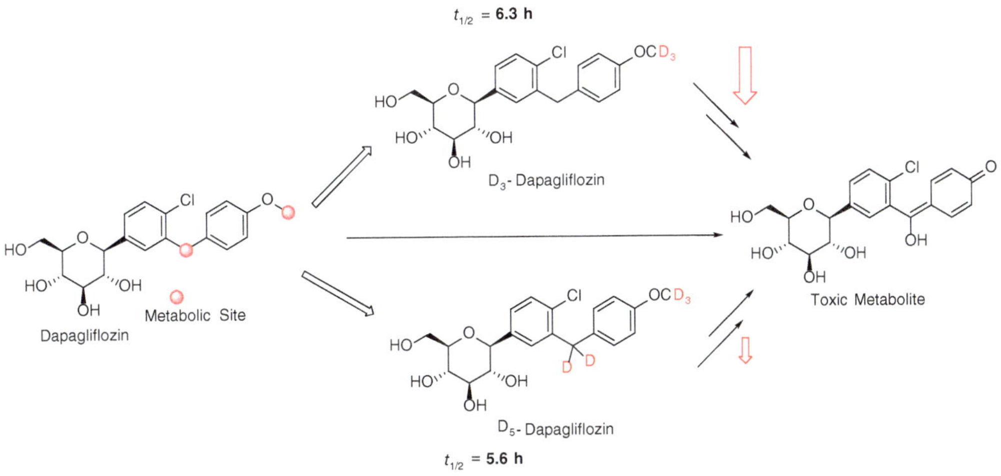

Figure 13.31 Chemical structures of Dapagliflozin and its deuterated derivatives. *Source:* [2] / American Chemical Society.

Due to the structural and metabolic properties of compounds, less than 10% of FDA-approved drugs can obtain more value after deuteration modification [7]. To avoid the uncertainty of deuterated drugs, it is first necessary to select the appropriate deuterated position based on the metabolic pathway. When the metabolism of a compound involves the cleavage of carbon–deuterium bonds and is dominated by oxidative metabolism, and the metabolic pathway affects the exposure and systemic clearance of the drug, deuteration of such a compound may improve its properties [8].

When a compound has multiple metabolic pathways that meet the above requirements, it is necessary to obtain compounds with different deuteration positions or numbers of deuterium elements, conduct simultaneous comparisons, and screen out candidate compounds with better DMPK properties. This requires high-throughput screening during the drug design stage.

13.5.6 Summary

The marketing of drugs such as deuterated Tetrabenazine and Donafenib has attracted much attention for deuterated drugs. However, it is difficult to simply imitate the original drug to meet clinical value. Only by obtaining better DMPK properties and compensating for the shortcomings of the original drug after deuteration, can increasingly stringent new drug application requirements be achieved. Therefore, in the early stages of drug design, it is critical to clearly define the metabolic pathways of the compounds to determine whether deuteration strategies can be applied. By selecting appropriate deuterated sites, maximum property optimization can be achieved. In addition, deuterated drugs also need to consider the limitations of patents, the cost of synthesis, and other factors to comprehensively determine whether the deuteration strategy has application value.

References

1 Qin, S., Bi, F., Xu, J. et al. (2020). P-86 comparison of the pharmacokinetics of donafenib and sorafenib in patients with advanced hepatocellular carcinoma: an open-label, randomized, parallel-controlled, multicentre phase II/III trial. *Annals of Oncology* 31: S117–S118.

2 Pirali, T., Serafini, M., Cargnin, S., and Genazzani, A.A. (2019). Applications of deuterium in medicinal chemistry. *Journal of Medicinal Chemistry* 62 (11): 5276–5297.

3 Atzrodt, J., Derdau, V., Kerr, W.J., and Reid, M. (2018). Deuterium-and tritium-labelled compounds: applications in the life sciences. *Angewandte Chemie International Edition* 57 (7): 1758–1784.

4 Xie, Y., Yin, W., Zhang, Y. et al. (2021). Design and development of an oral remdesivir derivative VV116 against SARS-CoV-2. *Cell Research* 31 (11): 1212–1214.

5 Zhang, Y.S. (2017). Past, present and future of deuterated drug research and development. *Progress in Pharmacy* 41 (12): 902–918.

6 Tung, R. (2010). The development of deuterium-containing drugs. *Innovations in Pharmaceutical Technology* 32 (32): 24–28.

7 Cargnin, S., Serafini, M., and Pirali, T. (2019). A primer of deuterium in drug design. *Future Medicinal Chemistry* 11 (16): 2039–2042.

8 Sharma, R., Strelevitz, T.J., Gao, H. et al. (2012). Deuterium isotope effects on drug pharmacokinetics. I. System-dependent effects of specific deuteration with aldehyde oxidase cleared drugs. *Drug Metabolism and Disposition* 40 (3): 625–634.

9 Gant, T.G. (2014). Using deuterium in drug discovery: leaving the label in the drug. *Journal of Medicinal Chemistry* 57 (9): 3595–3611.

14

Radiolabeled *In Vivo* ADME Studies

14.1 Placental Transfer and Milk Excretion Studies Based on Radioactive Tracer Technique

Mengjie Liu, Lin Li, Wei Lei, Lingling Zhang

14.1.1 Introduction

The placental transfer and milk excretion studies are important components of late-stage drug development, registration application, and post-marketing research. If a drug is developed for pregnant women, it needs to be assessed for potential exposure risks to fetuses and nursing infants. The use of the radioactive tracer technique to study the placental transfer and milk excretion of drugs in preclinical species can visually observe the distribution of drugs and quantify drug-related substances in fetal tissues and milk. This provides a strong basis for clinical research and guidance for clinical medication. This chapter elaborates on the current medication situation for pregnant and lactating women, as well as the mechanisms of placental transport and milk excretion. Through case analysis, it demonstrates the application of the radioactive tracer technique in researching placental transfer and milk excretion.

14.1.2 Overview of Medication Used in Pregnant and Lactating Women

In recent years, the prevalence rate of women of childbearing age has increased, and common diseases such as asthma, diabetes, and hypertension are often accompanied throughout pregnancy and lactation [1]. Based on statistics [2], about 60% of women worldwide take at least one prescription drug during pregnancy to treat chronic diseases or sudden diseases caused by pregnancy. Approximately 70% of women take the drug during lactation [3]. However, more than 90% of approved drugs lack guidance for pregnant and lactating women on their labels [4]. Due to the historical events of thalidomide and diethylstilbestrol from the 1940s to 1960s, drugs' clinical research in pregnant and lactating women in the following decades was subjected to many legal and ethical restrictions [5]. As a result, very few new drugs have been made available to pregnant women over the past half-century.

Considering the importance and necessity of special populations in the process of new drug development, the FDA) issued an industrial guidance "Pharmacokinetics in pregnancy—study design, data analysis, and impact on dosing and labeling" [6] in 2004. If a drug is developed for use

Drug Metabolism and Pharmacokinetics: Frontiers, Strategies, and Applications, First Edition.
Edited by Liang Shen and WuXi AppTec DMPK.

in pregnant women, PK studies in the pregnant population, especially during the middle and late pregnancy, need to be conducted to fully understand the PK differences in special populations. In 2014, the FDA issued the "Pregnancy and Lactation Labeling Rule, PLLR or final rule" [7]. The guidance recommended conducting animal studies, special population research, and clinical risk assessments to help researchers better understand drug information.

14.1.3 The Transport Mechanism of Drugs in the Placenta and Mammary Gland

Any drugs that enter the fetus or milk need to pass through the placental barrier or the blood–milk barrier. The placental barrier is a material exchange place between the fetus and the mother, through which the mother can transport nutrients to the fetus, while preventing the entry of some harmful substances. The blood–milk barrier is a place for nutrient exchange between the milk and blood to protect the integrity of the milk and the health of the newborn. It helps the mammary gland to defend against pathogenic microorganisms. The transport of drugs through the placenta and mammary gland is influenced by various factors, including the physicochemical properties of drugs and the activity of transporters.

14.1.3.1 Placental Barrier

As shown in Figure 14.1, drugs can cross the placental barrier through passive diffusion, facilitated diffusion, and active transport.

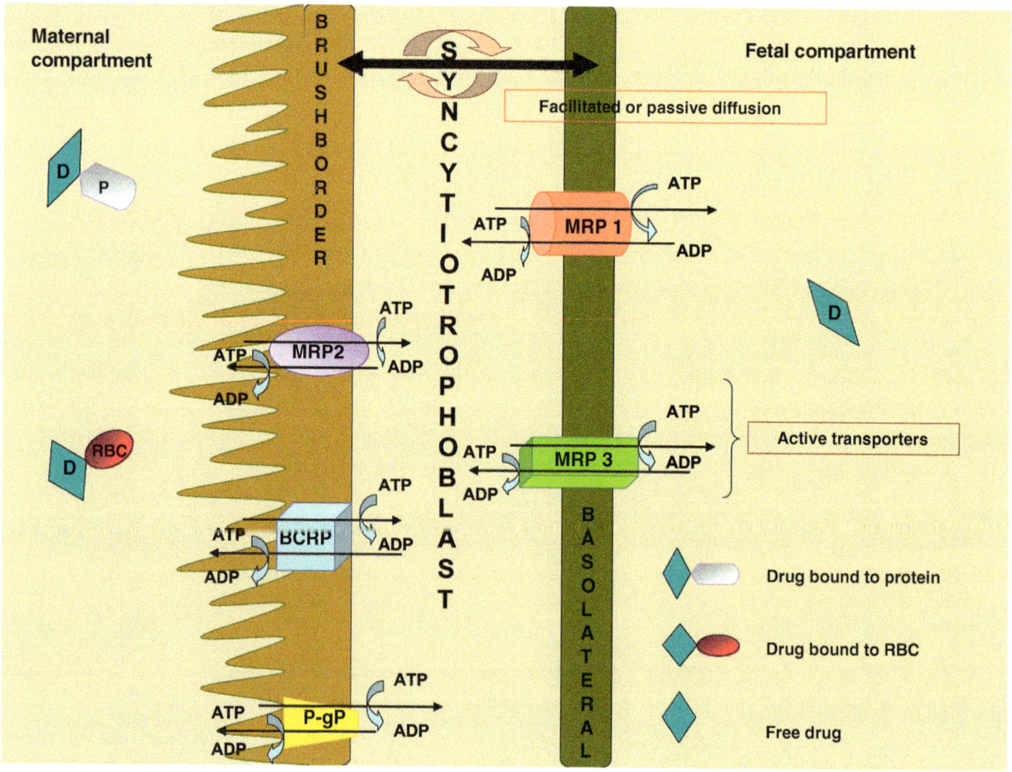

Figure 14.1 Mechanisms by which drugs cross the placental barrier. *Source:* [8] / ELSEVIER.

1) **Passive diffusion**
Passive diffusion is the main way in which most small molecules cross the placenta, and the extent of passive diffusion is influenced by the physicochemical properties of the drug. As shown in Table 14.1, drugs with smaller molecular weights, higher lipophilicity, lower ionization, and lower plasma protein binding are more likely to cross the placental barrier.

2) **Facilitated diffusion**
Drugs that are structurally associated with endogenous substances often cross the placenta by facilitated diffusion, which requires carrier proteins in the placenta to facilitate transport, such as glucose transporters (GLUTs).

3) **Drug transporters**
Whether a drug can pass through the placental barrier is also influenced by drug transporters. Maternal-side transporters include BCRP, p-glycoprotein (P-gp), and multidrug resistance protein 2 (MRP2). As the most expressed efflux transporter, BCRP can transport a variety of drugs and exogenous substances, which are bound to it from the fetal compartment back to the maternal circulation and protect the fetus. Fetal-side transporters include MRP1 and MRP3, which actively excrete drugs into the fetus and increase their exposure in the fetus [10].

14.1.3.2 Blood–Milk Barrier

Drugs pass through the blood–milk barrier mainly with paracellular transport and transcellular transport mechanisms. Paracellular transport refers to the process by which some substances are absorbed through the intercellular space. There are gaps between mammary epithelial cells during the first 3 to 4 days postpartum, through which relatively large molecules and cells can pass into the milk [11]. Transcellular transport refers to the transport process across the cell membrane, including passive diffusion, active transport, etc. The unbound drugs with low molecular weight and high lipid solubility in plasma are easier to enter the milk by passive diffusion. In the studies of mice, cows, and humans, BCRP is strongly induced in the apical epithelial cells of the mammary gland in the third trimester of pregnancy and during lactation [12]. It can actively efflux the drug bound to the breast cells into the milk, resulting in the accumulation of the drug at a higher concentration in the milk.

Considering the unstable composition of milk during the colostrum stage, the FDA recommends milk collection in clinical studies should be performed after the formation of mature milk (approximately 10 days postpartum) [13].

Table 14.1 Physicochemical properties of drugs affecting placental transport. Data from reference.

	Effect on crossing the placental barrier			
Physicochemical properties	**Very easy**	**Easy**	**Difficult**	**Very difficult**
Molecular weight (Da)	<350	350–500	500–1000	>1000
Lipophilicity (logP)	>2	1–2	0.1–1	<0.1
Ionization (%)	<50	50–90	90–98	>98
Protein binding (%)	<1	1–10	10–90	>90

Source: [9] / ELSEVIER.

14.1.4 Application of Radioactive Tracer Technique to Study Placental Transfer and Milk Excretion

As a valuable tool for studying drug absorption, distribution, metabolism, and excretion (ADME), the radioactive tracer technique has been widely used in PK studies. QWBA can be used for placental transport, directly obtaining drug-related substance distribution in the fetus. The total radioactivity concentration in milk can be determined by LSC, which is fast and accurate.

Nonclinical studies on placental transport and milk excretion are usually conducted in pregnant and lactating female rats [14]. Combined with nonclinical PK guidelines, the experiments are usually designed as in Table 14.2.

4) **Case**

Apixaban, a potent and highly selective factor Xa inhibitor, was approved by the FDA in 2012 for use in patients with nonvalvular atrial fibrillation to reduce the risk of stroke and blood clots. In 2014, it was approved for the treatment of deep vein thrombosis and pulmonary embolism and became one of the most widely used oral anticoagulants in the world. It is known that the risk of venous thromboembolism (VTE) in pregnant women is about 4–5 times that of nonpregnant women at the same age, which is one of the major causes of maternal death [15].

The physical and chemical properties of apixaban are as follows: small molecular weight (459.5), high protein binding (87%), medium lipophilic ($\log P = 2.23$), and non-ionization at physiological pH. Based on these properties, apixaban can be predicted to cross the placental barrier and blood–milk barrier easily. Then, the placental transfer and milk excretion properties of [^{14}C]-apixaban were investigated in pregnant and lactating rats.

14.1.4.1 Placental Transfer Study

Six pregnant female SD rats were used for the placental transfer study [16]. Each rat received a single oral dose of [^{14}C]-apixaban at 5 mg/100 μCi/kg, respectively. QWBA was performed to determine the radioactivity distribution in maternal and fetal tissues at 0.5, 1, 4, 8, 24, and 48 hours post-dose.

The results showed that the distribution of [^{14}C]-apixaban-derived radioactivity was extensive in the tissues of pregnant rats, while only low levels of radioactivity were detected in fetal tissues (Table 14.3). The C_{max} of radioactivity in fetal tissues was reached at 4 hours post-dose; then the radioactivity concentrations in tissues of fetuses declined quickly and were below the limit of quantitation at 24 hours post-dose. At 48 hours post-dose, the radioactivity in the maternal rats was below the quantitation limit. This may be due to the fact that apixaban acts as a substrate of the efflux transporters, P-gp and BCRP. Therefore, these efflux transporters may play an important

Table 14.2 The design for nonclinical placental transfer and milk excretion studies.

Study content	Placental transfer	Milk excretion
Study phase	Second and third trimesters	Lactation 10–14 d
Sample collection	Maternal and fetal blood, tissues, fetal appendages (placenta, amniotic membrane, etc.)	Plasma, milk
Analytical methods	Quantitative whole-body autoradiography (QWBA)	Liquid scintillation counter (LSC)

Table 14.3 Tissue distribution study data after oral administration of 5 mg/kg [^{14}C]-apixaban to pregnant rats.

	Apixaban					
Tissue	**0.5 h**	**1 h**	**4 h**	**8 h**	**24 h**	**48 h**
μg Eq/g						
Adrenal gland	1.49	1.90	2.30	0.18	N.S.	N.S.
Blood	1.43	1.44	1.30	0.09	N.D.	N.D.
Bone marrow	0.50	0.33	0.46	0.06	N.D.	N.D.
Brain	0.05	0.06	BLQ	BLQ	N.D.	N.D.
Brown fat	1.08	0.95	0.87	0.06	N.D.	N.D.
Harderian gland	1.00	1.09	1.10	0.19	N.D.	N.D.
Heart	0.68	0.93	0.83	0.06	N.D.	N.D.
Kidney (cortex)	2.11	2.03	1.93	0.21	0.05	N.D.
Liver	6.77	7.38	6.13	0.65	0.12	N.D.
Lung	1.65	1.56	1.33	0.15	BLQ	N.D.
Muscle	0.37	0.46	0.50	0.05	N.D.	N.D.
Pancreas	0.89	0.75	0.99	0.09	N.D.	N.D.
Pituitary	0.72	0.93	0.75	0.19	N.S.	N.S.
Salivary gland	0.73	0.95	1.03	0.08	N.D.	N.D.
Spleen	0.46	0.51	0.47	0.05	BLQ	N.D.
Thymus	0.38	0.58	0.54	0.05	N.D.	N.D.
Amnion	1.45	2.12	11.24	2.12	3.52	N.S.
Placenta	1.02	0.74	1.06	0.10	BLQ	N.D.
Fetal blood	0.27	0.31	0.51	0.07	N.D.	N.D.
Fetal brain	BLQ	0.05	0.06	BLQ	N.D.	N.D.
Fetal kidney	0.24	0.20	0.46	N.S.	N.S.	N.S.
Fetal liver	0.16	0.22	0.32	0.06	N.S.	N.S.

ND: not detected; BLQ: below the lower limit of quantitation (0.037 μg Eq/g tissue); NS: tissue not collected, not observed by autography, considered below BLQ.

role in the limited transfer of apixaban to the fetus for maternally administered drugs. Subsequent animal studies have not found any direct or indirect reproductive toxicity, so physicians can assess whether apixaban can be used during pregnancy based on risks versus benefits [17].

14.1.4.2 Milk Excretion Studies

Nineteen lactating SD rats were used, one as blank control, and each of the remaining rats received a single oral dose of [^{14}C]-apixaban at 5 mg/100 μCi/kg [16]. Milk and blood were collected from three rats per time point at 0.5, 1, 2, 6, 12, and 24 hours post-dose (0.2 ml of Oxytocin was injected subcutaneously 15 minutes prior to milk collection to stimulate lactation). The radioactivity in the milk, blood, and plasma was analyzed using an LSC.

Table 14.4 Pharmacokinetic parameters in the blood and milk of rats after a single oral administration of [^{14}C]-apixaban (5 mg/kg).

Rats	Sample	T_{max}	C_{max}	$t_{1/2}$	$AUC_{(0-t)}$	$AUC_{(0-\infty)}$
Lactating		h	ng Eq/g	h	ng Eq·h/g	ng Eq·h/g
	Blood	0.5	514	4.2	1940	1960
	Plasma	0.5	1040	4.3	3420	3470
	Milk	6.0	8920	3.7	103,000	104,000

The results showed that [^{14}C]-apixaban-derived radioactivity was extensively excreted in the milk and was detected at all time points through 24 hours post-dose. The radioactivity concentration of apixaban-equivalent in milk at each time point was greater than that in blood and plasma (Table 14.4). The T_{max} value of radioactivity in milk was 6 hours, later than that in plasma and blood (0.5 hours), while the elimination $t_{1/2}$ of plasma and milk were similar. The milk/plasma concentration ratios ranged from 2.8 to 37. The milk $AUC_{0\sim inf}$ was approximately 30-fold greater than the plasma $AUC_{0\sim inf}$, indicating that active transport proteins were involved in the distribution of Apixaban into milk. Apixaban, as a substrate of BCRP, is known to be transported into breast milk by highly expressed BCRP, resulting in higher concentration in milk. Therefore, Apixaban's potential hazards to newborns and infants cannot be ruled out, and the use of Apixaban in lactating women is not recommended [17].

14.1.5 Summary

Scientific experimental design is one of the important factors to ensure the medication safety of pregnant and lactating women. The radioactive tracer technique is often selected for comprehensive placental transport and milk excretion studies for new drug development, especially when analyzing the mechanism of drug transport and elimination.

References

1 Admon, L.K., Winkelman, T.N., Moniz, M.H. et al. (2017). Disparities in chronic conditions among women hospitalized for delivery in the United States, 2005–2014. *Obstetrics & Gynecology* 130 (6): 1319–1326.

2 Zhang, C., Zhang, L.L., Chen, L. et al. (2012). Medications used in pregnancy: a systematic review. *Chinese Pharmaceutical Journal* 47 (11): 858–862.

3 Mazer-Amirshahi, M., Samiee-Zafarghandy, S., Gray, G., and van den Anker, J.N. (2014). Trends in pregnancy labeling and data quality for US-approved pharmaceuticals. *American Journal of Obstetrics and Gynecology* 211 (6): 690–e1.

4 Illamola, S.M., Bucci-Rechtweg, C., Costantine, M.M. et al. (2018). Inclusion of pregnant and breastfeeding women in research–efforts and initiatives. *British Journal of Clinical Pharmacology* 84 (2): 215–222.

5 Al-Enazy, S. (2017). Placental control of drug delivery. *Advanced Drug Delivery Reviews* 116: 63–72.

6 Food and Drug Administration, Center for Drug Evaluation and Research (2004). Draft guidance for industry: pharmacokinetics in pregnancy — study design, data analysis, and impact on dosing and labeling.

7 Food and Drug Administration (2014). Content and format of labeling for human prescription drug and biological products: requirements for pregnancy and lactation labeling. *Federal Register* 79 (233): 72064–72103.

8 Gedeon, C. and Koren, G. (2006). Designing pregnancy-centered medications: drugs which do not cross the human placenta. *Placenta* 27 (8): 861–868.

9 Evain-Brion, D., Berveiller, P., and Gil, S. (2014). Le passage transplacentaire des médicaments. *Thérapie* 69 (1): 3–11.

10 Eshkoli, T., Sheiner, E., Ben-Zvi, Z. et al. (2011). Drug transport across the placenta. *Current Pharmaceutical Biotechnology* 12 (5): 707–714.

11 Maliepaard, M., Scheffer, G.L., Faneyte, I.F. et al. (2001). Subcellular localization and distribution of the normal breast cancer resistance protein in human cancer tissues. *Cancer Research* 61: 3458–3464.

12 Jonker, J.W., Merino, G., Musters, S. et al. (2005). The breast cancer resistance protein BCRP (ABCG2) xenotoxins into milk drugs and carcinogenic concentrins. *Nature Medicine* 11: 127–129.

13 Food and Drug Administration (2019). Clinical lactation studies: considerations for study design guidance for industry [EB/OL]. [2021-04-27].

14 Solon, E.G. and Kraus, L. (2001). Quantitative whole-body autoradiography in the pharmaceutical industry: survey results on study design, methods, and regulatory compliance. *Journal of Pharmacological and Toxicological Methods* 46 (2): 73–81.

15 Abdul Sultan, A., Tata, L.J., Grainge, M.J., and West, J. (2013). The incidence of first venous thromboembolism in and around pregnancy using linked primary and secondary care data: a population-based cohort study from England and comparative meta-analysis. *PLoS One* 8 (7): e70310.

16 Wang, L., He, K., Maxwell, B. et al. (2011). Tissue distribution and elimination of [^{14}C] apixaban in rats. *Drug Metabolism and Disposition* 39 (2): 256–264.

17 Cada, D.J., Levien, T.L., and Baker, D.E. (2013). Apixaban. *Hospital Pharmacy* 48 (6): 494–511.

14.2 Visualization of Tissue Distribution: Quantitative Whole-Body Autoradiography (QWBA)

Mengjie Liu, Lin Li, Wei Lei, Lingling Zhang

14.2.1 Introduction

QWBA is a mature radiotracer technique. After administration of [^{14}C]/[^{3}H]-labeled test article to animals, QWBA can be applied to investigate the distribution of total radioactivity throughout the body. Complete and detailed tissue distribution data can be obtained. Based on the exposure and half-life of rodent tissues, the estimated effective dose of radioactivity to human subjects can be calculated. QWBA is widely used in both drug discovery and development stages.

14.2.2 Advantages of QWBA Over Traditional Tissue Distribution

The number of tissues quantitatively determined by QWBA is much greater than that by the traditional tissue dissection method.

For traditional tissue distribution, after the [^{14}C]/[^{3}H] test article is administered to the animals, the animals will be sacrificed at specific time points post-dose, and about 13 specific tissues will be harvested, followed by tissue homogenization and aliquots. Then, the concentration of the parent test article will be determined by mass spectrometry, or the concentration of total radioactivity determined by LSC. For the QWBA study, one animal/gender will be euthanized at a specific time point post-dose and the intact carcass will be frozen, embedded, and sliced. Then, the whole-body slices, which represent all tissues, will be quantitatively measured for radioactivity using the phosphor imaging technique and AIDA software. The distribution of multiple tissues (45+ tissues) throughout the body could be seen. The main equipment is shown in Figure 14.2.

For the traditional tissue collection method, if investigators increase tissues of interests after the project is finished, another study needs to be arranged. For the QWBA method, the images of tissues have been saved, investigators can obtain any added tissues at any time.

QWBA can cover more than 95% of tissues throughout the body, including many small tissues, such as cornea, iris, lacrimal gland, and bile duct. If the drug is distributed widely in these tissues, it may cause potential hazard or risks. QWBA is used to observe the whole-body distribution of the test article in various animal species (rodent, nonhuman primate, etc.), which can comprehensively predict the risk associated with conducting clinical studies and scientifically estimate the clinical radioactive dosage of human radiolabeled mass balance studies.

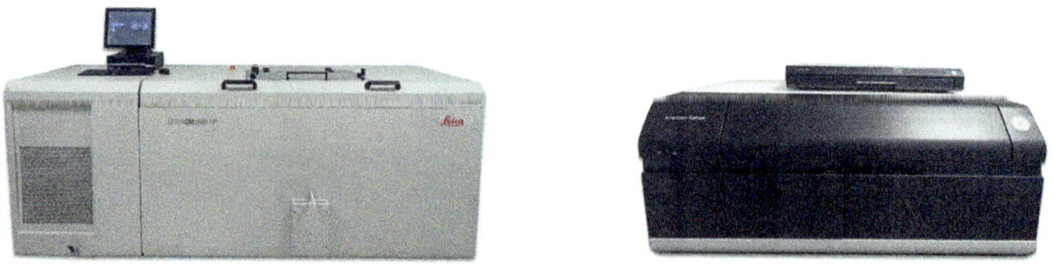

Figure 14.2 Main equipment for QWBA experiment.

14.2.3 When to Conduct a QWBA Study

PK studies of small molecule test articles in animals or humans, named ADME studies, include absorption, distribution, metabolism, and excretion studies. At the Investigational New Drug (IND) stage, a reasonable labeling position is decided based on the chemical structure, synthesis process, and early metabolite identification results of the test article. The [^{14}C]/[^{3}H]-labeled test article is used for QWBA study within the corresponding animal models, accompanied by the studies of mass balance, biliary excretion, and metabolite identification. At the clinical stage, especially after POC validation, a human radiolabeled mass balance study is required to provide information on the metabolism and excretion pathways of the test article in human beings. At that time, a QWBA study in pigmented rodents is conducted to predict human radioactivity dosimetry as the basis for obtaining successful clinical ethical approval.

14.2.4 How to Design a QWBA Study?

It depends on two scenarios. The first one involves using the pharmacodynamic and toxicological rodent model to conduct QWBA. Typically, an albino rodent or large animals such as monkeys, rabbits, etc., are used. Dissection tissue samples from 4~6 time points are collected after administration. In the other situation, melanin-containing rodents, such as Long Evans or C57BL/6 strains, are used to predict human radioactivity dosimetry. The carcasses are collected at 6~10 specific time points to obtain a relatively complete clearance profile. The dosage needs to be within the range of the effective dose, which is generally the same as the medium dosage in the IND package, while the radioactivity dosage is generally 100~200 μCi/kg for rodents and 10~40 μCi/kg for monkeys/dogs. The calculation of human radioactivity dosimetry is based on QWBA results from melanin-containing rodents because the human body contains melanin. Drug candidates such as tyrosinase inhibitors are likely to bind to melanin and accumulate, leading to various clinical adverse reactions, such as rash and pruritus. Prior to clinical trials, QWBA in melanin-containing rodents is necessary to assess whether the test article binds to melanin.

14.2.5 Application Cases for QWBA

Application Case 1
Albino rats and melanin-containing rats were administered with [^{14}C]-labeled test articles, respectively. The presence of melanin binding of the test article was observed based on the distribution of the radioactivity in the pigmented uveal tract, but it was not observed in albino rats.

The tissue distribution profiles of radioactivity after administration of [^{14}C]-test article showed a clear tendency of binding and accumulation of radioactivity to the skin in SD and LE rats. However, there still were some differences between the two rats, and the distribution and accumulation on the uveal tract of LE rats were obvious (c), indicating that the test article tended to specifically bind to the melanin-containing uveal tract (Figure 14.3).

Application Case 2
QWBA images clearly show the concentration differences between the outer and inner layers of the tissue through different colors corresponding to different concentration levels, providing incomparable advantages over traditional tissue distribution for assessing whether anticancer drugs can reach the tumor tissue and the depth and breadth of their distribution.

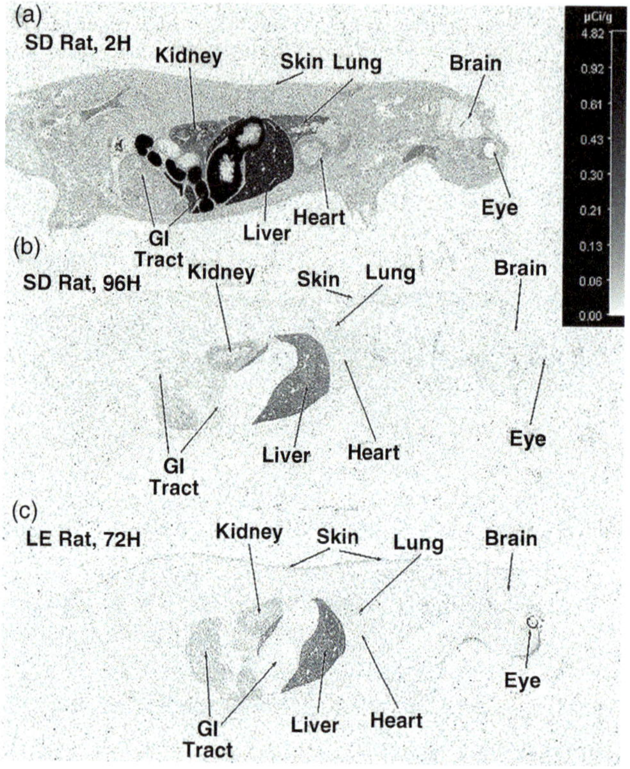

Figure 14.3 Melanin binding. Whole-body autoradiographs reveal tissue distribution of a ^{14}C-labeled drug in SD rats (a,b) and LE rats (c). *Source:* [1] Solon E G. (2015) / Springer Nature.

Using [^{3}H]-labeled folic acid bleomycin linkers ([^{3}H]-BMS-753493) to target the folate receptor (FR), after oral administration of [^{3}H]-labeled linkers to tumor-bearing mice, the radioactivity was extensively distributed in various tissues and organs, and only a very small amount of radioactivity detected in the brain. QWBA images revealed significant radioactivity distribution around the tumor containing FRs, with minimal distribution in the tumor without FRs. At 48 hours post-dose, although radioactivity was rapidly eliminated from various tissues, a relatively high radioactivity distribution was still observed in receptor-containing tumors, fully supporting the pharmacodynamics data (Figure 14.4).

Application Case 3
In phase III and IV clinical trials, investigational drugs need to be assessed for the effects on pregnant women and fetuses by investigating whether they can cross the placental barrier and blood–milk barrier to reach the fetuses and babies, providing pregnant women's medication guidance data in drug labels. After injection of [^{14}C]-bisphenol A into pregnant mice, the autoradiogram clearly showed, in addition to being distributed in various major organs of the maternal abdominal cavity, [^{14}C]-bisphenol A also crossed the placental barrier to reach the fetuses (Figure 14.5). Therefore, the use of products without bisphenol A is advocated during pregnancy.

14.2.6 QWBA Is the Gold Standard of Tissue Distribution Study

QWBA is a powerful tool for analyzing the tissue distribution of test articles, offering absolute advantages such as high image resolution, a broad tissue range, and quantifiability. QWBA can

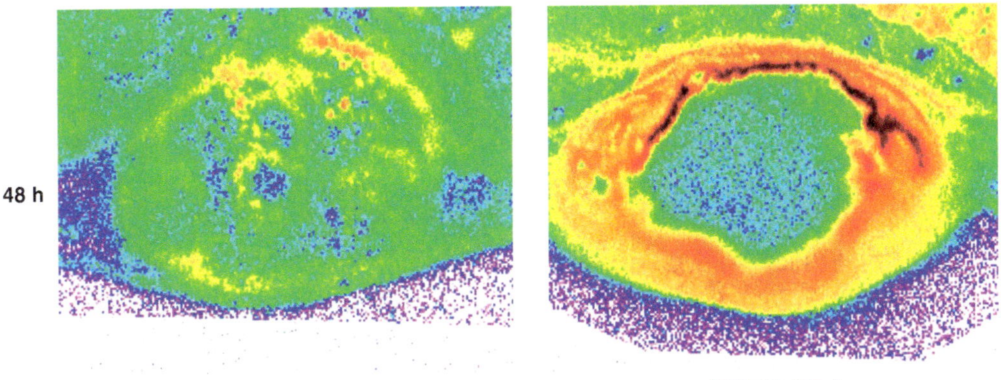

Figure 14.4 Distribution plots of radioactivity in M109 tumors with different FR (folate receptor) expression after administration of [^{3}H]-BMS-753493 to tumor-bearing mice. *Source:* [2] / ELSEVIER.

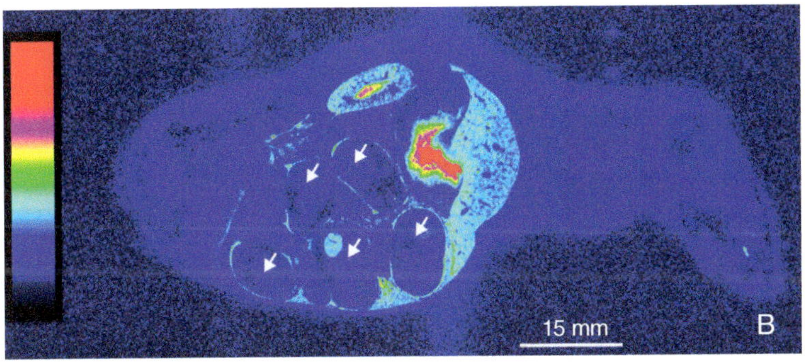

Figure 14.5 Distribution plot of radioactivity in pregnant and fetal rats following the injection of [^{14}C]-bisphenol A. *Source:* [3] / ELSEVIER.

be applied at any phase of drug development to timely and accurately discover the target organs associated with the test article. The QWBA data are commonly used to illustrate the results of pharmacodynamic and toxicological studies and to systematically interpret the transformation processes of drugs.

References

1 Solon, E.G. (2015). Autoradiography techniques and quantification of drug distribution[J]. *Cell and Tissue Research* 360 (1): 87–107.

2 Shen, H., Wang, L., Chen, W. et al. (2016). Tissue distribution and tumor uptake of folate receptor–targeted epothilone folate conjugate, BMS-753493, in CD2F1 mice after systemic administration. *Acta Pharmaceutica Sinica B* 6 (5): 460–467.

3 Tanakaa, M., Kawamotob, T., and Matsumotoa, H. (2010). Distribution of 14C-bisphenol A in pregnant and newborn mice. *Dental Materials* 26: e181–e187. 1–3.

14.3 Preclinical and Clinical Research Strategy for Human Radiolabeled Mass Balance Studies

Huan Li, Lian Guo, Lingling Zhang

14.3.1 Introduction

In May 2022, the FDA published the draft guidance of "Clinical Pharmacology Considerations for Human Radiolabeled Mass Balance Studies," which described recommendations regarding clinical pharmacology considerations for conducting human radiolabeled mass balance studies of investigational drugs, including: (i) whether and when to conduct the study, (ii) recommendations on study design, and (iii) result reporting [1]. Referring also to "Safety Testing of Drug Metabolites [2]," and nonclinical PK study practice in the industry, this chapter describes the strategies for designing preclinical and clinical mass balance studies for investigational drugs at different development stages.

Traditional techniques often fail to quantify all metabolites; therefore, drugs recovered in mass balance studies rarely achieve a recovery rate greater than or equal to 90%. Some metabolic pathways may be ignored, and the main metabolic enzymes may be inaccurate due to incomplete metabolite identification. Once some high-level metabolites are generated but cannot be identified by LC–MS, such as metabolites that do not have UV absorption or cannot be ionized, their potential safety risks cannot be fully evaluated.

Radioactive tracer technology can compensate for the shortcomings of traditional mass spectrometry technology and help in fully elucidating the process of drug disposition in the body.

14.3.2 Test Articles Requiring Human Radiolabeled Mass Balance Study

In general, human radiolabeled mass balance studies should be conducted for all new molecular entities. Unless clinical concerns suggest otherwise, a mass balance study might not be recommended in some circumstances. For example:

1) Drugs for which mass balance study results can be obtained from acceptable literature sources or FDA-approved drug labeling.
2) Drugs such as monoclonal antibodies, endogenous substances, and analogs (e.g., peptides, hormones, and oligonucleotide therapeutics) with known metabolism and elimination pathways based on basic pharmacology and nonclinical ADME information.
3) Drugs with the majority of the dose (i.e., greater than or equal to 90%) recovered in the urine as the unchanged parent drug with minimum metabolism.
4) Drugs with no or negligible systemic exposures.

14.3.3 Considerations for Designing Mass Balance Studies

14.3.3.1 Administered radioactivity dose and investigational drug dose

The dose of the non-radiolabeled investigational drug used in the mass balance study should be the final intended dose or at least within the anticipated therapeutic dose range (taking into account the safety profile of the drug in the study population). If the therapeutic range has not been identified at the time of conducting the mass balance study, the study should use a dose within the PK linearity range.

The absorbed dose of radioactivity should be estimated via dosimetry calculations based on data from animal radiolabeled mass balance studies and QWBA studies. Guidelines of other groups concerned with human safety (e.g., the International Commission on Radiological Protection (ICRP) and Advisory Committee on Radiological Protection (ACRP)) should also be considered, as applicable.

14.3.3.2 Route of administration and formulation of the investigational drug

The routes of administration for the mass balance study should include the final intended routes of administration unless precluded by practical considerations (e.g., inhalation products).

The formulation prepared at a small scale in the laboratory and used in the human radiolabeled mass balance study is different from the final intended clinical formulation manufactured in the Good Manufacturing Practice (GMP) workshop. Although formulation differences may cause some changes in ADME parameters (e.g., absorption), the formulation used in the study should still provide sufficient information to fulfill the objectives of the mass balance study.

14.3.3.3 Sample collection and handling

Plasma, urine, feces, and other matrices, as applicable, should be collected for quantitative analysis of total radioactivity and metabolite profiling. Sample collection should continue until: (i) the cumulative radioactivity exceeds 90% of the administered dose in urine and feces and (ii) the radioactivity in the urine and feces is less than 1% of the administered dose over 24 hours on two consecutive sample collection days. Therefore, it is necessary to detect samples as soon as possible during sample collection and determine the collection of endpoints on time.

14.3.3.4 Reporting requirements

The mass balance clinical study report should include the following:

1) Plasma and whole blood concentration-time profiles of total radioactivity, and descriptive statistics of PK parameters for total radioactivity.
2) The cumulative percentage of the administered radioactive dose recovered in urine, feces, and total excreta (urine and feces combined) concentration-time profiles.

For example, Figure 14.6 shows the total radioactivity in plasma/blood concentration-time profiles after single oral administration of [^{14}C]-test article to subjects; Figure 14.7 shows the cumulative

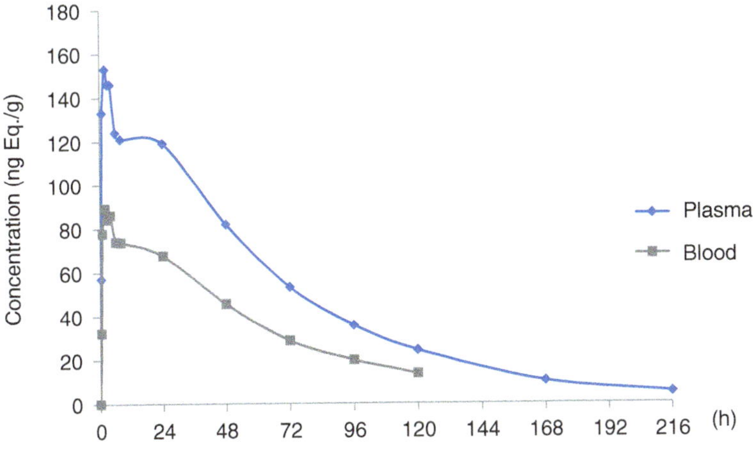

Figure 14.6 Total radioactivity in plasma/blood concentration-time profiles after single oral administration of [^{14}C]-test article to subjects.

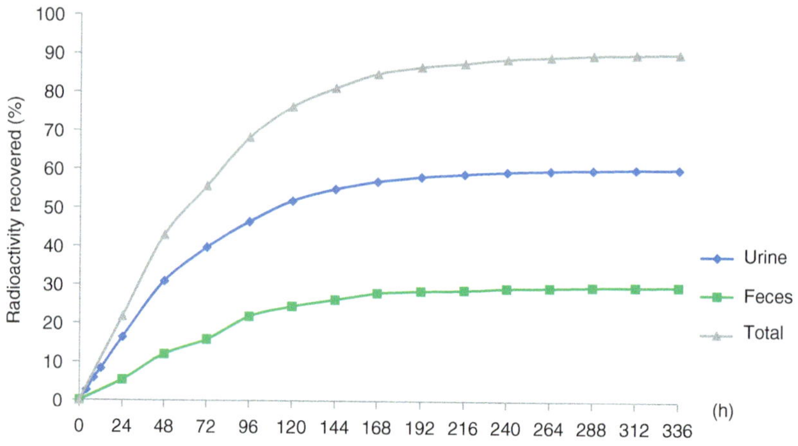

Figure 14.7 Cumulative excretion rates of the administered radioactive dose recovered in urine and feces from 0 to 336 h after single oral administration of [^{14}C]-test article to subjects.

excretion rates of the administered radioactive dose recovered in urine and feces from 0 to 336 hours, after single oral administration of [^{14}C]-test article to subjects.

3) Plasma concentration-time profiles for the non-radiolabeled moieties including the parent drug and metabolite (if possible), with descriptive statistics of PK parameters for the parent drug and metabolite (if possible).
4) Metabolite profiling combining radiolabeled and non-radiolabeled analytical techniques: for quantitative profiling in plasma, urine, and feces, samples should be analyzed separately for each subject and not pooled across subjects. Identification of metabolites is usually done after pooling of samples in the matrix of interest (plasma, urine, or feces) within each subject. In certain cases (e.g., low levels of metabolites), it may be helpful to pool samples across subjects.

Generally, if a metabolite accounts for more than 10% of the total drug-related exposure in plasma, the metabolite should be structurally characterized. Ideally, more than 80% of the radioactivity recovered in the excreta should be identified. Adequate justification should be provided if less than 80% of the recovered radioactivity is identified.

14.3.4 Considerations for Unconventional Mass Balance Studies

A multiple-dose study can be considered in some scenarios such as when the investigational drug and/or active metabolite exhibits time-dependent PKs or when the study is conducted in patients where a single-dose study is not feasible.

For drugs with a long half-life (parent or metabolites), when an extended stay in the clinic becomes impractical to achieve greater than 90% recovery, alternative sample collection strategies should be considered to estimate the final recovery.

The routes of administration for the mass balance study can be adjusted if precluded by practical considerations (e.g., inhalation products).

The labeled position of the radioisotope should be chemically and metabolically stable to ensure the radionuclide is not lost during metabolism, allowing for the detection and quantification of both the parent drug and metabolites. Two separate labeling positions can be used if needed.

When a human radiolabeled mass balance study cannot be conducted (e.g., safety concerns because of the potential for radiolabeled moieties accumulating in critical organs), the sponsor should consider alternative approaches. Sponsors should consult with the appropriate administrative review division before employing alternative methods.

14.3.5 Research Strategy at Clinical Stages

Phase I clinical trials are performed to assess the preliminary clinical pharmacology and safety. The final intended dose is identified in single/multiple ascending dose studies. It is recommended to collect plasma, urine, and feces samples from the therapeutic dose group for the non-radiolabeled metabolite identification studies. Radiolabeled mass balance studies should be conducted early in drug development, at the latest before initiating any late-phase clinical trials. The information from the mass balance studies can inform the recommendation for and the design of DDI studies specific to the pathways involved in metabolism and excretion. In addition, this can guide decisions for conducting renal and/or hepatic impairment studies. Figure 14.8 shows the research strategy of when to perform the radiolabeled mass balance studies at clinical stages.

Bioanalytical method development should be performed before quantification determination in clinical trials, focusing on aspects such as urine and plasma adsorption, as well as the stability of plasma, urine, and feces. The purpose of these studies is to ensure satisfactory cumulative percentage of the administered radioactivity and to establish a biotransformation scheme, including the structures of metabolites. When available. These findings can provide information for sample collection and pretreatment procedures.

Non-radiolabeled metabolite identification studies should be initiated as early as possible during clinical trials. The results will be used to assess the need for conducting safety testing for drug metabolites. Whether the PK characteristics of metabolites are time dependent can be assessed by comparing the results in humans following single and multiple administrations, or by observing changes in the abundance of metabolites relative to the parent compound or the total drug-related exposure. This information helps to provide information on the dose frequency in human radiolabeled studies. Figure 14.9 shows the research contents and significance of human radiolabeled mass balance studies.

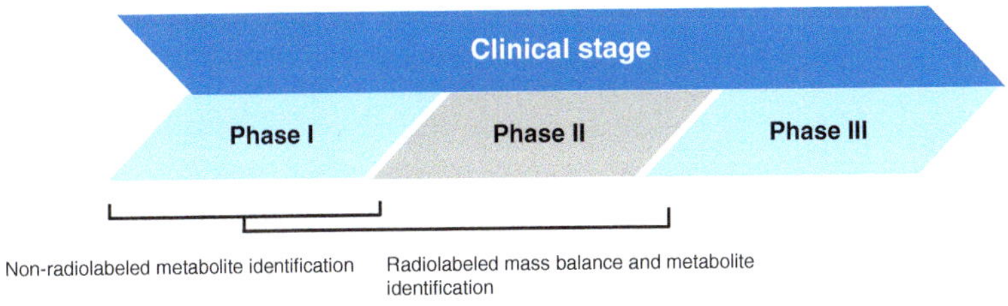

Figure 14.8 Research strategy of when to perform the radiolabeled mass balance studies at clinical stages.

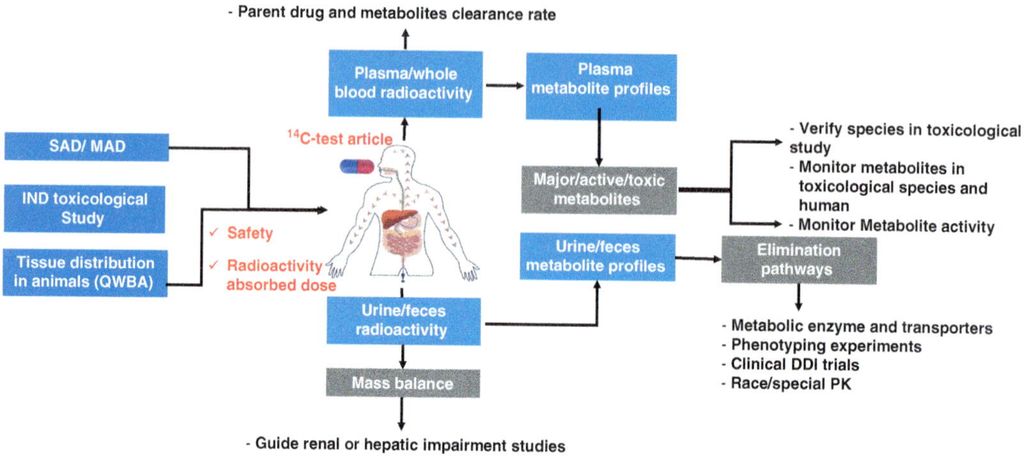

Figure 14.9 Human radiolabeled mass balance study.

14.3.6 Research Strategy at the Preclinical Stage

The overall pathways of metabolism and excretion of an investigational drug in preclinical species should be investigated to satisfy the high requirements of clinical trials. The cumulative percentage of the administered radioactivity recovered in preclinical studies is not suggested to be regarded as the only standard for mass balance, even if the cumulative percentage of the administered radioactivity recovered has exceeded 90%, safety risk may also be caused if the parent drug/metabolites are accumulated in some small tissues (such as uveal tract and thyroid gland). Therefore, the mass balance we focused on includes not only the regular urine and feces excretion data but also the tissue distribution data throughout the whole body.

The following radiolabeled studies are suggested to be conducted in preclinical species (Table 14.5).

Table 14.5 Recommended radiolabeled studies in preclinical species.

Test type	Description	Guidance
Plasma pharmacokinetics	Comparison of concentration-time profiles of total radioactivity and non-radioactivity following radiolabeled and non-radiolabeled test article administration at the same dose level or the dose within the pharmacokinetic linearity range.	Radioactivity concentrations significantly higher than parent concentration in plasma indicate a high proportion of metabolites are produced, and quantitative determination of some metabolites should be considered conducted in the subsequent PK, Toxicokinetics (TK), and clinical trials.
Mass balance	Biliary, urinary, and fecal excretion studies using pharmacodynamic or toxicological animal models. If the total radioactivity recovered is <90%, carcasses, syringes, recoveries of different sample pretreatments, or radioactive volatile substances will be analyzed.	Improve the clinical study design based on the preclinical study experiences, to avoid the total radioactivity recovered being <90%. Adequate justification should be provided when recovery is less than 90% based on the preclinical study results. If the total radioactivity recovered in rodent animals is <90% or non-rodent animals are closer to humans, mass balance studies in non-rodent animals may be considered.

Table 14.5 (Continued)

Test type	Description	Guidance
Tissue distribution using the QWBA method	Tissue distribution using the QWBA method in albino and pigmented rats and mice, such as Long Evans rats and C57BL mice.	Dosimetry calculation is conducted to estimate the human radioactivity absorbed dose based on total radioactivity concentrations in organ/tissue distribution and excretion data obtained after administration. PK studies are usually completed earlier than toxicological studies, and the tissue distribution data can predict which tissues may have abnormalities. Conducting QWBA studies in pregnant and lactating rats/mice can provide guided information on embryonic developmental toxicity and postpartum developmental toxicity.
In vitro hepatocyte/ liver microsome metabolite identification	Incubation of radiolabeled test articles with *in vitro* matrix from different species to obtain metabolite profiles for various species.	Assess interspecies metabolic similarity and if the metabolites in animals can cover metabolites in humans, then the main *in vitro* metabolic pathways are obtained. Compared with non-radiolabeled studies, radiolabeled studies can provide more accurate quantitative information on the proportion of metabolites.
In vivo metabolite identification	Obtain metabolite profiles of various matrices (e.g., plasma, bile, urine, and feces) in animals, following administration of radiolabeled test articles.	Compared with non-radiolabeled metabolite identification studies, radiolabeled metabolite identification studies can provide more comprehensive metabolite distribution in the systemic circulation of the test article and provide information for toxicological and pharmacodynamic mechanisms. Metabolites concentration as % Dose in various matrices (e.g., plasma, bile, urine, and feces) can be obtained based on the excretion of radioactivity, which can be used to determine the metabolic pathway of the investigational drug and to guide the corresponding phenotyping experiments. The results and methods of preclinical radiolabeled metabolite identification can also be used in radiolabeled clinical samples, which is conducive to accelerating the clinical process.

References

1 FDA (March 2022). *Clinical Pharmacology Considerations for Human Radiolabeled Mass Balance Studies Guidance for Industry*.

2 FDA (March 2020). *Safety Testing of Drug Metabolites Guidance for Industry, Revision 2*.

14.4 Key Considerations for Conducting Radiolabeled Human Mass Balance Studies

Huan Li, Lian Guo, Weiqun Cao, Lingling Zhang

14.4.1 Introduction

In May 2022, the FDA issued the draft guidance Clinical Pharmacology Considerations for Human Radiolabeled Mass Balance Studies [1], which gave the requirements and guidance for radiolabeled human mass balance studies. It is stated that in general, mass balance studies should be conducted for all new molecular entities, of which radiolabeled human mass balance studies are one of the most commonly used methods. If radiolabeled human mass balance studies are not performed, adequate justification should be provided. In this chapter, the current considerations on human mass balance in the process of new drug research and development are discussed and summarized.

14.4.2 What Is the Implementation of Mass Balance Studies for Marketed New Drugs?

66% ($n = 101$) [2] of the new small molecule drugs approved by the FDA from 2014 to 2018 had conducted human radiolabeled mass balance studies, while 81 new small molecule drugs approved by the FDA from 2017 to 2019 also conducted human radiolabeled mass balance studies [3]. The reasons for drugs approved without a mass balance study were usually due to endogenous substances, radiopharmaceuticals, drugs with existing study data, or drugs that are difficult to carry out but urgently needed in clinical practice. When human radiolabeled mass balance studies cannot be conducted (e.g., safety concerns), alternative approaches, such as *in vitro* assessment and animal radiolabeled mass balance studies should be used. However, extrapolation from animals to humans is challenging due to potential species differences in drug metabolism and excretion.

14.4.3 How to Select the Study Methods for Human Mass Balance?

Mass balance studies are typically conducted using radiolabeled methods (conventional and micro-radioactive tracer technique) or non-radiolabeled methods (e.g., fluorine-nuclear magnetic resonance [^{19}F-NMR] and mass spectrometry quantification). Among them, the human mass balance study using radioisotope (usually ^{14}C) tracer technology is a common method to obtain comprehensive and direct information about drug ADME in the human body, which is the "gold standard" for mass balance study.

14.4.3.1 Conventional radioactive tracer technique – liquid scintillation counter (LSC)

The radioactive tracer technique has been used in human mass balance studies for about 50 years [4]. The administered dose of radioactivity is usually set to range from 40 to 100 μCi/person [5]. LSC can directly analyze urine samples; for fecal samples, to avoid color quenching affecting the accuracy of the data, it is necessary to combust the fecal samples into [$^{14}CO_2$] and capture before determination. Compared with the other human mass balance methods, LSC assay is simpler in sample processing, and the less time-consuming detection method with the higher accuracy. According to statistics [5], 95% of the radioactive tracer technology used in mass balance studies is LSC assayed.

14.4.3.2 Micro-radioactive tracer technique – accelerator mass spectrometry (AMS)

AMS is an ultra-highly sensitive analytical instrument for analyzing isotopic abundance ratios with accelerators, and its detection limit can be more than three orders of magnitude lower than LSC [6, 7]. The administered dose of radioactivity by micro-radioactive tracer technique is usually set to 0.1 ~ 1 μCi/person [5], at which the conventional LSC cannot be accurately quantified, but AMS can meet the detection needs. For AMS, all samples must be graphitized and captured with $[^{14}CO_2]$, so sample preparation is much more time-consuming, cumbersome, and costly than LSC. According to statistics, when radioactive tracer technology was used in mass balance studies, 5% of the studies were AMS assayed [5]. Based on the recommendations of the guidelines, when the $[^{14}C]$-labeled radioactive dosage is less than 1 μCi, the radioactive dosimetry estimation based on animal data (mass balance and QWBA) may not be required, so it is beneficial to conduct an intravenous injection of $[^{14}C]$ compound in the early clinical trial stage (such as phase 0) of the drug to obtain the absolute bioavailability, PK, and metabolic characteristics.

14.4.3.3 Liquid chromatography–tandem mass spectrometry (LC–MS)

LC–MS analysis has the advantages of accuracy, sensitivity, and rapidity, but quantitative analysis of metabolites cannot be conducted without metabolite standards. Therefore, before mass spectrometry is used for mass balance studies, it is necessary to clarify the structure of metabolites and to obtain as many metabolite standards as possible, which is a great challenge for new drug development in competition with time. In addition, LC–MS quantitative analysis depends on the mass spectrometry response of compounds and is vulnerable to matrix interference. It is necessary to develop and validate bioanalytical methods for different metabolites in each matrix. This method is suitable for the test substances that exist in the unchanged form *in vivo* or have relatively single metabolites.

14.4.3.4 Fluorine-19 nuclear magnetic resonance (^{19}F-NMR)

The intensity of the signal observed in the ^{19}F-NMR is proportional to the number of fluorine nuclei in the sample. The quantitative results are not affected by chemical structure and matrix. Therefore, F-NMR is available in human mass balance studies that the test compounds are fluorochemicals, which can be completed by the drug itself (^{14}C-labeled drugs are not required), greatly reduce the research cycle and cost, are one of the effective methods of human mass balance research [8]. However, the sample processing procedure before F-NMR analysis is complex, including extraction, drying, and deuterium water reconstitution; meanwhile, the analysis time is affected by the concentration of F atoms, probe type, magnetic field intensity, etc., and the sample analysis time range is from about 6.5 minutes to 24 hours [8].

The characteristics of the above four detection methods are summarized (Table 14.6).

14.4.4 When to Carry Out Radiolabeled Human Mass Balance Studies?

Mass balance studies are usually conducted between clinical phases I and III of drug development. Radiolabeled human mass balance studies are encouraged to be conducted in the early stage of clinical development of pharmaceuticals and are recommended to be carried out during phase IIa to find out whether there is a low mass balance recovery, complex metabolites, and high proportion of circulating metabolites as early as possible, for a comprehensive understanding of clinical PK data.

Table 14.6 Comparison of detection techniques for human mass balance study.

	Radiolabeled methods		Non-radiolabeled methods	
	Conventional radioactive tracer techniques	**Micro-radioactive tracer technique**	**LC–MS**	**^{19}F-NMR**
Instrument	Liquid scintillation counter (LSC)	Accelerator mass spectrometry (AMS)	Triple quadrupole mass spectrometry	^{19}F-NMR
Typical radioactive dosage	40 ~ 100 μCi/person	0.1 ~ 1 μCi/person	None	None
Compound	Synthesis the [^{14}C] or [^{3}H] compound	Synthesis the [^{14}C] or [^{3}H] compound	Synthesis Standard	Contain fluorine
Sensitivity	Higher	Extra-high	High	Low
Detection limit	ng ~ μg	pg ~ fg	ng ~ μg	μg
Validation	Not required, verify instrument performance instead	Required	Required	Required
On-time monitor	Easy	Difficult	Not easy	Difficult

14.4.5 How to Select Nuclide and Administered Radioactivity?

[^{14}C] and [^{3}H] are the most commonly used nuclides for radiolabeled mass balance studies. The [^{3}H]-labeled drugs are relatively easy to synthesize, but they are easy to exchange with water molecules in the body to form tritium water, thus affecting the conclusion of mass balance and the accuracy of metabolite quantification; therefore, the more stable [^{14}C]-labeled drug is usually selected. For a drug with a larger molecular weight or a lower administered dosage, [^{3}H] labeling may also be used due to limitation in the specific activity of radioactivity (^{14}C: ~62.4 mCi/mmol; ^{3}H: ~28.6 Ci/mmol), but it needs to be fully verified that tritium–water exchange does not readily occur in the *in vivo* environment [9]. It has been reported [2] that of 104 human mass balance studies, 99% (n = 103) used [^{14}C], and only one study used [^{3}H] in the radiolabeled mass balance study.

The radioactivity administered to humans should follow regulatory requirements and be presumed to be acceptable through excretion and tissue distribution QWBA data obtained from animal studies. According to statistics, the administered radioactivity dosage used in 104 human mass balance studies ranged from 200 nCi to 500 μCi [2]. The detailed distribution data is shown in Table 14.7.

Table 14.7 The administered radioactivity dosage in the human mass balance study.

Administered radioactivity dosage	0.2–1 μCi	1–100 μCi	100 μCi	100–500 μCi
Distribution ratio	12%	27%	45%	17%

14.4.6 Considerations on Administration Frequency, Dosage, and Dose Route

Generally, a single-dose study is typically considered sufficient for a human mass balance study. It is reported [2] that 97.12% ($n = 101$) of the 104 human mass balance studies were single-dose studies, and only 3 were performed at the steady state (Figure 14.10a), that is, the volunteers administered the radiolabeled drug after the steady state reached using the non-radiolabeled compounds. Of these three drugs, two exhibited nonlinear PKs (Lemovir and Eligrusat). Another drug is gefitinib, which needs to be studied in cancer patients and a single dose is not feasible (ethical consideration).

The route of administration of the compounds in the mass balance studies was generally consistent with the final approved clinical route of administration. When it is not feasible to administer radiolabeled drugs by the intended route of administration, other routes of administration (e.g., intravenous) may be used. In 104 studies, it was reported that 98.08% ($n = 102$) used the same route of administration as the final approved route of administration (Figure 14.10b), and 1.92% ($n = 2$) used oral and intravenous routes of administration instead of inhalation for the treatment of pulmonary diseases.

Exclusion of 2 topical medications and 2 routes of administration inconsistent with the final route, in 100 human mass balance studies [2], 54% ($n = 54$) of studies selected doses within the final approved dose, 17% ($n = 17$) used doses lower than the final approved dose, and 29% used doses higher than the final approved dose (Figure 14.10c). Overall, 85% of the doses used in these studies were in the linear range of kinetics, and the results of the mass balance study can

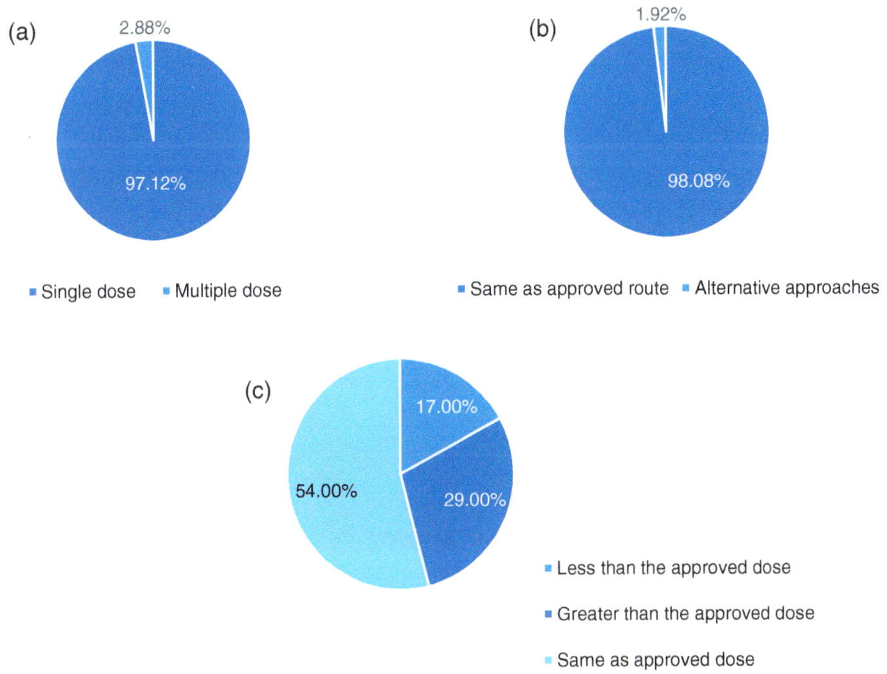

Figure 14.10 Design characteristics of human radiolabeled mass balance studies performed to support the initial approval of new small-molecule drugs approved between 2014 and 2018. (a) The mass balance study was performed as a single-dose study versus at the steady state. (b) Route of administration utilized. (c) Dosage of the investigational drug used in the study.

be extrapolated to other doses in this range. This may affect the credibility of the results of the mass balance study when the dose used is not the final approved dose and the clearance of the drug is nonlinear.

14.4.7 How to Select Volunteers?

Most mass balance studies are conducted in healthy male volunteers.

It is reported that [2] 90.38% (n = 94) of the studies were conducted in healthy volunteers (Figure 14.11a), and 9.62% (n = 10) of the studies were conducted in the target patient population (nine in patients with cancer and one in patients with endocrine-related diseases) due to risk factors such as expected adverse effects. 85.58% (n = 89) of the studies selected only male volunteers, 12.50% both men and women, and 1.92% selected only women (Figure 14.11b). Female-specific drugs or drugs that are toxic to men may be the reason for choosing women only, such as Olaparib, which is used to treat ovarian cancer, and Venetok, which has testicular toxicity.

The mass balance study is typically performed with six volunteers [2]. The sample mean, median, and mode size for the mass balance study were all 6 in 104 studies. 13% (n = 14) of the studies reported less than 6 subjects, and when more than 8 subjects, most of these studies had multiple objectives, such as evaluating the mass balance of drugs with different radiolabeled positions or simultaneously determining absolute bioavailability.

14.4.8 What About Insufficient Recovery?

The guidelines state that the total recovery for the mass balance study should reach 90% in principle. In 102 mass balance studies [2], only 62.75% (n = 64) reported recoveries ≥90%, 28.43% (n = 29) recovered 80%~90%, and 8.82% (n = 9) recovered <80% (Figure 14.12). There are many reasons for insufficient recovery, which include: (i) improper selection of radiolabel location, and loss of radionuclide; (ii) incomplete sample collection, especially for drugs with long half-life; and

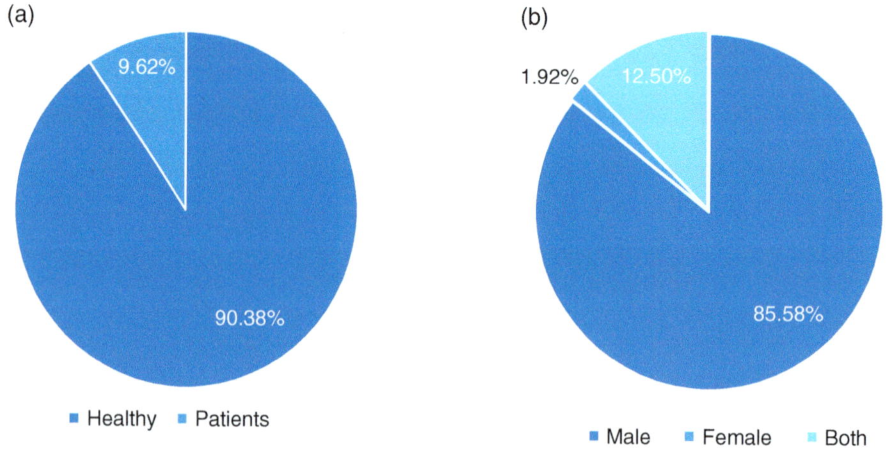

Figure 14.11 Volunteer characteristics in radiolabeled human mass balance study of new small molecule drugs approved from 2014 to 2018. (a) Healthy and patient proportion (n = 104); (b) gender composition (n = 104).

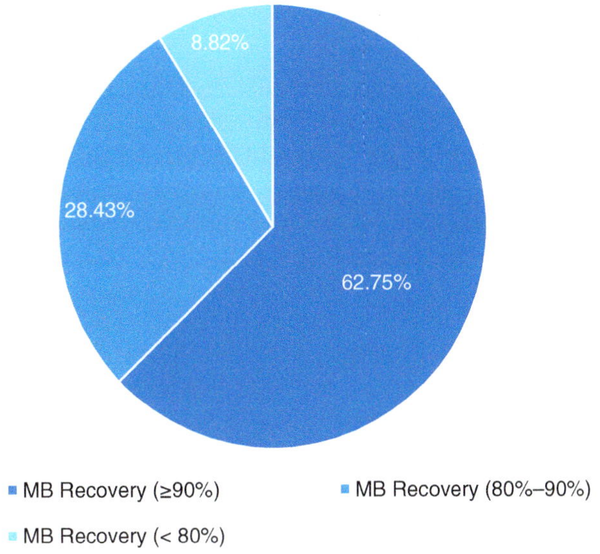

Figure 14.12 Recoveries in radiolabeled human mass balance of new small molecule drugs from 2014 to 2018.

(iii) inadequate or insensitive analytical methods used. In summary, it is necessary to fully describe the reasons for the insufficient recovery rate of test compounds.

14.4.9 What Are the Other Considerations of Guidelines?

In addition to the above concerns, the following key points should also be paid attention in the guidance of "Clinical Pharmacology Considerations for Human Radiolabeled Mass Balance Studies":

- The study is non-randomized and open-label.
- The position of the radioisotope is in a chemically and metabolically stable position, if necessary, two or more separate labeling sites may be used.
- Radioactivity dosage to be administered in humans can be predicted based on animal studies dosimetry calculations and follow applicable guidelines (e.g., ICRP and ACRP).
- Encourage the simultaneous design of absolute bioavailability and mass balance studies in one clinical protocol.
- Continuous sample collection to ensure maximum recovery of radioactivity in the relevant biological matrix and comprehensive data interpretation.
- Real-time monitoring of radioactive recovery of samples to determine volunteer departure time.
- For drugs with a long half-life, alternative sample collection strategies can be considered to get an estimate of the final recovery.
- For the quantitative analysis of biological samples, the samples of each subject shall be analyzed separately.
- For metabolite identification, samples from different time points of the same subject can be mixed for analysis.
- Metabolites in plasma greater than 10% of total exposure must be identified.
- The metabolites accounting for more than 80% of the administered dose (in excreta) must be identified.

References

1 Clinical Pharmacology Considerations for Human Radiolabeled Mass Balance Studies (Draft Guidance), Guidance for Industry (fda.gov).

2 Ramamoorthy, A., Bende, G., and Chow, E.C.Y. (2022). Human radiolabeled mass balance studies supporting the FDA approval of new drugs. *American Society for Clinical Pharmacology and Therapeutics, Clinical and Translational Science* 15: 2567–2575.

3 Wang, L., Zeru, Z., and Feng, S. (2021). 14C-labeled human mass balance study design and results introduction. *Chinese Journal Clinical Pharmacology* 37 (5).

4 Sheng, M., Ling, Y., and Liyan, M. (2020). Clinical research progress of isotope labeled drug mass balance and biotransformation. *Chinese Journal of Pharmacy* 55 (14).

5 Cerny, M.A., Spracklin, D.K., and Scott Obach, R. (2023). Human absorption, distribution, metabolism and excretion studies - origins, innovations and importance. *Drug Metabolism and Disposition* 51 (6): 647–656.

6 Gupta, N., Zhang, S., and Pusalkar, S. (ed.) (2017). A phase I study to assess the mass balance, excretion, and pharmacokinetics of [14C]-ixazomib, an oral proteasome inhibitor, in patients with advanced solid tumors. *Investigational New Drugs* 36 (3): 407–415.

7 Shaoyuan, W., Zheng Xin, H., and Bei, J.J. (2018). Current status of accelerator mass spectrometry application in early drug clinical research. *Chinese Journal of Clinical Pharmacology* 34 (5).

8 Haitao, H., Katyayan, K.K., and Czeskis, B.A. (2017). Comparison between radioanalysis and 19F nuclear magnetic resonance spectroscopy in the determination of mass balance, metabolism, and distribution of pefloxacin. *Drug Metabolism and Disposition* 45: 399–408.

9 Zhou Xiaowen, J., Wenzheng, Z.X. et al. (2020). Application of radionuclide tracer technology in human body mass balance of innovative chemical drugs. *Pharmaceutical and Clinical Research* 28 (2).

15

Bioanalysis

15.1 Application of High-Throughput Mass Spectrometry in *In Vitro* ADME

Yuanqiang Su, Xinxin Wen, Xiaotong Li, Lili Xing

15.1.1 Introduction

Data on *in vitro* ADME have played an increasingly important role in drug development over the past 10–15 years. The use of low-cost, high-throughput screening (HTS) technology enables rapid and effective elimination of interfering factors, making it possible to research the individual ADME properties of drug candidates. Combined with *in vivo* animal experimental data, the kinetic parameters of drug candidates in human can be more accurately predicted. Liquid chromatography–mass spectrometry (LC-MS/MS) technology, as a common analytical technology for *in vitro* ADME research platforms, has promoted the development of the platform in a faster and more accurate direction due to its high sensitivity, high specificity, and good reproducibility. However, millions of *in vitro* ADME HTS samples are produced every year, such as those for early metabolic stability assay, posing new speed challenges for traditional LC-MS/MS analysis [1, 2].

To address this challenge, this chapter will share several representative mass spectrometry (MS)-based HTS analysis techniques and their application cases.

15.1.2 Mass Spectrometry-Based High-Throughput Screening Analysis Technology

15.1.2.1 Cassette Technology

Cassette assay, also known as N-in-one, includes two types: (i) cassette incubation and analysis and (ii) single incubation and cassette analysis. It accelerates the drug candidate screening and optimization process by analyzing multiple compounds simultaneously, as shown in Figure 15.1. The advantages of cassette analysis include reducing the number of samples, improving analytical throughput, and saving time and cost while having less impact on the quality of the results. Jun Zhang et al. [3] compared the correlation of protein binding ratios between cassette analysis ($n = 4$, 8, 16, and 32, respectively) and individual analysis in plasma protein binding experiments, and the correlation coefficients were ≥0.94 for all data sets.

DDIs need to be considered when using cassette analysis. On the one hand, structurally similar compounds may cause saturation of metabolism enzymes if they share the same metabolic

Drug Metabolism and Pharmacokinetics: Frontiers, Strategies, and Applications, First Edition.
Edited by Liang Shen and WuXi AppTec DMPK.

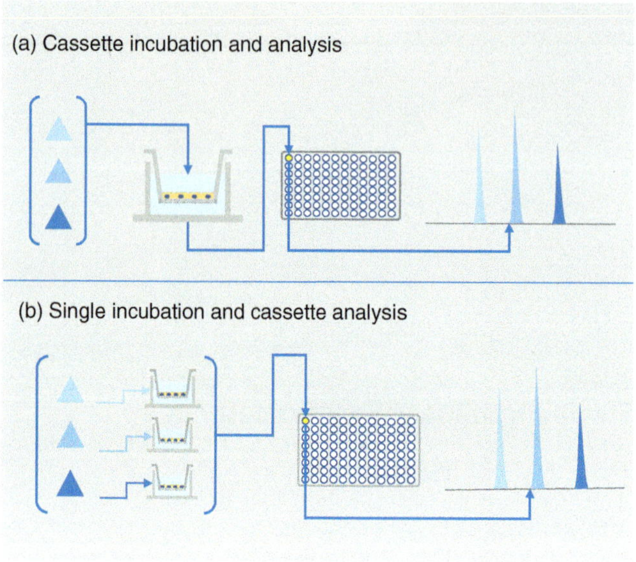

Figure 15.1 Cassette analysis. (a) Cassette incubation and analysis; (b) single incubation and cassette analysis.

pathway, thereby reducing drug clearance. On the other hand, when the molecular weight difference between structurally similar compounds is small, there may be issues of MS interference or suppression. Solving the latter problem usually requires the assistance of other equipment, such as introducing LC separation or using HRMS, to enhance the separation capability of the system. Despite these challenges, cassette analysis remains an effective screening technique in the early drug discovery stage.

15.1.2.2 Multiplex LC Technology

Multiplex LC-MS technology is achieved through the combination of multiple LC systems with a single mass spectrometer. Multiple independent LC systems are sequentially connected to the mass spectrometer via switching valves, which have the advantage that part of the mobile phase containing the target LC peak is introduced into the mass spectrometer and the rest is switched into the waste, thereby improving the effective detection of MS without affecting the LC separation efficiency. Multiple LC streams can be multiplexed depending on target compound retention time and total gradient time, such as two sets of alternating LCs that can theoretically increase throughput by a maximum of twofold, as shown in Figure 15.2.

Although multiplexing more LCs can save time, two limitations should also be considered. The first one is the MS acquisition window, where multiplexing more LCs results in narrower acquisition windows, potentially preventing compounds with large differences in retention times from entering the mass spectrometer. The second factor is the complexity of the system, and the introduction of more LCs requires more cooperation and coordination, making the system more complex. Therefore, the current two sets of LC systems are commonly used for staggered injections. Jingtao Wu et al. [4] used dual alternating LC to simultaneously analyze oxazepam and methotrexate in plasma samples, reducing the analytical time of each sample from six to three minutes without losing the LC separation, achieving a twofold increase in analytical throughput.

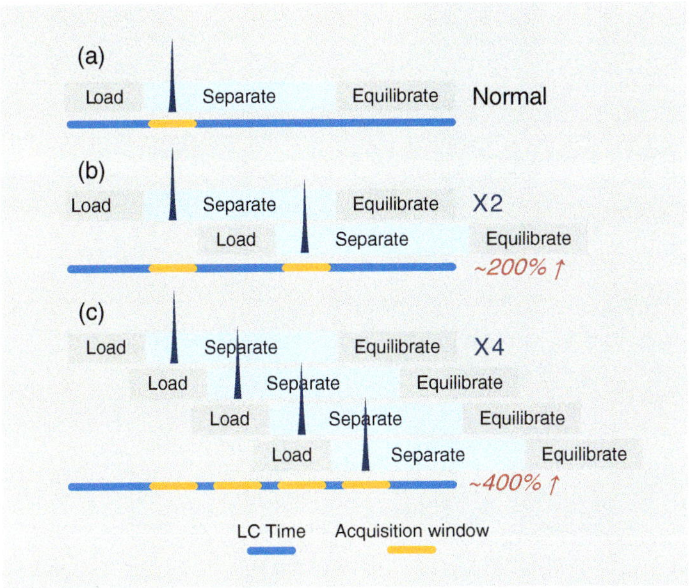

Figure 15.2 Multiplex LC system. (a) Traditional LC mode; (b) multiplex two sets of LC systems; (c) multiplex four sets of LC systems.

15.1.2.3 Online Solid-Phase Extraction (Online SPE) Technology

Online SPE technology, also known as the "Trap and Elute" mode, provides even faster analytical speed compared to previous multiplex LC technology. At its fastest, a sample can be analyzed every 10 seconds, but the chromatographic separation capability is reduced. Typical patterns of online SPE analysis techniques are shown in Figure 15.3. First, the compounds in the sample are carried through the weaker elution mobile phase into a 10 mm cartridge for retention, and the salts in the sample are washed into the waste with the weaker elution mobile phase. The compounds retained in the column are then eluted by switching to the stronger elution mobile phase through the switching valve and carried to the MS for detection. Online SPE technology has been widely used in HTS platforms, such as the RapidFire system, Apricot Designs Dual Arm (ADDA) system, and LeadSampler (LS-1) system. In particular, the LS-1 system, which can support a variety of LC applications and workflows, combines multiple LC technologies and microflow LC based on online SPE technology to meet the analysis needs under different application scenarios. John Janiszewski and Brendon Kapinos *et al.* [5, 6] successively validated and evaluated the ADDA system and LS-1 system for *in vitro* ADME assay such as DDI and metabolic stability.

15.1.2.4 Acoustic Ejection Mass Spectrometry

Acoustic ejection mass spectrometry (AEMS) innovatively couples acoustic droplet ejection (ADE) device and the standard electrospray ionization (ESI) source of a mass spectrometer with the open-port probe (OPP) sampling interface, mainly including three parts: (i) ADE: ADE device; (ii) OPI: open-port interface; (iii) ESI-MS: MS base on ESI ionization technology. In this system, ADE ejects the sample droplet (2.5 nL) from the well of the sample plate into the inverted OPI. The carrier solvent pump then advances the sample from the OPI through the transfer tube to the MS ion source at a rate of 200–500 μL per minute, which is then detected by MS. The main schematic is shown in Figure 15.4.

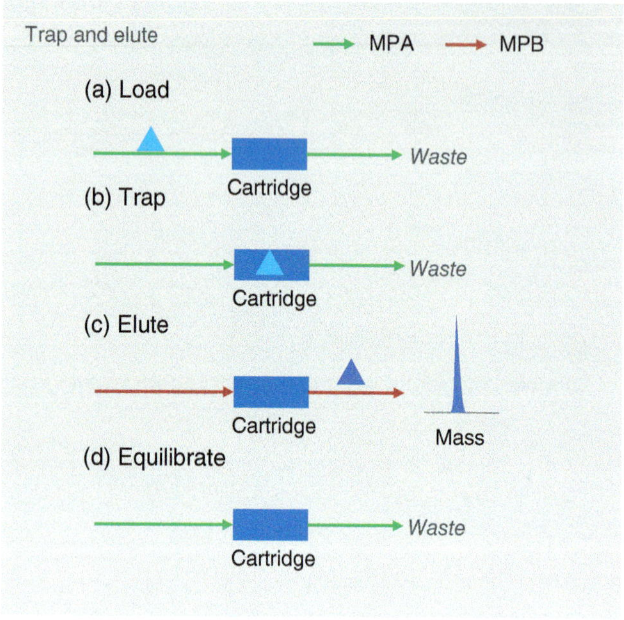

Figure 15.3 Online SPE technology. (a) Load; (b) trap; (c) elute; (d) equilibrate.

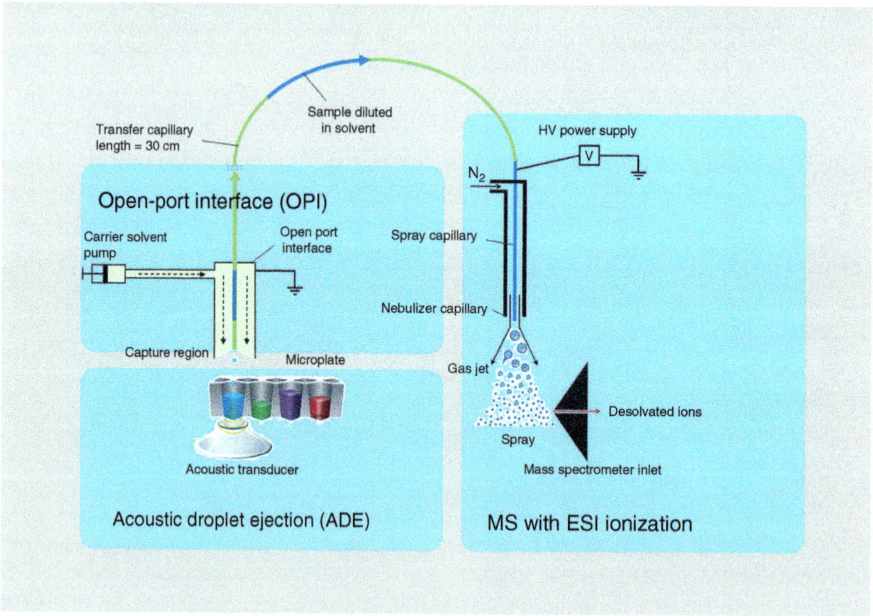

Figure 15.4 Schematic of an ADE-OPI-MS system. *Source:* [2] / Society for Laboratory Automation and Screening.

Based on the unique design, AEMS has the advantages of better matrix tolerance and lower carryover [7, 8]. In addition, AEMS increases analytical speed to a second level, up to six samples per second [9], which can promote the daily throughput of HTS MS to 100,000. The speed is comparable to fluorescence detection, greatly improving the existing MS analytical speed. This makes it more suitable for the future HTS development trend.

15.1.3 Practice of HTS MS in DMPK

Our team has rich experience in HTS platforms and has successively established HTS technology application platforms, including cassette analysis, multiplex LC-MS, rapid online SPE analysis, and AEMS, supporting early drug screening, *in vitro* ADME, rapid bioanalysis of biomarkers, and HTS compound quality control (QC), providing diversified and rapid bioanalysis solutions (Figure 15.5 and Table 15.1), delivering high-quality analytical data, and accelerating the process of drug research and development.

15.1.3.1 Case Study I: Online SPE-Based Rapid MS Analysis of 5-Aminolevulinic Acid (5-ALA)

5-ALA is a photosensitizer. In photodynamic therapy, 5-ALA is a precursor of protoporphyrin IX, which can produce cytotoxic reactive oxygen species when exposed to specific wavelengths of light and is used to treat a variety of cancers, including skin cancer, esophageal cancer, glioblastoma, and bladder cancer. 5-ALA is hard to retain in reversed-phase chromatography systems due to its strong polarity. Therefore, common analytical methods include derivatization and ion exchange chromatography, which usually require longer analysis time. In this study, the fast online SPE

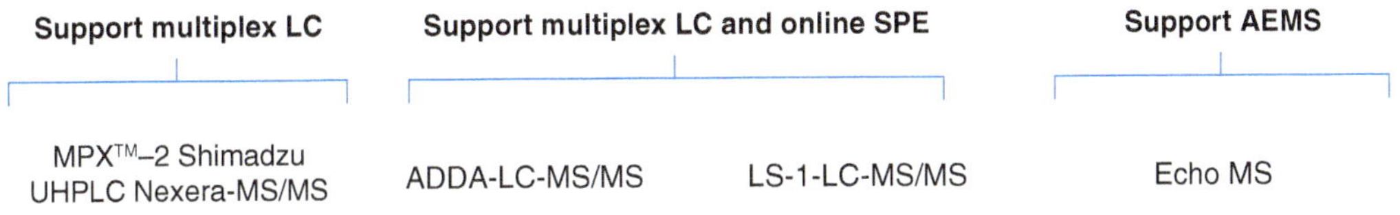

Figure 15.5 HTS MS application platforms.

Table 15.1 Characteristics of diverse HTS MS platforms.

Technology	Runtime[a] (s)	Throughput[b]	Features
Multiplex LC	15–30	~10^3	Flexible support of various LC gradients and reduced analysis time
Online SPE	10–15	~10^4	Fast analysis supports a large number of samples. It is one of the fastest online chromatography purification technologies for MS detection.
AEMS	1–2	~10^5	Fast supports ultra-high-throughput samples with high matrix tolerance and reproducibility

[a] Runtime: mass analysis time per sample.
[b] Throughput: number of samples analyzed daily.

technique was used for the quantification of 5-ALA in samples in 12 seconds per sample, which can support the detection of a large number of samples (Figure 15.6).

15.1.3.2 Case Study II: Library Compounds QC Based on AEMS

High-throughput compound QC has widespread application in drug discovery, ranging from the QC of assay-ready plates, to compound integrity checks in the screening collection, to reaction monitoring of chemical syntheses in microtiter plates. Due to the large number of samples (thousands per batch) that need to be monitored, these workflows can put a significant burden on typical LC-MS platforms. HTS platform based on AEMS provides feasibility due to the advantages of direct, noncontact liquid sampling, minimized sample consumption, and ultrafast analytical speed. We have used the AEMS technology (Echo MS) to complete QC analysis for nearly 5,000 compounds within four hours, while the traditional LC-MS analysis platform would take about 10 days.

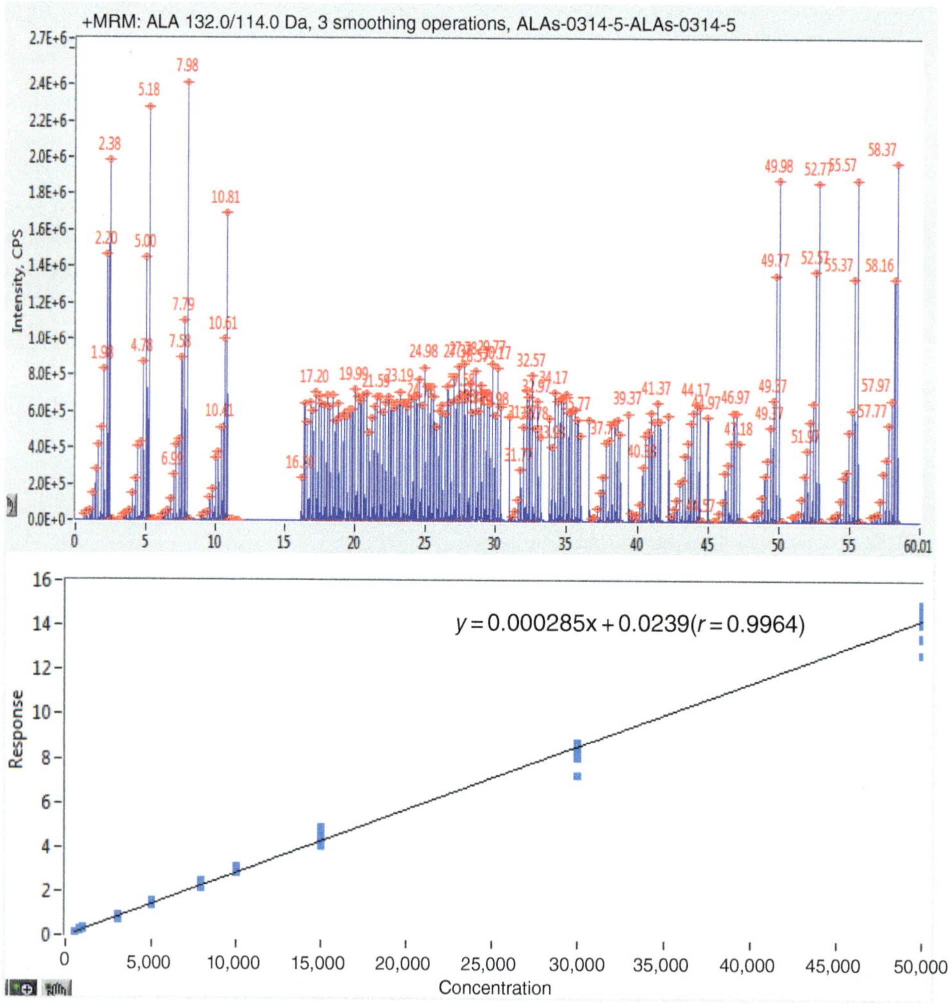

Figure 15.6 High-throughput quantification of 5-ALA based on online SPE MS technology. Top: 5-ALA LC-MS chromatogram; bottom: linear regression curve.

15.1.4 Summary

In retrospect, MS technology has played an increasingly important role in *in vitro* ADME high-throughput analysis platforms. With the continuous emergence of new technologies such as new modalities, new screening approaches, and new concepts (e.g., automation and digitization), high-throughput MS analysis is set to embrace more innovative applications and solution strategies. These innovations jointly drive continuous development and improvement in the field of HTS, thereby promoting the development of an *in vitro* ADME HTS platform in a more accurate and efficient direction, which will ultimately help shorten the drug research and development time, accelerate the marketing of new drugs, and thus better serve the cause of human health.

References

1 Kyranos, J.N., Cai, H., Wei, D., and Goetzinger, W.K. (2001). High-throughput high-performance liquid chromatography/mass spectrometry for modern drug discovery. *Current Opinion in Biotechnology* 12 (1): 105–111.
2 McLaren, D.G., Shah, V., Wisniewski, T. et al. (2021). High-throughput mass spectrometry for hit identification: current landscape and future perspectives. *SLAS DISCOVERY: Advancing the Science of Drug Discovery* 26 (2): 168–191.
3 Zhang, J., Maloney, J., Drexler, D.M. et al. (2012). Cassette incubation followed by bioanalysis using high-resolution MS for *in vitro* ADME screening assays. *Bioanalysis* 4 (5): 581–593.
4 Wu, J.T. (2021). The development of a staggered parallel separation liquid chromatography/tandem mass spectrometry system with online extraction for high-throughput screening of drug candidates in biological fluids. *Rapid Communications in Mass Spectrometry* 15 (2): 73–81.
5 Janiszewski, J., Schneider, R., Kapinos, B. et al. (2012). Development of a high-speed, multiplexed sample-delivery instrument for LC-MS/MS bioanalysis. *Bioanalysis* 4 (9): 1039–1056.
6 Kapinos, B., Liu, J., Piotrowski, M. et al. (2017). Development of a high-performance, enterprise-level, multimode LC-MS/MS autosampler for drug discovery. *Bioanalysis* 9 (21): 1643–1654.
7 Zhang, H., Liu, C., Hua, W. et al. (2021). Acoustic ejection mass spectrometry for high-throughput analysis. *Analytical Chemistry* 93 (31): 10850–10861.
8 Liu, C., Van Berkel, G.J., Cox, D.M. et al. (2020). Operational modes and speed considerations of an acoustic droplet dispenser for mass spectrometry. *Analytical Chemistry* 92 (24): 15818–15826.
9 Häbe, T.T., Späth, C., Schrade, S. et al. (2021). An ultrafast and flexible liquid chromatography/tandem mass spectrometry system paves the way for machine learning driven *in vivo* sample processing in early drug discovery. *Rapid Communications in Mass Spectrometry* 35 (12): e9096.

15.2 Application and Detection Techniques of Biomarkers in Drug Development

Hongfang Cui, Hongmei Wang, Honghong Zhang, Lili Xing

15.2.1 Introduction

Biomarkers are objectively measurable indicators that reflect physiological or pathological processes and the biological effects of medical interventions. According to their application value, they can be divided into diagnostic, monitoring, pharmacodynamic, predictive, prognostic, and risk assessment biomarkers. The exploration of biomarker identification commenced in 1847 with the discovery of urine periplasmic protein as a cancer biomarker. After more than 150 years of progress in biomarker development, the completion of the Human Genome Project marked the commencement of investigations into genetic biomarkers. The discovery and application of biomarkers have become a crucial research direction in the biotechnology and biopharmaceutical industries. In this section, we will discuss the application of biomarkers in drug development and, with experience from our department, elucidate state-of-the-art methods for biomarker detection.

15.2.2 Application of Biomarkers in Drug Development

In December 2021, the NMPA Drug Evaluation Center of China released the "Technical Guidance Principles for the Application of Biomarkers in the Clinical Development of Antitumor Drugs." This initiative represents a substantial regulatory step by the NMPA Drug Evaluation Center, indicating its emphasis on biomarkers. If a systematic evaluation process for the effectiveness of biomarkers is implemented during the preclinical development stage, including assessing drug-resistance-related biomarkers, efficacy-related biomarkers, and safety-related biomarkers, it will enhance transparency and facilitate the management of the development process. The clarity of the development goals will result in a substantial increase in the success rate of drug development during the later stages of clinical trials. Figure 15.7 presents the data regarding the reasons for the failure of clinical phase II drug development from 2008 to 2010 and clinical phase III drug development from 2007 to 2010. The data show that the primary reason for the clinical failure of

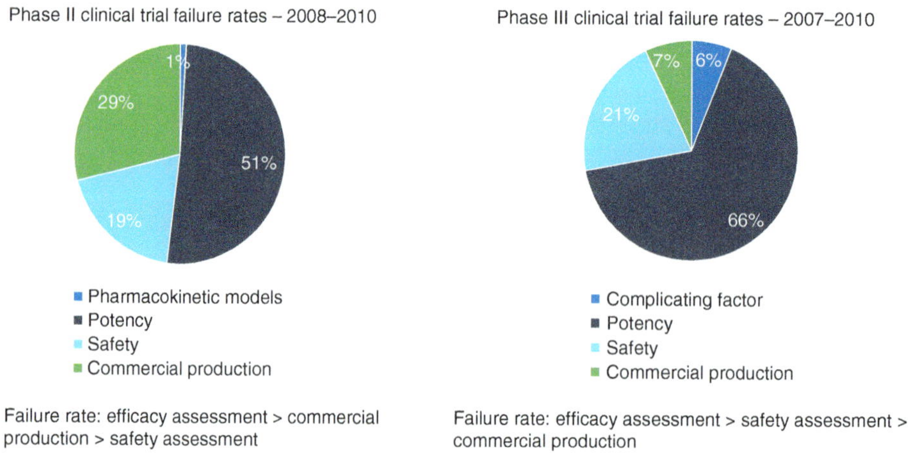

Figure 15.7 Factors influencing drug development failures. *Source:* Data from reference [1].

drugs is efficacy issues, accounting for more than 50% of the failures. Simultaneously, funding, safety, and PK impact the success rate of drug development to varying degrees [1].

Novel drug development can offer considerable advantages in treating diseases when applied to appropriate patients, with the correct dosage, time, and target. In recent years, biomarkers have been used to select appropriate targets and patients, and these are recognized as effective tools in drug development [2, 3]. As depicted in Figure 15.8, the rational utilization of biomarkers in preclinical and clinical studies accelerates drug development. It greatly assists in the applications of IND, new NDA, and biologics license applications (BLAs), encompassing the confirmation of drug action targets and candidate drugs, risk assessment, experimental design, dose optimization, patient stratification, and safety monitoring [4].

What is the impact on the success rate of drug clinical development when biomarkers are used? According to the "Clinical Development Success Rates and Contributing Factors from 2011 to 2020," jointly released by the Biotechnology Innovation Organization, Quantitative Life Sciences, and Informa Pharma Intelligence, the success rate of drug development, from phase I clinical trials to new drug applications, BLAs, and final approval for market launch, is more than twice as high when biomarkers are involved, reaching 15.9% (Figure 15.9) [5]. The involvement of biomarkers substantially enhances the efficiency of clinical drug development. Therefore, pharmaceutical companies can improve the efficiency of drug development by transforming the research and development model to monitor changes in biomarkers. Additionally, this shift in the development

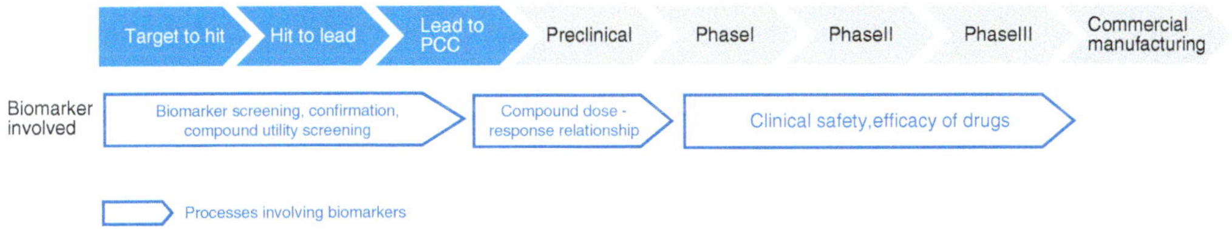

Figure 15.8 Application of biomarkers in drug development.

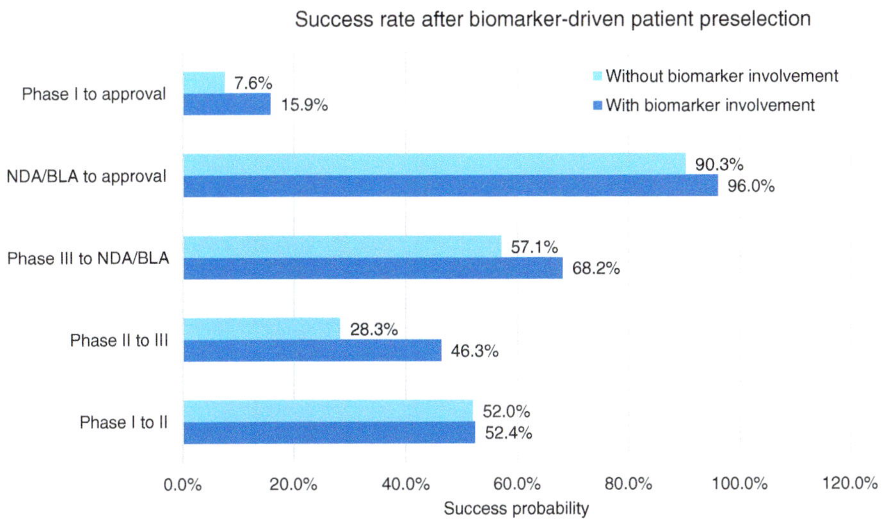

Figure 15.9 The success rate of drug development with and without biomarker involvement. *Source:* [5] / MDPI. / CC BY 4.0.

model has the potential to reduce the time taken to introduce drugs into the market, leading to cost savings during drug development.

15.2.3 Biomarker Detection Technologies and Testing Methods

15.2.3.1 Biomarker Detection Technologies

An appropriate detection platform can be selected based on the molecular weight of the biomarker, as illustrated in Figure 15.10.

NMR, nuclear magnetic resonance; GC-MS, gas chromatography-mass spectrometry; LC-MS, liquid chromatography-mass spectrometry; ELISA, enzyme-linked immunosorbent assay; MSD, mesoscale discovery; SIMOA, single-molecule array.

Nonprotein biomarker detection platforms include those for endogenous substances, such as lipids, polysaccharides, amino acids, fatty acids, and vitamins. Metabolomics, connecting genotype and phenotype, is the scientific study of the quantities and patterns of changes in metabolites (nonprotein endogenous metabolites) in an organism after a disturbance (such as changes in genetic or environmental factors). The main detection platforms for metabolomics-related tests of nonprotein biomarkers include nuclear magnetic resonance spectroscopy, LC-MS, and GC-MS, as shown in Table 15.2. Liquid chromatography-tandem mass spectrometry, with high sensitivity and a wide detection range, is the primary detection platform for metabolomics.

Protein biomarker detection platforms for free-form protein biomarkers include ELISA, ultra-sensitive multiplex electrochemiluminescence analysis (MSD), single-molecule array assays (SIMOA), and multiplex detection systems (Luminex). For non-freeform biomarkers, detection platforms include flow cytometry, fluorescence microscopy, and laser confocal microscopy.

Due to the size of hormone or peptide biomarkers, which falls between nonprotein and protein biomarkers, optimized ELISA or MS platforms can serve as detection platforms.

15.2.3.2 Biomarker Detection Methods

Depending on the specific experimental purpose, biomarker assays can be categorized into quantitative and semiquantitative assays.

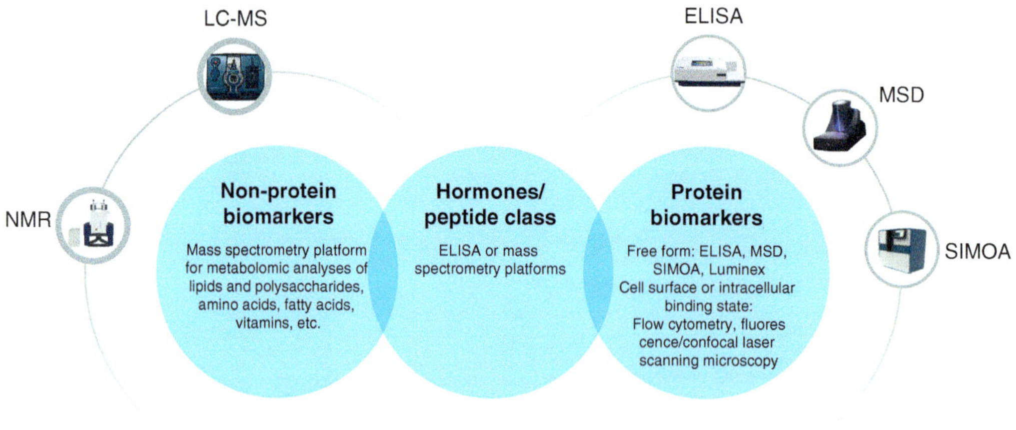

Figure 15.10 Biomarker detection platforms.

Table 15.2 Sensitivity and pros/cons of nonprotein biomarker detection platforms.

Analysis methods	Sensitivity (moles, mol)	Pros	Cons
Nuclear magnetic Resonance spectroscopy	10^{-6}	Nondestructive detection	Low sensitivity
		Good reproducibility	Limited dynamic range
		Easy sample handling	High cost
		Unbiased analysis	Poor repeatability for quantitation
		High-throughput	
		Small sample volume	
Gas chromatography-mass spectrometry	10^{-12}	High-resolution	Cumbersome derivatization steps
		Comprehensive database	Mainly analyzes volatile substances
			Unsuitable for thermo-unstable and nonvolatile compounds
		Good selectivity	Cumbersome sample handling
		Low cost	
		Good qualitative and quantitative results	
		Small sample volume	
Liquid chromatography-mass spectrometry	10^{-15}	High sensitivity	
		High resolution	Poor chromatographic separation compared to gas chromatography
		Wide dynamic range	Data parsing relies on databases
		Unbiased analysis	
		Good reproducibility	
		Broader spectrum of analytes	

Quantitative assays are capable of accurately obtaining the exact concentration of the analyte in the sample. For quantitative assays, it is critical to select the appropriate method for the preparation of calibration standards and QC samples. For biomarkers, however, the actual biological matrix usually contains unknown amounts of the target analyte, and therefore, conventional preparation of calibration standards and QCs is not feasible. The most used approaches for the quantitation of biomarkers are the surrogate matrix or surrogate compound methods, as shown in Figure 15.11.

(i) Here isotopically labeled compounds means the stable isotopes labeled (SIL) compounds. (ii) Internal standards (IS): target compounds labeled with stable isotopes used simultaneously as internal standards and surrogate analytes.

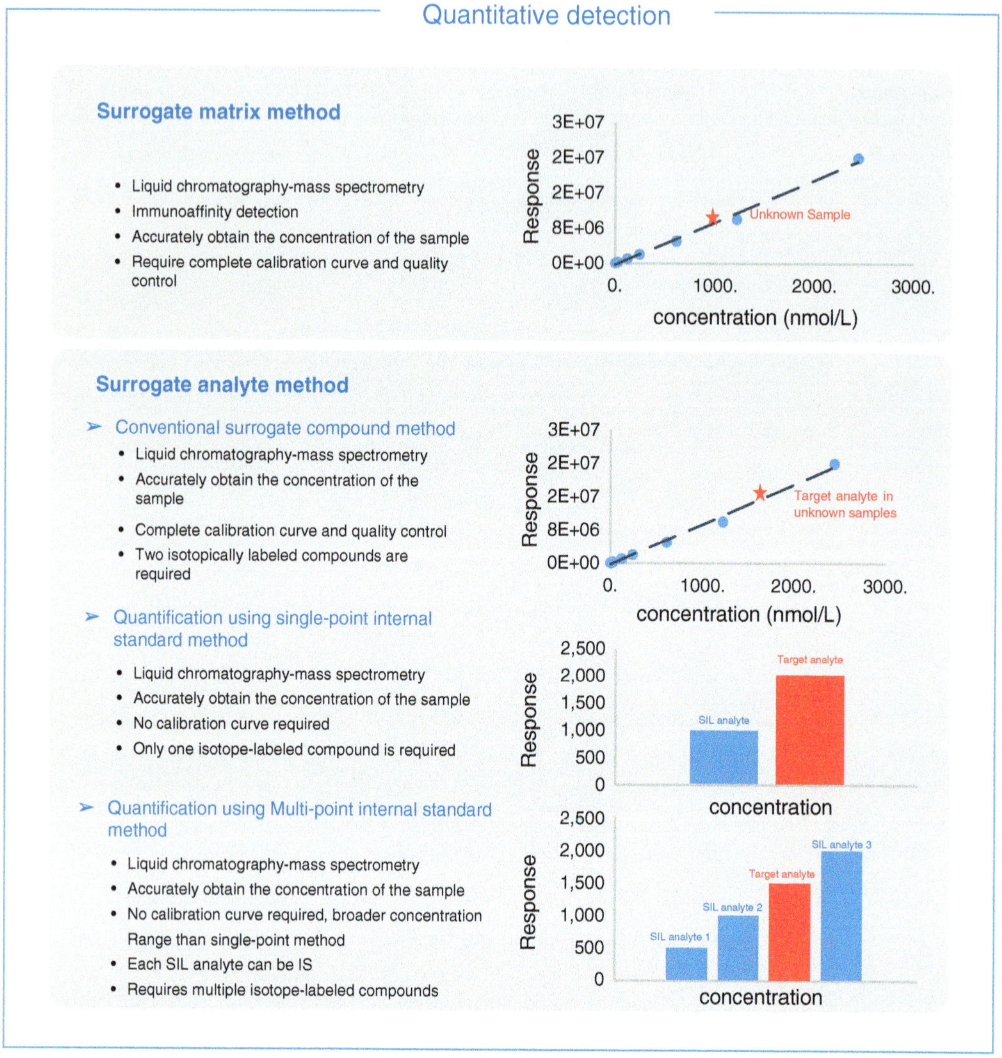

Figure 15.11 Quantitative detection of biomarkers.

The surrogate matrix method is widely employed for quantifying endogenous compounds such as biomarkers. This approach applies to experiments aimed at decreasing or increasing the target biomarker levels. When utilizing liquid chromatography-mass spectrometry or immunoaffinity detection of compounds, complete standard curves and QCs are necessary. However, this method introduces matrix effects and variations in recovery rates due to differences between the matrix used for the calibration curve and actual samples. Using the SIL target compounds as the IS can effectively correct the influence of different matrix and recovery rate disparities.

The surrogate analyte method uses the substitute analytes that are either absent or low concentration in the target matrix as the analyte, it should exhibit consistent signals with the target biomarker after correction. The SIL target biomarker is always used as the surrogate analyte, then the IS will not be as critical as the surrogate matrix method. This approach can use authentic matrices for the calibration curve, which is consistent with the unknown samples. The surrogate analyte method includes the complete calibration curve method, the single-point internal standard method, and the multipoint internal standard method. As for the complete calibration curve method, similar to the routine bioanalysis, the unknown sample concentration will be obtained via the calibration curve.

The single-point internal standard method requires one SIL analyte as the calibration point and facilitates the accurate determination of sample concentrations within a specified range without the need for a standard curve. However, the multipoint internal standard method necessitates multiple stable isotope-labeled compounds as surrogate compounds of the sample, each with different concentrations, forming a standard curve. The concentration of the target analyte can be determined by comparing the signals of the target compound with those of the surrogate analytes. This method has a broader detectable concentration range compared to the single-point internal standard method.

For experiments that only require an understanding of the relative trends in target biomarkers, a semiquantitative detection approach can be employed. This method characterizes the relative trends of target analytes in the samples by comparing changes in the peak area or peak area ratio to IS.

In the semiquantitative experiment, using liquid chromatography-mass spectrometry will be convenient with the known ion transition. In contrast, in immunoassay detection, semiquantitative detection can only be performed if the method has been validated, with experimental steps like quantitative detection. Compared to conventional tandem MS, which focuses on limited ion transitions, HRMS can utilize its ultra-high resolution characteristics, employing a full-scan mode to monitor signals of all the target compounds and potential metabolites. This approach allows for the maximization of sample information within a limited sample volume. Therefore, in cases where information about some target biomarkers is available and a more extensive understanding of sample information is required, semiquantitative detection can be performed using HRMS with a full-scan mode.

15.2.4 Summary

Biomarkers have emerged as crucial tools in drug research and development. With the wide application of biomarkers, an increasing number of signaling molecules and pathways are being discovered as therapeutic targets for drugs. Biomarkers serve as valuable indicators for the biological pharmacological effects of drugs, facilitating accurate assessment of their activity and exhibiting a high level of predictability for clinical trials. For monitoring biomarkers, it is crucial to select customized experimental designs, appropriate detection platforms, and technical methods based on the experimental objectives.

References

1 Khanna, I. (2012). Drug discovery in pharmaceutical industry: productivity challenges and trends. *Drug Discovery Today* 17 (19–20): 1088–1102.

2 Califf, R.M. (2018). Biomarker definitions and their applications. *Experimental Biology and Medicine* 243 (3): 213–221.

3 Kraus, V.B. (2018). Biomarkers as drug development tools: discovery, validation, qualification and use. *Nature Reviews Rheumatology* 14 (6): 354–362.

4 Lee, J.W. and Hall, M. (2009). Method validation of protein biomarkers in support of drug development or clinical diagnosis/prognosis. *Journal of Chromatography B* 877 (13): 1259–1271.

5 Emwas, A.H., Roy, R., McKay, R.T. et al. (2019). *NMR spectroscopy for metabolomics research. Metabolites* 9 (7): 123.

15.3 Analysis and Detection Strategies of Liposome Drugs in Biological Samples

Peiyun An, Jinlian Lu, Lili Xing

15.3.1 Introduction

Liposomes, which are commonly used nanocarriers for various potentially active hydrophobic and hydrophilic molecules, are mainly nanometer-sized, enclosed vesicles with a lipid bilayer composed of phospholipids and cholesterol. Hydrophilic molecules are encapsulated in the hydrophilic nuclear region of the liposome, and hydrophobic molecules are encapsulated in the hydrophobic region of the phospholipid bilayer, as shown in Figure 15.12 [1].

Depending on the purpose of the research, modifications can be made to the components of liposomes (such as different phospholipid and cholesterol content), surface charge (generated by the charge of the constituent phospholipids), etc. This allows for the preparation of long-circulating liposomes (such as PEG-modified) for extended circulation time, cationic liposomes (modified to carry a positive charge) for gene drug carriers, environmentally sensitive liposomes (changes of *in vivo* pH, enzymes, or physical conditions such as light or heat effects) that respond to specific sites for drug release, and immunoliposomes that improve targeting (e.g., by connecting antibodies) [1, 2]. Based on the above characteristics, H. Nsairat et al. [1] summarized the clinical applications of liposome drug products on the market. The primary application is in antitumor treatments, as detailed in Table 15.3.

Liposomal drugs, as carrier nanomedicines, compared to conventional dosage forms, have special nanoparticle size, phospholipid bilayer structure, and special surface modifications, which cause changes in the physicochemical properties and biological behaviors. The *in vivo* absorption characteristics of conventional dosage forms are mainly reflected by the determination of the concentration of free drugs in the systemic circulation, while the difference between liposome drug products and conventional dosage forms lies in the existence of their functional unit "drug particles." After entering the body, drugs exist in the form of free (or non-liposomal) drugs substance (the drugs released from liposomes mainly exist in two forms: free drug and plasma protein binding drug in the body) and liposome-associated drugs (the drugs contained within liposomes), as shown in Figure 15.13. Therefore, it is necessary to determine the concentration of non-liposomal and liposome-associated drugs in blood or plasma respectively when conducting PK studies [3].

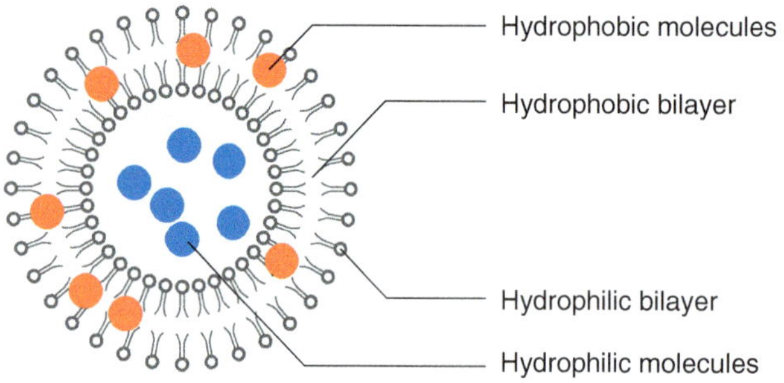

Figure 15.12 Liposome structure.

Table 15.3 Clinically used liposomes grouped by therapeutic usage.

Usage	Trade name	Active ingredient(s)	Liposome platform (molar ratio)	Manufacturer	Year approved	Administration route
Anticancer	Doxil®	Doxorubicin	HSPC:cholesterol:PEG 2000-DSPE (56:38:5)	Sequus Pharmaceuticals	1995	I.V
	DaunoXome®	Daunorubicin	DSPC:cholesterol (2:1)	NeXstar Pharmaceuticals	1996	I.V
	Depocyt®	Cytarabine	DepoFoam™	SkyPharma Inc.	1999	Spinal
	Myocet®	Doxorubicin	Cholesterol:EPC (45:55)	Elan Pharmaceuticals	2000	I.V
	Mepact®	Mefamurtide	DOPS:POPC (3:7) Multilamellar liposome	Takeda Pharmaceutical Limited	2004	I.V
	Lipodox®	Doxorubicin	DSPC:cholesterol:PEG 2000-DSPE (56:39:5)	Sun Pharma	2012	I.V
	Marqibo®	Vincristine	SM:cholesterol (60:40)	Talon Therapeutics	2012	I.V
	Onivyde™	Irinotecan	DSPC:cholesterol: MPEG-2000-DSPE (3:2:0.015)	Merrimack Pharmaceuticals	2015	I.V
	Lipusu®	Paclitaxel	NA	Luye Pharma Group	2006	I.V
	Vyxeos®	Cytarabine:Daunorubicin 5:1	DSPC:DSPG:cholesterol (7:2:1)	Jazz Pharmaceuticals	2017	I.V
Antifungal	Ambisome®	Amphotericin B	HSPC:cholesterol:DSPG (2:1:0.8)	Astellas Pharma	1997	I.V
	Fungisome®	Amphotericin B	PC:cholesterol (7:3)	Lifecare Innovations	2003	I.V
Photodynamic therapy	Visudyne®	Verteporphin	Verteporphin:DMPC&EPG (1:8)	Novartis AG	2000	I.V
Analgesic	DepoDur™	Morphine sulfate	DepoFoam™	SkyPharma	2004	Epidural
	Exparel®	Bupivacaine	DepoFoam™	Pacira pharmaceuticals	2011	I.V

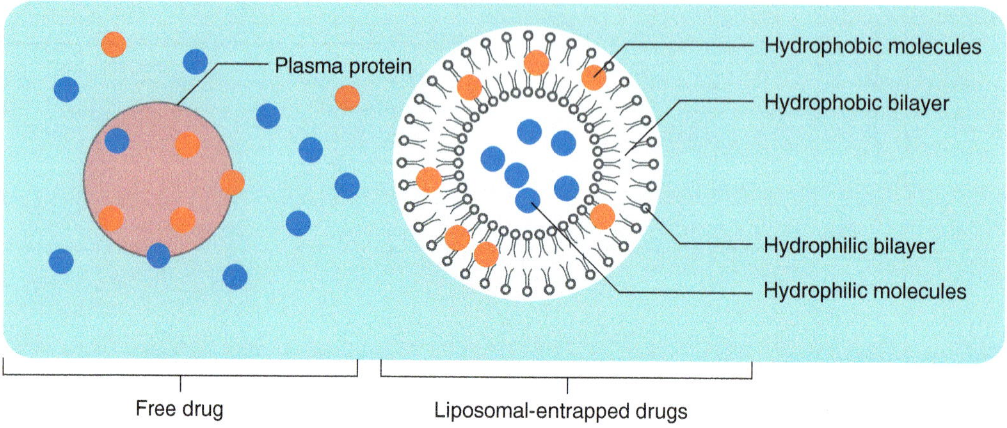

Figure 15.13 The existence form of liposomal drugs in plasma.

Due to the specific clinical advantages of liposomal formulations, as well as the unique PK characteristics, *in vivo* drug research on this formulation has become increasingly hot in recent years. As regulatory agencies become more mature for nonclinical PK of nanoparticle technology, it is necessary to establish a reproducible, robust method that can accurately determine free and liposomal-entrapped drugs in biological samples.

During the establishment of biological methods, the separation of free and liposomal-entrapped drugs and how to ensure the stability of liposomes in the biological matrix during sample storage and pretreatment are challenges during method development, compared with conventional dosage forms. The preparation of calibration standards and QCs, matrix effect, selectivity, recovery, stability, and even centrifugal speed involved in the development of methods have different considerations from those of conventional dosage forms. Based on the understanding of regulations and existing project experience, this section will introduce the key points in the detection process of biological samples of this type of drug.

15.3.2 Establishment of Determination Methods for Liposomal Drugs in Biological Samples

Compared with conventional dosage forms, the existence morphology of liposomal drugs in the body is complicated, so the following contents will introduce the critical points that need additional attention during the method development for bioanalysis.

15.3.2.1 Separation Method

For the determination of free and liposomal-entrapped drugs *in vivo*, whether it is direct method (determination of free and liposomal-entrapped drugs) or indirect method (determination of total and free drugs), to ensure accuracy, during sample processing, both methods need to ensure that the state of different morphological components such as loaded particles and free drugs does not change. Several methods have been proposed to isolate free (or non-liposomal) and liposomal drugs, such as equilibrium dialysis, ultracentrifugation, ultrafiltration, and SPE [3]. According to different experimental purposes, appropriate methods can be selected for separation and detection. We briefly introduce these types of methods, as detailed in Table 15.4.

Table 15.4 Separation method selection.

Separation method	Separation principle	Separation state	Application
Equilibrium dialysis	On both sides of the semipermeable membrane, only small molecules are allowed to pass through, while large molecules cannot	F-F permeates the membrane, and F-P is retained in the L section	Compounds that do not adsorb to the semipermeable membrane, and the liposomes are stable in this buffer solution
Ultracentrifugation	Different intensities of centrifugal force cause substances of different masses to separate in stages	F-F, F-P, and L can all be separated	Suitable for liposome formulations of hydrophilic compounds
Ultrafiltration	Under the action of centrifugal force, small molecule drugs can pass through the ultrafiltration membrane, while large molecules are retained	F-F permeates the membrane, and F-P is retained in the L section	Suitable for hydrophilic compounds with a lower protein binding rate in biomatrix
Solid-phase extraction	The electrostatic or hydrophobic interaction forces between the packing material and the compounds retain the free compounds on the packing material, while the loaded drugs pass directly through without being retained	F and L can be separated	Suitable for these compounds that can be adsorbed by the stationary phase of SPE for biomatrix

F, free drug; F-F, the free drug that is not bound to plasma proteins; F-P, the free drug that is bound to plasma proteins; L, liposomal entrapped.

The equilibrium dialysis method is time consuming, and for drugs with a high binding rate to proteins, the concentration of the free form in the free state is low, which requires a higher sensitivity for the detection method. The high centrifugal force of the ultracentrifugation method may lead to the agglomeration of particles, the destruction of the bilayer structure of liposomes, and the leakage of drugs. SPE requires that the test substance can be adsorbed by the solid-phase packing and that the components on the surface of the liposome bilayer do not interact with the solid-phase packing, flowing out smoothly through the gaps between the packing particles. Ultrafiltration is suitable for hydrophilic compounds with a low protein binding rate and complements the SPE method. Based on the above methods, it is possible to choose according to actual work requirements. SPE and ultrafiltration are commonly used separation methods for liposome-type drug bioanalysis (the operation steps can be referred to in Figures 15.14 and 15.15).

15.3.2.2 Preparation of Calibration Standard Samples

The "Technical Guidance for Non-clinical Pharmacokinetic Studies of Nanomedicine (Draft)" of China points out that, when establishing the *in vivo* analysis method of carrier nanomedicines, it is recommended to prepare the calibration standards and QCs by simulating the actual *in vivo* state of loaded particles, free drugs, liposomal-entrapped drugs, and carrier materials.

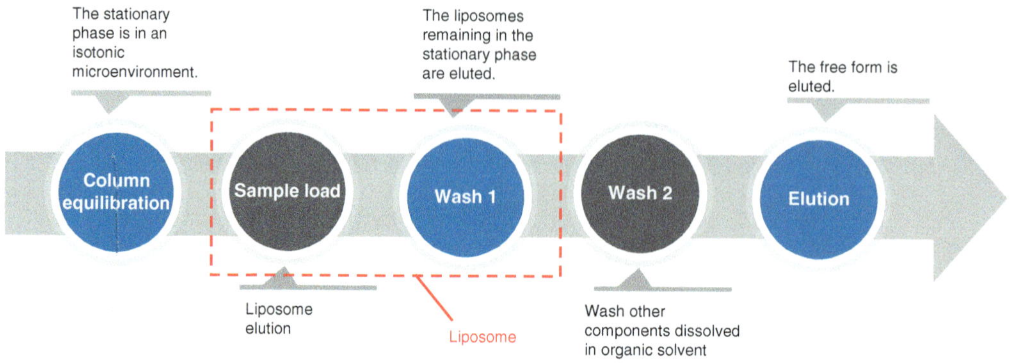

Figure 15.14 General process of SPE.

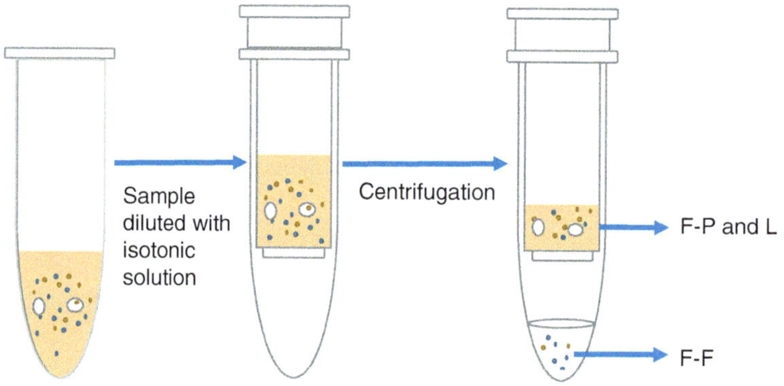

Figure 15.15 General process of ultrafiltration.

For total drug determination: to preliminarily understand the concentration differences of free and liposomal-entrapped drugs at different time points of *in vivo* samples according to the results of the preliminary test, it is recommended to simulate the *in vivo* state of free and liposomal-entrapped drugs after administration and prepare calibration samples according to a certain proportion.

For free drug determination: considering the low *in vivo* concentration of free drug (generally $< 20\%$ of the total drug in the concentration) and the possible influence of liposome encapsulation efficiency, it is generally not recommended to prepare the calibration sample by mixing the liposome and free drug.

When the free drug is used to prepare the calibration standards, due to the presence of phospholipids and charges on the surface of coexisting liposomes, nonspecific binding may occur with the drug, and it needs to be considered that the coexisting liposomes may cause the free drug to be eluted from SPE.

For liposomal-entrapped drug determination: it is required that different vial liposome formulations of the same lot have consistent encapsulation efficiency and good stability in the solution state. Liposomes can be directly used to prepare calibration standards into the matrix. Since it is necessary to simulate the *in vivo* state, and their concentration is generally several times or even tens of times of the free drug, then it is necessary to investigate the stability of liposomes during the pretreatment process, as well as the impact on the concentration of the free drug in case of leakage.

15.3.2.3 The Additional Focus of Validation Items

Refer to the validation items involved in the implementation recommendation of ICH M10 (Bioanalytical Method Validation and Study Sample Analysis of International Council for Harmonisation of Technical Requirements for Pharmaceuticals for Human Use), and according to the actual work requirements, we summarized the areas requiring additional attention in Table 15.5.

Because liposomal drugs contain both the test article and other formulation components in the sample after administration, it is necessary to consider the impact of the formulation component when preparing the simulating biological samples for verification. A plasma matrix containing blank liposomes (it is recommended to choose the amount of blank formulation at the total drug upper limit of quantitation (ULOQ) concentration, or the maximum concentration of blank liposome formulation from which the free drug has been eluted) can be used for the investigation of selectivity and matrix effects.

The special components of liposomal formulations lead to the liposome surface easily adsorbing hydrophobic compounds. When SPE is used for separation, liposomes may also compete with the stationary phase of the column to adsorb hydrophobic compounds, causing part of the free drug to enter the loading buffer along with the liposomes, resulting in inconsistent recovery of free drug at low and high concentrations. Therefore, the recovery of free drug in the presence of blank liposomes needs to be investigated during method development, meanwhile, the maximum blank liposome formulation concentration that causes free drug to be eluted should also be investigated.

Table 15.5 Additional validation considerations in method development.

		Liposomal drugs		
Validation items	**Conventional dosage forms**	**Total drug**	**Free drug**	**Liposome-associated drug**
Preparation of calibration standards and QCs	Matrix and free drug	Matrix, free drug, and liposome-associated drug	Matrix and free drug	Matrix and liposome-associated drug
Matrix effect	Six sources of blank matrix and one source of hemolytic matrix	Six sources of blank matrix and one source of hemolytic matrix	Extra attention should be paid to the impacts brought by the high concentration of blank liposome formulation	Six sources of blank matrix and one source of hemolytic matrix
Recovery	Normal preparation	Normal preparation	Extra attention should be paid to the impacts brought by the high concentration of blank liposome formulation	Normal preparation
Stability	Normal preparation	Normal preparation	Extra attention should be paid to the impacts brought by liposomes after storage and freeze-thawing	Normal preparation
Selectivity	Six sources of blank matrix	Extra attention should be paid to the impacts brought by the high concentration of blank liposome formulation		

Therefore, during the detection of plasma samples of free drugs after administration, the samples containing high-concentration liposomes (if the concentration of liposome samples is higher than the maximum concentration) should be diluted to the appropriate concentration range for detection.

In terms of stability investigation, it is necessary not only to consider the stability of free drug at different temperatures in the matrix but also to pay attention to whether the introduction of additives (stabilizers, anticoagulants, etc.) will cause the destruction of liposome structure. Therefore, when investigating the stability conditions of the free drug, attention should also be paid to the stability of the liposomes under the same conditions, to prevent the occurrence of "false-positive" conclusions about the stability of the free drug caused by minor leakage from the liposomes.

15.3.2.4 Centrifuge Speed

For nanoliposomes freeze-dried powder, typically forms a suspension after dissolution. For the blood samples collected after administration, stratification of the nanoliposomes may occur after a period of time. Then during the preparation of plasma samples by centrifugation, there may be uneven distribution of liposomes in the plasma. Therefore, the effect of centrifugal force on liposome delamination and leakage needs to be investigated.

The difference in the peak areas of liposomes and free drugs in plasma at different centrifugal forces can be used to determine whether the homogeneity of liposomes is affected so that the appropriate centrifugal force can be selected for the preparation of plasma samples.

In addition, the drug may leak from the liposome during vortex mixing, so it is necessary to gently mix the sample and working solution as low as possible during bio-sample preparation.

15.3.3 Summary

Liposomal drug product components are complex, and the physicochemical properties of different surface modifications vary significantly in *in vivo* studies; therefore, the considerations in the process of establishing bioanalytical methods still need to be "drug-specific". However, it is always inseparable from the principle of finding the appropriate separation method for the separation of free and liposomal-entrapped drugs, and it is necessary to ensure that the liposomes remain in a stable state from collection to treatment and do not affect the detection of free drugs.

References

1 Nsairat, H., Khater, D., Sayed, U. et al. (2022). Liposomes: structure, composition, types, and clinical applications. *Heliyon* 8: e09394.

2 Deb, P.K., Al-Attraqchi, O., Chandrasekaran, B. et al. (2019). Protein/peptide drug delivery systems: practical considerations in pharmaceutical product development. *Basic Fundamentals of Drug Delivery* 651–684.

3 Center for Drug Evaluation, NMPA. (2021, Aug 27). Technical Guidance for Non-Clinical Pharmacokinetic Study of Nano Drugs (interim). https://english.nmpa.gov.cn/2021-08/27/c_693010.html.

15.4 DMPK Research Strategies of Chiral Drugs Based on UPCC-MS/MS and UPLC-MS/MS Dual Analytical Platforms

Li Qu, Hongmei Wang, Lili Xing

15.4.1 Introduction

Chiral compounds exist in natural and biological environments. Natural sugars, nucleic acids, starch and cellulose, proteins, and amino acids are chiral [1]. The chiral compounds exhibit mirror symmetric in space but cannot superimpose by rotation in space (Figure 15.16). Many organic compound molecules possess "enantiomers" which means they have "chirality."

Significant differences may exist in the pharmacodynamic activity and PK properties of a pair of chiral drug enantiomers. To assess the activity, toxicity, and ADME properties of individual enantiomers, as well as possible chiral inversion in biological systems or organisms, analysis of chiral compounds in biological samples is crucial and necessary for the evaluation of individual enantiomers and metabolites.

15.4.1.1 Different Enantiomers, Different Functions

Chirality determines the biological interaction, safety, and efficacy. Different isomers of chiral drugs may have different effects, leading to different clinical applications. For example, the R-enantiomer of thalidomide has nausea and vomiting of pregnancy inhibitory and sedative effects, but the S-enantiomer is teratogenic. Despite having similar structures, it is recognized two chiral enantiomers can exhibit significant differences in their effects on organisms. With the in-depth research of their pharmacological mechanism of action, thalidomide has been confirmed to have strong immunomodulatory and inhibitory effects on angiogenesis. For safety considerations, this drug was used for the treatment of immune and tumor-related diseases in specific populations [3]. Only the S-enantiomer (levetiracetam) of etiracetam is biologically active, while the R-enantiomer is not [4]. At present, levetiracetam is mainly used for the treatment of partial seizures in adults and children over four years with epilepsy in clinical practice.

15.4.2 Regulations for Chiral Drugs

National drug regulatory authorities have developed policies and specific guidelines for the development of chiral drugs (Table 15.6) and pharmaceutical companies need to follow strict regulatory requirements to ensure the safety of drugs.

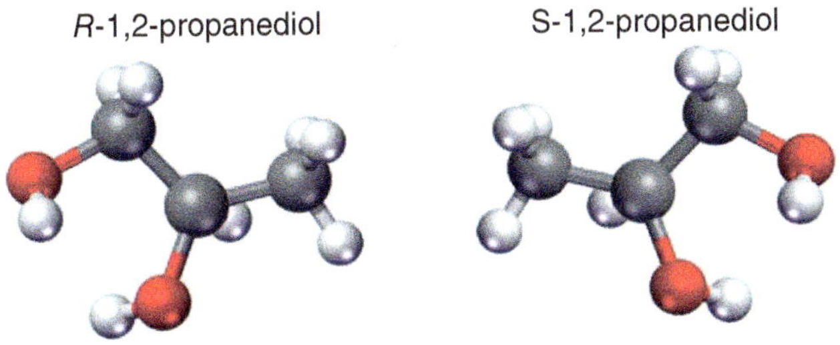

Figure 15.16 Two chiral enantiomers of 1,2-propylene glycol. *Source:* [2] / Springer Nature.

Table 15.6 Regulatory authority opponent drug requirements in various countries.

US FDA	Development of new stereoisomeric drugs [5] requires that drugs with a chiral structure require pharmacodynamic and toxicity tests before they are allowed to be marketed. Isomers of drugs must be separated and researchers must provide their pharmacological, toxicological, and kinetic results.
European EMA	European Pharmacopoeia (EP 7.0) [6]; there are a total of 1,350 compounds, of which chiral drugs account for about 50%, requiring the study of enantiomers of specific configurations.
Japan PMDA	In the Japanese Pharmacopoeia (15th Edition) [7], according to incomplete statistics, there are about 900 compounds, of which about half are chiral drugs with specific requirements for chiral configuration, the study of pharmacodynamics is required.
China NMPA	The Pharmacopoeia of the People's Republic of China (2010 Edition) [8], according to incomplete statistics includes 1,020 drugs with different structures, of which drugs with chiral centers were about 40%.

15.4.3 Chiral Drugs Pharmacokinetic Research Strategy

The stereoisomerism of chiral drugs may lead to big differences in their pharmacodynamic and PK properties. That means one enantiomer can work on the target, but the other enantiomer may be inactive or even harmful. Therefore, it is necessary to measure the concentration of individual enantiomers and chiral metabolites in biological samples to assess the absorption, distribution, metabolism, excretion, and toxicity of individual enantiomers. In addition, chiral assays are needed to assess the potential chiral inversion in organisms and to correctly assess the pharmacodynamic activity, kinetic properties, and associated toxicity of chiral compounds.

Method development and validation of chiral drug assays are more challenging compared to common drug assays. Therefore, based on the development strategy and molecular characteristics of the drug candidate [9], specific solutions should be implemented to meet the specific needs arising from different scenarios of chiral drug development, including method development for racemates and single enantiomers, racemate conversion, quantification of enantiomeric metabolites and appropriate chiral columns (Figure 15.17) should be selected for qualitative or quantitative analysis of the chiral drug.

At the same time, chiral methods allow rapid chiral enantiomeric conversion studies. For example, it is reported [11] that GNE-C was administrated to three relevant preclinical species and then quantified by using SFC-MS/MS. As shown in Figure 15.18, the enantiomer of GNE-C and GNE-C* was found in the mouse plasma, whereas no chiral inversion of GNE-C to GNE-C* was observed in either rat or dog plasma.

For the PK research of chiral compounds, *in vitro* chiral inversion studies can be performed to understand the transformation state of compounds in various biological matrices and environments, and then to evaluate *in vivo* characteristics and conversion. In an *in vivo* study, the chiral method can also be used to monitor the sample at various time points, thereby directly understanding the conversion and metabolic characteristics of compounds. The strategy for the study of chiral PK is shown in Figure 15.19.

15.4.4 Challenge for Chiral Drug Analysis in Biological Samples

In preclinical and clinical research, the detection of chiral compounds is crucial. Compared with common drugs, the PK research of chiral drugs is extremely challenging analytically. This is mainly reflected in the following aspects.

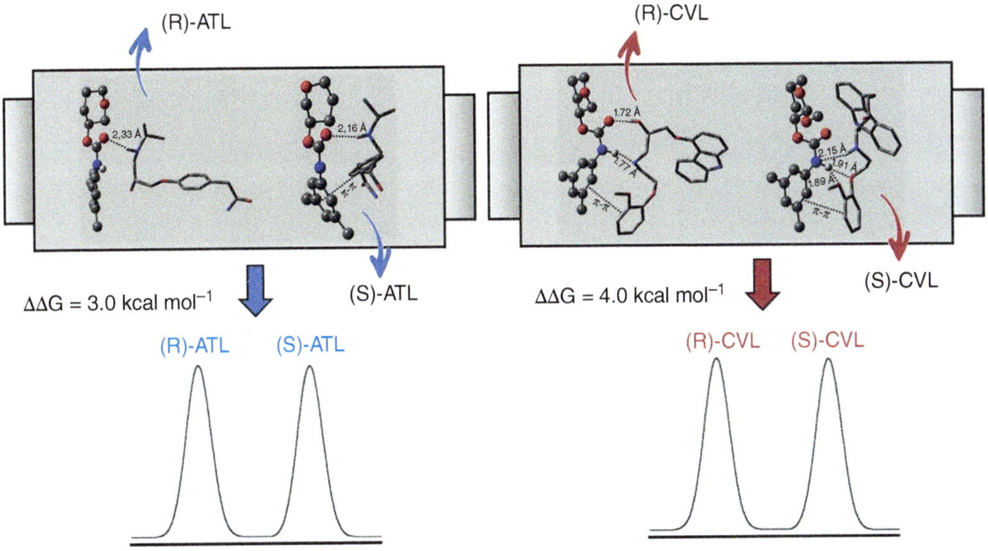

Figure 15.17 Chiral recognition process of polysaccharide-derived stationary phases with two β blocker drugs atenolol (ATL) and carvedilol (CVL). *Source:* [10] / ELSEVIER.

15.4.4.1 Enantiomers' Properties

1) Chiral drug enantiomers may interconvert under certain physical, chemical, or physiological conditions and require optimization of pretreatment methods.
2) The enantiomer may have different PD and PK properties, resulting in substantial differences in concentrations within the sample and varied sample handling.

15.4.4.2 Chromatographic Separation Difficulties

1) Due to the similar enantiomeric structure of chiral drugs, the parent ions and product ions are the same in the mass spectrum, so the enantiomers must be baseline separated for their accurate characterization and quantification.
2) It is necessary to screen suitable chiral columns and optimize the mobile phase.

15.4.4.3 Other Bioanalytical Challenges

1) Biological samples are small in volume, diverse, and complex in matrix (plasma, tissue).
2) Chiral separation also has high requirements for both sensitivity and detection speed.
3) The number of biological samples was large in some experiments, and the pretreatment process was needed to ensure that the compounds remained stable and untransformed.

15.4.5 Chiral Drug Bioanalytical Approach

With the continuous progress of analytical technology, the common separation methods for separating chiral compounds include GC, LC, supercritical fluid chromatography (SFC), and compatible chromatography (Figure 15.20).

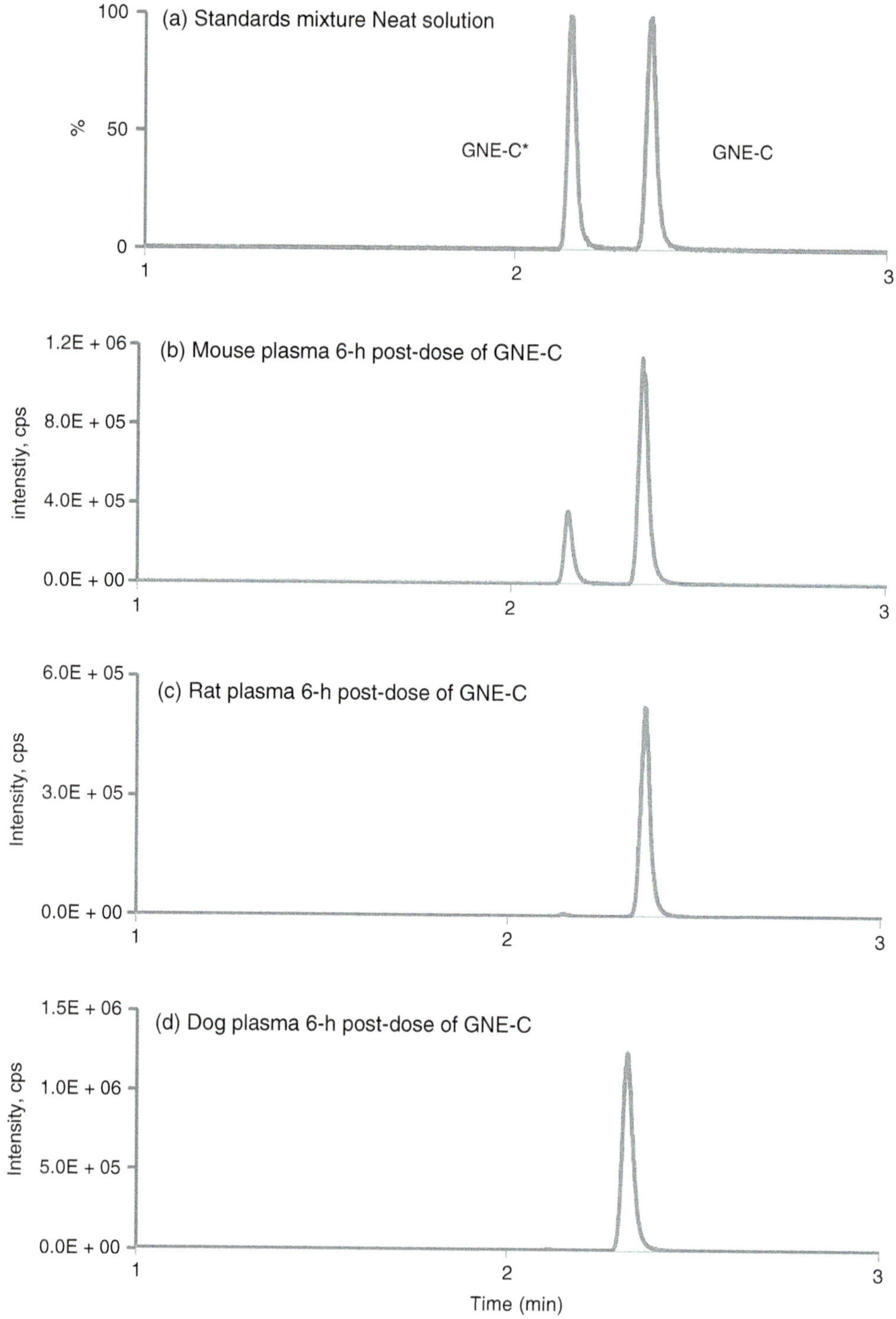

Figure 15.18 Chiral SFC-MRM chromatogram of GNE-C and GNE-C* in (a) standard mixture of neat solution, (b) mouse plasma after six hours post-dose of GNE-C, (c) rat plasma after six hours post-dose of GNE-C, and (d) dog plasma after six hours post-dose of GNE-C. *Source:* [11] / ELSEVIER.

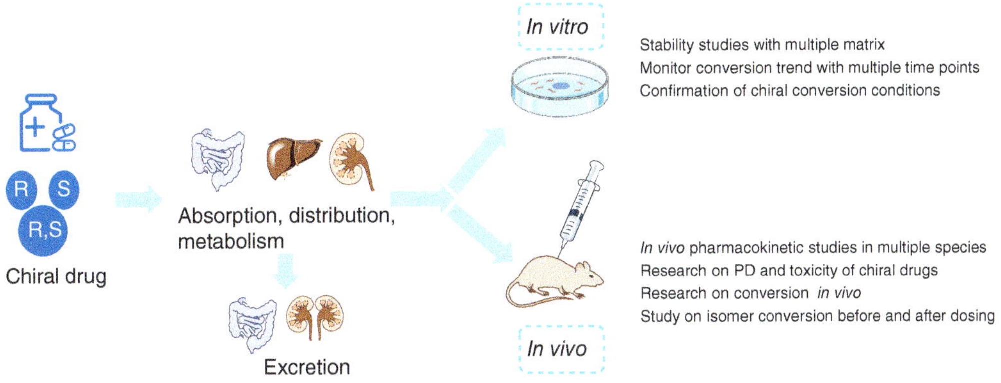

Figure 15.19 Chiral pharmacokinetic study strategy. *Source:* Adapted from References [12] Barreiro et al., 2021; [13] Calcaterra & Acquarica, 2018.

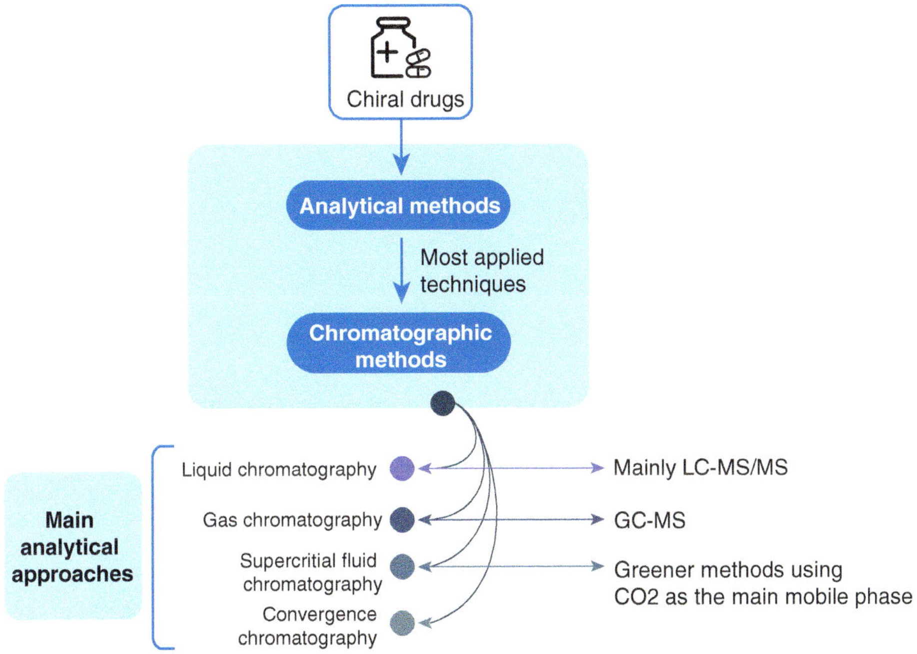

Figure 15.20 Major analytical approaches for chiral drugs. *Source:* Adapted from reference [14].

GC is suitable for chiral drugs with low boiling points and high volatility. However, compared with liquid phase enantiomeric separation techniques, GC has poor compatibility with aqueous samples and very limited chiral stationary phase selection, therefore it is less attractive in bioanalysis. LC offers a high degree of automation and good reproducibility. HPLC modes such as normal phase (NP), reverse phase (RP), and HILIC enhance the applicability of the technique for various types of chiral analytes. In SFC, a supercritical fluid is used as the mobile phase, and modifiers (such as methanol) can be added to adjust the polarity of the mobile phase to achieve the purpose of separating compounds. Convergence chromatography (CC) combines the advantages of low viscosity, high diffusivity of GC, and wide applicability of LC. It is an upgraded version of SFC with supercritical carbon dioxide fluid-bound partial solvents as the mobile phase and is separated according to the different partition coefficients of each compound between the stationary phase and the mobile phase.

15.4.6 Chiral Drug Bioanalytical Platforms

In the research and development of new drugs, the chiral analysis and research of compounds are imperative. The types of detection of chiral compounds are becoming increasingly complex, and the scope of assessment is also expanded from concentration detection to the identification of metabolites, especially the PROTACs and other biological products [15]. This presents new challenges to chiral separation and method stability for bioanalysis. Therefore, it is crucial to establish an accurate, reliable, and high-throughput chiral drug analysis technology platform (Figure 15.21).

For the separation and detection of chiral drugs, UPLC-MS/MS and UPCC-MS/MS analysis technology platforms can cover conventional small molecule drugs and PROTAC drugs. We have experience in chiral enantiomer separation, the detection of chiral drugs in biological samples, and chiral drug enantiomer conversion studies. We have rapidly separated and quantified chiral drugs and successfully performed ADME *in vitro* and PK *in vivo* assays (Figure 15.22).

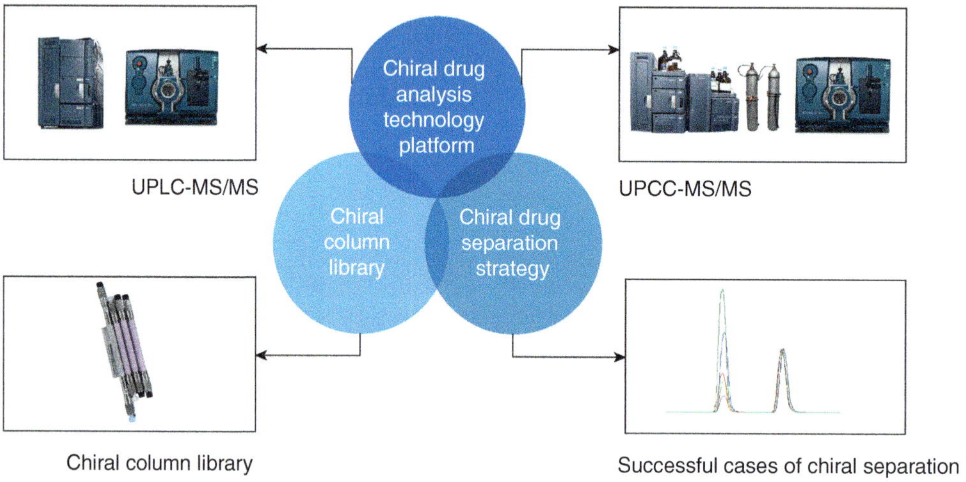

Figure 15.21 Chiral drug analysis technology platform.

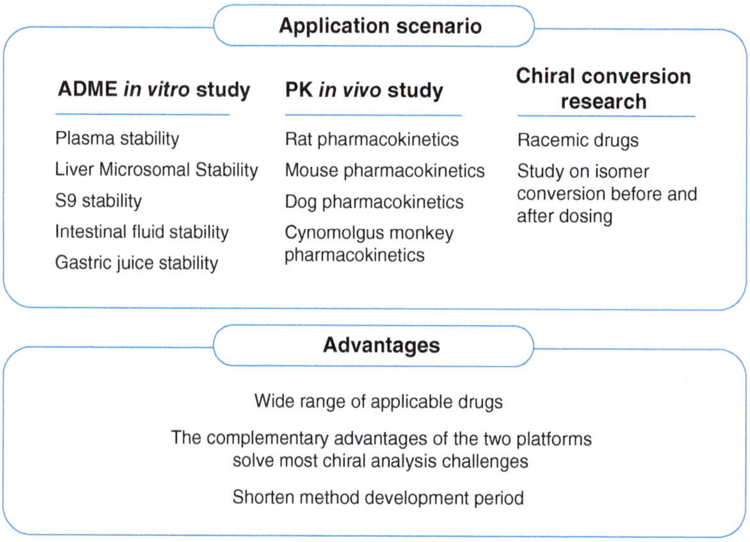

Figure 15.22 Advantages and application of two chiral drug analysis technology platforms.

15.4.7 Summary

With the increasing number of chiral drugs year by year and the escalating complexity in the development of multi-chiral center drugs and PROTACs chiral drug methods, it becomes crucial to separate chiral drugs quickly and efficiently. According to the requirements of the application, both racemates and chiral enantiomers need to be evaluated in terms of pharmacology, toxicology, and pharmacokinetics. Both UPCC-MS/MS and UPLC-MS/MS serve as chiral drug analysis platforms, each offering unique advantages. The suitable platform can be selected based on the compound types and analytical requirements.

References

1 Junli, P., Wang, H., Huang, C. et al. (2022). Progress of molecular imprinting technique for enantioseparation of chiral drugs in recent ten years [J]. *Journal of Chromatography A* 1668: 462914.

2 Patterson, D., Schnell, M., Doyle, J.M. et al. (2013). Enantiomer-specific detection of chiral molecules via microwave spectroscopy [J]. *Nature* 497 (7450): 475–477.

3 Raje, N. and Anderson, K. (1999). Thalidomide-a revival story [J]. *The New England Journal of Medicine* 341: 1606–1609.

4 Jian, Y., Guard, S., and Bookbook, H. (2009). Method Improvement for the Synthesis of Etiracetam. *Synthetic Chemistry* 17 (1): 118–119.

5 GUIDANCE DOCUMENT Development of New Stereoisomeric Drugs. MAY 1992 Final, U.S. Food and Drug Administration. https://www.fda.gov/regulatory-information/search-fda-guidance-documents/development-new-stereoisomeric-drugs [1992-05-01].

6 European Pharmacopoeia EP 7.0. European Medicines Agency. https://www.edqm.eu/en/european-pharmacopoeia.

7 The Japanese Pharmacopoeia 15th Edition. Pharmaceuticals and Medical Devices Agency. https://www.pmda.go.jp/english/rs-sb-std/standards-development/jp/0011.html [2006-05-31].

8 Pharmacopoeia of the People's Republic of China 2010 Edition &#32.

9 Jian, W., Edom, R.W., Huang, M.Q. et al. (2014). Bioanalysis of chiral compounds during drug development using a tiered approach [J]. *Bioanalysis* 6 (5): 629–639.

10 Maia, P., Nascimento, C.A., Silva, C.F. et al. (2022). Chiral separation study of atenolol and carvedilol β-blocker drugs by DFT calculations. *Computational and Theoretical Chemistry* 1213: 113741.

11 Chen, L., Dean, B., La, H. et al. (2019). Stereoselective supercritical fluidic chromatography-mass spectrometry (SFC-MS) as a fast bioanalytical tool to assess chiral inversion *in vivo* and *in vitro*. *International Journal of Mass Spectrometry* 444: 116172.

12 Barreiro, J.C., Tiritan, M.E., and Cass, Q.B. (2021). Challenges and innovations in chiral drugs in an environmental and bioanalysis perspective. *TrAC Trends in Analytical Chemistry* 142: 116326.

13 Calcaterra, A. and Acquarica, I. (2018). The market of chiral drugs: Chiral switches versus de novo enantiomerically pure compounds. *Journal of Pharmaceutical and Biomedical Analysis* 147: 323–340.

14 Ribeiro, A., Maia, A.S., Ribeiro, C. et al. (2019). Analysis of chiral drugs in environmental matrices: current knowledge and trends in environmental, biodegradation and forensic fields. *TrAC Trends in Analytical Chemistry* 124: 115783.

15 Chankvetadze, B. (2021). Application of enantioselective separation techniques to bioanalysis of chiral drugs and their metabolites. *TrAC Trends in Analytical Chemistry* 143: 116332.

15.5 Integrated Qualitative and Quantitative Bioanalysis of Intact Protein with High-Resolution Mass Spectrometry

Zhiyu Li, Lili Xing

15.5.1 Introduction

Intact protein profiling refers to an analytical technique that uses HRMS to skip enzymatic hydrolysis or fragmentation and directly characterize protein drugs at the intact level. The mature industrial application of the technology is to evaluate the purity and heterogeneity for batch release in the QC process of protein drug production. However, in recent years, with the rapid development of immunocapture technology and HRMS platforms, intact protein profiling has been used increasingly in *in vitro* and *in vivo* bioanalysis of large molecule drugs. Compared with conventional analysis methods, intact protein profiling can provide comprehensive structural information and accurate quantitative information for large molecule drugs from the background of complex components in *in vivo* samples [1]. This section will discuss the application of intact protein profiling techniques in preclinical bioanalysis as well as some practice cases.

The classical intact protein bioanalysis procedure for monoclonal antibody drugs is shown in Figure 15.23. First, immunoaffinity capture of the target antibody in the biological matrix is conducted, then purified samples are analyzed by LC-HRMS to obtain original data, and, finally, the obtained data is quantitatively or qualitatively processed.

15.5.2 Basic Techniques and Methods

15.5.2.1 Immunoaffinity Capture

For the biological analysis of proteins, the purification and enrichment with immunoaffinity capture is often the first step. Immunoaffinity capture technology has two key components: capture reagent and stationary phase. Capture reagents, as the name implies, are key reagents for affinity capture of target proteins from biological matrices, often generic antibodies, specific antigens, or anti-idiotype antibodies with specific binding ability to target proteins. The stationary phase is the surface used for the "binding" of the capture reagent (usually the high-affinity specific binding of streptavidin on the surface of the stationary phase to the biotinylated capture reagent). According to its form, in addition to the commonly used immunocapture plates in LBA, there are also magnetic beads, affinity tips, and affinity membrane/resin plates. With the help of IMCSTips and Hamilton automation device, we realize the complete automation of affinity immune capture, greatly improving the reproducibility and robustness of pretreatment of large molecule biological samples.

15.5.2.2 Front-End Separation Technique

15.5.2.2.1 Reversed-Phase Liquid Chromatography (RPLC)

RP chromatography is the most widely used front-end separation technique in the field of intact protein analysis. Compared with its great success in the field of small molecule and peptide analysis, there are still many challenges in the separation of intact proteins by RPLC, of which the most prominent is the peak shape (peak broadening and tailing) problem. Of course, the peak shape problem is closely related to the complex structure and heterogeneity of the protein itself. In addition, due to the nature of some intact proteins, nonspecific adsorption and carryover in

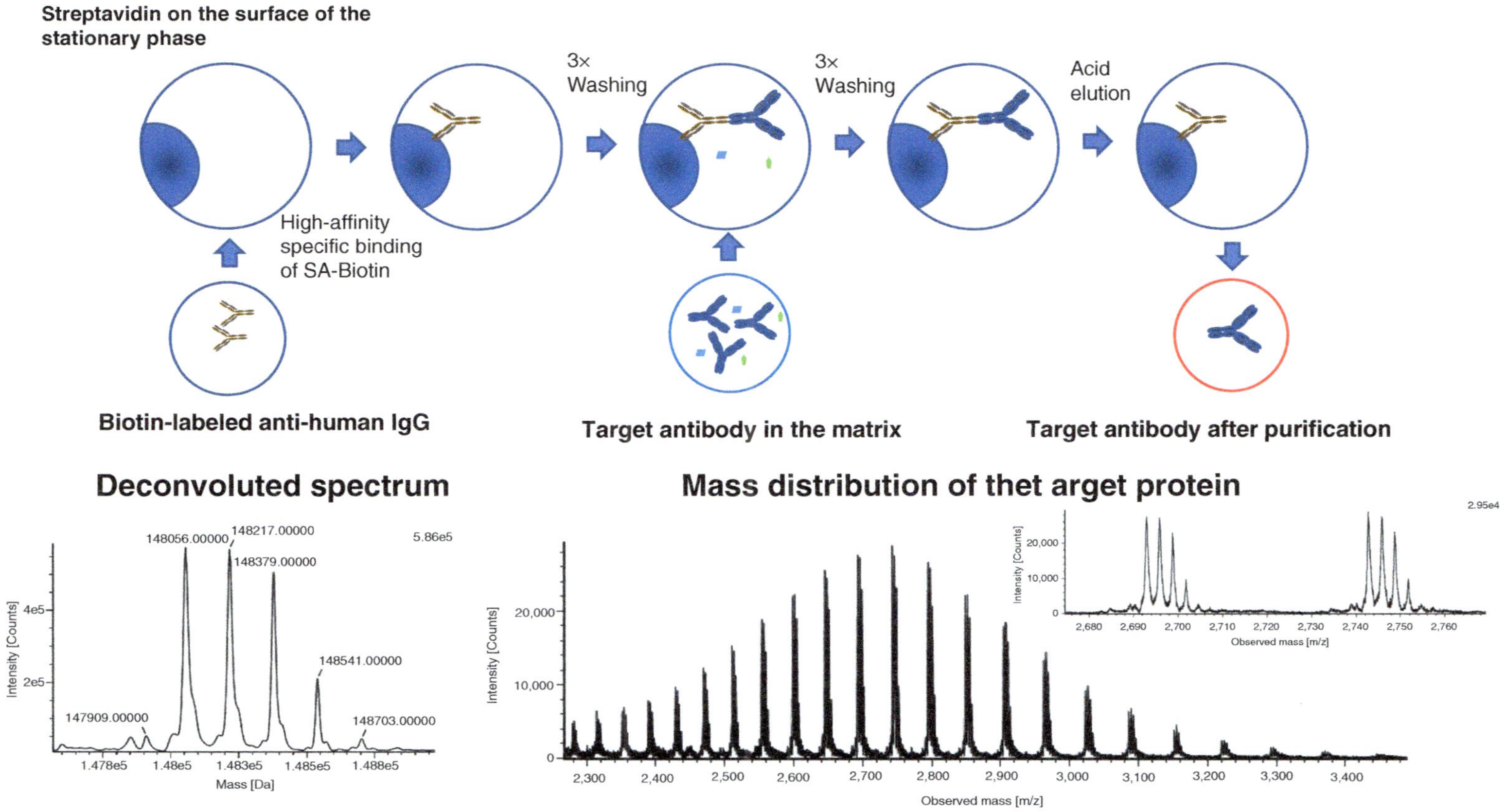

Figure 15.23 Classical intact protein analysis procedure with monoclonal antibodies: antihuman IgG is used to capture the target antibody (trastuzumab) in the matrix, followed by acid elution after washing. The eluted product is analyzed by LC-HRMS. The obtained mass spectrum can be directly quantitatively analyzed or deconvoluted to obtain the mass distribution of the target protein.

various parts of the liquid phase system often need to be addressed. We have rich experience in intact protein analysis for these potential challenges: (i) screen appropriate RPLC columns for intact protein analysis to obtain better peak shapes; (ii) specifically optimize the mobile phase and gradient to obtain the optimal peak shape and LLOQ; (iii) for proteins with severe nonspecific adsorption problem, add an appropriate amount of bovine serum albumin or surfactant in the eluent after immunocapture to inhibit adsorption of target proteins in the LC system; (iv) develop appropriate methods to remove glycosylation modifications for some proteins with too complex glycosylation modifications, simplify the sample structure and thus simplify the mass spectrum and improve the response.

15.5.2.2.2 Other Front-End Separation Techniques

In addition to RPLC, size-exclusion chromatography (SEC) and CE techniques have also been shown to be useful for intact protein separation and can be used in combination with MS.

In classical RPLC, the mobile phase is acidic with low pH, which means that the large molecules are already under denaturing conditions before entering the mass spectrometers for analysis, while SEC-MS and CE-MS can keep the protein in the native conform while entering the mass spectrometer with the help of neutral pH elution phase that mimics the physiological conditions. Both of them have unique applications in the field of intact protein profiling because of their different separation principles, such as native MS characterization with SEC, and capillary isoelectric focusing of CE-MS for charge heterogeneity analysis of antibody drugs [2]. In recent years, with the continuous development of related interface technology and MS technology, both SEC-MS and CE-MS have been used in *in vivo* PK studies of intact antibody drugs [3, 4]. The main drawback of these two methods compared to conventional RPLC is their poor sensitivity, which often needs to be overcome by increasing the sample volume or increasing the concentration of the sample at the time of injection (which can be achieved by decreasing the elution volume at the elution in the immunoaffinity capture process), but this is often difficult to implement for low-dose *in vivo* samples, thus limiting the application of these two techniques in the field of large molecule protein PK analysis.

15.5.2.3 Ionization Technique

The development of soft ionization technologies such as ESI and matrix-assisted laser desorption ionization (MALDI) is also one of the foundations for breakthroughs in intact protein profiling. Among them, ESI technology allows LC to be connected directly with MS, in this case, the protein can remain in the aqueous phase until entering the mass spectrometer.

15.5.2.4 High-Resolution Mass Spectrometry

Compared with low-resolution MS, where only a limited number of ion channels are often monitored, HRMS can use the ultra-high resolution of the detector and adopt the full-scan mode to comprehensively characterize the mass-to-charge information of the analyte and its potential metabolites, to make the integration of quantitative and qualitative analysis possible.

Compared to small molecules or peptides, due to their large and complex conformation, intact proteins often require higher declustering potential or cone voltage during ionization to reduce metal ion and solvent addition peaks to obtain cleaner and clearer mass spectra. However, higher cone voltage means that possible insource fragmentation may destroy the structure of the protein, thus giving the structural information of deviation, which requires us to optimize the MS parameters in combination with the actual situation, to obtain the most useful experimental results.

MS data analysis of intact proteins also requires tailoring. The original MS data of intact protein presents the distribution of multiple charge states. We can select one or more charge states for ion spectrum extraction, to realize the quantification of single or multiple components; we can also use analysis software to first deconvolute the original spectrum, remove the interference of multiple charge distribution, and use the deconvoluted spectrum for quantification. The method of processing raw data requires comprehensive consideration based on the purpose of the experiment and the nature of the analyte. In summary, for the data processing of intact protein analysis using HRMS, there is currently a lack of clear regulatory guidance in the industry, and we will continuously track and actively participate in the development of new rules in the industry.

Through the discussion in the above four parts, we can see that successful MS bioanalysis of intact proteins requires many indispensable prerequisite technologies on one hand; on the other hand, the requirements for laboratories and experimental operators are also comprehensive. First, laboratories need to be equipped with precise and expensive high-resolution mass spectrometers, and then laboratory personnel should be very familiar with the sample pretreatment of bioanalysis. In addition, the processing and interpretation of data are also more complex and troublesome than general PK analysis.

15.5.3 Case Sharing

15.5.3.1 Study on Pharmacokinetics of Monoclonal Antibody by Intact Protein Profiling

The PK study of antibody drugs usually depends on the LBA method or LC-MS/MS method based on MRM, which has been introduced in detail in the previous section. These two methods have advantages, but their common disadvantages include unclear the specific structure of the target protein to be quantified, such as whether it is a homogeneous single protein or a mixture of multiple protein structures formed after biotransformation of the original protein. With the help of the LC-HRMS platform, the quantification of antibody drugs at the intact level can solve the above problems, and the heterogeneity of antibody drugs (such as glycosylation and potential biotransformation) can be characterized.

We developed and validated a bioanalytical method of intact protein profiling for PK analysis of trastuzumab, a monoclonal antibody drug. Results are shown in Figure 15.24, with the help of a highly selective, high-recovery immunoaffinity capture method and carefully optimized

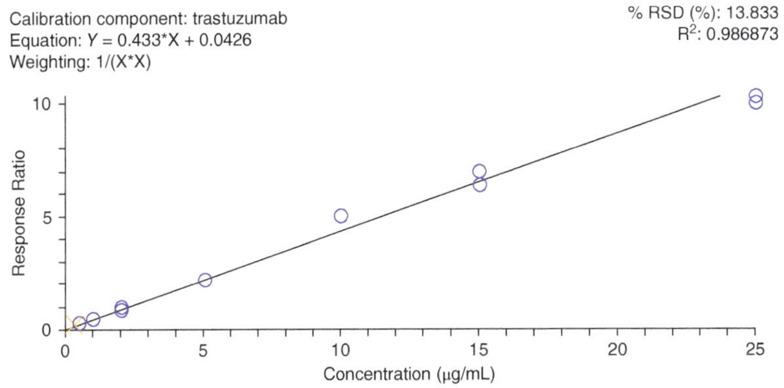

Figure 15.24 Quantitative standard curve for intact trastuzumab in rat plasma, linear range: 200–2,500 ng/mL.

RPLC-HRMS condition, the lower limit of detection of the method was 0.2 μg/mL, which was sufficient to support conventional PK analysis of monoclonal antibodies in the case of 20–30 μL of initial plasma samples. Afterward, we applied this method for PK analysis of *in vivo* rat samples and compared the results with those measured by the surrogate peptide method. When a single glycoform, G0F/G0F, was selected for quantification, intact protein profiling revealed that the G0F glycoform had a relatively faster PK clearance rate and a low half-life, as shown in Figure 15.25. To characterize the PK clearance rate analysis of different glycoforms, the traditional analytical method needs to first confirm the sites of glycosylation modification, then enzymatically hydrolyze the antibody, and then develop an MS quantitative method for the modified peptides [5]. The intact protein analysis can quickly and straightforwardly provide experimental results consistent with the literature reports, and the whole experimental process is much simpler and more direct.

In the processing of intact protein raw data, we specially selected only the peaks of a single type of glycosylation modification of G0F/G0F for peak extraction and quantification, and finally compared them with the surrogate peptide method to reveal the phenomenon of faster G0F/G0F clearance. Comparison of data at the initial time point (B) and collection endpoint (C) collected for the drug administration also revealed a decrease in the relative content of trastuzumab for the sugar form G0F/G0F.

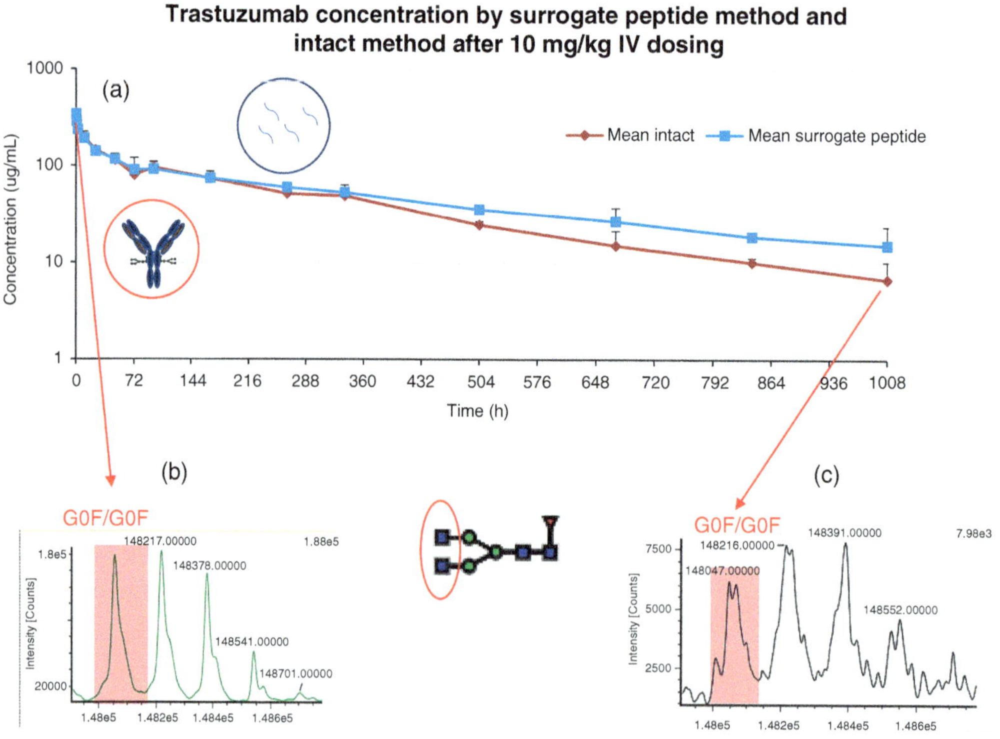

Figure 15.25 Comparison of trastuzumab concentration–time profiles in rat plasma using the surrogate peptide method and quantification of intact protein.

15.5.3.2 Intact Protein Profiling for Bioanalysis of Antibody-Conjugated Drugs

As a naturally heterogeneous molecule, intact protein analysis of ADC drugs has unique significance and importance. This importance is mainly reflected in multiple aspects:

1) Detection of DAR distribution versus mean DAR values. As shown in Figure 15.26, intact protein profiling of T-DM1 can reveal that the distribution of DAR values of T-DM1 gradually shifts to lower DAR values with prolonged *in vitro* incubation time. In addition, cysteine-conjugated ADC, due to its structural characteristics, will undergo random dissociation of light and heavy chains under classical RPLC acidic mobile phase, and for such ADC drugs, it is necessary to separate the light and heavy chains and detect their respective DAR value distributions, with representative results shown in Figure 15.27.
2) Detection of biotransformation. As shown in Figure 15.28, intact protein profiling of T-DM1 in rat plasma four days post-dosing revealed that the early *in vivo* biotransformation occurred as the release of small molecule toxins and cysteine exchange of small molecule toxins.
3) For ADC drugs with homogenous DAR values, PK analysis at the intact protein level is also possible. For example, the appearance of new connection technologies has reduced the heterogeneity of some ADCs to be close to or consistent with monoclonal antibody drugs [6, 7],

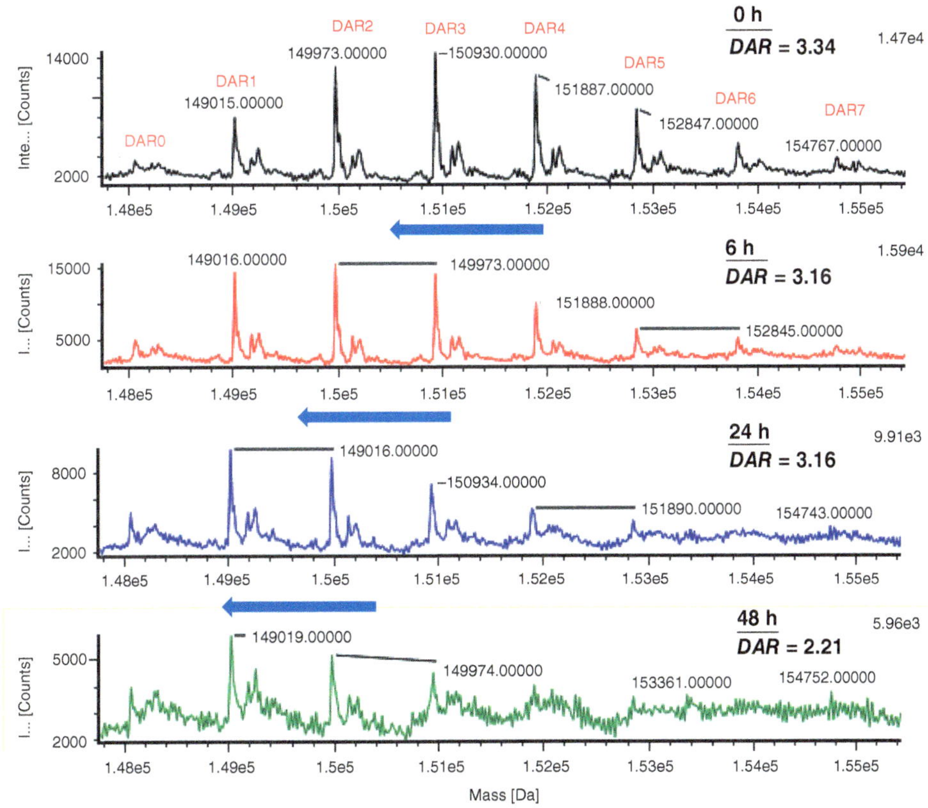

Figure 15.26 T-DM1 *in vitro* plasma stability test results. With the increase of incubation time, the content of ADC (DAR6 in the figure disappeared after 24 hours) with high DAR value gradually decreased, and correspondingly the mean DAR value of ADC decreased from 3.34 at T0 to 2.21 after 48 hours.

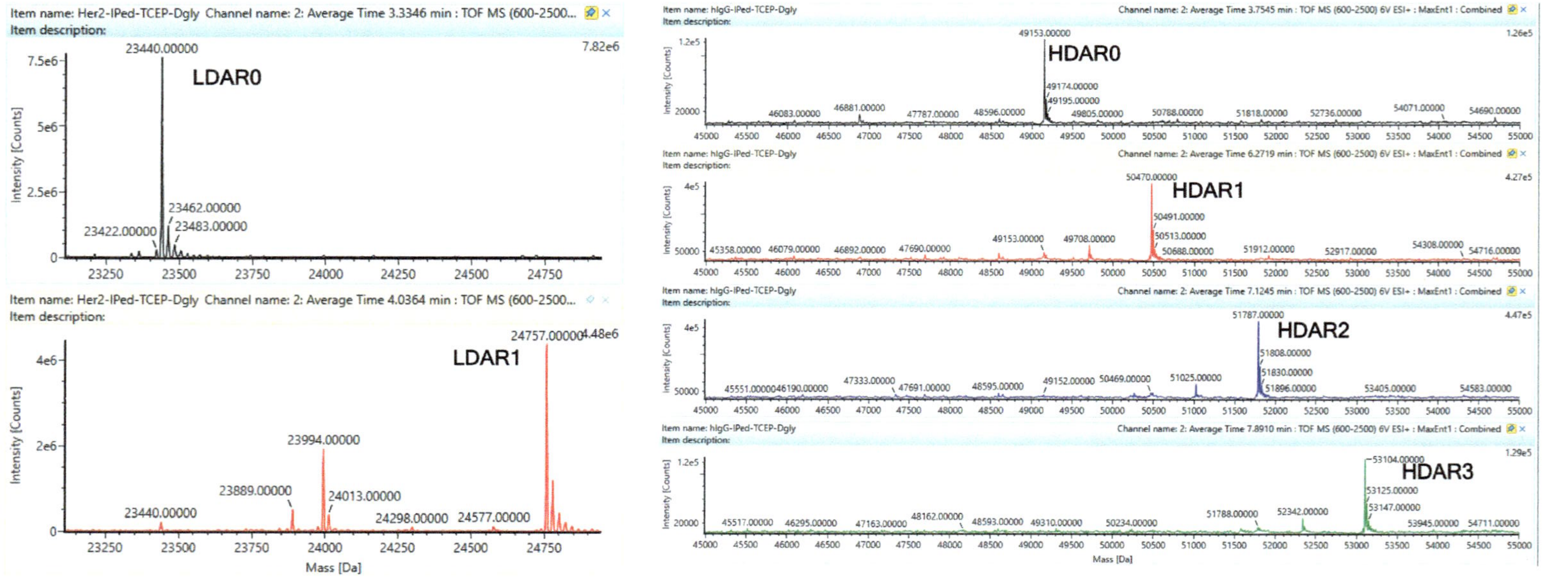

Figure 15.27 DAR value characterization of cysteine-conjugated ADC based on RPLC-HRMS. After a simple reduction of the ADC, RPLC-HRMS analysis was performed, and the obtained results were deconvoluted to obtain the distribution of DAR values for the light and heavy chains, respectively, and then the average DAR values of the original ADC were calculated.

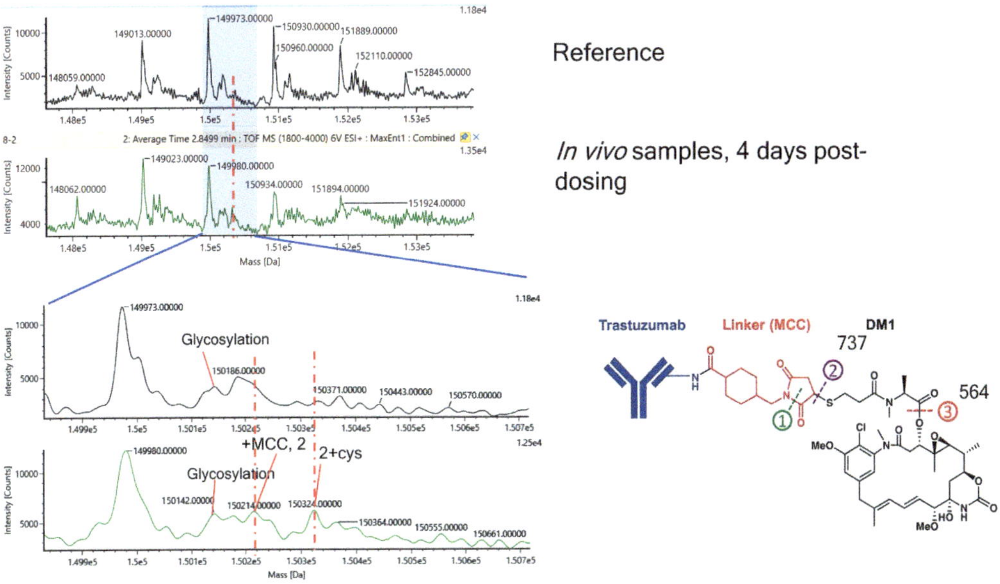

Figure 15.28 Identification of biotransformation products of T-DM1 in rats. Through the study of DAR2-related metabolites, it was shown that the main *in vivo* biotransformation was reflected as the shedding of small toxin molecules, and by comparing the blood samples collected from rats four days post-dosing versus the T0 samples, new peaks 150,214 and 150,324 corresponded to the shedding of toxin molecules and the cysteine exchange products, respectively.

which enabled PK analysis of the full ADC. As shown in Figure 15.29, from the literature [8] by Huang et al., through the synthesis of site-specific DAR2 and DAR8 ADC, with the help of intact protein analysis, different PK curves based on surrogate peptide and intact protein analysis were successfully plotted, and the possible *in vivo* biotransformation experienced by ADC was revealed.

15.5.4 Conclusion

Finally, we cite Figure 15.30 in the article published by Kellie et al. in 2021 as a summary to summarize the current classical application of intact protein analysis in bioanalysis and its implementation. It can be seen that although the authors of the article focus on antibody drugs, the decision-making idea can actually be extended to other proteins, such as small protein drugs with a molecular mass of ~10 kDa as well as oligonucleotide drugs.

Intact protein profiling has great potential in the field of bioanalysis, but due to the novelty of the method and the absence of relevant regulations, it is primarily being used as an alternative or supplement to classical methods. However, it cannot be ignored that intact protein profiling technology has unique advantages that LBA and surrogate peptide methods cannot match, that is, it can provide macromolecular drugs with metabolite identification and detection capabilities similar to small molecules, and can achieve biotransformation characterization for *in vivo* samples. With the increased structural complexity of large molecule drugs, especially the birth of ADC, a new molecular type of large and small molecule conjugation, this detection ability for biotransformation has become more and more important, which can help us to better study the stability and biotransformation of ADC drugs *in vitro* and *in vivo*.

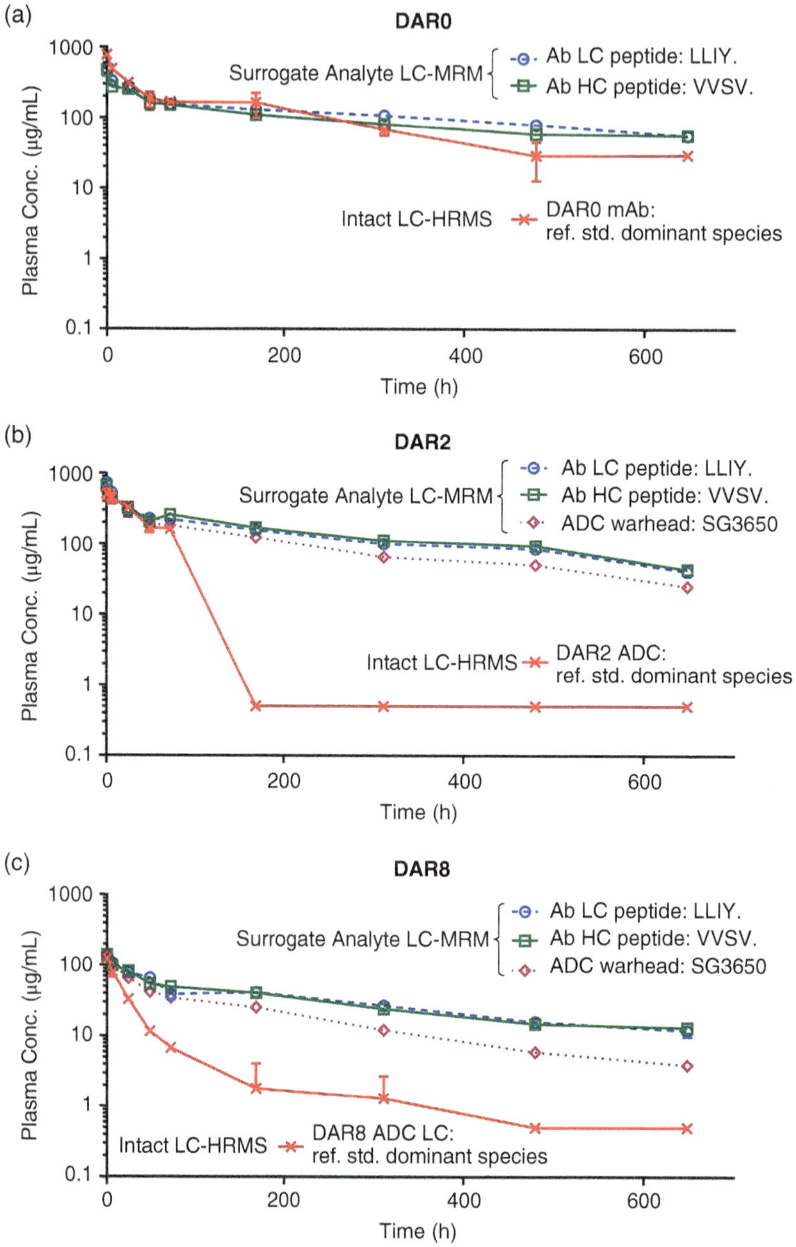

Figure 15.29 Site-specific ADC concentration–time curves for different DAR values reported in the literature. (a) DAR0 (25 mg/kg), (b) DAR2 (25 mg/kg), and (c) DAR8 (25 mg/kg). From the quantitative method based on surrogate peptides, the PK trends of antibody backbones of the three ADCs were similar, but the analysis based on intact protein revealed that the ADCs with toxin molecules were more likely to undergo *in vivo* biotransformation, and the specific analysis of the intact protein spectra after a period of administration showed that the toxin molecule SG3584 would be rapidly hydrolyzed *in vivo*. *Source:* [8] / American Chemical Society / CC BY 4.0.

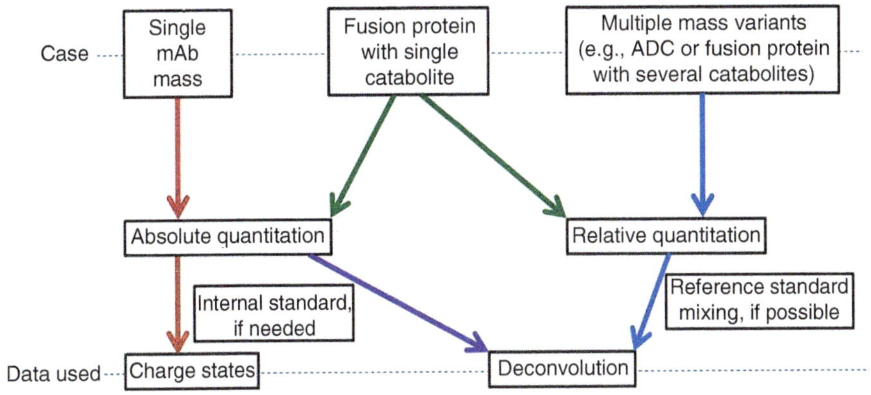

Figure 15.30 Representative case of intact protein analysis and its data analysis strategy. For single-component antibody drugs, absolute quantification can be achieved. For quantification, it can be based on extracted ion chromatographic peaks of single or multiple charge states, or based on the deconvoluted MS peaks. The addition of internal standards can improve the stability and reproducibility of the method. For fusion protein analytes with slightly complex components, such as single metabolites, we can use an absolute quantification method similar to the single antibody or semiquantitative analysis based on the relative intensity of deconvoluted peaks. For protein drugs with extremely complex components and ultra-multiple components or metabolites, such as heterogeneous and biotransformed complex ADC drugs, the implementation of deconvolution-based qualitative/semiquantitative bioanalysis is more realistic. *Source:* [9] / with permission of American Chemical Society.

References

1. Kellie, J.F. (2020). Intact protein LC-MS for pharmacokinetics. *International Journal of Pharmacokinetics* 4 (4): IPK05.
2. Dai, J., Lamp, J., Xia, Q., and Zhang, Y. (2018). Capillary isoelectric focusing-mass spectrometry method for the separation and online characterization of intact monoclonal antibody charge variants. *Analytical Chemistry Journal* 90 (3): 2246–2254.
3. Zhang, L., Vasicek, L.A., Hsieh, S. et al. (2018). Top-down LC–MS quantitation of intact denatured and native monoclonal antibodies in biological samples. *Bioanalysis* 10 (13): 1039–1054.
4. Kotapati, S., Passmore, D., Yamazoe, S. et al. (2020). Universal affinity capture liquid chromatography-mass spectrometry assay for evaluation of biotransformation of site-specific antibody drug conjugates in preclinical studies. *Analytical Chemistry Journal* 92 (2): 2065–2073.
5. Falck, D., Thomann, M., Lechmann, M. et al. (2020). Glycoform-resolved pharmacokinetic studies in a rat model employing glycoengineered variants of a therapeutic monoclonal antibody. *MAbs* 13 (1): 1865596.
6. Coumans, R.G.E., Ariaans, G.J.A., Spijker, H.J. et al. (2020). A platform for the generation of site-specific antibody–drug conjugates that allows for selective reduction of engineered cysteines. *Bioconjugate Chemistry Journal* 31 (9): 2136–2146.
7. Lee, B.I., Park, M.-H., Byeon, J.-J. et al. (2020). Quantification of an antibody-conjugated drug in fat plasma by an affinity capture LC-MS/MS method for a novel prenyl transferase-mediated site-specific antibody–drug conjugate. *Molecules* 25 (7): 1515.
8. Huang, Y., Mou, S., Wang, Y. et al. (2021). Characterization of antibody–drug conjugate pharmacokinetics and in vivo biotransformation using quantitative intact LC-HRMS and surrogate analyte LC-MRM. *Analytical Chemistry journal* 93 (15): 6135–6144.
9. Kellie, J.F., Tran, J.C., Jian, W. et al. (2021). Intact protein mass spectrometry for therapeutic protein quantitation, pharmacokinetics, and biotransformation in preclinical and clinical studies: an industry perspective. *Journal of the American Society for Mass Spectrometry* 32 (8): 1886–1900.

15.6 Nanobodies: Kinetics and Bioanalytical Strategies

Maotian Zhou, Zhiyu Li, Lili Xing

15.6.1 Introduction

In 1989, functional heavy chain antibodies were found in the sera of camelids including dromedaries, camels, llamas, guanacos, and vicuñas, and the variable regions of this heavy chain antibody had complete antigen binding capacity. Based on this heavy chain antibody, Ablynx (bought by Sanofi in 2018) created antibodies comprising just variable domains of heavy chain antibody fragments (variable heavy domain of heavy chain, VHH), which inspired Ablynx to propose the notion of "nanobody" due to its nanoscale microscopic size (Figure 15.31).

Relying on its structural and functional properties, nanobody has excellent penetration, stability, and solubility. These properties are currently garnering significant interest in the domains of medication detection, diagnosis, and therapy. Remarkably, the lack of conserved antibody sections and the small molecular weight (small size) of nanobodies pose a challenge to their bioanalysis. The universal sensitive competitive ELISA (USCE), a unique bioanalytical method developed in accordance with their characteristics, will be introduced in this chapter along with its structure, function, and kinetic characteristics. The sensitive ELISA method can be established with only one specific reagent, greatly reducing the difficulty of the development of bioanalytical methods in early development.

15.6.2 Structure and Function of Nanobodies

As illustrated in Figure 15.32, the structural properties of nanobodies are tightly linked to their functionalities. These are a few of its structural and functional attributes in brief.

a) Small molecular weight and small size: it contains only heavy chain variable regions in its structure, so it is characterized by small molecular weight (~ 15 kDa) and small size (~ 2.5 × 4 nm, conventional IgG 8.5 × 14.2 nm). Its compact size allows for improved tissue penetration and offers a special chance for druggability.
b) With **complementarity-determining region 1/2/3 (CDR1/2/3)** and **framework region (FR)**, as in Figure 15.33, the FR2 region has highly conserved hydrophilic amino acids. The heavy chain variable region of IgG (VH) with highly conserved hydrophobic amino acids (like V42, G49, L50, and W52 in VH), whereas VHH is on the more hydrophilic/smaller amino

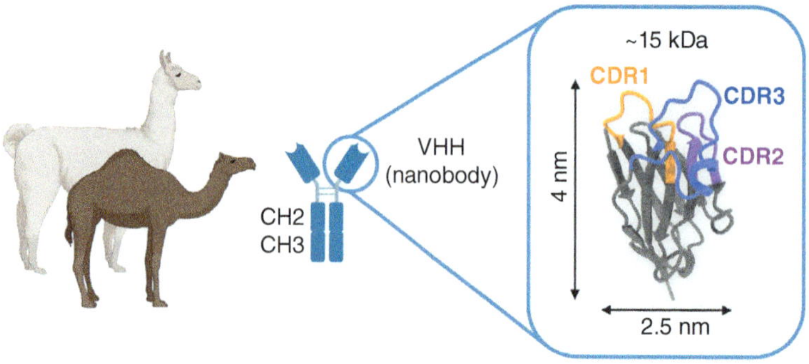

Figure 15.31 Acquisition and composition of nanobodies. *Source:* [1] / ELSEVIER.

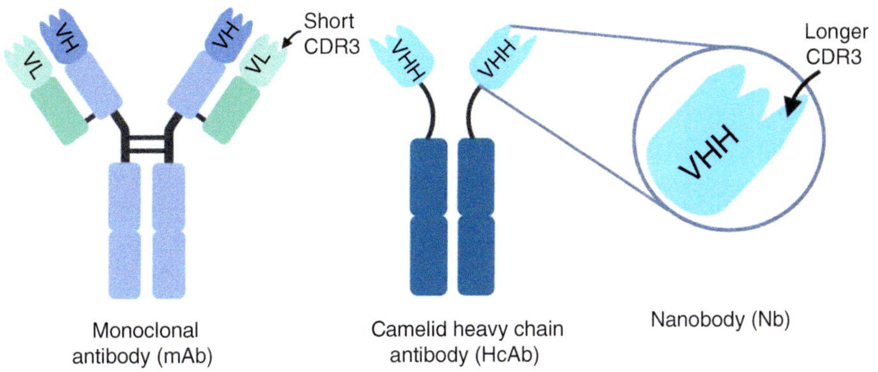

Figure 15.32 Basic structures of humanized monoclonal antibody, camelid heavy chain antibody, and nanobody. *Source:* [2] / Frontiers Media S.A. / CC BY 4.0.

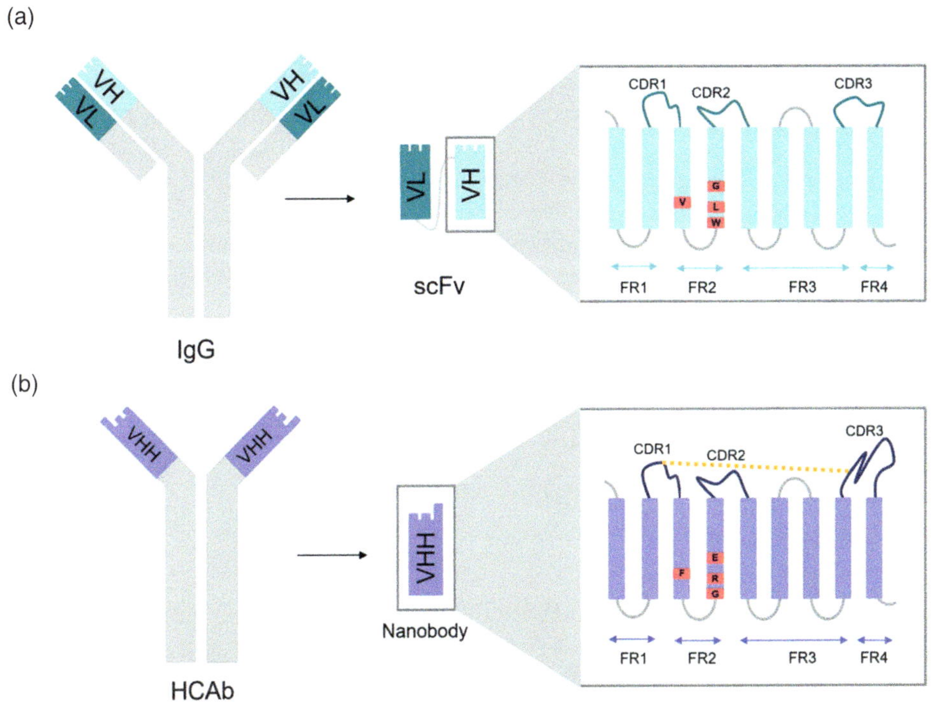

Figure 15.33 Schematic representation of CDR and FR regions of (a) IgG VH and (b) heavy chain antibody VHH. *Source:* [3] / Springer Nature / CC BY 4.0.

acids (F42 or Y42, E49, R50, and G52). These hydrophobic interactions between IgG VH and VL formed six pieces of CDR regions, stabilizing the conformation of the whole antibody. Unlike IgG, the hydrophilic structure of VHH helps it maintain conformational stability in the monomeric state and does not lead to aggregation. Due to the dependence on the hydrophobic structure to maintain the conformation, the fragment VL/VH (like scFv) of IgG may aggregate or mismatch at high concentrations.

c) Compared with the VH of IgG, the **CDR3 zone** of nanobody VHH is longer, which compensates for the absence of the VL domain. The CDR3 region mainly determines the specificity of the nanobody VHH and maintains strong binding to the antigen. Its finger-like extension can reach epitopes that cannot be bound by conventional antibodies.
d) Nanobody VHH has **intercyclic disulfide bonds** inside it, and these additional disulfide bonds help it to maintain conformational stability in harsh environments. Thus, compared to traditional antibodies, nanobodies are frequently more resilient to extreme temperatures, acid–base pH, and organic solvents.

The term "nanobodies" in this context refers to the heavy-chain variable region antibodies originating from camelids. Instead of VHH, antibodies derived from elasmobranchs, as shown in Figure 15.34, are one of the smallest antibody units. The smallest antibody form with antigen binding capacity is the variable new antigen receptor (VNAR) of Ig-NAR (new antigen receptor), which consists solely of CDR1 and CDR3 and has a substantially smaller molecular weight (~12 kDa).

The structural characteristics of nanobodies allow them to have unique properties. Thus, the advantages and disadvantages of some nanobodies in drug formation, biological detection, and other applications are listed in Table 15.7.

The benefits and drawbacks of nanobodies are strongly correlated with their structural features, as the table above illustrates. For instance, nanobodies can target solid tumors because of their small size, low molecular weight, and good tissue penetration. However, because of their tiny size and lack of Fc structure, they do not have antitumor effects such as ADCC/ADCP/CDC and have fast blood clearance (half-life typically varies from one to three hours), which further restricts the drug's use in the therapeutic field to some extent. It is important to note that the properties of nanobodies' quick clearing also allow them to have good application potential in the fast in-and-out detection field.

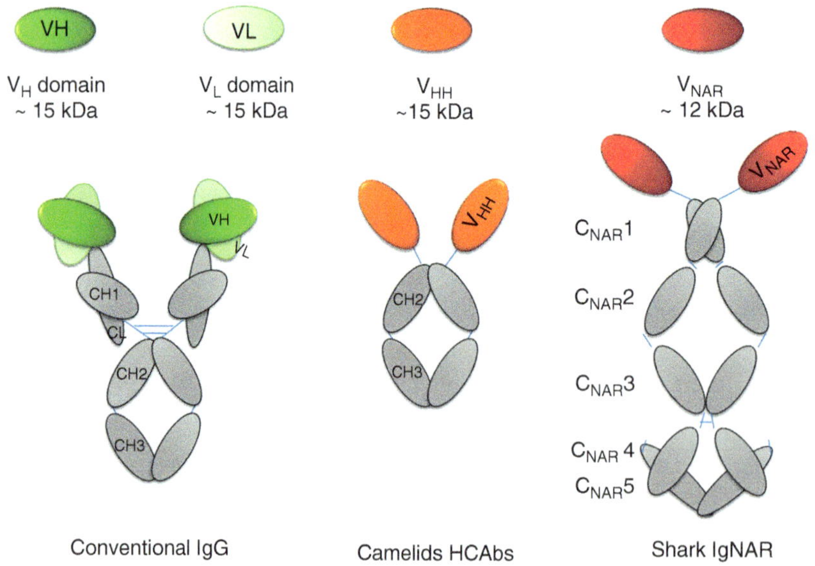

Figure 15.34 Comparison between the camelids-derived heavy chain of IgG antibodies and IgNAR antibodies produced by elasmobranchs such as sharks. *Source:* [4] / MDPI. / CC BY 4.0.

Table 15.7 Nanobody representative summary of advantages and disadvantages [5, 6].

Benefit	Disadvantages
Small size, strong tissue penetration	Fast blood removal speed (lack of Fc, quick renal filtration due to small molecular weight); short half-life, difficult to maintain effective concentration
High solubility and stability (hydrophilic backbone)	More off-target effects if the target has a large distribution of nontarget tissues due to the high tissue penetration
Identification of new epitope (small size and long CDR3)	Lack of Fc receptor-mediated ADCC/ADCP/CDC and other antitumor effects
Fast blood removal speed (advantages for scenarios requiring fast in and out, e.g., imaging tests with radionuclides)	Potential renal toxicity
Mature platforms for VHH screening; various well-developed and productive expression systems like *E. coli*, yeast, or mammalian cell lines (only ~100 amino acids and small size)	High potential immunogenicity (consisting of all variable regions and originating from camel); usually only partially humanized treatment: fully humanized treatment may cause problems such as aggregation due to hydrophobic changes in the exposed regions of FR. However, considering the rapid clearance rate of nanobodies, the overall immunogenicity situation is still debatable
Suitable for the synthesis of multivalent or multifunctional antibodies (the entire molecular size is not too large) because of the hydrophilic single-domain properties of VHH; relatively simple to produce; VL/VH mismatches can cause aggregation or reduced affinity in ScFv	

15.6.3 Kinetic Properties and Optimization Strategies of Nanobodies

Due to the structural similarity between nanobody monomers, they also share some commonalities in ADME characteristics, as shown in Figure 15.35.

In general, the absorption of nanobodies is characterized by **nonlinear kinetics**, with a rapid decline after the plasma drug peaks (IV, SC). After IV administration, C_{max} increased linearly with the dose, while AUC nonlinearly and less than the dose proportionally increased. Generally, nanobodies are mainly distributed in the systemic circulating blood, kidney, or liver (free targets-binding nanobodies usually have liver distribution). In terms of metabolism and excretion, nanobodies mainly rely on renal excretion, and it is also possible that there is target-mediated hepatic metabolic clearance or ADA-mediated metabolic clearance.

From these characteristics, nanobodies have their unique advantages as well as significant shortcomings. To better satisfy the requirement for druggability, nanobodies are typically optimized in terms of the following aspects:

1) **Humanization:** VHH antibodies generated from whole camels are typically humanized and immunogenic. The conserved hydrophilic amino acids in the FR2 region were converted into hydrophobic amino acids within the fully humanized VHH, leading to problems with solubility and stability. As a result, partial humanization is commonly used to preserve conserved

Figure 15.35 Summary of ADME characteristics for nanobody monomer.

hydrophilic amino acids in the FR2 region, which is mildly immunogenic and has high solubility and stability.

2) **Dimerization/multimerization**: nanobodies can be dimerized or multimerized to target different target proteins, epitopes, or even the same epitope within the same target protein, increasing the possibilities for more therapeutic mechanisms and prolonging their half-life.
3) **The conjugated half-life extender (half-life extender):** combined with particularly long half-life proteins like serum albumin, nanobodies can extend the half-life. Hence, HSA or a VHH that targets HSA are common conjugations for nanobodies [7, 8].

15.6.4 Industry R&D Progress of Nanobodies

The following table provides an overview of nanobodies that have entered phase 2/3 clinical phase by 2021, and nanobodies are primarily classified into the following groups based on distinct applications:

1) First, the role of neutralizing or regulating signaling pathways based on the mechanisms for therapy.
2) Second, conjugated nanobodies.
3) Third, as synthetic receptors for cell therapy.
4) Fourth, molecular imaging assays.

Up to now, only one nanobody drug, Caplacizumab [9], has been marketed, and the development and application of nanobodies are still in a relatively early stage (Table 15.8).

Here is a brief introduction to caracizumab, also known as ALX-0081 or its commercial name Cablivi. It was developed by Sanofi's subsidiary Ablynx, which received EMA approval on September 3, 2018, for the treatment of acquired thrombotic platelet purpura (aTTP) in adults. Cablivi became the first specific aTTP treatment and the first marketed nanobody drug. On February 6, 2019, the FDA authorized it for the treatment of aTTP when combined with immunotherapy and plasma exchange. The drug is structurally $(VHH)_2$ structured and the VHH is 90% humanized, retaining the key VHH conserved hydrophilic amino acid sequence. The two VHHs are linked by AAA chains with a total of 259 amino acids (molecular weight 28 kDa), as shown in Figure 15.36.

Von Willebrand factor (vWF) is a key protein in coagulation, and autoantibodies in aTTP patients may inhibit the activity of the vWF cleavage protease ADAMTS13, resulting in the adhesion of

Table 15.8 Nanobodies in current or phase 2/3 clinical phases.

Form	Antibody fragment	Target	Indications for use	Status
Neutralizing antibody	**ALX-0081** **Caplacizumab**	**Von Willebrand factor**	**Thrombocytopenia** **Purpura**	**FDA clearance**
	ALX-0061	**Interleukin 6 receptor (IL6R)**	**Systemic lupus erythematosus**	Clinical trial Phase II
	Vobarilizumab		**Rheumatoid arthritis**	Clinical trial Phase II
	ALX-0171	**Respiratory Syncytial Virus**	**RSV infection**	Clinical trial Phase II
	ALX-0761	**IL17A, IL17F, and IL17A/F**	**Psoriasis**	Clinical trial Phase II
	ATN-103 Ozoralizumab	**Tumor necrosis factor**	**Rheumatoid arthritis**	Clinical trial Phase II
	KN035	**PD-L1**	**Breast cancer hepatocellular carcinoma**	Clinical trial Phase II
	LMN-101	**Flagellin filament protein**	***Campylobacter* infection**	Clinical trial Phase II
	VHH batch 203027	**Rotavirus**	**Rotavirus diarrhea**	Clinical trial Phase II
	N3088 **N3130** **Colchicine**	**SARS Cov2**	**COVID-19**	Clinical trial Phase III
	BI 836880	**VEGF/Ang2**	**Tumor metastasis**	Clinical trial Phase II
Small molecule drug carrier	**L-DOS47**	**CEACAM6**	**Pancreatic carcinoma** **Lung adenocarcinoma**	Clinical trial Phase II
CAR-T-related synthetic receptor	**TC-210 T Cells**	Mesothelin	**Mesothelioma** **Cholangiocarcinoma** **Ovarian cancer** **Interstitial lung cancer**	Clinical trials I and II
Molecular imaging	**68GaNOTA** **Anti-HER2** **VHH1**	**HER2**	**Breast cancer**	Clinical trial Phase II
	68GaNOTA **Anti-MMR** **VHH2**	**Macrophages** **Mannose receptor (MMR)**	**Breast cancer** **Head and neck cancer** **Melanoma** **Carotid artery stenosis** **Atherosclerosis**	Clinical trial Phase II

Source: [3] / Springer Nature /

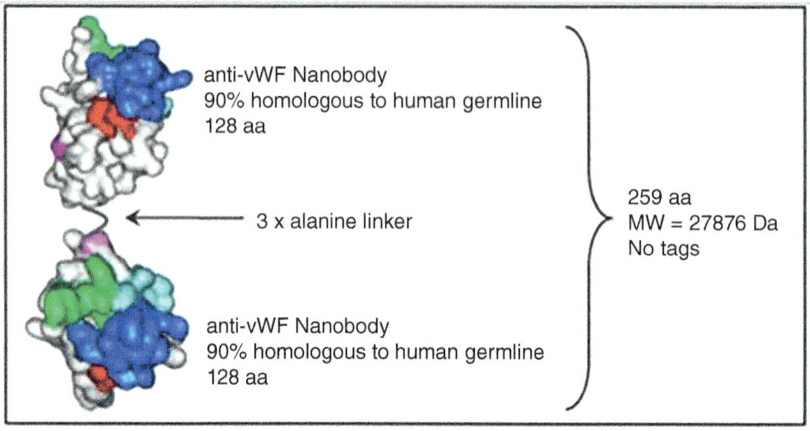

CDR-loops are colored as follow: CDR1 in green, CDR2 in cyan, and CDR3 in blue. Red and purple residues refer to as hallmark residues which are different for a VHH compared to a human VH. CDR: Complementarity Determining Region

Figure 15.36 Basic structure of carasitide. *Source:* [5].

vWF to platelets and the formation of microvascular thrombi. Concomitant tissue ischemia and multiple organ dysfunction may lead to the occurrence of thromboembolic events, which can be fatal in severe cases. Caplacizumab interferes with the binding of vWF to platelets via interacting with the A1 domain of vWF. Moreover, it can interact with both active and inactive forms of vWF, thereby preventing vWF-mediated platelet adhesion and thus treating coagulation abnormalities.

15.6.4.1 Bioanalytical Strategies for Nanobodies

Nanobodies are typically more difficult to bioanalyze than conventional IgG monoclonal antibodies due to their small size and lack of conserved region structure. **Special affinity critical reagents** are frequently required for this process. Using the bioanalysis of carassian antibodies as an example, the total karasidone antibody is studied in preclinical and clinical trials because it targets the vWF protein in the free form and is not expressed *in vivo* in low concentrations. Figure 15.37 shows the exact assay format. To obtain the analysis of total caracil antibodies, it is delicately designed to be captured by non-neutralizing anti-karasidone antibodies, followed by the addition of saturating levels of vWF and, lastly, the employment of antibodies vWF antibody detection.

In view of the above, the conventional sandwich method was mainly utilized to analyze nanobodies. Conventional sandwich method ELISA analysis is more suitable for nanobodies employing various types of modifications to prolong half-life, such as fusion of human origin IgG Fc or nanobodies fused with HSA protein. Critical reagents and ELISA assay format selection can be flexible and diverse according to the needs of analysis, such as ligands, specific antibodies, and ADA-positive control.

However, there are only a limited number of epitopes available for the analysis of nanobody monomers because of the small size of nanobodies. An appropriate sandwich ELISA has requirements for both the matching and targeting epitopes of key reagent pairs, which requires the screening or preparation of key reagents for capture/detection. Frequently, this preparation is time consuming, expensive, and difficult.

Is there a simpler or more versatile bioanalytical method that meets the need to establish appropriate nanobody analytical methods rapidly?

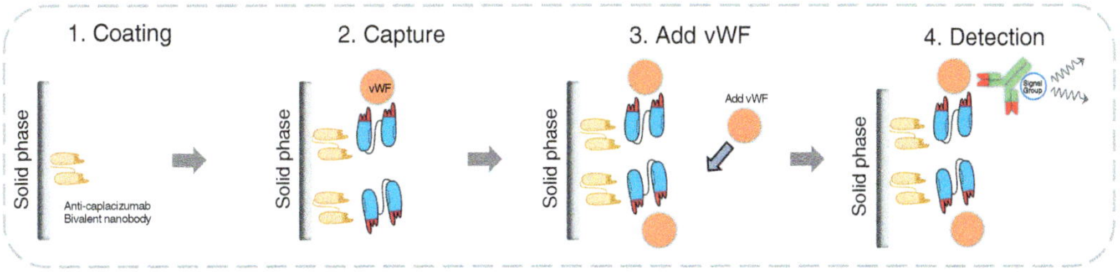

Figure 15.37 Schematic diagram of ELISA test form of total carasitide antibody.

We developed a set of **USCE systems** based on the competitive ELISA method. This system can adapt to a microplate reader or MSD detection platform as well as meet the high demand for sensitivity. This system requires only one suitable key reagent to complete the analysis of nanobodies, greatly shortening the reagent preparation time and method development cycle. With the USCE system, **lead time can be close to zero** for the bioanalysis of screening projects. Once the nanobody to be evaluated arrives, animal experiments can begin, and sample analysis can be done once the experiments are finished (Figure 15.38).

Each parameter of the method validation can be fulfilled by the bioanalytical methods based on the USCE system, applicable to a preclinical submission IND program or a clinical program. This is the case for the cynomolgus monkey IND project (Figure 15.39), where the nanobody is a monomeric structure that targets membrane receptor targets. To meet the requirements of the

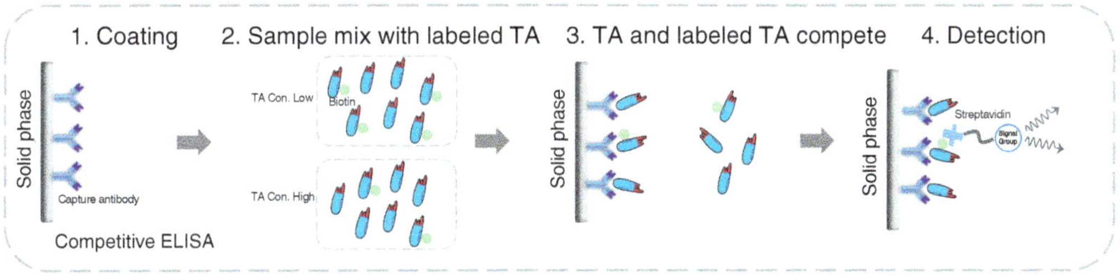

Figure 15.38 Schematic of USCE competitive ELISA assay format.

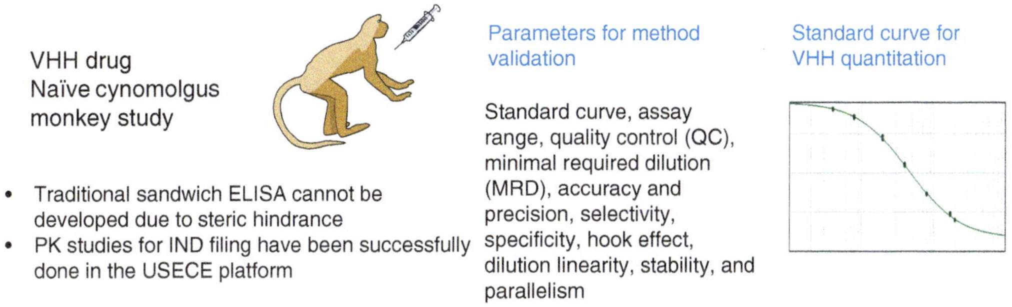

Figure 15.39 Content of method validation and quantitative standard curve diagram of nanobody application IND Project.

FDA/NMPA submission, we developed the bioanalytical method based on USCE and examined its parameters, including precision and accuracy, selectivity, hemolysis, dilution linearity and hook effect, robustness, short-term stability, room temperature 2°C–8°C, freeze-thaw stability, long-term stability, parallelism, and other factors.

15.6.5 Summary

Nanobodies are characterized by small size, strong penetration, stability, and good solubility in water and have received much attention in the fields of disease treatment, diagnosis, and substance detection. The unique structure of nanobodies gives them obvious advantages and disadvantages in drug formation, biological detection, and other applications. At present, the industry has further broadened its application prospects through a variety of modifications. Though bioanalysis of nanobodies is more challenging than conventional antibodies, the well-established USCE bioanalysis platform may assist producers of nanobody therapeutics in evaluating the *in vivo* PK features of their products at various phases of research more effective and precise way, laying a solid foundation for subsequent safety evaluation and clinical research.

References

1 Shen, Z., Sang, Z., and Shi, Y. (2022). Nanobodies as a powerful platform for biomedicine. *Trends in molecular medicine* 28 (11): 1006–1007.

2 Yang, E.Y. and Khalid, S. (2020). Nanobodies: next generation of cancer diagnostics and therapeutics. *Frontiers in Oncology* 10: 1182.

3 Asaadi, Y. et al. (2021). A comprehensive comparison between camelid nanobodies and single chain variable fragments. *Biomarker Research* 9 (1): 1–20.

4 Leow, C.H., Fischer, K., Leow, C.Y. et al. (2017). Single domain antibodies as new biomarker detectors. *Diagnostics* 7 (4): 52.

5 Committee for Medicinal Products for Human Use of European Medicines Agency (2018). Assessment Report: Cablivi. https://www.ema.europa.eu/en/documents/assessment-report/cablivi-epar-public-assessment-report_en.pdf (accessed 21 June 2024).

6 Bannas, P., Julia, H., and Friedrich, K.-N. (2017). Nanobodies and nanobody-based human heavy chain antibodies as antitumor therapeutics. *Frontiers in immunology* 8: 1603.

7 Karami, E., Mahdi, B., and Fatemeh, K.-L. (2020). Albumin nanoparticles as nanocarriers for drug delivery: using Focon antibody and nanobody delivery and albumin-based drugs. *Journal of Drug Delivery Science and Technology* 55: 101471.

8 Yang, E.Y. and Shah, K. (2020). Nanobodies: next generation of cancer diagnostics and therapeutics. *Frontiers in Oncology* 10: 1182.

9 Elverdi, T. and Eskazan, A.E. (2019). Caplacizumab as an emerging treatment option for acquired thrombotic thrombocytopenic purpura. *Drug Design, Development and Therapy* 13: 1251–1258.

Index

Note: The page numbers in *italics* denotes figures and in **bold** denotes tables.

Drug Metabolism and Pharmacokinetics: Frontiers, Strategies, and Applications, First Edition.
Edited by Liang Shen and WuXi AppTec DMPK.

d

m

p

t